P9-BYO-163

*Forest Ecology*

# FOREST ECOLOGY
## Third Edition

**STEPHEN H. SPURR**
The University of Texas at Austin

**BURTON V. BARNES**
The University of Michigan

KRIEGER PUBLISHING COMPANY
MALABAR, FLORIDA
1992

Third Edition 1980
Reprint Edition 1992

Printed and Published by
**KRIEGER PUBLISHING COMPANY**
**KRIEGER DRIVE**
**MALABAR, FLORIDA 32950**



**Library of Congress Cataloging-In-Publication Data**
Spurr, Stephen Hopkins.
Forest ecology / Stephen H. Spurr, Burton V. Barnes. -- 3rd ed.
p. cm.
Originally published: New York : Wiley, 1980.
Includes bibliographical references and index.
ISBN 0-89464-659-1 (acid-free)
1. Forest ecology. I. Barnes, Burton Verne, 1930-
II. Title.
QK938.F6S68 1991
574.5'2642--dc20 91-31799
CIP

10 9 8 7 6 5 4 3 2

# Preface

Forest ecology deals with forest ecosystems—assemblages of trees and their communities and environments in which they live. It is designed for use as a textbook in forest ecology, silvics, or principles of silviculture, for foresters, wildlife managers, and others interested in the ecology of forest land.

Since the appearance of the first edition in 1964, ecology has come into prominence popularly as well as scientifically. Increased ecological interest and concern present us with an opportunity and a challenge in managing our natural resources wisely. Ecological principles governing forest establishment, competition, succession, and growth are basic to managing trees and forests for whatever purpose and scale—from a city shade tree to a forested watershed. This book provides an understanding of the basic ecological relationships of individual trees and forest ecosystems.

The book has four major subdivisions. "The Forest Tree" considers the variations between individual trees, the causes of diversity within and between species, and selected aspects of the life history, structure, and function of trees. "The Forest Environment" treats *autecology*, the influences of solar radiation, atmospheric conditions, climate, soil, moisture, and fire on the individual forest plant. In "The Ecosystem," its components of the site and the biotic community are treated together with the study of whole ecosystems; *synecology* is also considered in this section. Finally, "The Forest" explores *phytogeography*. Here, the actual historical development and distribution of the North American forest are briefly developed.

In this edition we have tried to retain the readable qualities of the previous editions while adding new chapters and new material. Due to the enormous volume of ecological literature, a rigorous selection was necessary. Although over 1000 items of literature are cited, they constitute only a small sample of the many thousands of references in the authors' bibliographic collections. Preference has been given to major English-language works published since 1950. At the same time, however, an effort has been made to include a judicious selection of important earlier work, as well as representative papers written in languages other than English dealing with forests other than those in the United States.

We are indebted to many colleagues for their contributions and help in preparing the revised edition. In particular we acknowledge reviews by Terry L. Sharik, Paul E. Marshall, Bruce P. Dancik, Jean MacGregor, Warren H. Wagner, Jr., Willard H. Carmean, William J. Mattson, Richard E. Miller, Dale R. McCullough, John R. Bassett, and James R. Boyle. Scientists of the Coniferous Forest Biome and Deciduous Forest Biome contributed significantly to the ecosystem analysis chapter and other sections—in particular we thank Jerry F. Franklin and David E. Reichle. In addition, we thank especially text readers Frances G. Barnes and Lenora W. Barnes.

**Stephen H. Spurr**
**Burton V. Barnes**

*Austin, Texas*
*Ann Arbor, Michigan*

# Contents

## Part III The Ecosystem: Site, Community, and Ecosystem Analysis



## Part IV The Forest

# *Forest Ecology*

# 1
# Concepts of Forest Ecology

A **forest** is a biological community dominated by trees and other woody vegetation. **Ecology** is the science of the interrelationships of organisms in and to their complete environment. **Forest ecology**, therefore, is concerned with the forest as a biological community, with the interrelationships between the various trees and other organisms constituting the community, and with the interrelationships between these organisms and the physical environment in which they exist.

## ECOLOGY

The broader the field of scientific inquiry, the more difficult it becomes to limit and define that field. Ecology, as the broadest of the biological sciences, is also the most indistinct. Since Ernst Haeckel in 1866 proposed the term **oecology**, from the Greek **oikos** meaning house or place to live, the term has been applied at one time or another to almost every aspect of scientific investigation involving the relationship of one organism to another, or to the relationship of an organism to its environment. Ecology, however, clearly is confined to a study of life at the plane of the individual whole organism and not at levels of portions of plants. In contrast, the study of the life processes of the various portions of plants falls within the discipline of **plant physiology**; the investigation of events within the cell is termed **cytology;** while the analysis of ultimate chemical structure and life processes involves the science of **biochemistry.**

It is perfectly proper to investigate the life of an organism on any of these planes (Figure 1.1). The level-of-integration concept has great importance in the development of a logical approach to ecology (Rowe, 1961). As Decker (1959) illustrates the alternatives:

> *A life history can be described as a chain of events, with each event involving a whole plant. For example, a seed germinates, the seedling grows to be a sapling, the sapling is killed by fire. The same life history can be stated in terms of several chains of organ events, that is, in terms of what happens to leaves, roots and stem. Or it can be described in terms of a still*

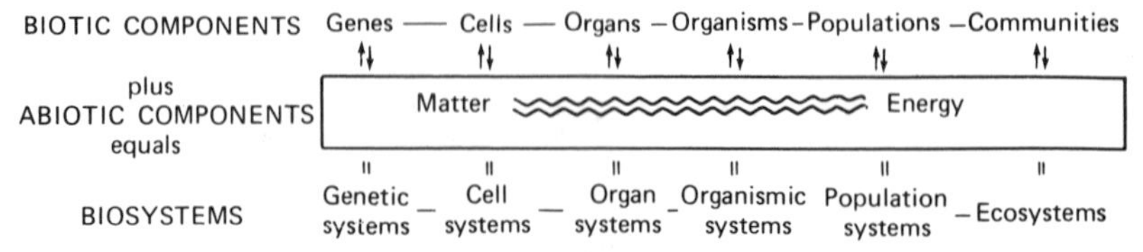


**Figure 1.1 Levels-of-organization spectrum. Ecology focuses on the right-hand portion of the spectrum, that is, the levels of organization from organisms to ecosystems. (After Eugene P. Odum, *Fundamentals of Ecology* © The W. B. Saunders Company, Philadelphia, Pa.)**

*larger number of tissue events. The resolving can be done repeatedly down through cellular events, subcellular events, colloidal events and molecular events. That is, if information were available, a life history could be described as an incomprehensibly large number of molecular events. Environmental factors can be resolved similarly; for example, illumination of a mesophyll cell, a proton striking a chlorophyll molecule.*

Investigations at each plane lead to an understanding of interrelationships at that plane. Ecological studies, therefore, may be expected to provide information as to observed correlations between whole organisms and their environment, and even as to cause-and-effect chains of events involving the whole plant or animal. Causality at other planes must be studied at the levels of other disciplines. It follows that ecological studies, involving as they do the complex of biotic communities, tend to lead to an understanding at the broad or integrative level; whereas the highly focused and localized attention of the biochemist, at the other extreme, proposes to understand the individual building blocks of life.

The term **ecology** was defined by Haeckel as the study of the reciprocal relations between organisms and their environments. A strict division of ecology into the study of individuals in relation to environment (autecology) and of communities in relation to environment (synecology) is widely held by European scientists. In their view the study of the development, composition, characteristics, and interactions of groups of plants or plant communities is defined as plant sociology, or phytosociology. The contrasting Anglo-American interpretation, and the one we follow, is that ecology is the study of the interrelationships of organisms to one another and to the environment (Hanson, 1962; Daubenmire, 1959, 1968a).

## THE FOREST

The forest is one of the basic physiognomic life forms by which biotic communities may be classified. Characterized by the predominance of woody vegetation substantially taller than humans, forests are widespread on land surfaces in humid climates outside of the polar regions. It is with forests in general, and with the temperate North American forest in particular, that the present book is concerned.

As with all biotic community types, the forest may be defined at several levels. Most obviously, it may be considered simply in terms of the trees, those plants which give the community its characteristic physiognomy. Thus we think of a beech-maple forest, a spruce-fir forest, or of other **forest types**, for which the naming of the predominant trees alone serves to classify the community.

A second approach to the definition of the forest takes into account the obvious interrelationships that exist between other organisms and forest trees. Certain herbs or shrubs are commonly associated with a beech-maple forest while others are concentrated in spruce-fir forests nearby. Similar interrelationships may be demonstrated for birds, mammals, arthropods, fungi, bacteria, and the like. The forest may be considered as an assemblage of plants and animals living in a **biotic** association, or **biocoenosis.** The **forest association**, or **forest community**, then, is an assemblage of plants and animals living together in a common environment, and is thus a more explicit and narrowly defined unit than the forest type, which is defined on the basis of the trees only.

A forest community exists in a physical environment composed of the atmosphere surrounding the aerial portions and the soil containing the subterranean portions. This environment is not static but rather is changing constantly due to the rotation of the earth, the fluctuations in solar radiation, the changing atmosphere, the weathering of the soil, and indeed the effects of the forest community itself upon both the local climate and the local soil. The forest community and its habitat together comprise an ecological system, **ecosystem**, or **biogeocoenosis**, in which the constituent organisms and their environments interact in vast and complex cycles of carbon, water, and nutrients. The **forest ecosystem**, taking into account both the organic and inorganic aspects of the cyclic processes of life, is at the same time the most precise and the least comprehensible definition of the forest community. The ecosystem is increasingly our focus of study whether we stress organisms or processes.

## AN APPROACH TO THE ANALYSIS OF FOREST ECOLOGY

The scope of forest ecology, then, may be defined as the analysis of the forest ecosystem. Such an analysis is facilitated first by the segregation of the ecosystem into its organic and inorganic aspects and then by consideration of forest communities and entire ecosystems. The sequence of analysis is (1) the forest tree, i.e., the variation, diversity, and life history of forest species, (2) the forest environment, (3) the forest ecosystem, and (4) forest history.

The forest tree owes its appearance, rate of growth, and size, in part, to the environment in which it has grown throughout its life. Put in modern biological language, the **phenotype**, the individual as it appears, is the product of the effect of the environment on its **genotype**, its individual hereditary constitution. This statement is basic to an understanding of the forest and, indeed, defines the scope of this book. An understanding of forest trees and communities in managed, exploited, or wilderness conditions is based in part upon the environment of the forest and the way it affects the trees making up the forest; but also in part upon the genetic nature of the trees themselves and the way in which this affects the response of trees to the environment they find themselves in. These genetic and environmental interactions are discussed in Part I.

The physical environment of the forest ecosystem together with biotic factors constitute the **habitat**, or **site**. Although the forest environment is the composite end product of many interacting factors, it is convenient to present and discuss separately each physical factor of the atmosphere surrounding the forest and of the soil in which the forest is grown.

The **habitat**, or **site**, as it is more commonly called in the case of trees, is the sum total of environmental conditions surrounding and available to the plant. These conditions are primarily the atmospheric and soil factors of the physical environment but also include important biotic influences of animals and plants, including the trees themselves. Among atmospheric factors, solar radiation, air temperature, air humidity, and carbon dioxide content all vary throughout the day, from day to day, month to month, and year to year, making up the complex we term **climate**. Below the surface, the supply of soil nutrients, the soil moisture regime, the physical structure of the soil, the soil fauna and flora, and the nature and decomposition pattern of the organic matter, that is, factors pertaining to the soil—**edaphic** factors—all affect the growth and development of plants. Furthermore, the growing vegetation itself affects the climate and soil so that the site changes as the plants themselves grow and change. The study of the environmental factors and their effects on plants constitutes the field of **autecology**. Part II summarizes those aspects of forest autecology most pertinent to an understanding of the forest ecosystem and its management.

The forest ecosystem is considered in Part III. First, the forest site quality, as determined by the integration of individual site factors, is examined in detail. The important roles of animals in the ecosystem are considered, and the forest community, its composition, and its competitive and dynamic relationships are described. The structure and composition of the forest community change from time to time and from place to place. **Dynamic plant ecology**, emphasizing forest succession, is concerned with changes in time, while **plant sociology** stresses variations in space. The term **synecology**, referring to the study of biotic communities and the interaction of the organisms which compose them, is inclusive of both.

The physical environment and the community are components of the forest ecosystem and are operationally inseparable. Thus one must ultimately deal with the ecosystem in emphasizing the interdependence and causal relationships of plants and animals in their physical environment. The understanding of functioning forest ecosystems and the method of systems analysis in their study is considered in the final chapter of Part III. Because of the size and complexity of forest ecosystems, it has only been in recent times that the biological and mathematical skills, data processing methods, and funds have been available to conduct the comprehensive programs capable of determining how an ecosystem works. Such investigations have led to the creation of an exciting and entirely new level of ecological science.

## PRESENT-DAY FORESTS

Although the principles of forest ecology may be presented by considering the forest as a complex ecosystem (Part III) arising out of the interactions of the forest trees and their environment (Parts I and II), so many factors are involved over so long a period of time that it is wise to conclude with a consideration of forests as they actually are and not necessarily as they should appear to be in accordance with our philosophy. Part IV brings together information on the present forest communities of the temperate portions of North America and on the past development of these communities as deduced from paleoecological studies or as recorded in the history of human exploration, land settlement, logging, and farm abandonment. The forest of today is affected most strongly by the conditions that existed at the time the present individuals became established on that site, but it is also affected by all that has happened since that time, including climatic fluctuations and soil profile development. In the forest, the history of logging, land clearing, urban development, fires, windstorms, insect and disease epidemics, and other happenings that affect the life and growth of the trees all will influence the present forest stand.

Here the biological historian must play the part of the detective, for written records are few and scanty, and precise measurements in the past are almost nonexistent. Fortunately, recent developments in the reconstruction of past forests through analysis of fossil pollen accumulations and other plant remains, in the dating of fossil organic matter through radioactive carbon analysis and related techniques, and in precise tree ring studies are all increasing our knowledge and understanding of the past.

## OTHER TREATMENTS OF FOREST ECOLOGY

Such a complex subject as ecology may be approached in various ways, and no one approach has any inherent virtues outside of the clarity of its logic and the resulting understanding. Furthermore, no approach has any particular originality in modern times, all having evolved from man's long curiosity about his surroundings and from the gradual evolution of natural history into the science we know now as ecology.

Toumey's classic *Foundations of Silviculture* (1947) is an American statement of the classic German approach, including both autecology and synecology, the latter influenced in his day by the theories of F. E. Clements on plant succession.

A physiological approach to the biological basis of silviculture also has its merits. An analysis based upon the life processes of the tree is just as valid as that based upon the organism itself. Thinking along this line has been greatly influenced by the English translation of Münch's revision of Büsgen's monograph on the *Structure and Life of Forest Trees* (1929), which remains a basic reference despite its age; the more recent and detailed revised German edition of this work (Lyr et al., 1967) continues the physiological approach. Kramer and Kozlowski (1979) and Kozlowski (1971) have summarized present-day information on tree physiology as such.

Recent and important European works include the two-volume revision of Dengler's famous treatment of the biological basis of silviculture (Bonnemann and Röhrig, 1971, 1972), the two-volume study of environmental factors in relation to tree and forest growth (Mitscherlich, 1970, 1971), and a concise treatment of the forest as a living community (Leibundgut, 1970). Significantly, the thought of the famous Russian ecologist Sukachev and his school is now available in English translation (Sukachev and Dylis, 1964).

## APPLICABILITY TO SILVICULTURE

Although knowledge and understanding for their own sake are sufficient justification for the ecologist, forest ecology is taught in the professional schools of forestry and natural resources because of its importance in influencing silvicultural practice. **Silviculture** is the theory and practice of controlling forest establishment, composition, and growth (Spurr, 1945). Man's conscious motives for control of forests have changed drastically since the turn of the century and continue to change from a primarily timber-oriented management to multiple-use management where human and water resources play an ever-increasing role. Whatever the purpose, obviously, management based on knowledge and understanding is better than that not so based. A knowledge of forest biology, most frequently organized and taught with an ecological point of view, is as important to the resource manager as is a knowledge of business economics and a competence in the mathematics of forest measurements and estimates. Beyond this, opinions differ.

At one extreme, many foresters follow the classic viewpoint of the agriculturist in choosing the desired crop species and then growing it in formally planned and obviously artificially-made communities. Pure, even-aged plantations of such trees as pine, spruce, rubber, apple, and the date palm represent the type of arboriculture where the forester and the horticulturist differ only in that one is primarily concerned with the growing of timber and wood fiber while the other is similarly involved in the growing of fruit. Since such plantation crops are frequently grown on limited areas of carefully chosen site conditions and since the value of the crops is usually high, management may be relatively intensive, and normal ecological trends may be countered successfully by fertilizers, herbicides, cultivation, and other measures. It does not follow, however, that a knowledge of these ecological trends and the consequences of herbicides, fertilizers, and cultivation is unnecessary to the plantation manager—quite the contrary.

At the other extreme, the "naturalistic" school of silviculture preaches that nature's ways are the best and the safest, and that the silviculturist should grow forests in harmony with natural trends.

While it is true that it is cheapest to sit back and allow the forest to develop in the absence of further interference, it does not necessarily follow that the greatest net return similarly will be realized either in tangible or in intangible values. The pines and Douglas-fir—indeed, most of our commercial timber species—require the aid of considerable effort by humans if high yields are to be maintained year after year despite continual harvest of the timber crop. The more drastic the departure from the patterns and processes of the natural ecosystem, the more knowledge and understanding one must have of the ecosystem.

So, as with so many other things, the middle course is generally the best. As Lutz (1959) sums the matter up:

> *In conclusion, it seems to me that the silviculturist should seek to understand ecological principles and natural tendencies as they relate to the trees and forest communities with which he works. I do not infer that he is bound to follow them blindly but neither do I think that he can safely ignore them. Between the two extremes of passively following nature on the one hand, and open revolt against her on the other, is a wide area for applying the basic philosophy of working in harmony with natural tendencies.*

To this we must also add that the consequences of our manipulation or noninterference must be considered by silviculturists, resource managers, and planners not solely on monetary or intangible benefits in the local ecosystem, but also on larger, interrelated terrestrial and aquatic systems. In most cases this means a change in our thinking to a new approach—considering not only the effects of our acts on a single stand but on the ecosystem of which it is a part and also on the larger systems that in turn may be affected. Our understanding of forests and streams as ecological systems and of the integrated functioning of all component parts must play a much greater role in the present and future than it has in the past.

# Part I
# The Forest Tree

# 2
# Forest Tree Variability and Diversity

Too often and for too long in the past the approach to the ecological foundations of silviculture has been strictly environmentalistic. Most treatments have completely ignored the genetic basis that is equally important in determining the nature of the forest. In a sense, forest ecology has been too much a direct descendant of pre-Darwinian environmental naturalism.

Life, whether in a forest tree or in any other organism, cannot exist without a governing biochemical control mechanism that can be passed on from generation to generation to perpetuate the species; and the organism cannot exist independently of an environment. It is futile to argue whether genetic or environmental factors control the form and development of an organism. Both always together determine the nature of the phenotype. Environmental factors, being generally visible and readily accessible, are the most obvious, and it is natural that most ecologists have been preoccupied with their study. Genetic factors have been fully appreciated only in recent years, and their assessment in forest trees has lagged far behind their study in smaller and more tractable organisms.

The ability of organisms to live and reproduce in a given range of environments, termed adaptedness (Dobzhansky, 1968), had been known since the days of Aristotle, and a convincing explanation was given by Darwin even though the causes of hereditary variation were then unknown. Darwin's theory of evolution by natural selection is an ecological theory based on ecological observations. It is important for the forest ecologist to consider adaptedness of forest species and their ability to adapt to changing environmental conditions. The long life and wide range of forest trees affect the adaptive strategies of many associated plants and animals.

In the present chapter, the primary concern is variation: the sources of variation; the kinds and extent of variation within and between individuals, populations, and species; and how the environment and biotic factors elicit adaptive changes in tree populations. Although the evolutionary framework is briefly described, no attempt is made to discuss in detail physiological and population genetics; they are well treated in a number of texts (Stebbins, 1950, 1970; Grant, 1963, 1971, 1977; Solbrig, 1970; Mettler and Gregg, 1969). Treatments of forest genetics and tree improvement include those of Syrach-Larsen (1956), Wright (1976), and Stern and Roche (1974).

## COMPONENTS OF PHENOTYPIC VARIATION

As indicated in Chapter 1, the ecologist may work with organisms at various levels of complexity—the individual, the population, the community, the ecosystem. In general, the individual is the least arbitrary of these units. The genetic constitution of an individual is termed the **genotype**. We can never see a genotype because from the moment of fertilization the genotype is influenced by the plant's environment, the internal environment of cells, tissues, and biochemical reactions, and the external environment of temperature, moisture, and light. We see only the result—the **phenotype**—the observable properties of an organism produced by the genotype in conjunction with its environment. We may express this relationship for the entire organism or for individual characters by the simple formula $P = G + E + GE$. The phenotype or phenotypic character *(P)* is the sum total of the effects of three components, the genetic information coded in the chromosomes *(G)* the environment *(E:* all nongenetic factors including those of the plant and its physical and biotic environment), and the interaction of the genotype and environment *(GE)*. Often the genotype-environment interaction is small and usually can be disregarded; in certain cases it is of special importance. Through complex pathways, the genes control physiological functions, and many of these influence the morphological and anatomical features of the plant. The interrelationship of the genotype, environment, and plant processes of the phenotype is illustrated in Figure 2.1.

A recurrent problem confronting the forest ecologist is the degree to which a phenotypic character is controlled by the genotype and by the environment. For example, to what degree are the cushion-like phenotypes of spruce trees at timberline due to wind, snow, and cold, and to what degree are they the result of genotypes that favor this growth form? If we set *P* in the formula equal to 100, do the environment and genotype contribute equally (50–50), or is one more important than the other? If so, how much? Or consider a dominant tree exhibiting excellent growth in a forest plantation and an adjacent suppressed and dying tree of the same species. The dominant has a superior phenotype, but to what degree is the genotype

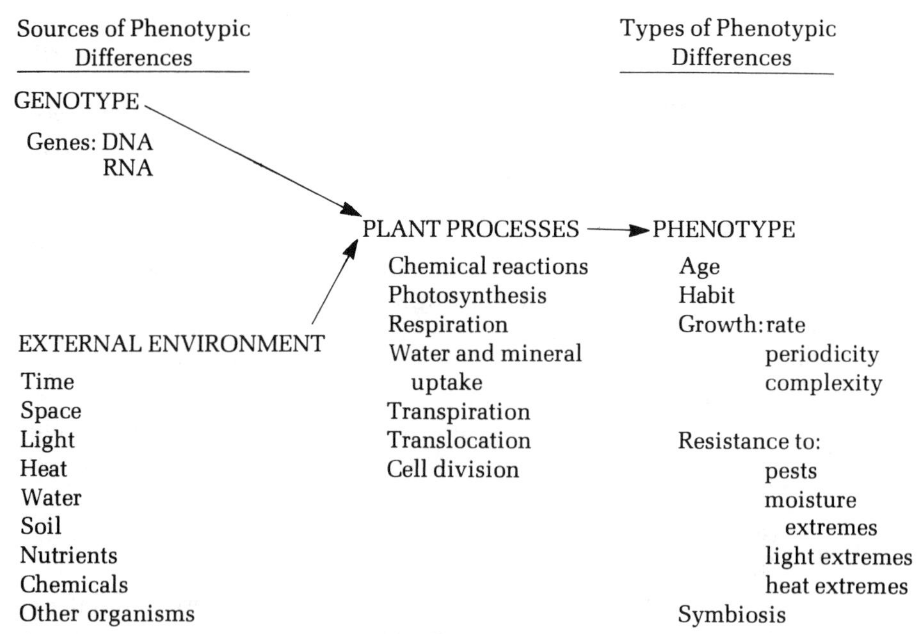


**Figure 2.1. Relationship of an individual's phenotype to its genotype and environment. Differences in plant phenotypes have their origin in the genetic constitution or genotype of an individual and in environmental factors. Phenotypic differences of the individual become apparent as physiological processes occur in the internal environment of plant cells, tissues, and organs. (Modified from a diagram by J.W. Hanover, Michigan State University.)**

responsible and to what degree does its microenvironment or such chance factors as better handling in the nursery account for the difference? Despite its poor phenotype the genotype of the suppressed tree may be superior to that of the dominant—or, of course, it may not.

The determination of the relative effects of genotype and environment is for the ecologist a recurrent and often unconscious feature in ecological studies. Both components are always involved and no two phenotypes are exactly alike. The important point is the relative degree of genetic and nongenetic influence on a given character.

Forest geneticists use statistical methods to compute genetic ($V_g$) and environmental ($V_e$) variances in the process of making a quantitative estimate of the strength of genetic control of a given character. Once these values have been determined, the total phenotypic variance ($V_p$) is equal to their sum ($V_p = V_g + V_e$). The strength of genetic control, termed **heritability**, of a character is then determined by using the ratio of genetic variance to the total phenotypic variance [heritability $= V_g / V_p$ or $V_g / (V_g + V_e)$]. If the ratio of the genetic to total variance is high (for example 80 percent), that is, a high heritability, it indicates a strong genetic control for the trait. Strong genetic control (high heritability) has been shown for bole straightness, stem-wood specific gravity, monoterpene composition in certain conifers, susceptibility to leaf rusts, and date of bud burst. Traits that are highly influenced by the environment, such as height (strongly influenced by soil fertility and moisture) and diameter (strongly influenced by density of the stand) are predictably under weak genetic control and usually have low heritabilities accordingly.

## Plasticity of the Phenotype

A given genotype may assume one set of characters (exhibit one phenotype) in one environment and exhibit a markedly different phenotype in another environment. The degree to which a character of a given genotype can be modified (a nongenetic change) by environmental conditions is termed its plasticity or, in general, **plasticity of the phenotype** (Bradshaw, 1965).

For herbaceous species Bradshaw cites as plastic characters: size of vegetative parts, number of shoots, leaves, and flowers, and elongation rate of stems. Nonplastic characters include leaf shape, serration of leaf margin, and floral characteristics. There is good reason to believe these findings hold true for most tree species. In general, characters formed over long time periods of meristematic activity, such as stem elongation, are more subject to environmental influences and are more plastic than characters like reproductive structures that are formed rapidly or than traits such as leaf shape, whose pattern is impressed at an early stage of development.

Plasticity may have substantial adaptive value to plants in general and tree species in particular since trees are rooted in their environment and have life spans typically longer than annuals or herbaceous perennials. An example of plasticity of adaptive significance is the rooting habit of individuals of many tree species, particularly Norway spruce, white spruce, and balsam fir. The roots may be either shallow or moderately deep depending on the soil environment, for example, a poorly drained swamp versus a sandy loam upland soil. If individuals of a species exhibit a high plasticity for certain establishment and growth characters, they may be able to establish and maintain themselves in a variety of habitats and as adult trees endure decades or even centuries of fluctuating climate. Such a mechanism tends to decrease the need for specially adapted genotypes or races (genetically distinct populations of a species), each one fit for a special microhabitat.

We can summarize the differences in phenotypes by examining three hypothetical situations involving individuals of a given species:

| Situation A | Situation B | Situation C |
|---|---|---|
| $P_1 = G_1 + E_1$ | $P_4 = G_1 + E_1$ | $P_7 = G_1 + E_1$ |
| $P_2 = G_2 + E_2$ | $P_5 = G_1 + E_2$ | $P_8 = G_2 + E_1$ |
| $P_3 = G_3 + E_3$ | $P_6 = G_1 + E_3$ | $P_9 = G_3 + E_1$ |

The phenotypes in A illustrate the typical situation in the field. All phenotypes have different genotypes, and the environments are also different enough to contribute to differences in the phenotypes. Situations B and C illustrate experimental situations where we can either hold constant the genotype (in B) or test different genotypes in a given environment (in C).

Situation B illustrates plasticity but would have to be related to a specific character. Different phenotypes of a single genotype ($G_1$) are the result of the environmental differences; modification has occurred. In nature, the degree of plasticity of a character cannot be measured precisely because each individual has a different genotype (as in A). The extent of environmental modification can only be inferred. For example, individuals of an even-aged stand in rolling terrain may occur from a dry ridge top to a moist, fertile valley. We observe a marked increase in height of the trees as we progress from ridge top into the valley. If it is unlikely that there are major changes in the genotypes along the gradient, we may infer that environment is the major factor controlling the phenotypic differences in height. To determine precisely the plasticity for representative genotypes we would have to conduct experiments based on model B.

In situation C we see that if the environment is the same for all individuals, phenotypic differences are due to differences among genotypes, and the amount of genetic variation can be estimated directly from the phenotypes. In practice the environment cannot be held constant. However, we may approach this ideal by using either growth chambers or relatively uniform field test plots and a replicated experimental design. This method is widely applied in determining genetic differences among selected individuals or populations.

## SOURCES OF VARIATION

As we have seen, variation among phenotypes is attributable partly to the genotype and partly to the environment. The major sources of genetic variation are mutation and recombination of genes. Mutation is the ultimate source of variation and in its broadest sense includes both changes in the molecular structure of genes at individual loci (gene mutation) and chromosomal aberrations such as duplications, deletions, inversions, and translocations. Gene mutations have the effect of adding to the pool of genetic variability by increasing the number of alleles (the different forms of a gene) available for recombination at each locus.

Continuous or polygenic variation is typical for most characters of trees. This is due to the simultaneous segregation and harmonious interaction of many genes affecting the character and the continuous variation arising from nongenetic causes. Only a few traits are controlled by a single gene with major effects. One prevalent trait controlled by one or a few genes is chlorophyll deficiency in seedlings of species of the Pinaceae (Franklin, 1970) and in many other conifers and angiosperms. The albino and yellow seedlings usually die soon after germination, but yellow-green types may turn green and survive in controlled environments.

Although mutation is the ultimate source of genetic variation, it is recombination that spreads mutations and extracts the maximum variability from them. Recombination is regarded as by far the major source of genetic variation of individuals in sexual systems. It makes available the raw material of variation which is acted upon by natural selection.

The exchange of genes between different populations is termed **gene flow** or migration and may also be considered a source of variation. Migrants, in the form of pollen and seed, bring to a population new genetic material from another population. When the populations involved are substantially different (such as species), the process is often termed **hybridization**.

The major sources of nongenetic variation are (1) the external or physical environment (biotic, climatic, and soil factors) and (2) the internal or somatic environment of the plant. Factors of the physical environment modify plants in many important ways and are discussed in Chapters 4 to 11. Much less appreciated is variation within individuals that is not directly related to factors of the external environment.

Although all cells of a tree have the same genetic constitution, the internal environment of the organism may affect the expression of genes and hence the traits we observe or measure. In the development of a seedling to an adult tree, striking physiological changes occur, and a series of developmental stages is recognized. Best known are the differences between the juvenile and adult stages. The characteristic features of these stages are apparently due to changes that take place in the apical meristems as they age. The most universal feature is the inability of trees in the juvenile stage to flower (See Chapter 3).

Schaffalitzky de Muckadell (1959,1962) investigated many other characters exhibiting differences in juvenile and adult stages. In European beech and oak, the brown and withered leaves are retained over winter by trees in the juvenile stage; they are not retained in the adult stage. This feature is also observed in American beech and many oaks. The entire portion of the lower trunk and branches even of very old trees may remain juvenile. Reciprocal grafting experiments of juvenile and adult branches show that the juvenile stage consistently leafs out later than the adult stage of European beech and ash. This juvenile trait may be of adaptive value since late spring frosts can pose a serious problem in young beech stands.

In many species, characters of adaptive importance are exhibited by juvenile forms. The presence of thorns protects young trees against animal damage, and greater shade tolerance increases the chances of a young plant staying alive in a shaded understory. Also, vegetative propagation is easier with juvenile material than with that from the adult stage.

## THE EVOLUTIONARY SEQUENCE

The starting point of evolutionary change is the formation of individuals with different genotypes. Without the heritable variation of these genotypes, natural selection cannot bring about evolutionary changes. Individuals, each usually a different genotype, may be grouped in populations for various reasons, and a general and nonrestrictive definition is preferred: a population is any group of individuals considered together because of a particular spatial, temporal, or other relationship (Heslop-Harrison, 1967).

In evolutionary biology, the basic unit in sexual populations is no longer the individual, which by itself has only a limited future; instead it is the local interbreeding population, a group of individuals that are potentially interbreeding and which constitute the gene pool for that population. Genes are sorted out and combined in genetically different male and female gametes at meiosis. Upon fertilization a new set of genotypes is produced, and the surviving phenotypes form the next generation having a new gene pool. Assuming the population is large enough (theoretically infinite), the frequencies of the respective genes will not change from generation to generation. They maintain an equilibrium or steady state, as shown by the Hardy-Weinberg law, and no evolution occurs. Evolution is the cumulative change in genetic makeup of populations of an organism during the course of successive generations—in simplest form a change in gene frequencies. Such changes are brought about by the guiding force of natural selection. Gene flow and genetic drift(chance effects that operate in small populations to make the new generation a random sample of the previous generation) may interact with natural selection to alter the way in which natural selection guides the course of evolution.

Natural selection is strongly associated with the nineteenth century naturalist's concept of differential mortality of individuals, Darwin's "survival of the fittest." Of the thousands or millions of zygotes of a species on a given site that might develop to maturity and contribute offspring to the next generation, only a few survive. Some individuals die as zygotes and fail even to develop as live embryos in seeds. Other ill-adapted plants fail to germinate or die soon after germination. Some individuals may die as a result of competition with existing vegetation for moisture and light. Diseases and insects may selectively take their toll. This is not to say that only the most highly adapted individuals survive, since chance factors are always operating. Some otherwise well-adapted seedlings may, by chance, become established in a moist microsite only to die later when the site dries out, whereas others may be eaten by animals along with less well-adapted types. Nevertheless, the continuous sifting and elimination of less well-adapted phenotypes occur throughout every phase of life, and are a very important part of natural selection. However, natural selection can take forms other than differential mortality of individuals, important as that may be. For example, surviving and otherwise fit plants may have differing abilities of reproduction. In certain tree species, vigorous individuals may be infertile year after year and may make little if any contribution to the succeeding generation. In addition, selection may act upon gametes as well as individuals. As a consequence the modern biologist defines **natural selection** as the differential and nonrandom reproduction of genotypes, stressing that ultimately it is the differential reproduction of genotypes, rather than just differential mortality of individuals, that brings about evolutionary change.

An example of the raw materials available to selection and the severity of selection comes from sugar maple. Curtis (1959) reported that of 6,678,400 potentially viable sugar maple seeds, 55.7 percent germinated, giving 3,673,100 seedlings per hectare. In late summer of the same year 198,740 remained, and two years later only 35,380 seedlings remained alive—less than 1 percent of those germinating. He estimated that the opening resulting from the death of a mature sugar maple tree (about 6 to 7 m in diameter) would initially support about 15,000 seedlings. They would be reduced to about 150 during the first three years, and eventually one or two trees would occupy the opening. Although selection acts in every phase of the plant's life, it is most effective on young seedlings and in many plant species eliminates all that are not well adapted to their immediate environment (Harper, 1965).

## Sexual and Asexual Systems

The sexual breeding system, dominated by cross-pollination, characterizes most woody species. Seedlings from self-pollination are consistently slower growing than outcross seedlings of the Pinaceae (Franklin, 1970) and are at a selective disadvantage in most natural environments. Evolution proceeds only in a sexual system where new genotypes are constantly generated. Nevertheless, most woody species have some form of asexual reproduction or **apomixis**—any means of reproduction, including vegetative propagation, which does not involve fertilization. In angiosperms, this includes sprouting from stems (oaks, chestnut) and roots (aspens, sweet gum, beech) following fire or cutting. Conifers rarely form sprouts (an exception is redwood), but the production of adventitious roots from branches (termed **layering**) in larch, white spruce, and firs is an important device in swamp or moist upland environments. Apomixis may even occur when seeds are formed asexually as in the hawthorns. Whatever the mechanism, asexual reproduction gives the plant immediate fitness in the prevailing environment; it is a uniformity-promoting device. A striking example is seen in the aspens of North America and Eurasia, which sucker from roots to form natural colonies termed **clones**. A clone is the aggregate of stems produced asexually from one sexually produced seedling. The clone is the typical growth habit of aspens throughout their worldwide range (Barnes, 1967), and their asexual proclivity may be the major factor in their ability to compete successfully in conifer-dominated landscapes. In parts of western North America, aspens occur in clones, some over 40 ha in size (Kemperman and Barnes, 1976), and they may live for indefinite periods by recurrent suckering, barring major climatic changes.

## GENECOLOGY

The foregoing review of the causes of changes in gene frequencies in populations leads to a consideration of genecology, a term the Swedish ecologist Turesson (1923) applied to the study of the variability of plant species and their hereditary habitat types from an ecological point of view. More specifically, genecology is the study of adaptive properties of any sexual population—race, species, subspecies, local interbreeding population—in relation to its environment (Langlet, 1971). Turesson, like others before him, demonstrated conclusively that ecologically correlated phenotypic variation among populations was typically genetically based rather than merely the result of environmentally induced modification of individuals. This concept has major practical implications for the silviculturist when introducing populations into a new environment. In tree-introduction attempts throughout the world, we have largely used the trial-and-error method—with some resounding successes and some dismal failures. Such a method is too arbitrary, costly, and time consuming as a general practice. Instead, we need to be able to predict how a given population will perform when grown in a new environment. To do this we need to know what environmental and biotic factors elicit a genetic response, how closely populations are adapted to these factors, and the patterns of adaptation along major environmental gradients. The basis of our knowledge of genecological adaptation and examples of it are presented below.

Comparative cultivation of seedling populations of forest trees, originating from environmentally different sites, was pioneered by Duhamel du Monceau about 1745 (Langlet, 1971), and the methods were continued and refined by other workers such as P. de Vilmorin (in the 1820s), Kienitz (1879a,b), Cieslar (1887–1907), Engler (1905–1913), among others (Langlet, 1971). The careful historical documentation by Langlet (1971) makes it clear that the concept of genecological diversity, originating from an adaptational response to the environmental factors of the particular habitats, was known to forest botanists long before Turesson's work with herbaceous species. For example, Cieslar in Austria (1895, 1899) and Engler in Switzerland (1905, 1908) determined experimentally that forest trees in the Alps were genetically adapted to the climatic conditions of their respective environments. In 1895 Cieslar published evidence of a continuous gradient of juvenile height growth for Norway spruce demonstrating its genetic adaptation to growing-season conditions grading from low to high altitude. These results served to document observations of the previous century, published in 1788, that seedlings of lowland provenances proved worthless on mountain sites.

Unlike Cieslar and Engler, who used seeds for their experimental work, Turesson transplanted whole individuals from markedly different habitats but, like his predecessors, grew them under standard conditions of cultivation. This method of comparative cultivation is often termed the **common garden technique**. The phenotypes Turesson observed in nature were usually different in habit of growth (procumbent or erect) and in various morphological characters. These differences were usually maintained in the garden and hence indicated genetic differences between the populations studied.

The causes of the adaptive differences may be inferred by correlating the performance of the populations in the garden with the environmental and biotic factors of the original sites. Presumably, natural populations have been exposed to the factors of their respective environments for generations. Forces of natural selection have guided the genetic differentiation of each population so that it is more or less adjusted to the daily, seasonal, yearly, and even longer-term climatic, soil, and moisture fluctuations of its respective environment.

Because of the problems of preconditioning of whole plants (Rowe, 1964) and the difficulty in transplanting whole forest trees, forest scientists typically collect seeds from the desired populations, termed **provenances**, raise the seedlings in a common garden, and study the differences among the provenances. Such experiments are termed provenance, or seed source, tests. This type of testing determines (1) if there are significant genetic differences between populations in the characters chosen for study, and (2) the amount of genetic differentiation among provenances under the environmental conditions of the common garden. Provenance testing does not directly indicate what mechanisms caused the differences, although these may be inferred. Nor does it indicate whether such differences would exist or be of the same magnitude at another test site.

## Patterns of Genecological Differentiation

A wealth of evidence has accumulated confirming that genetically based ecological differentiation or divergence of populations, termed genecological differentiation, is a recurrent feature of plants in general. However, controversy has arisen over the pattern of differentiation—whether it is discontinuous or whether it is continuous or clinal in nature. Huxley (1938, 1939) introduced the term **cline** to designate a gradation in measurable characters, which might be continuous or discontinuous, stepped or smooth, or sloping in various ways. The term itself, as the definition indicates, does not mean or necessarily imply a genetically based gradation in a character. It could refer to a gradation of phenotypic characters observed along a natural gradient. If, however, a gradation in characters were found for populations in a common garden test, an adaptive cline would be demonstrated.

Genecological differentiation is a multidimensional response of individuals of a population to their environment. Although the response is unique for each species and population, we present the following generalizations as best summarizing current understanding of differentiation in forest species.

1. The total range of a species, the distribution pattern (continuous, discontinuous, mosaic) of a species within this range, and the way in which the conditioning environmental factors vary are three major determinants of the differentiation pattern. If a species is distributed continuously over a wide range, particularly in latitude or elevation, it is subjected to more or less continuously varying climatic factors, and adaptive variation tends to be continuous. If discontinuities occur in the species distribution or if the conditioning factors are discontinuous and sufficiently distinct, a discontinuous pattern may result. The variation pattern may be visualized as a series of contour lines whose spacing reflects the rate of change in the conditioning factors.
2. The results of a given genecological study tend to be related to the scale in which it is conceived and conducted (Heslop-Harrison, 1964). A wide-ranging investigation of a species along a north-south gradient may expose a clinal variation pattern that may mask other clines associated with elevation at a given latitude, or local discontinuities that might arise from a major change in soil type or soil-moisture conditions.
3. However continuous a cline may be, it is usually possible to show a seeming discontinuity by incomplete sampling and certain methods of data analysis (Langlet, 1959).
4. The dominant pattern of genetically based variation is more or less continuous; the discontinuous pattern is usually the exception. The clinal pattern has a fundamental basis in the genetic system of many forest trees favoring a high degree of recombination and outbreeding (cross-pollination) and the associated features of (a) long life span of individuals, (b) greater habitat and community stability, (c) high and selective seedling mortality, and (d) high physiological tolerance in the adult to fluctuating environmental conditions. Discontinuous variation, being favored by inbreeding, low rates of gene recombination, short life span in ephemeral communities, and strongly fluctuating habitats, is more typical in herbaceous than in tree species.

These and other evolutionary aspects of differentiation are discussed by Heslop-Harrison (1964).

## The Ecotype Concept

Turesson (1922a,b) defined **ecotype** as the product arising as a result of the genotypical response of a population to a "definite habitat" or "particular habitat." Controversy and confusion have arisen over the distinctness of ecotypes and over the related feature of their size—whether they are local or regional phenomena. The term ecotype is not only abused when applied in situations where genetic differences have not been proved, but it is also used in a number of different senses, each having a different genecological significance. Thus the convenience of classification by types is offset by the inadequacy of types to treat continuous variation.

The problem is traceable to two features of Turesson's ecotype concept. First, the definition as a genetical response of a species to a particular habitat has been interpreted to mean both local differentiation in specialized habitats (meadows, swamps, shifting sand dunes) and large-scale differentiation, such as climatic races embracing large portions of a species' range. Examples of the latter type are those described by Wells (1964a,b) for ponderosa pine (Figure 2.2). Each regional ecotype occupies a large geographic area within which local differentiation undoubtedly occurs. For even a small part of the California ecotype, marked clinal differentiation along an elevational gradient has been demonstrated (Callaham and Liddicoet, 1961).

Turesson himself used "ecotype" not only to characterize local genecological differences but also in a much wider sense—"alpine ecotype," "coastal ecotype," and so on—applying to major habitat complexes. This usage implies that the term indicated populations adapted to types of environments and not to particular or special environments. Thus the term ecotype is ambiguous and fails to provide a useful concept except for the general fact of genetic adaptation to environment.

Second, Turesson's methods of sampling and cultivation culminated in his view of a species as a mosaic of populations, each adapted to a distinct habitat. Each ecotype was shown to be morphologically different and distinct from others. Turesson's work in demonstrating ecotypes naturally led him to sample populations from distinctly different habitats, and the genetically based differences led him to stress the discontinuity of ecotypes. Subsequent investigations have indicated that the discreteness of ecotypes was exaggerated (Gregor and Watson, 1961). In general, by incomplete

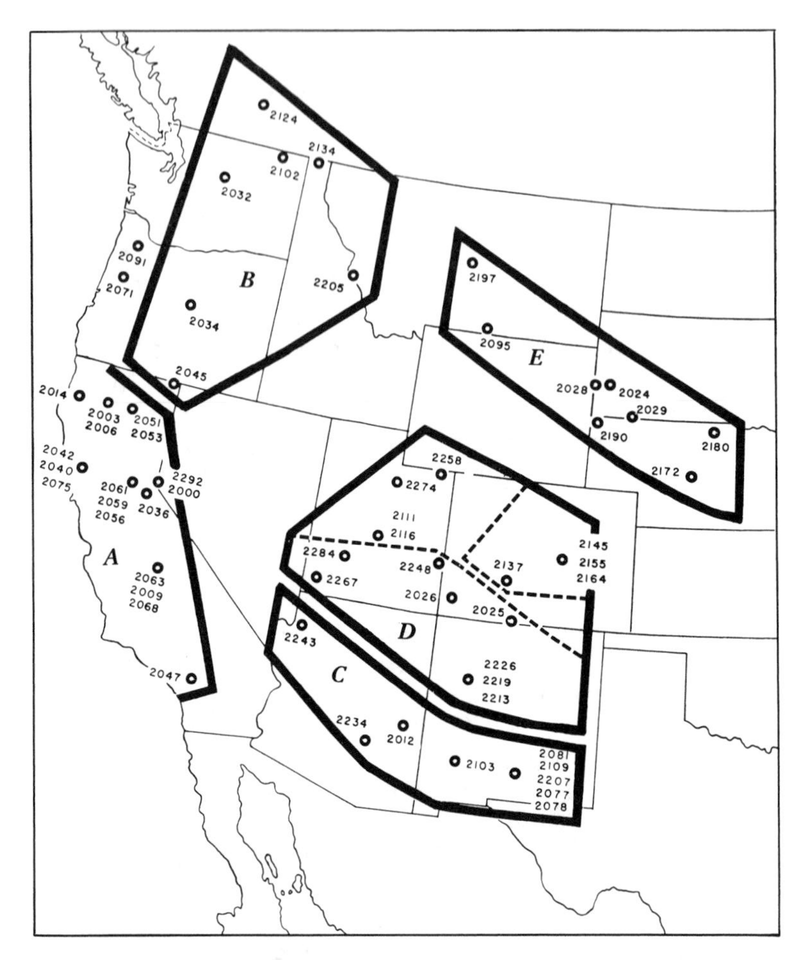


**Figure 2.2 Geographic ecotypes of ponderosa pine, based on 60 stand progenies grown for 2 years in East Lansing, Michigan: (A) California; (B) North Plateau; (C) Southern Interior; (D) Central Interior; (E) Northern Interior. (After Wells, 1964a. *Silvae Genetica* 13, Sauerländer, Frankfurt.)**

sampling or, as in Turesson's work, by ordering and analyzing the data in groups a seeming discontinuity may be demonstrated no matter how continuous the cline. Thus whenever the term ecotype is used, the user's precise definition of the term always should be sought. The trend today, however, is away from classification of populations by ecotypes and toward the study of the patterns of genecological differentiation in a species and of the mechanisms evoking genotypical response. Such patterns for the selected species described below illustrate the relative continuity and scale of adaptation and the kinds of physical and biotic factors of the environment evoking a genetic response.

## Examples of Genecological Differentiation

The woody-plant literature abounds in references citing genetic differences in morphological and physiological characters among populations in a great variety of species. Much of the information comes from provenance tests established to meet the practical objective of finding the most suitable provenances for planting in the general area of the test site. Many of these tests are therefore not designed and cannot be expected to answer, except in a very general way, genecological questions of how and why populations are adapted to their environments. The term geographic is almost invariably associated with the literature surrounding these tests. The provenances are usually stratified by geographical regions, political units (country, state, or county), or varietal rank; genetic distinctions are then shown between these subdivisions. Ecological data on the origins are often sketchy, and genecological interpretations, if present, are as a result incomplete.

Despite inadequacies of many tests from a genecological standpoint, far-reaching provenance studies have uncovered major adaptive responses, primarily along latitudinal and altitudinal gradients. Scots pine, the most wide-ranging pine, has been investigated in many studies in Europe (dating from 1745) and in North America. Genetic differences in such diverse characters as height growth, foliage color, dry-matter content of seedlings, stem form, rooting habit, resistance to insect attack, fruitfulness, tracheid length, and time of bud set have been demonstrated (Langlet, 1959; Troeger, 1960; Wright et al., 1966; Wright et al., 1967). A clinal pattern is evident for most of these characters.

Portions of a cline, sometimes designated as "ecotypes" or "varieties," may be useful as the basis for selecting seed-collection zones, particularly where major differences in tree characteristics are important. American Christmas tree growers, for example, prefer Scots pine varieties from Spain and elsewhere in southern Europe, which remain green in winter, to those of northern Europe which turn yellow (Wright et al., 1966). The use of "varieties" in specifying such seed-collection zones may be convenient but may also convey a false impression of uniformity within the named group. The fact is that populations within a specified geographic area may be quite diverse. Obtaining seed from within the geographical boundaries of the "variety" seemingly best suited to a planting area is satisfactory only if the populations within the "variety's" boundaries are reasonably uniform.

**Environmental Factors Eliciting Ecological Differentiation.** Marked genetic differences in growth and other characters usually are expressed when populations are grown at latitudes or elevations substantially different from that of their native habitats. Limiting environmental factors affecting the length and nature of the growing season in the native habitat (such as temperature, thermoperiod, photoperiod, and amount and periodicity of rainfall) are important selective forces acting on growth rate and related characters.

Associated tree species and animals may play an important role in the evolution of some characters. The crooked stem form of lowland Scots pine sources in Germany may be attributed in part to competition of pine with European beech and oak. The hardwood species are highly phototropic[1] and probably exert a much different selection pressure on juvenile pines than that encountered at high elevations with spruce and fir.

Coevolutionary systems have been reported for animals and reproductive traits of various woody-plant species. In studies of woody legumes in Central America, Janzen (1969) listed 31 traits that may act to eliminate or lower the destruction of seeds by bruchid beetles. The major defense mechanisms against these predators are deterrents, such as biochemical repellents (alkaloids and free amino acids), or an increase in the number of seeds to the point of predator satiation, probably requiring a decrease in seed size. Another important device is a dispersal system that effectively disperses seeds soon after maturation. In western North America, squirrels may act as important selective agents on the reproductive characters of conifers in the process of maximizing their own feeding efficiency (Smith, 1970, 1973). These studies emphasize the multidimensional nature of adaptation. Not only are characters influenced by many factors of the physical environment, but also by plant associates and the interrelated selective pressures of insects, mammals, and birds (see Chapter 13).

Survival and growth are closely related to the efficient use of the growing season. Plants must not grow too late in the fall or they will be damaged or killed by early fall frosts. They must not cease growth too soon because longer-growing individuals may overtop and suppress them. In their native habitats, species anticipate seasonal fluctuations by responding genetically to the more reliable factors of the

[1]Response of a plant shoot to turn or grow toward light, which is the orienting stimulus.

changing environment (such as day length) than to more variable factors such as frost and drought. In temperate regions, response to a photoperiodic signal of shortening days of summer and fall sets in motion a gradual and complex pattern of development which we call dormancy (Chapter 5). **Photoperiodism,** the response of plants to the timing of light and darkness (usually expressed as day length), is a biological clock enabling plants to adjust their metabolism to seasonal fluctuations. Unlike other environmental factors, day length changes everywhere in a regular yearly cycle, except at the equator. Plants apparently can measure these changes with remarkable precision.

Photoperiod largely controls the entrance into dormancy of many woody plants, particularly species with a northern range. These species are genetically adapted to a photoperiod that enables them to become dormant before the time when particular factors of their prevailing environment, such as cold or drought, become limiting. For trees in northern climates or at high elevations, early frosts in autumn and cold winters are factors of primary survival value. Hence a reliable mechanism, such as photoperiod, in triggering the dormancy sequence, may be highly developed. In other sites, moisture stress during the midst of the growing season may be a severe limiting factor. An adaptive system may evolve to promote dormancy and yet allow growth to resume again when moisture conditions are favorable.

In almost all genecological and provenance tests, populations are grown in day-length regimes different from that prevailing in their native habitats. In black cottonwood, for example, individuals of high-latitude provenances ceased height growth in June when planted at a low-latitude site near Boston, Massachusetts (Figure 2.3; Pauley and Perry, 1954; Pauley, 1958). Southerly provenances, moved north to the test site, continued height growth until September and October; some individuals ceased growth only when their terminal shoots were killed by the first severe frost. Generally, movement from the natural habitat northward, into longer days, prolongs the active period of growth and results in greater plant size. However, it renders the plant susceptible to early frosts. Movement southward, into shorter days, shortens the active growth period in comparison to plants native at or south of the test site. In the black cottonwood examples, individuals of high-latitude provenances grew only about 15 to 20 cm, whereas those from southern localities grew about 2 m (Pauley, 1958). When clones from the high latitude at the test site were given longer days by artificial light, they grew over 1.3 m (Pauley and Perry, 1954), indicating a strong influence of day length in regulating growth.

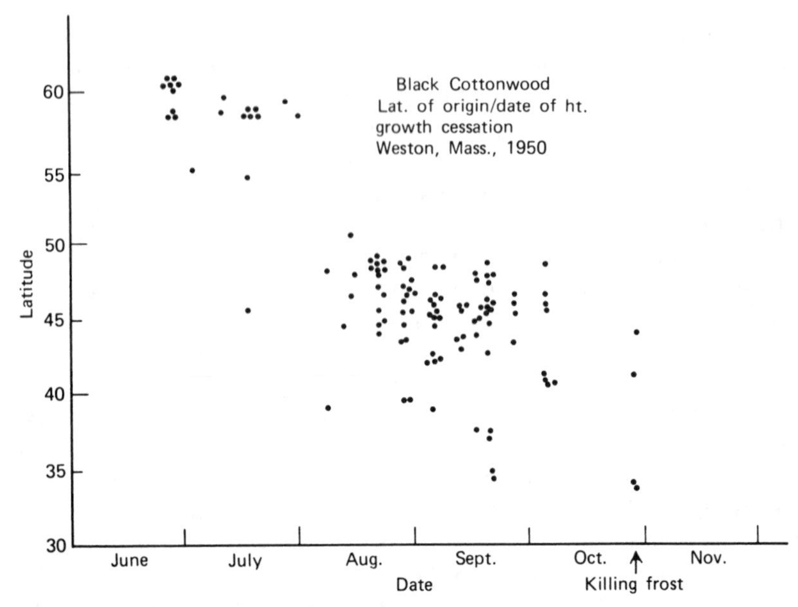


**Figure 2.3. Relation between latitude and time of height-growth cessation in black cottonwood. (After Pauley and Perry, 1954; from Kenneth V. Thimann, *The Physiology of Forest Trees* © 1958, The Ronald Press Company.)**

Although a significant, genetically based, clinal response was shown in relation to latitude, the response is not simple and direct, as evidenced by substantial variation among provenances from 44 to 48 degrees (Figure 2.3). This group included a variety of sources sampled from the Pacific coast to western Montana and over an elevational range from sea level to 1525 m. Although the difference in latitude is not great, there is known to be a marked difference in the length of growing season among these sources due to elevation, aspect, or microsite conditions. Clinal genetic adaptation to length of growing season was found within the narrow latitudinal range of 45–47 degrees (Figure 2.4) and probably explains much of the variability not accounted for by latitude. Thus populations at low and high elevations at a given latitude have growing seasons of different lengths and become adapted to different photoperiods accordingly, and in particular a photoperiod in autumn which is important in regulating their entrance into dormancy. High-elevation populations necessarily cease growth earlier than low-elevation populations due to earlier occurrence of killing frosts. Hence they adapt to relatively

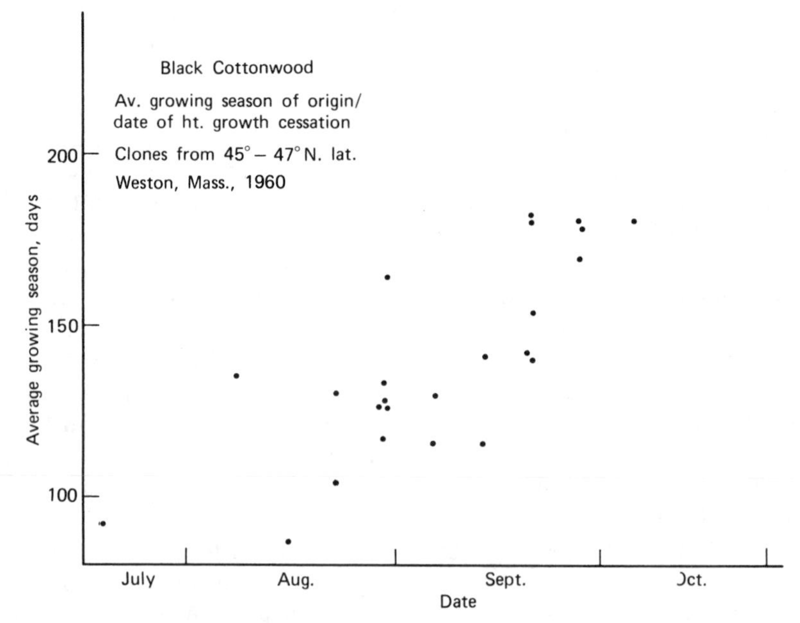


**Figure 2.4. Relation between length of growing season and time of height-growth cessation in black cottonwood. (After Pauley and Perry, 1954; from Kenneth V. Thimann, *The Physiology of Forest Trees* © 1958, The Ronald Press Company.)**

longer day-lengths (occurring earlier in the year) than those at low elevations.

Carrying this approach one step further, we see that a population at a high elevation may have the same length of growing season as a population at a lower elevation but several degrees of latitude farther north. Through equivalence in length of growing season, they would have a similar photoperiodic adaptation mechanism and, if interchanged, may show only negligible differences in date of growth cessation and growth rate.

This interrelationship between elevation and latitude has rarely been recognized in genecological studies. Almost without exception, correlations of cessation of growth or plant size and latitude of source are confounded by elevational differences. However, Wiersma (1962) modified a formula developed for Swedish conditions, using growing season (number of days ≥ 6°C) to relate latitude and elevation. He reported that a displacement of 1 degree of latitude north is equivalent to a displacement of 100 meters upward in altitude. Wiersma (1963) recomputed correlations of latitude of source and various characters from published papers using this adjustment and found greatly improved relationships. Sharik and Barnes (1976) computed a corresponding value based on length of growing season (number of days $\geq 0°C$) for the Appalachian Mountains from New Hampshire to North Carolina and found a relationship of 1 degree of latitude equal to a displacement of 189 meters. Adjustment of latitude by this factor substantially improved the correlation of latitude of origin and cessation of height growth for yellow birch and black birch populations, compared to the unadjusted relationships.

Genetic differences in plant size and growth cessation for numerous hardwood and conifer species have been related to latitude (Kriebel, 1957; Wright and Bull, 1963; Genys, 1968; Fowler and Heimburger, 1969; Sharik and Barnes, 1976; Wearstler and Barnes, 1977) and elevation of the source (Cieslar, 1895; Callaham and Liddicoet, 1961; Genys, 1968; Hermann and Lavender, 1968). These studies indicate that photoperiod is a timing device of major adaptive significance. The consistency with which given individuals of many species cease growth from year to year reinforces this conclusion. The clinal variation patterns of some growth traits are surprisingly regular despite interruptions of a species' range (Scots pine, for example) and apparently reflect the strong adaptation of populations to the highly precise photoperiodic cycle.

The close association of various adaptive responses with different limiting factors of the native environment has been shown for Douglas-fir seedlings grown near Corvallis, Oregon, by Irgens-Moller (1968). The late cessation of growth of coastal provenances (Vancouver Island) at the test site is related to the long growing season of their native habitat. In the northern Rocky Mountains, low summer precipitation and a short frost-free season apparently are responsible for early onset of dormancy of their provenances at Corvallis. Early dormancy was displayed, although soil moisture was kept in ample supply, indicating the lack of a direct effect by moisture stress. Photoperiod was again shown to be important since only long photoperiods could keep the plants actively growing. Sources from Arizona and New Mexico grew intermittently. They entered a short period of dormancy after which the majority resumed growth before entering winter dormancy. The distinct intermittent growth in the southwestern provenances and its lack in northern Rocky Mountain provenances may be explained by differences between the two areas in seasonal distribution of precipitation. A relatively high summer rainfall is received in Arizona and New Mexico as compared to Northern Idaho (64 percent of total annual rainfall compared to 29 percent). The intermittent growth may permit seedlings to go into early dormancy during periods of soil moisture stress and then resume growth quickly when moisture is abundant. Seasonal distribution of precipitation also may be an important factor affecting the adaptation patterns of ponderosa pine (Squillace and Silen, 1962) and slash pine (Squillace, 1966a).

The rate of change in a clinal variation pattern is well illustrated in an intensive study of the phenotypic and genetic variation of seedling slash pines by Squillace (1966b). This study revealed weakly defined or highly fluctuating gradients as well as distinct clinal trends. The reversal in the general cline, common to many of the 25 characters studied, is well demonstrated in the variation pattern of needle length (Figure 2.5). From a low of 16 cm in southernmost Florida, needle length of the progenies increased to its longest values, 19 to 20 cm, in south central Florida and then progressively decreased to the north.

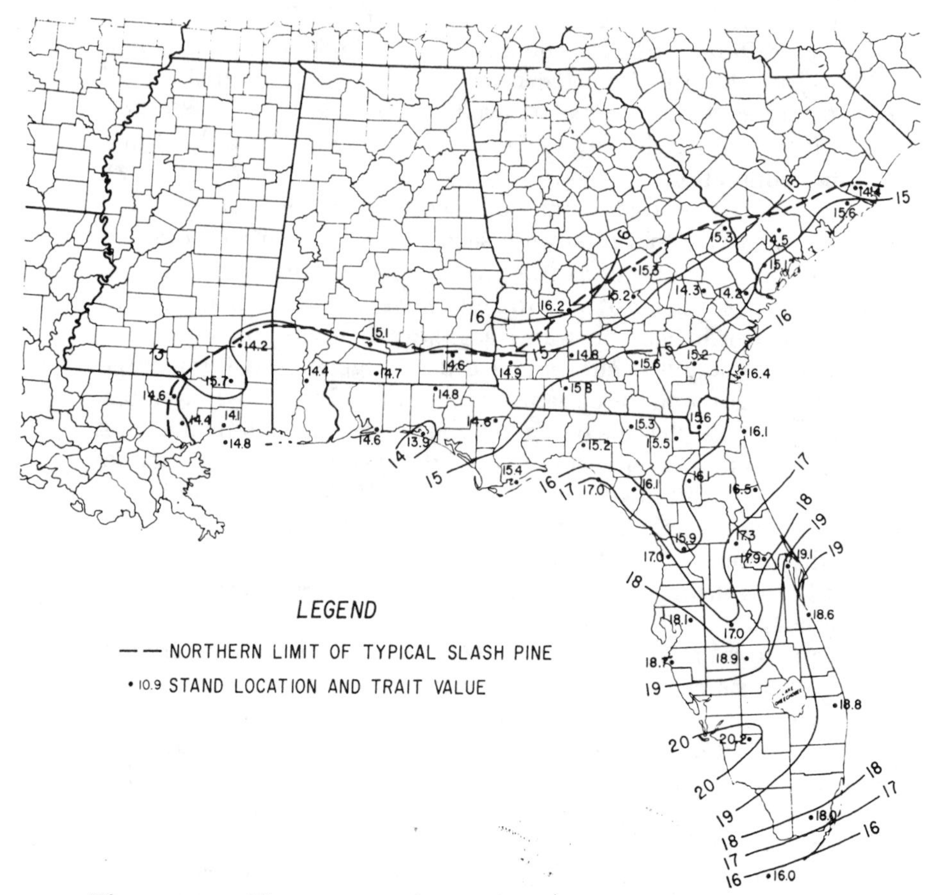


**Figure 2.5. The pattern of variation of needle length (cm) in seedling progenies of slash pine. (After Squillace, 1966b.)**

**Local Diversity.** Natural situations may present local variation in soil and topography that cause specialized environments. Local adaptation to soil and climatic conditions within major climatic regions has been demonstrated repeatedly for herbaceous species (Heslop-Harrison, 1964; Jain and Bradshaw, 1966; Kruckeberg, 1969). The physical and chemical properties of soils can elicit sharp discontinuities in plant distribution, and woody-plant species showing marked edaphic preferences are not uncommon. However, infraspecific genetic adaptation to the local soil type or to soil-moisture differences has rarely been demonstrated. Although Kruckeberg (1967, 1969) demonstrated genetically different races of several herbaceous species on serpentine and nonserpentine soils, such differentiation was not immediately apparent for knobcone pine and digger pine seedlings originating from parents growing on serpentine and nonserpentine soils (McMillan, 1956; Griffin, 1965).

Genetic differences have been demonstrated between a few populations of northern white cedar from well-drained upland sites and those from frequently inundated swamps (Habeck, 1958; Musselman et al., 1975), although they have not been proven for upland and lowland populations of black spruce (Fowler and Mullin, 1977). Habeck found that root length of seedlings of the upland provenances was twice that of swamp seedlings grown in well-drained soils, but only about half as long in a poorly drained soil. Besides illustrating genetic sensitivity to soil-moisture conditions, this differential performance illustrates a genotype-environment interaction. An interaction occurs when different genotypes (or populations) do not respond in the same manner in different environments. It may often mean that the best genotypes in one environment are not the best in another environment. Genotype-environment interactions have often been reported (Squillace, 1970) and are to be expected whenever genetically diverse populations are grown in sites markedly different from their native habitats.

Localized genetic differences were reported by Squillace and Bingham (1958) for western white pine trees growing on a dry and on a moist site (0.8 km apart, 15-m difference in elevation). Differential germination percentages of the respective populations under simulated dry and moist conditions prompted this conclusion. They speculated that differences in seedbed moisture occurring between opposing north and south slopes might cause differences in selection pressures strong enough to evolve local races. Such aspect differences have been demonstrated for Douglas-fir seedlings originating from north and south slopes in southern Oregon (Hermann and Lavender, 1968). In growth chamber experiments, north slope seedlings from 458-m, 915-m and 1525-m elevations had a longer growing season and greater dry weights of shoots and roots than seedlings from south-aspect parents of similar elevations. Campbell (1976) has also demonstrated substantial microenvironmental adaptations of Douglas-fir in the western Oregon Cascades. Additional and even more detailed investigations are needed to determine just how closely populations are adapted to aspect, soil-moisture differences, and soil type. Through longer-term field testing, we will also learn how small a difference in environment will require specially adapted genotypes.

## Factors Affecting Differentiation: Gene Flow and Selection

The extent of differentiation depends on the amount of gene flow via pollen, seeds, and other propagules between populations and on the intensity of selection. Gene flow is a cohesive force acting to keep populations from diverging, and its role in differentiation was reviewed by Ehrlich and Raven (1969). Isolating factors such as spatial distance between populations, or ecological isolation (south versus north slopes, swamps versus uplands) act to disrupt gene flow. Selection is also important. Populations will tend to remain similar if subjected to similar selection forces but will differentiate if they are not. Over short distances gene flow is likely to be great and differentiation seems therefore unlikely. When it does occur in these situations, it is brought about chiefly by the impact of high selection pressures. First let us consider gene flow and then an example of intense selection.

Three important factors acting to restrict gene flow are (1) limited number of breeding trees, (2) differences in flowering times of individuals, and (3) limited dispersal of pollen and seeds. It is well known that there are marked differences among species in the age of seed bearing and periodicity of seed crops (Daniel et al., 1979). Furthermore, within a species trees vary greatly in their reproductive capacity; some are highly fruitful, some moderately so, and others are completely barren year after year. Thus of the many trees in a population that could potentially exchange genes, only a few breeding trees may contribute appreciably to the next generation. For example, ple. Schmidt (1970) found that in an 89-year-old Scots pine stand 30 percent of the trees produced 71 percent of the female strobili and 64 percent of the male strobili. For sugar maple Wright (1962) estimated that if there were 62 large fruiting trees per hectare, only 5 to 7 on the average would be breeding trees.

The time of pollen release and female receptivity (phenology of flowering) is vital in pollination and is closely related to air temperature and humidity. Trees within a given population are more likely to be synchronized with one another than with trees progressively farther away. Ordinarily the time of flowering is earlier at more southerly localities than at northerly latitudes, and at lower altitudes compared to higher altitudes. However, high-altitude variations in regional and local temperature and the advent of spring make the situation quite different in different years. For example, overlap in flowering time of Scots pine between southern and central Finland was observed to occur in four years in the 10-year period 1957 to 1966; in one year overlap was found between central Germany and central Finland (Koski, 1970). We may conclude that, although the timing of flowering on the average favors local gene exchange, the possibility of gene flow from distant sources definitely exists.

Although viable pollen may be transported many miles, it may reach another stand too early or too late to compete effectively with local pollen. This becomes even more important in Scots pine and probably other pines and conifers since the capacity of the pollen chamber is limited (Sarvas, 1962), and all grains do not have an equal chance to fertilize the eggs. Of the many pollen grains reaching the micropyle only two, on the average, have the opportunity to fertilize the eggs of each ovule, one of which eventually develops into the embryo. Because of the greater probability of their being first to reach the micropyle, pollen grains of neighboring trees may have a higher probability of achieving fertilization than those of trees at more distant sites.

In forest trees, the range of seed dispersal is limited although there are exceptions in the willows, birches, and poplars. A mechanism favoring heavy seed has even evolved in island species, assuring that most of the seeds are not literally blown into the sea. In contrast, we know that pollen can be carried great distances (Lanner, 1966). Andersson (1963) reported that pollen was blown from Germany to southern Sweden, a distance of 72 km, and that in Sweden one year the pollen crop was so heavy that clouds of pollen were mistaken for forest fires. The pollen-dispersal distance of Scots pine is at least in the tens of kilometers, and its transfer 600 to 700 km in 10 to 12 hours has been reported (Koski, 1970).

The many reports of widespread dispersal might lead to the conclusion that populations over large areas are prevented from diverging because of widespread gene exchange. However, accumulating evidence indicates that in wind-pollinated species (most north temperate species) most individuals are pollinated and fertilized by their neighbors of the surrounding stand or from adjacent stands. A study of Norway spruce showing effective pollination and fertilization within a 40-m radius led Langner (1953) to conclude that fertilization of a given tree in a forest stand is effected by its immediate neighbors. However, detailed studies of pollen dispersal in extensive stands of Scots pine in Finland indicate that only half the pollen comes from trees growing less than 50 m away (Koski, 1970). The remainder may come from trees more than 50 m away in the same stand or from adjacent stands. Although long-distance transport of pollen occurs on a large scale, most of the pollen from afar ordinarily overflies the forests at an altitude of several hundred meters and only a small fraction of the grains that diffuse into the lower regions can participate in pollination in a stand. Due to the great difficulty of monitoring flow and destination of pollen from individual trees and from different stands, and considering the vagaries of environmental and biological factors, we can expect no generalized answer to the question of the extent of gene exchange. At times gene flow may be restricted to nearest neighbors, favoring inbreeding, whereas at other times some gene exchange may occur over considerable distances, thus enriching the gene pool of receptor populations. It is against this background of gene flow that natural selection guides the genetic make-up of populations.

The evidence available from many factors suggests that gene flow is limited in varying degrees in population systems. Thus selection pressures of the particular environment play a significant role and in herbaceous species may be effective over a distance of 2 to 4 m (Aston and Bradshaw, 1966). The effect of intense selection was strikingly demonstrated by clinal changes in glaucous and nonglaucous (green) phenotypes of *Eucalyptus urnigera* in Tasmania (Barber and Jackson, 1957). Green phenotypes are typical of low elevations and more sheltered habitats, whereas glaucous individuals become more frequent in increasingly exposed environments along a gradient from low to high elevation. The change from glaucous to green types was essentially complete over a vertical distance of 122 to 152 m (0.8−1.6 km ground distance) in the adult populations. Glaucous seedlings may be produced from nonglaucous mother trees and vice versa, indicating gene flow via insect and bird pollinators. Nevertheless, intense selection eliminates glaucous seedlings as the forest matures in the lower elevations. At the higher elevations green seedlings are eliminated. The authors emphasize that strong selection can build up great genetic diversity even in the face of considerable gene flow.

## ECOLOGICAL CONSIDERATIONS AT THE SPECIES LEVEL

As populations change through time and radiate into new areas, they typically become increasingly diverse in response to genecological differentiation. Also, due to geographic separation and other isolating mechanisms, some populations become less able to exchange genes with other populations; they become reproductively isolated in varying degrees. An important route of speciation involving geographic isolation followed by reproductive isolation is shown in Figure 2.6 (Stebbins, 1966). The isolating factors which act singly or, more usually, in combination, are presented in Table 2.1. The vari-

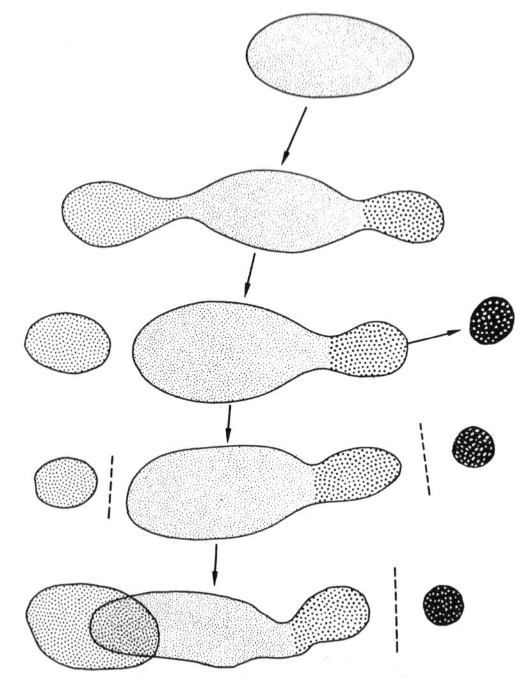


**Figure 2.6. Diagram showing the sequence of events which leads to the production of different races, subspecies, and species, starting with a homogeneous, similar group of populations. (After G. Ledyard Stebbins, *Processes of Organic Evolution* © 1966, p. 86. Reprinted by permission of Prentice-Hall, Inc., Englewood Cliffs, New Jersey.)**

ous degrees of population differentiation (Figure 2.6) have been classified in a hierarchical system. Populations exhibiting minor differences are termed races, local or geographic. Populations that have more substantial differences in morphology and physiology, and are reproductively isolated, are classified as subspecies or species. These two classes are formal taxonomic designations while race is not.

The assumptions that different species are of the same degree of differentiation and that individuals within a given species are uni-

**Table 2.1. Major Isolating Mechanisms Acting to Separate Plant Species**

---

*Prefertilization Mechanisms:* Prevent fertilization and zygote formation.

1. Geographical separation. Populations live in different regions (allopatric).
2. Ecological separation. Populations live in the same regions (sympatric) but occupy different habitats.
3. Seasonal or temporal separation. Populations exist in the same regions and may exist in the same habitat, but they have different flowering times.
4. Ethological separation. Pollination is accomplished by specific pollinators (as in some tropical species).
5. Gametic incompatibility. Pollination may occur but gametes are incompatible before or at fertilization.

*Postfertilization Mechanisms:* Fertilization takes place. Hybrid zygotes are formed, but they are inviable, or give rise to weak or sterile hybrids.

6. Hybrid inviability or weakness. Zygotes are formed but are unable to germinate or become established. If established, they break down before reproductive organs are formed.
7. Hybrid sterility. Hybrids are sterile because reproductive organs develop abnormally or meiosis breaks down due to chromosome incompatibilities.
8. $F_2$ breakdown. $F_1$ hybrids are normal, vigorous, and fertile, but $F_2$ generation contains inviable or sterile individuals.

---

SOURCE: Adapted from Stebbins (1966) and Solbrig (1970).

form pose problems in forest ecology and management. In analyses of communities over wide areas, species are often entered into formulas or manipulated as if they were uniform units of comparable divergence. In physiological studies, results may be generated based on a few trees whose individual variation or genecological differentiation may or may not permit extrapolation of the findings as a general rule for the species. In management practice, examples are legion in Europe and North America where seed has been collected from genetically inferior trees or plantations established from ecologically inappropriate seed sources.

Forest species were named and described on the basis of morphological characters by the early taxonomists. They believed that species were specially created and morphologically uniform. Again, it is important to reiterate that the recognition of a class, such as a species, may convey a false impression of uniformity within the class.

Species are not all of the same degree of divergence. This can often be seen in the degree of morphological difference and in their reproductive isolation. For example, there is no gene exchange whatsoever between species such as loblolly pine and southern red oak or ponderosa pine and white pine. In other recognized species, however, such as Engelmann spruce and white spruce, there is considerable intergrading of populations and there may be considerable gene exchange. "Good" species, those that have full species status from the standpoint of reproductive isolation, are those that exist in the same geographic area (**sympatric** species) and yet maintain their distinctness, even though individuals of each are within effective pollinating range of one another and interbreeding would be possible (Figure 2.6, fifth stage). Sympatric species that are reproductively isolated by one or more of the isolating mechanisms include red pine and white pine, white oak and black oak, yellow birch and paper birch, and loblolly pine and shortleaf pine. These "good" species remain distinct even though they may interbreed and form hybrid individuals such as in loblolly pine and shortleaf pine and yellow birch and white birch. Isolating mechanisms operating at a later stage (stages 6–8 in Table 2.1) render hybrids sterile or act against hybrid or backcross progeny to prevent gene flow.

Where individuals of two species come together and hybridize, the species distinctions may be dissolved or swamped out in a flood of intermediates. In this situation, complete reproductive isolation has not yet occurred, and the populations may be more appropriately regarded as subspecies. This situation is indicated in Canada where the morphologically similar white spruce and Engelmann spruce, and balsam poplar and black cottonwood, intermingle. In both cases the recognition of subspecies has been suggested (Taylor, 1959; Brayshaw, 1965; Viereck and Foote, 1970).

Biologists recognize that no single definition of a species is entirely applicable to classify the enormous diversity of organisms nor serve the various purposes desired by different scientists. However, it is important to remember that although a system of populations is recognized as a "species," this in no way indicates that it is directly comparable in degree of divergence or morphological distinctness to other species. The problem of speciation has been considered concisely by Heslop-Harrison (1967), Stebbins (1966), and Solbrig (1970), and in detail by Stebbins (1950 and 1970) and Grant (1963, 1971, 1977).

## Niche

Competition has resulted in marked genetic differentiation among species such that each occupies a different niche. Ecologists use the term niche to attempt to express in one word: where, when, and how a species is genetically adapted to compete with other species (for light, moisture, nutrients, etc.) in its ecosystem, that is, its site or habitat, its time of dominance in the successional sequence, and its functional (physiological) adaptations. The niche of a species is the result of the multidimensional specialization of that species in its ecosystem. By occupying different niches, species can coexist in an ecosystem with a minimum of direct competition.

For convenience in examining the niche differentiation of woody species we recognize three components: a spatial component (the physical site conditions to which the species is adapted), a temporal component (the relative time a species dominates in a sequence of vegetational development of an area), and a functional component (physiological or behavioral adaptations, sometimes termed natural history traits—number of seeds produced, dispersal time and mechanism, growth rate, tolerance of shade, drought, fire, flooding, etc.). The functional adaptations that a species evolves specialize it to occupy a characteristic geographic range and habitat within its range and a characteristic time in the course of succession when the species dominates the area. These three niche components identify where (spatial), when (temporal), and how (functional) a species competes in the ecosystem. We use the term niche as the most concise and appropriate formulation of this genetic specialization.

The **spatial component** of the niche may be illustrated by silver maple, which occupies river floodplain sites, whereas the related sugar maple thrives on upland sites. Their niche differentiation is primarily one of site conditions and the different functional adaptations of the species to establish and grow on the respective sites. The **temporal component** may be illustrated by paper birch and sugar maple on an upland site in the northern hardwood forest. The paper birch, a pioneer species,[2] dominates the site following fire early in the succession of biota on the area. Sugar maple seedlings enter the birch stand in the understory and dominate the site a century or more later as the birches decline and die. In this case, the species are niche differentiated, not by site conditions primarily, but by the physiological adaptations that enable them to dominate the site at different times. The **functional component** may be illustrated by the physiological adaptations of paper birch and sugar maple. Birches seed in the burned site quickly and their seedlings grow rapidly, being highly photosynthetically efficient at high light levels. In contrast, maple seedlings are photosynthetically efficient at low light levels and outcompete paper birch, and most other species, on the shaded forest floor.

## Hybridization

Hybridization, the crossing between populations (races, subspecies, species) having different adaptive gene complexes, is frequent in

[2]A pioneer species is one that is among the first to establish itself in an area.

natural populations of many woody-plant groups: in pines (Mirov, 1967), larches (Carlson and Blake, 1969), birches (Dancik and Barnes, 1972; Barnes et al., 1974), oaks (Tucker, 1961; Maze, 1968; Hardin, 1975), poplars and aspens (Brayshaw, 1965), and many others. The great number of reports of hybrids during this century probably reflects the widespread disturbance of forest communities providing open sites for the establishment of hybrids.

Interspecific hybridization is probably of minor evolutionary significance (Wagner, 1968, 1970) although hybrids may act as evolutionary catalysts (Stebbins, 1969, 1970). However, they are of major ecological and practical significance.

Artificial hybridization has been pursued widely among species and races (Wright, 1976). Many hybrids are important in forest management and horticulture due to rapid growth, good form, disease resistance, or frost hardiness (Duffield and Snyder, 1958; Wright, 1976; Nikles, 1970).

Natural hybrids often occur in zones of contact between species (Remington, 1968; Brayton and Mooney, 1966) and in disturbed habitats. The disturbed area may be an intermediate or hybrid habitat (Anderson, 1948) where neither parent is well adapted. The common denominator of environments where hybrids become established, whether disturbed or not, seems to be lack of competition.

It has been popular to report natural hybrids, usually based on morphological characters, perhaps due to their presumed rarity, or as Wagner (1968) relates, seeking hybrids "was all part of the 'game,' and added to the thrill of the chase, like adding a new stamp to the collection." However, little detailed ecological study has been devoted to the comparative establishment of hybrids and their parents or the presumed differences between the so-called hybrid habitat and that of the parents.

Furthermore, little critical genecological study has been given the popular concept of **introgressive hybridization** or **introgression**: the gradual infiltration of germ plasm of one species into that of another as a consequence of hybridization and repeated backcrossing[3] (Anderson, 1949). After examining the variation of some 30 genera of flowering plants in the subcontinental area between the Great Plains and the Atlantic (including species of junipers, oaks, maples, and redbuds), Anderson (1953) reported: "For each of these genera, all the readily detectable variation can be ascribed to introgression." Introgression has been widely accepted, probably more out of faith and intuition than from fact. For example, a detailed reevaluation of Anderson's classic example of introgression, an iris complex in Louisiana, revealed hybrids in the populations but "no statistically

[3]Backcrossing is the crossing of a hybrid with either of its parents.

significant effects of taxonomic or evolutionary importance attributable to introgression" (Randolph et al., 1967).

Introgression is presumably achieved in three phases: (1) initial formation of $F_1$ hybrids, (2) their backcrossing to one or other of the parental species, and (3) natural selection of certain favorable recombinant types (Davis and Heywood, 1963; Figure 2.7). This is simply gene flow between species. If hybridization occurs between two closely related species, the probability of gene flow is higher than when the species have diverged sufficiently to have well-integrated, but different, gene pools. Genes from one population will be incorporated into the gene pool of the other (regardless of rank—species, subspecies, etc.) if they improve the well-integrated harmony of the foreign gene pool. If they tend to disrupt the harmony, their frequency will be reduced; this is merely natural selection in action (Bigelow, 1965). Not surprisingly, we find frequent reports of introgression between the closely related spruces of western Canada (Horton, 1959; Daubenmire, 1968b; Ogilvie and von Rudloff, 1968; Roche, 1969; Hanover and Wilkinson, 1970) and particularly white spruce and Engelmann spruce (Nienstaedt and Teich, 1971).

In many instances introgression has been postulated or inferred rather than compellingly demonstrated. Often the range of variability of the parent species and the $F_1$ hybrids is not well known; $F_1$

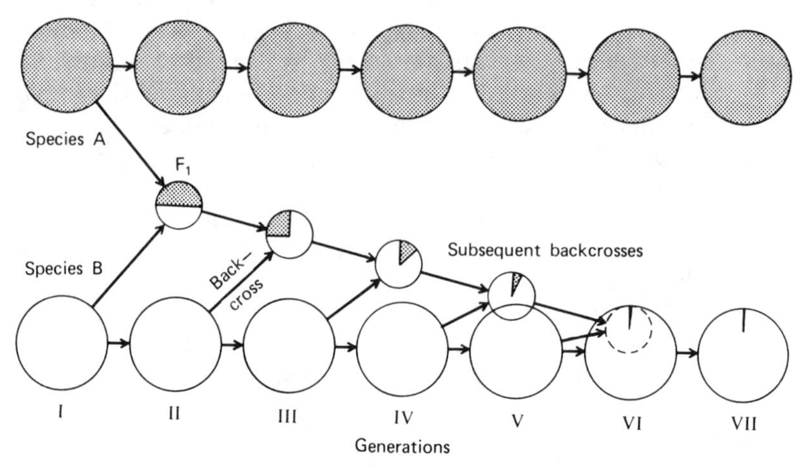


**Figure 2.7. Diagram illustrating introgression—the interbreeding of species, followed by backcrossing to one parent and ultimate absorption of some genes from one parent into at least some members of the population of the other. (After Lyman Benson, *Plant Taxonomy: Methods and Principles* © 1962. The Ronald Press Company.)**

hybrids may be mistaken for backcrosses (Dancik and Barnes, 1972) and backcrosses for a parent. The diagram in Figure 2.7 indicates that individuals of the first backcross generation are relatively distinct from the recurrent parent. This has not been borne out in research by the junior author and Farmer (1977) with the morphologically distinct species of bigtooth and trembling aspen in southeastern Michigan. Many backcross individuals, produced by controlled pollinations, could not be distinguished from $F_1$ hybrids using many leaf and bud characters; other backcross individuals could not be distinguished from the recurrent parent (either bigtooth or trembling aspen). There is enormous variation in morphological characters of woody plant populations, and detailed studies of variation of the parents and hybrids using standardized collections and many characters are needed to estimate the extent of hybridization and gene flow. In the case of black and red spruce of the northeastern United States and adjacent Canada, a detailed study of variation within both species and $F_1$ hybrids revealed much less hybridization than had previously been reported (Gordon, 1976). Introgression was very limited in occurrence.

In some cases, we suspect that divergence caused by intense selection, as cited by Barber and Jackson (1957), is misinterpreted as introgression. An example of the dilemma is a study of phenotypic variation of morphological traits of *Juniperus* (Van Haverbeke, 1968a,b). Upon finding two extreme types, Rocky Mountain juniper and eastern red cedar, connected by a series of intermediates in the Missouri River basin, the author first concluded that the data could be interpreted as evidence for introgression. He then stated that an even more tenable interpretation would be migration of junipers from west to east, bringing divergence of the populations.

In determining the amount of gene flow between species it is important to establish, through more than observations of morphological characters of phenotypes in nature, that (1) hybridization and backcrossing have actually taken place, and (2) increased variability of the parent species occurs outside the area of hybridization and is due to hybridization and not solely to intense selection.

## Polyploidy

Polyploids are organisms with three or more sets of chromosomes. The ploidy level of a species (triploid = 3 sets, tetraploid = 4 sets, etc.) is measured in relation to the base or **x** chromosome number established for the genus or family, usually the lowest haploid (gametic) number for the group. For example, the base number for the genus *Pinus* is 12 ($x = 12$) and all pine species have 24 chromosomes ($2x = 24$, the diploid number). In *Betula* the base number is 14, and black birch has the 2x number of 28 chromosomes. However, other members of the genus are tetraploids (bog birch and European white birch, 4x = 56), and yellow birch is a hexaploid (6x = 84). Paper birch is a complex population system with individuals having 28 (2x), 42 (3x), 56 (4x), 70 (5x), or 84 (6x) chromosomes (Brittain and Grant, 1967).

Conifers are almost exclusively diploid; only three species are known to be polyploids (Khoshoo, 1959), and the most important of these is redwood, a hexaploid with 66 chromosomes. Nearly 40 percent of the angiosperms are polyploids. Some woody angiosperm genera have no polyploid species (*Populus, Juglans, Robinia*), but many others each have species of various ploidy levels (*Prunus, Salix, Betula, Alnus, Magnolia, Acer*, and many others). Polyploidy in woody plants has been reviewed by Wright (1976), and the evolutionary aspects have been discussed by Stebbins (1950, 1970) and Jackson (1976). Polyploids apparently arise most often by hybridization of related species followed by doubling of the chromosomes of the hybrids.

Polyploids are of considerable ecological significance. Compared to related diploids, they are more frequent in higher northern latitudes than in lower latitudes. Polyploids apparently are better able to colonize and persist on sites in the colder climates, particularly those resulting from Pleistocene glaciation. A major reason for the success of polyploids compared to related diploids under these severe conditions is that polyploidy is a uniformity-promoting mechanism. In general, polyploidy acts like a sponge, absorbing mutations but rarely expressing them. In diploids, mutations or new recombinations are more easily expressed due to the lower chromosome number and fewer of each kind of chromosome. Polyploids, however, have many chromosomes, often four to six of each kind, and new gene combinations are not likely to produce a major change in the phenotype. Mosquin (1966) reasoned that polyploids represent an efficient buffering system, resisting the effects of natural selection on particular genes and promoting and preserving phenotypic uniformity. According to Mosquin, the narrow adaptational limits of high latitude and weedy polyploids are an adaptive feature corresponding to the narrow and relatively uniform environments of boreal, arctic, and disturbed or weedy habitats. The range of polyploids is often great, as in high-latitude birches and willows, because such habitats are themselves widespread.

Polyploids, however, are not confined to recently disturbed boreal habitats, and some appear just as variable as diploid species. Temperate species such as American basswood and tulip poplar are apparently polyploids of ancient origin that have outlived their ancestors. High levels of polyploidy have also been reported for certain tropical floras, some of which are of very ancient origin. According to Stebbins (1970), newly opened habitats were available in ancient times for evolving angiosperms, and increasing polyploidy accompanied the establishment and spread of new groups of angiosperms during the early stages of their history. The ability of polyploids to colonize newly opened environments is apparently the common denominator of their success in diverse regions of the world, whether the time of their origin was ancient or modern.

## The Fitness-Flexibility Compromise

To survive and persist through time a population must not only show adaptedness to its present environment, but have adaptability, the potential to change. Mather (1943) expressed this compromise between fitness for the environment as it exists and flexibility that will permit further adaptive change. Flexibility is favored by variability-promoting mechanisms such as cross-pollination and a high rate of recombination (Heslop-Harrison, 1964; Mosquin, 1966). Inbreeding, apomixis, polyploidy, and a low rate of recombination are uniformity-promoting devices that favor fitness in the given environment.

We have already cited polyploidy as a uniformity-promoting device, often in northern and glaciated or disturbed sites. Polyploidy may be coupled with cross-pollination, as in the willows and birches, adding to the flexibility side of the ledger. In most woody species, we see various mixes of uniformity- and variability-promoting devices giving the plants the best of both possible worlds. Selection through time has produced a different mix and different mechanisms, depending on the particular environmental situation. The proportion of each and the nature of the mechanism differ between species even within genera. The fitness-flexibility compromise is closely related to the limiting ecological factors of the species' environment.

For example, compare trembling aspen, widely distributed in northern, glaciated, and disturbed habitats, with eastern cottonwood, also wide ranging but primarily found in lower latitudes and growing in stream and river valleys. Both species are primarily dioecious (male and female flowers borne on different individuals), the most effective device for ensuring cross-pollination. Thus a great amount of variation is generated in these diploid species and then widely circulated through abundant seed production and excellent dispersal. Aspen has the remarkable ability of vegetative propagation by root suckers, which assures genetic uniformity; the clonal growth habit is pronounced throughout its range. Fire is probably the environmental factor that has most favored this adaptation. In contrast, cottonwood rarely produces root suckers in nature (in an essentially fire-free environment), but its branches and young shoots root easily in soil. This may be of considerable selective advantage and of immediate fitness in riverside sites subject to periodic disturbance through flooding. Thus both species have strongly developed mechanisms giving both fitness and flexibility.

Pines, typically cross-pollinating species, maintain a certain amount of self-pollinating ability in their breeding systems. In western white pine, for example, a great variation exists in self-fertility—from infertile individuals to highly self-fertile individuals (Bingham and Squillace, 1955). Since fire plays a major role in establishment of pines around the world, the ability of one or a few survivors of a severe fire to colonize the site, if necessary by self-fertilization, may be of selective advantage. Colonization is likely to be accompanied at first by an increase in the degree of inbreeding, but outbreeding will tend to be restored as the stand density increases (Bannister, 1965).

Selection pressures of this type have apparently been instrumental in promoting a very high degree of self-fertility in red pine (Fowler, 1965a). The species is highly self-fertile, and unlike most other pines, seedlings resulting from self-pollination are as vigorous as those from cross-pollination (Fowler, 1965b). However, a high degree of self-fertility has been achieved, seemingly at the cost of variability, since red pine is one of the most uniform of all woody-plant species.

## SUGGESTED READINGS


Barber, H. N., and W. D. Jackson. 1957. Natural selection in action in *Eucalyptus*. *Nature* 179:1267–1269.

Critchfield, William B. 1957. Geographic variation in *Pinus contorta*. Maria Moors Cabot Foundation Publ. No. 3. Harvard Univ., Cambridge, Mass. 118 pp.

Davis, P. H., and V. H. Heywood. 1963. *Principles of Angiosperm Taxonomy*. (Populations and the environment, pp. 387–416; Evolution and the differentiation of species, pp. 417–461.) D. Van Nostrand Co., Inc., New York. 558 pp.

Dobzhansky, Theodosius. 1968. Adaptedness and fitness. *In* Richard C. Lewontin (ed.), *Population Biology and Evolution*. Syracuse Univ. Press, Syracuse, New York. 205 pp.

Gordon, Alan G. 1976. The taxonomy and genetics of *Picea rubens* and its relationship to *Picea mariana*. *Can. J. Bot.* 54:781–813.

Heslop-Harrison, J. 1967. *New Concepts in Flowering-Plant Taxonomy*. (Ecological differentiation of populations, pp. 44–58; Geographical var-

iation and reproductive isolation, pp. 59–78.) Harvard Univ. Press, Cambridge, Mass. 134 pp.
———. 1964. Forty years of genecology. *Adv. Ecol. Res.* 2:159–247.
Kruckeberg, Arthur R. 1969. The implications of ecology for plant systematics. *Taxon* 18:92–120.
Langlet, Olof. 1959. A cline or not a cline—a question of Scots pine. *Silvae Genetica* 8:13–22.
———. 1971. Two hundred years' genecology. *Taxon* 20:653–721.
Libby, W. J., R. F. Stettler, and F. W. Seitz. 1969. Forest genetics and forest-tree breeding. *Ann. Rev. Genetics* 3:469–494.
Mirov, N. T. 1967. *The Genus "Pinus."* (Physiology and ecology, pp. 396–464.) The Ronald Press Co., New York. 602 pp.
Mosquin, Theodore. 1966. Reproductive specialization as a factor in the evolution of the Canadian flora. *In* Roy L. Taylor and R. A. Ludwig (eds.), *The Evolution of Canada's Flora.* Univ. Toronto Press. 137 pp.
Roche, L. 1969. A genecological study of the genus *Picea* in British Columbia. *New Phytol.* 68:505–554.
Squillace, A. E. 1966. Geographic variation in slash pine. *Forest Science Monog.* 10. 56 pp.
Stern, Klaus, and Laurence Roche. 1974. *Genetics of Forest Ecosystems.* Springer-Verlag, New York. 330 pp.
Wright, Jonathan W. 1976. *Introduction to Forest Genetics.* Academic Press, Inc., New York. 463 pp.

# 3
# *Life and Structure*

In this chapter we consider the life history of forest trees, with special attention to particular environmental factors that influence reproduction and growth of individuals and populations. In addition, it is important to discuss selected aspects of the structure and function of shoots, roots, and stems. Excellent treatments of the physiology of forest trees are found in the books of Büsgen and Münch (1929), Kramer and Kozlowski (1979), Kozlowski (1971), and Zimmermann and Brown (1971). Tree growth has been the subject of shorter works by Wilson (1970) and Morey (1973). The life history of plants has been comprehensively examined from the viewpoint of structure and function by Büsgen and Münch (1929) and from the viewpoint of population biology by Harper and White (1974) and Harper (1977).

Physiological processes occur in organs and tissues of individual plants but they affect the growth and form of the whole plant and, furthermore, the associated populations of plants and animals. We need to consider the life history and structure of woody plants not only from the viewpoint of the individual but also from the standpoint of the population. Harper (1977) has pioneered the area of plant population biology, and his work provides the forest ecologist with a useful conceptual model of the essential features of the life cycle. In Figure 3.1 we observe individuals of a species population recruited and established from a "bank" of seeds stored in the soil or on the forest floor (Phases I and II). Growth in height and mass, Phase III, requires space, nutrients, moisture, and other environmental resources that may be insufficient to allow vigorous growth of all individuals. Some plants die (unbranched stems, ┬) while others thrive (as shown by the branched systems). Although potentially capable of branching indefinitely, the individuals are sooner or later restrained by physical and biotic limits of their environment resources. The environmental constraints of limited resources are shown by the vertical bars on either side of the population. Development of the living individuals brings changes to the environment of the understory and forest floor (this feedback shown by the dashed arrow) that in turn affects the recruitment of new

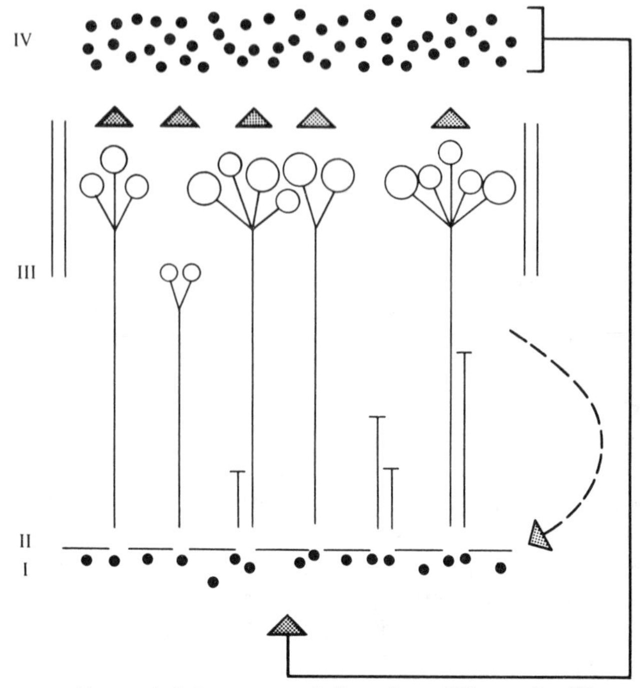


Figure 3.1. **Essential features of the plant life cycle, illustrating a tree as a series of repeating modular units (the shoots). (After Harper, 1977; © by John L. Harper.)**

**Phase Feature**

**I the bank of seeds of the forest floor**

**II the recruitment and establishment of seedlings**

**III growth in height, mass, and in number of modular units (vertical bars represent the limiting resources of the environment on growth; the dashed line indicates the influence of the overstory on the recruitment and establishment phase.)**

**IV seed production and dispersal**

individuals from the seed bank. Reproduction occurs in Phase IV and seeds are dispersed to the forest floor. Our consideration of the life history of trees emphasizes several features of this model: reproduction, dispersal, establishment, and some basic aspects of tree structure and growth in relation to environmental factors.

## REPRODUCTION

Sexual reproduction is the basic mode by which plants maintain their populations, adapt to changing environmental conditions, and thus persist through time. Male sperm and female egg cells unite to form a zygote, which is genetically different from either parent and other offspring. Asexual reproduction in plants is actually a growth process whereby genetically identical stems or **ramets** are derived from a sexually produced plant (the **genet**) and form a spontaneous clone (Chapter 2, p. 19). The multistemmed clone best illustrates the concept of a tree as a population of modular units, the shoots, where each shoot becomes a ramet of the original plant. Almost all woody plants are capable of some form of cloning. By fragmenting its genotype, such a plant gains growing space, water, and nutrients and eventually increases its capacity for sexual reproduction. As far as we know, a species' vigorous asexual reproduction does not lead to a diminished sexual reproductive ability in producing flowers and seeds. The balance of fitness and flexibility traits, in part determined by asexual and sexual reproduction, was discussed in Chapter 2.

### Sexual Reproduction

**Ability to Flower.** Individuals of forest trees progress from a juvenile stage, characterized by no flowering,[1] to the adult stage in which flowers, fruits, and seeds are produced. The duration of the juvenile period varies markedly among species. Fast-growing, shade-intolerant species, such as paper birch and Virginia pine, flower sooner than slow-growing, shade-tolerant trees such as beech and hemlock. For example, the juvenile period is estimated to last 5 to 10 years in Scots pine whereas in European beech it is 30 to 40 years (Wareing, 1959).

Attainment of the adult stage is more closely related to tree size than to age. Usually, however, height and diameter are highly correlated with age so that either size or age may be used to predict when flowering may commence. Nevertheless, small trees in the forest understory, suppressed by the overstory canopy of the local population, may not flower—even at ages of 50 to 100 or more years. For many species the attainment of a certain minimum size, rather than the number of periods of growth and dormancy, is the critical factor

[1]Angiosperms produce flowers whereas gymnosperms do not bear flowers but produce cones (strobili) that bear naked ovules. For simplicity we will use the term flowering to denote the reproductive process of both groups.

in attaining the adult or flowering stage. For example, European larch normally remains juvenile for 10 to 15 years but was observed to flower in just 4 years when seedlings were grown to the minimum size for flowering using warm temperatures and long days in a greenhouse (Wareing and Robinson, 1963). Flowering itself is not directly triggered by the attainment of a critical size, but through the activation of "flowering genes" by internal growth regulator controls. Gibberellins apparently play an important role in inducing flowering in some juvenile conifers, especially for members of the Cupressaceae and Taxodiacae (Pharis, 1974).

Trees in natural stands exhibit great variation in their genetic disposition to flower, particularly in number of reproductive buds produced and the ratio of female to male buds. Some trees of adult stage and favorably situated in the overstory canopy do not flower or rarely flower whereas others are highly fruitful. Some trees that bear both male and female unisexual flowers on the same tree (**monoecious** condition) are predominantly male whereas others are predominantly female.

Species of several genera (*Acer, Ailanthus, Diospryos, Fraxinus, Gymnocladus, Maclura, Populus, Sassafras*) bear unisexual male and female flowers on different individuals (**dioecious** condition). The dioecious trait promotes outcrossing between genetically different individuals and precludes self-pollination. In tree species, selfing, and inbreeding generally, lead to growth depression (red pine is a notable exception; see Chapter 2, p. 45). Selfed or inbred seedlings are typically eliminated at an early stage in natural populations through competition. Angiosperms rarely produce viable selfed seeds. Conifers, however, are more likely to produce viable self-pollinated seeds (especially pines), and this may be of ecological significance when isolated trees survive fire or other disturbances (see Chapter 2, p. 45, and Chapter 16, p. 423).

**Increasing Seed Production.** The most effective way to reduce the length of the juvenile stage is to grow seedlings in such a way that they attain a large size as rapidly as possible. This is accomplished by applying fertilizers or using other cultural methods. The use of gibberellins, together with cultural measures such as fertilization, appears to be a promising development for increasing flower production of members of the Pinaceae as well as in the Cupressaceae and Taxodiaceae.

Once trees are in the adult stage, they may be stimulated to increase flower and seed production by a variety of cultural methods. Many investigators have demonstrated such increases over untreated controls in a great variety of species using many different methods (see reviews by Matthews, 1963 and Kozlowski, 1971).

Considerable success may be achieved by fertilizing, irrigating, and releasing trees from competition by cultivation or thinning. Methods that injure the tree, such as girdling, root pruning, grafting, and bending or knotting branches, may be temporarily successful with individual trees. However, because of the potentially harmful effects they are not recommended in large-scale practice.

**Reproductive Cycles.** The entire reproductive cycle of forest species is closely adapted to the complex of environmental factors of the place where it grows, i.e., its site or habitat. Many river floodplain and other wetland species (most willows, cottonwoods, elms, silver maple, river birch) and other trees, such as red maple and bigtooth and trembling aspen, flower in the early spring and disseminate seeds 4 to 6 weeks thereafter. The seeds are dispersed long distances by strong spring winds or fall short distances into flooded or recently flooded lowland sites where they germinate readily. In these instances the period between pollination and seed dispersal is only a few weeks, and large seeds are not usually developed. Instead, thousands of small seeds are produced, some of which find a favorable seedbed for establishment and growth so that a large food reserve in the seed is unnecessary.

In contrast, in most other North Temperate Zone trees, fruits and seeds develop throughout the growing season of 2 or 4 months and are disseminated in the fall or winter. The relatively large seeds that are produced typically lie dormant over the winter and germinate in the moist forest floor the following spring. Seedlings of these species (such as pines and upland oaks) are soon subjected to drying soils and moisture stresses; the large amount of stored food in the seed is therefore needed to develop a root system that can cope with the decreasing moisture supplies (see discussion below). Space limitations preclude a description of the details of the reproductive cycle for each species; such descriptions are found in two USDA Handbooks (USDA 1965; USDA 1974). However, it is important to note that the great physiological complexity of the cycle through many stages of development has been selected for and is closely adapted to the habitat conditions in which the seeds germinate and the seedlings establish and grow.

In virtually all tree species, reproductive buds or **primordia** are initiated during the growing season of the year before the opening of the flowers (anthesis). Usually the flower buds are visible along the current year's shoots of conifers and in the axils of leaves of hardwoods in the fall of the year prior to flowering. In most cases, reproductive buds cannot be distinguished at this time from vegetative buds without examination by hand lens or microscope. The reproductive cycle of Douglas-fir, shown in Figure 3.2 (Allen and

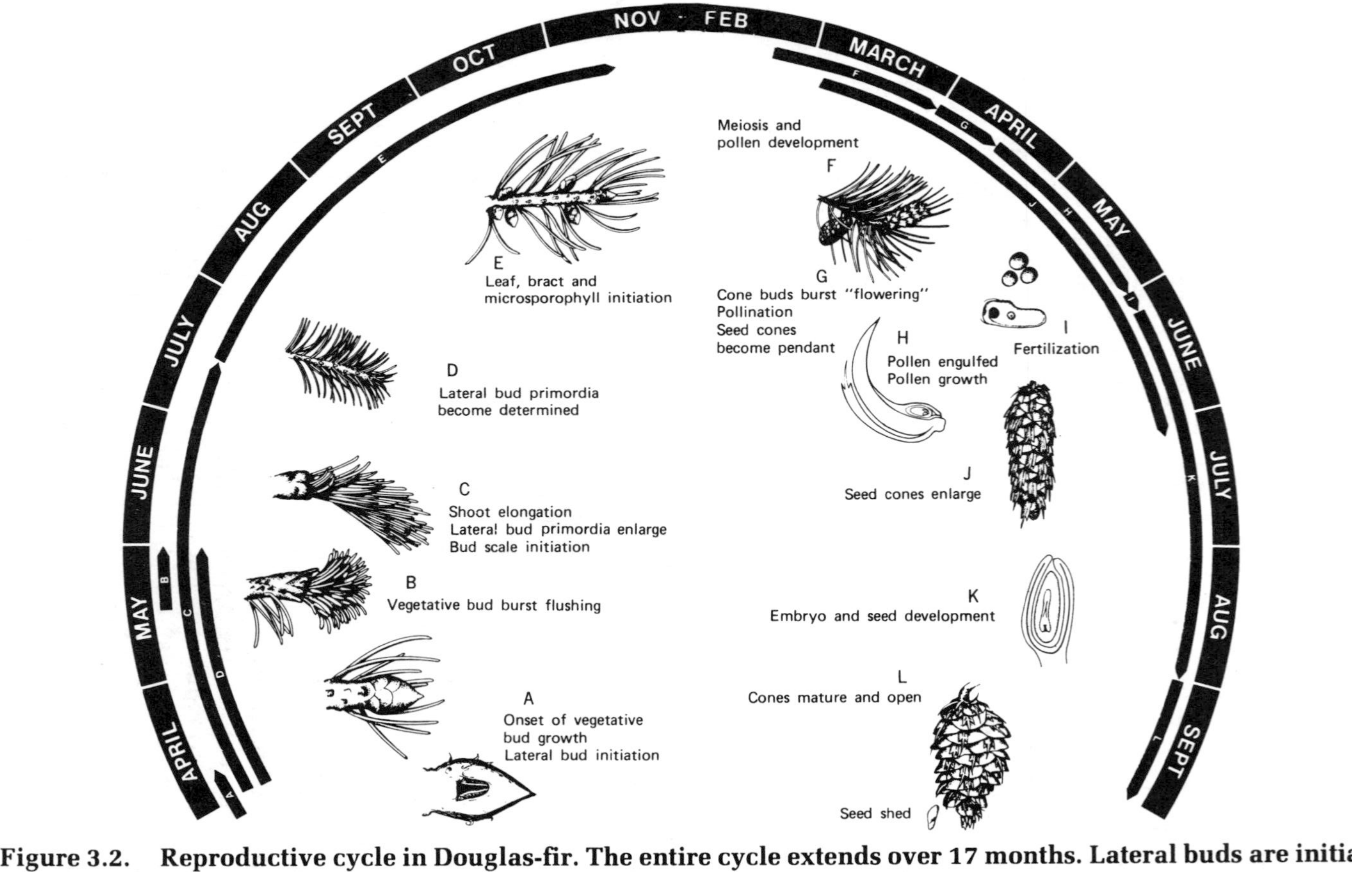


**Figure 3.2.** **Reproductive cycle in Douglas-fir. The entire cycle extends over 17 months. Lateral buds are initiated in April and differentiate into vegetative, pollen, or seed cone buds during the ensuing 10 weeks. Pollination of the seed cones occurs the following April and the mature seeds are shed in September of the second year. The various stages are identified by letters A–L and are briefly described. The approximate length of each stage is shown by the arrows. (After Allen and Owens, 1972. Courtesy Canadian Forestry Service.)**

Owens, 1972) illustrates that lateral bud primordia are initiated in April along the vegetative shoot that develops in the terminal vegetative bud (Figure 3.2A). Some or many of these primordia (young lateral buds) may become pollen or seed cone buds provided the internal environment is favorable. As the vegetative bud bursts, needles flush out (Figure 3.2B), the shoot elongates (Figure 3.2C), and the lateral buds become visible (by late July or August) along the young shoot (Figure 3.2D,E). These buds will enlarge and burst the following spring to produce branches or male and female cones. By October one can determine whether the lateral buds are vegetative or reproductive. Lateral buds at the base of the young shoot tend to become pollen cones, whereas buds toward the tip of the shoot become either seed cones or vegetative shoots.

The alternative pathways of lateral bud primordia development are shown in Figure 3.3. Some bud primordia abort early, degenerate, and leave no trace. Others form bud scales and then cease to develop; they are termed latent buds. If the terminal bud is removed, latent buds are usually stimulated to develop into vegetative buds. The remaining primordia develop into pollen cone buds, seed cone buds, or vegetative buds.

The internal nutrition and hormonal relations in the shoot and tree largely determine the disposition of lateral bud primordia. Although the same number of primordia may be initiated in two consecutive years, the proportion of vegetative and reproductive buds may be quite different. For example, the number of lateral primordia initiated in Douglas-fir study trees did not differ significantly in 1967 and 1968 (Allen and Owens, 1972). However, abundant seed cones developed in 1967 but none developed in the same trees in 1968. Thus the marked variation in seed cone production in Douglas-fir (and probably many other forest trees) is the result of the

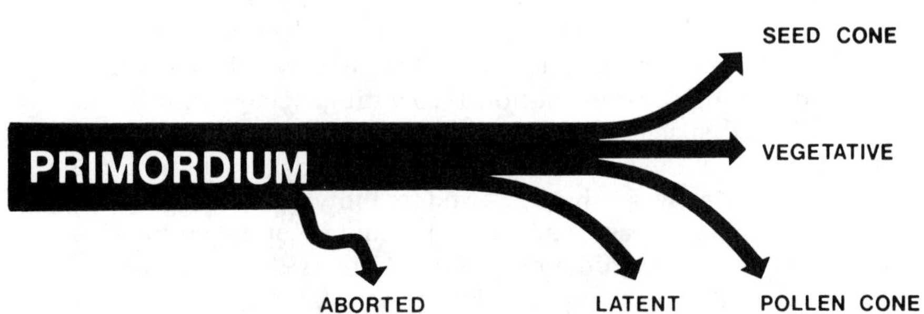


**Figure 3.3.** **Alternative pathways of lateral bud primordial development. (After Allen and Owens, 1972. Courtesy Canadian Forestry Service.)**

proportion of primordia that develop into cones rather than variation in the number of primordia originally initiated. In Douglas-fir, pollination and fertilization take place the same year; cones mature during the summer, and seeds are released in September. This cycle is typical of most conifers and many hardwoods as well, although details of the reproductive process vary from species to species.

Despite the fact that the life cycle of the genus *Pinus* is often used to illustrate the conifer reproductive cycle, pine species are atypical in that fertilization occurs 12 months after pollination, and their cycle is a full year longer. The reasons for this are not clear. However, it does provide a period for cones to abort and to be destroyed by frost or insects before fertilization occurs. For example, a loss of 60 percent of the cones between the first and second year of development was reported for red pine in northern Wisconsin (Lester, 1967). Thus the photosynthate, water, and nutrients that are needed the second year to bring about cone and seed maturity are conserved by being channeled only into those cones and seeds that have survived the selective process.

**Pollination.** The timing of pollen release and female receptivity in deciduous species is closely related to the mode of pollination, whether by wind or animals. Wind-pollinated species (aspens, elms, red maple, birches) flower in the early spring before the leaves form. Insect-pollinated species, such as basswood, yellow-poplar, and black cherry, typically flower later when the leaves are forming. In the tropical rain forest, pollination by animals is the rule (see Chapter 13). Conifers are exclusively wind pollinated. Wind pollination is promoted by the production of enormous amounts of pollen (much more than in insect-pollinated deciduous species) and the positioning of the female cone buds at the ends of shoots in the top third of the crown.

Climatic conditions, especially temperature, markedly affect the shedding of pollen and the receptivity of female flowers. Flowering of female and male buds is synchronized but does not precisely coincide for female and male flowers on the same tree, thus reducing the likelihood of self-pollination. Flowering occurs rapidly in wind-pollinated species. For Scots pine in Finland, Sarvas (1962) found that most of the pollen for individual stands is shed within 3 to 7 days, and even over a shorter period for individual trees. In years of abundant pollen release, one day in the middle of the period (the day of maximum shed) often exceeded all others in the amount of pollen shed. Pollen discharge was highly correlated with high temperature and low humidity. The day of maximum shed coincided with the warmest day of the flowering season in most years. Even in unfavorable springs, pollen discharge was delayed until the occurrence of several favorable days (or even one) which were adequate for pollen shed and spreading. Sarvas cites the rapidity of pollination as one of the distinct advantages of wind pollination. In 14 years of monitoring pollination of Scots pine and European white birch, not once was the major part of the pollen crop destroyed by unfavorable weather.

**Periodicity of Seed Crops.** Abundant flowering and seed production occur irregularly in natural stands of forest trees. The cyclic nature of seed production is one of the most important traits of the life history, greatly promoting the establishment of tree seedlings. In general, fast-growing, shade-intolerant species (aspens, birches, cherries, cottonwoods, junipers, and maples except sugar maple) tend to have fewer years between good seed crops than slower growing, more shade-tolerant species (beeches, sugar maple, red and white oak, hemlocks).

The abundance and periodicity of flowering is not only controlled by the internal environment of the plant but is strongly influenced by external environment as well, particularly light, temperature, and moisture. Large, open-grown trees, being well lighted throughout the crown and well supplied with moisture, flower more frequently and abundantly than equally large members of the same species in the forest stand. Such open-grown trees also typically produce large quantities of seed each year, provided they are well pollinated. However, in isolated, open-grown conifers, many seeds may be produced by self-pollination, typically increasing the number of unfilled seeds, decreasing the viability of filled seeds, and causing reduced seedling growth. Lack of pollen results in collapsed ovules in pines and some other tree species. In the closed forest stand, the large dominant trees that have crowns well exposed to sunlight are the primary seed producers; smaller trees with narrow and suppressed crowns yield few if any seeds.

If flowering and seed production are studied over a long time, a typical cycle of seed production is evident, although some seed is usually produced each year by the dominant trees. For example in western white pine, good cone crops occur about every 3 to 4 years (Figure 3.4; Rehfeldt et al., 1971; Eis, 1976). Cone production tends to drop markedly in years following peak production (1952, 1960, and 1963 in Figure 3.4). Individual trees tend to be on a similar cycle although there is considerable variation in inherent productive ability. For example, tree 58 is highly productive compared to tree 17; low years for tree 58 are higher than most years for tree 17. In the genus *Pinus*, good cone and seed crops typically occur at 3- to 7-year intervals. However, they may occur nearly every year in some species (Virginia, Monterey, and lodgepole pines). A 3-year cycle of

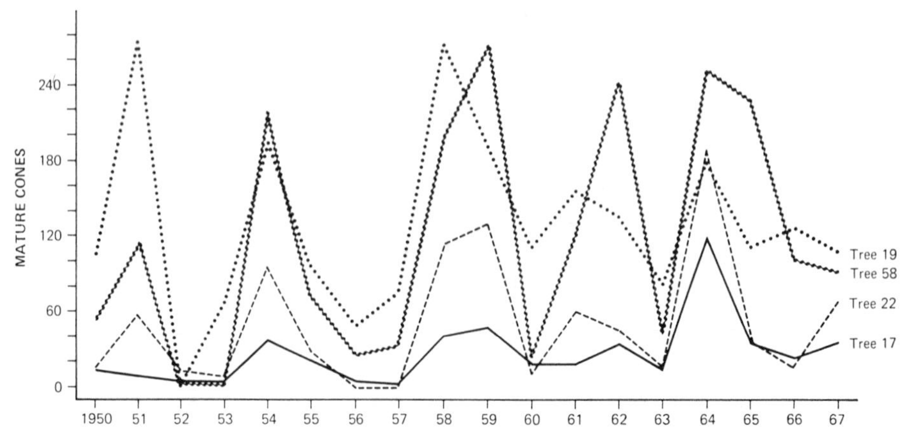


**Figure 3.4. Periodicity of cone production of four western white pines in northern Idaho over 18 years. (After Rehfeldt et al., 1971.)**

good cone crops has been reported for several northwestern fir species (Franklin, 1968), and in Douglas-fir the cycle is about 5 years. The periodicity of flowering and the minimum seed-bearing age for most North American trees and shrubs have been summarized by the U.S. Department of Agriculture (1974).

The production of seed is governed by two major sets of factors: those that influence the initiation of flower primordia and those that cause the loss of flower buds, fruit or cones, once they are initiated (assuming pollination and fertilization proceed efficiently). Little is known of the relative effect of the nutritional, hormonal, and the external climatic factors determining the initiation of flower buds. In deciduous fruit trees, flower bud production is promoted by high light intensity and is related to high rates of photosynthesis and a high level of nutrition (Kozlowski, 1971). Other woody plants are probably not markedly different from fruit trees in which a supply of metabolites, in excess of shoot, root and developing fruit requirements, is needed and mediated by a control system of growth regulators. In general, a very small seed crop occurs in the year following a very heavy seed crop, but the strong biennial bearing pattern of many fruit trees is not typical of forest trees.

Little information is available on the factors of the physical environment that actually determine the number of lateral buds initiated or the number of flower buds that eventually develop from these lateral bud primordia. Because of the relative ease of determining the number of mature cones or fruits, researchers have attempted to relate climatic and other site factors to the abundance of the seed crop rather than the number of flower or cone primordia initiated. Although no single factor or general relationship has emerged for all species, high temperatures and dry conditions in the summer of the year before flowering and pollination often have been associated with good seed crops, as in ponderosa pine (Daubenmire, 1960), European beech (Holmsgaard, 1962), red pine (Lester, 1967), and western white pine (Rehfeldt et al., 1971; Eis, 1976). The relationship is particularly complex because two and sometimes three (in pines and oaks) reproductive cycles are proceeding simultaneously on the mature tree. In addition, because female reproductive buds are initiated at different times of the year in different species, a particular combination of physical factors is unlikely to affect all species in the same way. Furthermore, the magnitude of the maturing crop strongly influences the effect of climatic and other site factors on the reproductive structures that are beginning to develop or those just being initiated.

Generally, the more favorable the site conditions (climate, soil moisture and nutrients), the greater the flower and seed crop. For example, Sarvas (1962) found that on poor sites in southern Finland (height of dominant trees, 16 m at 100 years) fewer than 30 seeds were produced per square meter. On sites of medium fertility (height of dominants, 23 m) and high fertility (height of dominants 27 m) 60 seeds $m^{-2}$ and 90 $m^{-2}$ were produced. Similarly, pollen yields were much higher on fertile sites (35 kg $ha^{-1}$) than on infertile sites (9 kg $ha^{-1}$).

Once flower buds have been formed, various physical and biotic factors cause mortality to the flowers and the seed crop. Spring frosts may kill newly formed reproductive buds or injure immature fruits and cones. Exceedingly high temperatures or severe droughts may cause abortion of immature fruits and cones or cause a marked decrease in seed size. Strong winds and hail may mechanically damage the seed crop. Cone insects and seedbugs may also cause severe losses. In general, the entire reproductive cycle is adapted to the prevailing regime of site factors, and a major departure from such conditions tends to disrupt the reproductive cycle, but only temporarily. The development of cones and fruits and the factors affecting their development are considered in detail by Kozlowski (1971).

Considerable loss may occur from biotic agents after seed dispersal, and these are considered together with the benefits of animal dispersal in Chapter 13. Trees and shrubs are coadapted to animals. The massive seed crops of certain years satiate seed consumers and make available large numbers of seeds for establishment. The most remarkable flowering cycle in woody plants is that of bamboo. Bamboo clones of many species in the Indian-Asian tropics grow vegetatively for 12 to 120 years, flower and fruit synchronously over very large areas, and then die (Janzen, 1976). According to Janzen, animals eliminated bamboo plants that were out of phase and, through such selective action, were responsible for the synchronized reproduction that is of adaptive advantage to bamboo. The massive seed crops produced are large enough to satiate all seed consumers and still provide sufficient seeds for successful regeneration.

**Effects of Reproduction on Vegetative Growth.** Flowering and the production of fruit and seed crops reduce vegetative growth. Many years of research and experience with fruit trees have demonstrated a strong decrease in shoot, cambial, and root growth with increasing fruit productivity (Kozlowski, 1971). In forest trees, too, heavy seed crops markedly decrease both height and radial growth (Morris, 1951; Blais, 1952; Eis et al., 1965; Tappeiner, 1969). In mature balsam fir, the weight of new foliage produced in a cone year was only 27 percent of that in a noncone year (Morris, 1951). The developing cones or the fruits of angiosperms are major reservoirs or metabolic "sinks" to which photosynthetic materials (photosynthate) of the current year are allocated. Not only is current shoot and radial growth depressed, but terminal and lateral buds may receive less photosynthate and nutrients so that immature shoots telescoped within them are reduced in size. This in turn affects the shoot growth the following year. In heavy seed-crop years, the radial growth of European beech over 100 years old was markedly reduced to about one-half of that in years unaffected by seed crop; and it was also considerably reduced in the following 2 years (Holmsgaard, 1955).

In yellow birch, heavy fruit crops can lead to crown dieback as well as reduced radial growth. For example, the enormous seed crop of 1967 in Ontario was eight times greater than the previous year, which had been regarded as a good seed year (Gross and Harnden, 1968; Gross, 1972). Fruits were produced at virtually every bud site, and three to five male catkins were present on nearly every shoot terminal (Figure 3.5). The nutrient demand for flowering and fruiting suppressed early leaf expansion, and leaves in heavily fruiting crown areas rarely developed to more than one-quarter of normal size. The small leaves contributed to poor vegetative growth and small buds or lack of buds on associated shoots. Marked dieback was evident in 1968 (Figure 3.6), and a new crown developed below the bare branches. Radial increment in 1967 was reduced to 47 percent of the annual average increment, and diameter growth of current shoots (Figure 3.5) was less than 5 percent of that in the previous year. Negligible fruit crops were produced the two following years.

**Figure 3.5.** **Yellow birch shoots and fruiting catkins. A. A typical shoot terminal from a heavily seeded upper crown as it appeared in December 1967. *(a)* Dead terminal showing staminate catkin scars. *(b)* Fruit attached to a budless short shoot. Note the small leaf scars indicative of dwarfed leaves in 1967. *(c)* Latent bud developed in 1966, which remained dormant in 1967. B. Part of a shoot terminal showing *(a)* catkin scars, *(b)* buds rated as small, and *(c)* a long shoot typical of the type present in heavily seeded upper crowns. (After Gross, 1972. Reproduced by permission of the National Research Council of Canada from the *Canadian Journal of Botany*, Vol. 50, pp. 2431–2437, 1972.)**

Because flowering is under at least moderate genetic control, a dilemma presents itself in the genetic improvement of forest crops. Early and abundantly flowering genotypes may be highly advantageous in the seed orchard where high seed production is desired from selected trees. However, their progeny may exhibit the same abundant flowering—hence reduced vegetative growth—when planted in the field. Here increased wood volume rather than seed production is sought.

**Figure 3.6. Marked dieback in the upper crown of a yellow birch photographed in 1968. This tree is typical of many that developed extensive dieback in those parts of the crown that bore heavy seed crops in 1967. (After Gross, 1972. Reproduced by permission of the National Research Council of Canada from the *Canadian Journal of Botany*, Vol. 50, pp. 2431–2437, 1972.)**

**Dispersal.** Seeds are dispersed by wind, water, animals, and by combinations of these agents to microsites where they may germinate and become established seedlings. The mechanism of seed dispersal is one of the functional traits of a woody species that enables it to compete in a particular spatial and temporal niche. In general, most tree species disperse their seeds locally, with a high proportion falling within 40 to 50 m or so of the parent. However, a few seeds of many species are carried long distances, and many seeds of a particular group of species are widely distributed by water and wind. At this latter extreme are pioneer trees such as willows, cottonwoods, sycamores, birches, and aspens. Although most of the seeds of even these species may be dispersed locally, many are blown or carried by water to disturbed habitats where establishment is far more probable than in the locality of the parent. Seedlings of these light-seeded, shade-intolerant, short-lived pioneer species are less likely to reestablish in the same locality where their parents have matured sexually than in freshly disturbed sites some distance away. In the presettlement forest, disturbance was so common that widespread dispersal was a successful means of reproducing pioneer species by sexual reproduction.

The majority of trees and shrubs of northern latitudes, however, have a more or less local dissemination pattern. Heavy-seeded species, such as walnuts and oaks, are distributed in the general vicinity of the parents, primarily by animals. It is of selective advantage for their seeds to be dispersed relatively close (10 to 30 m) to the parent where soil–site conditions are likely to be similar to those where the parent tree became established. This may be of such overriding importance that these species have evolved various other adaptations that allow them to establish and compete with the plants and animals of this habitat. Black walnut, for example, is highly sensitive to site conditions, failing to grow where the soil is too wet, too dry, or too shallow. Being relatively site specific, its heavy seeds are an adaptation for ensuring that they reach the ground under the trees. Squirrels and other animals then disperse the seeds to microsites away from the parent tree but similar in soil–site conditions.

Most pines have moderately heavy, winged seeds, and most seeds are wind disseminated within 50 m of the source. In longleaf pine, the maximum amount of seed per square meter falls within 10 m of the base of the tree (Buttrick, 1914). In one study, about 80 percent of the seeds caught in traps (placed at increasing distances from an isolated mature tree) fell within 40 m of the tree. Similarly, for shortleaf pine in Arkansas, Yocom (1968) reported that one-half of the seeds trapped in a forest opening fell 10 m from the edge of the stand; 85 percent fell within 50 m of the stand edge. Because pines are typically regenerated by fire, and because chances are good that fire will occur within the lifetime of the parent tree, regeneration is likely to occur in the vicinity of the parent.

Bird and mammal dispersal for the stone pine and pinyon pine groups are much more important than wind. Seed wings are absent or rudimentary in stone pines (limber, whitebark, and Swiss stone, among others), and seed wings of the pinyon pines remain attached to the cone scales as the cones open and seeds are shed. Bird and mammal dispersal is extremely important for many forest species and is considered in Chapter 13.

The typical exponential dispersal pattern of many tree species may be characterized by the relationship shown in Figure 3.7 (Roe, 1967a). The dispersal of Engelmann spruce seeds of "bumper" seed years from four stands into adjacent openings illustrates that most seeds fall near the parent. In these four stands, approximately 70

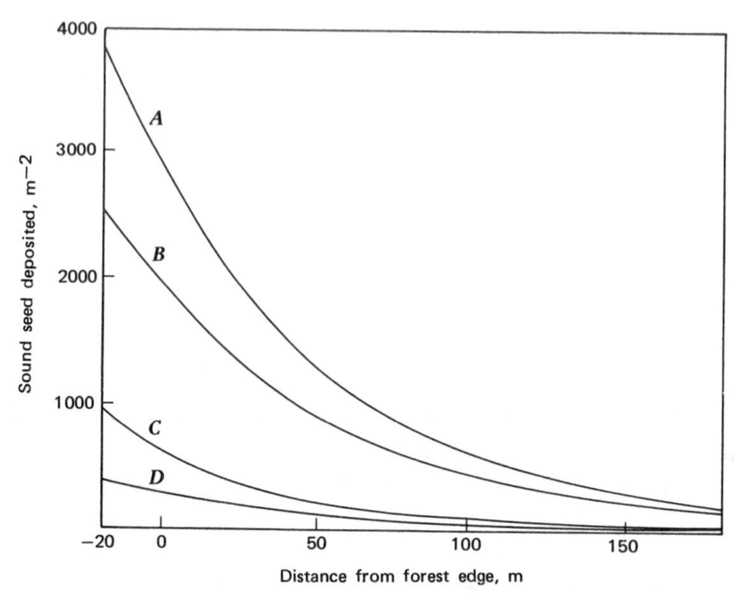


**Figure 3.7. The distribution of seeds of Engelmann spruce in good seed years from the edges of forest stands into adjacent openings. *A*—Togwotee Pass (1964), Teton National Forest, Wyoming; *B*—Falls Creek (1952), Flathead National Forest, Montana; *C*—Griffin Top (1964), Dixie National Forest, Utah; *D*—Fisher Creek (1964), Payette National Forest, Idaho. (After Harper, 1977, © John L. Harper; data from Roe, 1967a.)**

percent of the seeds fell within 50 m of the edge of the standing timber. However, in continuous forest stands, tree crowns of the overstory and understory intercept many seeds. Thus dispersal distances into openings are greater than may be expected within the stand itself.

Dispersal of seeds may occur over a short or long time period, and this depends in part on the establishment adaptations of the seedlings. Seeds of elms, aspens, and silver maple, all spring germinators, are dispersed over a relatively short period in the spring. In marked contrast, some pine species (jack, lodgepole, Monterey, Virginia) have cones that remain closed (termed **serotinous** cones) up to 5 years or indefinitely. These species are adapted to wildfires, whose heat opens the cones, and their seeds are dispersed on a fire-prepared seedbed (Chapter 11). In most other pines, cones open shortly after ripening and seeds are rapidly dispersed over a 4- to 8-week period.

The effect of weather conditions on duration of seedfall of eastern white pine is illustrated in Figure 3.8 (Graber, 1970). In 1965, 75

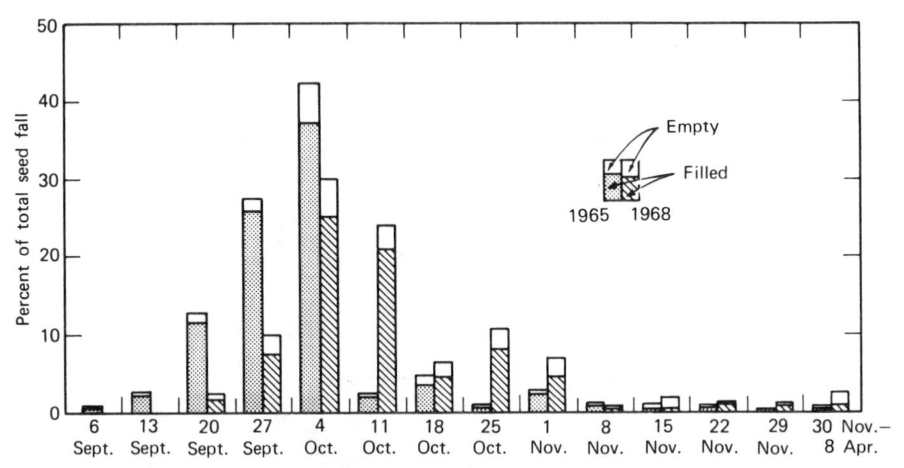


**Figure 3.8. The time of seed fall of eastern white pine in 1965 and 1968, southwestern Maine. (After Graber, 1970.)**

percent of the filled seeds were shed over a 3-week period under warm and dry conditions. In contrast, cool, moist conditions of 1968 delayed and lengthened the dispersal period; approximately the same amount of seed was shed over a 7-week period.

In moist, cool climates seeds of conifers, notably hemlocks, spruces, and firs, are typically dispersed over many months. In fir species, dispersal usually begins in September or October and continues over the winter as the cones disintegrate and release the seeds. In black spruce, cones are retained on the tree for several years in a semiserotinous state. The cones open when fire occurs or when the weather is warm and dry; they close when cool and moist conditions prevail. A continuous seed supply is thereby assured. Another adaptive advantage is that the seeds are not all on the forest floor where a wildfire would destroy them.

**The Seed Bank, Dormancy, and Germination.** Seeds disseminated to the forest floor are stored there, in what Harper (1977) terms the seed bank, for a shorter or longer time until they germinate (Figure 3.9). Seeds of many forest trees arrive at the forest floor in a nongerminable or dormant state. They remain in the dormant seed bank until internal and external conditions are favorable for germination. Seeds of some tree species are capable of germinating immediately following dispersal, and they enter directly the active seed bank (Figure 3.9). Compared to seeds of grasses and herbaceous plants, tree seeds are relatively uncommon in the dormant seed bank (Harper, 1977). The period of residence of viable tree seeds in the forest floor is relatively short; being relatively large, they are eaten by animals, and they decompose easily.

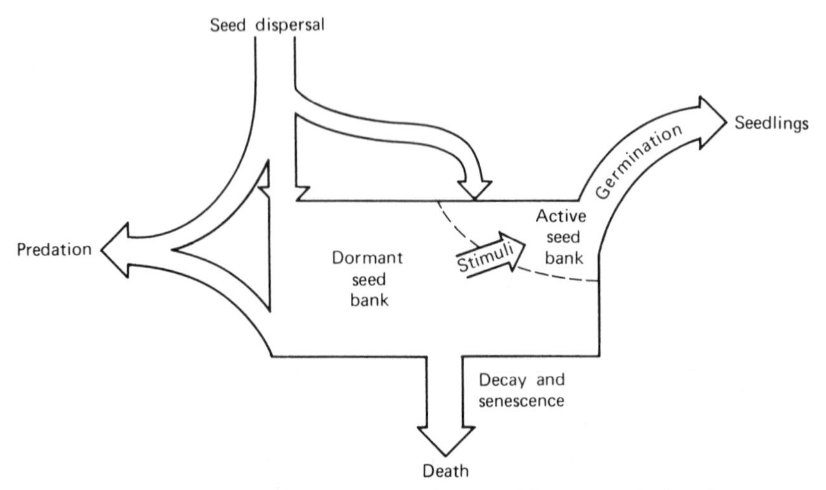


**Figure 3.9. Conceptual model of the seed bank and the dynamics of the population of seeds. (Modified from Harper, 1977, © by John L. Harper.)**

Seeds of some species, particularly those disseminated in the spring (e.g., eastern cottonwood and silver maple) germinate readily a few days or weeks after being dispersed. To germinate, viable seeds (1) imbibe water, (2) activate metabolic processes, and (3) initiate growth of the embryo. Seeds of species which are unable to satisfy these requirements (i.e., blockage of any of these functions) are said to be in a state of dormancy.

In species with nondormant seeds, the seeds are typically adapted to certain favorable germination conditions immediately following dispersal. This is the case for seeds of spring-disseminated, river floodplain species. Also, seeds of many species of pines, especially the hard-pine group (including jack, lodgepole, Virginia, ponderosa, pitch, and longleaf), lack a dormancy period. Being strongly adapted to fire, their seeds are ready to germinate on the fire-prepared seedbed. Other pines, including all pines of the soft-pine group (white, sugar, stone pines, among others), exhibit a pronounced period of seed dormancy, even though they are also fire-dependent species.

During the breaking of dormancy, morphological and physiological changes must occur before the seed is capable of germinating. For seeds that overwinter in the forest floor, these changes take place slowly, and the seeds typically germinate in the following spring. Seeds of some species, however, including basswood, white ash, and black cherry, may require much longer periods before their dormancy requirements are satisfied. Thus their seeds are stored in the dormant seed bank for varying periods and germinate 1, 2, or even 3 years after dissemination. Some seeds of many northern and high-elevation conifers (soft pines, Douglas-fir, some firs and spruces) may also germinate in the second year after dissemination. Seeds of many shrub species (*Ribes*, *Rhus*, *Ceanothus*, and many chaparral species) may be stored for several years in the soil and may germinate vigorously after fire.

At this point we must emphasize the great variation in various factors, particularly seed size and seedbed environment, that affect the dormancy of overwintering seeds such that they germinate over a period of weeks or even years. Seed size and weight vary greatly among different individuals of a species and even markedly within a single tree. For example, weight variation per seed as much as two- to threefold is not uncommon in pine species. Besides affecting dispersal distance, weight and size may be related to predation by animals, penetration of the forest floor, and internal requirements for breaking dormancy and for early growth. Seedbed conditions and the microenvironment where the seed resides are extremely varied in nature, and no two seeds in the dormant seed bank reside in exactly the same conditions. Thus the time of germination in the spring may vary over several weeks, and in many species some seeds may not germinate until the following spring.

Adaptive selection pressures have favored various kinds of mechanisms that cause dormancy in one form or another in as many as two-thirds of American trees (USDA, 1974). In satisfying dormancy requirements so as to germinate under favorable environmental conditions for establishment, seeds undergo various morphological and physiological changes. Imbibition of water and oxygen may be prevented by a hard, impermeable seed coat as in many legumes (black locust, for example) and basswood. Cracking or dissolving of the pericarp facilitates germination. Immature embryos may result from unfavorable conditions during maturation, and morphological changes must take place to complete the embryo or enlarge it before germination may occur. In addition, physiological changes are usually required in the embryo, cotyledons, or endosperm before metabolic processes are activated. In nature, the moist, cool conditions of the forest floor over a period of weeks or months act to decrease germination inhibitors, increase germination-promoting growth regulators, and create favorable internal conditions for germination. To hasten the germination process artificially, many treatments and techniques have been employed, including mechanical abrasion or chemical treatment of seed coats, and stratification—the placement of seeds at low temperatures (1–5°C) and moist conditions from a week to four months. A comprehensive treatment of the requirements for breaking dormancy and stimulating germination for North American and many introduced species is given by the USDA (1974).

Filled, viable seeds are a prerequisite for germination. Many seeds of normal size may contain only a shriveled endosperm and embryo. These so-called "empty" seeds often result from either lack of pollination or the failure of fertilization to occur after pollination, or in pines and probably other conifers, the union of lethal or sublethal genes at the time of fertilization (Sarvas, 1962). In addition, filled or sound seeds may be nonviable, that is, they are deeply dormant or are inherently incapable of germinating. The highest percentages of filled seeds apparently occur in the most productive seed years. For example Curtis (1959) found 37 percent filled seeds of sugar maple in good seed years (average of 9.5 million seeds $ha^{-1}$) whereas only 11 percent were filled in the intervening average and poor seed years (average of 770,000 seeds $ha^{-1}$).

The degree of dormancy varies among individual trees of a local population and among races of a species. In general, more northern races require a longer period to satisfy dormancy requirements than southern races. This has been demonstrated for sugar maple (Kriebel, 1958), sweetgum (Wilcox, 1968), and sycamore (Webb and Farmer, 1968).

**Establishment.** Upon germination, woody plants put out roots, stems, and leaves, and the seedlings cope with their environment. Because thousands or millions of seedlings perish soon after germination, the few seedlings that survive and exhibit vigorous growth are regarded as established. The period of establishment typically lasts for 1 to 3 years or more depending on the species and site conditions.

Establishment is the single most critical stage in the life history of an individual. As indicated in earlier sections, a great many adaptations have occurred that enable the species to produce and disperse seeds in such a way that many seedlings are likely to encounter favorable conditions and survive. Each species has its own establishment ecology. Seedlings must not only survive under the particular physical site conditions of their seedbed, but compete with seedlings of their own species as well as plants already established on the area. These interactions bring about the apparent order and organization that enable us to recognize particular forest communities characteristic of different habitats.

Although each species exhibits different establishment adaptations, two general patterns of germination and early seedling development may be contrasted: species with **epigeous** and those with **hypogeous** germination and development. In the epigeous condition, the cotyledons, often still with the pericarp (the fruit wall) attached, are elevated above the surface by the elongating hypocotyl (Figure 3.10, *A* and *B*). This is the typical pattern of nearly all conifers and most angiosperms. In hypogeous species (oaks, hickories, walnuts, buckeyes), the cotyledons remain below the surface (remaining attached to the seedling for weeks or months) while the epicotyl grows upward and develops true leaves (Figure 3.10, *C* and *D*). Species with epigeous development store relatively little food in endosperm and the cotyledons; they rely strongly on the cotyledons for photosynthesis to stimulate early root development. Four stages in cotyledon development are recognized (Marshall and Kozlowski, 1977):

- **Storage.** Cotyledon cells are packed with stored foods (fats, carbohydrates, proteins) and mineral nutrients. These foods and nutrients are utilized in the first days of seedling growth.
- **Transition.** When exposed to light, a series of changes takes place: chloroplasts develop and chlorophyll synthesis begins, stomates develop, epidermal cells expand and large intercellular spaces form in the mesophyll.
- **Photosynthesis.** Major contributions in photosynthesis occur to further development of the shoot and tap and lateral roots. In black locust, red maple, and American elm, appreciable photosynthesis begins 4 to 6 days after radicle emergence (Marshall and Kozlowski, 1976). Photosynthesis peaks 8 to 15 days after radicles emerge and continues for about 4 weeks.
- **Senescence.** Dry weight decreases and some mineral nutrients are translocated back into the seedling as the cotyledon function declines.

The cotyledons are extremely important to the development of seedlings during the first few weeks. Any damage to the cotyledons, such as might be caused by animals or frost, inhibits seedling growth.

In contrast, hypogeous seedlings have large fleshy cotyledons that remain below ground during seedling development and are enclosed in the pericarp (Figure 3.10, *C* and *D*). The large amount of stored food favors extensive root development prior to the development of a transpiring shoot and leaf system. The cotyledons also store considerable water and have enough food reserves to reestablish the epicotyl should it become damaged. Furthermore, being underground and inside the pericarp, the cotyledons are protected from browsing animals.

The hypogeous system is characteristic of the large-seeded tree species whose seeds are typically buried by squirrels and rodents

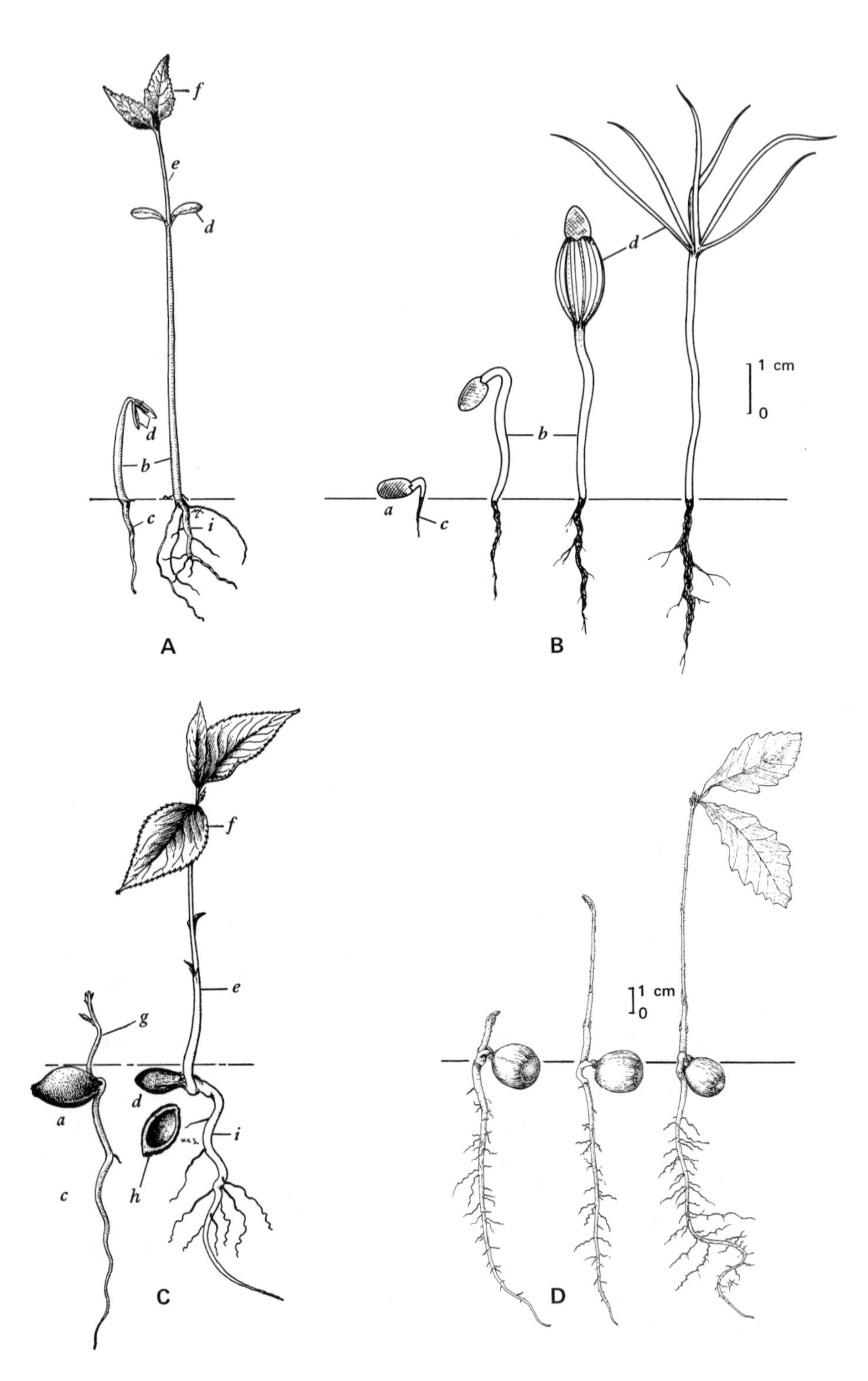
f
e
d
d
b
c
i
A
d
1 cm
0
b
a
c
B
f
e
g
a
d
i
c
h
C
1 cm
0
D

and distributed not far from the parent tree. Many oaks and hickories grow in dry sites characterized by summer drought. Their seedlings also typically establish themselves under overstory trees whose crowns shade the forest floor and whose roots compete for moisture and nutrients. Thus production of relatively few but well-provisioned seeds has proved of adaptive advantage. The production of many smaller seeds with less stored food and epigeous development also has proved of selective advantage for many other species for their sites and methods of establishment. In mixed hardwood forests of the central and eastern North America, species with both systems often grow together under the same general site conditions: shagbark hickory, red and white oak (hypogeous) and sugar maple, white ash, and beech (epigeous).

Following germination, young seedlings pass through a succulent stage during the first several weeks. At this time tissues are soft and highly susceptible to fungal infections (especially damping-off), damage by animals, smothering, and desiccation. Tissues soon begin to harden, and there follows a juvenile period when the seedling becomes increasingly hardy but still subject to mortality. Seed germination and survival of Douglas-fir seedlings throughout the most critical part of the growing season are illustrated in Figure 3.11 (Lawrence and Rediske, 1962). A major reason for the low accumulative germination and survival is that 46 percent of the seeds were destroyed by fungi and animals prior to germination; another 27 percent failed to germinate the first year. The diagram illustrates the timing of factors causing mortality and the relatively few seedlings left at the end of the first growing season. A detailed discussion of the period of seedling establishment and the factors affecting seedling survival is presented by Baker (1950) and Kozlowski (1971).

Of particular interest to the forest ecologist are the kinds of habitats and seedbeds where seedlings of different species become

---

**Figure 3.10. Epigeous and hypogeous seed germination and development. A—epigeous germination of pin cherry seedlings at 1 and 10 days; B—epigeous germination of red pine at 1, 2, 6, and 10 days; C—hypogeous germination of Allegheny plum seedlings at 1 and 9 days; D—hypogeous germination of bur oak seedlings at 1, 5, and 12 days. *(a)*, seed; *(b)*, hypocotyl; *(c)*, radicle; *(d)*, cotyledons; *(e)*, epicotyl; *(f)*, leaves; *(g)*, plumule; *(h)*, pericarp; *(i)* primary root. The plumule consists of the epicotyl and the emerging leaves. (A, C, and D after USDA, 1974; B by W. H. Wagner Jr.)**

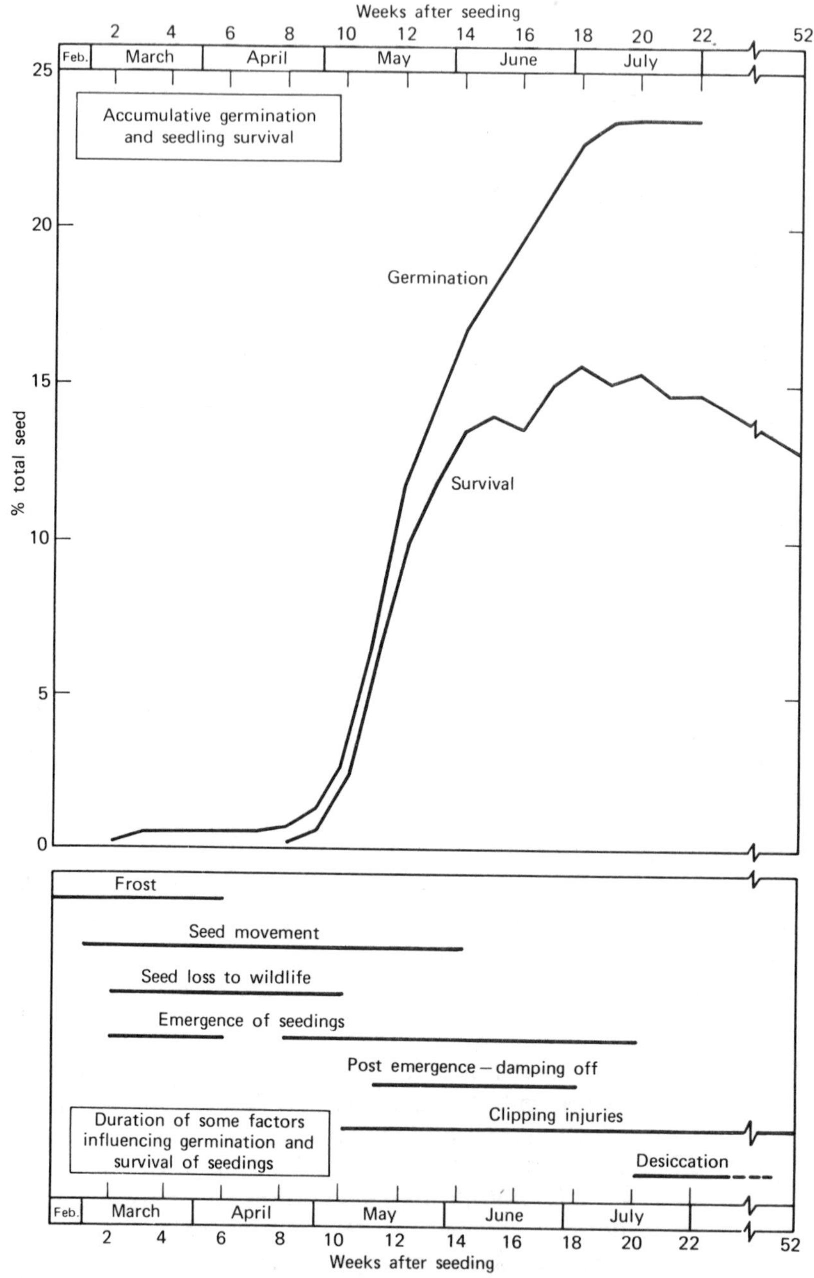


**Figure 3.11.** **The cumulative germination and survival of Douglas-fir seeds during the first growing season. 440 Scandium$^{46}$-tagged seeds were seeded by hand in February and their fate carefully followed. (After Lawrence and Rediske, 1962.)**

established. The vast majority of tree species are dependent either on fire, flooding, or windthrow to provide suitable seedbeds for their establishment. Fast-growing, light-demanding species (all pines, willows, cottonwoods, aspens, paper birch) are adapted to catastrophic disturbances of fire or flooding, and they can thrive in these open and extreme sites. The adaptations of fire species are presented in Chapter 11.

While the establishment of tree species is favored or not severely limited by partial shade of an overstory, few species can establish and persist in a heavily shaded understory. Similarly, although most seedlings of tree species can become established under continuously moist forest soils, only a few are adapted to establish and persist as soils dry out and moisture stress becomes severe. Thus each species has a unique set of adaptations that facilitates establishment under certain physical and biotic conditions. Although it is difficult to generalize, we have split the continuum of establishment systems into three groups of species that differ in establishment pattern in relation to the amount and kind of disturbance.

1. **Pioneer species** seed into open areas following major disturbances such as fire and flooding. Little existing tree competition is present, and the environment is harsh or extreme (hot, dry, wet, exposed). Germination and growth are rapid. Roots soon penetrate deep enough into the soil, as in the case of pines, to enable the seedlings that eventually become established to withstand summer drought year after year. Floodplain species tolerate variable water levels and have the ability to generate adventitious roots from their stems if the stems are covered with silt.

   Fire-dependent pioneer species, such as the pines of southeastern North America (loblolly, longleaf, shortleaf, and slash) typically establish in dense, even-aged stands on areas that have been recently burned. Fire prepares the seedbed in various ways (Chapter 11), and the seedlings develop rapidly during the first year (Figure 3.12; Wakely, 1954). Tap-root development is rapid because considerable moisture is needed to maintain the water balance in the needles and shoots that develop rapidly in the ensuing juvenile period.

2. **Gap-phase species** germinate and establish under the existing forest canopy and are shade-tolerant enough as seedlings so that a small number exist until a local disturbance enables them to penetrate a gap in the canopy (see Chapter 14). White ash, black cherry, white and red oak, red maple, yellow birch, basswood, black walnut, slippery elm, white spruce, and eastern and western white pine, are representative species of this group.

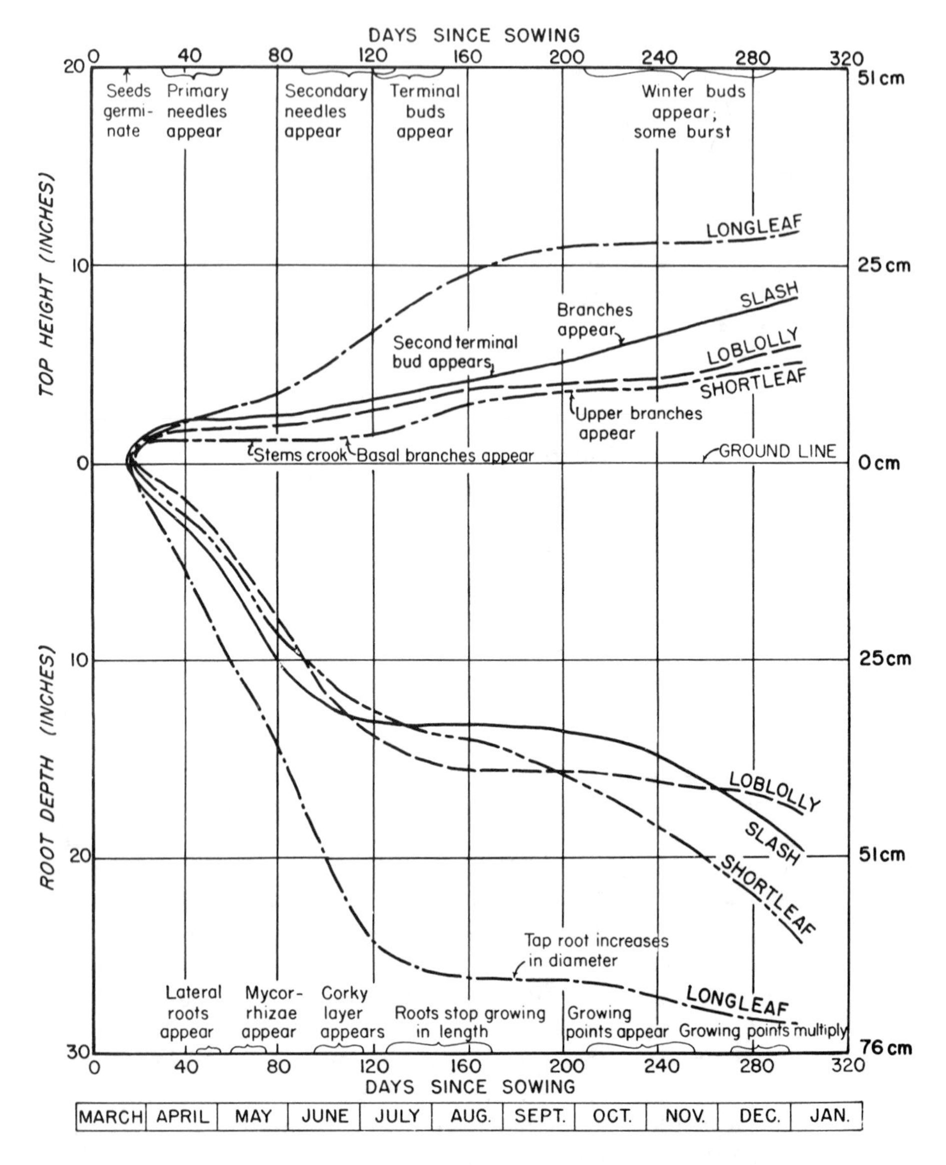


**Figure 3.12. The course of first-year shoot and root development of four species of pines under favorable conditions for establishment (in a forest nursery) in Louisiana. (After Wakeley, 1954.)**

3. **Extremely shade-tolerant species** establish in the shaded understory and persist for long periods. They gradually penetrate the canopy as overstory trees die or windthrow provides openings for them. Sugar maple, American beech, hemlocks, many true firs, and western red cedar belong to this group.

Groups 2 and 3 are similar in that seedlings are established under a forest canopy and must persist in the understory until conditions are favorable for growth into the overstory. Many different kinds of adaptations have evolved, enabling the various species to persist in the understory and respond to release depending on (1) the physical and biotic conditions of the understory, (2) the species' intrinsic growth rate, and (3) the nature of disturbance that sooner or later releases the young trees so that they may reach the overstory (e.g., fire vs. windthrow). In both groups (but much more so in group 3), there is an accumulation or buildup of a few to hundreds or millions of seedlings per hectare in the understory.

Species of the two groups have different functional adaptations that enable them to tolerate and survive the physical rigor (low light and moisture stress) and biotic hazards (herbivores and diseases) of the understory environment. Gap-phase species typically require higher light irradiance to survive than the extremely shade-tolerant species, but they can better tolerate moisture stress. In addition, the deciduous gap-phase species have the marked ability to sprout after fire or animal damage and continue to persist in the understory. In contrast, we have constituted group 3 with the most shade-tolerant and slow-growing species; as a group they require more moisture and have less sprouting ability. They may accumulate large populations and experience high mortality. However, they endure to shade out any seedlings that might try to establish under them. Sometimes these seedlings build up for 20 to 30 years or more. For example in spruce-fir forests, Ghent (1958) found that a severe attack of the spruce budworm, that seriously damaged the overstory, released balsam fir seedlings. Most of them (75 percent) were 1 to 25 cm high and were about 12 years old on the average. In extreme cases, understory hemlocks may be 100 or more years old. Thus in these two groups, the trees eventually reaching the overstory have been released from a "bank" of established and persisting seedlings.

The establishment ecology of two associated species of the northern hardwood forest, yellow birch (group 2) and sugar maple (group 3), provides an example of a representative of each of the two nonpioneer groups. Although these two species may grow side by side in the northern forest, their establishment ecology and realized niche are markedly different. In contrast to pioneer species, members of the gap-phase and shade-tolerant groups become established on a forest floor that is more or less shaded by an existing overstory.

Sugar maple disperses seeds in the early fall (before snowfall) during leaf drop. Although yellow birch seeds also mature in the early fall, they are gradually dispersed throughout the winter. They may be blown for great distances on top of the snow crust that forms after mild thaws (Figure 3.13*a*; Tubbs, 1965). Sugar maple seeds germinate in early spring under the snow and a layer of leaves where temperatures are only slightly above freezing (about 1°C) (Figure 3.13*b* and *c*). At the same time yellow birch seeds still rest on top of the snow (Figure 3.13*a*). As the snow melts, birch seeds come to rest

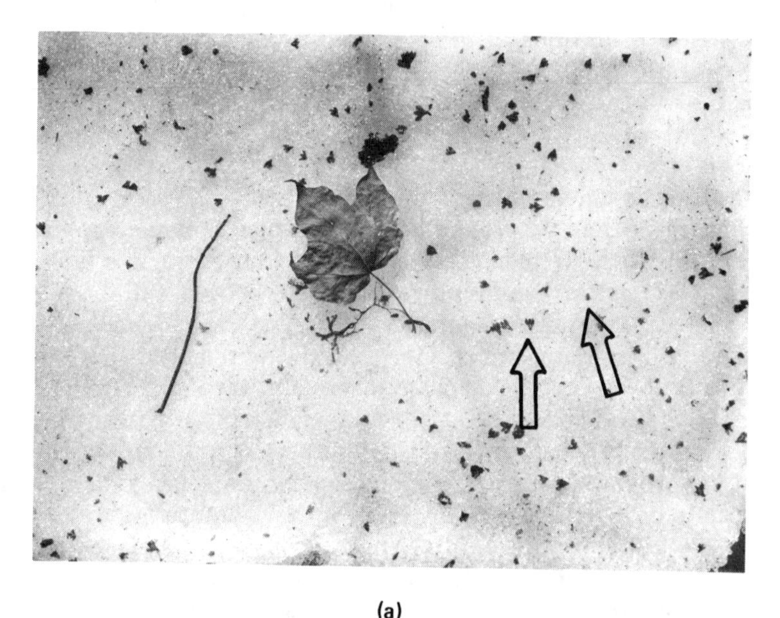

(a)

**Figure 3.13. Sites of seed dispersal and germination of sugar maple and yellow birch in the northern forest. (After Tubbs, 1965; U. S. Forest Service photos.) *(a)*—Yellow birch bracts and seeds litter the surface of the melting snow on April 30 in a northern hardwood stand on the Upper Peninsula Experimental Forest. The overstory is composed primarily of sugar maple of seeding age, but seldom are maple seeds observed on top of or within the snow cover even after a bumper seed crop. Right arrow indicates yellow birch seed; left one points to bract. *(b)*—Removing the snow from the exact spot shown in *(a)* to the top of the previous fall's leaf layer reveals no seed of any species. The leaves are compressed into a soggy mat, which is often partially frozen. Spring ephemerals have pushed through the mat (arrow). *(c)*—When the**

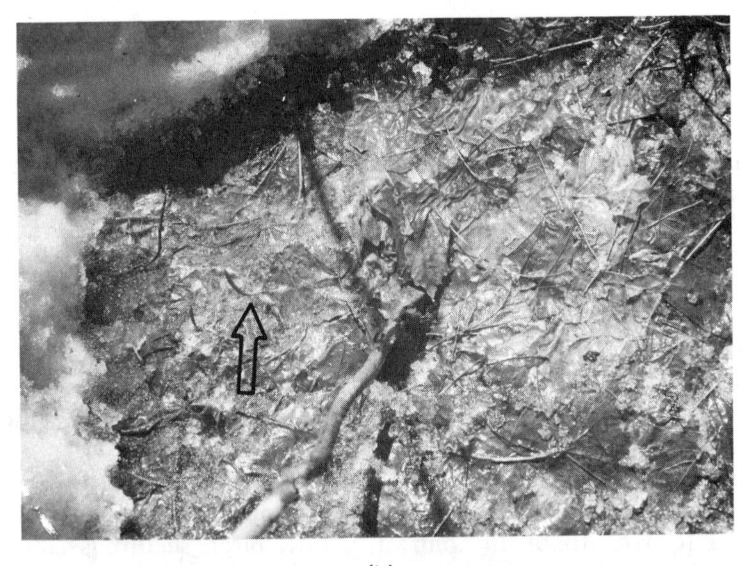

(b)

(c)

**top leaf layer shown in *(b)* is removed, the ability of sugar maple to germinate underneath a snow cover in early spring is revealed. Arrows point to germinated seed. In those areas sampled on the Experimental Forest, the bulk of the sugar maple seed was found under a layer of leaves whereas yellow birch seed occupied the top of the snow as illustrated in *(a)*.**

on the forest floor and germinate in late spring at higher temperatures (about 10°C). Radicles of maple seedlings penetrate the wet leaf mat and, following a good seed year, establish by the millions, often forming a carpet of first-year seedlings on the forest floor. The tiny radicles of the small birch seeds (over 50 times lighter than sugar maple seeds) are unable to penetrate the thick leaf mat, which tends to dry out rapidly. Most yellow birch seedlings soon desiccate and die. However, some yellow birch seeds come to rest by chance on rotten logs, moss-covered rocks, mineral soil of mounds that have formed as a result of a tree being blown down, and in shallow depressions that stay moist throughout most of the growing season. In these microsites, birch seedlings may become established provided sufficient overhead light is available.

Sugar maple seedlings are highly shade tolerant and can survive the shaded understory environment better than any other tree species (Curtis, 1959). In contrast, yellow birch seedlings cannot endure heavy shade. However, at higher light levels yellow birch inherently grows faster than sugar maple. Thus when a gap in the canopy occurs, an appropriately positioned yellow birch is able to outcompete the slower-growing maples for a place in the overstory canopy. Similarly, many other gap-phase species are rapidly able to colonize openings in the canopy; but they must be able to endure the rigors of the understory (low light, high root competition, smothering by leaves and woody debris, and the attacks of insects, disease, and herbivores) until a gap develops. Over much of the upland northern forest, sugar maple is able to dominate both understory and overstory because of its effective seed production, dispersal, germination, and establishment adaptations.

**Postestablishment Development.** Under favorable environmental conditions, tree species exhibit rapid growth in the juvenile stage. This grand period of growth is followed by a leveling off of the growth curve and a declining rate in old age. Different rates of growth are typical for different species. They are generally correlated to the shade or understory tolerance of the respective species (Chapter 14). For convenience, silviculturists recognize a sequence of size classes as trees increase in diameter: seedling, sapling, pole, small sawtimber, large sawtimber. The high mortality of the establishment period, in terms of number of stems, decreases with time. Trees increase in size and their density (number of stems per unit area) decreases. Competition, however, does not diminish. Competition in the understory and overstory of the forest community is considered in Chapter 14.

Radicles of maple seedlings penetrate the wet leaf mat and following a good seed year - establish by the millions - often forming a carpet of first year seedlings on the forest floor.

The tiny radicles of the small birch seeds (over 50 times lighter than sugar maple seeds) are unable to penetrate the thick leaf mat.

which tends to dry
out rapidly. Most yellow
birch seedlings soon
desicate and die. However
some come to rest
by chance on rotten logs,
moss covered rocks, mineral
soil of mounds that
have formed as a
result of a blown down
tree and in shallow
depressions that stay
moist. In these micro sites
yellow birch might become
established provided enough light is
available

### Vegetative Reproduction

In many situations, particularly following disturbances of various kinds, reproduction by vegetative means is more important for the survival of woody plant populations than sexual reproduction. All woody angiosperms are able to reproduce themselves vegetatively once they have become established; it is a major fitness trait (Chapter 2). Conifers are less adapted to reproduce by vegetative means. This advantage of angiosperms may be one of the major reasons for the dominance of angiosperms over conifers in diversity and abundance since the Cretaceous Period.

Vegetative reproduction enables the plant to survive and reestablish itself in place after disturbance and often expands the portion of the site it occupies. Sprouting and the rooting of branches are the primary types of vegetative reproduction common in woody plants. The specific methods are briefly described below:

- **Root-collar sprouting.** New shoots develop from dormant or adventitious buds from the root-collar at the base of an established plant. They form multiple-stemmed shrubs or trees (oaks, hickories, basswoods, ashes, walnuts, birches, and species of many other genera).
- **Lignotuber sprouting.** New shoots arise from a buried mass of stem tissue termed the lignotuber; it is characteristic of eucalypts.
- **Root sprouting.** New shoots arise from adventitous buds on roots (or rhyzomes) as in *Vaccinium*, *Symphoricarpos*, and many other shrubs. Root sprouting is characteristic in clone-forming trees and shrubs: aspens, beeches, sassafras, sumacs, and sweetgum, among others.
- **Fragmentation.** Branches that are broken off willow trees by wind often take root after being buried in soil of the stream bank.
- **Stolons.** Arching branches of shrubs such as red-osier dogwood (*Cornus stolonifera*) take root when they come in contact with the soil surface.
- **Runners.** Procumbent stems or runners take root at various points along their length as they come in contact with the soil (creeping strawberry bush, *Euonymous obovatus*, and many woody vines).
- **Layering.** Lower branches of cold-climate conifers are often pressed into the soil by the weight of snow. They take root and form colonies around the parent tree (spruces, firs, larches).
- **Tipping.** Northern white cedar trees in swamps are often tipped by wind and go down slowly, eventually coming to rest on the

peat surface (Curtis, 1959). The ascending branches of the tree develop into independent trees as moss covers the old tree, and roots form at the base of each branch.

Besides maintaining individual genotypes for long periods of time in disturbance-prone environments, vegetative reproduction expands the area occupied by many shrubs and trees: blueberries (*Vaccinium* spp.), dogwoods (*Cornus* spp.), sumacs, sassafras, beech, aspens, sweetgum, and sycamore. Trembling aspen clones, consisting of many genetically identical ramets up to 0.1 ha in size, are very common in burned-over forests, and clones of very large size occur in the Interior West of North America. Figure 3.14 illustrates a clone in south-central Utah 43 ha in extent (see Chapters 2 and 16).

**Figure 3.14. Large clone of trembling aspen, 43 ha in extent, with the boundary outlined; south-central Utah. Note the smaller clones around the large clone; the differences in tone indicate differences in fall coloration of clone foliage. (After Kemperman and Barnes, 1976. Courtesy Canadian Forestry Service. Reproduced by permission of the National Research Council of Canada from the *Canadian Journal of Botany*, Vol. 54, pp. 2603–2607, 1976.)**

Vegetative reproduction plays a significant role in the juvenile period following establishment of deciduous tree seedlings. Although fire or a herbivore may damage or destroy a portion or all of the above-ground part of an established seedling, it can sprout back vigorously. Many species are able to withstand repeated browsing or defoliation and still resprout for many years. Sprouting ability is greatest in the juvenile stage (when it is of greatest adaptive significance) and declines as the plant matures. All deciduous trees and shrubs are capable of sprouting from roots, root-collar or stem, and some are particularly vigorous: oaks, hickories, aspens, white and red ash, red alder, red maple, paper birch, basswood, yellow-poplar, and eucalypts.

Conifers rarely reproduce vegetatively. Even in the northern, boreal, and alpine areas where reproduction by layering is common, conifers reproduce primarily by sexual means. Only a few species are able to sprout vigorously and regenerate themselves after fire, grazing, or cutting: redwood sprouts vigorously from root-collar, stump, and stem; and various hard pines, including pitch pine, Virginia pine, and pond pine sprout from dormant buds along the stem after fire or cutting (Kozlowski, 1971). Near the upper limit of tree growth in the southern Rocky Mountains, Marr (1977) reported "tree islands" of alpine fir and Engelmann spruce that moved along the ground by repeated layering and growth to leeward (Figure 3.15). Movement of 5 m in 11 years was common, and some clonal islands apparently have moved at least 15 m. Seedlings apparently become

**Figure 3.15. A tree island of subalpine fir with the typical mid-winter leeward snow drift in Niwot Ridge, Colorado Front Range. Dead windward branches (left) form an interconnected meshwork and some are connected with currently live stems with needles. (After Marr, 1977. © by the Ecological Society of America; photo by John W. Marr.)**

established in sheltered microsites, and the clones expand and colonize adjacent microsites that are inhospitable to seedling establishment.

## STRUCTURE AND GROWTH

The development of the seedling into a large tree is a complex process involving three dimensions of growth: extension of each growing point forming the shoots of the crown and the roots (**primary growth**), and expansion of stem and root diameter (**secondary growth**). A combination of these growth processes gives each species its characteristic aerial structure and form. We have selected a few features of woody plant structure and growth, and discuss them from an ecological perspective. For a comprehensive treatment of the physiology of tree structure and function, the reader is directed to books by Büsgen and Münch (1929), Kozlowski (1971), Zimmermann and Brown (1971), and Kramer and Kozlowski (1979).

### Tree Form

In northern and alpine environments, trees (primarily conifers) are genetically adapted to grow in a conical form (Figure 3.16*a*), apparently due to strong selection pressures of snow, ice, and wind. That this trait is under strong genetic control is demonstrated by the maintenance of the conical form when spruces, larches, and firs from high altitudes are grown in parks or arboreta far from their native site. In these species, the terminal shoot or leader grows faster than the lateral branches below it in such a systematic way that a central stem and a conical crown are the result. This is termed the **excurrent** tree form; it is typical of northern conifers and a few deciduous trees such as yellow-poplar, sweetgum, and white ash. The excurrent form is an expression of strong apical control, whereby the terminal leader maintains control year by year over the laterals below it (Brown et al., 1967; Kozlowski, 1971; Zimmermann and Brown, 1971).

Not all conifers exhibit the excurrent form, and the change to a more rounded and spreading form is associated with the species' climatic region. Pines of the southeastern United States, such as loblolly and longleaf, exhibit a strong central stem, but a more rounded crown than spruces and firs of the far north. In the arid Southwest, pinyon pines and junipers tend to have short, compact, rounded, and bushy crowns. This form is related to the high temperatures and moisture stress of their environment; tall excurrent growth would severely expose the crown to strong, dry winds.

**Figure 3.16. Examples of excurrent *(a)* and decurrent or deliquescent *(b)* forms of trees. (After Hosie, 1969. Courtesy Canadian Forestry Service.)**

In many deciduous species, the lateral branches grow nearly as fast or faster than the terminal leader, and the main stem may fork repeatedly giving rise to a spreading form (Figure 3.16*b*). This is termed the **decurrent** or **deliquescent** form; it is typical of elms, oaks, maples, and many other species. In these species, weak apical control operates, and a multiple terminal leader system is often developed; it is most pronounced in American elm and various oaks. The broad, spreading crowns may have an adaptive function in spacing out the relatively large deciduous leaves compared to the densely packed needles of many excurrent conifers. The spreading form allows light to reach the leaves, and the growth rate of the shoots acts together with branch angle, gravity, and other factors to control the density or compactness of the crown.

The kinds of shoots and the patterns of shoot growth are important in determining tree form. Before discussing shoot growth it is important to consider the importance of shoots in plant population biology. In the study of plant populations, two levels of population structure are important (Harper, 1977). One level is the number of different genetic individuals present (the number of original zygotes or genets), and another level is the number of modular units of plant construction. In trees, the convenient unit is the annual shoot (Büsgen and Münch, 1929), the collection of leaves and buds attached to the stem. For example, Figure 3.17 shows the characteristic sigmoid growth curve, here applied to the number of modular shoot units on individuals of staghorn sumac (*Rhus typhina*) of different ages. In studying the demography of tree populations in relation to environmental factors, the variations observed may be related to changes in the number of individuals or the number of shoot modules or some combination of them. Knowing only the actual number of individuals often gives us little information about the population. We will examine below the kinds of shoots that make up the modular units and also the patterns of shoot growth that affect tree structure and function.

**Short and Long Shoots.** Trees usually have two morphologically distinct kinds of shoots: short and long shoots. The short shoots may be called dwarf shoots in conifers and spur shoots in deciduous trees. The boundaries between consecutive annual shoots are marked by groups of ringlike scars left by the bud scales. Annual short-shoot growth is a few millimeters or less so that the annual ring scars lie virtually next to one another; one 28-year-old short shoot of yellow birch measured only 7 cm. In contrast, the annual growth of long shoots is measured in several to many centimeters. Long-shoot growth of over 3 m in one growing season is not uncommon in eastern cottonwood on alluvial sites. All trees have long shoots; short shoots are characteristic of pines, larches, gingko, beeches, birches, and maples. The frequency of short shoots tends to increase with age. For example, Wilson (1966) reported that short shoots accounted for 90 percent of the shoots on red maple trees over 30 years of age. Short shoots bear "early" leaves that are preformed in the overwintering bud. Long shoots typically have early leaves along their basal portion and "late" leaves, those formed during the current growing season, along the upper portion.

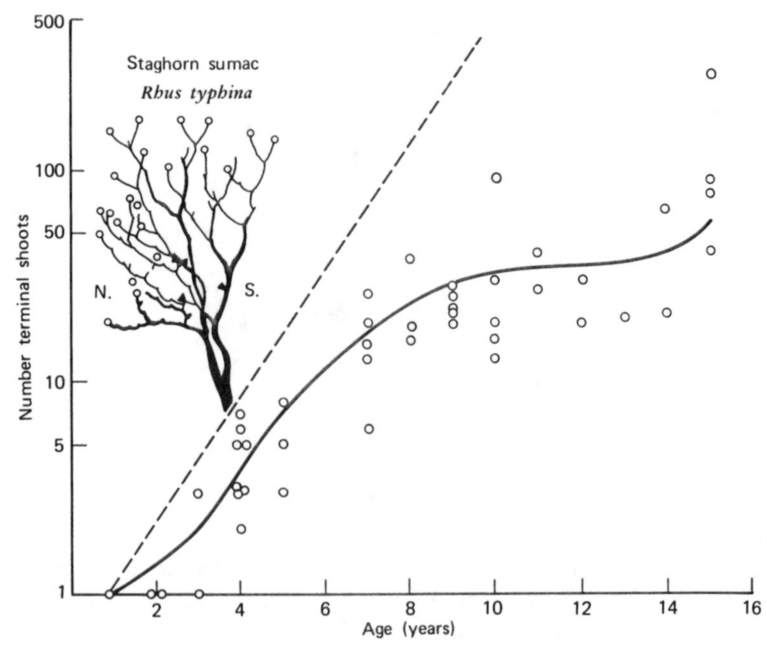


**Figure 3.17.** **Tree growth as the development of a population of shoot modules. Each point (○) on the graph represents the number of terminal growing points on a tree of a particular age. The species shown is staghorn sumac, *Rhus typhina.* The broken line indicates the number of shoots expected if each growing shoot would give rise to two new shoots in the next growth period. The terminal meristem in this species aborts. On the tree diagram ▲ indicates dead branches and ○ indicates inflorescences. (Courtesy of James White; modified from Harper, 1977; © by John L. Harper.)**

**Patterns of Intermittent Growth.** The growth of long shoots of trees is highly variable, depending both on the species and the environment. This variation markedly affects crown form as well as the nutritional relations of the species.

For example in *Populus*, the late leaves that are produced on leader and terminal shoots after the early leaves are formed bear little resemblance to early leaves (Figure 3.18). The different forms are all expressions of the same genotype and are not directly attributable to the external environment. Instead, they are due to the internal environment at the time and place of primordia formation and develop-

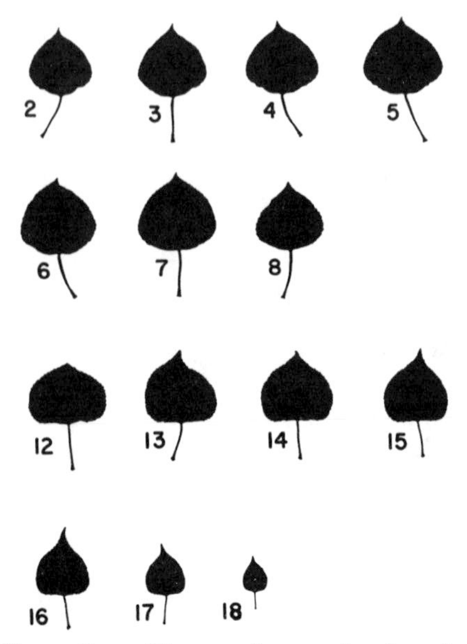


**Figure 3.18. Silhouettes of leaves from a leader shoot of trembling aspen. Leaves 2–7 are typical early leaves; 8 is a transition form to late leaves; and leaves 12–18 are typical late leaves. (After Barnes, 1969. *Silvae Genetica* 18, Sauerländer, Frankfurt.)**

ment. Morphological analyses may be useful in ecological and taxonomic studies, but they may be misleading unless this kind of variation is taken into account.

Zimmermann and Brown (1971) cite 4 patterns of long-shoot extension:

1. A single flush of terminal growth followed by formation of resting bud.
2. Recurrent flushes of terminal growth with terminal bud formation at the end of each flush.
3. A flush of growth followed by shoot-tip abortion.
4. A sustained production of leaves including early- and late-leaves up to the time of terminal bud formation.

**Type 1.** Many northern species, including white pine, red pine, oaks, hickories and buckeyes, make a burst of growth in spring or early summer and then form terminal buds (Figures 3.19 and 3.20). Extension growth of the shoots terminates before moisture stress of midsummer becomes severe. During the remainder of the growing season, a portion of the plant's resources is utilized in developing

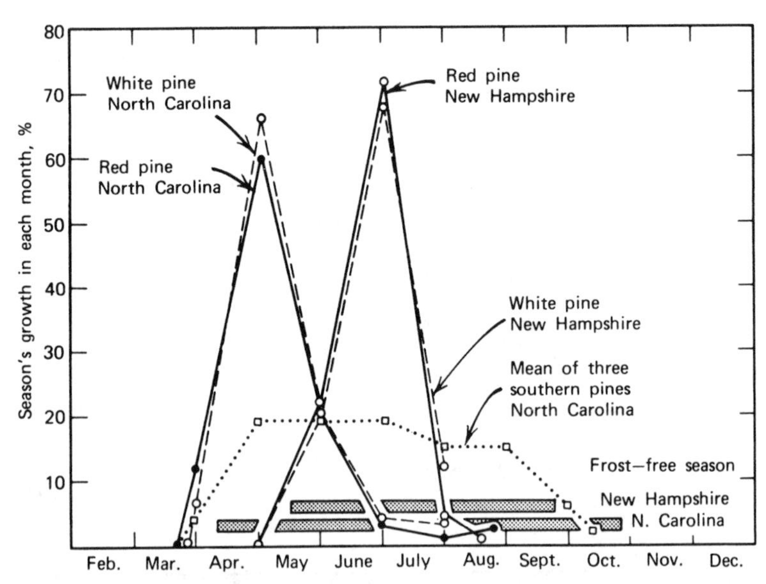


**Figure 3.19. Variations in seasonal height growth patterns of red pine and white pine in North Carolina and in New Hampshire, and of three southern pines (loblolly, shortleaf, slash) in North Carolina. The northern pines have preformed shoots and usually have one annual growth flush whereas the southern pines grow in recurrent flushes. (After Kramer, 1943. Reproduced by permission from *Plant Physiology*, Vol. 18, pp. 239–251.)**

shoots and foliage inside the protective bud scales of the newly formed terminal buds. The bud contents expand the following year. Therefore, the current year's shoot growth is primarily dependent on environmental conditions of the previous year (when leaf primordia formed). Red and white pines make a single seasonal flush such that groups of branches (false whorls) are separated from one another by long branch-free internodes. The marked nodal occurrence of branches makes aging relatively easy and facilitates the estimation of site quality based on the amount of growth between the nodes (Chapter 12).

Ecological conditions of the native habitat largely control when the burst of growth begins. This is shown in Figure 3.19 by the marked difference in timing of shoot growth of red and white pines in North Carolina as compared to New Hampshire. Except for the oaks, the structure of the crowns seems to be more regular than in the following types.

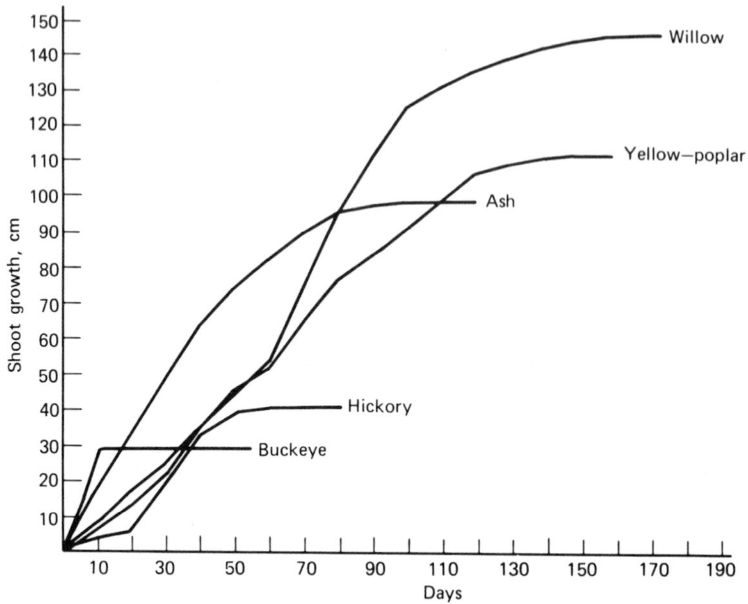


**Figure 3.20.** **Rate and duration of shoot growth among several woody species in the Georgia Piedmont. Measurements of shoot elongation were made biweekly on nine trees ranging from 8 to 15 years in age: painted buckeye, mockernut hickory, red ash, yellow-poplar, and black willow. (After Zimmermann and Brown, 1971.)**

**Type 2.** Trees in regions in which conditions for growth may be periodically favorable, such as the pines of the southeastern United States, typically exhibit recurrent flushes of growth during the growing season. This type of growth is the most typical one among conifer and deciduous trees in the subtropical and tropical regions (Zimmermann and Brown, 1971). The environmental conditions of the site, especially moisture, determine the number of flushes of growth.

**Type 3.** In this common type, the shoot tip aborts following the extension of the preformed leaves (beech) or considerably later in the season after many late leaves have been produced (black willow, birches, mimosa). In many cases, the shoot-tip abortion (and, in the case of birches, shoot tip and terminal bud) is associated with shortening days of summer. In the following spring, shoot extension proceeds from the last fullyformed lateral bud. Zimmermann and Brown list 16 genera in which shoot-tip abortion occurs. The ecological significance is unclear. However, it may be an effective mechanism to prolong shoot extension (avoid setting the terminal bud earlier than necessary) yet cease growth prior to significant frost damage.

**Type 4.** In a few species, some shoots make continuous and prolonged shoot growth, including both early and late-leaf development, before setting a terminal bud. Examples of species with southerly distributions include yellow-poplar, sweetgum, and eastern cottonwood. Within their range the potential of severe frost damage is low, and selection pressures for shoot-tip abortion may not be strong. They also occupy moist sites so that the pressure for early cessation of growth to avoid drought also may not be great. This type of shoot extension and true terminal bud formation lead to a more symmetrical crown and more pronounced main stem than in trees exhibiting shoot-tip abortion.

Since long shoots are responsible for height growth and crown extension, it is of selective advantage to use the growing season effectively. The above types of long-shoot extension illustrate various ways trees make full use of favorable and potentially favorable parts of the growing season. While long shoots extend the crown, short shoots bear leaves that contribute photosynthate to satisfy current growth needs.

**Lammas Shoots.** Favorable growth conditions in autumn, especially moisture, can stimulate trees of many species to produce abnormally late-season bursts of growth. **Lammas** shoots[2] result from shoot growth from terminal or false-terminal buds on leader or main branches; the flush of growth at the same time from lateral buds situated at the base of the terminal buds (as in pines and oaks) is termed **prolepsis**. Lammas shoots may cause false rings to form. They sometimes do not harden off sufficiently and are susceptible to winter damage. Proleptic shoots may cause forking in conifers and cause branchiness because of the development of many secondary lateral branch buds.

## Roots

Tree roots perform many important functions. Two important ones are the firm anchoring of the tree in the soil and the absorption of water and nutrients. Anchorage is accomplished by the whole root

[2]Late-season shoots apparently occurred about the time of old European holidays, Lammas Day (August 1) in England and the festival of St. John in Germany. Thus they are termed lammas shoots in English and **Johannistriebe** in German.

system, whereas absorption is mainly effected by the root tips of the many fine nonwoody roots. Other functions include storage of carbohydrates and other materials, synthesis of organic compounds, secretion of chemical substances, and generation of vegetative shoots (as in aspens, beeches, and sweetgum). The growth and functioning of roots has received attention from many authors, and the major review papers are those of Röhrig (1966), Lyr and Hoffman (1967), Sutton (1969), and Kozlowski (1971).

Important differences exist in the root systems of angiosperms and conifers. Deciduous trees, evolving later than conifers, have generally developed root systems that are more extensive and efficient than those of conifers (Voigt, 1968). The roots of the more primitive conifers probably have a greater ability to extract nutrient ions from primary soil materials. In addition, conifers such as spruce and larch absorb quantities of silica from soil minerals, and this may contribute to the slow decomposition of their foliage (Viro, 1956). Leaves of most deciduous trees decompose rapidly, and these species may have evolved extensive and finely divided root systems to intercept and absorb the nutrient ions released from the leaves upon decomposition. In general, root tips of deciduous species are smaller in diameter than those of conifers (Voigt, 1968). Thus nutrient absorption may be more efficient because absorption per unit of surface area increases as root radius decreases (Nye, 1966). The relative capacity for root development of several conifer and deciduous trees was studied by Kozlowski and Scholtes (1948). Several of their comparisons are presented in Table 3.1 demonstrating that seedlings of the deciduous species developed more extensive root systems than the conifers in greenhouse and forest environments.

**Table 3.1. Comparative Root Development of Seedlings of Deciduous and Coniferous Species.**

| SPECIES | AGE IN MONTHS | NUMBER OF ROOTS | TOTAL ROOT LENGTH meters | GROWING CONDITIONS |
|---|---|---|---|---|
| Black locust | 4 | >7000 | 326 | Greenhouse |
| Loblolly pine | 4 | 419 | 1.6 | Greenhouse |
| Flowering dogwood | 6 | 2657 | 52 | Greenhouse |
| Loblolly pine | 6 | 767 | 4 | Greenhouse |
| White oak | 12 | 196 | 23 | Forest |
| Loblolly pine | 12 | 148 | 10 | Forest |

After Kozlowski and Scholtes, 1948.

**Kinds, Forms, and Occurrence.** The tree root system is characterized by relatively large, woody, long-lived roots supporting a mass of small, short-lived, nonwoody absorbing roots, many of which are associated with fungi (these mycorrhizal relationships are discussed in Chapter 9). In many species, seedlings develop tap roots that provide early stability and survival, especially on dry sites. Tap roots are common in pines and in species that have large amounts of reserve food in their seeds (oaks, hickories, chestnut, walnuts). Lateral roots gradually increase in size and complexity. The lateral system may become widespread and at intervals develop vertical sinker roots that reach lower soil horizons.

As in shoots, long and short roots may be distinguished for many species. The long or main roots have the capacity for prolonged extension growth, either horizontally or vertically (as in tap or sinker roots). At varying distances along the main roots, lateral roots are formed that may become long roots or remain short. The root system is thus characterized by several orders of woody, perennial long roots and one or more orders of small, nonwoody roots.

The form and structure of the root system is to an important degree genetically controlled by the species and the individual. Nevertheless, site conditions markedly influence the form and pattern of root development. Species adapted to sites characterized by summer droughts may possess deep-penetrating tap roots, as found in pines and upland hickories and oaks. Where the water table is near the surface as in swamps or peat bogs, the root systems of woody species are shallow. The uprooting of trees by wind (i.e., windthrow), is common on such sites where rooting is shallow. A sample of the great variety of rooting forms that develop in relation to different soil and water table conditions is presented in Figure 3.21. Although many species exhibit deep rooting in sandy soils, their rooting depth may be severely curtailed where impermeable or poorly aerated layers occur near the surface (see Chapter 8). Detailed illustrations of root systems of individual European species were presented by Köstler et al. (1968), and racial differences in root morphology of Scots pine were described by Bibelriether (1964).

Not illustrated in Figure 3.21 are aerial roots that are formed by some tropical trees. In eastern North America, yellow birch is noted for its aerial rooting habit. Birch seeds germinate on decaying logs and stumps, and their roots grow downward and into the soil. The roots enlarge and are exposed as the woody debris decays and slumps away. Yellow birch also germinates in the moss on boulders. Descending roots penetrate the soil and eventually a large tree is perched on the boulder or outcrop of bedrock. In humid environments such as the Southern Appalachian Mountains, large, internally decaying yellow birch trees have the ability to generate roots

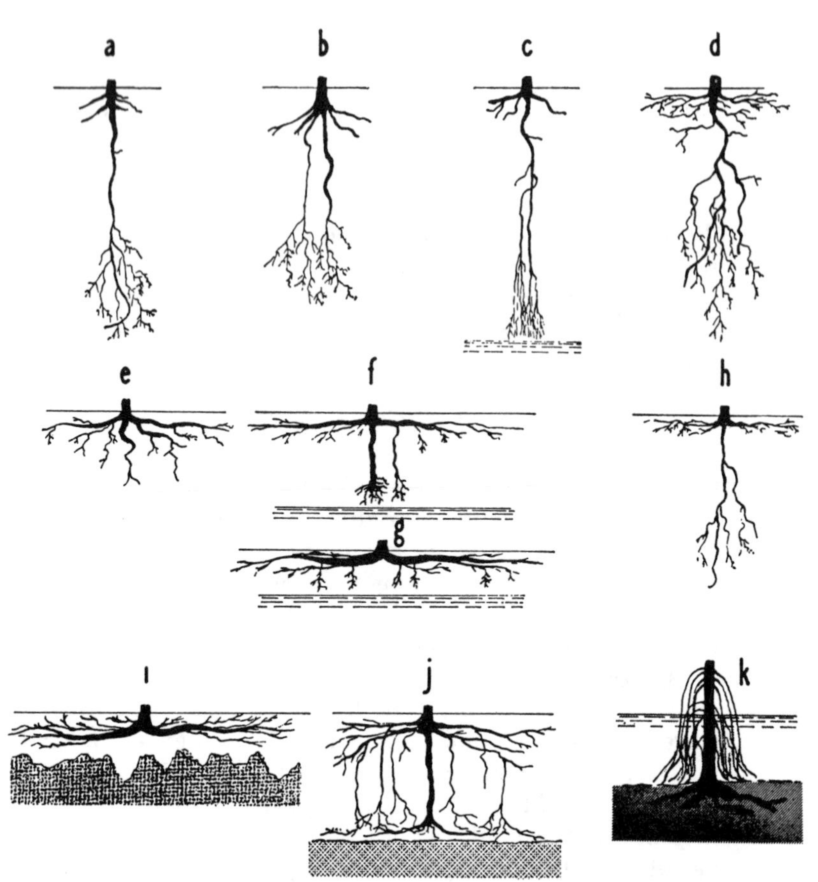


Figure 3.21. Modification of root systems of forest trees by site. *(a,b)* Taproots and heartroots with reduced upper laterals: patterns found in coarse sandy soils underlain by fine-textured substrata. *(c)* Taproot with long tassels, a structure induced by extended capillary fringe. *(d)* Superficial laterals and deep network of fibrous roots outlining an interlayer of porous materials. *(e)* Flattened heartroot formed in lacustrine clay over a sand bed. *(f)* Plate-shaped root developed in a soil with a reasonably deep ground water table. *(g)* Plate-shaped root formed in organic soils with shallow ground water table. *(h)* Bimorphic system of platelike crown and heartroot or taproot, found in leached soils with a surface rich in organic matter. *(i)* Flatroot of angiosperms in strongly leached soils with raw humus. *(j)* Two parallel plate-roots connected by vertical joiners in a hardpan spodosol. *(k)* Pneumatophores of mangrove trees in tidal lands. (After S. A. Wilde, *Forest Soils*, 1958, The Ronald Press Company.)

that grow down through the decaying heartwood of the standing tree. (Figure. 3.22).

**Figure 3.22. Aerial roots growing inside a living yellow birch tree. Several roots, as large as 9 cm, were found growing in the hollow lower trunk of a yellow birch tree, 60 cm in diameter, in the Great Smoky Mountains National Park, Tennessee.**

The majority of roots, especially the small absorbing roots, are located in the upper soil horizons where favorable aeration, nutrients, and moisture conditions occur. Throughout much of eastern North America and northern and central Europe, considerable precipitation falls during the growing season. Following a rain, the wetting front may saturate the upper 5 to 10 cm or more of soil. This is a major source of much of the moisture required by woody vegetation during the growing season. Decomposing organic matter of the surface layers holds this moisture effectively and in addition provides nutrients for small roots. Numerous studies have reported a high concentration of roots in the surface layer (Kozlowski, 1971). For example, in two pine and two oak forests of the Piedmont of North Carolina, from 94 to 97 percent of the roots less than 2.5 mm in diameter were in the upper 13 cm of soil (Coile, 1937).

The number of root tips per unit volume varies greatly during the growing season and at different soil levels. For example, in Scots pine stands in sandy soil in Finland, Kalela (1957) found a markedly greater number of root tips (expressed by Kalela as roots $m^{-2}$ surface area) in the early summer than in the late summer and in the organic (humus) layer than in the deeper mineral soil layers (Figure 3.23). He also determined that the variation in number of root tips ($m^{-2}$ surface

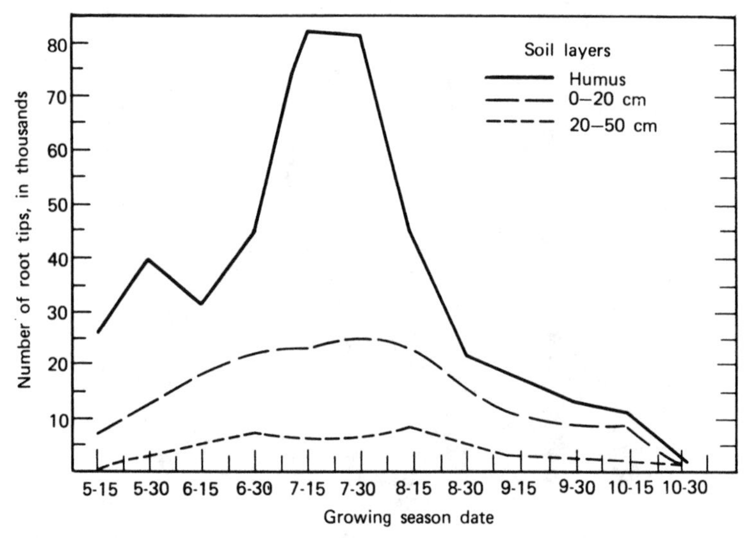


**Figure 3.23. Number of root tips ($m^{-2}$ surface area) in different soil layers in Scots pine stands in Finland during one growing season. (After Kalela, 1957.)**

area) during the growing season was almost entirely due to the amount of fine roots (< 1 mm diameter). There was a continual flux of dying and replacement of fine roots. He concluded that the amount of small roots was closely related to the favorableness of local ecological conditions. Under unfavorable conditions (drought, cold, lack of photosynthate) fine roots died en masse, but they were rapidly formed again once conditions became favorable. Mortality of fine roots is high in winter due to low temperatures and lack of photosynthate. The massive and rapid fine-root turnover described by Kalela has been confirmed in studies of the deciduous forest biome (see Chapter 18).

Moisture may also be obtained from deeper soil layers and it may be of survival value for many species in times of drought. For example, prairie-inhabiting species of *Quercus*, *Gleditsia*, *Juglans*, and *Maclura* are able to survive droughts by their inherent deep-rooting ability (Albertson and Weaver, 1945). Although rooting depths of from 2 to 3 m are not uncommon for many forest trees, deep-rooting species (rooting depths of 3 to 6 m or more) include red cedar, hackberry, black and honey locust, bur oak, mulberry, and osage orange (Kozlowski, 1971). Lyr and Hoffman (1967) cite various studies reporting rooting to 10 m for apple, 15 m for *Prosopis* (mesquite), 20 m for *Robinia*, and 30 m for *Tamarix* (tamarisk). In southern Arizona, roots were found 53 m below the surface and were thought to be those of the desert shrub mesquite (Phillips, 1963).

The horizontal development of perennial long roots is often considerable. In sandy sites, lateral roots of pines, birches, and black locust may extend 10, 20, or even 40 m as they continually occupy new volumes of soil (Lyr and Hoffman, 1967). Lateral spread is related to the nature of the rooting medium, being more extreme in sandy soils than in clay. For example, in a sandy Wisconsin soil, roots of 20-year-old red pine trees extended about seven times the average height of the trees, and those of eastern cottonwood for more than 60 m from the source (Kozlowski, 1971). Excavations of roots in stands of black pine, growing in sandy soil at the University of Michigan's Stinchfield Woods, showed that the major lateral roots of dominant trees extended over 20 m or about eight times the crown radius of the trees. These reports suggest that in many conifer stands a given tree competes with the roots of many individuals, not just with those of adjacent stems. A cultural practice such as fertilization will not be effective when broadcast only under the crowns of selected crop trees. As in thinning, the entire stand must be treated to achieve satisfactory results.

**Periodicity of Primary Root Growth.** Extension growth of roots, like that of shoots, is intermittent. For many trees of temperate latitudes, roots typically begin growth earlier in the spring and continue growth later in the fall than the shoots. However, species vary, and in larch species short-shoot needles are expanded before root growth begins (Lyr and Hoffman, 1967). Root growth of many trees, particularly angiosperms, exhibits a peak period of activity in the spring and fall when moisture and temperature conditions are especially favorable. Roots typically show reduced activity during summer drought periods, and, with the onset of winter and low soil temperatures, root activity declines markedly. The course of root growth and the marked difference in root and shoot growth of eastern white pine are illustrated in Figure 3.24. Spring and fall peaks of activity are interrupted by a summer rest (Figure 3.24*a*). Root growth continues long after shoot growth has ceased (Figure 3.24*b*).

With the onset of winter and low soil temperatures, roots tend to cease growth. However, unlike shoots of species of temperate latitudes that remain dormant over the winter period, roots of some species may grow during the winter provided temperature, moisture, and aeration are favorable (Lyford and Wilson, 1966; Lyr and Hoffmann, 1967). Winter root growth has been reported from regions with mild winter temperatures and frost-free soils, such as in the southern United States, coastal British Columbia, the Crimea, and in parts of Europe. For example, Woods (1957) reported considerable growth of longleaf pine roots in the surface layer of sandy soils of the coastal plain of the southeastern United States. The roots of scrub oaks and pine were concentrated in the upper 12 cm of soil. Warm temperatures of 16 to 22°C in the upper 8 cm (6° warmer than soil at depths of 20–30 cm) apparently stimulated the root development.

Although seasonal periodicity of root growth is not as distinct as that in shoot growth, quiescent periods may be induced by factors such as temperature, drought, and aeration (Zimmermann and

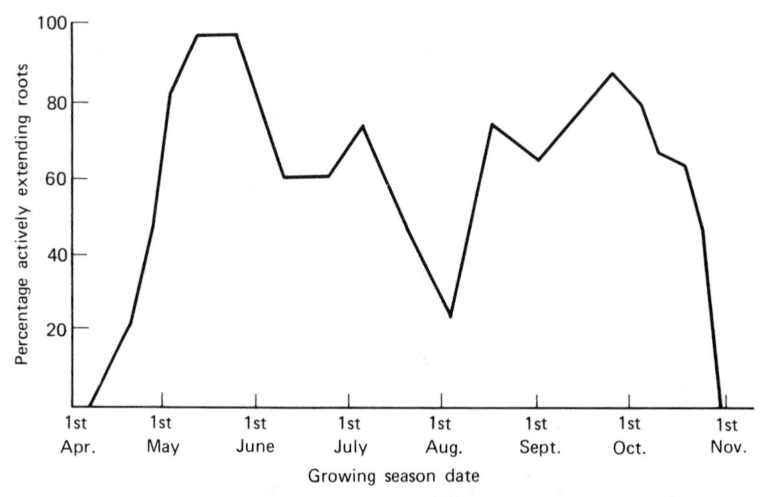


**Figure 3.24*a*. Seasonal course of root growth in eastern white pine. The percentage of roots that are actively extending is high in spring and again in autumn. Most roots are inactive during midsummer. (Modified from Stevens, 1931.)**

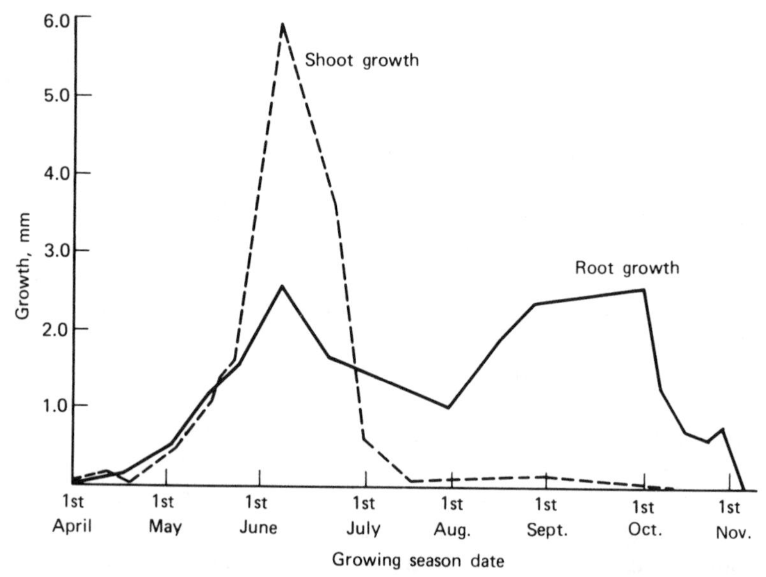


**Figure 3.24*b*. Comparison of average daily extension for shoots and growing roots. Root and shoot growth follow distinct but different seasonal patterns. The growth rate of active roots declines in mid-summer. (Modified from Stevens, 1931.)**

Brown, 1971). Thus individual roots may undergo alternating periods of growth and rest independent of seasonal activity. Roots of small seedlings are much more sensitive to the aerial environment than those of large trees because they are in close proximity to the shoots. For example, root growth in silver maple seedlings ceased after leaf fall and did not resume until the chilling requirement of buds had been satisfied (Richardson, 1958).

Diurnal growth rhythms also exist; more rapid growth occurs during the night than the day. Lyr and Hoffman (1967) found the following percentages of greater night than day growth: black cottonwood—60 percent; red oak—37 percent; Scots pine—36 percent; Norway spruce—30 percent.

**Root Grafting.** Natural root grafting among individuals of a species is relatively common in woody plants (Graham and Bormann, 1966). It is probably most frequent in pure stands of species (especially pines, spruces, and other conifers) that exhibit a widespreading system of roots concentrated in the surface soil layer. Living stumps and girdled trees in forest plantations provide striking evidence for intertree grafts. Grafting is accomplished when a union forms between the cambium layers of two closely associated roots. The phloem and xylem tissues connect so that food, water, growth substances, and pathogens may be transported from one tree to another.

Root interconnections are frequent and effective in red pine stands. Stone (1974) studied red pine plantations in New York in which approximately 30 percent of the stems had been girdled in thinning operations. Many of the trees were still alive up to 18 years after girdling, and completely girdled stems accounted for about one-third of the live basal area of the stands 5 to 10 years after thinning. Water transport of the girdled trees still functioned and food was supplied to their roots through phloem connections with intact trees. Dye transfer studies in one stand indicated that each girdled pine was grafted to an average of three nongirdled trees up to a distance of 4.5 m. Radial growth of girdled trees stabilized at about two-thirds that of intact trees. Growth of intact trees was not affected by the increased drain on their resources. Thus evidence did not support the existence of a communal root system that would allow unrestricted sharing of food, growth regulators, and other phloem-transported substances. The work of Stone and others with red pine indicates that up to 90 percent of stems in plantations may be grafted to at least one other individual. The physiological and ecological implications of root grafting are discussed by Bormann (1966) and Stone (1974).

Roots play an important but different role in the transmission of systemic wilt diseases such as Dutch elm disease and oak wilt (caused by *Ceratosystis ulmi* and *C. fagacearum* respectively) and in root rots caused by *Armillaria mellea*, *Fomitopsis* (*Fomes*) *annosus*, and *Phellinus* (*Poria*) *weirii*. Research on root grafting in American elm has shown that nearly all trees within 2 m of one another and about 30 percent of those within 8 m are connected by root grafts (Verrall and Graham, 1935; Himelick and Neely, 1962). This interconnection permits rapid transmission of the fungus from infected to uninfected trees. Spores move passively through the vessel elements of the roots, then germinate and grow through impediments such as pits. In the root rots, however, this vascular transmission is not found. Instead, the fungus (1) grows and decays its way between adjacent trees via root grafts (as in *F. annosus*); (2) spreads over the outer surface of roots, occasionally penetrating the root and causing a new infection (as in *P. weirii*); (3) grows from plant residue, including old dead roots, to the roots of an actively growing plant that happens to come into contact with it (as in *A. mellea* and others).

Another site-related root feature is the formation of root projections, knees or pneumatophores (roots that function as a respiratory organ), by various species that grow in periodically flooded river valleys or subtropical and tropical swamps. Kozlowski (1971) lists 14 genera for which knee-roots or pneumatophores have been reported. In bald cypress, knee-roots, like buttresses, may be formed as a response to the air-water interface. The upper side of the root, exposed to the air, exhibits greatly increased cambial activity compared to the submerged underportion of the root. Knee-roots in bald cypress apparently are not important in root aeration but may act to strengthen the roots to give a firmer anchorage in yielding soil medium (Mattoon, 1915; Gill, 1970). However, some pneumatophores may contain lenticels and act to facilitate exchange of gasses with the roots. The precise role of knee-roots and pneumatophores is not known.

**Specialized Roots.** Buttresses, flattened triangular plates that develop between the trunk and lateral roots, prop or guy tree crowns under site conditions unsuitable for extensive vertical root growth. They are typical of many tropical rain-forest trees but also occur in temperate latitude trees, especially American elm, and even the Lombardy poplar (Senn, 1923). Although buttress formation is related primarily to flooding and shallow soil conditions (Richards, 1952), buttresses also occur on aerial portions of trees (Johnson, 1972). In bald cypress trees of the southern United States, buttresses develop as a response to the level of flooding. The kind of buttress formed (shallow, cone, bottle, bell) corresponds to a different water-level regime (Kurz and Demaree, 1934). The trunk produces a marked increase in the number or size of individual cells (hypertrophy) at the air-water interface where the trunk is neither completely aerated or completely saturated. This is caused apparently by the inhibition of growth-regulator translocation at the floodline or by buildup of ethylene. Thus the stem diameter at any height up the stem is directly proportional to the frequency of occurrence of that height of flooding.

Mechanical stresses may also affect buttress development. Based on studies of 191 trees of *Triplochiton scleroxylon* in Ghana, Johnson (1972) reported that buttressing was closely associated with site conditions. Small buttresses were found on well-drained upland soils, whereas larger buttresses occurred on mid- or lower-slope areas of imperfectly drained, shallow, gravelly soil. He suggested that buttresses developed in response to mechanical stresses occurring at the juncture of the lateral roots and the stem. Soil conditions that permit deep rooting allow a strong vertical root system to absorb stresses that would otherwise concentrate at the lateral root-stem juncture. If the soil is effectively shallow, due to impermeable layers or a high water table, vertical root development is impeded and stresses will concentrate at the stem base, thus inducing greater buttress growth.

## Stems

The ability to form consecutive layers of structural tissues to the primary stem distinguishes woody species from all other plants. This secondary growth strengthens the stem and increases the transport of food and water between the shoots and the roots. The meristematic sheath of cells that surrounds the stem, shoots, and roots is the **cambium**. It originates the successive layers of secondary growth and is functionally a single cell layer, although a zone of actively dividing cells is typically observed. Evolution of the cambium was a significant event because the cambium provides the structural modifications required to withstand the stresses of aboveground and belowground environments of terrestrial habitats.

Toward the center of the stem, the cambium gives rise to the water-conducting cells, the **xylem**, and to the outside the cambium generates the food-conducting inner bark or **phloem**. The xylem cells become lignified and form the dead, woody axis of the tree. The phloem cells serve as transport channels for photosynthetic products, growth regulators, and many other substances within the crown, from crown to other parts of the tree, and from the roots to the crown. Phloem cells are not lignified and must be regularly differentiated because they function for only a few years before collapsing and becoming part of the outer bark. The anatomy and physiology of the xylem, phloem, and bark have received considerable attention, and the reader is directed to Esau (1977) for anatomical relations and to Zimmermann and Brown (1971) and Kozlowski (1971) for physiological considerations. A few points concerning stem growth and environmental factors are discussed briefly below.

**Xylem Cells and Growth Rings.** Diameter growth of trees is primarily due to the periodic formation of layers of xylem cells by the cambium. In conifers, the primary water-conducting cells are the vertically oriented, overlapping tracheids (Figure 3.25). These thick-walled, tapering cells are 3 to 5 mm long and up to 100 times longer than wide (Kozlowski, 1971). Conifer wood is relatively homogeneous and composed primarily of tracheids that are arranged in uniform radial rows. Other cells, parenchyma and ray tracheids, are oriented horizontally and function in the lateral transport of water, food, and other substances. Perforations in the walls of the tracheids provide the means for transfer of water and other substances to adjacent cells. Conifer growth rings are distinguished by differences in cell diameter and cell-wall thickness of the tracheids

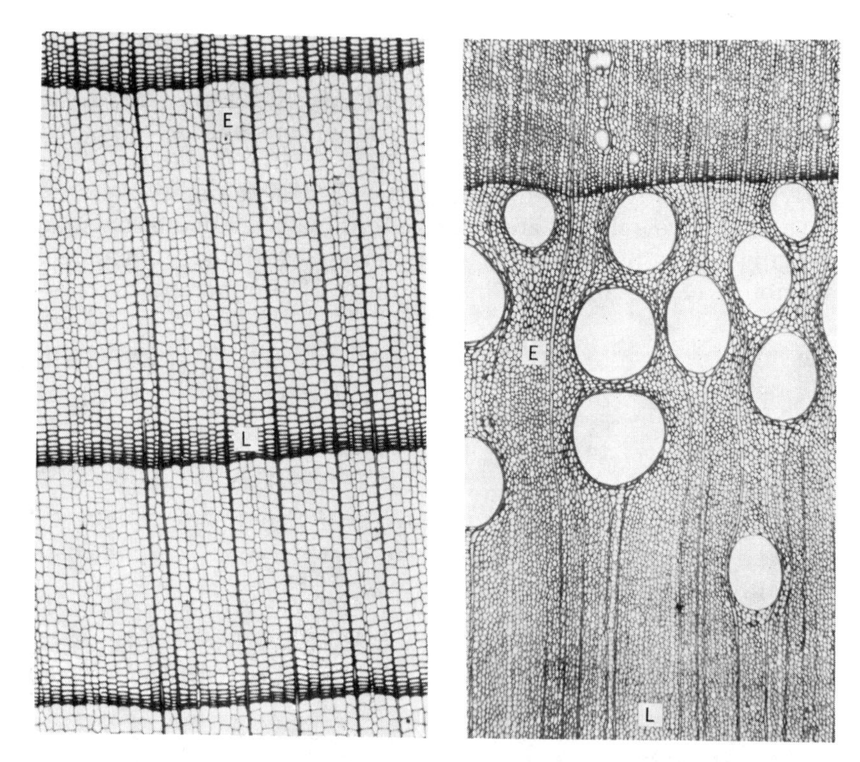


**Figure 3.25. (left)** **Transverse section of wood of ponderosa pine showing transition in size and shape of earlywood (E) and latewood (L) in successive growth rings. (After Zimmermann and Brown, 1971; Photo by Claud L.Brown.)**

**Figure 3.26. (right)** **Transverse section of wood of a ring-porous black oak showing extremely large vessels in the first formed earlywood (E) and the preponderence of fibers in the latewood (L). (After Zimmermann and Brown, 1971; Photo by Claud L. Brown.)**

produced during the early (earlywood) and late (latewood) parts of the growing season (Figure 3.25).

Xylem cells of angiosperms are more specialized, and their longitudinal components include vessels, tracheids, fibers, and parenchyma. Vessels are the primary water-conducting elements although the majority of the xylem cells are fibers. Vessels are composed of single cells whose end walls have disintegrated, thus creating tubes that may be up to several meters long. Therefore, they may rapidly transport water and nutrients to the shoot system. Vessels vary greatly in size, and, in certain genera (*Fraxinus*, *Carya*, and *Quer-*

*cus)*, vessels that form early in the growing season may have diameters 100 times as great as those formed late in the growing season (Figure 3.26). Growth rings of such **ring-porous** species are easily distinguished because of the concentration of large vessels in the earlywood. Many other deciduous trees, **diffuse-porous** species, have smaller vessels that are more uniformly distributed throughout the growth ring (Figure. 3.27). Growth rings of diffuse-porous trees are not as clearly distinguished as those of ring-porous trees, al-

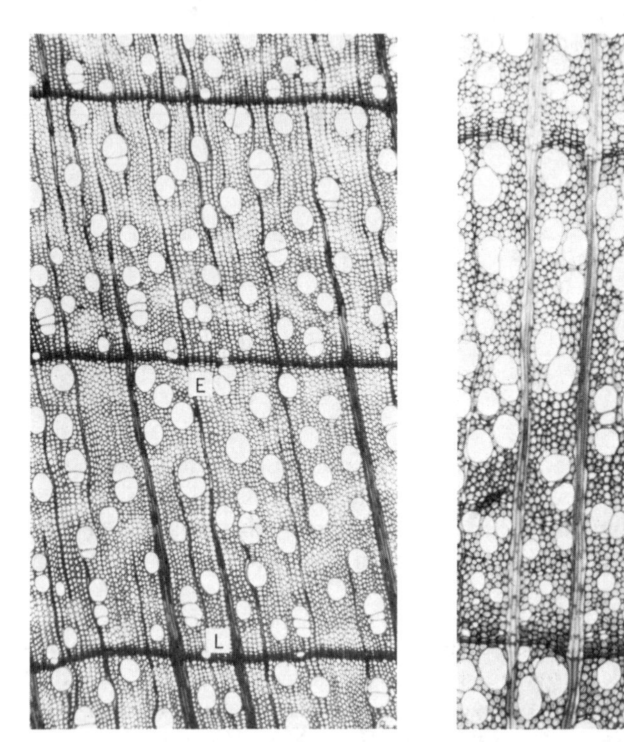


**Figure 3.27. (left)** **Transverse section of a diffuse-porous hardwood, yellow-poplar, showing fairly uniform distribution of vessels throughout the growth ring and radial flattening of the last formed latewood cells. (After Zimmermann and Brown, 1971; Photo by Claud L. Brown.)**

**Figure 3.28. (right)** **Transverse section of a diffuse-porous hardwood, American beech, showing a higher proportion of vessels in the earlywood (E) than in the latewood (L) and a corresponding increase in the proportion of thick-walled fibers in the latewood. (After Zimmermann and Brown, 1971; Photo by Claud L. Brown.)**

though in some species (beeches) the vessels become progressively smaller and the proportion of fibers and tracheids increase toward the end of the growing season (Figure 3.28).

**Periodicity and Control of Secondary Growth.** Diameter growth of North Temperate Zone trees is typically characterized by a single annual ring. In addition to variation caused by species, age, and soil conditions, the width of the growth ring depends largely on weather conditions of the current year. The number and size of xylem cells of a given species is directly related to the size of the crown and the amount of growth regulators and photosynthetic material produced by the foliage. Although diameter growth generally continues longer into the growing season than shoot extension, most of it occurs in the spring and early summer when moisture conditions are favorable. However, species vary greatly in the duration of the growth period. For 21 tree species on one site in the Georgia Piedmont, the most rapid period of growth varied from 70 to 209 days (Jackson, 1952). In a study of European species, ash, Norway spruce, and larch formed the growth ring rapidly, with little wood laid down after July (Ladefoged, 1952). Species of birch, beech, alder, maple, and oak laid wood down from May until early September, with up to one-third of the ring being formed in the late summer. The width of the growth ring and the proportion of earlywood to latewood cells are closely related to the amount of growth regulators produced by the foliage and more indirectly to the availability of soil moisture. The effects of moisture deficits on tree growth are considered in Chapter 10.

Multiple growth rings during a growing season are common wherever multiple flushes of shoot growth occur during the season. In Temperate Zone trees, multiple rings occur following defoliation, drought, fire, frost, wind damage, and heavy fruiting, provided the tree reflushes. They are prevalent in subtropical and tropical trees that exhibit multiple shoot flushes. This would include the pines and some deciduous trees (water and willow oak) of the southern United States. Multiple rings are more prevalent in young than in old trees because young trees are more vigorous and more capable of multiple flushes of shoot growth. Missing or discontinuous rings may be common in heavily defoliated trees, suppressed trees of the understory, overmature trees, and drought-stressed trees of the American southwest.

The initiation of cambial growth in the spring is closely associated with renewed bud activity and leaf development. Growth regulating hormones, auxin, and other substances (promoters and inhibitors) from the expanding buds and leaves trigger cambial activity (Zimmermann and Brown, 1971). In conifers, cambial activity begins at the base of actively developing buds and leafy shoots (before shoot extension begins) in the crown and moves rapidly downward throughout the twigs, branches, and the main stem. In diffuse-porous hardwoods, the process is similar except that the wave of activity proceeds rather slowly. In contrast, in ring-porous species cambial activation is almost simultaneous throughout the entire stem. This is an important adaptation because many of the large earlywood vessels of the previous year are nonfunctional, and the newly forming vessels are vital in supplying water to the foliage that will soon develop (see following discussion).

**Control of Earlywood and Latewood Formation.** Differences in the properties and physiology of formation of earlywood and latewood have attracted interest for over a century. Latewood cells have a higher specific gravity than earlywood cells because of a greater proportion of cell wall substance per unit volume. Thus in conifers the specific gravity of latewood is about two to three times that of earlywood (Kozlowski, 1971). Because the amount of latewood in structural timber and paper pulp affects the properties of these materials, considerable research has been marshalled to determine the factors controlling early and latewood formation with the aim of manipulating the amount of latewood cells. Research has demonstrated that the balance of growth regulators controls the kind of cells produced (earlywood or latewood) and that environmental factors, such as soil moisture stress, act indirectly to influence earlywood or latewood production through their influence on the balance of growth regulators. The results of many investigations are summarized succinctly by Zimmermann and Brown (1971):

> *. . . any condition that enhances bud break, rapid shoot growth, and continued leaf development, results in high levels of auxin production and large diameter cells of the earlywood type. Conversely, low temperature, drought, or short photoperiods which adversely affect shoot extension and leaf development, lowers the levels of diffusible auxin, and brings about the formation of smaller or radially flattened tracheids of the latewood type.*

**Winter Freezing and Water Transport.** The freezing of water in xylem cells in winter causes problems in water conduction for trees of northern latitudes. Dissolved gasses in the xylem water causes bubbles to form upon freezing; upon thawing, gaps develop in the water columns thus breaking continuity, forming an air lock. The problem is potentially most serious for the ring-porous species that have very large earlywood vessels. Zimmermann and Brown (1971) cite three ways trees may cope with this situation:

1. Small-diameter xylem cells. In species with small conducting cells, the bubbles may be so small that they redissolve when the water thaws, and the vertical water columns rejoin before transpiration begins. Conifers and certain diffuse-porous trees with very small vessels (willows, trembling aspen, paper birch, alders) can easily survive the extreme freezing and thawing regime of the boreal or alpine forest.
2. Root pressure. Enough pressure may be developed by roots of certain diffuse-porous species (birches) that the gaps in the vessels can be filled.
3. Spring formation of large vessels. Many ring-porous species with large earlywood vessels conduct water mostly in the last-formed (youngest) growth ring. Possessing very rapid cambial activation, they are able to produce some large earlywood vessels in the spring before the leaves flush out. Upon flushing these large vessels quickly transport quantities of water as it is required by the developing and transpiring leaves.

The mechanism of ring-porous species is especially interesting because some species of this group are particularly late flushing—oaks, hickories, walnuts, and ashes. Flushing is delayed until the xylem system is ready to supply the water required by the foliage. Delayed flushing has the disadvantage that the leaves develop very rapidly in the warm temperatures of late spring when transpiration is high. However, the large vessels are highly efficient in rapidly supplying large amounts of water to the crown.

The system is advantageous in that these late-flushing species may have a greater probability of escaping late frost than earlier flushing species. However, by specialization for water conduction in the youngest ring, such species are susceptible to severe damage if significant injury befalls the newly formed ring. Two North American trees, the American chestnut and the American elm, have been dramatically affected by diseases that block the water vessels (see Chapter 16).

The distribution of tree species is at least partially related to the structure of the xylem cells controlling water conduction. Boreal forest and alpine trees (including certain spruces, firs, willows, and birches) are adapted to maintain water conduction, in spite of freezing and thawing, by their small xylem cells. No ring-porous trees have a boreal distribution, and many oaks, hickories, walnuts, and other ring-porous trees are most abundant in the central or southern portions of eastern North America.

## SUGGESTED READINGS


Allen, George S., and John N. Owens. 1972. The life history of Douglas-fir. Environment Canada, For. Serv., Ottawa. 139 pp.

Gill, C. J. 1970. The flooding tolerance of woody species—a review. *For. Abstr.* 31:671–688.

Graham, B. F., Jr., and F. H. Bormann. 1966. Natural root grafts. *Bot. Rev.* 32:255–292.

Harper, John L. 1977. *Population Biology of Plants.* Academic Press, Inc., New York. 892 pp.

Harper, J. L., and J. White. 1974. The demography of plants. *Ann. Rev. Ecol. Syst.* 5:419–463.

Kozlowski, T. T. 1971. *Growth and Development of Trees. I. Seed Germination, Ontogeny, and Shoot Growth,* 443 pp. II. *Cambial Growth, Root Growth, and Reproductive Growth.* Academic Press, Inc., New York. 520 pp.

Lyr, Horst, and Günter Hoffmann. 1967. Growth rates and growth periodicity of tree roots. *Int. Rev. For. Res.* 2:181–236.

Matthews, J. D. 1963. Factors affecting the production of seed by forest trees. *For. Abstr.* 24(1):i–xiii.

Morey, Philip R. 1973. *How Trees Grow.* Inst. of Biology, Studies in Biol., No. 39. Edward Arnold, Ltd. 59 pp.

U.S. Dep. Agr. 1965. *Silvics of Forest Trees of the United States.* USDA For. Serv. Agr. Handbook No. 271. Washington, D.C. 762 pp.

U.S. Dep. Agr. 1974. *Seeds of Woody Plants in the United States.* USDA For. Serv. Agr. Handbook No. 450. Washington, D.C. 883 pp.

Wilson, B. F. 1970. *The Growing Tree.* Univ. Mass. Press, Amherst. 152 pp.

Zimmerman, Richard H. 1972. Juvenility and flowering in woody plants: a review. *Hort. Sci.* 7:447–455.

Zimmermann, Martin H., and Claud L. Brown. 1971. *Trees, Structure and Function.* Springer-Verlag, New York. 336 pp.

# Part II

# The Forest Environment

# 4
# Solar Radiation

## AUTECOLOGY

A facet of the forest community can no more be attributed solely to the forest environment than solely to the genetic or physiological makeup of the trees. It is not a question of one or the other, for the forest community is always the product of the two working together.

The forest environment divides itself naturally into the physical environment surrounding the aerial portions of the trees and that surrounding the subterranean portions of the trees. The sum total of these factors is the **forest site** or **habitat.** In addition, biotic factors, the plants and animals occupying the site, locally influence the physical environment of atmosphere and soil.

These chapters on the forest environment fall within the subdivision of ecology known as **autecology**, the study of the organism—in this case the forest tree—in relation to its environment. Earlier coverages of forest autecology may be found in the writings of Toumey (1947), Daubenmire (1959), and Lundegardh (1957). German-language texts dealing with forest ecology include Rubner (1960), Leibundgut (1970), and Bonnemann and Röhrig (1971).

### Site Factors

It is fairly easy to enumerate the factors making up the environment upon which the life and growth of the forest tree depend. It is exceedingly difficult, however, to understand and evaluate the sum total of the interactions among these environmental factors that make up the complex we term "site" or "habitat."

The tree grows with its crown in the atmosphere and its roots in the soil. To the crown come the warmth, light, carbon dioxide, and oxygen; and to the roots come the mineral nutrients and water necessary for photosynthesis and the other life processes. These are the basic factors which the site must supply. Their availability to the tree, however, depends upon an endless system of changing climate, day length, and soil development—changes that are in part related to the developing vegetation itself.

## Classification of Site Factors

The single site factors, therefore, may be divided into broad groups, which may be considered separately. The site factors interact to yield the light, heat, water, etc., that are directly available and used by the plant (Figure 4.1). Climatic factors (Chapters 4–7) are those relating to the atmosphere in which the aerial portions of the trees grow. These include solar radiation, air temperature, air humidity, wind, lightning, and carbon dioxide content of the air. Precipitation, whether rain or snow, is usually not a major direct climatic factor as it affects plant growth primarily through its indirect effect on soil moisture.

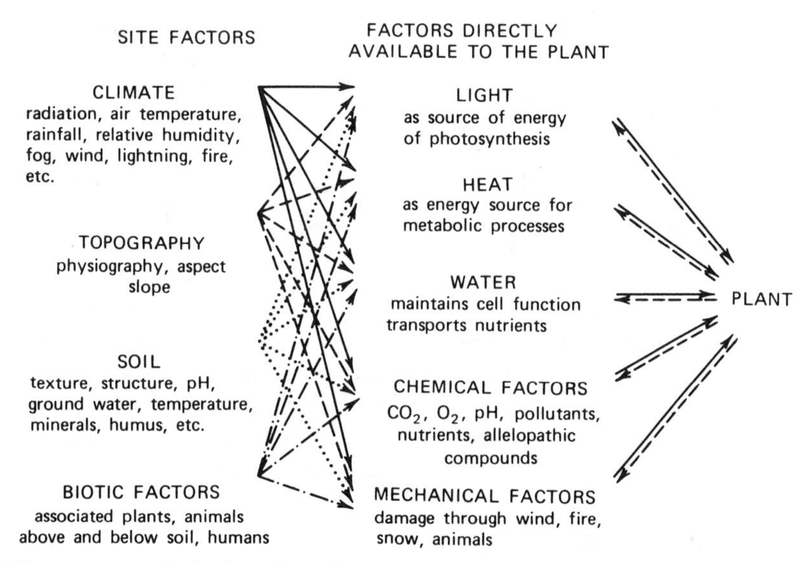


**Figure 4.1. Relationships between site or habitat factors and factors directly responsible for plant life. (After Ellenberg, 1968.)**

Soil factors (Chapters 8, 9, and 10) include all the physical, chemical, and biological properties of the soil. The nature of the parent material, the soil profile, the physical properties of the soil, the soil fauna and flora, the nutrient cycle between the soil and the trees, and soil-moisture and soil-air relationships, are particularly important in determining site quality.

Biological factors include the effects of plants and animals, both visible and microscopic, on the climate and the soil—and therefore on site quality. Large organisms (such as trees, grazing animals, and humans) create the most obvious changes in the microclimate and the soil. Small animals and plants occurring in great numbers (including fungi, bacteria, earthworms, rodents, and many others) can also bring about substantial changes in the site. Because the influence of plants and animals on site is exerted through changes in the climate and soils, a discussion of various biotic factors is integrated into the chapters on climatic and soil factors. In addition, the roles of animals in the forest ecosystem and the effect of grazing animals and humans on site quality are discussed in Chapter 13.

Fire is also a most important factor affecting forest species and site quality (Chapter 11). The burning of soil organic matter and the heating of the top layers of the mineral soil result in changes in the physical and chemical properties of the soil and the soil biota. The great importance of fire in affecting the composition of the forest community is considered in Chapter 16.

Once the individual factors affecting forest site have been considered, it becomes necessary to integrate them into a whole if the forest site is to be characterized simply and accurately. Forest productivity and the determination of forest site quality are considered in detail in Chapter 12.

### Interrelations Between Site Factors

While it is both conventional and convenient to discuss the various factors of the forest environment one at a time, a word of caution must preface such a discussion. The presentation of the principal factors one by one may create a dangerous tendency to think of each factor as an independent force affecting the tree. We tend to think and talk in terms of simple direct relationships, and to make such statements as "at a temperature of 55°C, the plant suffers direct heat injury" or "precipitation is the limiting factor in determining the distribution of this species." Such statements have their value and even truth, but they *do* tacitly ignore the fact that the plant lives in the total complex of the environment—and a change in any one factor of this complex may well bring about a changed requirement of the plant for other factors. In other words, these individual environmental factors are not isolated independent forces operating on the plant, but rather interdependent and interrelated influences that must in the last analysis be considered together.

Much of this philosophy is embodied in the changing concept of the **law of the minimum**. As originally enunciated by Liebig in 1855, this law holds that the rate of growth of an organism is controlled by that factor available in the smallest amount. It has been demonstrated repeatedly, however, that such a law does not hold true—but that the rate of growth of a plant or tree can be increased by changes in the supply of more abundant factors that can compensate for the so-called "limiting factor." The law still retains much value and is part and parcel of the thinking of the forest ecologist. In its modern form, though, it is usually restated to take into account the interactions existing between the various single factors. We may thus say with authority that "whenever a factor approaches a minimum, its relative effect becomes very great."

The changing concept of the law of the minimum has arisen from the natural development of ecological research methods. In early years, most ecological theories were based upon observations of trees and other plants under natural growing conditions. For instance, observations that seedlings die under dense forests may lead to conclusions that low light intensity is responsible for the mortality. On reflection, however, the ecologist will realize that many environmental factors are affected by a dense forest cover. Light intensity, of course, is low, but so also is the supply of moisture in a soil permeated by the roots of the many trees present. Under a dense forest, the temperature regime and wind speed are greatly changed from that in the open or even that under an open forest. The humus type and the soil organisms too, will be greatly affected. In short, the whole environmental complex is related to forest cover density, and it is misleading if not downright incorrect to attribute changes in plant response to any single factor.

## Importance of Site in Forest Ecology

There is good reason for considering the site before one considers the vegetation. In the forest community, the site is more concrete, more stable, and more easily defined than the animals and plants that occupy it. In a sense, therefore, the site constitutes a better basis for the description of the forest community than do the trees, other plants, and animals.

By site is meant the habitat of the forest community—a more or less homogeneous area as regards soil, topography, aspect, and climate; and one on which a more or less homogeneous forest type may be expected to develop. This homogeneity is strictly relative and must be defined with respect to the tree rather than to the lesser plants. A given forest site will inevitably contain many different microsites where more restricted plant and animal communities will be found. The forest floor, the tree trunks, the crown, even the different levels within the forest all may contain specialized "layer communities" or synusia—subcommunities within the forest.

The forest site occupies a given geographical area and is capable of fairly precise definition. Depending upon chance, upon past history, and upon changing environmental conditions within that site, various types of forest communities may develop on that particular site. The superiority of site to plant species as a basic unit of plant community classification becomes apparent when consideration is given to the number of quite different communities that may be characterized by a given species or even by a given group of species. Ideally, it is the combination of site and the biota, the ecosystem itself, that is necessary in describing, classifying, and more significantly, understanding the forest.

## SOLAR RADIATION AND THE FOREST ENVIRONMENT

Our climate arises from the interaction of solar radiation and the atmospheric blanket that surrounds the earth. From the sun come, directly or indirectly, the light that makes possible photosynthesis and the heat that warms the air and the soil to the point that permits the life processes of the plant to continue. From the atmosphere come oxygen, the carbon dioxide required for the photosynthetic process, and much of the moisture needed by the tree. Gases absorb very little solar radiation. Solids and liquids absorb many times more. Thus most heating of the atmosphere comes by conduction from the ground to the air at their boundary. Movement of the atmosphere profoundly affects the distribution of this heated air and the active gases—and wind is of course a prime factor governing both tree development and life span.

The sum total of these factors is climate. Climatic data obtained from standard weather stations define the macroclimate of a given region. Many valid generalizations can be drawn between the coincidence of broad climatic boundaries and the observed presence or behavior of forest trees. The climate in which a given tree or stand lives, however, may be entirely different from the regional macroclimate. In recent years, much has been learned about the microclimates—the climate near the ground, surrounding the tree, or affecting a critical part of the tree.

### Nature of Solar Radiation Reaching the Earth

The energy making possible the growth of trees and other plants comes either directly or indirectly from the sun. Thus the nature and the amount of solar radiation received on the surface of the earth will affect the distribution and the growth of the forest. Photosynthesis, being a reaction that takes place only in the presence of light, is obviously affected by the quantity and quality of light. In addition, though, the structure, growth, and even the survival of the tree are affected in other ways by the light factor. The relative lengths of the day and night, too, are frequently prime factors in determining which plant can prosper best in a given environment. Finally, solar radiation ultimately governs air temperature and thereby indirectly determines the thermal conditions around and within the plant. The biological effects of radiation in many of its phases have been summarized by Duggar (1936). Other major works treat energy exchange in the biosphere (Gates, 1962, 1968) and solar radiation in relation to forests (Reifsnyder and Lull, 1965; Reifsnyder, 1967).

The solar radiation that reaches the outside of the earth's atmosphere is dissipated in many ways, being in part reflected and in part absorbed by the atmosphere itself, by clouds and other masses of solid particles in the atmosphere, by the vegetation, and finally by the earth itself (Figure 4.2). The radiation reflected back into space amounts to about one-third of the total amount received, and this **albedo** gives a brightness to the earth that is similar to that of the planet Venus. Clouds, of course, reflect the most light, but atmospheric scattering and reflection from the surface of the earth are also important. Of the radiation absorbed by clouds, vegetation, and the earth, a substantial portion is reradiated. On cloudy days much of this radiation is retained by the cloud layer so that the days remain relatively warm. Night temperatures on the earth's surface also remain relatively mild, because of the thermal radiation of cloud layers.

Solar radiation (total energy) reaches the outside of the earth's atmosphere at an average rate of about 1.95 gram-calories per square centimeter per minute. Actually this value, the so-called **solar constant**, is not constant, but varies about 1½ percent with variation in the activity of the sun, and about an additional 3½ percent depending upon the distance from the sun to the earth (List, 1958).

The wavelengths of the bulk of solar radiation are divided into three regions, termed the **ultraviolet**, the **visible**, and the **near infrared** (Figure 4.3). The shorter ultraviolet rays are almost completely absorbed by the atmosphere. Radiation with wavelengths ranging from approximately 0.4 to 0.7 micrometers[1] is visible to the unaided human eye and is termed **light** (Figure 4.3). Approximately one-half of the total solar energy reaching the earth's surface is visible. The farther the sun is positioned from the zenith, however, and the greater the amount of atmosphere through which the sun's radiation must pass, the higher is the percentage of the solar radiation that

[1]One micrometer ($\mu$m) = 1000 nanometers (nm) = 1 micron ($\mu$).

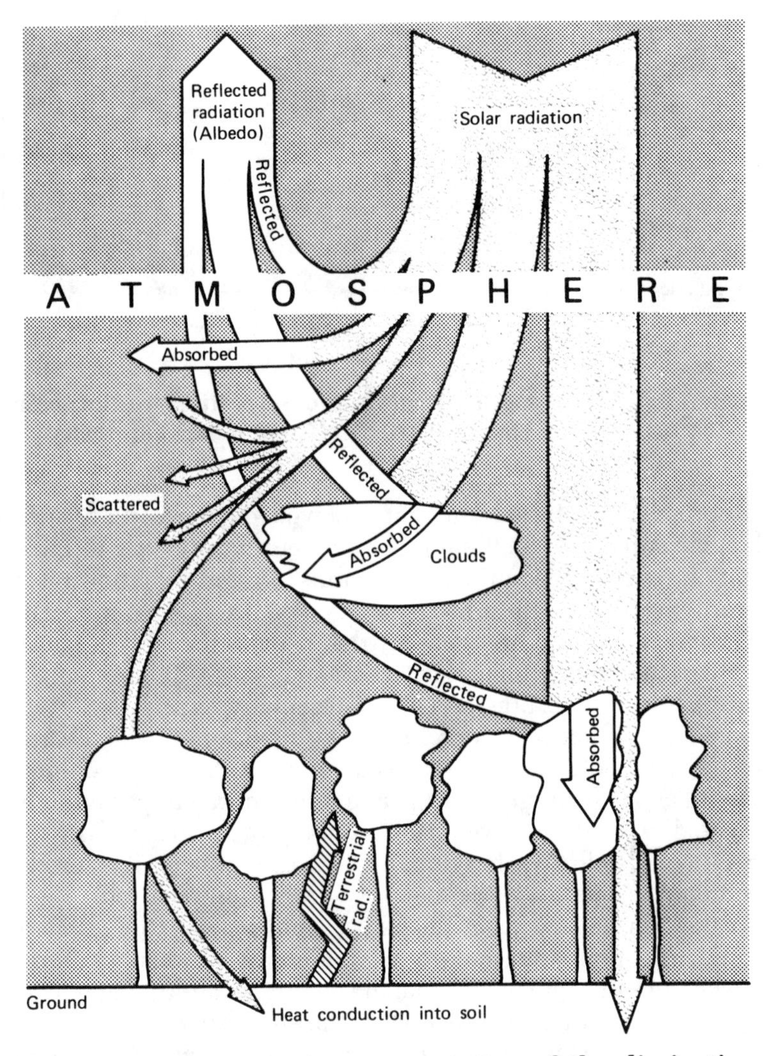


**Figure 4.2.** **Diagrammatic representation of the dissipation of solar radiation. The widths of the lines approximate relative amounts of energy. (Adapted from Geiger, 1950.)**

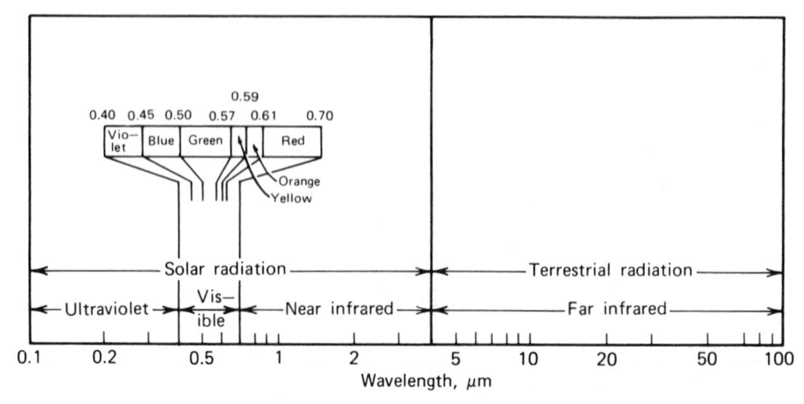


**Figure 4.3. Wavelength of solar and terrestrial radiation and spectral bands. (After Reifsnyder and Lull, 1965.)**

is absorbed or scattered and not received on the earth's surface. When the sun is low on the horizon, therefore, very little ultraviolet radiation reaches the surface, and more infrared radiation reaches the earth than does light.

The total amount of solar energy that reaches the earth is substantially less than that which reaches the atmosphere (Figure 4.4). The atmosphere itself absorbs much. Representative values reported by Abbot (1929) for clear sky conditions with the sun directly overhead give the total solar energy received on Mt. Whitney in California (4418 m) as 1.75 gram-calories per square centimeter per minute, as compared to 1.45 to 1.62 on Mt. Wilson (1737 m), also in California, and 1.15 to 1.45 for Washington, D.C.—near sea level. The lower the elevation, the smaller is the amount of radiation received.

As the sun departs from the zenith and the radiation is slanted through the atmosphere, the total amount is even further diminished. At an angle of 20° from the zenith, an additional 6 percent is absorbed over that absorbed when the sun is at zenith; at 40°, the loss is 30 percent more than the loss at zenith; at 60°, the loss is doubled; and at 88°, the loss is increased by 20 times.

Clouds, ozone, smoke, and other atmospheric impurities, of course, may intercept or reflect most if not all of the direct solar radiation. Values obtained for radiation on overcast days commonly range from one-quarter to one-half of those for cloudless days under the same conditions. Most of the radiation under such overcast conditions is scattered radiation—solar radiation that has been scattered by the atmosphere and other interceptors, and then reaches the ground more or less circuitously. Even on clear days, scattered radiation may amount to 10 to 15 percent of the total.

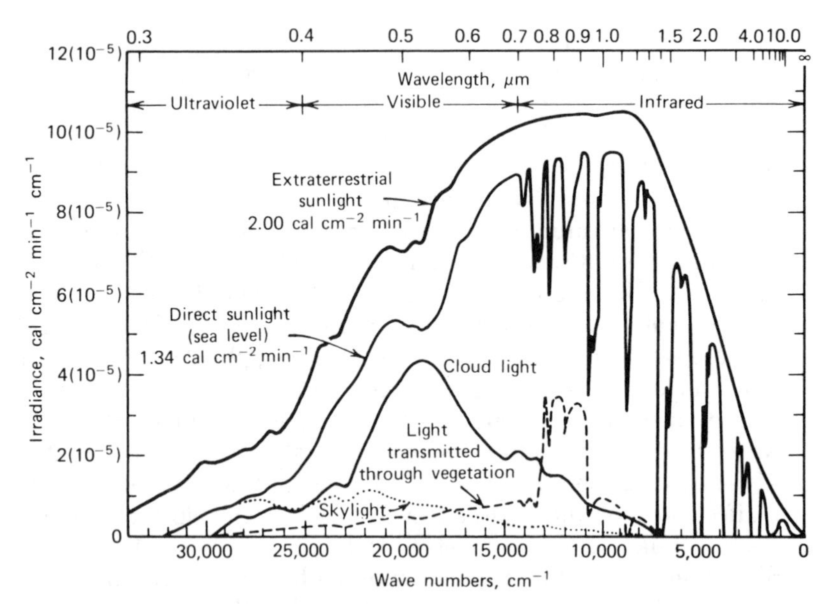


**Figure 4.4. Spectral distribution of the irradiance of extraterrestrial sunlight, direct sunlight at the earth's surface, scattered skylight, cloudlight, and light transmitted through a vegetation canopy. (Reprinted with permission from David Gates, "Energy exchange between organism and environment," in Biometeorology, © 1968 by Oregon State University Press.)**

A distinction must be made at this point between total incident solar radiation per unit surface area and radiation only in the visible range, or **light**. These terms denote radiation measured in watts per square meter or in gram-calories per square centimeter per minute (**langleys per minute**). Many past ecological studies, however, have been based on measurements that represent the stimulation of the human eye by radiant energy. This measure of radiation is called **illuminance** and is measured as lumens per square foot or square meter. Illuminance is commonly expressed in the English system in foot candles and in the metric system in luxes. One foot-candle (ft-c) is equal to 10.764 lumens $m^{-2}$ (lux). More and more studies measure **photosynthetically active radiation** (PAR) of the 0.4- to 0.7-$\mu$m waveband using quantum or **photon flux density**, expressed in microeinsteins per square centimeter per unit time ($\mu$e $cm^{-2}$ $s^{-1}$). This measures the total number of photons within the visible spectrum striking a surface and is a better measure of photosynthetically active radiation than illuminance.

Although illuminance is not a reliable indication of the energy in radiation in the visible range for biological processes (such as photosynthesis) other than vision, it has been widely used because of the availability of photoelectric light meters which are calibrated to measure visual response. Total solar radiation and illuminance, however, are related and approximate conversions are possible. Using an average value of 1.42 gram-cal $cm^{-2}$ $min^{-1}$, the maximum solar illuminance at sea level attained on a clear day with the sun at the zenith would be approximately 9500 ft-c. The figure 10,000 ft-c or approximately 108,000 lux is commonly used for full sunlight. This value is lower than would be obtained at high altitudes with the sun overhead, but is higher than true values at low elevations or whenever the sun is at a considerable distance from the zenith.

All of the solar radiation values given above apply to radiation received by a horizontal surface. The slope and aspect of the ground also affect the amount of radiation received. In the Northern Hemisphere, south-facing slopes receive more radiation per unit area than north-facing slopes, the highest amounts being received by slopes most nearly facing the sun at its highest elevation during the day. The relative irradiance on a slope of any given degree and aspect may be approximated from standard formulae. The effects of these factors may be illustrated by curves for north-, south-, and east-facing 100

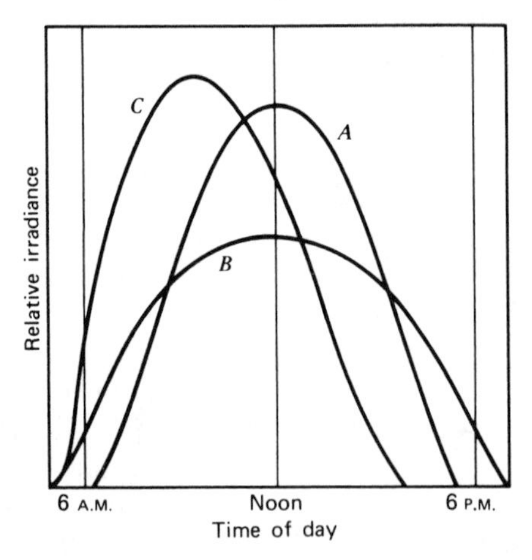


**Figure 4.5. Relative irradiance received throughout the day on a 100-percent slope in the southern Appalachians. *(A)* South. *(B)* North. *(C)* East. The west slope curve is a mirror image of that for the east slope, the maximum occurring in the afternoon. (After Bryam and Jemison, 1943.)**

percent (45°) slopes at north latitude 35°30' on June 21 (Byram and Jemison, 1943). They show lower maximum irradiance levels on north slopes as opposed to south, a morning maximum for east and an afternoon maximum for west slopes (Figure 4.5). Tables of direct solar radiation for various slopes and latitudes are given by Buffo et al. (1972).

## The Energy Budget

The environment, through climatic factors, transfers energy to all living things. This flow of energy, which determines the energy budget of the plant and affects plant temperature, is accomplished primarily by solar and terrestrial radiation, convection, and transpiration. Each process by which energy is transferred between a plant and the environment can cause loss or gain of energy, but the sum total of all energy transferred must balance. Energy gained by the plant from the environment may be stored as heat or converted into photochemical energy by photosynthesis; it may be lost to the environment by radiation from the plant, by heat conduction or convection, or by evaporation and transpiration (evapotranspiration). The remainder of the arithmetic sum of energy flow to (+) and from (−) a surface is termed **net radiation.** The energy budget per unit area of an organism may be written as follows when net energy is neither gained nor lost:

$$Q_{abs} - \epsilon\sigma T^4{}_1 \pm C \pm LE \pm M = 0 \qquad \text{(Gates, 1968)}$$

where $Q_{abs}$ = the total incident radiation of short- and long-wave radiation (always positive).

$\epsilon\sigma T^4{}_1$ = outgoing radiation dependent on the fourth power of absolute temperature and emissivity of the surface.

$C$ = energy flow by convection and conduction.

$LE$ = energy flow by water conversion (energy loss by evaporation or gain by condensation).

$M$ = metabolic rate of the organism (negligible in plants).

If the plant is to survive, it can neither gain nor lose net energy over an extended period of time. If the net radiation at a point on the leaf surface is negative, then the energy arriving at the point is exceeded by the outflowing radiation, typical of forest conditions on cold, clear nights. Severe short-term energy deficits may lead to freezing injury of leaves and other plant parts. With a positive net radiation, the energy may heat the leaf surface (leading to injury if in excess), heat the air adjacent to the leaf, be transformed into latent heat through transpiration, or converted into photochemical energy by photosynthesis. About 4 to 5 percent of gross solar energy is absorbed and converted to chemical energy, but only 1 to 2 percent of total solar energy is used in photosynthesis.

Plant foliage is well adapted to handle solar radiation through its absorptance, transmittance, and reflectance. These properties are remarkably similar for many species (Reifsnyder, 1967) and are illustrated for a leaf of eastern cottonwood (Figure 4.6). The curves indicate high absorptivity in the ultraviolet and the visible range, strikingly low absorptivity in the near infrared (the 0.7−1.5-$\mu$m range, high in energy—see Figure 4.4), and high absorptivity of far-infrared radiation where solar irradiance is very low (Gates, 1968).

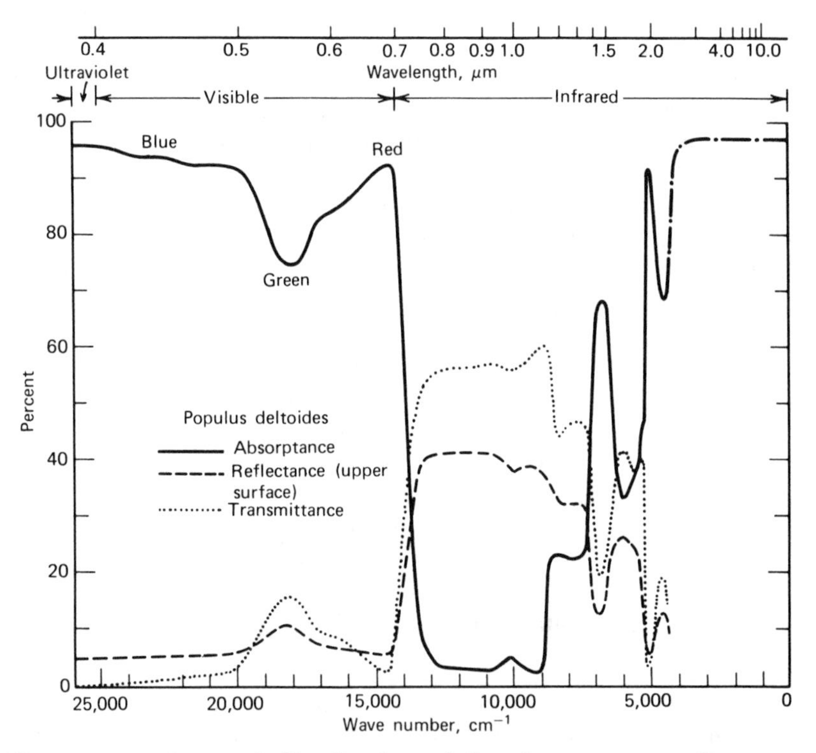


**Figure 4.6. Spectral distribution of the absorptance, reflectance, and transmittance of a leaf of eastern cottonwood, *Populus deltoides*. (Reprinted with permission from David Gates, "Energy exchange between organism and environment," in *Biometeorology*, © 1968 by Oregon State University Press.)**

## Interception of Radiation

The effect of the forest itself in intercepting radiation is quite obvious. Only a small percentage of the incident sunlight reaches the floor of a dense forest. Many determinations have been made of the **relative illumination** (RI) within the forest, expressed as a percentage of total solar illumination.[2] Under leafless deciduous trees, the relative illumination may be as high as 50 to 80 percent of full sunlight; under open, even-aged pine stands, 10 to 15 percent represents a common range; under temperate hardwoods in foliage, values from 1 to 5 percent are common; while beneath the tropical rain forest (Carter, 1934), the relative illumination may be as low as ¼ to 2 percent. Among the densest temperate forests are those formed by pure stands of Norway spruce. Under high closed spruce canopies in Switzerland, total solar radiation may average only 2½ percent of that in the open (Vézina, 1961).

The general relationship between percent forest cover and percent radiation transmission based upon conditions existing under conifers at high elevations in the Sierra Nevada of California during the spring melt season indicates a marked dropping off of relative illumination as crown density increases from zero to about 35 percent, with a more moderate decrease with further increases of density (U.S. Corps of Engineers, 1956). In this particular example, the relative illumination was 18 percent for 0.5 (i.e., 50 percent) density and only 6 percent for full crown density (Figure 4.7).

For western white pine, Wellner (1948) has developed a similar curve relating relative light irradiance to basal area[3] per unit area. At 92 $m^2$ $ha^{-1}$ (400 square feet of basal area per acre) the mean light irradiance was 10 percent of that in the open. In a similar study for shortleaf pine in Georgia, Jackson and Harper (1955) found a strong inverse correlation between the logarithm of stand basal area and the logarithm of relative illumination.

It is somewhat misleading, though, to express radiation beneath the forest in such relative terms. Actually, much of the forest floor is in heavy shade most of the time, with relative illumination values under dense forest ranging from ¼ to 1 percent. Sunflecks sweep over the ground as the day progresses, however, bathing areas

[2]Ecologists have used relative illumination either to express the relative amount of radiation received in the visible spectral range (= **light irradiance**, measured in gram-calories per unit area) or the visual response of radiation (= **illuminance**, measured in luxes or foot candles).

[3]Basal area is the cross-sectional area of a tree at 1.3 m above the ground. It may be expressed on a per-tree basis or summed for all trees giving an amount per unit area, for example, 80 $m^2$ $ha^{-1}$.

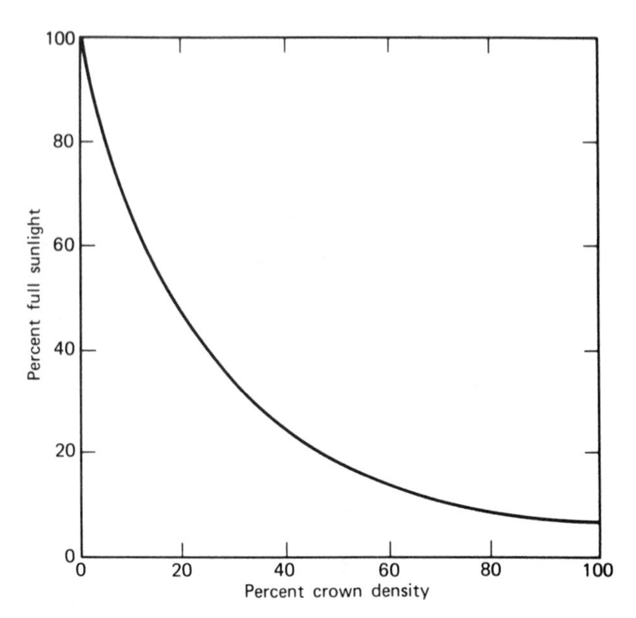


**Figure 4.7. Effect of crown density on penetration of solar energy into conifers in the California Sierra Nevada, during spring snow-melt conditions. (After U.S. Army Corps of Engineers, 1956.)**

momentarily with radiation perhaps half as much as sunlight in the open (Figure 4.8). In a study of the Nigerian rain forest, Evans (1956) has estimated that from 20 to 25 percent of the ground is occupied by sunflecks at noon—and that these sunflecks account for 70 to 80 percent of the total solar energy reaching the ground. Under such conditions, it is likely that photosynthesis may occur only during the times that a leaf is in the sunfleck. Some tree seedlings such as sugar maple and white ash that open stomates rapidly and exhibit slow stomatal closure with increasing light may be able to make use of short-lived sun flecks (Davies and Kozlowski, 1974). This efficiency in response to changing light conditions may explain in part their ability to endure shaded conditions of the forest floor.

## Light Quality Beneath the Forest Canopy

Leaves of the canopy transmit from 10 to 25 percent of the visible radiation they receive. The quality of radiation reaching the understory depends on the optical properties of leaves as well as incoming direct and scattered light penetrating openings in the canopy.

**Figure 4.8. Sunlight penetrates the dense redwood forest canopy and sunflecks sweep over the ground, creating an important source of radiation for understory vegetation. (U.S. Forest Service photo.)**

The broader transmission ability of conifers compared to the more selective absorption and transmission by deciduous canopies has been observed by many workers. In comparisons of light beneath sugar maple and red pine canopies Vézina and Boulter (1966) found far-red radiation to predominate in the sugar maple understory on a clear day, with lesser amounts of green, blue, and red light and ultraviolet radiation (Figure 4.9). Sunflecks were strikingly evident in the morning and early afternoon in Figure 4.9, and during these periods red light increases sharply. The red pine canopy was much less selective in transmission than sugar maple foliage, and for red

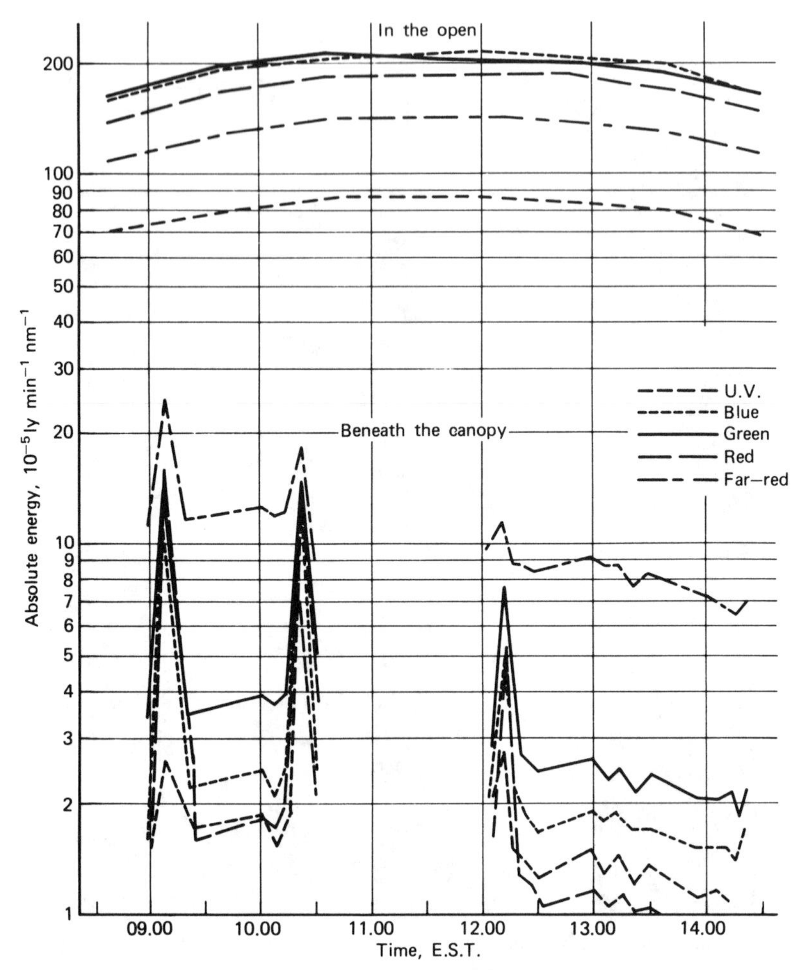


**Figure 4.9. Radiation quality in the open and beneath the sugar maple canopy. (After Vézina and Boulter, 1966. Reproduced by permission of the National Research Council of Canada from the *Canadian Journal of Botany*, Vol. 44, pp. 1267–1284.)**

pine no significant differences were found among the transmission values of the various wavelengths in the visible spectrum. On cloudy or overcast days, the relative percent of radiation transmitted by forest canopies is higher in general than on clear days, and the sugar maple canopy was not as selective as it was on clear days. Similar relationships for conifer and deciduous species and for cloudy and clear days were reported by Federer and Tanner (1966) in studies of scattered shade light. Under sugar maple, clearly defined minimums were found at 0.47 and 0.67 μm (probably resulting from high absorption by chlorophyll) on clear days but were not as well defined under overcast conditions.

The differential absorption and reflectance of wavelengths by forest species have been utilized in aerial photography and are the basis for remote monitoring of the environment, the subject of the following section.

## Remote Sensing

Decreased leaf absorption and increased leaf reflectance in the wavelength region between 0.7 and 1.0 μm are more striking with broadleaved than coniferous species. This fact is the basis for the use of infrared-sensitive films for forest aerial photography when separation of broadleaved and coniferous species is desired. Figure 4.10 demonstrates that reflection from broadleaved and coniferous species is essentially the same in the visible spectrum (Figure 4.10*a*), but that the broadleaved species are significantly more reflective than conifers in the near-infrared spectral band (Figure 4.10*b*; spectral range 0.7–0.98 μm). While aerial photography has been an indispensable part of forestry for many years (Spurr, 1960), devices for recording energy over a broad spectrum of wavelengths now make it possible to monitor or "sense" environmental phenomena from remote stations, including aircraft and spacecraft. Thermal remote sensors capable of recording infrared energy with wavelengths too long to be recorded photographically (4.5–5.5 μm) can detect small fires from altitudes as high as 7,000 m. For example, the eight bright spots in Figure 4.10*c* were caused by small charcoal fires with a surface area of approxiately 80 $cm^2$ each. These hot spots are apparent in the 4.5-to-5.5-μm band because this is the wavelength of maximum emittance for an object whose temperature is approximately 300°C.

Remote sensors operating at still longer wavelengths (8.0–14.0 μm) can detect the effects of several kinds of insect and disease attacks, in some instances before visual symptoms appear. The affected trees are consistently 1–3°C warmer than unaffected trees (Weber, 1971). Once the visual symptoms appear, such insect-infested trees can easily be identified using color or black-and-white film, as illustrated in Figure 4.11 by a lighter tone than that of the surrounding healthy individuals. Remote sensing has been successfully used in the inventory of big game animals whose body temperatures are several degrees warmer than their background (McCullough et al., 1969). Techniques and applications of remote sensing in forestry and ecology are presented in the comprehensive treatment by Reeves (1975).

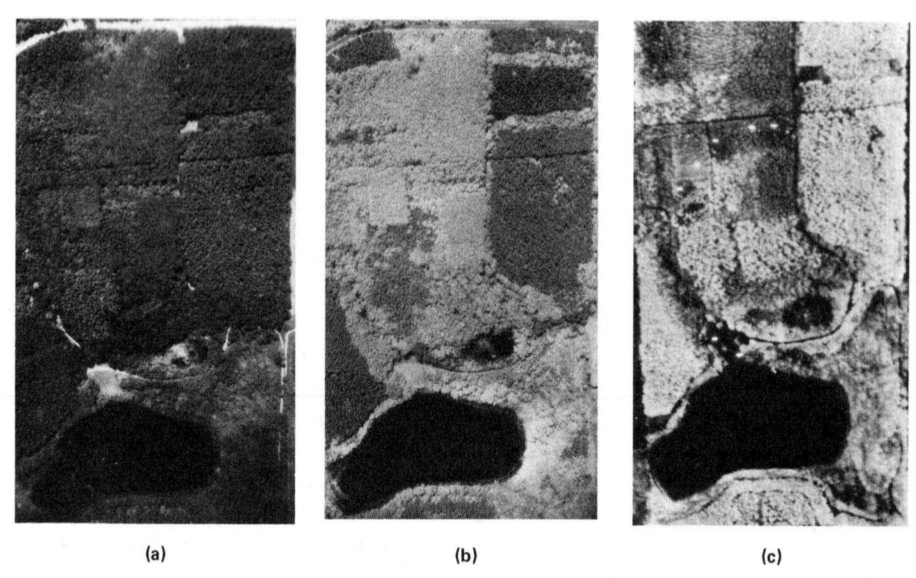


**Figure 4.10. Differences of reflectance of conifers and hardwoods in the visible and near-infrared spectral bands and emittance in the thermal infrared band, Saginaw Forest, Ann Arbor, Michigan. (Courtesy of University of Michigan.) *(a)* Visible spectral band (0.4–0.7 $\mu$m); hardwood plantations not distinctly different in tone from conifers in midsummer. *(b)* Near-infrared spectral band (0.7–0.98 $\mu$m); hardwoods markedly lighter in tone than conifers in midsummer. *(c)* Thermal infrared band (4.5–5.5 $\mu$m); small charcoal fires detected; conifers warmer (lighter tone) than leafless hardwoods in midwinter.**

## Light and the Growth of Trees

The most obvious importance of solar radiation to forest trees lies in the dependence of life upon photosynthesis and the dependence of photosynthesis in turn on light. The term **light** is used here, because it is quite well established that it is the solar radiation in the visible bands of the spectrum that affects the photosynthetic process. We use **light irradiance** to express the amount of radiation received per unit area in the visible spectral band.

Since chlorophyll is green, it follows that green foliage reflects a higher percentage of the green wavelengths than the blue-violet or the longer yellow-red wavelengths. Such indeed is the case (Figure 4.6). A corollary is that the blue-violet and the yellow-red

**Figure 4.11. Living (dark tone) and dead (light tone) ponderosa pine trees in a small stream drainage in the Black Hills of South Dakota. Trees recently killed by the mountain pine beetle (*Dendroctonus ponderosae* Hopk.) appear white whereas the older kills are grayish. (U.S. Forest Service photo. Courtesy of Philip Weber.)**

wavelengths, being absorbed by the plant instead of being reflected, would exert a relatively greater influence on photosynthesis. This concept is confirmed in a classic study by Hoover (1937) with wheat, in which he found that the effectiveness of solar radiation on photosynthesis was almost entirely confined to the visible spectrum and that the most effective bands were the violet-blue and the orange-red. Similar relationships apparently hold for most plants and forest trees, although Burns (1942) found that, for Norway spruce, eastern white pine, and red pine, the orange-red wavelengths were most effective in stimulating photosynthesis, while the shorter violet-blue waves were relatively unimportant, a finding contrary to the earlier study of wheat by Hoover. Linder (1971), working with Scots pine seedlings, found absorption maxima in the red and the blue regions, but the photosynthetic efficiency of blue light was less than that of red light. In general, absorption in both regions is needed for maximum efficiency.

Since the growth rate of trees is obviously closely related to their rate of photosynthesis, and since the rate of photosynthesis can be measured under controlled conditions by measuring the uptake of carbon dioxide from the air by the plant, there have been many studies of the effect of varying amounts of light upon the rate of photosynthesis. At very low irradiance levels, photosynthesis takes place at such a slow rate that it fails to utilize all the carbon dioxide evolved by the plant in respiration. Under such conditions, carbon dioxide is actually given off by the plant rather than being absorbed by it from the atmosphere. The **light compensation point** is that level of photosynthesis where carbon dioxide is neither given off nor taken up. In other words, it is the point at which photosynthetic gains balance respirative losses.

For forest trees under otherwise optimal conditions, the light compensation point appears to occur at from 1 to 2 percent of full sunlight (RI). For example, Grasovsky (1929) found that this point for well-watered eastern white pine occurred at approximately 1830 lux. It thus appears that, in very dense forests, **net photosynthesis** (i.e., more photosynthesis than respiration) occurs only when the leaves are bathed in sunflecks. Similarly, many understory species in dense deciduous forests can make net gains in growth only during the leafless period of the overstory trees. Under dry conditions, the compensation point occurs at relatively higher light irradiance, so that more light is needed to permit an understory plant to hold its own where root competition limits its supply of soil moisture than when the plant is well watered (Takahara et al., 1955).

The minimum light requirements vary for different species. In general, trees which are seldom found under dense canopies require more light at the compensation point than do those that commonly occur in the understory. In a test of 14 species, Burns (1923) found that ponderosa pine and Scots pine required light irradiances at a compensation point three times as high as that of eastern white pine. At the other extreme, hemlock, beech, and sugar maple required the least amount of light to maintain themselves. Similar results were obtained for 12 species in the West, with redwood and Engelmann spruce having the lowest light requirements, and limber pine and pinyon the highest (Bates and Roeser, 1928). Redwood has one of the lowest light requirements of all species that have been studied, redwood seedlings having shown 90 percent survival and 66 percent increase in dry weight when grown for 55 days at 320 lux (Shirley, 1929). For shade and sun plants in general, approximately 550 and

1600 lux (0.5 to 1.5 percent RI) respectively, are required at the compensation point under natural conditions (Perry et al., 1969).

Plants may live for a considerable length of time on stored food after environmental conditions develop to the point at which respiration exceeds photosynthesis. For example, Grasovsky found that white pine seedlings remained in good condition throughout the growing season under 700 lux, even though 1830 lux is seemingly required at the compensation point. Sooner or later, though, such plants are bound to die. In studies with seedlings of several species, Grime (1966) observed the greatest height growth and least mortality in shade for species having large seeds. Mortality of black birch seedlings (0.7 mg dry weight of seed reserve) was four times that of red oak (1969.2 mg dry weight of seed reserve). The occurrence of one- or even two-year-old seedlings under dense forest stands does not demonstrate that light under those conditions is sufficient for growth; the seedlings may still be living on stored food material from the seed.

As the irradiance is increased above the compensation-point level, photosynthesis is increased proportionately. After reviewing the evidence of other workers as well as carrying out tests of his own, Shirley (1945a) concluded that, in the range from 1 to 15 percent of full sunlight, photosynthesis is directly proportional to irradiance if other factors are favorable. The increase in photosynthesis will continue until other factors combine to bring growth to a halt. At very high irradiances, such secondary factors as high respiration, water deficit causing stomatal closing, and overaccumulation of photosynthate in the leaves may result in decreased photosynthesis.

The relative effect of illumination upon photosynthesis in different species has been studied by a number of investigators. For example, both Kramer and Decker (1944) and Kozlowski (1949) have shown that seedlings of loblolly pine show increased photosynthesis for increased illuminance at all levels up to full sunlight, while associated hardwood species reach their maximum photosynthetic rate at 30 percent or less relative illumination. Such a finding tends to explain the better relative competitive ability of the hardwoods under open pine stands where the relative illumination is commonly in the neighborhood of 30 percent full sunlight (Figure 4.12).

Another factor affecting the relative ability of different species to compete under given light conditions is the color, shape, and arrangement of the leaves—factors that regulate the amount of light actually reaching the chlorophyll. Pines, for instance, characteristically have rounded needles in dense clusters. Such an arrangement results in much scattering of light and much mutual shading (Kramer and Clark, 1947). Many hardwoods associated with pine in

**Figure 4.12. A conspicuous hardwood understory in an open 80-year-old loblolly pine stand, North Carolina. (U.S. Forest Service photo.)**

the eastern United States, on the other hand, have broad, thin leaves, arranged perpendicular to the direction of incident radiation and spaced by their branching habit so that shading of one leaf by another is at a minimum. It follows that the pine chlorophyll receives a much smaller fraction of the light incident to it than does the dogwood, to cite a specific example. This fact is at least one of the reasons why pine can maintain a high net photosynthetic rate at high light irradiances, which actually inhibit the growth of dogwood, while dogwood can carry on photosynthesis efficiently in low light conditions under a loblolly pine stand.

The failure of seedlings in shade is often associated with fungal attack. Compared to tree species intolerant of shade, shade-tolerant species are less susceptible to infection both above and below the compensation point (Grime and Jeffrey, 1965). The predisposition of shade-intolerant species to fungal attack may be due to characteristics arising from their adaptation for survival in habitats of high light irradiance. These characteristics are (1) attenuation of the stem and mechanical collapse when shaded, (2) inherently high rates of respiration, and (3) marked rise in respiration with increasing temperature (Grime, 1966).

The experiments detailed above have dealt with the rate of photosynthesis as measured by the uptake of carbon dioxide from the atmosphere. It is also possible, of course, to measure the actual height growth and weight growth of trees over a period of years under different degrees of irradiance. This has been done by a number of investigators, who for the most part have reduced the possible effect of soil moisture by watering the plants grown under all irradiances to keep the soil at a more or less optimum soil-moisture level.

In general, at least 20 percent of full sunlight is required for survival over a period of years, the actual amount varying with the species and with growing conditions. For species that are usually found growing in full sunlight, growth commonly increases with irradiance up to full sunlight. Pearson (1936, 1940) grew ponderosa pine under various shade conditions in Arizona for periods up to 10 years and found the best growth under full sunlight, although the trees grown at 50 percent showed only slightly less height growth and about one-half the diameter growth. Shirley (1945b) and Logan (1966a) obtained similar results for red pine and jack pine.

Many other species, however, make as much or more height growth under partial light as under full light. Among conifers, white spruce, white pine, and Douglas-fir provide examples (Gustafson, 1943; Shirley, 1945b; Logan, 1966a; Lassoie, 1980). Among hardwoods, this relationship seems to be the rule. Seedling height growth data for six hardwood species in Missouri (McDermott, 1954) showed that five of the six grew tallest in either 50 or 33 percent relative illumination (Table 4.1). For five- to six-year-old hardwood seedlings in Ontario, Canada, Logan (1965, 1966b) reported the maximum height of white birch, yellow birch, and American elm was at 45-percent RI, whereas it was at 25 percent for basswood and sugar maple (both highly shade-tolerant). Few species, conifers or hardwoods, however, make better weight growth under partial light than under full light.

Regardless of whether or not height growth is increased or decreased by shaded conditions, there seems little doubt that root development of seedlings is sharply impaired. The lower the irradiance, the greater is this impairment, so that, at low irradiance levels, seedlings of most forest trees have relatively shallow and poorly developed root systems. The data of McDermott (Table 4.1) serve to illustrate this point. American elm may make the tallest shoot growth at 33 percent RI, but the top is 3.2 times as heavy as the roots under these conditions. The shorter plant grown under full sunlight has a relatively better-developed root system, the top being only 1.7 times as heavy under these circumstances. Similar results have been reported for many other North American species (Baker, 1945; Shirley, 1945b; Logan, 1965, 1966b) and for European species (Harley, 1939; Brown, 1955; Leibundgut and Heller, 1960).

**Table 4.1 Seedling Height and Top Root Weight Ratio of Newly Germinated Seedlings as Influenced by Amount of Sunlight**

| | 20% RI | | 33% RI | | 50% RI | | 100% RI | |
|---|---|---|---|---|---|---|---|---|
| SPECIES | HT(cm) | T/R | HT(cm) | T/R | HT(cm) | T/R | HT(cm) | T/R |
| American elm (*U. americana*) 13 weeks | 73 | 3.1 | 81 | 3.2 | 69 | 3.0 | 37 | 1.7 |
| Winged elm (*U. alata*) 13 weeks | 59 | 5.2 | 71 | 3.9 | 76 | 4.4 | 28 | 1.9 |
| Sycamore (*P. occidentalis*) 15 weeks | 43 | 4.6 | 41 | 3.5 | 40 | 4.1 | 33 | 2.0 |
| River birch (*B. nigra*) 10 weeks | 26 | 13.4 | 33 | 9.3 | 52 | 8.3 | 41 | 3.6 |
| Red maple (*A. rubrum*) 13 weeks | 26 | 4.1 | 27 | 3.7 | 29 | 3.5 | 26 | 2.0 |
| Alder (*A. rugosa*) 14 weeks | 21 | 3.4 | 20 | 3.5 | 38 | 4.7 | 30 | 2.8 |

Source: After McDermott, 1954.

The marked effect of reduced light in reducing root growth is of major importance in explaining the growth and survival of plants in the understory. Under a growing forest, root competition substantially reduces the amount of soil moisture available to the seedlings (Chapter 10). The combination of reduced root size due to low light irradiances and reduced soil moisture due to root competition is frequently fatal to the seedlings. Frost-heaving, too, is increased for seedlings grown under low light irradiance (Hagem, 1947).

The fact that, under the forest canopy, both light and soil moisture are reduced makes meaningless any attempt to relate relative light irradiance to survival and growth of seedlings under uncontrolled conditions. Curves showing the effect of light irradiance upon seedling development ignore the existence of variations in soil moisture and other environmental factors which also vary with different canopy densities. Nevertheless, such efforts have their value if the light measurements are considered to indicate merely the variation in the crown density and are not interpreted to mean the effect of light alone. For example, Shirley (1932) found that red pine reproduction was uncertain under forest stands sufficiently dense to allow only 17 percent of the light to reach the ground, but that satisfactory reproduction became established when the stand was opened up to the point that the relative illumination on the ground was 35 percent.

In studies where establishment and growth involve several environmental factors, compensating effects may be expected. In a study of yellow birch regeneration, Tubbs (1969) employed 64 different combinations of three major factors—light irradiance, soil moisture, and seedbed type—to assess germination, survival, and height growth. On well-drained sites good germination occurred under full sunlight on mineral soil, but on heavy and organic soils heavy shade was required. On mineral soil height growth was best in full sunlight on the well-drained site, whereas shade was necessary for maximum growth on the imperfectly drained site. Although partial shade has been found generally to promote better seedling development of yellow birch, full sunlight may be required under certain conditions. The study also demonstrated the remarkable phenotypic variation of the seedlings, all from one mother tree, in germinating and surviving over a wide range of environmental conditions.

## Light and Tree Morphology

Plants grown under shade develop a structure and appearance different from the same plants grown under full sunlight. These morphological changes are of importance ecologically in understanding the capacity of a given species to become adjusted to shaded conditions and the reaction of such a plant when suddenly released, as by cutting of the overstory.

Since the leaf is the principal photosynthetic organ of the tree and therefore presumably most affected by changes in radiation and light, it has been the subject of many detailed investigations. The structure of leaves of typical understory plants has been compared with those of typical overstory plants; and sun leaves have been compared anatomically with shade leaves of the same plant. The findings are in essential agreement (Büsgen and Münch, 1929).

Some species exhibit little plasticity in leaf anatomy with the result that there is little difference between their sun leaves and their shade leaves. Such plants are usually found only in the overstory or only in the understory—their anatomy is suited for their survival under only one set of environmental conditions. They are either obligate sun plants or obligate shade plants.

Most forest trees, however, have the faculty of developing different anatomical structures in leaves grown in the shade than in those developed in the sun (Hanson, 1917; Büsgen and Münch, 1929; Jackson, 1967). Typically, shade leaves are thinner and less deeply lobed, and have a larger surface per unit weight, a thinner epidermis, less palisade, more intercellular space and spongy parenchyma, less supportive and conductive tissue, and fewer stomata than comparable sun leaves off the same tree. Similar differences in leaf structure are found between species that characteristically grow in shaded conditions and those that grow under fully exposed conditions.

The typical shade leaf is adjusted to carry on photosynthesis when efficiently protected from the detrimental effects of too much light. Furthermore, shade leaves require less light at the compensation point, so they typically survive and show net photosynthesis with very little light.

Despite the assumption that shade leaves develop in response to a reduced light, it must be remembered that other factors may also be involved. The temperature is obviously different in sun and shade situations. Furthermore, leaves within the crown and in its lower portions are under less water stress on clear days than are the sun leaves at the top of a tree.

When shade leaves are suddenly exposed to full light, as commonly happens after partial cutting, they are frequently unable to survive. This failure may be due in part to excessive moisture loss and in part to excessive light reaching the chloroplasts. Under these conditions, the shade leaves commonly die and drop off, even in the case of conifers such as Norway spruce (Stålfelt, 1935). Survival of the branch depends upon the development of a new crop of leaves with anatomical characteristics suited to the new set of environmental conditions.

Much less work has been done on the effect of light upon bark thickness, the stimulation of adventitious buds, and upon root growth. Certain facts, however, seem to emerge from careful observation. For example, there is evidence that the bark of eastern white pine is substantially thinner, smoother, and more vulnerable to insolation when grown in the shade. Sunscald, a winter damage phenomenon apparently related to rapid temperature changes of the cambium (Huberman, 1943), occurs chiefly in thin-barked white pine whose shade bark is exposed to direct solar radiation by the removal of adjacent trees.

Adventitious buds in the bark may be activated to form **epicormic branches** when the bark is exposed to direct solar radiation. It does not follow, though, that there is a necessary cause-and-effect relationship between light and bud stimulation. The removal of surrounding trees, for example, will affect the environment of the remaining trees in many ways, and carefully controlled experimentation is required to determine which factor or combination of factors actually initiates the chain of physiologic circumstances that lead to the development of epicormic branches. It is known that the crown exerts a substantial control on sprouting; decapitation of the crown (Books and Tubbs, 1970) and pruning of live branches (Kormanik and Brown, 1969) stimulate sprouting. It is generally agreed that sprouting is triggered by a change in growth regulators within the tree. All that can be deduced from field observations is that epicormic branches form on the exposed boles of many trees, particularly those with poorly developed crowns, and that European foresters have been able to reduce the amount of epicormic branching on such species as oak by keeping the bole clothed with the foliage of understory trees such as beech and hornbeam. The occurrence of epicormic branches after thinning is particularly serious with many of the oaks, maples, birches, and other hardwoods where stem quality is of paramount economic importance. Under extreme conditions of change from shade to light, epicormic branches may develop in most forest trees, both conifers and hardwoods.

## Photocontrol of Plant Response

The growth and development of all parts of the plant, including stem elongation, root development, dormancy, germination, flowering, and fruit development are subject to the same causal photocontrol whereby light is absorbed by a reversible pigment system in the plant. This system is the physiological basis of photoperiodism, which has been repeatedly demonstrated to affect seasonal rhythm, timing, and amount of tree growth. **Photoperiodism** is the response of the plant to the relative length of the day and night and the changes in this relationship throughout the year.

The term photoperiodism was proposed by Garner and Allard

(1920, 1923), who carried on extensive tests which demonstrated that both vegetative development and the initiation of reproductive processes in plants could be greatly affected by varying the relative length of day and night. Although many plants were not found to be especially sensitive to day length, others could be characterized as short-day plants because flowering could be induced under suitable conditions by exposure to days shorter than a certain critical length, while others could be termed long-day plants as these responded to days longer than a given critical length or even to continuous illumination. Responses to differing day lengths are clearly different for different species.

As we have seen in Chapter 2, photoperiodic requirements are not constant within a species, but vary with the latitude and altitude of the source. Thus genecological differentiation is characteristic of many tree species, having arisen as a result of natural selection for a particular photoperiod associated with the limiting factors of the growing season.

When plants respond to light they do so because light is absorbed by the phytochrome pigment system of the plant. The bluish protein phytochrome has been detected in all parts of higher plants and exists in two forms:

$$\text{Pr} \underset{\text{far-red light } 0.73\ \mu\text{m}}{\overset{\text{red light } 0.66\ \mu\text{m}}{\rightleftarrows}} \text{Pfr} \xrightarrow{\text{darkness}} \text{Pr}$$

One form, Pr, has an absorption peak (0.66 $\mu$m) in the red region of the spectrum; the other form, Pfr, biologically active, has an absorption peak (0.73$\mu$m) in the far-red region. Absorption of light at the appropriate wavelength readily converts one form to the other; when the red-absorbing form receives red light, it is changed into the far-red absorbing form (Pfr), and vice versa. Growth responses elicited by red light are reversed by far-red light. Besides reversibility, the system is characterized by a slow drift in darkness from the far-red absorbing to the red-absorbing form. The detection of phytochrome, its properties, and effects in morphogenesis of plants are reviewed by Siegelman (1969), Mohr (1969), and Vince-Prue (1975).

Growth cessation under short-day conditions and continuous growth under long days may be explained by the phytochrome system. At the end of the daily light period, in which red light predominates, more than 70 percent of the pigment is in the Pfr form. During the dark period that follows, the pigment slowly reverts to the Pr form, and if the dark period is long enough less than 10 percent remains in the Pfr form (Downs, 1962). If the pigment remains in the

Pr form for a substantial time during each dark period, woody plants cease growth and enter dormancy, because the amount of biologically active Pfr form is inadequate. Thus it is the length of the night, rather than that of the day, that initiates through phytochrome the biochemical reactions controlling growth. Under long days or continuous light the pigment is in the Pfr form for an appreciable time and growth is promoted. This physiological system is responsible for the development of a continuous sequence of photoperiodic races of North Temperate Zone trees throughout their ranges (see Chapter 2).

Unlike growth cessation, bud-burst or leaf flushing in the spring is not appreciably controlled by photoperiod. Once the plant's chilling requirement has been met, flushing is primarily dependent on temperature. The photoperiodic effect on the flowering of trees is not well understood. It has not been possible to establish clearly long-day and short-day flowering types as has been done in herbaceous species. This has led to the conclusion that most trees are day-neutral. The extreme difficulty of experimentation with large trees capable of flowering has restricted our knowledge.

## SUGGESTED READINGS


Downs, Robert Jack. 1962. Photocontrol of growth and dormancy in woody plants. *In* Theodore T. Kozlowski (ed.), *Tree Growth*. The Ronald Press Co., New York. 442 pp.

———, and H. A. Borthwick. 1956. Effects of photoperiod on growth of trees. *Bot. Gaz.* 117:310–326.

Gates, David M. 1962. *Energy Exchange in the Biosphere.* Harper & Row, Inc., New York. 151 pp.

———, 1968. Energy exchange between organism and environment. *In* William P. Lowry (ed.) *Biometerology*. Oregon State Univ. Press, Corvallis. 171 pp.

———. 1970. Physical and physiological properties of plants. *In Remote Sensing, with Special Reference to Agriculture and Forestry.* National Academy of Sciences, National Research Council, Washington, D. C. 424 pp.

Lowry, William P. 1969. *Weather and Life, an Introduction to Biometeorology.* (Chapter 2, Energy and ecology, pp. 9–12; Chapter 3, Radiation, pp. 13–30; Chapter 7, The energy-budget concept, pp. 113–121; Chapter 8, Energy budgets of particular systems, pp. 122–151.) Academic Press, Inc., New York. 305 pp.

Reifsnyder, William E. 1967. Forest meteorology: the forest energy balance. *Int. Rev. For. Res.* 2:127–179.

———, and Howard W. Lull. 1965. Radiant energy in relation to forests. USDA Tech. Bull. 1344. 111 pp.

Shirley, H. L. 1945. Reproduction of upland conifers in the Lake States as affected by root competition and light. *Amer. Mid. Nat.* 33:537–612.

Treshow, Michael. 1970. *Environment and Plant Response.* (Chapter 8, Light mechanisms, pp. 99–115; Chapter 9, Light stress and radiation, pp. 116–124.) McGraw-Hill Book Co., Inc., New York. 422 pp.

# 5 Temperature

Solar radiation is the source of the heat that controls the temperature regime near the ground of the earth. The great importance of terrestrial radiation and of air movements in affecting the level and distribution of temperature, however, makes it desirable to discuss solar heat and air temperature separately from solar radiation and light.

Basically, the mean annual temperature at any given spot on the earth's surface is a function of the incoming solar insolation at that spot modified by secondary heat transfer arising from terrestrial radiation and air movements. Heating of the surface layers of air during the day is greatest under conditions where the greatest amount of infrared radiation is received, i.e., in tropical latitudes, at high elevations, and where the air is freest from water vapor, clouds, and atmospheric impurities.

Temperatures during the night, on the other hand, depend largely upon the amount of heat absorbed by terrestrial objects and atmosphere during the day and the rate at which this heat is given off as terrestrial thermal radiation. To cite an important example, the great capacity of large bodies of water to absorb solar energy and to reradiate it slowly and steadily results in greatly modified temperature extremes on lands subject to winds that have passed over the water. Thus the coastal areas of western North America and western Europe show little variation in temperature between day and night and even between winter and summer. This is the characteristic of maritime climates as contrasted with continental climates characterized by extreme changes in temperature between day and night, and between summer and winter.

## TEMPERATURES AT THE SOIL SURFACE

The focal plane of temperature variation is the line of contact between the atmosphere and the ground (or, more accurately, the surface exposed to the sun). Both the surface of the ground and the adjoining surface layer of air heat up greatly under direct sunlight and cool off greatly at night as heat is lost through terrestrial thermal radiation. Direct solar insolation commonly produces a surface temperature at midday of at least 70°C, even in northern latitudes in the summertime (Vaartaja, 1954; Maguire, 1955).

The exact temperature reached at the soil surface depends upon the rate of absorption of solar energy and the rate at which it is dissipated once absorbed, which in turn is dependent primarily upon the amount of vegetation and litter cover, and only secondarily upon the color, water content, and other physical factors of the soil itself, if exposed.

The importance of soil color has been demonstrated by Isaac (1938). He found that charcoal-blackened soils will reach a temperature of 73°C when the air temperature is 38°C, at which time comparable grey mineral soil heated up only to 64°C and yellow mineral soil to 67°C. These values were obtained on bare soil in the Douglas-fir region of southern Washington.

The rate at which surface materials dissipate heat received was found to be highly important in determining surface soil temperatures by Smith (1951), who worked with white pine in Connecticut. A surface of white pine needle litter in a small clearing heated up to 68°C on a day when the air temperature reached only 24°C. In contrast, the surface temperature of bare mineral soil reached 46°C and that of polytrichum moss reached only 39°C. The specific heat and the conductivity of the surface materials are apparently the important physical factors involved. Wind speed, however, is most important of all.

Much of this variation in surface temperatures is due to the water content of materials. Moist material has a much greater capacity to dissipate heat through evaporation of water, while the evaporated water itself tends to reduce the amount of incoming solar energy. Thus Maguire (1955), in California, sprinkled bare mineral soil with the equivalent of 25 mm of rain on May 17 and found that the watered soil was 23°, 14°, and 9°C cooler on the succeeding three days than the comparable unwatered soil.

Temperature variations drop off sharply within the soil. Diurnal variations may disappear within 30 cm of the surface and annual variations within several meters. In fact, Shanks (1956) has argued that soil-temperature data from a 15 cm depth measured at weekly intervals serve as a useful integration of radiation and air temperature conditions during the preceding week. Indeed, temperatures measured deep in the ground, such as may be measured in caves, frequently remain constant throughout the year and may well measure the mean annual temperature of the area (Poulson and White, 1969). Such measurements may prove useful in measuring the extent of climatic fluctuation.

## TEMPERATURES WITHIN THE FOREST

Within the forest, light crown cover and trees without foliage, as in the case of deciduous trees during the leafless season, tend to reduce air movement while allowing solar radiation to penetrate the canopy. Under such conditions, the mean air temperature may be higher within the forest than outside it. For example, Pearson (1914) found that the mean annual air temperature within the open ponderosa pine forest of northern Arizona was 1.5°C higher than in adjoining open parks, a phenomenon partly attributed to lower radiation losses from the forest and partly to the effect of the forest in deflecting cold air masses moving down from surrounding mountains. In the Copper Basin of Tennessee, Hursch (1948) found that air temperatures were 0.3° to 1.1°C higher in the deciduous forest than in the open during the winter months, although they ranged from 1.2° to 1.9°C lower during the summer months.

When trees are in full leaf, the extremes within the forest are generally less than outside, and the diminution of radiation within the forest may result in lower mean annual air temperatures. To cite one of many examples, data collected within and adjacent to a white pine plantation by Spurr (1957) gave a summer air temperature range within the forest of 15.9°C compared to 21.6°C in the open, and a winter range of 19.4°C within the forest compared to 23.5°C in the open (Table 5.1). Throughout the year, at a height of 1 m from the ground, the maxima and means were lower and the minima higher within the forest as contrasted to the open station.

**Table 5.1 Mean Weekly Maxima, Minima, and Mean Temperatures (°C) in the Open and Under a Dense 20-year-old White Pine Plantation [a]**

| | WINTER | SPRING | SUMMER | FALL |
|---|---|---|---|---|
| *Open* | | | | |
| Maximum | 5.1 | 22.8 | 29.7 | 14.2 |
| Minimum | −18.4 | −2.2 | 8.1 | −7.3 |
| Range | 23.5 | 25.0 | 21.6 | 21.5 |
| Mean | −6.7 | 10.4 | 18.9 | 3.4 |
| *Under forest* | | | | |
| Maximum | 2.7 | 19.9 | 25.6 | 11.0 |
| Minimum | −16.7 | −2.8 | 9.7 | −5.7 |
| Range | 19.4 | 20.2 | 15.9 | 16.7 |
| Mean | −7.1 | 9.8 | 17.7 | 2.7 |

[a] Petersham, Mass.; by 13-week quarters, 1943–44.

Higher up in the forest stand, temperature variations are less than they are near the ground. Fowells (1948) measured the temperature profile up to a height of 37 m in a mature Sierra Nevada mixed conifer forest ranging up to 60 m in height. He found that temperature variation decreased with increased distance from the ground. At 37 m, minimum temperatures ranged up to 2°C higher than at 1.4 m while maximum temperatures were similarly lower. At the top of the tree crown itself, however, surface conditions similar to those described for the forest floor exist. Consequently, extreme variation in temperature may be expected and has been described for the beech forest of Ohio by Christy (1952).

## TEMPERATURE VARIATIONS WITH TOPOGRAPHIC POSITION

The effects of local topography upon local temperature have been studied by a number of investigators (Pallman and Frei, 1943; Hough, 1945; Wolfe et al., 1949; Spurr, 1957, among others). All have found that low concave landforms tend to radiate heat rapidly on still cold nights and to accumulate cold air, which flows in from surrounding higher land. As a result, such sites will frequently have air temperatures near the ground as much as 8°C lower than that of the surrounding terrain. The resulting condition is known as **inversion** in that the temperature increases with height in the layer of air near the ground in contrast to the usual decrease in temperature with height. The concave areas are known as **frost pockets** because the temperatures commonly occurring near the ground result in late spring frosts, early fall frosts, and consequent short growing seasons. The concave area need not be deep. Spurr (1957) found that minimum temperatures in a small depression only 1 m deep were comparable to those in a nearby deep valley 60 m below the general land level.

In contrast, relatively high convex surfaces will tend to drain off cold air as radiation from the ground proceeds, so that night minima remain fairly high. During the day, the low concave landforms tend to accumulate radiant energy and reach high maximum temperatures near the ground, while mounds, ridges, and other convex surfaces tend to remain cooler during the day.

A few examples will suffice. In a study of temperature variation 1 m off the ground in the open, Spurr (1957) found that the weekly maximum temperatures on slopes with good air drainage averaged 3°C lower throughout the year than in nearby frost pockets. The weekly minimum temperatures, however, averaged 5°C higher on the well-drained slopes. The effect of this range on the growth of plants is indicated by the fact that on May 19, in the middle of the spring growing season, a frost dropped temperatures to −1°C at the top of the highest hill (400 m above mean sea level), while temperatures reached a low of −8°C in a small frost pocket. The effect of inversion in the frost pockets resulted in frost-free summer periods of only 77 and 104 days as compared to 161 days at stations with good cold air drainage. Despite the greater extremes in the concave areas, however, the mean annual temperatures remain essentially the same as on the upland sites with better air drainage.

The major influence of local topography on temperature is illustrated by the southerly distribution of eastern hemlock in Wisconsin (Adams and Loucks, 1971). Hemlock occupies cool sites of lower valley slopes (similar to sites in its northern range) where low temperatures and moist soils favor net photosynthesis compared to warm upslope sites.

Although variations in temperature due to local topography undoubtedly exert a strong influence on the distribution of plants, and almost certainly play a part in inhibiting tree growth in frost pockets, it must be remembered that soil water drainage is frequently an interacting factor. The same factors—especially the existence of convex surfaces—that make a soil very well drained are apt to insure good drainage of cold air and to indicate a site with lower maximum and higher minimum temperatures than would be indicated by a regional climatic average. Thus a very well-drained site is apt to have a moderated climate suitable to plants of a generally southern distribution. Similarly, a very poorly drained soil is apt to result from a concave land surface that inhibits the drainage of cold air as well as of water. Thus the very poorly drained sites are apt to be characterized by temperature extremes and a short growing season, and might prove suitable for plants of a generally northern distribution.

Finally, it should be realized that frost pockets can be created by the forming of a small clearing in a forest stand, the surfaces of the tree crowns of the surrounding forest functioning to channel cold air into the clearing as terrestrial radiation chills the surface layers of the air on still, clear nights.

## EFFECTS OF LATITUDE AND ALTITUDE ON AIR TEMPERATURE

In general, temperatures decrease with increasing distance from the Equator toward the Poles and with increasing altitudes.

Many factors affect temperature other than latitude and altitude. Local topography, the direction and character of prevailing winds, storm patterns, and location relative to water bodies all contribute to the climate of a given place. It is both instructive and useful, however, to generalize the effect of latitude and altitude on temperature, holding these other factors as constant as possible.

Along the eastern seaboard of the United States, the mountains and uplands are oriented parallel to the coast. We would expect, therefore, a relatively uniform gradient in temperature from the crest of the Appalachians down to the Atlantic and from Canada south to Florida. Such indeed is the case. In a study of weather bureau data, mean annual temperature at sea level varied from 21°C interpolated at latitude 30° North to 3°C at latitude 50° North. Mean annual temperatures for these latitudes and altitudes are given in Table 5.2. In this study (Spurr, 1953), mean temperature was used, as it tends to average out differences in local climate due to topography. It was estimated that 1 degree of latitude (111 km; 69 mi) represents an average change of about 1°C, and that 300 m of elevation represents a change of 1.7°C (1000 feet change = 3°F).

In the mountains of the American West, it is possible to utilize the climatic summaries of Baker (1944) and others to make similar generalizations. From the Pacific Ocean to the crest of the Cascade–Sierra Nevada chain, the prevailing on-shore winds are the dominant influence in the climates, and greatly diminish temperature variation from north to south. On the west side, there is a difference in mean annual temperature of less than one degree for each degree of latitude (0.4°C or 0.7°F; and about 0.3°C per 100 km or 1°F per 100 miles). This is only about one-third of the change found on the east coast. However, east of the Cascade–Sierra Nevada barrier continental conditions generally prevail. The overall change here in the interior is nearly 0.8°C (1.4°F) per degree of latitude, or 0.7°C per 100 km (2°F per 100 miles).

Temperature variation with altitude is similarly affected by prevailing wind patterns. In the maritime climate of California, the temperature decrease with each 300 m (1,000-foot) rise in elevation

**Table 5.2 Relation of Mean Annual Temperature to Latitude and Altitude in Eastern United States**

| | MEAN ANNUAL TEMPERATURE (°C) | | |
|---|---|---|---|
| LATITUDE | Sea level | 600 m (2,000 ft) | 1200 m (4,000 ft) |
| 35° | 18 | 14 | 11 |
| 40° | 12 | 9 | 6 |
| 45° | 7 | 4 | 0.6 |

is between ½ and 1°C for the coastal ranges and 1.4 to 2°C for the Sierra Nevada. In the interior of the American West, the decrease for each 300 m averages about 2°C.

## TEMPERATURE VARIATION WITH TIME

It must also be realized that temperatures change with time, not only in geologic terms but also from year to year and from decade to decade. This fluctuation may or may not be cyclic—occurring at regular intervals or with a regular rhythm—but it does occur. The evidence comes in part from actual temperature records. The oldest and most complete set of data—from New Haven, Connecticut—dates from the late eighteenth century. From this it can be seen that the mean annual temperature was substantially lower in the early nineteenth century than at present, a fact confirmed by other meteorological records throughout the world and by indirect evidence such as the retreat of glaciers, the thinning of arctic and antarctic ice, and the changing distribution of plants and animals.

In New Haven, the coldest 10-year period since measurements were initiated in 1780 was that between 1812 and 1821, and the coldest 5-year period was between 1814 and 1819. The mean annual temperature there has risen from 8.7°C (47.6°F) for the 10-year period ending in 1821 to a high of 10.9°C (51.6°F) for the 10-year period ending in 1931, a rise of 2°C (4°F). Even a smooth curve drawn through the data indicates a rise of about 2½°C. (Figure 5.1).

Kincer (1933) has demonstrated that this rise in temperature is significant, that it is duplicated in records from weather stations throughout the world, and that it is equally apparent at rural as well as at urban stations, obviating the possibility of city development having caused the trend. Recent studies further substantiate the warming trend of up to 1½°C in the eastern United States from the middle of the nineteenth century to the 30-year period 1931–1960, but show a minor cooling trend in the western mountain states and Great Basin (Wahl, 1968; Wahl and Lawson, 1970). The general cooling trend in recent years suggests that we have passed the height of the warming episode (1880–1950). However, some scientists believe a new warming trend is likely due to human intervention with the global carbon dioxide cycle (Chapters 6 and 7).

The long-term effect of climatic change upon forest migration and development is discussed in Chapter 16. It is worthy of note here, however, that the rise in temperature from 1817 to 1950, nearly 1.7°C, represents a movement of the climate about 300 m up in elevation, or about 160 km north at the same elevation in terms of the

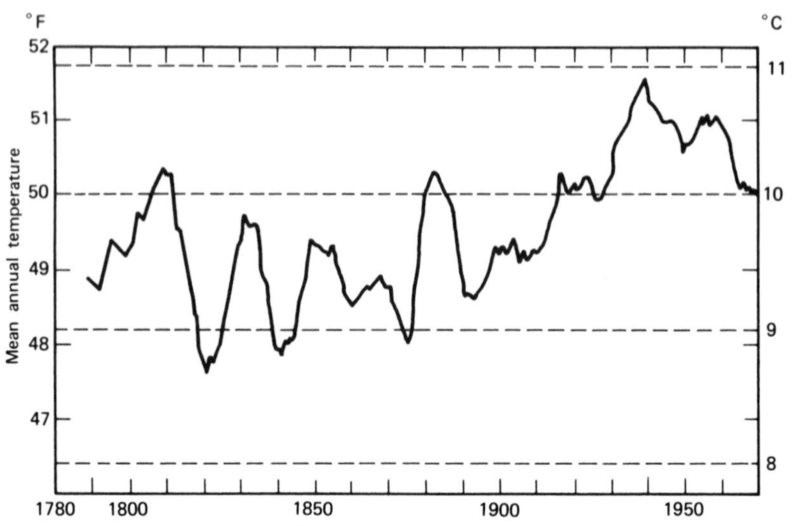


**Figure 5.1. Mean annual temperature at New Haven, Connecticut (10-year moving average).**

temperature gradient on the eastern seaboard of the United States (Table 5.2).

Quite obviously, such a change in temperature may exert a real influence upon the occurrence, natural regeneration, and growth of forest tree species, especially those growing near the limits of their natural range. Under far-northern conditions, temperature is the decisive climatic factor affecting the radial growth of trees (Mikola, 1962). For Scots pine in Finland the influence of summer temperature (mean July temperature) on tree growth increases from south to north, although the correlation is quite evident in both south and north. Radial growth increment increased sharply in the first part of the twentieth century, peaked in the mid-1930s, and declined with temperature thereafter.

## TEMPERATURE AND PLANT GROWTH

Plants regulate their temperature by dissipating part of the energy they absorb and thus preventing death due to excessive temperature. Of the three major mechanisms, reradiation, transpiration, and convection, dissipation of energy by reradiation accounts for one-half of the energy plants absorb (Gates, 1965). Transpiration accounts for an additional heat loss as the leaf is cooled when energy is expended in changing water to water vapor. In addition, convection across the thin air zone that surrounds all surfaces in still air, the **boundary**

**layer**, acts to transfer heat from the leaf to the cooler air. When the leaf is cooler than the air, as is typical at night, heat is transferred to the leaf from the air by convection and conduction. Through the interaction of these and other adaptations described below, the plant maintains a heat balance with its environment. These interactions tend to favor overall plant efficiency, tending to keep leaves warmer than cool air but cooler than warm air.

Plant processes function in a broad range of tissue temperatures, generally 0 to 50°C, as long as living cells and protein compounds are biologically stable and enzymatically active. As the seasons change, the foliage becomes conditioned and functions well at temperatures associated with the season. In forest trees, photosynthesis can take place at air temperatures below freezing, down to about −8°C, although at such temperatures tissues are usually warmed to near or above freezing by solar and terrestrial radiation. Low temperatures depress the rate of photosynthesis; nevertheless appreciable photosynthesis in conifers may take place in winter. For example, although the optimum temperature range for Douglas-fir is between 10 and 25°C, net photosynthesis at 0°C is still 70 percent of the amount occurring at 10°C (Lassoie, 1980).

As temperature increases, plant activities increase up to an optimum temperature and then decrease until at very high temperatures death occurs. The processes influenced most strongly by temperature are (1) the activity of enzymes that catalyze biochemical reactions, especially photosynthesis and respiration, (2) the solubility of carbon dioxide and oxygen in plant cells, (3) transpiration, (4) the ability of roots to absorb water and minerals from the soil, and (5) membrane permeability. Because different growth processes may require different optimum temperatures, one simply cannot characterize the growth or dry-weight production of a species by a certain optimum temperature. Actually, various phases of the temperature regime—day temperature, night temperature, heat sums, and the difference between day and night temperature (thermoperiod)—all affect growth. Also, optimum growth requirements vary between species and populations within species and vary in a way related to the environmental conditions under which a population has evolved.

The experimental work of Hellmers and associates (1962, 1966a,b, 1970) and of Kramer (1957) has shown that seedlings of tree species may respond to one or more of the following temperature conditions. Night temperature elicited the greatest growth response of Engelmann spruce and digger pine whereas redwood responded mainly to day temperature, reaching its maximum growth in the moderate range of 15−19°C. The finding that survival and growth of Engelmann spruce increase with increasing night temperature (Figure 5.2) seems to confirm Wardle's (1968) suggestion that night temperature during the summer was an important factor in the survival of Engelmann spruce in Colorado. Although redwood and Engelmann spruce grew best with the same day temperature (19°C), spruce grew better under warmer nights and tolerated warmer days better than did redwood. Redwood grew best with only slight differences in day-night temperature or at constant temperature, apparently reflecting the lower diurnal changes in its native coastal climate. In contrast, Engelmann spruce is apparently adapted to the greater diurnal changes of continental, mountain environments.

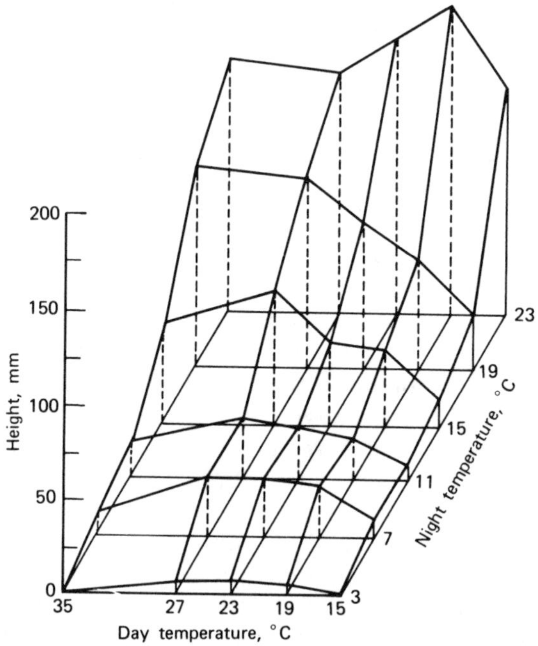


**Figure 5.2. Average height of 36 Engelmann spruce seedlings from each of 30 combinations of day and night temperature where the plants were grown for 24 weeks. (After Hellmers et al., 1970.)**

Several species (Jeffrey pine, erectcone or Calabrian pine, eastern hemlock) show a marked growth response to total heat received during a day, irrespective of the time of application. For example, Jeffrey pine exhibited the most growth when 300 to 400 degree-hours (hours × temperature = heat sum) of heat were received, regardless of what different day and night temperatures were used.

Although some species show a primary response to day temperature, night temperature, or total daily heat, many plants require nights considerably cooler than days for best growth. This difference between day and night temperature is termed **thermoperiod**. Low night temperatures coupled with moderate day temperatures are important in the flowering and fruit set, flavor, and quality of various crop plants and fruit trees (Treshow, 1970).

Seedlings of some forest trees respond strongly to thermoperiod. Maximum top growth of loblolly pine seedlings occurred under thermoperiods of 12°C, with night temperature colder than day temperature (Kramer, 1957). A similar response has been consistently reported for Douglas-fir, although the temperature differential has varied among the studies conducted and the provenances tested (Helmers and Sundahl, 1959; Lavender and Overton, 1972).

Two effects of thermoperiod were observed for red fir (Hellmers, 1966a). First for maximum height growth to occur, a warm day must be augmented with a cool night while a cool day must have a cold night. Second, maximum height growth was obtained when the thermoperiod was 13°C. Although maximum growth under a 17°C day and a 23°C day was nearly equal, this growth occurred only when the cooler day was augmented by a 4°C night and the warmer day with a 10°C night, i.e., a 13° thermoperiod in both cases.

In contrast to the three species described above requiring cold nights and warm days, seedlings of ponderosa pine grow best with warm days and warm nights of 23°C (Larson, 1967), or with warm nights and cool days (Callaham, 1962). In the latter study the optimum growth of six provenances was with a thermoperiod of 5°C, with nights warmer than days. Furthermore, pines from diverse parts of the range showed significantly different responses to the temperature regime (Figure 5.3). Seedlings from east of the Rocky Mountains (Figure 5.3c) required a high night temperature for optimum growth. Seedlings from the Southwest grew remarkably fast under cold days and hot nights. Seedlings from the west slope of the Sierra Nevada Mountains in California grew well with lower night temperature (14°C). Root growth of ponderosa pine is more dependent on soil temperature and top growth more dependent upon air temperature (Larson, 1967). Optimum root growth occurred in 15°C air and 23°C soil. Optimum root growth for red oak, basswood, and ash also occurs at relatively high soil temperatures (Larson, 1970). Low soil temperatures tend to reduce metabolic activity and reduce membrane permeability so that uptake of water and nutrients is limited. Temperature of the crown also affects root growth. As crown temperature increases so does respiration and transpiration; carbohydrates and water are utilized in the crown and become less

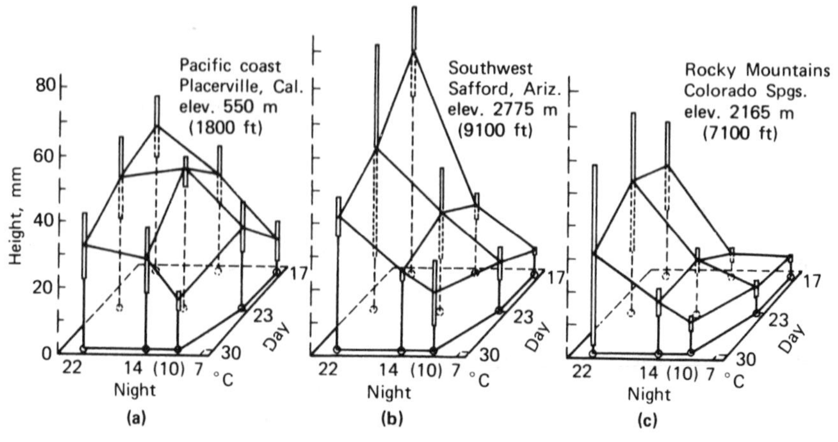


**Figure 5.3. Mean height growth of ponderosa pine progenies from three provenances in three regions. Treatment included 16-hr days and nine combinations of three day and three night temperatures. (The vertical bars show the range of the mean ± standard errors of the mean.) (After Callaham, 1962; from Theodore T. Kozlowski, *Tree Growth* © 1962, The Ronald Press Company.)**

available for use by the roots. For example, root growth of silver maple seedlings decreased markedly as temperature of the crown increased from 5 to 30°C (Richardson, 1956). Thus much root growth takes place at night when crowns are relatively cool and other factors are favorable.

The range of response in net photosynthesis of woody species to temperature has been demonstrated by Pisek and associates (Pisek et al., 1969). Species were chosen to represent alpine, montane, and Mediterranean climates of central Europe. The high-elevation and alpine species exhibit lower temperature optima for net photosynthesis than Mediterranean and low-elevation species. The photosynthetic apparatus of the alpine species is more efficient in cold temperatures and less efficient in high temperatures than that of the Mediterranean species.

Studies with various tree species have shown that light interacts with air temperature in controlling growth rates. For example, the optimum temperature for photosynthesis is known to vary with light irradiance. Increasing temperature increases photosynthesis up to a point, and thereafter light irradiance must be raised if photosynthesis is to increase.

With increases of temperature above the optimum, the photosynthetic rate lessens. At about 30°C many enzymes tend to become disrupted and if high temperature persists enzymes become nonfunctional, acting to slow growth processes. Furthermore, respiration increases greatly until at very high temperatures (perhaps 50°C) respiration exceeds photosynthesis. Finally, at even higher temperatures, death of cells occurs. A number of studies (cf. Baker, 1929; Lorenz, 1939) seem to indicate that the lethal point occurs at about 55°C.

The temperature values referred to are those of the plant tissues themselves and not of the outside atmosphere. It stands to reason that air temperatures must be even higher to warm up the plant tissues to the lethal point. Thus prolonged exposure to air temperatures of 50°C may eventually induce death while the same plants might be able to withstand brief periods of exposure to outside temperatures as high as 66°C. This time factor has been demonstrated by a number of workers and is of considerable importance in seedling survival in that extremely high temperatures may be interrupted by even the lightest shade, such as is the case when cotyledons of the seedlings shade the basal portion of the seedling stem. High-temperature stress and heat injury of plants is discussed in detail by Levitt (1972).

Leaf arrangement and orientation, a response to light irradiance, may act to reduce the amount of solar energy absorbed and hence be of survival value by preventing overheating of the leaf. For example, when red maple seedlings are subjected to high light intensities the leaf blades are deflected downward until they hang in a vertical position; when shaded they return to the horizontal (Grime, 1966). Such a mechanism may explain in part why red maple is a versatile species, colonizing both dry exposed sites and shaded habitats.

Furthermore, seedlings may be able to maintain lower temperatures through the effects of transpiration and thus avoid injury or maintain a better metabolic balance under very high outside temperatures. Shirley (1929) found that, for northern conifer seedlings in the Lake States, the killing temperature was higher in dry air, which favored seedling transpiration, than in moist air, which reduced it. In addition to the mechanisms of reradiation, transpiration, convection, leaf arrangement, and orientation already cited, leaf morphology, coloration, cuticle, pubescence, and degree of protein hydration may also act to enable the plant to function effectively at high temperatures (Treshow, 1970; Levitt, 1972).

Since temperatures above 50°C are largely confined in forested regions to the ground-air boundary, direct heat injury in forest trees is most significant in its effect on small seedlings, which have relatively unprotected live tissues in this critical zone. However, leaves of many mature hardwoods and conifers suffer leaf damage due to water deficiency in cells, particularly along the leaf margin and the tip of conifer needles. For example, widespread leaf damage to a wide variety of woody species occurred in northern California when temperatures suddenly rose above 38°C in areas that had had an exceptionally cool spring (Treshow, 1970). Conifer needles may also exhibit a needle curl which has been reported for southern pines and lodgepole pine (Parris, 1967) and attributed to short periods of high temperatures, 37–43°C, in the absence of moisture stress.

## COLD INJURY TO PLANTS

Unlike lethal high temperatures, lethal cold temperatures occur periodically throughout the entire zone of tree growth in the temperate and boreal regions of the earth. Thus they affect to a greater or lesser degree the distribution and growth of trees in these zones.

Death of plant tissues, particularly of actively growing plants and succulent tissues, may occur from rapid freezing and formation of ice crystals within the protoplasm. Rapid thawing is very harmful, causing sudden turgor changes that disrupt cell membranes. Slow freezing also may kill many semihardy plants at temperatures of −15° to −45°C at rates of cooling that commonly occur in nature (Weiser, 1970). According to Weiser, water between cells freezes, dehydrating cell contents until a point is reached when only "vital" water remains in the protoplasm. As temperature decreases further, vital water is pulled away from the protoplasm, initiating denaturation and ultimately causing death. In tropical plants, death may occur at above-freezing temperatures ranging from 0 to 10°C.

Most trees in the temperate and boreal zones, however, become increasingly inactive as the day length shortens and temperatures drop at the end of the growing season. As dormancy sets in, the water content of protoplasm is reduced and many species are able to survive subfreezing temperatures without damage. The more hardy the species, the greater the capacity of its cells to tolerate dehydration. **Frost hardiness** is a function of this decreased moisture content and is developed by natural selection, so that local races of a given tree species are usually able to withstand normal cold periods in their locality. For example, woody plants of the boreal forest and arctic (paper birch, willow, trembling aspen) survive prolonged subzero temperatures. Moreover, if the freezing rate to −30°C is relatively slow and plants are fully acclimated, these and numerous hardy species are known to withstand experimental freezing to −196°C (Weiser, 1970). The extreme of cold resistance is exhibited by the seeds of certain pine and spruce species, which cannot be frozen at any normally occurring temperature when in a dry, dormant condition. A detailed consideration of plant tissue freezing and freezing injury and resistance is given by Levitt (1972), and winter dormancy has been considered in detail by numerous authors (Romberger, 1963; Vegis, 1964; Wareing, 1956, 1969; Perry, 1971).

Acclimation or hardening are terms used to describe the change of plants from a succulent (tender) to a hardy or dormant condition. Three stages of dormancy are ordinarily recognized (Vegis, 1964; Perry, 1971): early rest (predormancy), winter rest (true dormancy), and afterrest (postdormancy). Short days act as an early warning system triggering growth cessation and initiating metabolic changes characteristic of the early rest or first stage of acclimation (Weiser, 1970; Figure 5.4). These changes facilitate further plant response in

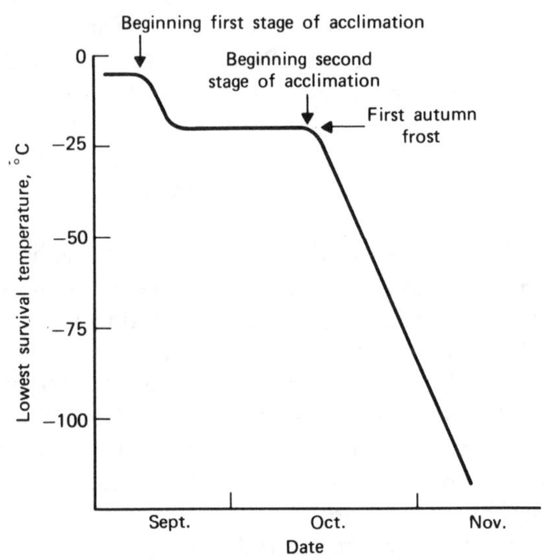


**Figure 5.4.** **A typical seasonal pattern of cold resistance in the living bark of red-osier dogwood *(Cornus stolonifera)* stems in Minnesota. In nature, acclimation in this hardy shrub and in a number of other woody species proceeds in two distinct stages as shown. The beginning of the second stage of acclimation characteristically coincides with the first autumn frost. (After Weiser, 1970. © 1970 by the American Association for the Advancement of Science.)**

the second stage of acclimation (also accompanied by metabolic changes) that is triggered by frost (Figure 5.4). The third stage of acclimation and attainment of true dormancy is a purely physical process and is induced by low temperatures of −30° to −50°C. Truly dormant buds and seeds cannot be induced to immediate normal growth by any means. However, some species, birches, European beech, and some oaks, do not show true dormancy and hence pass readily from early rest to afterrest.

## WINTER CHILLING AND GROWTH RESUMPTION

Buds and seeds of most woody species of the Temperate Zone require a period of winter chilling or stratification (seeds) before growth is resumed in the spring. Temperatures near 5°C are most effective (Perry, 1971). The nature of dormancy and the chilling requirement may vary with the local race of the species. For example, red maples from New York State attain true dormancy and require a month or more of chilling before resuming active growth (Perry and Wang, 1960). Red maples in south Florida cease active growth, drop their leaves but are unable to withstand freezing temperatures, and have no chilling requirement. As the afterrest period proceeds, physiological changes take place enabling growth to resume once minimum-temperature and sometimes day-length requirements are met.

Once the winter-chilling requirement has been satisfied and dormancy is broken, temperature primarily regulates the time of bud bursting (leaf flushing), bud-scale initiation, and shoot axis elongation. Vegetative bud dormancy is broken when mitotic activity begins in the buds. It precedes flushing of these buds, usually by several weeks in conifers (Owens et al., 1977). After winter dormancy is broken, **temperature sums**[1] may be used to predict accurately when a stage of development, such as flushing, will occur or how rapidly the shoot will elongate.

Seeds of many temperate forest species require a cold and moist period over winter in the forest floor before germinating the following spring. The cold treatment acts to alter the balance between growth inhibitors and promotors to favor spring germination when conditions are ideal for growth. Other species, including trees that fruit in the spring (elms, cottonwoods, aspens, willows) and many fire species (including most hard pines) lack a cold requirement.

[1]A **temperature** or **heat sum** is the product of temperature above a certain base or threshold level (such as 0° or 5°C) and the time duration of that temperature. It may be expressed in degree-hours or degree-days.

They germinate quickly following dispersal, once sufficient moisture is available.

## LIFE FORMS

Besides classifying plants taxonomically by family, genus, and species, plants may be ordered ecologically by characters of their structure and function in relation to important site factors. The basic classification into evergreens, deciduous plants, perennials, and annuals is so well accepted that we almost forget it is a classification at all, so thoroughly have the terms been integrated into our thinking. The importance of survival during unfavorable periods initially led to ecological classification of life forms based upon the dormant condition of the plant under climatic conditions unsuitable for growth.

This classification was elaborated and systematized in 1907 by Raunkiaer (1937), who has suggested terms that have become widely accepted by ecologists. **Phanerophytes** include all seed bearing plants, both evergreen and deciduous, in which the dormant buds are in the atmosphere. All trees, of course, fall within this category. In ***Chamaephytes***, the surviving buds are situated close to the ground; in ***Hemicryptophytes***, the buds pass the unfavorable season in the soil surface, protected by the soil itself and by litter; while in ***Cryptophytes***, the buds survive completely concealed in the ground or at the bottom of the water. Finally, the **Therophytes**, or annuals, complete their life cycle within a single favorable season and remain dormant in the form of seed throughout unfavorable periods.

This system has been modified and expanded to 23 major life forms and includes characteristics of the favorable season as well as structural and seasonality features of the crown, foliage, and shoot systems (Mueller-Dombois and Ellenberg, 1974). Communities may be contrasted with such a classification, and the system forms a basis for classifying forests and vegetation of the world (Chapter 21).

## SUGGESTED READINGS

Burke, M. J., L. V. Gusta, H. A. Quamme, C. J. Weiser, and P. H. Li. 1976. Freezing and injury in plants. *Ann. Rev. Plant Physiol.* 27:507–528.

Gates, D. M. 1965. Heat transfer in plants. *Sci. Amer.* 213:76–84.

George, M. F., M. J. Burke, H. M. Pellet, and A. G. Johnson. 1974. Low temperature exotherms and woody plant distribution. *Hort. Sci.* 9:519–522.

Hellmers, Henry. 1962. Temperature effect on optimum tree growth. *In* T. T. Kozlowski (ed.), *Tree Growth*. The Ronald Press Co., New York.

Levitt, Jacob. 1972. *Responses of Plants to Environmental Stress*. Academic Press, Inc., New York. 697 pp.

Lowry, William P. 1969. *Weather and Life, An Introduction to Biometeorology*. (Chapter 3, Environmental temperature, pp. 31–63.) Academic Press, Inc., New York.

Parker, J. 1963. Cold resistance in woody plants. *Bot. Rev*. 29:123–201.

Threshow, Michael. 1970. *Environment and Plant Response*. (Chapter 5, Temperature mechanisms, pp. 51–62; Chapter 6, Disorders associated with high temperatures, pp. 63–76; Chapter 7, Disorders associated with low temperatures, pp. 77–98.) McGraw-Hill Book Co., Inc., New York.

Vegis, A. 1964. Dormancy in higher plants. *Ann. Rev. Plant Phys*. 15:185–224.

Villiers, T. A. 1972. Seed dormancy. *In* T. T. Kozlowski (ed.), *Seed Biology*. Academic Press, Inc., New York.

Wareing, P. F. 1956. Photoperiodism in woody plants. *Ann. Rev. Plant Phys*. 7:191–214.

———. 1969. Germination and dormancy. *In* Malcolm B. Wilkins (ed.), *Physiology of Plant Growth and Development*. McGraw-Hill Book Co., Inc., New York.

Weiser, C. J. 1970. Cold resistance and injury in woody plants. *Science* 169:1269–1278.

Wilson, B. F. 1970. *The Growing Tree*. (Elongation and dormancy, pp. 47–67.) Univ. Mass. Press, Amherst. 152 pp.

# 6
# Atmospheric Moisture and Other Factors

Water is the inorganic substance most needed by plants and is present in plants in large amounts. The principal source of the water that reaches the tree is, of course, the soil, and soil moisture is treated in Chapter 10. The amount of moisture in the atmosphere is important, however, because it is the source of most soil moisture, and in addition it affects the rate of water loss from the leaves through transpiration. Also, water supplied by dew, rain, or fog may be absorbed directly by the foliage.

Precipitation as such is not a factor directly influencing the growth of the plant under most conditions. Rain, snow, and other forms of precipitation are indirectly of the greatest importance in recharging soil moisture and thus influence plant growth indirectly through their effect on the soil. Precipitation also carries nutrient ions that may be utilized by the tree via the nutrient cycle (Chapter 9). Situations do arise, though, where precipitation directly influences plant growth, and various types of precipitation—particularly snow, glaze, and sleet—are important causes of damage to trees.

## NOMENCLATURE OF WATER VAPOR

Water is always present in the atmosphere in the form of water vapor. The actual weight of water present per unit volume of air is termed the **absolute humidity**, while the percentage of vapor present relative to the maximum quantity that the air can hold is termed **relative humidity.** The vapor-holding capacity of the air is greatly affected by temperature. At 27°C, the air can hold twice as much water as it can at 16°C. In other words, the absolute humidity at 27°C is twice that at 16°C when the relative humidity is 100 percent at both temperatures.

Since the weight or pressure exerted by a given mass of air is increased as water vapor is added to it, absolute humidity can be measured in terms of **vapor pressure,** usually expressed in millimeters of mercury or millibars. Thus saturated air at 16°C will elevate a mercury column 13 mm over its height in vapor-free air; while at 27°C, with twice as much water being held by the saturated air, the elevation will be 26 mm, or twice as much as at 16°C.

The difference between the actual vapor pressure and the vapor pressure of saturated air at the same temperature is termed the **vapor pressure gradient** (or **vapor pressure deficit**), and is a measure of the absolute amount of moisture that can be taken up by the air at that temperature. Thus when the relative humidity is 70 percent and the air temperature is 27°C, the vapor pressure is $0.70 \times 26$, or 18 mm, and the vapor pressure gradient is $26 - 18$, or 8 mm. The vapor pressure gradient is the primary mechanism controlling the upward movement of water in the plant and the movement of water from soil into plant roots (see Chapter 10).

## EXCHANGE OF WATER VAPOR BETWEEN THE PLANT AND ATMOSPHERE

From the standpoint of tree growth, atmospheric moisture is important in changing the water relations of the plant tissues. Moisture will move from the plant into the atmosphere when the vapor pressure of the plant is greater than that of the atmosphere. This is the normal daytime situation when it is not raining. Moisture may also move from the atmosphere to the plant when the vapor pressures are reversed, as during a rain or when dew covers a plant that is not fully turgid. There is no exchange of water vapor between the plant and the atmosphere when the vapor pressures are equal as occurs when the air in both is saturated with water and the temperatures are the same within the plant and in the atmosphere, often the situation at night.

The actual rate of water movement is proportional to the vapor pressure gradient between the plant and the atmosphere. This, in turn, depends in large measure upon a thin layer of undisturbed air, the **boundary layer**, surrounding the leaves, the thickness of which depends upon the wind velocity. Since the air within the leaf is normally saturated under growing conditions, vapor will normally move out from the leaves into the surrounding air unless the outside air is also saturated at the same or higher air temperature. **Transpiration** results. Even if the air is at 95 percent relative humidity, transpiration takes place. In the absence of actual precipitation, arid conditions and even deserts can develop in regions of characteristically high humidities. Braun-Blanquet (1964) cites as an example the dry cactoid-euphorbia-scrub along the coast of southern Morocco where the summer relative humidities average 90 percent.

The rate of transpiration, however, is directly dependent upon plant and air temperatures, relative humidity of the air, and air movement affecting the thickness of the boundary layer. The drier the air and the greater the air turbulence, the higher will be the rate of water loss, assuming that soil moisture is available.

Since the epidermis of many leaves and most fruits, stems, and other aerial organs is cutinized to a greater or lesser extent, most of the exchange of water vapor between the plant and the atmosphere takes place through the stomata. The rate of exchange, therefore, is further modified by the opening and closing of these stomata. Transpiration is similar to evaporation except that the movement of water vapor from plant cells is controlled to a substantial degree by **leaf resistances** that are not involved in evaporation. The stomata and the cuticle are the chief sites of leaf resistances.

Although most of the movement of water vapor is from the plant out into the atmosphere, the direction of movement may be reversed when the plant air has a lower vapor pressure than the outside air. This condition occurs during rains when the plant previously has been subjected to desiccating conditions, or more commonly occurs when the atmosphere is warmer than the plant tissues, resulting in the formation of dew on plant surfaces (Stone 1957; Stone et al., 1956).

Transpiration is the dominant process in water relations of plants because it provides the energy gradient that causes the movement of water into and through plants. Furthermore, transpiration causes water deficits to develop in the plant that may result in growth reduction or death. These and other aspects of transpiration are discussed in Chapter 10.

## PRECIPITATION OF WATER VAPOR

The discussion of atmospheric moisture to this point has dealt with water vapor and its movement between the plant and the air. Equally important is precipitated water vapor—the main source of moisture upon which plant growth depends.

Precipitation occurs when warm moist air is cooled below the point of its water-holding capacity. This cooling may result from currents of air rising to higher elevations as occurs when cold air masses wedge under warmer air, or when warm air rides up over cold air (warm front); it occurs when moist air rises over warm land surfaces (convectional precipitation) and when air currents rise over elevated land masses (orographic precipitation).

If the condensation takes place below the freezing point, **snow** is formed; above the freezing point, **rain** ensues. Rain that freezes as it falls through subfreezing air layers becomes **hail** or **sleet**; if it freezes as it falls on a subfreezing surface, it forms **ice** or **glaze**.

Condensed water vapor in particles sufficiently small to remain suspended in the air forms **clouds** above the ground or **fog** in the layer of air near the ground. When droplets of undercooled clouds and fogs touch obstacles, white layers of ice crystals or **rime** forms. Finally, condensation of water vapor directly onto cooled surfaces near the ground gives rise to **dew**.

As has been pointed out previously, the greatest importance of precipitation to trees is not a direct one, but rather the indirect dependence of soil moisture supply upon replenishment by precipitation. The physical impact of precipitation, particularly glaze and snow, upon trees is a principal source of damage to forests. Indeed, in many areas, it is the most important factor determining the practicality of growing timber.

In a forest, much of the rainfall is intercepted by the crowns, whence it may drip to the ground, fall to the ground in the case of rime (Gary, 1972) or glaze, flow down the stems to the ground, be evaporated into the air, or be absorbed by the leaf surfaces. Reviews of interception studies conducted in the United States (Zinke, 1967) and elsewhere (Penman, 1963; Molchanov, 1963) indicate that forests often intercept a significant proportion of the annual precipitation ($\cong$ 20 percent). A few examples will indicate the results from typical interception studiès. In second-growth ponderosa pine in California studied for a 6-year period, the average annual precipitation was 1200 mm, of which 84 percent reached the floor by penetration or drip and 4 percent through stem flow, leaving 12 percent that was intercepted and either absorbed by the foliage or evaporated (Rowe and Hendrix, 1951). In a young loblolly pine plantation in South Carolina, rainfall under the forest was only 86 percent of that in the open (14 percent being intercepted), with about one-fifth of the rainfall reaching the ground as stem flow (Hoover, 1953). In very light rains, as much as 93 percent may be retained by the crowns, while in heavy downpours as little as 6 percent may be intercepted (Ovington, 1954).

Deciduous species are reported to be similar in their ability to intercept rainfall during the growing season; conifers intercept more precipitation than deciduous species (Zinke, 1967). In a German example, the mean annual interception for European beech and Norway spruce was 7.6 and 26.0 percent, respectively (Eidman, 1959). In this study, more summer rainfall reached the forest floor through stem flow in beech stands (16.5 percent) than in spruce stands (0.7 percent).

There are few data available as to what proportion of the rainfall intercepted by the crowns actually enters the foliage and is utilized. Much evidence has demonstrated, on the other hand, that the rain hitting the tree crowns actually removes substantial amounts of mineral nutrients and organic substances from the foliage and twigs (Chapter 9).

## RAIN AND SNOW AND THE FOREST

The amount and distribution of rainfall is related to meteorological conditions, season of year, topography, and many other factors. It may even be related in some local circumstances to the presence and structure of forests. For example, in the fume-denuded Copper Basin of eastern Tennessee, four-year (1936–1939) records were kept at two stations each in forest clearings (mean annual precipitation 1460 mm), grassland (1340 mm), and denuded land (1280 mm). Hursch (1948) concluded that the forests in the area had a slight but statistically significant effect in raising the amount of precipitation. However, meteorologists now believe forest cover cannot exert any significant, large-scale influence on the amount of precipitation (Penman, 1963); increases brought about by the presence of forests are smaller than the errors in rainfall measurements (Stanhill, 1970).

The problem of measuring rainfall in the forest is complicated by the variability of precipitation from place to place, especially on rugged terrain (Hamilton, 1954), the effect of the clearing in which the rain gauge is placed upon local precipitation and catch, and the necessity of shielding rain gauges to reduce air eddies around the orifice. Because of the high variability in rainfall from place to place and from time to time, precipitation records from standard meteorological stations will only roughly approximate the rainfall in a given forest.

In the case of precipitation occurring as snow rather than as rain, interception is apt to be greater, but much of the snow intercepted will eventually drop off onto the ground (Hoover and Leaf, 1967), thus increasing the total amount of water that reaches the soil under the forest. For example, Kittredge (1953) found that from 13 to 27 percent of the seasonal snowfall on the west slope of the Sierra Nevada was intercepted by the forest canopies; but Rowe and Hendrix (1951), also working in California, found that an average of 4 percent more precipitation reached the forest floor during snow than during rain storms. In any event, it may be assumed that little snow moisture is directly absorbed by tree foliage.

## FOG AND DEW AND THE FOREST

The interrelationship between condensed moisture and trees has been the subject of much speculation. Fog and dew may well be quite important in determining the growth and distribution of forests, but the extent and mechanism of the effect have proved difficult to demonstrate in precise experimentation. In addition to reducing transpiration losses, there can be no doubt that forests can condense appreciable amounts of moisture from fog—and also that appreciable amounts of moisture can be condensed on tree foliage during the night through dew formation. Whatever portion of this condensed moisture falls to the ground through fog drip or dew drip may be added to the soil and thus indirectly benefit the forest.

There is an obvious correlation in many parts of the earth between the distribution of certain trees and the presence of fog belts. Perhaps the most spectacular example occurs along the Pacific Coast of the United States and Canada where Sitka spruce (from Oregon north) and redwood (in California) characterize the very dense and fast-growing temperate rain forest (Figure 6.1).

The fact that the coastal zone of heavy summer fog more or less coincides with the Sitka spruce and redwood ranges, however, does not prove that a cause-and-effect relationship necessarily exists between the fog and the tree distribution, although such a relationship has been categorically stated on many occasions. The coastal zone differs climatically in many ways from adjacent forest zones, not only for its heavy summer fogs, but for its heavy winter rains and the general oceanic climate with little cold or hot weather but rather equable temperatures throughout the year. Furthermore, the summer fog itself may affect the tree climate in many ways. It may well have a direct effect in supplying summer moisture. In addition, however, it also has indirect effects in reducing the hours of summer sunshine, in reducing the summer daytime temperatures (Byers, 1953), and in increasing the supply of carbon dioxide (Wilson, 1948). However, the relationship between fog and tree growth is too complex to conclude that redwood and Sitka spruce must owe their survival to the moisture-giving powers of summer fog along this coast.

Let us examine the actual evidence on hand. In the first place, forests can remove moisture from the air through condensing it from incoming fog. Fog may be collected and condensed by tree crowns and dripped to the ground in appreciable quantities when no moisture is caught by rain gauges in the open. For example, Isaac (1946) found that, on a ridge 3 km from the Pacific Ocean in the Oregon fog belt, precipitation was one-quarter greater under trees than in the open because of fog drip; while in a valley 8 km inland, precipitation

**Figure 6.1. Redwood forest along California's coastal fog belt with luxuriant understory vegetation. (U.S. Forest Service photo.)**

was one-third less under the forest due to interception. In the San Francisco peninsula, Oberlander (1956) recorded up to 1500 mm of fog drip under five trees during a rainless period of 40 days in mid-summer. In detailed studies of the "fog-preventing forest Zone" on the southeast coast of Hokkaido, Japan (Hori, 1953), it was found that the forest could remove 3400 liters of moisture per hectare of surface per hour under standardized conditions of fog density and wind movement. Under similar conditions, grassland captured not more than one-sixth to one-tenth this amount. The windward vertical edge of the forest captured as much fog water as a horizontal surface three times as large in area.

In the high-altitude conifer forests around the world, much of the total precipitation is received as fog, rime, or hoar-frost. At 1150 m in the Bavarian Mountains of southern Germany, Baumgartner (1958) reported that 42 percent of the yearly total precipitation of 2000 mm was received as fog. In the Green Mountains of northern Vermont, a screened gauge simulating the effect of coniferous needles in intercepting and collecting cloud droplets, collected 1.7 times more water than an unscreened gauge (Vogelmann et al., 1968). The needlelike foliage and twiggy character of the spruce-fir forests serve as effective mechanical collectors of wind-driven cloud droplets. At a high altitude in New Hampshire (1400 m), where the vegetation is a mixture of balsam fir krummholz and alpine tundra, artificial foliage, structurally resembling that of balsam fir, caught 4.5 times more water than paired open bucket collectors (Schlesinger and Reiners, 1974). This is admittedly an extreme situation where high interception is expected due to high cloud frequencies and high wind speeds. However, it and the previous example demonstrate the adaptation of conifer needles to intercept moisture (and nutrients) and the inadequacy of open bucket collectors placed in a forest clearing to measure adequately the total precipitation available to forest stands. A detailed review of mist precipitation and vegetation was given by Kerfoot (1968).

In arid and semiarid regions, foliar absorption of water condensed as dew occurs primarily at night (Gindel, 1973). Stomata are present on both sides of leaves of the majority of indigenous species of arid regions. One group of xerophytic trees studied by Gindel (1966) obtained 80 percent of the water necessary for transpiration from dew and mist during the critical summer months of moisture shortage. In some studies, it has been shown that water is taken up by foliage in appreciable amounts, particularly through dew formation at night. For example, Gates (1914) found that severed black spruce and tamarack branches when exposed to moist air in Michigan summer conditions could take up moisture amounting to a 6 percent increase in weight in 4½ hours. Later experiments by others confirm that considerable moisture may enter plants through the condensation of either fog or dew.

## GEOGRAPHICAL VARIATION IN PRECIPITATION

As with air temperature, precipitation varies with latitude and altitude. The distribution of precipitation over the face of the earth, however, depends primarily upon the interrelationships between air currents and large water bodies. The heaviest precipitation occurs when air currents passing over water bodies come into contact with cooler land masses or are caused to rise by elevated land masses, becoming cooler in the process. On windward coasts, over 2500 mm of rainfall per year are normal in such forest regions as the Pacific Northwest of the United States and Canada, Hawaii, New Zealand, and India. To the leeward of major mountain ranges, **rainshadows** occur over which warming air currents depleted of moisture drop little on the land surface.

Elevated mountain masses, by causing air to rise, cause it also to cool, thereby increasing precipitation. Such **orographic** rainfall is essentially local, falling immediately on the windward side of the elevated land mass.

Precipitation, therefore, is not closely correlated with elevation above sea level as such. A major topographic feature such as the Mogollon Rim in Arizona may receive considerably more precipitation than is indicated by its altitude of 2100 m, while points on the Mogollon Mesa at the same elevation but remote from the rim may receive substantially less precipitation. An isolated mountain induces relatively slight orographic precipitation.

In very high mountains, the maximum precipitation is commonly reached somewhere along the slope, the air currents becoming too depleted of moisture to provide as much precipitation at higher elevations. Precipitation patterns are particularly affected by air currents and mountain barriers. Along the Pacific Coast, mean annual precipitation at the lower elevations increases sharply with altitude, at rates varying from 13 to 17 mm per 100 m in the coastal range of Washington to 7 to 8 mm in the Sierra Nevada. The maximum precipitation occurs at the middle elevations—ranging from perhaps 900 m in the Olympics to 1500 m in northern California and 2400 m in the southern Sierra Nevada. Above this elevation, precipitation decreases with altitude. East of the coastal mountain barrier in the mountains of the interior, the effect of elevation on rainfall is much more predictable, being approximately 3 to 4 mm per 100 m rise in elevation (range 900 to 1500 m). These values seem to hold from the east side of the Cascade–Sierra Nevada east through the Rockies, and from Canada south to Mexico.

## CARBON DIOXIDE

Although solar energy, including light and heat, and atmospheric moisture are the principal climatic factors affecting the distribution and growth of forest trees, other climatic factors must also be considered. Among these, the carbon dioxide content of the atmosphere, wind, atmospheric pollutants, and lightning exhibit the most important direct effects upon forest trees and are treated in the present chapter.

Carbon dioxide is present in low concentrations in the atmosphere, is required in large amounts for photosynthesis, and is given off in both plant and animal respiration, in forest fires, and in large quantities in the burning of fossil fuels. Because of these factors the amount of $CO_2$ in the air at any one time varies from time to time and from place to place. On the average, the air contains about 0.03 percent $CO_2$ by volume (300 ppm), but daily variations of 10 to 400 percent of this amount occur commonly. Rain and fog substantially increase the $CO_2$ content of the air (Wilson, 1948; Selm, 1952). During periods of fog and rain, low light conditions reduce photosynthesis and hence the uptake of $CO_2$. Respiration of $CO_2$ continues, and its concentration increases. In addition, low air movement during fog and rain prevent the rapid transfer of $CO_2$ out of the forest.

The supply of $CO_2$ in the atmosphere is increasing and is of worldwide concern (Baes et al., 1976; Bolin, 1977; Woodwell et al., 1978; Stuiver, 1978; Olson et al., 1978). The amount of $CO_2$ has increased from the mid-19th Century, preindustrial value of an estimated 270 ppm to the present value of 330 ppm. During the early part of the period 1850–1970, the increase was mainly due to deforestation: carbon was transferred out of the organic reserves in standing forests and soil organic matter into the atmosphere. More recently, the burning of fossil fuels has been the primary source. Although the primary source of $CO_2$ released into the atmosphere in the near future will come from the burning of fossil fuels, forest removals, especially in the tropics, are likely to account for 20–40 percent of the total. Although the increased $CO_2$ in the atmosphere is available to plants, and the amount of carbon fixed by plants may increase, $CO_2$ also may cause a rise in global average air temperature that can have undesirable effects (Chapter 7).

Carbon dioxide is being evolved continually by respiration of plant roots and organisms decaying organic matter. For example, in a yellow-poplar stand in Tennessee, 35 percent of the $CO_2$ evolved from the forest floor was accounted for by respiration of live roots, 42 percent from root decay, and 21 percent from the $O_1$ and $O_2$ litter layers (Edwards and Harris, 1977). The $CO_2$ level is therefore usually high in damp, calm air near the forest floor, where organic-matter decay by soil organisms is actively taking place. In such situations, the $CO_2$ percentage may rise to 0.18 percent by volume of the air (Fuller, 1948). Because respiration takes place during the night when photosynthesis ceases, and because of the prevalence of moist and calm air conditions during the night, $CO_2$ concentrations in the forest reach their highest levels toward the end of this period. However, the natural variation of $CO_2$ concentration at the crown level during the daylight hours, at most ±25 ppm, is not enough to influence photosynthesis significantly (Koch, 1969).

Photosynthesis requires large quantities of $CO_2$ that must be assimilated by trees from the atmosphere. Baker (1950) estimates that over 27 metric tons of the gas are needed to generate the annual dry-matter production on an average productive hectare (12 tons per acre) of forest, more $CO_2$ than is present above that hectare at any one time. It is obvious, therefore, that the $CO_2$ requirements of trees can be met only through the circulation of $CO_2$ in the ecosystem from oceans and land sources far removed from the local forest together with the rapid return of the gas from organic respiration and the circulation of this to the leaves through air movement. It follows that the $CO_2$ level in the forest will be lowest during the periods of maximum photosynthesis, that is, at midday during the height of the growing season. Such appears to be the case (Mitscherlich et al., 1963).

The amount of carbon dioxide in the air surrounding tree crowns (from 0.03 to 0.04 percent) definitely limits photosynthesis. At such relatively low concentrations, photosynthesis tends to vary linearly with the $CO_2$ concentration (Decker, 1947). As $CO_2$ concentration is increased under artificially controlled conditions, the rate of increase of photosynthesis becomes less. After reaching a peak at from 5 to 8 times the normal concentration, photosynthesis decreases for various tree species, including Norway spruce, Scots pine, European silver fir, European beech, and European cottonwood. At the optimum $CO_2$ concentration photosynthesis is increased up to 3½ times the normal rate (Stålfelt, 1924; Koch, 1969). According to Koch, closure of stomata at increasing levels of $CO_2$ above the normal concentration eventually causes photosynthesis to decrease.

Despite the fact that carbon dioxide content in the atmosphere apparently limits photosynthesis in tree leaves, it is difficult to see how this factor can be improved by silvicultural management. However, the $CO_2$ content is often increased in greenhouses where seedlings are grown for forest outplanting.

## WIND

The turbulence and movement of air has many effects on the distribution and growth of tree species: (1) Air movement regulates in large measure evapotranspiration from leaf surfaces, decreasing the thickness of the boundary layer and increasing gas exchange. Thus wind exerts a major influence on the water regime of the plant and at the same time helps cool the leaves. (2) Air movement circulates the small quantities of carbon dioxide in the air to the leaf surfaces, making possible photosynthesis, and may also circulate salt and atmospheric pollutants that may damage or destroy forest and agricultural crops. (3) A minor biological effect of wind is the increased illumination within the forest resulting from the twisting and turning of foliage. (4) Continued winds from a single direction will exert a strong influence upon the morphology and size of tree crowns. (5) Sway caused by the wind will influence both the form of the tree, making for short boles and excessive taper, and the quality of the wood (Jacobs, 1955). (6) Wind causes several types of root injury, including breaks, tears, and abrasions (rubbing of roots across rock surfaces) that provide entry courts for decay fungi (Stone, 1977). (7) Wind is a major cause of stem breakage, windthrow, and subsequent mortality. (8) Finally, wind is essential in the dissemination of pollen and seed for many forest species.

Air movement is deflected to a large extent by the exterior surfaces of a forest stand. Furthermore, movement near these surfaces is slowed by friction. It follows that wind velocity decreases with decreasing height above the stand and becomes markedly less within the stand. Wind velocity as affected by forest stands has been summarized for four separate studies by Reifsnyder (1955; Figure 6.2). Wind velocity is slowed in the vicinity of the crown and becomes negligible within the stand. In the case of forest clear of understory, however, wind velocity is somewhat higher in the trunk zone than

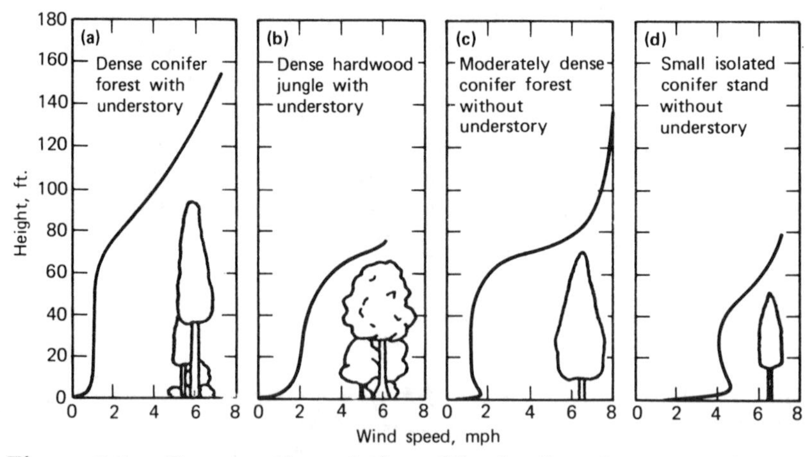


**Figure 6.2. Comparative wind profiles for four forest stands. (After Reifsnyder, 1955.)**

near the ground or in the region of the tree crowns. In comparison, grass and brush vegetation affect wind velocity only through air friction on their surfaces, and this effect is marked only in the zone 3 to 6 m above the vegetation (Fons, 1940). The small amount of air movement within a stand is indicated by a study in a dense aspen stand in Utah, 4 m high (Marston, 1956). Over the summer, the mean wind velocity was only 3 km per hour in comparison with 7 km per hour in a clearing 20 by 60 m in size.

Biologically, the chief importance of wind is its effect on evapotranspiration. When a plant is exposed to wind, leaf-water deficits develop more readily than in sheltered locations. A tree making all its development in an environment leading to water deficits will differ physiologically, morphologically, and anatomically from a tree sheltered from the exposure effects. The desiccating effect of wind on forests is demonstrated vividly in the dwarfed and deformed vegetation along windward ocean shores or in windswept mountain areas. The effects may be important economically within the forest. For example, on the north coast of Puerto Rico, the average growth is nearly twice as rapid on the west-facing leeward slope as on the east-facing windward slope (U.S. Tropical Forest Experiment Station, 1951).

It should not be assumed, however, that the drying effect of wind is the sole reason for such conditions. The mechanical shearing action of snow is a chief factor in the development of the flag form of many trees at timberline, where the prostrate portions are protected by a cover of snow in the winter months (Figure 6.3). The prostrate tree form, termed **krummholz**, is a type of forest characteristic of alpine regions.

In the critical stage of forest establishment, wind may be a crucial factor, particularly for broadleaved species, for example black walnut seedlings (Schneider et al., 1970; Heiligmann and Schneider, 1974, 1975). Seedlings grown on level, windswept, open-field sites had less vigorous growth, reduced leaf area, and greater foliage damage than those in forest openings or protected fields where similar soil conditions prevailed. When seedlings were protected by a wooden lath snowfence, height, diameter, leaf area, and shoot dry weight all increased markedly compared to the exposed plants. The favorable microclimate, including reduced wind velocity and lengthened growing season, was primarily responsible for the improved growth.

From a forest management viewpoint, wind is chiefly important because of its great destructive potentialities. Few forests can stand the devastating wind velocities and low air pressures of a hurricane or tornado. In fact, such cyclonic storms have affected the forests of

**Figure 6.3. Prostrate and erect forms of Engelmann spruce on the Snowy Range, Medicine Bow National Forest, Wyoming. Snow driven by wind above the winter snow line is responsible for keeping most trees sheared. (U.S. Forest Service photo.)**

much of the eastern half of the United States. In New England, occasional hurricanes and tornados have periodically blown down much of the older forests of the region.

Over and above the great destruction from major windstorms, however, is the immense amount of breakage and fall due to lesser winds, particularly around the edges of small holes, clearings, and cuttings in the forest. Local acceleration of wind due to the conformation of topography, the forest profile, and cutting boundaries may result in windfall under conditions where the unobstructed wind velocity is only moderate. Through the proper planning of harvest cuttings, windfall from these causes may be minimized. Indeed, much of the reason for the different silvicultural systems evolved in Germany and in the United States lies in the necessity for making partial cuts that will leave the residual stand as windfirm as possible. Often this involves cutting successive strips to eliminate vulnerable leeward boundaries by progressive cutting into the wind.

The prevailing direction of the high velocity winds must be

known before intelligent planning of cutting can be done. This is not necessarily the prevailing direction of all winds, and frequently cannot be determined from meteorological records. It can be determined, however, from existing windfalls. For instance, a tally of windfalls in the Fraser Experimental Forest in the Rocky Mountains of Colorado (Alexander and Buell, 1955) showed that most trees fell in an easterly direction, indicating that the destructive prevailing westerly winds were but little affected by the high mountain range 6 km to the east or by local topography within the experimental forest.

The practical problems of wind behavior in mountainous forested country have been summarized by Gratkowski (1956) and Alexander (1964), who found that much windfall is due to local acceleration of wind currents by either topography or forest borders. Such acceleration can take place where the wind is speeded up (1) on ridge tops, upper slopes, and the shoulders of a mountain, (2) on gradual and smooth lee slopes during severe windstorms, (3) in gaps and saddles of main ridges, (4) in narrow valleys or V-shaped openings in the forest that constrict the wind channel, and (5) where forest borders and cutting edges deflect wind currents, resulting in increased velocities where the deflected currents join others (Figure 6.4). The size of the clearing in the forest seems to be less important than its configuration.

Windthrow is most apt to occur where the concentration of air currents causes high velocities at a particular spot. It occurs primarily in shallow-rooted species on shallow soils and those with impeded drainage. Thus it is possible to identify high-risk species, black spruce and Norway spruce for example, and high-risk sites (Godwin, 1968; Pyatt, 1968).

## ATMOSPHERIC POLLUTANTS

Civilization and industrialization have brought ever-increasing destruction of vegetation through diverse kinds of air pollutants. For at least 2000 years sulfur dioxide has been evolved from the smelting of iron and copper ores, and emissions from coal burning (now accounting for approximately 60 percent of the pollution, Rohrman and Ludwig, 1965) have added to the problem. Extensive vegetation injury and soil degradation due to heavy metal accumulation in the soil (nickel, copper, lead, zinc, cadmium, arsenic) have been reported around the world (Hocking and Blauel, 1977). In some cases, metal contamination may cause more damage than sulfur dioxide. The effects of air pollutants on forests have been reviewed by Miller

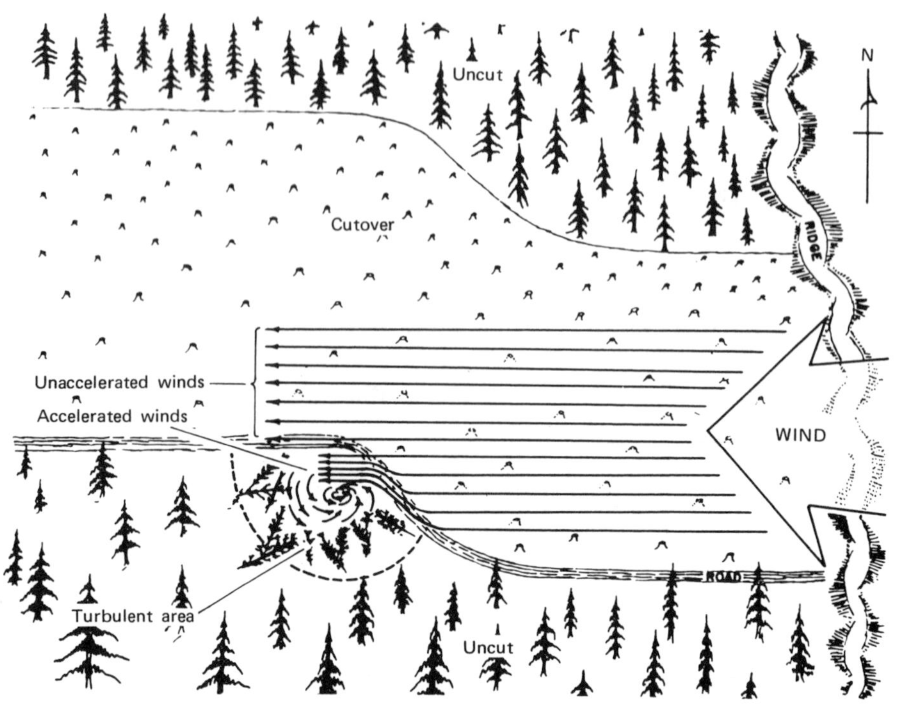


**Figure 6.4. Windthrow of standing timber following a harvest cut. A change in the direction of the cutting boundary that was parallel to the prevailing winds acts to funnel wind into standing timber, causing a pocket of blowdown. Routt National Forest, northwestern Colorado. (After Alexander, 1964.)**

and McBride (1975) and general reviews are also available (Treshow, 1970; Mudd and Kozlowski, 1975; Guderian, 1977).

Smelter-fume pollution may cause striking patterns of vegetation damage, and sulfur fallout may change markedly the acidity of lakes near the source (Gorham and Gordon, 1960a; Gordon and Gorham, 1963). In a northern Ontario locality, vegetation was severely damaged by sulfur dioxide emissions downwind of an iron smelter (Figure 6.5). The number of species per 40-m quadrat declined from 20–40 to 0–1 species between 16 and 3 km from the pollution source (Figure 6.6). Along a northeast transect, seedlings of eastern white pine were not observed within 50 km of the plant and those of white spruce, black spruce, and trembling aspen were not found within 25 km.

Emissions from a zinc smelter, combined with fires and erosion, have caused severe vegetation damage and denudation at Lehigh Gap, near Palmerton, Pa. (Jordan, 1975). High concentrations of zinc

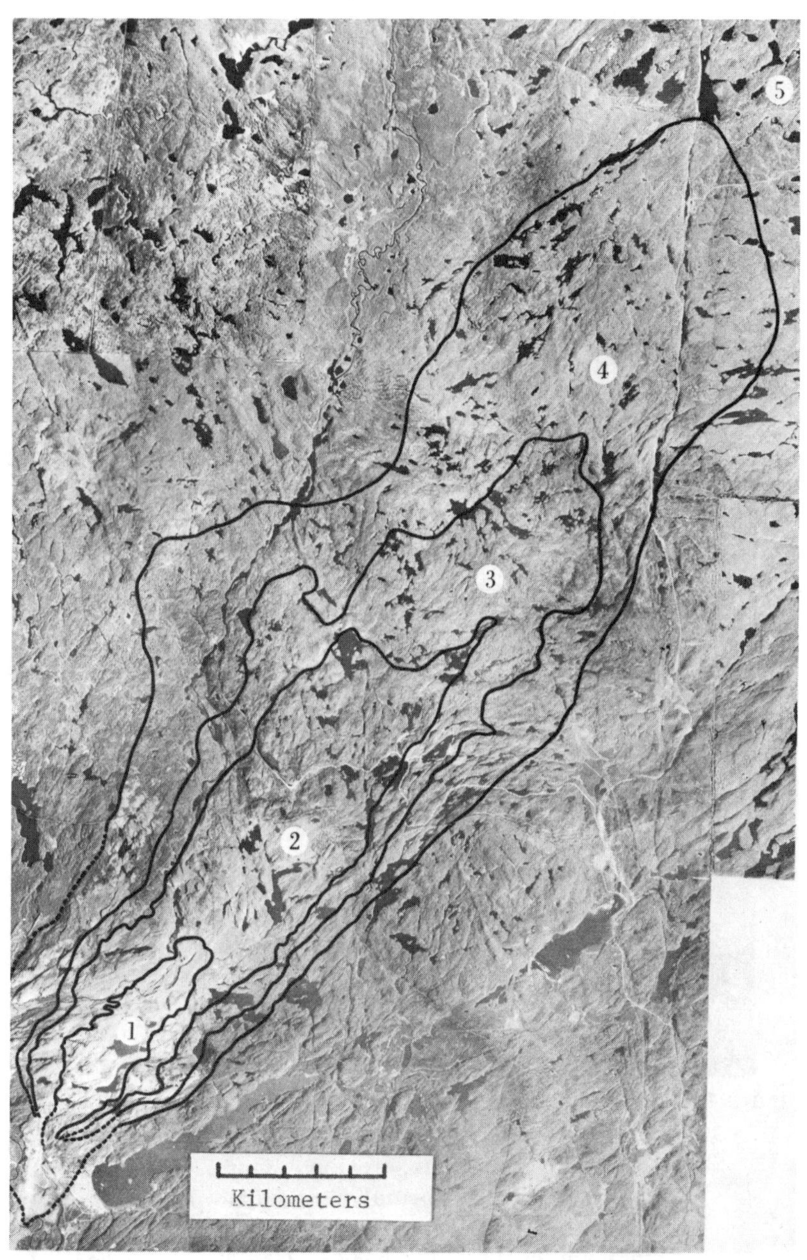


Figure 6.5. Aerial mosaic northeast of Wawa, Ontario, Canada, showing emission source and boundaries of the categories of vegetation damage superimposed. 1—Very severe; 2—severe; 3—considerable; 4—moderate; 5—not obvious. (Courtesy of Dr. Alan G. Gordon, Research Branch, Ontario Ministry of Natural Resources.)

Figure 6.6. Examples of damage categories shown in Fig. 6.5. 1—Very severe: overstory, shrub, and ground vegetation nearly all gone; erosion evident. 5—Not obvious: overstory with normal closed canopy. Normal understory, shrub layer, and ground flora. White spruce, balsam fir, white birch stand. (Courtesy of Dr. Alan G. Gordon, Research Branch, Ontario Ministry of Natural Resources.)

in the soil have apparently inhibited sexual reproduction and prevented rapid revegetation following local fires. Without vegetation erosion followed and revegetation was further limited, thus generating a cycle that will continue indefinitely if not ameliorated by humans.

Broadleaved species are generally more fume resistant than conifers. In their studies in Ontario, Gordon and Gorham (1963) found red maple, sugar maple, mountain maple, and red-berried elder (*Sambucus pubens*) quite fume tolerant. Highly sensitive conifers include western larch, Douglas-fir, ponderosa pine, and Engelmann spruce, while junipers are resistant (Katz, 1939). Rankings of the relative tolerance of woody plants to air pollution are available (USDA, 1973; Jensen et al., 1976).

Eastern white pine is highly sensitive to air pollutants, and serious disorders—chlorotic dwarf (Dochinger and Seliskar, 1970) and postemergence tipburn (Berry and Hepting, 1964)—are caused by ozone, sulfur dioxide, and fluorine. The level of resistance of both disorders is apparently under strong genetic control (Dochinger and Seliskar, 1965), and selection of resistant phenotypes and breeding for genetic resistance are recommended. Pollutants typically occur in combinations. The chlorotic dwarf decline of white pine is such an example; a mixture of sulfur dioxide and ozone are responsible.

Fluorides have been responsible for widespread mortality of coniferous trees near smelters, refineries, and power plants in the United States and Europe (Treshow, 1970). For example, needle burning and death of ponderosa pine over a 130 $km^2$ area near Spokane, Washington (Adams et al., 1952), were caused by fluoride emissions of a local aluminum plant. More recently, fluoride emissions from an aluminum plant were shown to cause widespread foliar burn and terminal dieback in forests of northwestern Montana and to predispose conifers to insect attack (Carlson and Dewey, 1971; Carlson et al., 1974).

A widely known example of air pollution is that in forests of the San Bernardino Mountains of California, which receive ozone, PAN (peroxyacetyl nitrate), and other harmful components of smog, blown 130 km from the Los Angeles basin by westerly winds. Ozone is a natural component of the upper atmosphere, where it acts to absorb harmful ultraviolet radiation. It also forms and accumulates in urban environments, due to the many sources of combustion that emit nitrogen oxide and hydrocarbons. Ozone is known to cause a major needle disease of ponderosa pine (chlorotic decline) resulting in decreased growth, needle drop, and death. In the San Bernardino National Forest, it is estimated that 1.3 million trees are affected by smog on an area covering more than 40,000 ha (Wert et al., 1970). In ponderosa and Jeffrey pines, the most sensitive of the dominant forest species, a yellow mottle of the normally green needles is followed closely by extensive needle drop, often leaving only the youngest needles on the tree (Miller, 1969). Chronic ozone levels bring about increased respiration and impaired photosynthesis; root growth declines and photosynthesis is further limited (Tingey et al., 1976). Although most pines are chronically affected, some trees remain green and healthy, suggesting a resistance of certain phenotypes to the disease. Most pines that are progressively weakened by smog injury are not killed by the disease itself, but by their natural enemy, the pine bark beetle (Cobb and Stark, 1970). Ponderosa pines are dying at a rapid rate (8 percent mortality in a 2-year period), and forest composition may shift to favor regeneration of shrubs (Miller, 1973). The resulting forest may become similar to other nearby areas converted to brush by fire or logging.

## LIGHTNING

No discussion of atmospheric effects on the forest can ignore lightning because of its great importance in terminating the life of trees, directly or indirectly, and in igniting fires, which play a dominant role in forest ecology (Taylor, 1974a, 1977; Chapters 11 and 16). The earth experiences 40,000 thunderstorms per day producing about a half million lightning discharges that strike forests (Taylor, 1974b). Thus it is not surprising that the effects of lightning on forests are widespread. Lightning effects on trees range from no obvious injury, to the usual stripping of bark, to the outright reduction of the tree to slabs and slivers. It can also kill branches, resulting in partial crown mortality, often termed "snag top" or "crown die back." It may also dig trenches and excavate soil at the base of trees, injuring and killing the roots. Lightning has caused substantial mortality to ponderosa pine, Douglas-fir, eastern hemlock, and loblolly pine (Taylor, 1969); Reynolds (1940) reported that it was directly or indirectly responsible for 70 percent of the total volume loss in pine forests in southern Arkansas.

Exceptionally large trees, especially in exposed locations, become the ground terminals of lightning discharges. Not only are individual trees struck, but a few to a hundred or more trees in compact, often circular groups, may be damaged or killed (Taylor, 1977). In one extreme case, four or more flashes from one storm affected 3255 trees on 52 hectares in one citrus grove. Six months later over 70 percent of the affected trees were dead or dying.

Species apparently differ in their predisposition to lightning damage (Treshow, 1970). For example, beech wood contains large amounts of oil while that of oak is free of it but high in water content. The high degree of hydration in oak may predispose it to lightning damage. Furthermore, since beech bark is smooth and constantly wet in thunderstorms, the electric current may be effectively conducted through the external water stream, minimizing injury.

Lightning indirectly influences forest trees by igniting fires and increasing their vulnerability to insects, disease, and windthrow. Lightning-damaged conifers attract bark beetles whose attacks on single trees often lead to mass attacks on surrounding healthy trees. In loblolly pine, lightning-struck trees are more susceptible to attack because of reduced oleoresin pressure, and they offer a more favorable brood environment for southern pine and Ips beetles (Hodges and Pickard, 1971). Direct damage by lightning and wounds caused by lightning-induced fires present potential courts of disease, which may be of major significance in trees of all ages. Trees mechanically weakened by lightning are susceptible to windthrow or wind breakage. More important, however, lightning-caused openings in the forest may serve as epicenters for extensive windthrow.

Most important of all, lightning ignites fires, which are a major agent of forest succession and regeneration (Chapter 15). Worldwide, lightning is estimated to cause about 50,000 wildland fires each year. Finally, lightning is known to cause some of the nitrogen of the atmosphere to form nitrogen oxides, the first of several steps ultimately making nitrogen available to plants.

## SUGGESTED READINGS


Alexander, Robert R. 1964. Minimizing windfall around clear cuttings in spruce–fir forests. *For Sci.* 10:130–142.

Carlson, Clinton E., and Jerald E. Dewey. 1971. Environmental pollution by fluorides in Flathead National Forest and Glacier National Park. USDA, Forest Service, Northern Region Headquarters, Div. State and Private Forestry. 57 pp.

Gordon, Alan G., and E. Gorham. 1963. Ecological aspects of air pollution from an iron-sintering plant at Wawa, Ontario. *Can. J. Bot.* 41:1063–1078.

Jensen, K. F., L. S. Dochinger, B. R. Roberts, and A. M. Townsend. 1976. Pollution responses. *In* J. P. Miksche (ed.), *Modern Methods in Forest Genetics*. Springer-Verlag, New York.

Kerfoot, O. 1968. Mist precipitation on vegetation. *For. Abstracts* 29 (1):8–20.

Palmer, R. W. V. 1968. Wind effects on the forest. Suppl. to Forestry. Oxford Univ. Press, London. 93 pp.

Stanhill, Gerald. 1970. The water flux in temperate forests: precipitation and evapotranspiration. *In* David E. Reichle (ed.), *Analysis of Temperate Forest Ecosystems.* Springer-Verlag, New York.

Taylor, Alan R. 1974. Ecological aspects of lightning in forests. *In Proc. Annual Tall Timbers Fire Ecology Conference.* 13:455–482. Tall Timbers Res. Sta., Tallahassee, Fla.

Taylor, Alan R. 1977. Lightning and trees. *In* R. H. Golde (ed.), *Lightning.* Academic Press, New York.

Treshow, Michael. 1970. *Environment and Plant Response.* (Chapter 10, Climatic extremes: lightning, hail, ice, and snow, pp. 125–137; Chapters 15–18 on atmospheric pollutants, pp. 245–353.) McGraw-Hill Book Co., Inc., New York.

# 7 Climate

The sum total of all climatic factors constitutes climate. We all can characterize climate in broad terms. When it comes to a precise description, however, our best efforts to date have left much to be desired.

Any usable climatic description must be based upon meteorological records which are generally available for most densely settled localities. This limits us pretty much to precipitation and air temperature. Such data as we have on solar radiation, cloudiness, wind, and evaporation are available from relatively few stations.

## CLIMATIC SUMMARIES

Actually, it is possible to characterize a climate reasonably well if the daily, monthly, and seasonal patterns of rainfall and air temperature are known. The difficulties arise from attempts to generalize from averages.

Climatic data for the United States are summarized in atlases compiled by Thornthwaite (1941) and Visher (1954) as well as in the 1941 U.S. Department of Agriculture *Yearbook*. Baker (1944) has brought together information available at that time on mountain climates of the western United States, while Pearson (1951) has compared climatic data in four widely separated portions of the ponderosa pine forest.

Let us consider a specific case. The meteorological station of the Southwestern Research Station of the American Museum of Natural History is located in a broad valley within the Chiricahua Mountains of southeastern Arizona at an elevation of 1650 m. The vegetation consists of southwestern pines, junipers, and oaks. In 1957 rainfall records covered a 7-year span and temperatures a 3-year span. The monthly temperatures and rainfall are summarized in Figure 7.1.

The quickest way to characterize the climate is in terms of the mean annual temperature and the mean annual precipitation. For this station, the former is 13°C and the latter is 400 mm. These figures convey a general picture of a warm climate with barely enough rain to support tree growth. Yet the mean temperature and

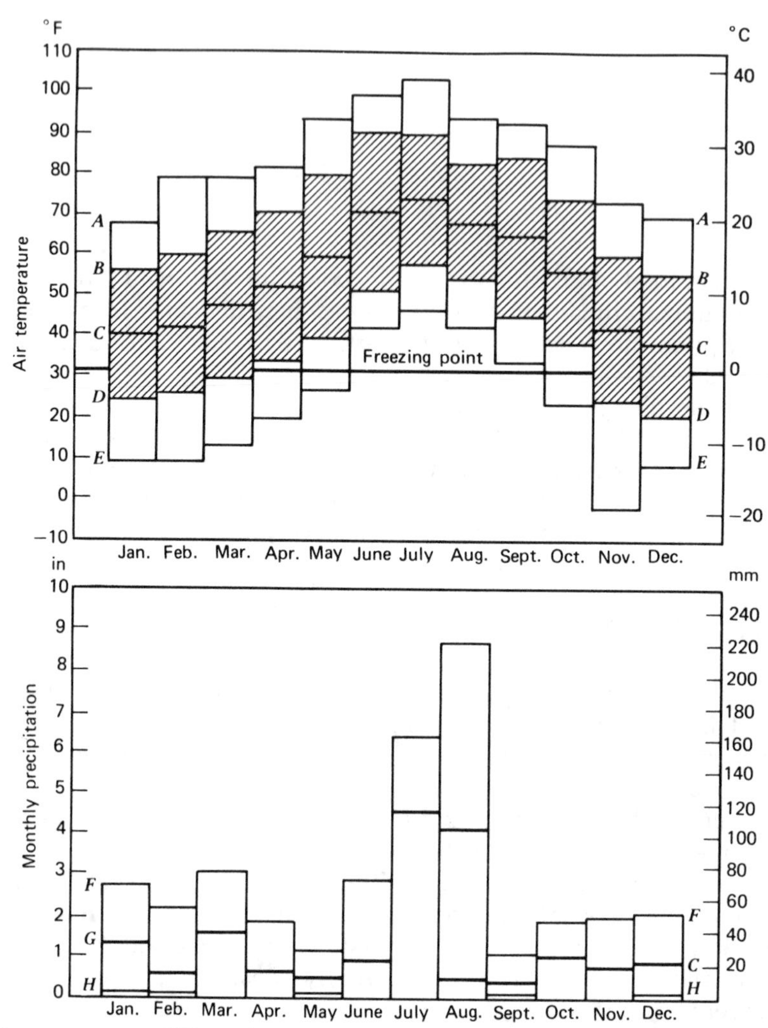


**Figure 7.1.** Climatic summary of Painted Canyon Basin, Chiricahua Mountains, Arizona. Temperatures (1955–1958) include *(A)* extreme high recorded for month, *(B)* mean daily maximum for month. *(C)* mean daily, *(D)* mean daily minimum, and *(E)* extreme low recorded. Precipitation (largely rainfall, 1951–1957) includes *(F)* extreme high, *(G)* mean for month, and *(H)* extreme low.

rainfall for the San Francisco peninsula in California are almost the same as those for the Chiricahua Mountains of Arizona, while the climate is completely different.

These mean values do not indicate many important facts about the climate: in the case of the Chiricahua Mountain station, that the daily fluctuation in temperature is extreme, averaging 19°C; that killing frosts are common throughout all but the summer months; and that over half of the annual rainfall (200 out of 400 mm) falls during the growing season. In contrast, the San Francisco climate shows but little daily and seasonal difference in temperatures (an average of only 5 degrees difference between January and July), virtually no killing frosts, and virtually no rain during the summer months. In other words, San Francisco has a typical maritime climate with a summer dry season, while the Chiricahua Mountain station has a typical continental climate with a summer wet season.

Many attempts have been made to summarize climatic values graphically, in order to permit a visual comparison of the climates in different parts of the country. Obviously, the rather complete graphing of temperature and precipitation as in Figure 7.1, showing the monthly variability as well as the monthly averages, is too complex to permit the visual comparison of a large number of climatic stations.

Considerable simplification can be achieved by working with the mean monthly temperatures rather than with the average daily maximum and minimum temperatures during the month. Actually, "mean monthly temperature" is a misnomer. It is really the average between two temperatures only: the average daily high and the average daily low. The value does not take into account intermediate temperatures at all. Even so, and despite the fact that it fails to give any indication of the daily fluctuation in temperatures, the mean monthly temperature is a reasonably reliable guide to the combined effects of solar radiation and air movement. It also has the virtues of both simplicity of concept and ease of calculation. A simple and meaningful technique is to graph the monthly trend in mean annual temperature and precipitation. In Figure 7.2, our two illustrative climates with the same mean annual temperature and precipitation are plotted on a monthly basis, and the differences between the two in temperature fluctuation throughout the year and in the pattern of rainfall are strikingly apparent.

## Climatic Classifications

Many attempts have been made to classify climates. None has proved satisfactory in explaining to any degree the distribution of forests and other vegetation types. Yet the terminologies of several of

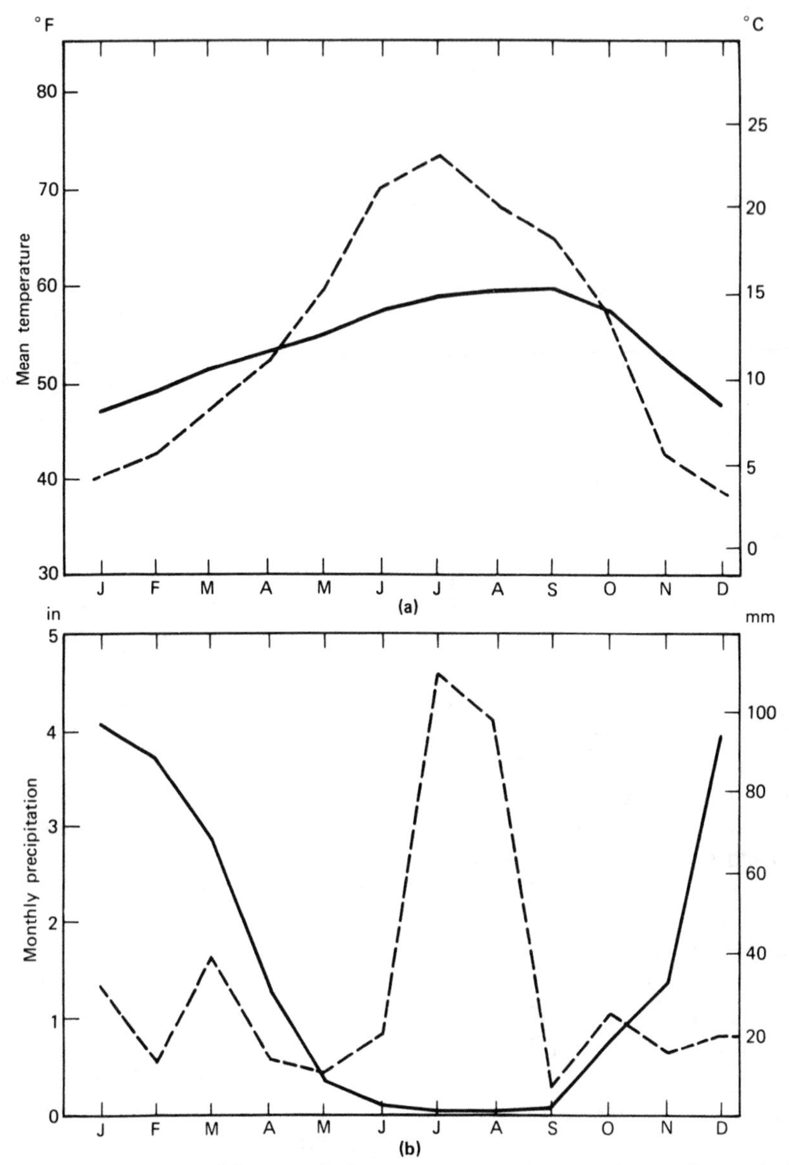


**Figure 7.2. Monthly trends in mean temperature *(a)* and precipitation *(b)* for San Francisco (solid line) and Chiricahua Mountains (dashed line).**

our climatic classifications have crept into our literature and have, indeed, a certain value in characterizing a broad and approximate climatic zone.

**Merriam's Life Zones.** In 1898, C. H. Merriam published a classification of life zones that is still used by biologists, especially in the western United States. Theoretically based upon the summations of daily temperatures for the period of the year when the mean daily temperature exceeds 6°C, this basis has proved impractical in the field. As used at the present time, Merriam's life zones are not based upon climate but upon broad vegetational zones. Thus the "Lower Sonoran" zone is a euphemism for the southwestern desert characterized by cacti, creosote bush, and the like. Similarly, the "Upper Sonoran" simply means that the vegetation consists of a woodland-savannah complex. Sagebrush, pinyon pines, desert junipers, oak savannah, and chaparral all fall within this zone. The "arid transition zone" refers to the ponderosa pine forest, and the "Pacific transition zone" to the Douglas-fir forest of the Pacific Northwest. Finally, the "Canadian zone" includes the higher-altitude conifers, and the "Hudsonian zone" the alpine and boreal spruce-fir types. Although the terminology of Merriam is useful in briefly characterizing broad vegetational zones, there is little basis for attributing climatic implications.

**Köppen's Climatic Provinces.** Among geographers, the classification scheme of this German scientist has achieved considerable use. Basically, the climate is classified by three code letters, the first referring to one of five zones of winter temperature, the second to seasonal rainfall pattern, and the third to one of three zones of summer temperature. As with other worldwide schemes, this system is broadly related to major vegetation types, but shows little correlation with the actual distribution of the vegetation in a specific area. At least, that has been the finding in attempts to relate it to the distribution of American forest trees.

**Holdridge's Life Zones.** Building on past attempts of climatic classification, Holdridge (1967) developed a system of life zones based on specific ranges of temperature and precipitation. This life zone system differs from Merriam's and other previous attempts in that the significant climatic factors of heat and precipitation are displayed in logarithmic progressions. The heat factor used, termed **biotemperature**, is the average annual temperature between 0°C and 30°C, which according to Holdridge is a measure of only the heat that is effective in plant growth. The resulting classification identifies a series of broadly defined life zones, such as rain forest, wet forest, dry forest, tundra, and steppe. Within each of the zones, plant and animal communities and their associated environmental features, such as topography, soils, and precipitation distribution patterns, are used to identify ecosystem units termed associations. Holdridge's life zones have found wide application in Central and South American countries in studies of land use planning.

**Thornthwaite's System.** It has been known that the balance between water gain and water loss plays a dominant role in both the development of soils and vegetation. Where the annual addition of water to the soil through precipitation exceeds the annual loss through evaporation, a humid climate exists. Soils tend to be leached downward and a forest vegetation tends to develop in both temperate and tropical zones. Conversely, where the evaporative potential is greater than the supply of water from precipitation, arid conditions exist. Soil chemicals tend to move upward toward the surface and desert or grassland normally is found.

In the United States, a formalization of this concept by Thornthwaite has found widespread acceptance. The original basis of this scheme involved the recognition of five major climatic regions, based upon the ratio of precipitation to evaporation (superhumid, humid, subhumid, arid, and superarid). These zones are in turn subdivided. In more recent writings, Thornthwaite, recognizing the difficulty of amassing and analyzing actual evaporation data, has redefined his system in terms of potential evapotranspiration, a value that can be computed from temperature data (Thornthwaite, 1948; Thornthwaite and Mather, 1955).

By computing potential evapotranspiration from air-temperature data throughout the year, he has been able to express climate in a compound graph in which the contrast between the precipitation curve over time and the potential evapotranspiration curve over time indicates seasonal water surpluses or deficits. Thornthwaite's graphical method has found many applications in forest climatology, as illustrated by representative curves for a forest in southern Arkansas (Figure 7.3; Zahner, 1956). The Thornthwaite approach may be used, as in Figure 7.3, to analyze seasonal variation in climate or to show broad climatic relationships (Patric and Black, 1968). Patric and Black reported close relationships of potential evapotranspiration with broad vegetation types of Alaska, by Thornthwaite's method. They found that, even though temperature is recognized to govern growth of far-northern forests, i.e., precipitation is everywhere adequate to supply the needs of the growth possi-

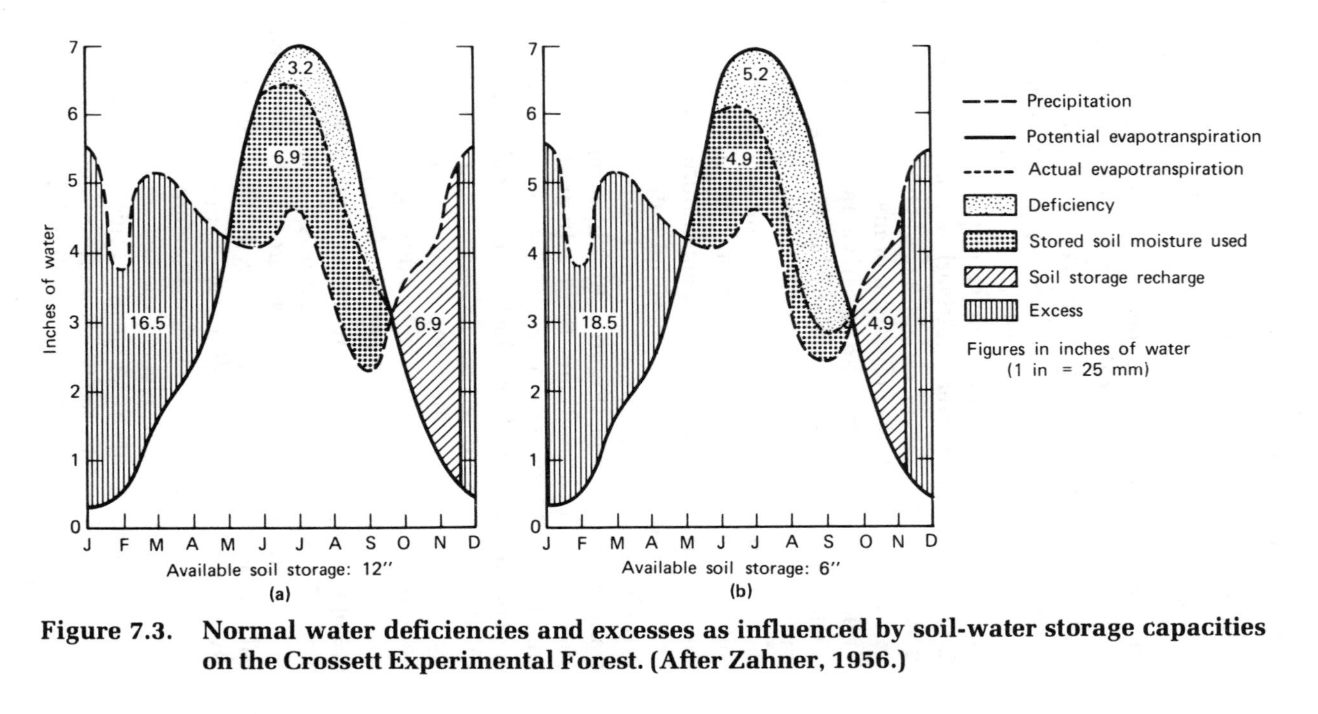


**Figure 7.3. Normal water deficiencies and excesses as influenced by soil-water storage capacities on the Crossett Experimental Forest. (After Zahner, 1956.)**

ble under such cool conditions (Hare, 1950), potential evapotranspiration reflects forest distribution better than temperature data alone.

### Multivariate Methods of Climatic Classification

Simple graphical methods, Köppen's classification, and multivariate statistical methods have been used to characterize the climates of British Columbia. Graphical or pictorial methods are satisfactory for demonstrating differences between a limited number of climates having major distinguishing features. For example, the precipitation, temperature, and growing season data clearly distinguish the climates of western British Columbia (*Cfb* of Köppen, marine west coast climate, Station 7), from that of the interior (*Dfb* of Köppen, humid continental cool summer climate, Station 59) (Figure 7.4; Krajina, 1959). However, multivariate methods offer the possibility to use a large number of variables to obtain the maximum differentiation and quantitative comparison of many climates (stations) simultaneously. By using the multivariate method of principal-components analysis, a large number of variables measured at each station may be replaced by one, two, or more synthetic variables, each of which is a combination of some or all of the original variables. For example, Newnham (1968) obtained climatic data for 19 variables (including seasonal temperature and precipitation data, length of growing season, elevation, and so on) for 70 stations in British Columbia (Figure 7.5) and subjected them to principal-components analysis (Figure 7.6). Three new variables were generated, the first, second, and third principal components, which accounted for 58, 29, and 5 percent, respectively, or together 92 percent of the total variation. The first principal component was primarily a combination of several factors related to length of the growing season while the second component was mainly determined by temperature and precipitation factors during the growing season.

The climates of the stations, as classified according to Köppen's system (Chapman, 1952), were reasonably well separated in the analysis (Figure 7.6). As in Figure 7.4, stations 7 and 59 and their respective *Cfb* and *Dfb* climates were widely separated in the principal components analysis. However, the Csb (humid continental cool, dry summer climate) stations were not well separated from one group of *Cfb* stations and station 18 seems to be misclassified.

The distribution of selected species, including coastal and interior races of Douglas-fir, Sitka spruce, Engelmann or white spruce, and ponderosa pine, was relatively well defined by the first two components of climate. Species distribution, however, was only generally related to climate.

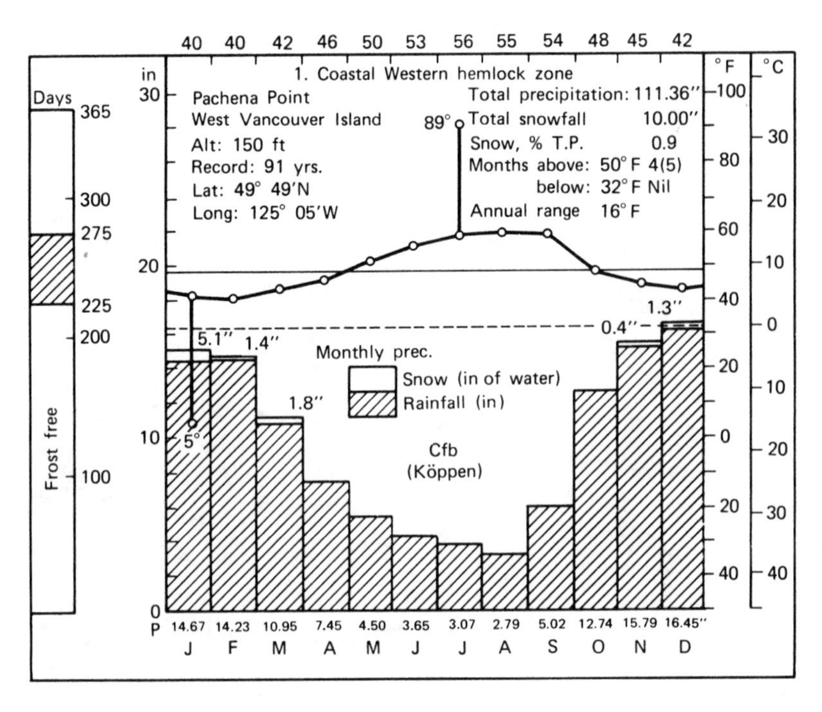


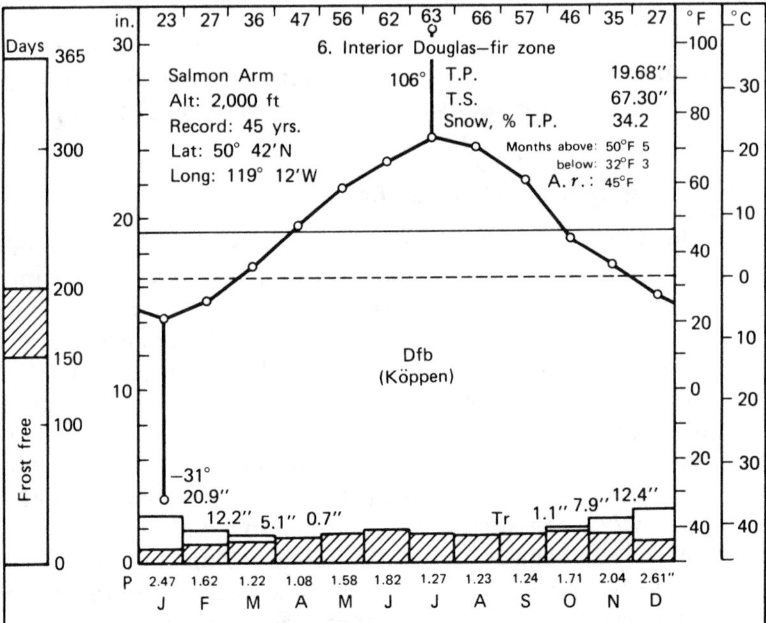


**Figure 7.4. Climatic characterization of stations representing coastal (1) and interior (6) climate. (After Krajina, 1959.)**

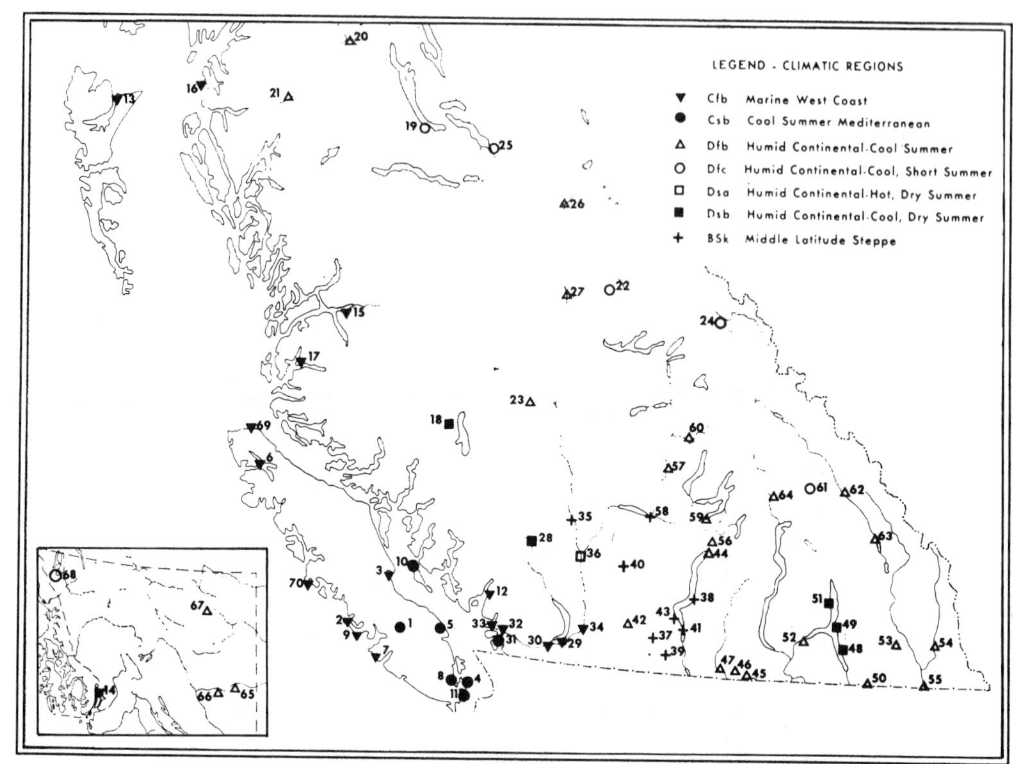


**Figure 7.5. Map of British Columbia showing the locations of the weather stations. (After Newnham, 1968.)**

## Climate and Distribution of Vegetation

Climate is a major determinant of the distribution of vegetation on a broad or regional scale, and microclimate may significantly determine the local distribution of species and communities. The importance of climate as a determinant of vegetation distribution is closely related to the time and space scale under consideration. Viewed over thousands or millions of years or over a wide latitudinal gradient at a given point in time, climate is the predominant factor. Within a given region and point in time, however, climate may be less important than soil conditions and stand history. Thus no single formula or universal classification scheme is adequate to explain or predict the existing vegetation distribution at regional and local levels. It is generally agreed that heat and water available to plants and the daily and seasonal fluctuations in these are the most critical climatic factors determining plant distribution. Daubenmire (1956) reviewed these factors in relation to the many attempts to combine them into

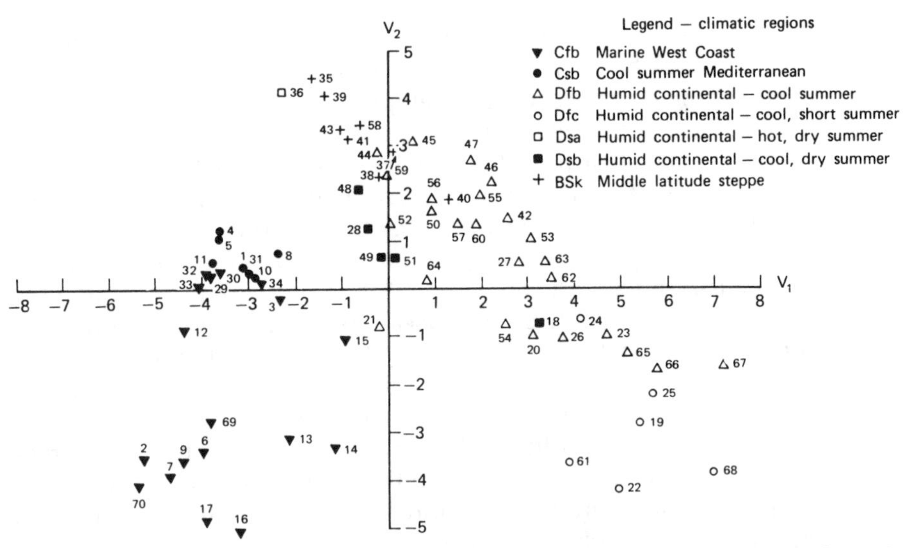


**Figure 7.6. The relationship between Chapman's (1952) climatic regions and the first and second principal components of climate. The coastal and inland stations of Figure 7.4 are 7 and 59, respectively. (After Newnham 1968.)**

formulas for classifying climates, classifications that might define the distribution of vegetation types. He found the classifications of Thornthwaite, Köppen, and Swain of little value in distinguishing areas of similar vegetation in the forests and adjacent grasslands of the northern Rocky Mountains. Furthermore the distinctions between grassland and forest climates could not be defined consistently by these classifications. However, a relatively high degree of correlation was achieved using graphs of mean monthly temperatures against median monthly precipitation.

In rare instances, climate may be the dominant factor responsible for the predominance of coniferous forest to the virtual exclusion of deciduous forest. In the Pacific Northwest of North America, forests of massive, long-lived conifers of 10 different genera dominate coastal and mountain forests (Waring and Franklin, 1979). Physiologically the conifers are highly adapted to the wet-winter, dry-summer environment. Evergreen foliage is capable of photosynthesis during much of the mild winter and is able to cope with the evaporative losses of warm, dry summers. The conifers are low in nutrient requirements, an advantage in the relatively young soils of the recent mountain ranges. In addition, nutrients (particularly nitrogen) taken up from the forest floor in the wet season when decomposition and nutrient availability are high, can be internally translocated within the tree from older tissues to developing shoots and needles. Most of the conifers are also adapted to fire, which periodically occurs in dry summers and is instrumental in their establishment.

Climatic factors may act directly to increase or decrease tree growth, which may change species distribution. Climate also may act indirectly by predisposing trees to insect and disease attacks and by intensifying an epidemic once it is initiated. For example, pole blight of western white pine, a disease reducing growth and causing substantial mortality in pole-sized stands in northern Idaho, has been linked to adverse temperature and moisture conditions for growth from 1916 to 1940 (Leaphart and Stage, 1971). These drought conditions on sites with shallow soils and severe moisture-stress characteristics led to root deterioration and rootlet mortality, crown decline, reduced growth, and ultimately the death of many trees. Recurrence of an adverse climatic pattern would increase the probability of the disease and would act to favor more drought-resistant species over western white pine on the sites with high moisture-stress characteristics.

The present distribution of species and communities is only generally related to climate. It also depends on physiography, on the age of the land and the nature of its soil, on the history of migration, on competition and disturbances, and on chance (Rowe, 1966). Obviously in specific situations climate is never the sole determinant, nor does one factor of climate always play the major role. The potential range of most species, based on climate and soil tolerances, is typically much greater than the actual range. Interspecific competition greatly limits the range. Many examples may be cited demonstrating that, once established by humans, woody species can survive and grow in climates markedly different from their native habitat. The microclimatic, soil, and biotic factors (including competition) impinging on seed germination and survival in the early years are neglected aspects in our understanding of species' distribution in nature. This, however, does not negate the fact that through climate's effect on plant processes and soils, its effect on genetic differentiation and interspecific competition, and as a causal factor of fires and windthrow, it is a major determinant of plant distribution (see Chapter 21, p. 574).

## Weather and Weather Modification

Weather is the current state of the atmosphere, the net effect of numerous and variable meteorological factors. We have seen how these factors influence tree growth, and that species distributions bear a general relationship to climate, which is the long-term average of weather conditions. Being long-lived perennials, woody species must not only be adapted to the mean climate, but must be able to withstand great fluctuations in weather from day to day, month to month, and year to year. Genetically evolved timing mechanisms, based on light, temperature, and moisture, enable them to withstand these fluctuations and yet maintain enough variability to meet the rate of long-term climatic change encountered in their respective environment.

Technology may soon enable man to realize an age-old dream of modifying weather to prevent monetary loss and personal suffering from severe storms, frosts, and droughts. In their detailed study of the ecological effects of weather modification, Cooper and Jolly (1969) point out that modification will be desirable if, among other reasons, significant options previously enjoyed by humans are not lost. These losses would include extinctions of plant and animal species, ruin of irreplaceable ecosystems, and destruction of valuable genetic material. If an increase of 10 percent in mean annual precipitation is achieved, changes in plant and animal communities are likely to be slow, but they could culminate in a rather extensive alteration of the original condition. No mass migration of communities is foreseen; rather, some plants will become regularly associated with species among which they were not commonly found in the past.

Humans are unwittingly changing the climate by burning fossil fuels and deforesting wildlands, which cause a gradual increase in the global $CO_2$ supply. This may raise mean annual temperature (Baes et al., 1976). The accompanying changes in regional climates are uncertain, but where climate becomes drier, lake and ground water levels might drop with significant effects on agricultural and forest production. Loss of coastal land might also occur due to rising seas fed by melting glacial ice. Although the oceans are the major reservoirs for $CO_2$, forests also incorporate carbon into their structures and accumulate it in the forest floor and soil. Thus deforestation and the expansion of agriculture are being monitored along with global changes in the $CO_2$ content.

## SUGGESTED READINGS

Daubenmire, R. 1956. Climate as a determinant of vegetation distribution in eastern Washington and northern Idaho. *Ecol. Monogr.* 26:131–154.

Fritts, H. C. 1976. *Tree Rings and Climate*. Academic Press, Inc., New York. 567 pp.

Krajina, Vladimir J. 1959. Bioclimatic zones in British Columbia. Botanical Series No. 1, University of British Columbia. 47 pp.

Leaphart, Charles D., and Albert R. Stage. 1971. Climate: a factor in the origin of the pole blight disease of *Pinus monticola* Dougl. *Ecology* 52:229–239.

Newnham, R. M. 1968. A classification of climate by principal component analysis and its relationship to tree species distribution. *For. Sci.* 14:254–264.

Patric, James H., and Peter E. Black. 1968. Potential evapotranspiration and climate in Alaska by Thornthwaite's classification. USDA For. Serv. Res. Paper PNW–71. Pacific Northwest For. and Rge. Exp. Sta., Portland, Ore. 28 pp.

Waring, R. H., and J. F. Franklin. 1979. The evergreen coniferous forests of the Pacific Northwest. *Science* 204:1380–1386.

Whittaker, R. H. 1967. Ecological implications of weather modification. *In* Robert H. Shaw (ed.), *Ground Level Climatology*. AAAS, Washington, D.C. Publ. 86. 395 pp.

# 8
# Soil

Soil is a natural body on the earth's surface in which plants grow; it is composed of organic and mineral materials. It is fruitless to argue whether climate or soil is more important in governing tree distribution and growth. Both are important. The individual forest tree requires favorable soil conditions just as it needs favorable climatic conditions. From the soil through the tree roots come the water and the mineral nutrients required for life processes. The soil is also the medium of support that holds the tree upright and in place—at least most of the time. The roots themselves can thrive only under favorable conditions of air supply, water supply, mineral nutrition, and warmth. Furthermore, certain soil fungi must infect the roots of many forest trees before the tree can grow normally and compete successfully.

Our concern here is with forest soils, but only those aspects of soils that are related to forest site quality. An understanding of forest ecology presupposes that the student has had, or will have had, exposure to the science of forest soils. We cannot here summarize this subject, but rather must confine ourselves to the relationship between certain soil characteristics and the definition of the forest site or habitat. Foremost among these are parent materials and the development of the soil profile (Chapter 8), the nutrient cycle (Chapter 9), the plant-soil water cycle (Chapter 10), and fire as a factor affecting forest site quality (Chapter 11).

Contrasting American approaches on forest soil relationships are presented by Lutz and Chandler (1946), Wilde (1958), and Armson (1977). General soils textbooks by Brady (1974) and Foth (1978) are available as well as an excellent reference text by Russell (1976). Publication of the proceedings of the North American Forest Soils Conferences (Youngberg, 1965; Youngberg and Davey, 1970; Bernier and Winget, 1975) offers a wide array of papers summarizing recent information on soil, site, and tree growth relationships. For northern Europe, excellent treatments are those of Aaltonen (1948) and Tamm (1950). Tropical soil problems are summarized by Mohr and Van Baren (1954).

## PARENT MATERIAL: GEOLOGY AND SOILS

The parent material from which soil develops may still reside at the original site or it may have been transported by gravity, water, ice, or wind (Figure 8.1). The importance of the underlying parent material in determining the distribution of vegetation types is often obvious. When two completely different vegetations abut, investigation will usually disclose that each is growing on a different geologic material of differing mineral composition and origin.

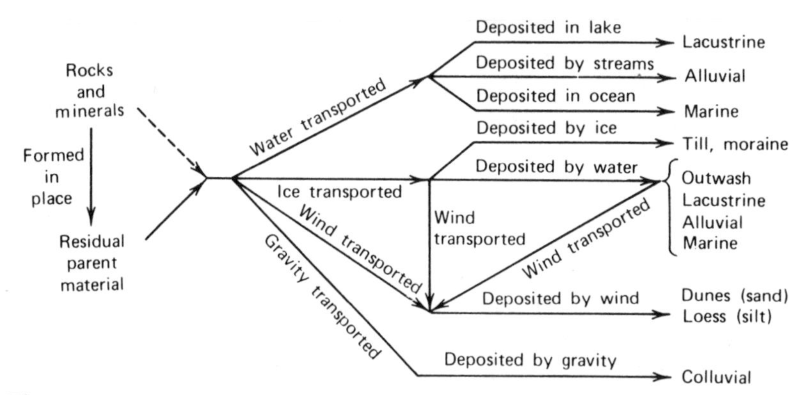


**Figure 8.1. How various kinds of parent material are formed, transported, and deposited. (Modified from Brady, 1974. Reprinted with permission of Macmillan Publishing Co., Inc. from *The Nature and Properties of Soils*, 8th edition, by Nyle C. Brady. Copyright © 1974 by Macmillan Publishing Co., Inc.)**

### Residual Soils

These plant-rock correlations are more apparent when the soils in which the plants grow are derived directly from the underlying rock. In such cases, distinct bands of vegetation often serve to delineate equally distinct bands of parent material. Many examples may be cited.

In the Appalachian and Ozark Mountains, soils derived from underlying sandstones support trees entirely different from those growing on adjacent soils weathered from limestones. The former material results in a relatively coarse, porous, and acidic soil in which pines have a great competitive advantage. The latter rock weathers

**Table 8.1 Tree Species in the Ozarks Ranked by Dominance Index on Soils Derived from Four Different Parent Materials**

| RANK | CHERT | LIMESTONE WITH CHERTY SURFACE SOIL | LIMESTONE | SANDSTONE |
|---|---|---|---|---|
| 1 | Black oak | Northern red oak | Eastern red cedar | Black oak |
| 2 | White oak | White oak | Northern red oak | White oak |
| 3 | Flowering dogwood | Black oak | Winged elm | Black gum |
| 4 | Mockernut hickory | White ash | Chinquapin oak | Flowering dogwood |
| 5 | Black gum | Winged elm | Shagbark hickory | Ozark chinquapin |

Source: Read, 1952.

into a finer, richer, and neutral or basic soil suitable for the more demanding hardwoods—walnut, beech, ash—and for such conifers as red cedar. A study by Read (1952) of forest composition in relatively undisturbed forest sites on northerly slopes in northern Arkansas (Table 8.1) is illustrative of the situation. Red cedar is concentrated on the limestone soils, with elm and ash and other demanding species also indicating the presence of limestone either at the surface or in the subsoil. In contrast, the sandstone and chert soils were dominated by black and white oak, mockernut hickory, and black gum and other less abundant species. In the mountain country of the southeastern United States, a band of pine forest lying parallel to a band of hardwoods mixed with red cedar will commonly connote parallel layers of sandstone and limestone (Figure 8.2).

Across the country in the mountains of the Pacific Coast, as in many other parts of the world, soils derived from serpentine support an entirely different vegetation than do surrounding areas (Whittaker, Walker, and Kruckeberg, 1954; McMillan, 1956). Serpentine is a mineral high in magnesium and iron and low in calcium, potassium, and sodium. Soils composed of its weathered fragments support only a poor and open plant life because of limiting nutrient conditions. In the ponderosa pine—mixed conifer forests, for example, serpentine soils are characterized by Jeffrey pine, a species otherwise restricted to higher elevations, and a distinctive group of shrubs and other lesser plants. One subspecies of lodgepole pine is almost completely restricted to serpentine soils. These "serpentine barrens" play a similar role in restricting the distribution of many other species and may elicit genetic differentiation in some species.

**Figure 8.2. Pines at higher elevation over hardwoods on lower slopes denote different geological strata in the Cumberland Plateau of Kentucky. (U.S. Forest Service photo.)**

## Transported Soils

Many of our forest soils, however, are developed from weathered rock fragments that have been moved to their present site by gravity, water, wind, or ice (Figure 8.1). During transport, a good deal of mixing usually occurs so that such soils are apt to contain fragments of many kinds of rocks and minerals. As a result, soils that have been transported are apt to differ more strikingly in their physical properties as these are related to the method of transport and the place of deposition rather than in their chemical properties.

Transported soils are characteristic of the Coastal Plains, the Great Plains, the valleys between the western mountain ranges, and the glaciated portions of Canada and the United States. Wherever differences in the mode of transportation and deposition occur, differences in composition of the vegetation may be expected.

For instance, in central New England, the native pines—white, red, and pitch—can usually compete successfully with hardwoods and form a major part of the plant community on sandy soils laid down by melt-water from the melting glaciers. On such sand **outwash plains** are found the best-developed and most persistent pine types. In contrast, **till**—nonstratified materials of all sizes rolled out by the advancing glaciers—has evolved into a finer-textured and more fertile soil where the various hardwood species normally outgrow and suppress the pines. The latter grow on these till soils only when farming, fire, or other disturbances have held back hardwood development and growth.

In the northern tip of the Lower Peninsula of Michigan, the direction of the glacier advance is reflected in the vegetation. The next to the last ice sheets of the Wisconsin glaciation advanced from the north, across the often-scraped and infertile Canadian shield of the Algoma section of Canada. Soils derived from the material carried by this ice are apparently coarser and less fertile than those in the same neighborhood derived from the last advance, which came from the northwest across the limestones and other rocks of the Upper Peninsula of Michigan.

In the Coastal Plain of the southeastern United States, finer-textured soils richer in organic matter are found in former lagoons rather than on the old beaches and beach dunes. The lagoons will often support a hardwood forest in localities where the latter grow longleaf pine. Other pine species may occupy the intervening sites.

Very commonly, when two distinctly different forest or other natural vegetation types abut, and there is no obvious difference in their topographic situations, soil differences resulting from the abutting of two types of parent materials will be found to be responsible.

## Physical Properties of Soils

Important physical properties of soils, particularly **texture** and **structure**, will be considered briefly as a basis for the discussion of the soil profile. Thereafter, texture and structure are each examined in relation to their influence on tree growth and development.

Soils are made up of four major components: minerals and organic matter, which together form the solid portion, and the soil solution and air, which occupy the pore space. Texture, the relative proportions of mineral soil particles of various sizes, and soil structure, the arrangement of particles into groups or aggregates, largely determine the physical properties of the soil.

The basic textural classifications are **sand, silt**, and **clay**. Sand particles are from 2.0 mm to 0.02 mm in diameter, silt particles from

0.02 to 0.002 mm, and clay particles less than 0.002 mm. Several kinds of clay minerals occur in forest soils, and silicate clays, composed of aluminum, silicon, and oxygen molecules, are important in the temperate regions (Brady, 1974). The mineral soil of the upper layers is also mixed with organic matter derived from decaying leaves, roots, and other plant tissues. The organic matter is important in providing water, nutrients, and aeration to the inorganic soil body. The clay particles and the very finely divided organic matter are all in the colloidal state and are characterized by a large surface area per unit weight. They are of great importance because water and nutrient ions, termed bases ($Ca^{++}$, $Mg^{++}$, $K^{+}$, etc.) are associated with their charged surfaces (Chapter 9).

The combination of the individual soil particles of different sizes into aggregates gives the soil its structure. The nature of soil structure affects the amount and size of the pores, which are filled by either water or air (Chapter 10). In addition to texture and structure, other important physical properties include soil air, water, and temperature.

## SOIL PROFILE DEVELOPMENT

As the blanket of rock detritus weathers and is occupied by vegetation and animals, it differentiates into more or less distinct horizontal zones, giving rise to a **soil profile**. The type of profile that develops depends upon the interaction of (1) climate, (2) parent material, (3) plants and animals occupying the soil, (4) relief of the land, and (5) the amount of time that has elapsed. Soil formation is in part a geological process in that it results from the normal weathering of rock fragments exposed to air and water, and in part a biological process, since the presence of plants and animals in and on the soil affects the nature and the pattern of the weathering.

### The Forest Soil Profile

Forests form the natural vegetation in many of the moister parts of the world—parts where the precipitation supplies more water than can be evaporated from the soil surfaces over the normal year. Under such conditions two factors dominate: the normal course of rain water is downward, thus tending to leach[1] the upper soil horizons of the more soluble chemicals; and the deep-rooted trees remove both

[1]To leach is to dissolve out the soluble constituents of a body (in this case soil) by the action of a percolating liquid.

water and nutrients from the root zone, transpiring most of the former and eventually returning most of the latter to the soil in leaf-, twig-, and fruit-fall and root decay.

In temperate regions, the typical forest soil, therefore, can be roughly subdivided into four zones, or **horizons** (Figure 8.3). The O or organic horizon is formed above the mineral soil and consists of an $O_1$ zone, containing recognizable litter from leaves, twigs, fruits, and dead plants and animals, and a lower $O_2$ zone, characterized by **humus** (decomposed litter) in which the original state of the organic material is unrecognizable. A similar and somewhat more discriminating method is to classify the organic horizon into three layers: **litter** (L), defined as the unaltered dead remains of plants and animals; **fermentation** (F), partly decomposed organic matter such that the source can still be identified; **humification** (H), well-decomposed amorphous organic matter. As decomposition proceeds

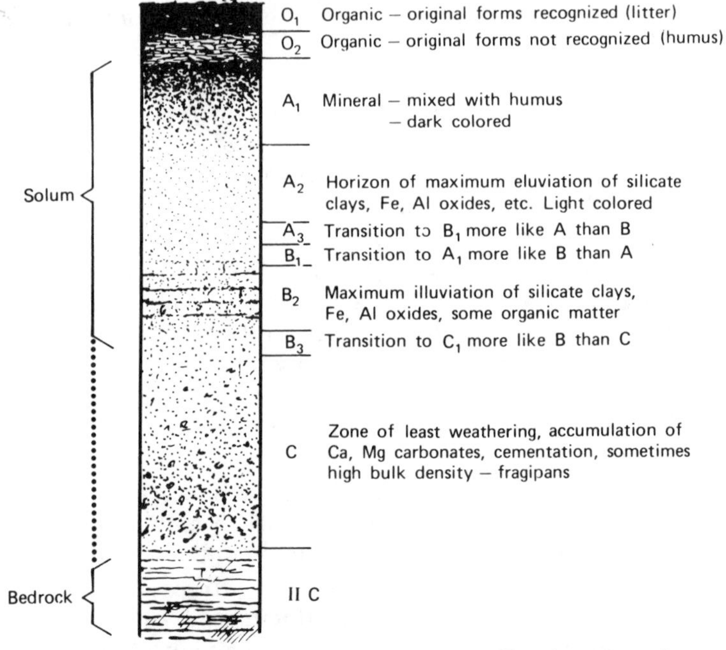


**Figure 8.3. A theoretical mineral soil profile showing the major horizons that may be present. (After Buckman and Brady, 1969. Reprinted with permission of Macmillan Publishing Co., Inc. from *The Nature and Properties of Soils*, 7th Edition, by Harry O. Buckman and Nyle C. Brady. Copyright © 1960, 1969 by Macmillan Publishing Co., Inc.)**

from identifiable organic tissues to minerals, $CO_2$, and water, an intermediate product is the dark amorphous residue termed **humus**.

The A horizon is the surface layer of mineral soil that is leached (eluviated) of its nutrients by the downward-moving water and inorganic and organic acids. It, in turn, is subdivided into an upper $A_1$, in which organic matter is constantly being added to mineral soil through litter decomposition and mixing by animal activity, and a lower $A_2$ horizon, which is merely leached. The term **melanization** is sometimes applied to the process of organic matter incorporation into the surface of mineral soil.

Below is the zone of accumulation (illuviation), the B horizon, characterized by the deposition of iron and aluminum oxides and minute clay and organic particles, all derived from the leaching of the minerals and organic matter above. As a result the B horizon is usually finer-textured and darker in color than the parent material, while the A is coarser-textured and may be either darker or lighter in color. Occasionally the B horizon is so packed with clays and iron oxides that it forms a hardpan and becomes quite impermeable to water and even air movement.

Below the zone of clay deposition is the C horizon, composed of the same parent material as that from which the A and B horizons are formed. Commonly, the upper part of this zone is somewhat weathered, as air in the pore spaces will have resulted in oxidation of at least some of the rock minerals and as downward-percolating water may have leached calcium carbonates and others of the more easily soluble minerals. The soils specialist can often recognize several minor strata in the soil, characterized by minor differences in weathering, leaching, or accumulation. A subsoil horizon composed of a parent material different than that from which the A and B horizons are formed is termed the IIC rather than the C horizon. An example of this case is soil formed in a thin layer of glacial deposits over residual bedrock.

The typical layered forest soil profile is developed only under conditions of good drainage, when water can move down through the soil. The presence of a high water table in the soil, or of rock or soil layers impervious to water movement, will result in a truncated and modified profile. In the zone of a fluctuating water table, oxidation is periodically restricted, giving rise to a mottled soil profile of yellow or gray (unoxidized) and brown or reddish (oxidized) minerals. A permanent or stagnant water table will prevent any soil oxidation and may even result in the extraction of oxygen (i.e., **reduction**) from previously oxidized soil minerals, forming a **gley** horizon.

It should be noticed that the terms A, B, and C horizons refer to zones which have been leached, enriched, and unaffected, respectively. It does not necessarily follow that the upper mineral horizon is always the A. Following sheet erosion, the B or C horizon may be exposed at the surface. On the other hand, a soil may be superimposed on a fossil profile consisting of A, B, and C horizons. These and many other combinations are possible.

The development of a forest soil profile can be followed with particular ease when a sand dune composed of uniform unstratified sand grains is stabilized by the planting of forest trees. An example is provided by the afforestation of coastal sand dunes with Corsican and Scots pines at Culbin on the north coast of Scotland (Wright, 1955, 1956). By the time the pines had established a closed forest (22 to 80 years), their roots had dried out the sand considerably; but litter added to the surface had increased materially the moisture-holding capacity of the upper soil. A soil profile had begun to form in this brief period, as indicated by the downward distribution of the major nutrients through leaching, their recirculation through root absorption, and subsequent leaffall.

Soil profile development may continue for many thousands of years. In the northern portion of the lower peninsula of Michigan, similar parent materials (both tills from advancing glaciers and sands from glacial melt-waters) can be dated from their position in relation to old lake levels of the Great Lakes (Spurr and Zumberge, 1956). One series, exposed for 3500 years, shows only slight weathering and profile development. Even on sands, lime is available within 30 cm or so of the surface to the extent of favoring the growth of northern white cedar. On similar sand soils exposed for 8000 and 10,000 years, the lime has been leached to a depth of about 1 m and clearly segregated A and B horizons have been formed. On adjacent clay loams originating as till, however, soil profiles are still relatively immature after 10,000 years. Reduced water percolation and leaching in the clay soil, compared to the sand soil, are largely responsible for this difference.

## Nonforest Soils

Quite different soil profiles normally develop under grass or brush vegetation than have been described for the forest. Ploughing, of course, mixes surface layers of the soil into a homogenized plough layer, below which the remains of the truncated soil profile may be detected. Under natural grassland, the dense fibrous root mass near the surface results in a higher surface content of incorporated organic matter and the retention of clays in this zone. Grassland soils, even in forested regions, develop a dark and rich topsoil, which can be detected long after the natural vegetation has been destroyed. In lower Michigan, extensive areas of prairie soils stretching across the lower part of the state bear witness to an extension of the prairies existing there prior to settlement.

Grassland, brushland, and desert soils in regions of low rainfall differ from forest soils to an even greater degree. Under conditions where precipitation is insufficient to support a closed forest canopy, moisture entering the soil is pulled toward the surface and evaporated, leaving a layer of precipitated chemicals in its wake. The surface layer thus tends to become enriched in calcium and other readily soluble nutrients. Deeper in the soil there is little leaching and the soil is only gradually changed in depth by geological weathering. The result is that the vegetation often seems to be growing on a mass of weathered rock fragments rather than on a soil with a well-developed profile. In the mountains of southern Arizona, for example, the open woodlands are formed by the live oaks, junipers, and pinyon and other pines of the semiarid country. The trees grow in rock detritus (colluvial slopes) at the base of outcropping bedrock.

## Podzolization

Although any given soil profile is the product of many different weathering and other reaction-producing processes, two general types of soil formation have attracted particular attention from soil scientists because of the prevalence of these processes over large areas, and their importance in changing the capacity of the soil to grow vegetation. **Podzolization** refers to the process of acid leaching whereby clay and organic particles and mineral ions, primarily iron and aluminum, are carried downward, a process that, brought to its logical conclusion, will leave a leached $A_2$ horizon composed predominantly of silica and from which the iron and aluminum minerals have been removed. The term is derived from the Russian *zola* meaning, ashes, and has become a part of basic soils vocabulary.

True podzolic profiles, termed **spodosols**, with a thick and impoverished $A_2$ horizon, are restricted to cool, damp climates and occur primarily under trees and shrubs that produce an acidic foliage low in content of calcium and other bases. The spruces and firs throughout Canada, northern Europe, northern Siberia, and in the high mountains of the North Temperate Zone generally create spodosols. So do heather, blueberry, and similar shrubs. For instance, in England, strong podzolic profiles under oak forest have been shown to have developed when the areas had been heathland prior to the invasion of the oak (Dimbleby and Gill, 1955).

Elsewhere in the Temperate Zone, **podzolic soils** tend to develop under the forest, the adjective implying that the process of podzolization is dominant, but that leaching has not or will not proceed to the point of reducing the $A_2$ horizon to a grayish and impoverished sand. Under pines and hemlocks, true spodosols may develop over a 50- to 100-year period, but if the forest is converted to hardwoods, the spodosol condition will often disappear in a similar span of time. When eastern white pine is planted on old agricultural land in central New England, a thin, leached podzolic horizon will normally appear in 40 to 50 years and may reach a breadth of 2 to 5 cm within 100. After the pines are harvested and are succeeded by red oak, red maple, and other hardwoods on till soils, the chemical effect of the hardwood foliage will gradually change this grayish-white horizon until, after 30 to 50 years, it has been recolored and partially recharged with nutrients (Griffith et al., 1930). In England, it is estimated that 60 to 100 years will elapse after birch has invaded heath moorland before the spodosol profile formed under the acid raw humus of the moorland is destroyed (Dimbleby, 1952b).

Podzolization is dominant in virtually all forest soils in moist temperate climates, and should be considered as a natural and inevitable soil-forming process under such climates. Under certain conditions, intense podzolization may result in lowered productivity of the forest. The $A_2$ layer itself, severely leached of nutrients, especially of calcium, may act as an effective barrier to root penetration.

More importantly, the extreme leaching characteristic of spodosol formation may result in the iron and aluminum minerals and other clay-forming materials effectively cementing the B horizon. Hardpans may develop that are impervious to air and water and that effectively seal off the tree roots from the soil moisture and nutrients below. Under such circumstances, the soil available to vegetation is the shallow zone above the hardpan, and this zone is very low in its supply of plant nutrients.

The classic example of soil deterioration due to podzolization is the heath and moorland of Great Britain and other parts of western Europe. Here, clearcutting and fires in bygone centuries have converted land that was originally forest to heather and related ericaceous species. Under the influence of the acid and indigestible foliage of these plants, podzolization was carried to the point of sealing off the B horizon with an iron pan. As a result, trees can no longer be grown to merchantable size on these sites. Yet, by breaking the iron pan with deep ploughs, by adding fertilizers, and by establishing trees and shrubs producing more nutritive foliage, the site can eventually be reclaimed for timber production. Oaks and birch are more effective in rehabilitating the soil than beech or Scots pine.

As early as the sixteenth century, and reaching a climax during the nineteenth century, many hardwood forests in central Europe were converted to stands of pure Norway spruce. Spruce was easier to regenerate than were hardwoods and became increasingly desirable because of its greater timber yields. Yet, in the second and third rotations of spruce, the yields in many cases fell off sharply. Although the decrease was attributed to the development of spodosols, such a generalization is unacceptable. The spruce monocultures do bring about marked physical and chemical changes in the topsoil, and on particular soils a combination of these, instead of or in addition to podzolization, may have evoked the decline.

The changes caused by spruce—even after one generation—include a reduction in the rate of organic matter decomposition, decreased earthworm and microorganism activity, decreased melanization, reduced water availability, and changed nutrient status in comparison to topsoil conditions under native hardwoods (Schlenker et al., 1969). Under these conditions reductions in nutrient cycling may have in part led to the observed growth decline. In addition, on heavy clay soils spruce roots exist as a flat latticework in the topsoil, for they are much less effective in penetrating the lower horizons than are the roots of the native oaks and beech. In the first spruce generation following hardwoods, the soil is permeated by the old root canals of the hardwoods, and spruce roots may take advantage of these in penetrating deeper horizons and extracting quantities of water and nutrients. In succeeding generations of spruce, ever-shallower root systems are developed. The decreased ability of the root system to reach water and nutrients in lower horizons may therefore in part reduce growth in successive generations. Finally, on some sites spruce may accelerate the natural tendency of the soil to become waterlogged (Werner, 1964). However, spruce monocultures may suffer much greater damage from windthrow, disease, insects, and snow breakage than from the effects of podzolization or soil deterioration. For example, heart rot, caused primarily by *Fomes annosus*, may cause severe growth losses in spruce stands of the first and successive generations on specific sites formerly occupied by hardwoods (Schlenker, 1976). Current forest practice in central Europe shows a strong concern for identifying and mapping the kinds of sites (Chapter 12) where pure spruce may be grown with the least risk of injury or soil deterioration, sites where spruce should be grown in mixtures with other conifers (primarily silver fir) and hardwoods (primarily European beech), and sites where spruce should be excluded entirely.

So much has been written and argued about this particular soil problem that there is a tendency to think of podzolization as a "bad" or undesirable process. Actually, podzolization is the inevitable process of soil formation wherever moist, cool conditions exist. Furthermore, in most instances, there is no evidence that podzolization decreases timber growth rates or otherwise affects undesirably the characteristics of the forest. In some special situations, podzolization has resulted from mismanagement of the vegetation and has lowered the productive capacity of the soil. The danger of podzolization, however, has been overemphasized. Podzolization definitely benefits the water-holding capacity and nutrient buildup in B horizons of extremely sandy soils. For example, in northern lower Michigan, Hannah (1969) found 29 percent more volume and 20 percent more dry weight of stemwood in 35-year-old red pines growing on soils with moderate to well-developed, spodosol $B_2$ horizons than that of pines growing on soils in which podzolization was absent or weakly developed.

**Laterization**

Whereas podzolization reaches its greatest development in cool or cold climates, **laterization** characterizes the tropics and is primarily a process of hot, wet climates (Prescott and Pendleton, 1952; McNeil, 1964). Through weathering and intense leaching under such conditions, the iron and aluminum minerals are changed to insoluble compounds, which remain in the A horizon after the silicas and other minerals are carried downward, giving rise to **oxisols**. The B horizon is either thin or completely missing. The soil is thus converted into a red or yellow body, a deep **oxic** subsurface horizon dominated by insoluble oxides of iron and aluminum, but impoverished of all else. Carried to the extreme, these minerals harden into a red rock or **laterite** (from the Latin word for brick or tile).

Laterization is essentially a geological weathering process in that it occurs with or without vegetation and that it may extend 30 or more meters deep, far below the influence of plant roots. Vegetation, however, does influence the extent and nature of the process. Oxisols may develop under grassland, savanna, or forest. Under the forest, however, the concentration of iron and aluminum compounds occurs at some depth from the surface, while under grassland it occurs at the surface. Clearcutting the forest and converting the area to grassland, though, may result in accelerating the hardening of lateritic soils into laterite rock, a process which greatly lowers the productive capacity of the site (Griffith and Gupta, 1948). For this reason, the conversion of natural forest into teak or other tree-plantation crops is not recommended when the forest soils are highly lateritic.

Oxisols developed under tropical rain forest are subject to excessive leaching when the forest cover is removed. Frequently, virtually all nutrients are leached out by the heavy rains to a depth of 2 m or more in 6 months or less. Only through the maintenance of a continuous cover of tree crops can soil nutrients be returned to the surface through litterfall and the fertility of such soil be maintained.

## SOIL TEXTURE AND STRUCTURE

### Soil Texture

By soil texture is meant the size distribution of the soil particles, which range from gravel to sand, silt (microscopic and floury to the touch), and clay (submicroscopic and plastic to the touch). Near-equal mixtures of sand, silt, and clay are termed **loams**, with a modifying term indicating the dominant fraction (e.g., **silt loam**).

Sandy soils (less than 15 percent silt and clay) are "light" in texture and can hold less water and minerals but more air than clays. In contrast, clay soils (more than 40 percent clay particles and less than 45 percent sand or silt) are "heavy" in texture and can hold more water and minerals but less air than sand soils. Soils containing generous proportions of silt particles, or about equal amounts of sand, silt, and clay, such as loam and silt loam soils, generally have the best balance between moisture, nutrients, and air.

The texture of the soil parent material is determined in part by the mineralogy of the soil-forming minerals and in part by the degree of weathering. A coarse-grained granite will tend to weather into a soil composed dominantly of coarse sands, while a fine-grained diorite will tend to weather into fine sands and silts. In the case of transported soils, the mode of transport is all-important (Figure 8.1). Rapidly moving waters will deposit only gravel or sand, while silt and clay will precipitate from very still water bodies. Ice, however, can move and leave all sizes of particles, so that glacial deposits may contain anything from large boulders to fine clays (i.e., till). Wind transports fine sand to form **dunes** and transports silt-size particles over much greater distances to form deposits termed **loess**.

As the soil profile develops, the different horizons tend to develop different soil textures. With podzolization, the leached A horizon becomes progressively coarser-textured (except at the surface, where organic matter may serve to help maintain a fine texture) and may eventually become composed entirely of sand grains. In contrast, the B horizon receives deposits of clays, silts, and organic matter and becomes increasingly finer-textured, even to the extent of becoming cemented and impervious to water and air movement. Under arid and semi-arid climates, where evaporation moves water toward the surface, the surface horizons may become finer-textured than those lower down. In the process of laterization, the residual weathered material becomes progressively coarser-textured as time elapses.

Soil texture is extremely important in affecting site quality. It influences the chemical properties of the soil, soil moisture and air relations, and root development.

From a chemical standpoint, the minute clay particles are the most active and present the largest surfaces from which nutrients may be released to roots. The fertility of a forest soil is commonly correlated with the percentage of **fines** in the soil. These include the mineral silt and clay particles and similarly sized organic matter. Electrostatic properties of clay-size particles, both mineral and organic (humus), are responsible for the phenomenon of cation adsorption in soils, a process that retains nutrient bases ($Ca^{++}$, $Mg^{++}$, $K^{+}$, $Na^{+}$) against leaching and thus enhances soil fertility (Chapter 9). Soils high in sand content are normally low in fertility.

Physically, the texture of the soil regulates the pore space and consequently both the water and the air-holding capacity of the soil. Coarse-textured soils have much large pore space. They are easily drained, though, and are apt to be excessively dry. Fine-textured soils have very few large pore spaces but hold much water on the large surface area (i.e., their capillary pore space is large). Heavy clays are but little pervious to air and water. The best conditions for both absorption and retention of both water and air are mixtures of sands, and fines, and organic matter. Well-balanced loams represent the most favorably textured soils under most forest climatic conditions (Figure 8.4).

Soil texture affects root development primarily through the influence of texture on aeration and nutrient retention. Different species vary markedly in their ability to penetrate soils. The roots of red pine in the Lake States, for instance, can only penetrate soils with a high sand or gravel content. In medium sands, extensive and deep root systems are developed. Satisfactory root systems can also be developed in rocky soils where the interstices among the rocks are filled with relatively coarse soils. In poorly aerated tills and other heavier soils, however, red pine root development is largely confined to the A horizon that has been rendered coarse by leaching. Spruces, particularly black and Norway spruce, are widely known as shallow-rooted, blowdown-prone species on heavy and poorly drained and aerated soils. However, Norway spruce develops extensive and deep root systems in sandy and sandy loam soils.

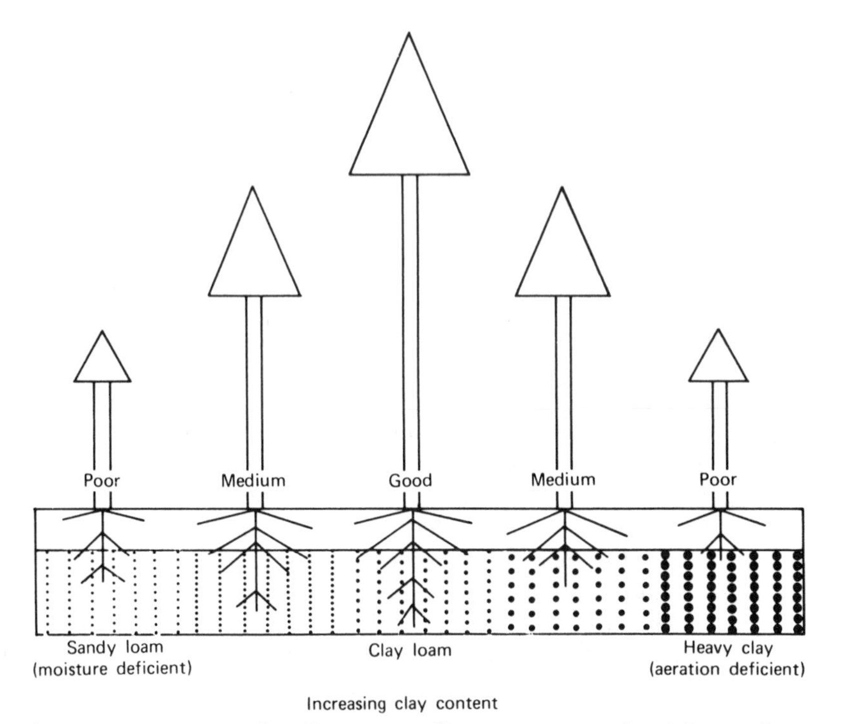


**Figure 8.4. How subsoil texture affects tree growth. (After Zahner, 1957.)**

## Structure

Whereas texture refers to the actual size of soil particles, structure refers to the mode of grouping of these particles into aggregates. Thus a fine-textured soil may actually appear coarse—and may function in many ways as a coarse soil—because of the coalescing of particles into larger granules. Structure is most important in soils high in silt and clay content. The development of a granulated structure in such soils permits good percolation of both water and air, reduces erosion, and results in a soil that has many of the desirable physical and chemical properties of a good natural loam.

Favorable soil structure is promoted by the incorporation of organic matter deep into the surface horizons. This generally favors aggregation of the fines into crumbs or other types of larger particles. Earthworms are particularly beneficial in accomplishing this.

Favorable soil structure developed under forest conditions may be destroyed by removing the forest and exposing the soil surface to the direct striking of raindrops and to intensive leaching. In the Coweeta Experimental Forest in the southern Appalachians, a mountain slope cleared of its natural hardwood forest vegetation and put into agricultural use showed little erosion for the first 2 years because of the soil aggregates that permitted the rain to penetrate the soil rather than to run off the surface. After exposure and farming had reduced the number and size of the surface aggregates, however, more than 180 metric tons of soil were washed off the 9-ha tract in the following 11 years.

## TOPOGRAPHIC POSITION

The treatment of soil factors would not be complete without considering the topographic position of the site, its slope, and its aspect. Soil and topography are closely related to one another. Topographic position affects soil depth, profile development, and the texture and structure of the surface soil and subsoil, hence influencing the composition, development, and productivity of the forest.

### Position on Slope

The location of the forest site with regard to surrounding land joins in importance the soil geology in determining the physical properties of the soil. High convex surfaces tend to be exposed to high winds, subject to erosion and weathering, and to be drier than is average for the region. At the other extreme, low concave surfaces tend to be sheltered from high winds, subject to accumulation of soil rather than to erosion, subject to cold-air drainage, and to be moister than is average for the region. Midslopes are generally intermediate in their characteristics. Level surfaces are very stable, their site properties being determined by the climate and the soil conditions.

The classification of forest site upon a basis of relative elevation (i.e., high slope, midslope, low slope, basin, high level, and low level) is frequently one of the most useful and meaningful criteria of site classification. Practically, it has the advantage of being capable of classification on aerial photographs viewed stereoscopically. Many studies have found position on slope to be the single most useful factor in evaluating growth potential of forest trees (Ralston, 1964; Carmean, 1975, 1977).

### Slope and Aspect

The angle of repose of the soil is usually measured in terms of percentage (the vertical rise expressed as a percentage of the horizontal distance) or in terms of degrees. The greater the slope, the greater also is the surface area per hectare or other area measured horizontally. For this reason, good forest sites of moderate slope usually contain more trees and produce greater yields per hectare (measured horizontally as it always is) than do comparable level sites.

The great importance of slope, however, is in orienting the site with regard to the sun and wind. In the North Temperate Zone, the sun is to the south during the warmth of the day, and south-facing slopes receive more intense sunlight than any other. At any given latitude, then, the hottest and driest sites are those which most nearly face the sun during the middle of the summer day. Both steeper and more gradual slopes receive less insolation. The amount of insolation received on a site governs other related factors, air and soil temperature, precipitation, and soil moisture, all of which are important for establishment and growth of plants. In a detailed study of eight environmental factors of four southerly aspects in desert foothills of Arizona, Haase (1970) found the sequence of warmest and driest to be S, SSW, SW, and SSE. However, the warmest and driest aspect may change when the time of year is considered. For example, the SW aspect exhibited drought extremes during the arid spring but had the mildest drought conditions during all other seasons.

North slopes, on the other hand, receive less sunlight and are invariably cooler and moister in the Northern Hemisphere (these relations are, of course, reversed in the Southern Hemisphere). East and west slopes show similar but less extreme variation. East-facing slopes are exposed to direct sunlight in the cool of the morning and are normally somewhat cooler and moister than west-facing slopes. In mixed upland oak forests of the Appalachian Mountains northeast aspects are the most productive, being approximately 15 percent better than the south and west aspects, which were the least productive (Chapter 12).

## SOIL CLASSIFICATION

The classification of soils presents many of the problems of classification of climates. So many separate factors enter into the formation of a given soil and so many different physical and chemical properties can be measured and evaluated that it is difficult to choose two or three values that adequately characterize the soil. Furthermore, all gradations of soils can be found between almost any two extremes, and any effort to divide the range into a set of classes becomes arbitrary and never more than partially successful.

Soils, however, are concrete in that they can be seen and described far more readily than the ephemeral and variable qualities of the atmosphere that we call climate. Furthermore, the importance of soils for agriculture and engineering is such that much attention has been given to the development of soil classification schemes, and much field work utilizing these schemes has been carried out.

A taxonomic system of soil classification, based largely on soil properties (soil morphology) rather than soil-forming processes (soil genesis), is widely accepted in North America. This system was developed by the soil survey staff of the U.S. Department of Agriculture. It is described in detail in Agriculture Handbook 436 (Soil Survey Staff, 1975) and is summarized by Brady (1974). The primary advantage of the system is that the soil profile itself, rather than a soil-forming process or presumed process, is classified. The system is modeled after the plant taxonomic system, with categories from "order" (the broadest group) to the "series" (roughly equivalent to a "species").

The main soil orders supporting forests in North America are **Entisols, Spodosols, Alfisols, Ultisols**, and **Inceptisols. Oxisols** are found in extensive areas of tropical and subtropical forests in Central and South America, southeast Asia, and Africa. Oxisols (from French *oxide*, and the Latin word for soil, *solum*) are highly weathered soils in which the subsoil is strongly oxidized. Entisols (recent soils) are mineral soils without, or with only the beginnings of, natural horizons. Examples of forested entisols are common: talus slopes, flood plains (alluvium), sand dunes, and shallow bedrock soils. Spodosols (from Greek *spodos*, wood ash) are characterized by acid leaching in cold, temperate climates and reach their best development under boreal forests. Alfisols (from the chemical abbreviations for aluminum and iron) are characterized by less pronounced podzolic leaching and found mostly in humid regions under hardwood forests. Forest soils of the eastern and central United States, as well as those of the Rocky Mountains, the Cascades, the Sierra Nevada Mountains, and the Pacific Northwest are primarily Alfisols. The Ultisols (from Latin *ultimus*, last) are developed by podzolic processes in warm to tropical climates and are more highly weathered and acidic than Alfisols but not as acid as Spodosols. Ultisols are formed on old land surfaces, usually under forest vegetation, and most of the soils of the southeastern United States are of this order. Inceptisols (from Latin *inceptum*, beginning) are young soils not characterized by extreme weathering or major accumulations of clay or iron and aluminum oxides. Prominent among soils of this group are large areas of Oregon, Washington, and Idaho, where productive soils are derived from volcanic ash and loess.

Soil series are the basic taxonomic units used in field classification by the U.S. Soil Survey, and about 10,500 series have been described. In many cases subdivisions or "phases" of a series, based on specific factors such as stoniness, soil depth, or slope, are described and mapped. For example, the soil "Miami loam" is given a series name, Miami, for the Miami River Valley in Ohio, and a type name, loam, describing the texture of the surface soil. All soils of a given series are developed from similar parent material, by similar soil-forming processes, and have horizons of similar arrangement and general features. Thus Miami loam refers to a soil with general characteristics of the Miami series and a loamy surface soil and not necessarily to one with a loamy parent material, or C horizon.

Soil classification systems are developed by soil scientists primarily for describing and mapping agricultural soils. They are the basis of systematic soil surveys of agricultural, urban, and forested lands of the United States. The system in current use provides a taxonomic classification of forest soils but cannot be readily used without interpretation to characterize the productivity of forest stands.

## SUGGESTED READINGS

Armson, K. A. 1977. *Forest Soils.* Univ. Toronto Press, Toronto. 390 pp.

Brady, Nyle C. 1974. *The nature and properties of soils,* 8th ed. (Chapter 3, Some important physical properties of mineral soils, pp. 40–70; Chapter 11, Origin, nature, and classification of parent materials, pp. 277–302; Chapter 12, Soil formation, classification, and survey, pp. 303–352.) The Macmillan Co., New York. 639 pp.

McNeil, Mary. 1964. Lateritic soils. *Am. Scientist* 211:96–102.

Olson, Gerald, W. 1976. Criteria for making and interpreting a soil profile description. *Kansas Geol. Surv. Bull. 212,* Univ. Kansas Publ., Lawrence. 47 pp.

Stålfelt, Martin G. 1972. *Stålfelt's Plant Ecology.* (Chapters 10–14.) Halsted Press, New York. 592 pp.

# 9
# The Nutrient Cycle

The forest and the soil together constitute a system in which each element of the community, both organic and inorganic, nurtures the others and in turn is affected by the others. The basic mineral nutrition of the vegetation is provided by the weathering of the soil minerals, and the soil itself is recharged and changed by the organic products of the vegetation. Other organisms form essential parts of the ecosystem. Bacteria are essential for the fixation of nitrogen, fungi are essential for the absorption of nutrients by tree roots, and a whole complex of soil biota is needed to effect the decomposition of organic debris to a state where it can be utilized over again by the vegetation. The cycling of organic matter, water, and chemical elements is so basic that quantitative ecologists in recent years have developed concepts of the ecosystem based on them (Ovington, 1962, 1968).

To summarize such a complex ecological system requires a simplifying logic based upon a consistent point of view. One such view is the chemical approach, based upon following the movement of the basic mineral nutrients through the cycle from atmosphere and soil to the vegetation and back again. This is the theme of the present chapter. This cycling, however, is dependent in large measure upon the water cycle, which controls the availability of nutrients to tree roots, their rate of movement through the tree, the conditions under which the tree litter is decomposed, and the development of the soil profile which in turn affects the availability of nutrients to the tree roots as recycling is initiated. The water cycle is treated in the following chapter. Both approaches are valid and necessary. Only through an understanding of the ecosystem of the forest, involving both the chemical approach typified by the nutrient cycle and the physical approach illustrated by the water cycle, can the importance of soil factors in the forest environment be evaluated.

## NUTRIENT UPTAKE

Although soil formation is in part a weathering process, it is also greatly affected by the circulation of soluble chemicals through the roots of plants into the stems and foliage; back to the ground in the form of leaves, fruits, twigs and roots; and into soluble compounds again by the decomposition of this litter through the combined action of bacteria, fungi, and soil animals. The relative amounts of the different nutrients taken up into the trees play a large part in determining the relative growth and competitive ability of the different species. Furthermore, the decomposition process will affect (and will also be affected by) the nature of the soil development and through this will exert an influence on succeeding vegetation.

## Nutrient Sources and Cation Exchange

Trees, like all other higher plants, require many chemical elements to live and grow. These include the gaseous elements (H, O, C), the macronutrients (Ca, K, Mg, N, P, S), and the micronutrients (B, Cu, Fe, Mn, Mo, Zn). The elements used by plants are derived ultimately from rock minerals, atmospheric gases, and water. The elements actually used in plant processes may come directly through rock weathering, in precipitation, or in the fixation of nitrogen from the atmosphere. They may also come secondarily processed or recycled through decomposition of organic matter (from aboveground and belowground plant remains), by being washed or leached from plant surfaces (throughfall and stem flow), by dry particulate fallout (dryfall), and by retranslocation within the plant.

Carbon, hydrogen, and oxygen are obtained in large part from water and carbon dioxide. Phosphorus, calcium, potassium, and the other nutrients, except nitrogen, weather from rock material. Nitrogen is fixed from the atmosphere by certain trees, shrubs, herbs, lichens, and bacteria. Nitrogen fixation is the combination of nitrogen gas with oxygen or hydrogen to form either nitrate-N ($NO_3^-$) or ammonium-N ($NH_4^+$).

In the soil, the available nutrients are found adsorbed on colloidal clay and humus particles and in the soil solution as nutrient salts. The positively charged ions or **cations**, including $NH_4^+$, $Ca^{++}$, and $K^+$, are mostly adsorbed on the colloids. The negatively charged ions or **anions**, such as $NO_3^-$, $Cl^-$, and $SO_4^=$, and some of the cations are found in the soil solution.

The minute clay and humus particles, termed **micelles** (microcells), have enormous surface areas and are negatively charged. Thousands of nutrient cations, often termed bases, and hydrogen ions are adsorbed to each colloidal particle. The silicate clay crystal (micelle) of Figure 9.1 illustrates both the external and internal surface area to which cations are adsorbed. Humus colloids are formed and destroyed (decomposed) much faster than the clay colloids, but they have much more capacity per unit weight in adsorbing cations. Thus the clay and humus colloids of forest soils are important "storehouses" for nutrient cations which might otherwise be moved out of the system by percolating water. The colloids are surrounded by films of water (soil solution) and the cations are taken up by plant roots when water is absorbed.

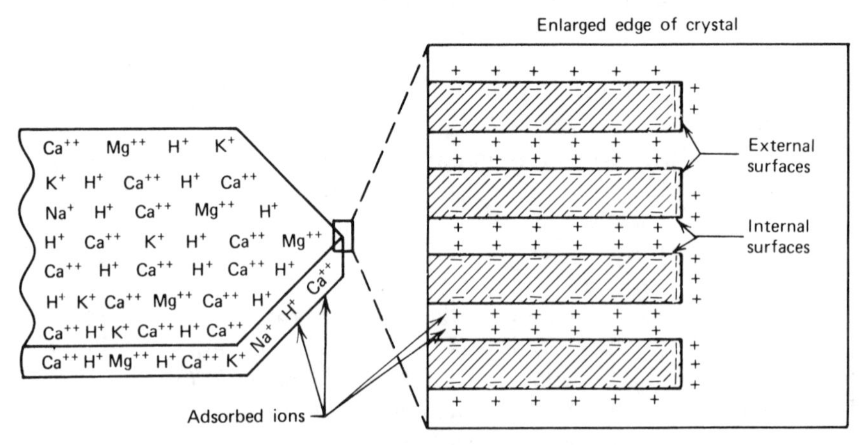


**Figure 9.1.** **Diagrammatic representation of a silicate clay crystal (micelle) with its sheetlike structure, its innumerable negative charges, and its swarm of adsorbed cations. An enlarged schematic view of the edge of the crystal illustrates the negatively charged internal surface of this particular particle, to which cations and water are attracted. Note that each crystal unit has definite mineralogical structure. (Modified after Brady, 1974. Reprinted with permission of Macmillan Publishing Co., Inc. from *The Nature and Properties of Soils,* 8th Edition, by Nyle C. Brady. Copyright © 1974 by Macmillan Publishing Co., Inc.)**

The cations are not all adsorbed with equal strength on the micelle. The order of tightness of adsorption, when they are present in similar amounts, is Al > Ca > Mg > K > Na (Brady, 1974). The hydrogen ion is also tightly held, more tightly than Ca. Cations in the soil solution are able to replace those already adsorbed on colloidal micelles (1) if the entering cation is held more tightly than one already adsorbed, and (2) by mass action. This replacement or **cation exchange** is illustrated by the replacement of one adsorbed ion of calcium by hydrogen ions (Brady, 1974):

$$\text{Ca}\,\boxed{\text{Micelle}} + 2\text{H}^+ \rightleftharpoons {}^{\text{H}}_{\text{H}}\,\boxed{\text{Micelle}} + \text{Ca}^{++}$$

The cation exchange capacity of a soil can be easily determined and is a measure of soil fertility. As expected, sandy soils have very low cation exchange capacities, whereas those of clay loams and clays are very high. Precipitation and organic matter decomposition provide $H^+$ ions which tend to replace less tightly held nutrient cations. The replaced cations move into the soil solution where they may be taken up by plant roots, or where they may be moved out of the system. In regions of high rainfall, they tend to be leached below the root zone and into the aquatic system, where they affect the water quality.

Rainfall, however, is not the only regulator of the leaching of cations; mobile anions in the soil solution also play an important role. In temperate forests of moderately acid soils, carbonic acid is a major leaching agent (McColl and Cole, 1968; Cole et al., 1975). Carbonic acid ($H_2CO_3$) is formed when $CO_2$, respired by soil organisms decomposing organic matter, dissolves in rain water moving into the upper soil layers (Figure 9.2). Carbonic acid then disassociates to the hydrogen ($H^+$) and bicarbonate ($HCO_3^-$) ions. The $H^+$ ion replaces a nutrient cation on a colloidal micelle and the replaced cation associates with a bicarbonate ion. The resulting bicarbonate salt now moves in the soil solution and may be removed from the rooting zone into the ground water. In strongly acid soils, for example cold region spodosols, leaching by the bicarbonate anion is reduced because disassociation of carbonic acid does not occur readily. However, organic acids (primarily fulvic), derived from decomposition of conifer foliage and other vegetation, apparently take over the leaching role played by carbonic acid in less acid soils. The major nutritional significance of this process is that $H^+$ ions may replace nutrient bases on micelles, leaving fewer bases to be absorbed by the root system. Although the nutrient bases in the soil solution may be absorbed, they also may be removed relatively quickly from the profile by percolating water.

## Nutrient Requirements

Nitrogen, phosphorus, potassium (these three constitute the NPK of the agricultural chemist), and calcium (commonly supplied in cultural practices as lime) are the chemicals most often in short supply in the soil. Sulfur, magnesium, and iron are seldom limiting in their availability in the soil. Minor elements are needed in minute amounts, and rarely, a complete absence of boron, manganese, zinc, copper, or molybdenum may cause malnutrition of forest trees. For example, in parts of western and southern Australia, the addition of slight quantities of zinc—even the driving of a galvanized nail into a tree or the erecting of a galvanized rabbit fence—is sufficient to

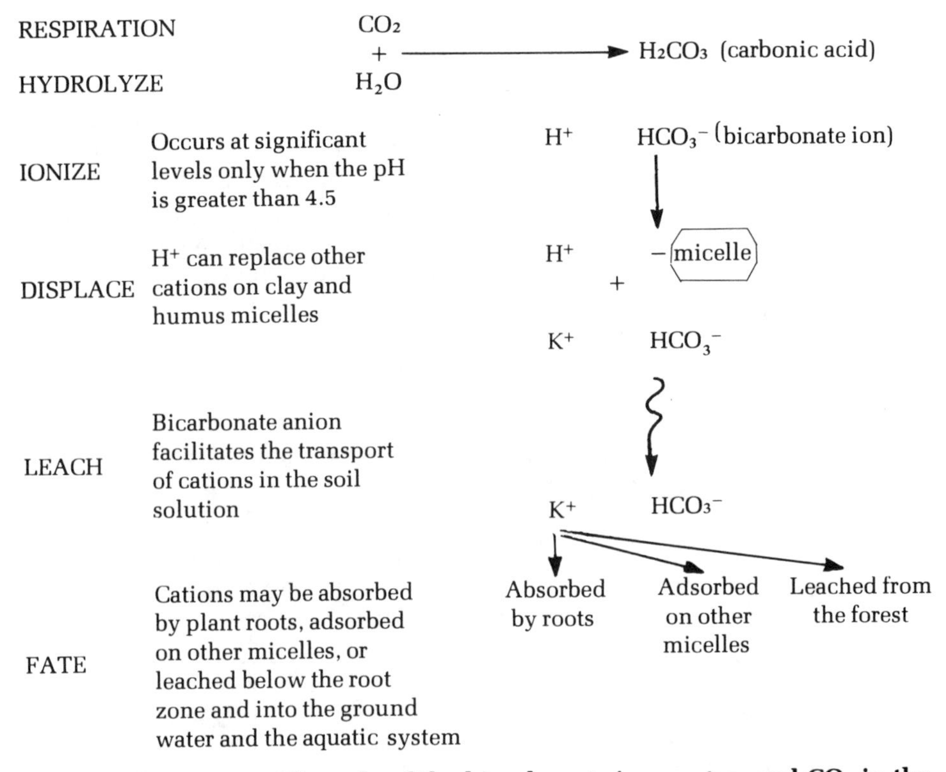


**Figure 9.2. The role of the bicarbonate ion, water, and $CO_2$ in the leaching process of forest soil. (Modified from Cole et al., 1975. Excerpt from *Forest Soils and Forest Land Management,* edited by B. Bernier and C. H. Winget, 1975, published by les Presses de l'Université Laval and reproduced with their permission.)**

correct stagnation and dieback in Monterey pine and *P. pinaster* (Stoate, 1951). Stone (1968) observed that Stoate's finding of a kangaroo skeleton beneath the single green pine in an otherwise deficient plantation has somewhat limited implications for silvicultural practice. Over a wider area in Australia and New Zealand, as well as in flatwoods soils of the southeastern United States, phosphate additions have been demonstrated as essential for normal growth of pines, while additions of lime may be necessary for the establishment of conifers on many acid heaths in western Europe.

For optimum growth, trees require a balanced supply of these various nutrients. Trees do differ from most agricultural crops, though, in growing quite well on relatively small amounts of nutrients. Many forest sites with adequate tree growth test relatively low in nitrogen, phosphorus, potassium, or calcium. Wood has a low ash content and can be produced by trees using only small quantities of nutrients (Table 9.1). The leaves, fine roots, and fruit, wherein much of the mineral ash is concentrated, are annually returned to the soil. As a result, the recycling of a small quantity of mineral nutrients together with those retranslocated within the trees themselves is often sufficient to keep the forest growing well. Conifers in general, and pines in particular (and even some evergreen angiosperms), require markedly fewer nutrients than most deciduous trees. Conifers are typically found on nutrient-poor sites such as sandy soils and in cold, wet climates where nutrients are relatively unavailable due to high acidity and slow decomposition.

**Table 9.1 Nutrient Demands of Forest Compared with Agricultural Crops in Western Europe**

| TYPE OF MANAGEMENT | NUTRIENT REMOVAL FROM SITE DURING 100-YEAR ROTATION OR CROPPING (kg $ha^{-1}$) | | |
|---|---|---|---|
| | Ca | K | P |
| Pines | 502 | 225 | 52 |
| Other conifers | 1082 | 578 | 101 |
| Deciduous trees | 2172 | 556 | 124 |
| Agricultural crops (oats, grass, potatoes, and turnips in rotation) | 2422 | 7413 | 1063 |

Source: Rennie, 1955.

## Accumulation of Nutrients by Trees

By analyzing the mineral content of the tree it is possible to determine the amounts of the various elements that have been taken up. The leaves are particularly responsive to nutrient supply, and foliar analysis has long been popular as a means of assessing soil fertility. In particular, the current year's foliage of the terminal shoot is especially indicative of soil conditions (Leyton and Armson, 1955; Leaf, 1968). For some species and nutrients, however, foliage from other portions of the crown, or other tissues such as bole wood, bole bark, live branches, or buds, may prove of greater diagnostic value (Leaf, 1968). The season of the year must also be considered, since the mineral content of the leaves changes as the season progresses.

Plants tend to take up soluble minerals as they are supplied to the roots. Nevertheless, the amounts of a given element in the leaves at a given time are not clearly related either to the amount available in the soil or to the nutritive requirements of the species. Trees take up what they can and not necessarily what they need. For example, trees are not known to require large amounts of sodium and do not accumulate it in their foliage. However, uptake of sodium is often substantial. The amount of sodium in aboveground biomass of a northern forest ecosystem at Hubbard Brook, New Hampshire, was 1.6 kg $ha^{-1}$, compared to 351 kg $ha^{-1}$ for nitrogen and 383 kg $ha^{-1}$ for calcium (Likens et al., 1977). Although annual uptake of sodium by the vegetation was considerable (34.9 kg $ha^{-1}$), 98 percent was returned to the soil in root exudates. In contrast, only 1 percent of the nitrogen and 6 percent of calcium uptake was returned by root exudates.

Nevertheless, there have been many comparisons between the mineral contents of the foliage of trees of the same species growing on different soils. In a study of white oak growing on different soil types in Illinois (McVickar, 1949), differences in the chemical composition of the leaf were significant only when trees growing on the poorest soils were compared to those growing on the best. Again, in New York State, when several hardwood species were compared on three soil types of varying limestone content, the calcium content of the foliage was found to be more dependent upon the inherent capacity of each species to absorb the nutrient rather than upon the calcium level of the soil itself (Bard, 1946).

The relative ability of different species to absorb nutrients has been studied in many parts of the world, particularly with reference to calcium uptake. In the northeastern United States, basswood, yellow-poplar, dogwood, and red cedar are among the trees that concentrate large amounts of calcium (and also phosphorus and potassium) in their foliage; while beech, red spruce, the pines, and hemlock are low in their uptake.

By putting together data for the mineral content of roots, stem wood, bark, branches, leaves, and fruit, it is possible to determine the annual uptake of nutrients by forest trees. The accumulation and distribution of nutrients in three major kinds of European forests, deciduous, nonpine coniferous, and pine, show that calcium is always accumulated in the greatest quantities, especially in deciduous forests (Figure 9.3; Rennie, 1955). The total calcium uptake for the three forests in 100 years was 2170, 1080, and 500 kg $ha^{-1}$. Pines have lower requirements than the other species. The magnitude of accumulation by the three types seems closely related to the site conditions on which each naturally grows, deciduous forests on the more fertile sites and pines on infertile sites.

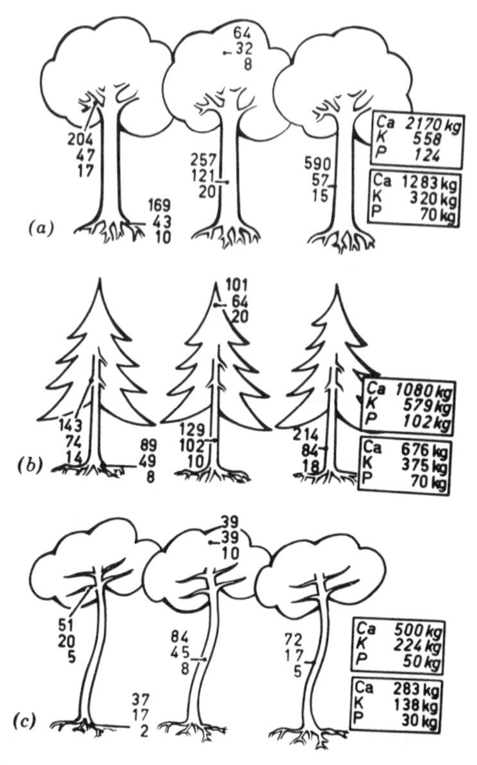


**Figure 9.3. Distribution (kg ha$^{-1}$) of calcium, potassium, and phosphorus in three major types of exploited forests. Total content in standing crop of 100-year-old forests (normal type, bottom box) and total uptake (including output by thinnings) of the same elements over the century (italics, top box). Values in boxes are for the total tree stand; various subtotals for leaves, branches, bark, wood, and roots are shown on the diagrams. (a) Deciduous hardwoods; (b) Conifers other than pines; (c) Pines. (Diagram from Duvigneaud and Denaeyer–De Smet, 1970; data after Rennie, 1955.)**

## Effect on Tree Growth

When nondemanding trees are grown on their natural sites, little correlation is generally found between their nutrient uptake, as assessed by chemical analysis of the foliage, and their growth rate. For example, a Finnish study by Aaltonen (1950) showed no correlation between the chemical constituents of Scots pine and Norway spruce needles and four forest sites of different fertilities from which they were collected, and upon which these species demonstrated markedly different growth rates.

With more demanding species, however, good correlations may be found. In the same Finnish study, European white birch leaves showed higher calcium and potassium contents (but lower phosphate contents) on the better sites than on the poorer. In a series of studies of coniferous plantations on poor heathland sites in Great Britain, Leyton (1955) found high correlation between the mineral content (nitrogen, phosphorus, and potassium) of the current year's foliage in the terminal shoot and the growth rate of the tree. On these degraded soils, Sitka spruce and Japanese larch seem to be more affected by the low mineral content of the soil than Corsican or Scots pine.

The differential nitrogen requirements of hardwood species native to northeastern and north central United States and Canada were demonstrated by Mitchell and Chandler (1939) in fertilizer experiments. Some species (red and white oak, red maple, and trembling aspen) were tolerant of nitrogen deficient soils (termed "N-tolerant" species) and grew poorly at the lowest levels of nitrogen in the soil. They responded well as N supply increased, but only up to a point where their growth leveled off. In contrast, "N-demanding" species (white ash, basswood, and yellow-poplar) grew even more poorly at the lowest N levels, but showed a much greater response to high levels of nitrogen supply. As borne out by field observations and site index comparisons (Stone, 1973), the N-tolerant species are typically superior competitors at low N sites and the N-demanding species superior on sites with abundant available N.

The effect of nutrients on tree growth has been intensively studied. Reviews of the nutrient requirements of forest stands (Tamm, 1964), the mineral nutrition of conifers (Morrison, 1974), the deficiencies of potassium, manganese, and sulfur in forest trees (Leaf, 1968), and the microelement nutrition of trees (Fortescue and Marten, 1970; Stone, 1968) provide entry into the vast literature. Besides foliar analysis, other major methods of diagnosing the nutrient status of forests include experimental field, greenhouse, or hydroponic studies (Ingestad, 1962), visual deficiency symptoms (Hacskaylo et al., 1969), and soil analyses (Tamm, 1964; Leaf, 1968). Despite the work involved, use of all the techniques may be justified in view of the large investments in forest fertilization that may be made on the strength of the diagnosis.

A useful analysis in field surveys to determine the nutrient status of stands involves sampling and analyzing the upper 4 cm of soil after removing the litter ($0_1$ horizon) (Evers, 1967). Since this soil zone is strongly influenced by the litter, this technique provides in effect a combination of soil and foliage analyses. Nutrient ratios, particularly C/P, C/K, and C/N, are then computed to describe the nutrient status of the stand. Ratios of C/P = 112, C/K = 92, and C/N = 20 characterized a Norway spruce stand of maximum growth in southwestern Germany, whereas nutrient disorders can be expected in pure and mixed stands of spruce with ratios higher than C/P = 400, C/K = 450, and C/N = 25. The method has proved inexpensive and simple, having no seasonal restrictions on sampling as with foliar analysis. Furthermore, the ratios are highly correlated with growth rate (Evers, 1971) and with the amount, availability, and cycling of nutrients (Evers, 1968).

## Soil Acidity

The measurement of the available nutrient content of the soil or of the plant is a long and tedious operation. In the soil, the problem is further complicated by the fact that only those nutrient ions in solution or capable of moving from colloids into solution can be absorbed by the plant. Much of the chemical content of the soil is held too tightly by the soil particles to be released to plant roots. Furthermore, the nutrient availability of the soil is constantly changing. Changes in the water content of the soil, the removal of nutrients by plant roots, and the chemical activities of soil organisms all constantly change the base-exchange balance.

To avoid both the tedious chemical analysis and the complexities of actual soil chemistry relationships, many workers with forest soils have emphasized the importance of soil acidity (Small, 1954). Acidity is measured in terms of pH, which depends on the concentration of $H^+$ and $OH^-$ ions in the soil solution. pH values are determined by taking the logarithm of the recriprocal of the $H^+$ ion concentration. Although the pH scale is expressed in whole numbers and extends from about 3 (extremely acid) through 7 (neutral) to about 11 (extremely basic), the logarithmic scale indicates a 10-fold difference in acidity between each integer. For example, a soil solution of pH 6 is 10 times more acid than that of pH 7, and pH 5 is 100 times more acid than pH 7.

Forest soils normally range from very acid, pH 4, up to slightly acid, pH 6.5. Only on calcium-rich (calcareous) soils does forest litter render soils neutral, pH 7, or slightly basic, pH 7.5. The pH can be measured with reasonable accuracy in the field with a portable potentiometer or can be approximated with simple colorimetric tests. Many efforts have been made to relate soil pH to the distribution and growth of trees and other plants. Certain species are typically associated with highly acidic soils (black spruce, Norway spruce, hemlocks, rhododendrons), whereas others thrive on basic soils (northern white cedar, eastern red cedar, beech, rock elm). However, in general tree distribution is more closely related to macroclimate and moisture conditions than to pH. Often it is competition among species that relegates some to extremely acid soils and others to basic soils.

Actually, soil acidity is far too complicated to be measured by a simple pH scale, and this measure of acidity must be accepted merely as an empirical number which provides a rough index to certain of the chemical properties of the soil. Approximate relationships may therefore be expected between soil pH and plant behavior, but cause-and-effect conclusions should not be drawn. The soil pH reflects both the chemical properties of the mineral soil as well as the chemical properties of the organic residue deposited as litter. Optimum growth of a tree may be associated with a narrow range in acidity. For instance, Sitka spruce in England has been demonstrated to grow better between pH 4 and 5 than under more acid or more alkaline conditions (Leyton, 1952). In general, lime-loving plants such as yellow-poplar and northern white cedar in the United States are found in neutral or alkaline soils, while acid-tolerant plants such as balsam fir and eastern hemlock are usually confined to highly acidic soils. Under actual soil nutrient conditions, however, almost any species can be grown successfully under a wide range of soil pH conditions (Wilde, 1954).

The forest tree itself influences soil acidity. The general trend is for conifers such as the pines, spruces, hemlocks and Douglas-fir to intensify the increased acidity of the upper soil to a greater extent than hardwoods or cedars. Individual species, however, vary widely in their effect. An example is provided by soil pH on old fields in Connecticut where eastern red cedar and common juniper were growing side by side (Spurr, 1941). Under the red cedar, soil pH was lowered in the root zone and raised in the topsoil where litter accumulated, indicating the high capacity of this species to remove calcium and other bases from the soil. In contrast, the juniper foliage increased the acidity of the topsoil, apparently because of the low base content of its foliage. Similar results have been reported elsewhere for other species.

## Mycorrhizae

When seedlings of forest tree species are grown in water cultures, root hairs develop. In the forest, however, root hairs are less conspicuous. Instead, the roots of forest trees are commonly ingrown with certain species of soil fungi. The association of plant root tissues and fungal mycelia is known as a **mycorrhiza** (meaning a fungus-root organ). In pines, spruces, firs (and all other genera of the Pinaceae), birches, beeches, oaks, basswoods, and willows, **ecto-mycorrhizal** fungi form a sheath or mat surrounding the rootlets, giving them a characteristic swollen appearance. The hyphae penetrate between the outer root cortical cells but do not enter the cells. Over 2400 species of fungi are known to form ectomycorrhizae on North American trees (Marx and Beattie, 1977). Ectomycorrhizae of loblolly pine, Douglas-fir, and western hemlock are illustrated in Figure 9.4.

A less conspicuous group, **endomycorrhizal** fungi, form no sheath, but the filaments grow within and between epidermal and cortical cells and extend into the soil. Although many species may exist, only about 20 endomycorrhizae have been identified. They are associated with various cultivated crops, grasses, and trees, including redwood and many hardwoods: maples, ashes, yellow-poplar, sweet gum, sycamore, black walnut, and black cherry (Harley, 1969; Marx and Beattie, 1977).

The presence of mycorrhizae is essential to successful growth of many species, particularly on drought and nutrient-poor sites. Mycorrhizae tend not to form in trees growing in nutrient-rich soil, even when the soil is inoculated with the fungus (Harley, 1969). Pines typically grow on nutrient-poor sites, and they have an obligate requirement for ectomycorrhizae; they do not grow normally without them. In many areas of the world, ectomycorrhizal trees and their associated fungi do not occur naturally (Marx, 1975), including the former treeless areas of the United States (Hatch, 1937). In these areas, afforestation attempts were either total or near failures until ectomycorrhizal infections occurred. In the case of many nurseries situated on nonforest soils and of many plantations situated on nonforest land, it has proved essential to introduce forest soil or litter infected with mycorrhizal-forming fungi. This has been necessary for all the extensive plantations of Monterey pine and other northern conifers in the Southern Hemisphere.

A true symbiotic relationship exists between tree feeder roots and the mycorrhizae. It is exceedingly complex and is as yet poorly understood. Whatever the mechanism may be, it is quite clear that the relationship is mutually beneficial (Kelley, 1950; Harley, 1969). A major benefit of mycorrhizae to forest trees is that they are highly efficient accumulators of nutrient ions and water in the **rhizosphere**, the zone immediately surrounding the root. The nutrients and water become available to the host tree and to the fungus. The ability of mycorrhizae to supply nitrogen to trees from nitrogen-poor soils has been heavily stressed. Investigations of beech in England (Harley, 1950 et seq.) and of Monterey pine in California (Stone and McAuliffe, 1954) also suggest that much of the benefit is due to the

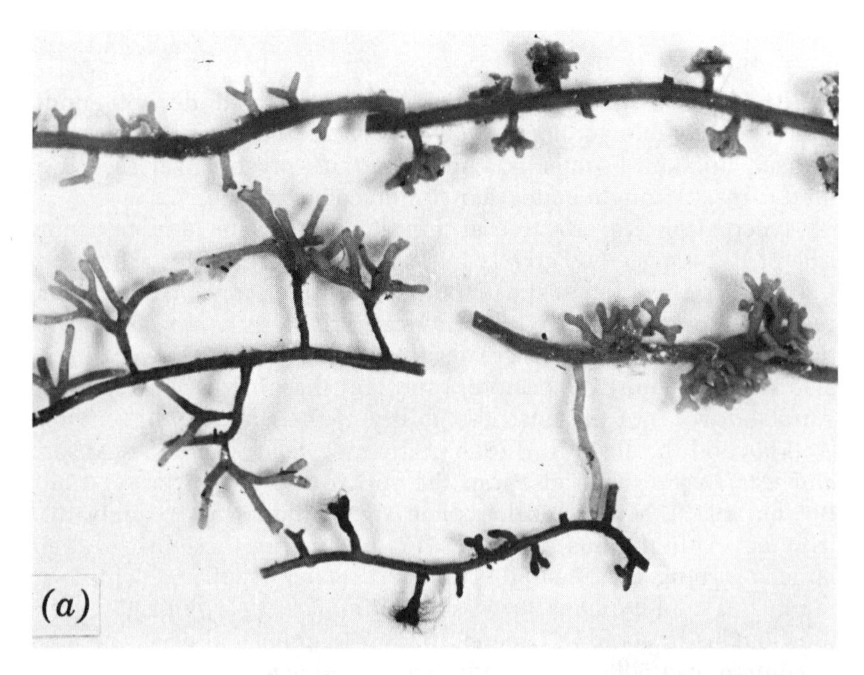

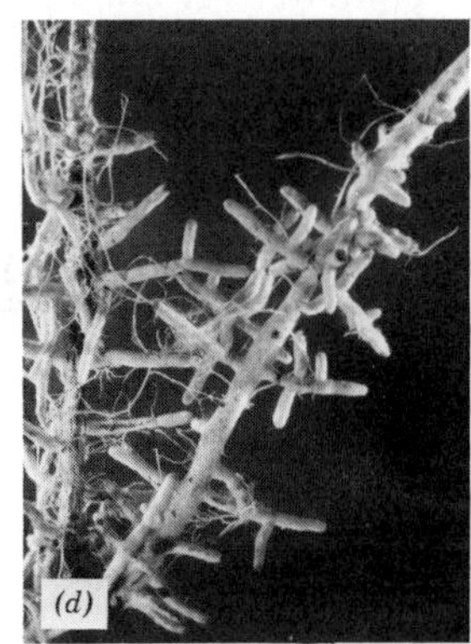

Figure 9.4. Ectomycorrhizae of loblolly pine, Douglas-fir, and western hemlock.

*(a)*—Five morphologically different forms of ectomycorrhizae of loblolly pine.

*(b)*—Smooth-mantled Douglas-fir ectomycorrhiza formed with *Lactarius sanguifluus.*

*(c)*—Douglas-fir ectomycorrhiza formed with *Corticium bicolor* with a dense covering of mycelium and strands of aggregated hyphae.

*(d)*—Ectomycorrhiza of western hemlock formed with *Byssoporia terrestris.*

(U.S. Forest Service Photos; *(a)* Courtesy of Donald H. Marx. *(b,c,d)* Courtesy of James M. Trappe.)

ability of mycorrhizal trees to take up phosphorus under soil conditions unfavorable to non-infected trees of the same species. Increased uptake of water is also important; mycorrhizal seedlings tend to resist drought better than nonmycorrhizal ones.

Mycorrhizae are effective accumulators of water and nutrients first of all because they greatly increase the absorbing surface area of the root system of the host. As compared with nonmycorrhizal roots, roots with mycorrhizae exhibit increased rootlet size, live and function longer, and respire at greater rates (Harley, 1969). Mycorrhizae also share a feature even more important than increased absorbing surface of roots of the host: the ability of their hyphae to function well beyond the absorbing zone of the root itself where they absorb and translocate materials from the soil to the host (Harley, 1969; Bowen, 1973). Mycorrhizal hyphae are abundant and widely distributed in the humus and soil, thus enhancing the effect of their large absorbing area. A single cubic millimeter of soil can contain as much as 4 m of hyphae (Burgess and Nicholas, 1961). In a 450-year-old Douglas-fir stand in Oregon, the top 10 cm of soil was estimated to contain over 5000 kg $ha^{-1}$ (dry weight) of ectomycorrhizae, about 11 percent of the total root biomass (Trappe and Fogel, 1977). Hyphae can fuse and create large mycorrhizal networks greatly increasing absorption and permitting translocation of nutrients and carbon between host trees sharing a mycelial system. For example, a single *Cenococcum* mycorrhiza can form from 200 to over 2,000 individual hyphae (Trappe and Fogel, 1977). A single hyphae emerging from a Douglas-fir + *Cenococcum* mycorrhiza in a rotten log extended over 2 m, including more than 120 lateral branches or fusions with other hyphae. At least 43 of the branches connected to other mycorrhizae, 34 to mycorrhizae on the same tree, and 9 to mycorrhizae of a western hemlock growing in the same log.

Mycorrhizae are important in nutrient cycling. They absorb and translocate nutrient elements and serve as reservoirs for nutrients that might otherwise be leached from the soil. Their fruiting bodies, sporocarps, contain substantial amounts of nutrient elements. They decay quickly or are eaten by insects or larger animals. Hence they provide an ephemeral but concentrated source of nutrients for soil organisms that are decomposing organic matter and for consumer animals. Mycorrhizae are not known to fix nitrogen, but they are associated with and may stimulate other organisms that fix nitrogen. They also can convert complex organic substances and minerals into usable nutrients.

Mycorrhizae are also beneficial in that they protect the host tree from root-rot pathogens by their antibiotic exudates (Marx, 1969 et seq.; Marx, 1973). The fungal sheath physically bars invading parasites, and the antibiotics deter root pathogens and the attack of root aphids (Zak, 1965).

Because the many benefits of mycorrhizae are known to increase the ability of trees to withstand drought and adverse soil conditions, research has been directed to developing methods of inoculating seedlings with mycorrhizae. This is accomplished by artificially introducing mycorrhizae into nursery soils or in containerized potting mixes (Mikola, 1970; Marx et al., 1976, 1977; Trappe, 1977). For example, pines infected with the ectomycorrhizal fungus *Pisolithus tinctorius* may be expected to exhibit increased tolerance to extremes of soil acidity, high sulfur and aluminum levels, and high soil temperatures of strip-mined land (Schramm, 1966; Marx 1975). *Pisolithus tinctorius* has nearly a world-wide distribution, occurring in 33 countries representing six continents. It is found on severely eroded soils and strip-mined coal spoils as well as in urban, orchard, and forest environments (Marx, 1977). It probably has the capacity to form ectomycorrhizae on the roots of most of the tree species in the world that form ectomycorrhizae. Thus with just this species and the techniques of inoculation currently available, there exists a tremendous potential for improving the survival and growth of many of the world's trees, especially in reforesting adverse sites. Significant advances also seem possible for growing hardwoods such as red maple, black locust, and alder on soils of low fertility and anthracite spoils using endomycorrhizae (Daft and Hacskaylo, 1976, 1977).

Excellent books on mycorrhizae are available (Harley, 1969; Marks and Kozlowski, 1973), and important review papers have appeared (Marx, 1975; Trappe, 1977; Trappe and Fogel, 1977).

## Fertilization of Forest Soils

Assuming that a forest soil contains mycorrhiza-forming fungi, growth may in many cases be stimulated by adding the necessary fertilizers. Occasionally, fertilization may be essential for survival of trees in nutrient-deficient soils, as in the case of adding lime to certain heath soils being reforested in England and super-phosphate or zinc to deficient soils in South Australia and being afforested to Monterey pine. More commonly, application of standard NPK (nitrogen, phosphorus, and potassium) fertilizers and lime in needed quantities and proportions will stimulate growth. Nitrogen has proved the most important single element generally in short supply in forest soils, and consequently addition of nitrogen fertilizers almost always stimulates forest growth.

In the past, fertilization of forest stands has been given only passing attention. However, under the economic conditions of the 1960s and early 1970s, which were more favorable for intensive forest management, fertilization became more practicable for general use. It was carried out increasingly in the Pacific Northwest (Gessel, 1968; Steinbrenner, 1968; Miller et al., 1976), the South (Pritchett and Smith, 1975), and in the northern forests of the United States and Canada (Northeastern Forest Exp. Sta., 1973; Canadian Forest Serv., 1974; Leaf et al., 1975; Armson et al., 1975). Fertilization became operational in the Pacific Northwest where approximately 250,000 ha were fertilized between 1966 and 1975 (Miller et al., 1976). In the South, fertilization is an accepted management practice where lack of one or more nutrients seriously curtails pine growth and where moderate additions of such nutrients markedly increase yields (Shoulders and McKee, 1973).

Fertilizers are used in hastening forest establishment, increasing forest growth in conjunction with thinning, stimulating seed production in tree seed orchards, and countering depletion of nutrients caused by intensive harvesting under short rotations. They are also used to improve food and cover for wildlife, in the production of specialty crops such as Christmas trees and maple sugar products, and to protect, repair, and enhance vegetative cover on exposed soil, burned or degraded sites, and heavily used recreation areas. Because of the increasing cost of fertilizers, priorities for their use will come into much sharper focus and research will intensify in determining the amounts, kinds, and schedules of fertilizer application for individual soils and species.

A significant feature of forest fertilization is its complex effects on the forest ecosystem and the adjacent aquatic ecosystems of streams and lakes (Hilmon and Douglass, 1968). The effects of forest fertilization on water quality are under study. For example, applications of nitrogen in the Douglas-fir region of western Washington and Oregon cause increases of nitrogen in stream water shortly after application, but peak concentrations of various soluble forms of N do not approach toxic levels (Fredriksen et al., 1975). The stream bank vegetation and stream organisms rapidly utilize the nitrogen entering forest streams and downstream movement further decreases the concentration. In upland forests, careful application of fertilizers (particularly avoiding direct aerial application to streams) apparently does not constitute a pollution hazard if water containing nutrients reaches the streams by moving through the soil.

## NUTRIENT INPUT, ACCUMULATION, AND RETURN

The minerals that are taken up into forest trees are eventually returned to the forest soil except for the amount carried out of the forest in logs and other forest products. Minerals are returned to the surface of the soil by litterfall and through the washing and leaching effects of rain on tree foliage and stems. Minerals are also added to the soil by rainfall and dryfall and by the belowground dying and sloughing of roots. On the forest floor and in the soil, a myriad of mammals, insects and other arthropods, earthworms, fungi, and bacteria attack the accumulating organic material, decompose it, and render it reavailable for plant nutrition.

## Litter

Litter is the organic remains of plants and animals that is found either on the soil surface or buried in the mineral soil itself. This includes large woody debris, such as whole tree trunks, in addition to the leaves, twigs, fruits, and the like, that may occur in abundance on the soil surface. Belowground litter includes dead roots, large and fine. Roots decompose and are an important source of nutrients although we have tended to think only in terms of the surface litter layer. A comprehensive treatment of plant litter, and the processes and organisms involved in its decomposition, is available in the two-volume set by Dickinson and Pugh (1974).

**Litterfall.** Leaves, small twigs, bark, and fruits add from 1,500 to 5,000 kilograms of oven-dry organic material to the surface of a fully stocked hectare in a single year. Leaf litter accounts for roughly 70 percent of the total (Bray and Gorham, 1964). Under open conifer stands the weight of litterfall may drop below 1000 kg $ha^{-1}$, and in the tropical rain forest as much as 10,000 kg may be accumulated (Lutz and Chandler, 1946; Bray and Gorham, 1964).

Climate exerts a predominant effect on litterfall as seen in the annual production figures (kilograms per hectare) in 4 major climatic zones: arctic-alpine—900, cool temperate—3100, warm temperate—4900, and equatorial—9700 (Bray and Gorham, 1964).

Considering a worldwide range of sites, total litter production of evergreen forests exceeds that in deciduous forests by about 13 percent (Bray and Gorham, 1964). However, considerable variation exists for specific areas. European beech in Germany, for example, shows consistently greater litter production than pine or spruce, apparently reflecting the predominance of beech on the more productive soils. Substantial annual fluctuations in leaffall are possible under evergreen forests of temperate climates. Although evergreen conifers typically hold their needles for two years or more, the oldest needles normally drop off each year. Under the influence of drought or other unfavorable conditions several years' accumulation of leaves may be dropped.

**Understory Litter.** Understory vegetation plays an important role in the circulation of nutrients that often has been ignored. Its contribution tends to be strongest in the early and late stages of stand development, when the amount of light reaching the understory is greatest. Under relatively open conditions understory vegetation may contribute up to 28 percent of the total litter (Bray and Gorham, 1964). Under white pine and mixed hardwood forests in Connecticut, Scott (1955) found that subordinate vegetation accounted for about 15 percent of the annual weight of the litter. The shrubs and herbs, however, contained higher percentages of many nutrient elements than did the tree foliage, so that as much as one-quarter of the annual return of nutrients to the soil came through the lesser plants. In an English sessile oak woodland, bracken fern (*Pteridium aquilinum*) contributed in both its litter and rainfall leachates nearly one-third of the total potassium reaching the soil annually (Carlisle et al., 1967).

**Large Woody Debris.** Large woody debris, such as decaying tree trunks and stumps, constitute a major component of the organic matter of the forest floor. In yellow birch–red spruce forests of the Adirondack Region of New York, such residue accounted for about 18 percent of the humus layer in a stand over 300 years old and 26 percent in a stand 135 years old (McFee and Stone, 1966). This residue of red spruce, yellow birch, white pine, hemlock, and balsam fir is that which remains after initial decay by brown rot fungi. White rot fungi decompose and remove the sapwood of these species and generally destroy the wood of hardwoods such as beech, red and sugar maple, and paper birch. Decayed wood is an important substrate for the establishment of yellow birch and coniferous seedlings because of its high moisture content and the ability of these species to extract nutrients that are available in the wood and bark.

In old-growth conifer forests of the Pacific Northwest, large amounts of woody debris litter the forest floor and the small streams (Figure 9.5; Swanson et al., 1976; Franklin et al., 1979; Chapter 18). Decomposition of large trunks is extremely slow, and the decaying material provides a reservoir of nutrients that slowly releases the nutrients over time. Nitrogen is known to be fixed by bacteria associated with fungi in decaying parts of living trees (Aho et al., 1974), and it also is fixed in woody debris on the forest floor (Sharp, 1975). Although litterfall of foliage is the most important source of recycled nutrients in the forest floor of young or intensively managed plantations, its role probably diminishes with increasing stand age while that of woody debris increases.

**Figure 9.5.** **Large masses of decaying logs are a dominant feature of old-growth forests. This 450-year-old stand of Douglas-fir, western hemlock, and Pacific silver fir (900 m, H. J. Andrews Experimental Forest, Oregon) illustrates maximal accumulations. (U.S. Forest Service Photo. Courtesy of Jerry F. Franklin.)**

**Belowground Litter.** Another kind of litter is the root material that dies each year and decays in the soil. Because the amount of root mortality is difficult to determine, especially the abundant fine roots $< 0.5$ cm, the significance of its contribution is just becoming appreciated. In a yellow-poplar forest in Tennessee, annual root mortality amounted to 9000 kg $ha^{-1}$ (Harris et al., 1977). This is about 75 percent of the total annual return of organic matter to the soil from both aboveground (litterfall) and belowground (root mortality) sources. Of the total annual return of nitrogen to the soil by litterfall and root mortality (95 kg $ha^{-1}$), root mortality contributes 67 kg $ha^{-1}$ or 70 percent (Burgess and O'Neill, 1975). Such root matter, occurring primarily in the upper 30 cm of soil, acts as a substrate for soil organisms, aerates the soil, holds moisture, and may contribute significant amounts of nutrients to the ecosystem. However, tree species vary in the amount of annual root mortality. White oak, for example, appears to exhibit significantly less annual root mortality than yellow-poplar.

## Rainfall and Dryfall

Atmospheric moisture plays a very important part in adding to the supply of soil nutrients. Nutrient input to the ecosystem is measured as the dissolved form in incident precipitation (**wetfall**) and as dry particulate fallout (**dryfall**) or as a combination of both, termed **bulk precipitation**. Precipitation of moisture dissolves nitrogen and precipitates dust and gaseous pollutants directly from the atmosphere. About 5 kg of nitrogen per hectare are added annually in the Temperate Zone, although the range is typically from about 1 to 10 kg $ha^{-1}$.

In addition, rain washes materials from surfaces of plants and leaches from foliage substantial quantities of mineral nutrients that may be rapidly recycled through ecosystems. Moisture dripping through the crown canopy is termed **throughfall**, and that flowing down the stems is termed **stem flow**. Washed-off materials include impacted airborne dust particles and aerosols, pollen, natural exudates, and materials released by the activities of damaging animals.

Nutrients leached most easily from foliage include potassium, sodium, calcium, and magnesium (Turkey, 1962). Unlike most nutrients that are taken up and incorporated into organic compounds, potassium remains in ionic form and is thus more easily leached from plant surfaces. In a mixed hardwood and natural loblolly pine stand in the North Carolina Piedmont, more potassium was contributed to the forest floor by throughfall than by litterfall (Wells et al., 1972).

In England, annual throughfall (in kg $ha^{-1}$) was found to have from two to eight times more sodium, potassium, calcium, and magnesium than rain water collected in adjacent open locations (Madgwick and Ovington, 1959). The amounts of these nutrients contained in the throughfall were the same as or greater than the amounts of these nutrients taken up permanently in the tree crops. Phosphorus was present in rainwater only in very small amounts; a similar finding has been reported in various North American regions (Wells, et al., 1972; Verry and Timmons, 1977; Likens et al., 1977). In an English oak forest, the contribution of precipitation (including throughfall, stem flow, and incident rainfall) to the total of nutrients reaching the soil exceeded that of litterfall for potassium, magnesium, and sodium (Carlisle et al., 1967).

Dryfall from windborne dust, pollen, and fly ash from forest fires and coal-burning factories is also a source of nutrients. Depending on the nutrient, from 5 to 30 percent of the atmospheric input normally may come from dryfall. However, at sites near coal-burning plants, the contribution may be much higher. At the Walker Branch watershed near Oak Ridge, Tennessee (site of electric generating plants), dryfall accounted for 28 to 64 percent of the cation input to the system (Swank and Henderson, 1976). Airborne dust, ash, and gaseous aerosols may also become attached to or impacted on tree surfaces and carried to the soil as throughfall and stem flow. The magnitude and significance of impacted nutrients, compared to that of ions leached from the plant surfaces, are poorly understood.

### Acid Precipitation

Special mention must be made of the increasing acidity of precipitation over the past 20 or more years. Normally, precipitation is acid because $CO_2$ gas dissolves in water to produce a slightly acidic solution having a pH of about 5.6. Acid precipitation (rain and snow with pH values less than 5.6) is caused primarily by strong acids such as sulfuric, nitric, and hydrochloric. It originates mainly from human-made pollutants (Galloway et al., 1976; Likens et al., 1977). Southern Norway and Sweden are severely affected by acids derived from industrial areas in England and Central Europe (Odén, 1976). Large emission sources can cause significant effects 800 to 1600 km away. Since the mid-1950s, acidity has greatly increased, and acidification of rivers and lakes has brought about a dramatic decline and disappearance of some fish populations (Wright et al., 1976).

In North America, high acidity was widespread in the northeastern United States and adjacent Canada by the mid-1950s, and the mean annual pH of precipitation now averages between 4.0 and 4.2 (Likens and Borman, 1974; Cogbill and Likens, 1974; Likens, 1976). Even in rural New Hampshire at the Hubbard Brook Experimental Forest, over 100 km from any industrial areas, annual pH values from 1964–1974 averaged just over 4 (Likens et al., 1977). Noticeable short term effects on forests are much less than on aquatic ecosystems where increases in pH over a critical level stops fish reproduction. Forested ecosystems effectively buffer acid precipitation and serve to protect rivers and lakes. In doing so, however, forests receive acid rain and snow that can cause increased leaching of substances from foliage, affect mycorrhizae and other soil organisms, increase leaching of nutrients from the soil, and possibly affect the important processes of flowering, fertilization, and seed germination. The effects of acid precipitation on the growth and composition of forests are not immediately evident due to the complexity of the forest ecosystem. Thus studies of entire ecosystems (Chapter 18) over time are needed to monitor the cumulative effects of acid precipitation. The proceedings of an international symposium (Dochinger and Seliga, 1976) provide a comprehensive examination of the problem.

## The Forest Floor

The leaffall and other litter gradually accumulate on the forest floor until decomposition begins. Initially litterfall may exceed decomposition but, sooner or later, an equilibrium will be reached between the yearly additions of organic matter and the yearly rate of decomposition. Later, in old, open forests litter decomposition may exceed litterfall, and the surface layer becomes less. Under optimum conditions for soil biotic activity—with the forest floor being warm, moist, and well-aerated for much of the year—decomposition will keep pace with additions and no organic matter is accumulated. Where soil biotic activity is inhibited by cold, acid conditions, insufficient moisture, or insufficient oxygen, however, litter will accumulate indefinitely. Thick accumulations of raw (i.e., little decomposed) humus, peat, muck, and, eventually, coal beds thus arise. The amount of time for the forest floor to reach near-equilibrium conditions of organic matter accumulation ranges from less than 10 years in fast-growing tropical forests to well over 100 years in the conifer forests of the western United States.

Under such conditions, the forest floor will normally contain at any one time the equivalent of several years' litterfall. Under upland oak forests in eastern Tennessee, where conditions are favorable for litter decomposition, the annual leaffall in one stand was found to be 2,915 kg $ha^{-1}$ as compared to a forest floor which contained 12,100 kg of organic matter after leaffall in the early winter and 9,420 kg at the end of the summer season of active decomposition (Blow, 1955). In a 27-year-old red pine plantation in Connecticut, annual needlefall was 4,030 kg $ha^{-1}$ compared to the forest floor, which contained 31,170 kg $ha^{-1}$.

In contrast to eastern American forests, where the forest floor seldom contains more than the equivalent of 5 to 10 years' leaffall, the coniferous forests of the Pacific Coast may hold as much as the litterfall of 50 years. When the woody material from tree crowns is considered, the total litter production in Douglas-fir forests increases with stand age and no equilibrium is reached (Long, 1980) unless periodic fires consume some of it. The woody component of total litter increases from about 6 percent in 20-year old stands to over 50 or 60 percent in stands over 70 years as large logs are added to the forest floor. The heavy accumulations of organic matter under the mixed coniferous forests of California and the Pacific Northwest result from the summer drought, which inhibits soil biotic activity. Most decomposition takes place in the wet and relatively mild "dormant" season.

In Britain, conifer forests frequently develop a thick layer of organic matter, about three to five times the annual litterfall, and Ovington (1962) assumes the annual litterfall takes 3 to 5 years to decay. In deciduous forests on similar sites there is often little accumulation from year to year. Decomposition is frequently complete within 12 months and often virtually complete in 6 to 9 months.

### The Soil Biota

The surface of the forest soil supports one of the richest faunas and floras of all ecological niches—both in terms of numbers of species and their weight per unit volume of space. Plant litter forms the basic food supply, but interactions between plants that carry on photosynthesis, saprophytes that live on dead organic material, parasites, and predators are exceedingly complex. In a few paragraphs we can only indicate the complexity and some of the principal types of organisms responsible for the decomposition of organic matter. From the standpoint of forest site, their importance is their combined effect on the reduction of the litter into soluble compounds that can be taken up by the roots of the forest.

Under favorable soil conditions characterized by sufficient warmth, moisture, and oxygen, bacteria are not only the most numerous of all soil organisms but are also the heaviest. The bacteria inhabiting forest soil may weigh as much as 1680 kg $ha^{-1}$. In such quantities they are capable of processing much of the organic matter that is added annually to the soil. Bacteria are more abundant where other soil animals, particularly earthworms, are present. The main role of bacteria in decomposition is that of further breakdown of materials already ingested and excreted by macroorganisms. Soils worked by earthworms and millipedes contain about two-thirds more bacteria than unworked soils (Kollmansperger, 1956; Went, 1963). Acid soils limit bacterial populations probably because acid soils limit macro-organisms such as earthworms.

The importance of fungi in forming mycorrhizae has already been discussed. The same fungi and many others are present in the soil in large amounts. Under conditions unfavorable to bacterial growth, fungi are better able to survive, grow, and become the dominant element of soil life. With the cool, wet, acid conditions characteristic of the forest floor under spruces and firs in boreal and high-altitude forests, fungi are the chief agents of humus decomposition. The fungi in such a soil may weigh 2 metric tons per hectare.

The capacity of earthworms to cultivate the soil has been well known ever since the exploratory studies of Darwin. Earthworm populations in the temperate forest reach the hundreds of thousands per hectare (e.g., 728,700 per hectare under mixed hardwoods in central Germany). Under the Nigerian rain forest, it has been estimated that 4480 kilograms of soil per hectare are cast up upon the surface by earthworm activity.

Earthworms show marked preferences in their choice of litter for food supply. In central Europe, for example, they prefer alder and elm to oak and beech and cannot thrive at all on Norway spruce. In southern Germany, earthworms were studied in a series of paired plots stocked with hardwoods and Norway spruce, respectively (Schlenker, 1971). Stands of each pair grew on similar soils. Five times as many earthworms were found under hardwoods as under spruce, even in first-generation spruce stands following hardwoods. In the United States, earthworms show similar preferences for leaves rich in minerals (such as ash), a reluctant ability to handle the tough, leathery leaves of oak and beech, and a distaste for acid conifer needles.

Among the arthropods, insects (particularly springtails), spiders, crustaceans, and mites are the most important (Birch and Clark, 1953; Murphy, 1953). Mites are microscopic or nearly microscopic relatives of the spiders that infest the surface layers of forest soils in immense numbers (Wallwork, 1959). Springtails are the most abundant of the many soil insects; as many as 60 species have been found in a single young white pine stand in Connecticut (Bellinger, 1954). Springtails live and feed in part on decaying plant material of the surface and in part on deeper-living microorganisms (Zachariae, 1962). Their role in the decomposition of organic matter is obviously great (Edwards et al., 1970).

Mammals, too, play a major role in the decomposition of organic matter. The smaller ground-inhabiting species—including the moles, shrews, voles, mice, and chipmunks—live upon plant materials and small animal life in the top soil layers. The tunnels and burrows they make aid materially in loosening compact soils, in incorporating organic matter in the deeper levels, and in providing passages for the downward movement of rain water containing soil nutrients. Because of their high metabolic work, their importance is great in proportion to their total weight.

## Decomposition of Organic Matter

The total effectiveness of the bacteria, fungi, and animals of the soil in decomposing litter is shown by the fact that, sooner or later, the litter disappears from the soil in some forests as fast as it is added from above. Direct leaching by rain water must be added to the soil biota as an important factor in decomposition. Complete decomposition may take from a few weeks to many years.

The nature of the decomposition process and the time it takes depend largely upon the forest tree species and upon the climate in which they grow. If the foliage is palatable to the soil organisms—and that condition seems to imply that it is rich in calcium and other nutrients and not excessively woody or leathery in structure—and if the forest soil is warm, well watered, and well aerated, organic matter is returned rapidly to the soil and litter does not accumulate. Such conditions characterize well-drained soils in the tropical rain forest and often occur in temperate zones in deciduous forests of the eastern United States on well-drained sites. Although more litter is deposited in tropical forests, the rich microflora and microfauna decompose it at a rate 6 to 10 times as fast as in temperate forests (Madge, 1965, 1966). Also, in the pinyon-juniper woodlands of the southwestern United States, where summers are both hot and wet and litterfall is relatively light, there is little or no humus accumulation.

Soft tissues of plants and animals are usually decomposed by soil microflora alone, while woody materials are typically broken down by the complex interaction of soil animals and plants. At the time of leaffall, leaves are already infested with an extensive external and internal microflora, and once on the forest floor they are also invaded by litter fungi. For example, in loblolly pine stands and adjacent upland hardwoods in northern Mississippi, 38 species of fungi were found on living pine and hardwood foliage, and about 50 species were found in the decomposing litter (Watson et al., 1974). Leaf litter darkens as it weathers and the water-soluble substances, chiefly sugars, organic acids, and polyphenols, are leached out. Fragmentation of litter by soil animals, mainly earthworms in the temperate regions and termites in the tropics, provides a desirable physical substrate for microfloral growth, and the litter is invaded by omnipresent microorganisms. If fragmentation is retarded experimentally, the whole decomposition process is slowed (Witkamp and Crossley, 1966). Tissue breakdown by microflora, in turn, favors attack by soil microfauna, and the cycle continues. Thus organic matter is gradually decomposed and incorporated into the soil through an intricate sequence of feeding by soil animals and growth of microorganisms.

If the foliage is palatable to soil organisms but the decomposition process is slowed down, the top layer of the mineral soil will gradually become mixed with finely divided, decayed organic matter. This mixture of mineral soil and humus creates what is known as a **mull** organic layer, a term taken from Danish, meaning "dust," (Müller, 1887). A mull humus layer is characteristic of much of the hardwood forest in the temperate zones.

If, on the other hand, the foliage is highly acid or is otherwise detrimental to soil biotic activity, raw humus tends to accumulate on top of the mineral soil. Rainfall then becomes a major decomposing agent, and the acids dissolved from the foliage by the downward-percolating water have a high capacity for leaching the upper layers of the mineral soil of nutrient ions. This **mor** type of humus thus tends to favor the development of a podzolic soil condition. Mor humus conditions are typically found in cold climates in the boreal zones or at high altitudes, especially under spruces and other conifers that produce an acid foliage. The nutrient supplying capacity (cation exchange capacity) of mor humus is substantially less than that of mull humus.

In the Temperate Zone forests of the northern United States and southern Canada and in similar zones in Europe and Asia, one tree species may favor the development of a mull condition while another may favor the mor. In central New England and Great Britain, natural hardwood stands fall in the former category and pine plantations in the latter. In the central hardwood region of the United States, oak litter is the slowest and elm the quickest to decompose (Kucera, 1959). In England, European beech litter has proved the most resistant to decomposition among major hardwood species; oak is relatively resistant, whereas elm, birch, and ash decompose completely within one year (Heath et al., 1966).

The same species may favor the development of quite different humus types on different soils. Thus European beech in England and France forms a mull on calcareous soils in all conditions and also on noncalcareous soils, if not too acid, in a mild, moderately humid climate. In colder and wetter climates, or on poor and acid soils, the same species causes a mor humus layer, accentuated acidity, and accelerated leaching (Duchaufour, 1947; Ovington, 1968).

## The Nitrogen Cycle

In considering the cycling of nutrients from the soil to the tree and back again, a special word must be said for nitrogen, as this element comes largely from the air and not from the decomposition of soil minerals. The part that rainfall plays in the addition of nitrogen to the soil from the atmosphere has already been mentioned. This input by precipitation occurs as inorganic ammonium, nitrate nitrogen and organic nitrogen.

Nitrogen-fixing bacteria living symbiotically in root nodules also contribute to supplying nitrogen directly from the air to plants.

Legumes are the most important plants with nitrogen-fixing bacteria. Such species as black locust, honey locust, acacias, and mesquite are trees of the legume family in the North Temperate Zone. There are hundreds of similar species in tropical climates. Also many herbs and shrubs of this family characteristically inhabit the forest floor throughout much of the world. In addition, over 300 nonleguminous plants have been shown to increase soil nitrogen supply by their presence through the medium of nitrogen-fixing bacteria in their roots (Stewart, 1967; Bond, 1967). Among these are the alders (both in Europe and North America), casuarinas in the tropics, and shrubs of the genera *Ceanothus* and *Myrica*. Young, dense stands of red alder of the Pacific Northwest are estimated to accumulate annually as much as 300 kg $ha^{-1}$ (Zavitkovski and Newton, 1968). Most of this is incorporated rapidly in the soil and only 20 percent is tied up in plant biomass. Annual litterfall contributes about 100 kg $ha^{-1}$.

In many forests, free-living soil bacteria of the genera *Azotobacter* and *Clostridium* are the most important agents fixing atmospheric nitrogen into water-soluble compounds. In boreal forests with acid soils and no legumes, the sources of nitrogen are apparently precipitation, nitrogen-fixing plants (such as *Sheperdia* and *Alnus*), lichens, and nitrogen-fixing organisms associated with mycorrhizae.

Blue-green algae, such as *Nostoc* and *Anabaena*, also fix atmospheric nitrogen. *Nostoc* is one of two algae present in the epiphytic lichen *Lobaria oregana* that is found in the crowns of old-growth Douglas-fir trees (Denison, 1973). *Lobaria* is estimated to fix approximately 8 to 10 kg of nitrogen per hectare per year (Chapter 18).

At any one time, the total amount of nitrogen in usable form in the soil is relatively small compared to that of other soil nutrients. Only the ammonium form, $NH_4^+$, and the nitrate form, $NO_3^-$, are taken up by higher plants. By far the greatest amount of nitrogen in any soil at any one time is in an organic form and not available in the soil solution. Furthermore, nitrogenous compounds are subject to losses by leaching and fire. The nitrate anion, $NO_3^-$, being very soluble, is subject to major losses by leaching, whereas the ammonium cation, $NH_4^+$, is much less affected, being retained by soil colloids through cation exchange. However, under favorable conditions $NH_4^+$ may be oxidized to the nitrate anion (termed **nitrification**), which then is subject to leaching. Being flammable, all forms of nitrogen are subject to loss by fire, and a hot fire can volatilize 50 kg $ha^{-1}$ or more of organic nitrogen to $N_2$ gas.

Successful tree growth, therefore, depends in large measure upon quick rotation of nitrogen. Much of the nitrogen taken up by the tree is returned to the soil, and under favorable conditions the nitrogen-

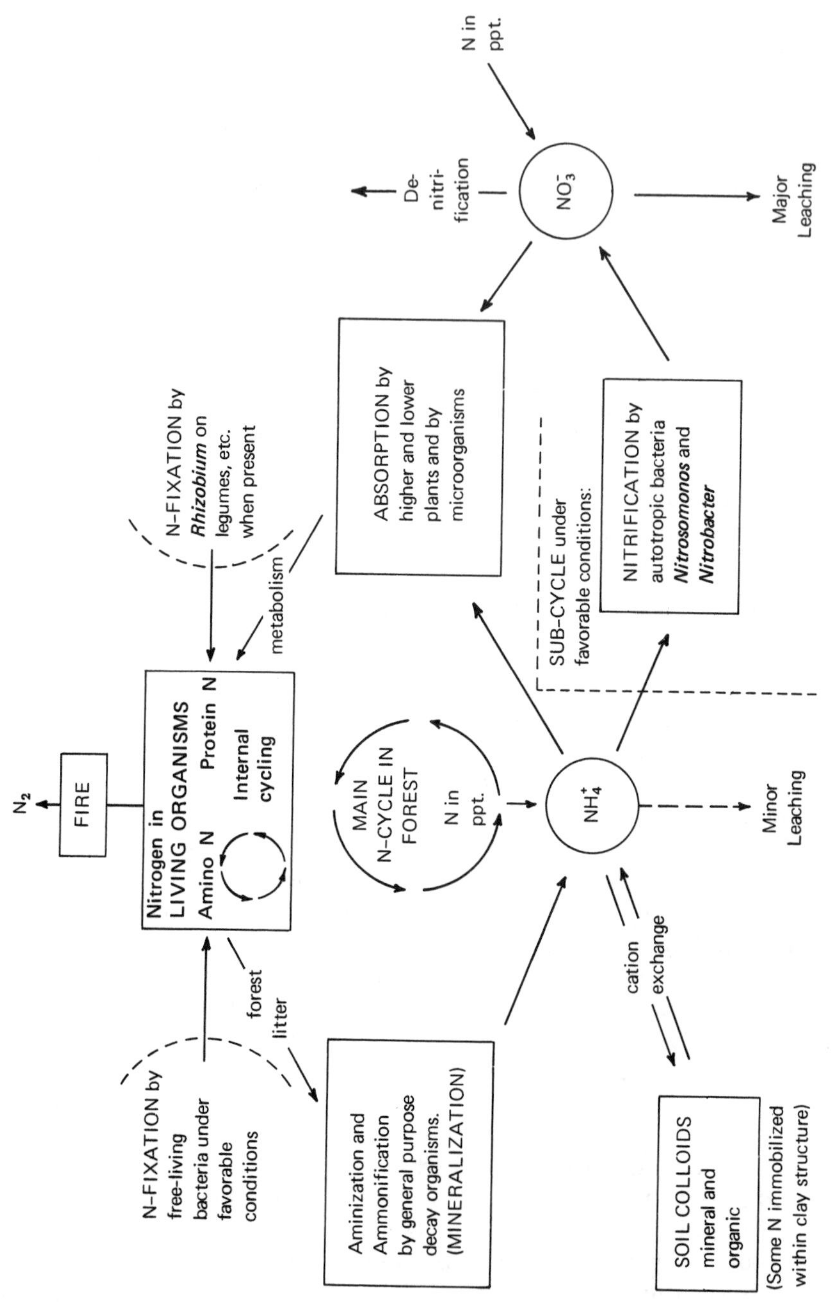


**Figure 9.6. Nitrogen balance in a stable forest community. (Diagram by R. Zahner.)**

ous litter is quickly decomposed by bacteria and other soil organisms. Nitrogen constitutes from less than 1 to nearly 3 percent of the soil organic matter, and this small amount is used over and over again.

Retranslocation of nitrogen within the tree itself (from senescing foliage to developing shoots and foliage) is also an important source of nitrogen for conifers such as loblolly pine (Switzer and Nelson, 1972) and Douglas-fir (Johnson et al., 1980) and deciduous trees (Duvigneaud and Denaeyer-DeSmet, 1970; Henderson and Harris, 1975). In a 20-year-old loblolly pine plantation, 39 percent of the nitrogen requirement was met by internal transfers. This illustrates the conservation of a nutrient in short supply and also an important adaptive mechanism of pines, enabling them to thrive on sites relatively low in nitrogen.

On sites low in nitrogen, such as infertile old fields, roots of some conifers have the ability to mineralize or otherwise extract from the soil nitrogen that is unavailable to the preceding herbaceous vegetation (Fisher and Stone, 1969). Ectotrophic mycorrhizae of the conifer roots may account in part for the plant's ability to obtain nitrogen. This may not only explain why conifers, such as larch and pine, exhibit rapid juvenile growth on infertile soils, but why hardwoods beneath these conifers, or succeeding them on the site, grow better than hardwoods without benefit of the conifer crop. For example, Carmean et al. (1976) found that hardwoods (yellow-poplar, sweet gum, red oak) planted on an old-field site formerly supporting a 23-year-old shortleaf pine plantation were about 4 times taller at age 10 years than those planted in an adjacent open field covered with herbaceous plants.

The major features of the nitrogen cycle of stable forest communities are diagrammed in Figure 9.6; a detailed flowchart is presented by Patric and Smith (1975). The main cycle involves the production of $NH_4^+$ from decomposition of litter and its absorption again by the vegetation. The subcycle involving $NO_3^-$ may be important under certain favorable circumstances when exchangeable bases are abundant.

Any condition that causes raw humus to accumulate on the forest floor rather than decompose quickly tends to result in a shortage of nitrogen for tree growth. Under the cold, wet conditions in which peat forms, nitrogen is in particularly short supply. It is on such sites that pitcher plants and other plants thrive, being able to obtain nitrogen by capturing insects. An excessive development of the mor type of raw humus usually indicates poor nitrogen supply in the soil. Such a condition will result in lowered growth of many forest trees but usually not of the spruces and those other conifers for which a mor humus layer is a normal soil development.

## OUTPUT OF NUTRIENTS

### The End Products of Decomposition

As the decomposition of organic matter proceeds, mineral nutrients are gradually transformed to ions that can be absorbed by tree roots. Losses from leaching, fire, removal from the forest in the form of forest products and animals, and other sources must in the long run be balanced by additions from the weathering of soil minerals, additions in rainfall and dryfall, and by the fixing of atmospheric nitrogen. As a by-product, carbon dioxide is respired by soil organisms and adds to the atmospheric supply.

### Potential Losses in Drainage Water and Timber Harvest

Evidence from deciduous and coniferous forests of the eastern and western United States, that are relatively undisturbed by humans, indicates that small net losses of nutrients in stream water and deep seepage, when they occur, are counterbalanced by weathering of parent material or by incoming precipitation (Likens et al., 1967, 1977; Johnson and Swank, 1973; Grier et al., 1974; Fredriksen et al., 1975; Henderson and Harris, 1975; Patric and Smith, 1975). In many natural systems, the amounts of some nutrients, nitrogen and phosphorus for example, often show an annual net gain. Precipitation inputs of inorganic nitrogen and phosphorus generally exceed losses in stream water (Likens et al., 1977).

Nutrient losses from conventional timber harvesting, when properly conducted and where only the tree boles are removed, typically are small on an annual basis over the rotation period. However, losses vary from slight to substantial during the first 1 to 2 years or so after harvest, depending on the severity of the cutting and soil-site conditions. The losses are replaced over the ensuing 50 to 100 years by soil supplies and inputs from precipitation, activities of microorganisms, and weathering. However, conventional harvesting, particularly clearcutting, disrupts the annual circulation of nutrients: (1) more organic matter becomes available for decomposition than in the undisturbed stand, and the rate of decomposition and mineralization is increased due to a warmer and moister forest floor; (2) root mortality of many species following cutting results in less nutrient absorption; (3) trees no longer absorb or intercept as much precipitation, thus reducing transpiration, increasing stream flow, and increasing the leaching of nutrients by percolating water; and (4) soil erosion may increase. As a result, the amount of dissolved nutrients in stream water following harvesting by clearcutting typically increases. However, it tends to return in a relatively short time to the preharvest level (Johnson and Swank, 1973; Fredriksen et al., 1975; Hornbeck et al., 1975). Nitrate nitrogen, because it is easily leached, is typically found in relatively high concentrations soon after harvesting. Rapid revegetation of the cut area and nutrient uptake by this new growth minimizes loss and acts to restore nutrient cycling to the preharvest level (Marks and Bormann, 1972; Likens et al., 1978). Regrowth also tends to reduce the rate of decomposition by shading the forest floor (hence reducing its temperature) and by intercepting precipitation. Although mycorrhizae and other soil organisms increase mineralization after cutting, they also immobilize large amounts of nitrogen and other nutrients that gradually become available to the regrowth.

An excellent yardstick for measuring potential losses of nutrients was provided by the denudation of a northern hardwood forest at the Hubbard Brook Experimental Forest in New Hampshire (Likens et al., 1970). On a small watershed all trees and shrubs were cut but not removed. Regrowth was inhibited by herbicide spray in each of 3 years following cutting. Elimination of vegetation was the purpose of the treatment. The nutrient cycle was disrupted, annual stream runoff increased, and percolating rain water flushed substantial amounts of nutrients from the system during the 3 years following cutting while vegetation was suppressed (Figure 9.7). Average stream water concentrations increased over 4 times for calcium and magnesium and over 15 times for potassium. Nitrate concentration increased 41-fold the first year and 56-fold the second, above undisturbed conditions. Nitrification of $NH_4^+$ by bacteria increased manyfold as absorption of this cation by higher plants ceased (see Figure 9.6). This led to a marked increase in nitrate and hydrogen ions. The nutrient ions, Ca, Mg, K, and Na, were released into solution as hydrogen ions replaced them on colloidal micelles. Two years following cutting, the total net export of inorganic substances from the denuded system was 14 to 15 times that of the undisturbed watersheds. The vegetation was allowed to regrow during the fourth growing season after deforestation. In 2 years, stream flow and nutrient export returned to levels similar to that of an undisturbed forested watershed (Figure 9.7). This experiment is of great value in illustrating the rapid nutrient loss after deforestation and the rapid recovery of the system following 3 years of denudation. However, the experimental denudation differs substantially from conventional clearcutting and should not be equated with this silvicultural practice.

Two recent developments, complete tree harvesting and ultrashort rotations (2 to 8 years), mean much greater nutrient removals than conventional harvesting procedures. The new logging sys-

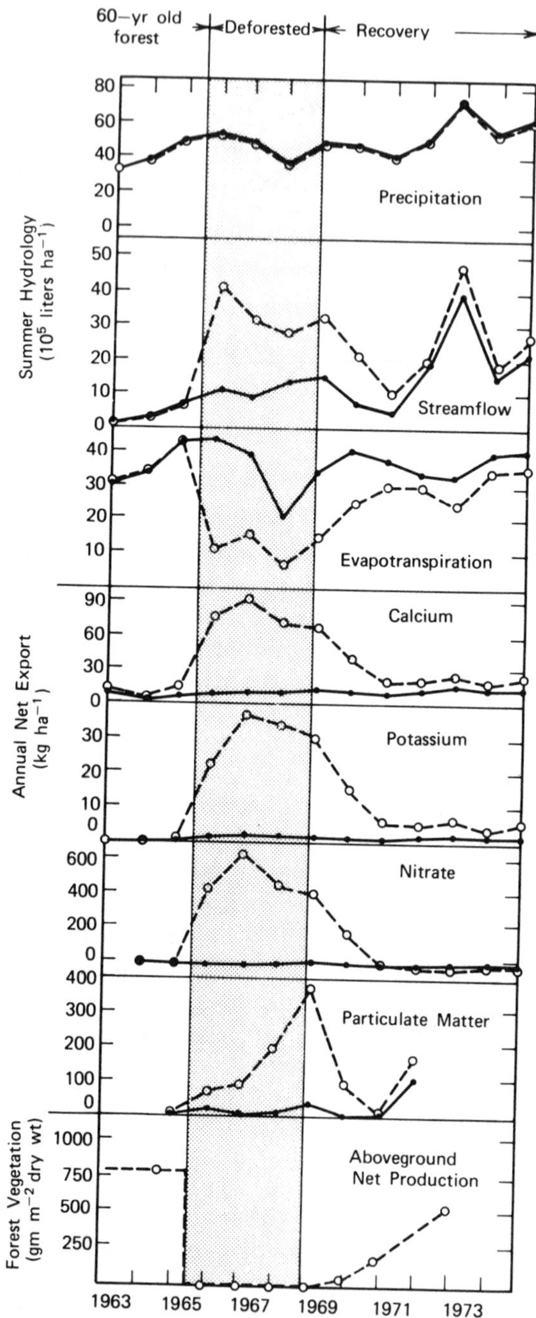


**Figure 9.7. Effects of deforestation on a northern hardwood forest ecosystem, Hubbard Brook Experimental Forest, New Hampshire. The hydrology, annual nutrient export,**

tems remove the whole tree to a central point for delimbing and chipping so that what residue remains is not spread evenly over the site. Besides the stem, branches, leaves, and roots may be utilized, and all standing trees and brush are often removed from the site. Under these conditions nutrient removals may be two or more times that taken out in conventional harvesting. A number of investigators have considered this problem (Weetman and Weber, 1972; Boyle et al., 1973; White, 1974; Jorgensen et al., 1975, Boyle, 1975; Hornbeck, 1977), and more research is needed on wildland sites where artificial replacement of the nutrients removed is not now anticipated.

Ultrashort rotations utilize fast-growing hardwoods that sprout from the stump or the root system following cutting. They cause a large nutrient drain on the site because nutrient uptake and accumulation is most rapid in youth and declines with increasing age. Thus five cuttings at 7-year intervals of coppice cottonwood or sweet gum would remove much more nutrients than one cutting at age 35 years. Such intensive culture, like modern agriculture, will undoubtedly necessitate fertilization to maintain continuous productivity at a high level.

## MINERAL CYCLING IN THE ECOSYSTEM

The components of the nutrient cycle previously discussed may be combined to provide a general picture of the annual circulation of chemical elements in ecosystems (Figure 9.8). Nutrient cycling is essentially a polycyclic phenomenon. It consists of three major cycles: the external geological cycle of nutrient inputs from the atmosphere, geological weathering of parent material, and losses through leaching; the biological cycle of plant-soil exchanges; and the within-tree cycle of uptake and retranslocation. Within each are short-term (daily, seasonal) and long-term subcycles.

The uptake by the forest (sum of the retained and returned elements) comes partly from the products of organic matter decomposi-

---

**and aboveground net production of an experimentally deforested ecosystem (○——○, W-2) are compared with a forested reference ecosystem (●——●, W-6). A 60-year-old forest (W-2) was experimentally devegetated during the autumn of 1965, maintained bare of vegetation for 3 growing seasons, and then allowed to re-vegetate during the growing season of 1969. (Likens et al., *Science,* Vol. 199, Issue 3, 1978, pp. 492–496 © 1978 by the American Association for the Advancement of Science.)**

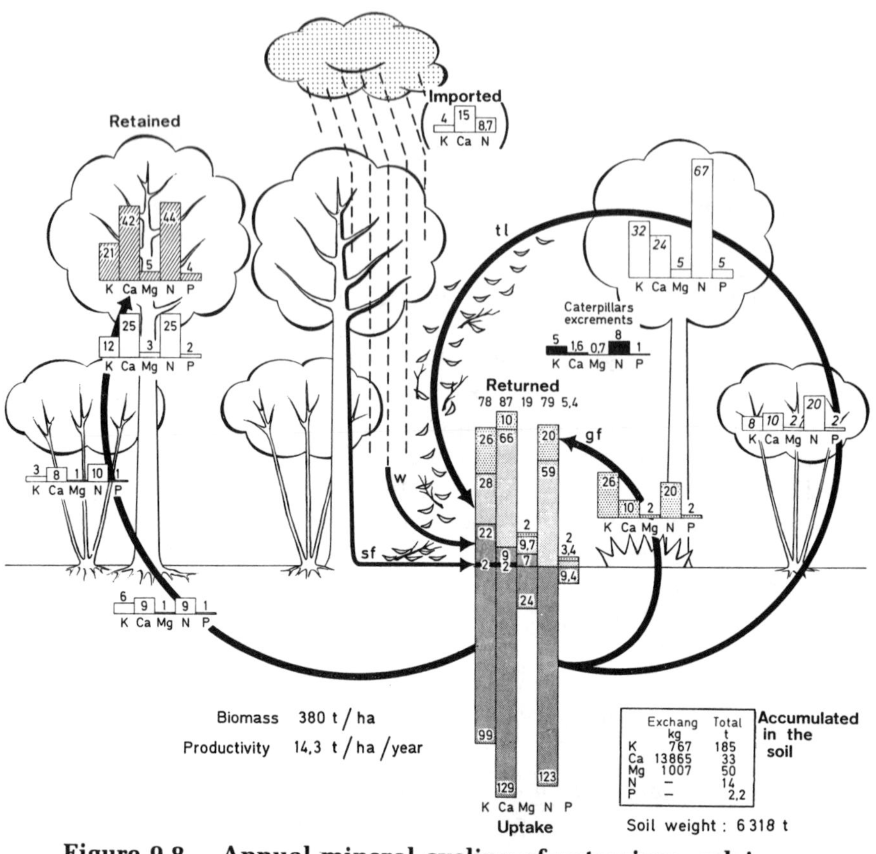


**Figure 9.8.** **Annual mineral cycling of potassium, calcium, magnesium, nitrogen, and phosphorus (in kg ha$^{-1}$) in a *Quercus robur—Fraxinus excelsior* forest with coppice of *Corylus avellana and Carpinus betulus* at Wavreille-Weve, Belgium. Retained: in the annual wood and bark increment of roots, 1-year-old twigs, and the above-ground wood and bark increment. Returned: by tree litter *(tl)*, ground flora *(gf)*, washing and leaching of canopy *(w)*, and stem flow *(sf)*. Imported: by incident rainfall (not included). Absorbed (uptake): the sum of quantities retained and returned. Macro-nutrients contained in the crown leaves when fully grown (July) are shown on the right-hand side of the figure in italics; these amounts are higher (except for calcium) than those returned by leaf litter, due to reabsorption by trees and leaf-leaching. Exchangeable and total element content in the soil are expressed on air-dry soil weights of particles <2 mm. (After Duvigneaud, 1968, and Duvigneaud and Denaeyer-De Smet, 1968.)**

tion, from atmospheric inputs, and in part from minerals weathered directly from the parent materials. Not all the nutrients needed for annual growth and maintenance of tissues come from annual uptake from the soil; considerable amounts may be circulated within the trees themselves. The important role of retranslocation of nitrogen in certain conifers has already been mentioned. Soluble compounds of nitrogen and phosphorus are probably responsible for their efficient retranslocation in northern hardwood species, whereas calcium, and to a lesser extent magnesium, are structural components and not soluble (Gosz et al., 1976). Internal cycling in deciduous trees is indicated by the seasonal variations found in the mineral content of leaves (Duvigneaud and Denaeyer-De Smet, 1970). Young leaves are richer in nitrogen, potassium, and phosphorus and poorer in calcium than mature leaves. In autumn nitrogen, potassium, and phosphorus contents of foliage decrease due to nutrient transport to other plant parts and to rainwater leaching, but calcium increases. Nutrient content of xylem sap of deciduous forests also varies seasonally, being higher in spring and fall than in summer and winter, reflecting the seasons of greatest root absorption.

Nutrients used by the forest stand may be stored in the soil, the litter, or the trees themselves. In temperate forests accumulating litter contains a high proportion of the nutrients. In tropical regions of high rainfall, where evergreen forests occur on soil low in nutrients and where decomposition of litter is extremely rapid, storage of nutrients in the vegetation is necessary if excessive leaching is to be avoided. In a study of nutrient contents of a tropical forest in Ghana, Greenland and Kowal (1960) found most of the nutrients in the living trees. In the Belgian deciduous ecosystem over two-thirds of the nutrients, except phosphorus, are stored in the litter.

An important nutrient transfer not illustrated in Figure 9.8 is that from root to soil, that is, the minerals released by root exudation and annual mortality. For example, annual root mortality in a deciduous forest in Tennessee exceeded litterfall in transferring nitrogen to the soil (Henderson and Harris, 1975; see Chapter 18).

Comparisons of ecosystems by their nutrient cycles show both similarities as well as striking differences (Duvigneaud and Denaeyer-De Smet, 1970). Oak ecosystems of Belgium and Russia were similar in their luxury consumption of calcium compared with the frugal turnover of calcium in spruce and pine ecosystems. The oak systems have significantly greater nutrient requirements, especially potassium and nitrogen, than beech, spruce, and pine forests of temperate Europe, and approach those of cultivated crops. Many differences between ecosystems in nutrient cycling, even forests of similar composition, not surprisingly, stem from basic differences in the chemical soil properties of the site. Considerable information is now available on nutrient cycling and budgets, but comparisons among the many ecosystems is beyond the scope of this treatment. However, further discussion of nutrient cycling is presented as a part of ecosystem analysis in Chapter 18.

### Litter Removal

Just as the fertilization of forest soil may improve site quality, the removal of tree litter may lower it. Litter removal is essentially defertilization in that the nutrients in the litter are taken out of the nutrient cycle.

Since nutrients in the litter and even in the semidecomposed humus are not in a form immediately available to trees, the removal of litter or humus may have no immediate effect upon site quality. In parts of Europe where these have been removed repeatedly for livestock bedding, mulching, and other purposes, however, litter removal resulted in severe deterioration of the site (Lutz and Chandler, 1946). German investigations indicate that the deficiency of nitrogen in soils subject to litter removal plays an important part in the impoverishment of the soil. In eastern North America, potassium deficiency commonly limits the growth of planted conifers on spodosols (Stone, 1967). Without exception the original surface humus layer of these soils has been destroyed, although usually by past cultivation, being cleared for pastures, or severe burning rather than just removal of the litter.

In rare circumstances, the losses in nutrients resulting from litter removal may be more than counterbalanced by improved soil properties. Thus the removal of upland peat in northern Europe may result in warming the soil, reducing soil acidity, stimulating soil biota, and generally improving site conditions.

## ALLELOPATHIC RELATIONS

Site quality and plant growth are not only affected by removals of chemical substances in the litter and harvested trees, but by additions of chemical compounds to the ecosystem. **Allelopathy** (derived from Greek words meaning "mutual harm"), any direct or indirect harmful effect of one plant on another by the production of chemical substances that escape into the environment, has long been observed and discussed (Tukey, 1969; Whittaker, 1970; Whittaker and Feeny, 1971; Rice, 1977). It is of widespread occurrence among woody plants and agricultural and wild species of many kinds. Comprehensive reviews are available in the books by Sondheimer and Simeone (1970) and Rice (1974).

Allelopathic effects have been observed for rain-forest and temperate-forest species and for shrubs of the desert and the northern forest. Among forest tree species, the black walnut has long been known to affect markedly other plants through chemical exudate of its foliage and roots (Bode, 1958). Many plants, including forest tree species, cannot grow if their root system comes into contact with that of a black walnut tree (Brooks, 1951). Others, such as Kentucky bluegrass, thrive and apparently grow better adjacent to the tree. Reduced growth of understory vegetation under cherrybark oaks was observed by Hook and Stubbs (1967) and confirmed by DeBell (1969) (Figure 9.9). The primary inhibitory substance in leaf extracts of cherrybark oak was found to be salicylic acid (DeBell, 1971). Leaching of this substance from oak crowns by rain presumably causes inhibition of vegetation beneath cherrybark oak. Similarly, herbaceous plants fail to grow under sycamore (Al-Naib and Rice, 1971), sugarberry, hackberry, and red and white oaks (Lodhi and Rice, 1971; Lodhi, 1976, 1978) due to toxins produced in the leaves of these trees. Other tree species known for allelopathic effects are ailanthus, eucalypts, Japanese red pine, Utah juniper, sassafras, and sugar maple. In addition, nine woody species of western Washington, including Pacific madrone, western red cedar, Engelmann spruce, vine maple, and grand fir, were found to have a definite allelopathic potential (Del Moral and Cates, 1971).

Among the forest shrubs, particular attention has been given to the effect of bracken and heather on the growth of tree seedlings. Both are detrimental. For example, in Norway, roots of bracken were dried, ground, and an extract prepared with cold distilled water (Torkildsen, 1950). This extract, after being put through a bacterial filter, was used to water newly germinated Norway spruce seedlings growing in sterile sand. The plants so watered either died or were dwarfed. Three-year-old seedlings, however, were not affected by the treatment. Similarly, a water extract of heather has been shown to cause yellowing and abnormal growth of Norway spruce.

Various grasses have long been known to hold back the growth of many tree seedlings, particularly hardwoods. That the effect is in part chemical has been demonstrated by watering tree seedlings with the water extract of living grass plants, a procedure which reduces the growth of the seedlings as compared to others treated with water only. For example, in northwestern Pennsylvania black cherry seedlings occurring in low-density stands, characterized by a dense ground cover of grass and **forbs** (nongraminoid herbaceous plants), grow slowly and soon die (Horsley, 1977). The allelopathic effect was demonstrated when water extracts of the grass and forbs were found to inhibit the germination and the shoot and root growth of the black cherry seedlings.

Allelopathy is a major factor in succession in infertile old fields (Rice, 1974). Toxic compounds of early successional old-field plants

Figure 9.9. Vigorous development of vegetation beneath a sweetgum *(a)* in contrast with retarded development under a nearby cherrybark oak *(b)* four years after a seed-

inhibit nitrogen-fixing and nitrifying bacteria, thereby giving a competitive advantage to plants having low nitrogen requirements. Old-field herbs may also inhibit mycorrhizal development and thus the entire supply of nutrients may be limited. By restricting nitrifying bacteria, ammonium-N becomes the chief available form, and its accumulation in time enables the higher nitrogen-requiring species (shrubs and trees) to dominate the old-field site.

Substances that may potentially be involved in allelopathy are released from plants via litterfall, leaching of foliage, volatilization from foliage, and root exudation. For example, leachates from eucalyptus leaves are known to inhibit the growth of grass and herbaceous species under natural conditions (Del Moral and Muller, 1970). In California shrub communities, volatile terpenes are released into the air from shrub species. They accumulate in the soil during the dry season and inhibit growth of herbs causing bare belts, 1 to 2 m wide, devoid of herbs (Muller, 1966; McPherson and Muller, 1969). Root exudates of sugar maple have been demonstrated to inhibit root growth of yellow birch (Tubbs, 1973), and the growth of various northern conifer seedlings (Tubbs, 1976).

It is obvious that both the plants within the forest and the lower plants within the soil may directly affect the growth of forest trees, especially in the seedling state. They may do this by increased competition for water and nutrients. In addition, however, many plants actually contain harmful chemicals which, in minute amounts, may change the growth pattern of other plants that absorb them. Allelopathic agents may act in many ways, directly and indirectly to modify plant growth. They may inhibit cell division and elongation, inhibit hormonal relations, modify mineral uptake, retard photosynthesis, inhibit protein synthesis, change permeability of membranes, inhibit specific enzymes, and affect respiration and stomatal opening (Rice, 1974). Indeed, as Whittaker and Feeny (1971) conclude, chemical agents are of major significance in the adaptation of species and the organization of communities.

## SUGGESTED READINGS

Bond, G. 1967. Fixation of nitrogen by higher plants other than legumes. *Ann. Rev. Plant Physiol.* 18:107–126.

**tree cut; coastal South Carolina. A sharp boundary between the affected and unaffected areas is evident in (b) (U.S. Forest Service photo. Courtesy of Dean S. DeBell.)**

Bray, J. Roger, and Eville Gorham. 1964. Litter production in forests of the world. *Adv. Ecol. Res.* 2:101–157.
Dickinson, C. H., and G. J. F. Pugh. 1974. *Biology of Plant Litter Decomposition.* Vol. 1, Part I. *Types of Litter.* Vol. 2, Part II. *The Organisms.*, Part III. *The Environment.* Academic Press, Inc., New York. 775 pp.
Duvigneaud, P., and S. Denaeyer-De Smet. 1970. Biological cycling of minerals in temperate deciduous forests. *In* David E. Reichle (ed.), *Analysis of Temperate Forest Ecosystems.* Springer-Verlag, New York.
Edwards, C. A., D. E. Reichle, and D. A. Crossley, Jr. 1970. The role of soil invertebrates in turnover of organic matter and nutrients. *In* David E. Reichle (ed.), *Analysis of Temperate Forest Ecosystems.* Springer-Verlag, New York.
Likens, Gene F., F. Herbert Bormann, Noye M. Johnson, D. W. Fisher, and Robert S. Pierce. 1970. Effects of forest cutting and herbicide treatment on nutrient budgets in the Hubbard Brook watershed-ecosystem. *Ecol. Monogr.* 40:23–47.
Likens, Gene E., F. Herbert Bormann, Robert S. Pierce, John S. Eaton, and Noye M. Johnson. 1977. *Biogeochemistry of a Forested Ecosystem.* Springer-Verlag, New York. 146 pp.
Marks, G. C., and T. T. Kozlowski (eds.). 1973. *Ectomycorrhizae—Their Ecology and Physiology.* Academic Press, Inc., New York.
Marx, Donald H. 1973. Mycorrhizae and feeder root diseases. *In* G. C. Marks and T. T. Kozlowski (eds.), *Ectomycorrhizae—Their Ecology and Physiology.* Academic Press, Inc., New York.
Marx, Donald H. 1975. Mycorrhizae and establishment of trees on strip-mined land. *Ohio J. Sci.* 75:288–297.
Patric, James H., and David W. Smith. 1975. Forest management and nutrient cycling in eastern hardwoods. USDA For. Serv. Res. Paper NE–324. Northeastern For. Exp. Sta., Upper Darby, Pa. 12 pp.
Rice, Elroy L. 1974. *Allelopathy.* Academic Press, Inc., New York. 353 pp.
———. 1977. Some roles of allelopathic compounds in plant communities. *Biochem. Syst.* and *Ecol.* 5:201–206.
Stone, Earl L. 1968. Microelement nutrition of forest trees: a review. *In Forest Fertilization.* Tennessee Valley Authority, Muscle Shoals, Alabama.
Switzer, G. L., and L. E. Nelson. 1972. Nutrient accumulation and cycling in loblolly pine (*Pinns taeda* L.) plantation ecosystems: the first twenty years. *Soil Sci. Soc. Am. Proc.* 36:143–147.
Tamm, Carl Olof. 1964. Determination of nutrient requirements of forest stands. *Int. Rev. For. Res.* 1:115–170.
Trappe, James M., and Robert D. Fogel. 1977. Ecosystematic functions of mycorrhizae. *In* J. K. Marshall (ed.), *The Belowground Symposium: A Synthesis of Plant-Associated Processes.* Range Sci. Dept. Sci. Series 26. Colo. State Univ., Ft. Collins.
Whittaker, R. H., and P. P. Feeny. 1971. Allelochemics: chemical interactions between species. *Science* 171:757–770.

# 10
# The Soil—Plant Water Cycle

The nutrient cycle, important as it is, is made possible only by the circulation of water from the soil through the roots to the foliage, to the atmosphere, and from the atmosphere back to the soil. Most soil nutrients must be in an ionic state before being absorbed, and this requires the presence of soil water. The water column extending from the roots to the leaves carries the nutrients in solution and provides the means of transport within the tree. Transpiration of moisture from the leaf surfaces makes possible the movement of water to the tops of tall trees and in turn plays an important part in recharging the atmosphere with moisture. Finally, precipitation of moisture from the atmosphere is either directly or indirectly responsible for the recharge of the soil with moisture.

The water cycle thus constitutes a part of the forest ecosystem just as does the nutrient cycle. Furthermore, because the greater part of world forests is under moisture stress for at least part of the year, water often is a limiting factor in determining the distribution and growth of forests. In those areas of evenly distributed high rainfall or of shallow water tables where water is not a limiting factor, soil air frequently may be limiting because of its displacement by the excess water supply.

The water cycle is thus basic to an understanding of forest ecology. It affects the behavior and growth of trees to a very great extent and thus is of importance to the silviculturist concerned with the growing of these trees. It also affects stream flow and ground water supply and is thus of interest to the watershed manager. These two groups differ only in their outlook. To the silviculturist, water is either used by the trees or lost to the ground. To the forest hydrologist, water is either lost to the trees or used by humans, who obtain it indirectly from the ground. The basic data and principles, however, apply to both fields. Many aspects of the forest–water relationship are reviewed by Penman (1963) and Sopper and Lull (1967), and an integrated treatment of plant and soil-water relationships is presented by Kramer (1969). Hydrology texts by Hewlett and Nutter (1969) and Satterlund (1972) are also available.

## SOIL WATER AND AIR

The supply of moisture to tree roots is inversely related to the supply of air (oxygen) to them. The pores in the soil may be filled with either air or water so that, as the supply of one increases, that of the other automatically decreases. For optimum tree growth, both air and water must be available to the roots at all times. This means that the larger soil pores should be occupied by air and the smaller by water. It is convenient to discuss and emphasize the water-holding capacity of soils. At the same time, though, we should remember that factors affecting the air-holding capacity are essentially the same, for any space in the soil not occupied by water will normally be occupied by air.

### Water-Holding Capacity

Soil minerals have an average specific gravity of 2.65. An undisturbed soil sample, however, contains water, air, and organic solids, as well as minerals. Thus weight of solids (minerals and organic matter) per unit volume of soil is called **bulk density,** and varies between 1.0 and 1.6 depending on the amount of pore space. If the specific gravity of soil minerals is 2.65 and the bulk density 1.325, one-half of the oven-dry soil is composed of air. The pore volume is thus 50 percent. Most forest soils have pore volumes between 30 and 65 percent. The pores are occupied by both air and water when the soil is in its natural state in the field. Air and water relationships for soils of different textures are illustrated in Figure 10.1.

When soil is completely saturated, the contained water falls into three well-defined categories. **Free water** fills the larger pores (larger than about 0.05 mm diameter) and drains off readily under the influence of gravity (Figure 10.2). This gravitational water is available but briefly to plant roots, and its direct effect on plant growth is transitory. It is, however, largely responsible for soil leaching.

After the saturated soil has been drained of gravitational water, it is said to be at **field capacity**. It holds all the water it can against the force of gravity. This force can be approximated in the laboratory by subjecting the saturated soil to a pressure of about ⅓ atmosphere over normal air pressure (Figure 10.2).

At field capacity, the water remaining in the soil is known as **capillary water** and is held largely as thin moisture films around individual soil particles and aggregates. Capillary water will move very gradually from moister to drier portions of the soil. Thus if tree roots and surface evaporation remove capillary water from the top soil layers, a certain amount of capillary water may move up into these layers from any moister soil beneath. The zone of capillary

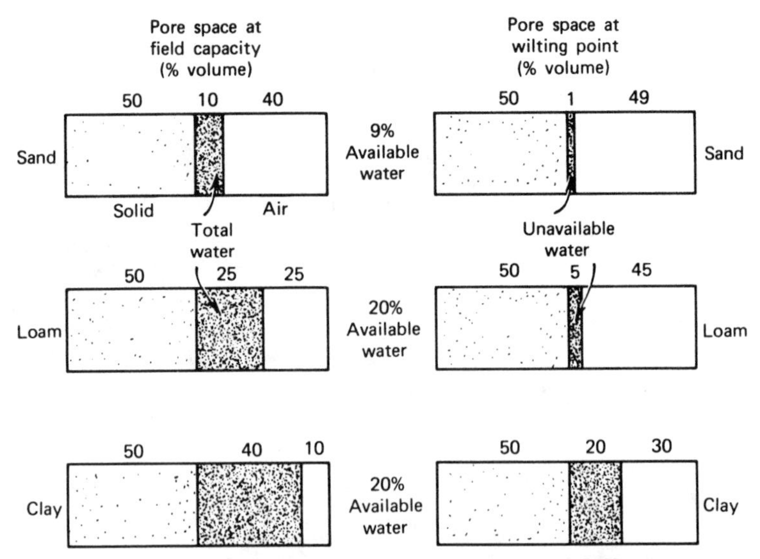


**Figure 10.1. Available water and aeration in different-textured soils, assuming 50 percent of the soil is mineral solid. Addition of organic matter increases capillary space in sand soils and air space in clay soils.**

movement beneath tree roots, however, is not great, varying from virtually nothing in coarse sands to a meter or so in clay soils. Therefore, to meet water requirements tree roots penetrate new volumes of soil.

Finally, yet additional water is held quite strongly by soil particles and can be removed only by prolonged heating above the boiling point. Such **hygroscopic water** of the soil is unavailable to plant roots.

## Water Available to Plants

Of the water that enters the soil, the only portion available to plants, then, consists of (1) gravitational water that comes momentarily into contact with roots on its downward journey, and (2) that portion of the capillary water held by forces slighter than hygroscopic water.

The mechanisms of water absorption and rate of water flow from soil to plant are primarily controlled by water evaporating from the leaves (i.e., along the vapor pressure gradient). In transpiring plants, water moves from the soil → root epidermis → root cortex and free space → root xylem → stem xylem → leaf veins → leaf mesophyll and free space → through stomates → atmosphere along gradients of decreasing **water potential**. Water in the plant-soil system has a

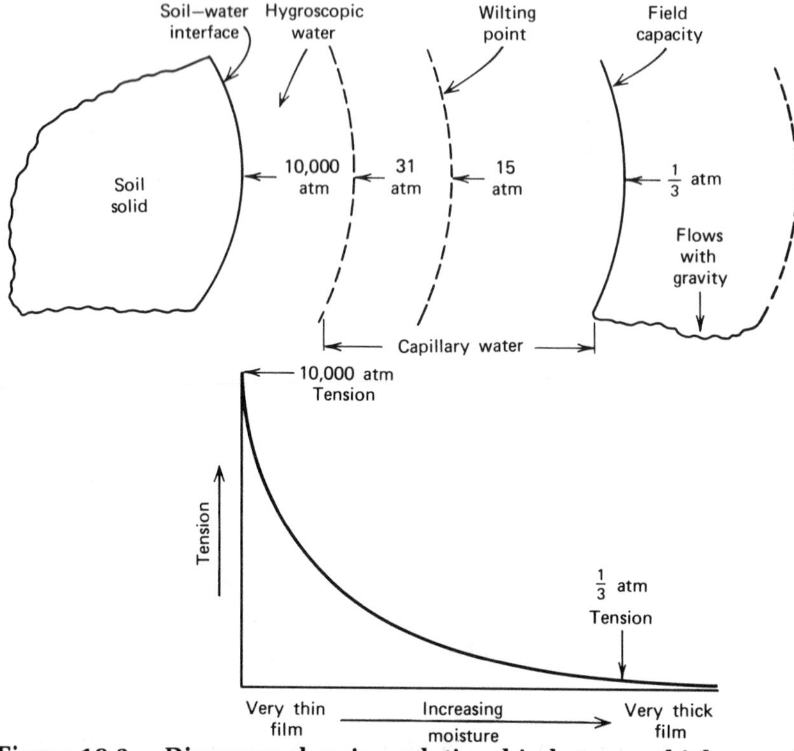


**Figure 10.2. Diagrams showing relationship between thickness of water films and the tension with which the water is held at the liquid-air interface. The tension is shown in atmospheres. *(Upper)* Sketch of water film thickness at several moisture levels. *(Lower)* Logarithmic change in tension with increase in thickness of moisture film. (After Buckman and Brady, 1969. Reprinted with permission of Macmillan Publishing Co., Inc. from *The Nature and Properties of Soils,* 7th Edition, by Harry O. Buckman and Nyle C. Brady. Copyright © 1960, 1969 by Macmillan Publishing Co., Inc.)**

chemical potential, a capacity to do work. By converting units of work to equivalent units of pressure, water potential may be expressed in bars or atmospheres. Pure free water is defined as having a potential of zero. The presence of solutes in plant cells or soil water reduces the chemical potential of the water in solution below that of pure free water. Likewise the adhesive forces holding water to the surfaces of soil particles reduce the potential of soil water far below that of free water.

A large water-potential gradient exists from the soil to the air surrounding the crown, and water moves toward the more negative water potential. Thus a sandy loam soil at field capacity might have a water potential ($\Psi$) of $-0.5$ bar, the $\Psi$ of roots might be $-3$ bars, the $\Psi$ of the midtrunk $-5$ bars, and the $\Psi$ of transpiring foliage $-15$ bars. This gradient is responsible for pulling water from the soil to the top of trees.

The movement of water through the soil-plant-atmospheric system is a continuous process, and the flow is determined by the difference in water potential and resistances in the plant and the boundary layer. (Kramer, 1974):

$$\text{Flow} = \frac{\text{Difference in water potential}}{\text{Resistances (root, leaf, boundary layer)}}$$

Resistance is highest in the roots where water must cross several layers of living cells in passing from the root surface to the xylem. Movement then takes place through the continuous column of the xylem elements along an increasingly negative water potential gradient described above. Resistance to evaporation also occurs in the leaf where water must pass through substomatal cavities and particularly through the stomata themselves. The degree of opening of the stomates provides the major leaf resistance to water movement; resistance is virtually complete if stomates are closed. Only minute amounts of water may pass through the epidermis and cuticle. If the stomates are open, the boundary layer surrounding the leaf provides a final resistance. The water vapor concentration (vapor pressure) is typically higher in the boundary layer than in the bulk air outside so that the rate of evaporation through the stomates is reduced. In quiet air, this resistance is higher than when wind is continually decreasing the thickness of the boundary layer.

In 1912, Renner described two water absorption mechanisms: so-called active absorption (metabolic energy required) operating in slowly transpiring plants and passive absorption (no metabolic energy required) when transpiration is rapid. When plants are growing in warm and moist soil the active accumulation of salts in the sap of root epidermis and cortex lowers the water potential below that in the soil. The resulting inward, osmotic movement of water (which is itself passive) produces the "root pressure" that is responsible for guttation and exudation from wounds, such as sap flow from grape vines and birch trees (Kramer, 1974). Maple sap flow, however, is not due to this mechanism. It is caused by localized stem pressures due to sugars that leak into xylem vessels.

Most of the water taken up by trees is absorbed passively through roots when trees are rapidly transpiring. The roots act as passive absorbing surfaces across which water is pulled from the soil by the water potential gradient generated in the plant by the transpiring foliage. Although we may describe the absorption process in terms of the plant exerting tensions to overcome those of the water held in the soil, water transport, whether in rapidly or slowly transpiring plants, is controlled by differential water potentials developed in the soil and in various plant tissues. Although most of the water is transpired out of trees, a small amount is used to circulate food materials in the phloem to growing parts. The mechanism of water absorption, the factors affecting it, and water movement in plants and soils are described in detail by Kramer (1969).

Water at field capacity can be easily absorbed by tree roots as it is held by only slight forces. As the soil dries out, however, the remaining water is more and more tightly bound to the soil particles, and can be removed with increasing difficulty by roots. Eventually, the point is reached at which root water potential is equal to soil water potential and no more water enters the roots. This is termed the **permanent wilting point**, as plants maintained in such soil will wilt beyond recovery and die (Figure 10.2).

Originally, the permanent wilting point of a soil was determined by finding the lowest moisture level at which plants will remain alive. This method works only for succulent plants, however (sunflower is the classic test plant), as woody plants possess structure permitting the dried plant to hold its form; also, many woody plants can endure drought by going into dormancy. The permanent wilting point cannot be determined in this way for most forest tree species.

A great many studies have been directed at the problem of determining minimum soil moisture for the maintenance of plant life. In general, it seems that roots of most higher plants can extract water that is held by the soil at water potentials less negative than −15 bars. For example, Richard (1953) in Switzerland found no significant difference between the permanent wilting points of Scots pine, European alder, and dwarf sunflower. It follows that the soil-moisture content at the permanent wilting point can be best measured by subjecting the soil to an artificial water potential of −15 bars. The resulting moisture content will vary widely for different soils, ranging from perhaps 2 percent by volume for coarse sandy soils to as high as 35 percent for compact clay soils.

Although plants can remove water from the soil down to levels of about −15 bars and below, most water actually taken up is that portion that is more readily available. Many field studies have shown that water uptake by forest trees drops sharply when soil water potential is more negative than −1 bar. Water uptake in drier soils continues at a very slow rate (Zahner, 1967).

Due to the resistance of water movement through the soil and into the roots, daily absorption, even in soil at field capacity, tends to lag behind transpiration. The resulting temporary, midday, internal water deficits tend to curtail growth but are not critical as the water in the tree is replaced overnight by continued absorption from the soil. The observation that growth of some crops and trees is greater at night than during the day (Kramer, 1969) is a response to this pattern of water availability. In midsummer, when absorption is markedly reduced by lack of soil water and overnight absorption fails to regain turgor of the tree, serious water deficits develop in tissues of trees and cause major reductions in forest growth.

In arid regions, even when moisture is unavailable from the soil, some may be obtained at night from dew and mist (Gindel, 1973, Chapter 6). Some desert xerophytes find deep sources of moisture which are tapped by roots penetrating to a depth of over 50 m below the surface. Evergreen xerophytes may maintain their green foliage and full turgor in leaves and roots for 3 to 5 months until rainfall starts and absorptions resume.

## Evapotranspiration by the Forest

Evapotranspiration is the general term widely used for the transfer of vapor from land and water surfaces to the atmosphere. For evapotranspiration to occur from the forest both water and energy must be available at one or more of the following surfaces: external surfaces of leaves and plant stems, internal plant surfaces connected to the external atmosphere via the stomata, soil and litter surfaces, and snow pack surface (Goodell, 1967). Thus the entire water vapor exchange of a forest is linked to evaporation, interception, and transpiration. Evaporation is used here for the physical process of vapor transport from soil or surfaces of vegetation and litter, and transpiration is used for the plant physiological process of water loss through the stomata. Interception originates on wet surfaces of plants and is principally derived from rain and snow (Chapter 6). Not all intercepted water may be evaporated to the atmosphere. Some may be absorbed by vegetation and utilized to recharge the internal water balance of the plants (Zinke, 1967). In addition intercepted water, while evaporating, may reduce the rate of transpiration, thus saving stored water in the soil. However, this effect appears to be slight (Hewlett, 1967; Rutter, 1968).

The soil moisture that enters tree roots passes up the stem into the foliage, bringing the nutrients to the growing organs of the plant.

However, much more water is lost in transpiration than is needed for nutrient transport. Only a small amount of water is actually used in the photosynthetic process: about 1 mm per year of water actually enters into the manufacture of cellulose and other organic products of the tree. The rest of the water that is taken up serves as a mode of transport for nutrients, maintains cell hydration, moves into the phloem to aid food translocation, leads to some cooling of the leaves, and passes on into the atmosphere.

Allowed free access to unlimited water supplies, trees can transpire immense quantities of water. Willows, cottonwoods, and other ***phreatophytes***[1] growing on the banks of permanent watercourses and reservoirs have been shown to be capable of transpiring as much as 1000 mm of water during the course of a single growing season. For a tree obtaining its moisture from 0.004 ha of land (assuming a fully stocked forest of 250 trees per hectare), this amounts to a transpirational use of 40,000 liters (10,000 gal).

Without access to the water table, forest trees will maintain maximum transpiration rates only as long as the soil water supply is excellent. Available water is first extracted from the zone of high root concentration in the upper layer of soil. Later in the growing season extraction becomes nearly equal with depth, even down to 6 m (20 ft) (Figure 10.3; Patric et al., 1965). When soils are recharged periodically by frequent rains, most water lost in transpiration by vegetation will come from the densely rooted surface soil.

With some exceptions, the upland forest, regardless of type, receives enough radiant energy during the growing season to deplete through evapotranspiration most, if not all, of the available moisture (Anderson et al., 1976). Zahner (1955), working in pine-hardwood forests of southern Arkansas, estimated a mean soil-moisture depletion of about 6 mm of water per day for the period in early summer when soil moisture was adequate. This estimate agreed closely with the computed theoretical evapotranspiration of the locality (Zahner, 1956). Similarly, a maximum soil-moisture depletion rate of 5 mm per day was reported from a loblolly pine plantation in South Carolina (Metz and Douglass, 1959).

In well-drained soils, the annual transpiration of forest trees is limited by the amount of annual precipitation that reaches the ground and the amount of water that moves upward into the root zones (especially from shallow ground water sources) less the amount that is lost to the trees by surface runoff and through gravitational drainage below the root depth.

In the hardwood forest of the southern Appalachians at the Cow-

[1]Phreatophytes are plants that derive their moisture supply from ground water and are more or less independent of rainfall.

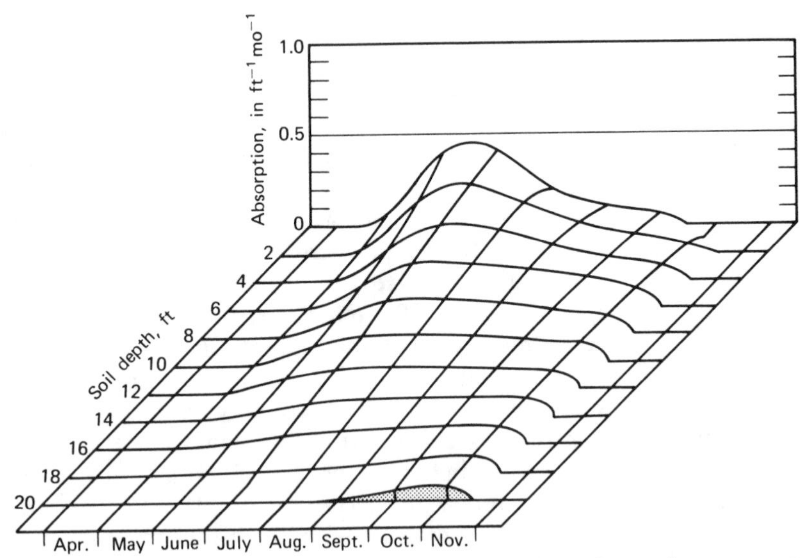


**Figure 10.3. The pattern of water absorption from the upper 20 feet (6 m) of soil at a southern Appalachian mountain site. Total absorption peaks in June and thereafter extraction of water comes more and more uniformly from the entire profile. (After Patric et al., 1965. Reproduced from *Soil Science Society of America Proceedings*, Vol. 29, p. 305, 1965, by permission of the Soil Science Society of America.)**

eeta research station, transpiration of forest trees has been determined on entire watersheds of 13 and 16 ha through measuring the effects on stream flow of clearcutting the forest. Of the annual rainfall of about 2000 mm, about 500 (range 430–560) mm was shown to have transpired through mature undisturbed hardwoods. The phreatophytic undergrowth of laurel and rhododendron along the major streams alone accounted for about 50 mm of transpiration (Johnson and Kovner, 1956). In pine stands, the understory vegetation may account for as much as one-quarter of the total evapotranspiration loss (Zahner, 1958b).

In southeastern Ohio, white oak forests growing on loam soils transpired from 300 to 580 mm of water during the growing season. The annual rainfall here is about 1000 mm (Gaiser, 1952). In cooler climates, transpiration is much less. For instance, aspen forests both in Utah (Croft and Monninger, 1953) and Russia transpired only 200 mm of water under the cooler conditions where these species occur.

The water-use figures here presented actually include transpiration from the trees and in most cases also evaporation from the forest floor. The latter is, however, only slight where the soil is covered with tree litter and is shaded by tree crowns. The great preponderance of evapotranspirational loss in dense forests is attributable to transpiration. In open forests with exposed forest floors, evaporation from the soil and litter is also a major factor.

Marked differences are observed in total evapotranspiration among cover types. Listed in decreasing order of water consumption, they are: wet meadows, open water, forests, grasslands, vegetable crops, and bare soil (Baumgartner, 1967). Differences between the cover types may be explained quantitatively in part by energy considerations. Forests, due to their greater absorption of radiation, vaporize water more readily than other types of vegetation. Major differences in rooting habit and depth also account in part for the observed differences between forests and grasslands and also for differences among forest trees. The consumption by trembling aspen of 150 mm more water than Engelmann spruce from the surface 2½ m of soil in Colorado was attributed to rooting differences (Rocky Mountain Forest Experiment Station, 1959).

## Plant Control of Transpiration

For the plant, transpiration is unavoidable. Stomates must be open during the day for plants to obtain $CO_2$ for photosynthesis, and simultaneously water evaporates into the atmosphere. The water deficits resulting from this water loss reduce growth and often kill plants. However, plants respond to the strong selection pressures of the diurnal and seasonal moisture regime of their respective habitats by adapting to cope with the stresses. For example, mechanisms have evolved enabling plants to control leaf resistance to water loss. A thick and impermeable leaf cuticle and external cuticular waxes are important either directly (in reducing water loss from epidermal cells and in plugging stomates) or indirectly (by increasing reflectance and thereby reducing leaf temperature) in reducing water loss. An even more important mechanism may be the opening and closing of stomates; this markedly affects transpiration rates. A dry site species such as black oak, for example, exhibits a sharp decrease in transpiration (primarily by stomatal closure) as temperature increases, whereas sugar maple, a mesic site species, maintains its relatively high transpiration rate (Wuenscher and Kozlowski, 1971a). Black oak is able to fix $CO_2$ rapidly while minimizing its loss of water in transpiration (Wuenscher and Kozlowski, 1971b). This may be one reason why black oak is highly competitive on hot, droughty sites, and sugar maple seedlings compete best on sites with low evaporative stress, such as the forest floor of a mesic site.

Another mechanism for reducing transpiration is reduced leaf surface area. In general, leaf area (leaf surface area per unit land area in $m^2\ m^{-2}$ is termed **leaf area index**) decreases with decreasing precipitation, and more significantly, with increasing moisture-stress of the site. For example, the leaf areas of diverse conifer communities of western Oregon are strongly related to the water balance of the site (based on temperature, precipitation, and evaporation; Grier and Running, 1977). Leaf area of a coastal, humid Sitka spruce forest was over five times greater ($38\ m^2\ m^{-2}$) than that of the dry, interior western juniper community ($7\ m^2\ m^{-2}$). Even within stands dominated by a single species, Douglas-fir, leaf area increases markedly from xeric to cool-moist habitats (Gholz et al., 1976).

## Seasonal Course of Water Uptake

The pattern of water uptake by trees varies widely. Water will be taken up in large quantities only when the soil is moist, the trees have transpiring foliage, and the atmosphere is warm and dry.

A simple situation is represented by pines on deep sand dunes in north temperate climate. Here, transpiration is great only during the growing season immediately after a rain. Large amounts of water are taken up immediately after a summer storm, water use tapering off and ending within a few days after the combined effects of gravity and root absorption have reduced the sands to near the permanent wilting point. Tree growth is largely concentrated in these brief periods.

In most forests, however, water is stored in the soil for longer periods of time. Wherever snow accumulates on the ground during the winter, and elsewhere in the temperate zones where precipitation is high during the dormant season, it is normal for the well-drained soil to be at or near field capacity at the beginning of the growing season. As the weather warms, growth of trees accelerates. Within a few weeks, the soil-moisture supply is depleted by transpiration to the point that tree growth begins to slow down. In the northeastern United States, growth is pretty much concentrated in the spring months—May and June—with growth tapering off in July and being maintained in August and September only in exceptionally wet years.

In the southern pine region, similar conditions prevail. With the onset of warm weather in the spring, transpiration reaches high levels until, in June, 5 to 6 mm of water are removed from the soil daily, as has already been seen. This maximum rate of transpiration loss, which continues as long as the soil can supply the moisture, agrees closely with the potential evaporation from the area as calculated with Thornthwaite's formula.

In the coniferous forests of the West, tree growth is largely concentrated in the spring months. The summer drought that characterizes the Douglas-fir region and much of the ponderosa pine region results in the soil becoming too dry for substantial tree growth during the summer months. A secondary growing season, however, sometimes occurs in the early fall, if substantial rainfall occurs while the weather is still warm.

From the evidence that has been gathered, it is apparent that forest soils reach field capacity or complete capillary water recharge only during dormant periods or exceptionally wet growing periods. Normally, during the growing season, light rains are largely intercepted by tree crowns, and the small proportion that reaches the ground goes to replenish the moisture in the surface soil only and is soon lost through absorption by surface roots and direct evaporation. Substantial rainfall is required to stimulate tree growth. Almost continual rain (or nearly 50 mm per week) is needed to maintain the soil at its field capacity during warm weather in temperate climates.

## Ground Water and Trees

The presence of a water table available to tree roots has both beneficial and detrimental aspects. On the plus side, it provides additional water for transpiration. Often, though, transpirational use is partly luxury consumption as the tree may be already absorbing all the moisture it needs to transport the optimum quantity of nutrients and required water to the foliage. On the minus side, a permanent water table will normally prevent the downward development of roots because of insufficient oxygen below the water line for root growth. It may even result in the development of a gley horizon in the subsoil, creating a nutrient-poor ground-water podzolic condition.

Where ground water is within reach of forest roots, the level of the water table is usually variable through the season, being highest just prior to the initiation of the growing season, and becoming depressed as transpiration by the forest depletes the water supply. The effective level of the water table, however, is pretty well marked by deoxidation or reduction that changes the affected soils to a grayish or whitish-gray color. This gley horizon is frequently mottled and easily detected. It is the upper limit of the reduced or mottled zone that should be taken as the effective height of the water table—not the actual height of the water at the time of measurement.

The great importance of the water table in affecting tree distribution and growth has led to the recognition of several different soil water conditions varying from completely drained to completely undrained conditions. A series of soils, alike in all respects except the position of the water table and its resulting effect on the profile, constitutes what is known as a **moisture catena** (from the Latin, meaning "chain"). The different members of the catena are arbitrarily defined, there actually existing a complete series of intergrading conditions. Since these classes can be defined to show a high correlation with forest site, however, it is worthwhile to consider them in some detail.

1. **Well-drained soils** are formed when the water table is permanently below the zone of tree roots and the A and B horizons. Since it is difficult to sample below 1.5 m either with a shovel or an auger, the absence of mottling to a depth of 1.5 m is normally considered evidence that the soil is well drained.
2. **Imperfectly drained soils** show evidence of a water table in the lower part of the B horizon or the upper part of the C. The A horizon and enough of the B horizon is well drained, however, to permit the normal development of the roots of shallow-rooted species and the normal cycling of nutrients in this upper zone. In imperfectly drained soils, mottling is normally found at least 60 to 90 cm below the surface of the mineral soil.
3. **Poorly drained soils** are affected by a high water table into the lower part of the A horizon. Root development is consequently much restricted, and nutrients leached out of the surface zones may be lost to the forest. Mottling may occur within 15 cm or so of the surface.
4. Finally, in **undrained soils**, free water stands above the surface of the mineral soil for much of the year. As a result, tree litter is trapped in more or less stagnant water and tends to accumulate rather than to decompose completely. Organic layers of markedly decomposed plant material are termed **muck**, while layers only slightly decayed are known as **peat**. In the boreal forests of Canada, northern Europe, and northern Asia, peat lands are a major feature of the landscape (Heikurainen, 1964; Heinselman, 1970; Moore and Bellamy, 1974). Peat may also accumulate on uplands where wet and cold summer weather normally inhibits the decomposition of certain types of litter. The formation of upland peat constitutes a serious forest problem in parts of Scotland and northern Europe.

Since forest trees may remove 500 mm or more of water a year from the soil, clearcutting the forest will result in raising a water table close to the surface. A rise in the water table of from 30 to 60 cm has been reported for such diverse forest types as aspen in the Lake States and loblolly pine in the southeastern coastal plain (Wilde et al., 1953; Trousdell and Hoover, 1955). Such a rise will often convert an imperfectly drained soil into a poorly drained soil or even into an undrained condition.

The detrimental effects of high soil water tables arise in part from their capacity for dissolving and removing soil nutrients and in part from the shortage of oxygen in stagnant or anaerobic soil water. Moving soil water that carries substantial quantities of dissolved air and nutrients is hence far more favorable to forest growth than stagnant soil water low in oxygen and nutrient ions. Thus in the Lake States, the presence of northern white cedar in wet sites is indicative of seepage conditions where the water table is moving and relatively high in oxygen. With completely stagnant and oxygen-poor water, only black spruce and associated ericaceous species can grow.

Inadequate aeration due to flooding or a high water table may cause decreased water absorption (Kramer, 1969). Many swamp and river floodplain species (such as bald cypress, swamp tupelo, willows, cottonwoods, red ash, and many others) are adapted to these conditions. However when upland species are flooded, they may suffer serious water deficits that lead to decreased growth and death. Under anaerobic flooded conditions many fine roots die, and the absorbing root surface is greatly decreased. The effects of high water levels and inadequate oxygen supply are usually more injurious during the growing season than during the dormant season. This is, at least in part, due to a reduced absorbing root surface and hence the failure of water uptake to meet transpirational requirements. In addition, the internal growth regulator relationships of flooded plants are often altered, resulting in altered plant metabolism that affects growth and survival.

## WATER DEFICITS AND TREE GROWTH

Tree growth responds more to water stress than any other perennial factor of the forest site. Thus soil water is the key to forest site productivity for many species in many parts of the world. Apical, radial, and reproductive growth of trees, as well as seedling germination and establishment, are highly correlated with environmental moisture stress and have been reviewed by Zahner (1968). Moisture relations in arid regions are considered by Gindel (1973).

In temperate climates moisture deficits during the middle of the growing season directly affect growth during both the current and succeeding growing seasons. Indirect evidence of the marked effect of moisture stress on height growth is seen in the low heights that trees attain in dry climates or on dry sites, compared with trees growing on moist sites. Direct measurement of shoot growth of seedling and sapling trees under moisture stress confirms that growth is closely correlated with water potential within the plant and to environmental soil-moisture deficits.

The effects of moisture stress for a given species depend partly on the species' seasonal pattern of shoot flushing. Some species, such as birches, yellow-poplar, and loblolly pine, are capable of maintaining shoot elongation during the complete growing season and hence are affected by late season droughts. Many other species, however, including red pine, eastern and western white pine, white ash, sugar maple, beech, and red and white oaks, complete height growth and set buds by midsummer, and their current year's growth is unaffected by late season water deficits. Their current height growth is, however, affected by late summer drought of the preceding year. For example, Zahner and Stage (1966) accounted for 72 percent of the variation in annual shoot growth in five stands of young red pine by water deficits of the previous and current growing seasons together. The water deficit of the previous summer (June 15 through October) accounted for as much reduction in annual height growth as the deficit for May 1 to July 15 of the current year. As expected, tree species exhibiting continued shoot flushing (yellow-poplar) or recurrent flushing (loblolly, pitch, and Monterey pines) show little correlation between total annual height growth and the previous year's rainfall (Tryon et al., 1957; Zahner, 1968).

Moisture stress plays an equally important role in the radial growth of trees. It affects the size of the annual ring, the proportion of earlywood and latewood, and various wood properties, particularly wood specific gravity. In conifers, the transition from earlywood to latewood (from large, weak, thin-walled xylem cells to small, strong, thick-walled cells) may be directly affected by water deficits in the cambium (Zahner, 1968), as well as markedly affected through reduction of auxin levels in the crown (Larson, 1963a, 1964; Zimmermann and Brown, 1971; Chapter 3).

Periodic changes (daily and annual) in radial growth of conifers and angiosperms have been repeatedly shown to be directly correlated with soil water availability (Figure 10.4). For several deciduous species over a 5-year period in Ohio, high air temperatures during periods of moisture stress in midsummer were consistently associated with temporary and sometimes permanent cessation of radial growth in all species (Phipps, 1961). Similarly, Fraser (1956, 1962) found the radial growth of the lower bole of northern hardwoods and conifers in Canada to be closely related to soil moisture. Zahner (1968) summarizes the typical response of trees growing in fully stocked stands on upland sites in temperate climates in this way:

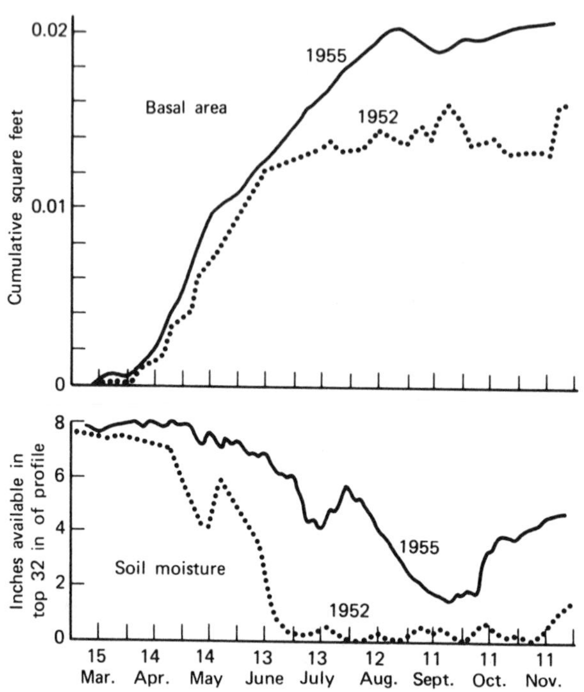


**Figure 10.4. Basal area growth per tree of shortleaf pine and trends of available soil moisture for relatively wet (1955) and dry (1952) growing seasons. Note that growth rate slowed in mid-June of 1952 but not until mid-August of 1955, at the time in each year when available soil water had been depleted to about 2 in (50 mm). (Redrawn from Boggess, 1956; after Zahner, 1968.)**

> . . . *water stresses are not serious prior to midsummer, at which time it is normal for absorption by roots to lag far behind transpiration, and the resulting dehydration of tissues in the crowns and stems causes important limitations in growth below the potential for that time of year. If the water stress is alleviated by late-season rains, radial growth usually resumes if the mid-season water deficit has not been severe.*

Many studies demonstrate that water stresses normally occurring in fully stocked stands may be alleviated by silvicultural practices such as thinning or wide spacing of trees. Radial growth of residual trees is faster and more prolonged in thinned than in unthinned stands (Figure 10.5; Zahner and Whitmore, 1960). Furthermore.

heavy thinning of loblolly pine stands may alleviate summer moisture stresses such that residual trees may continue to grow longer throughout the season (Bassett, 1964).

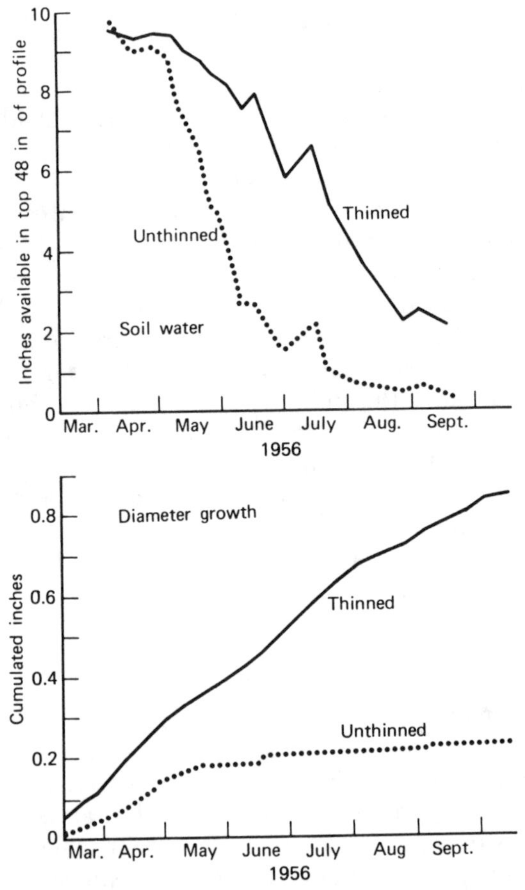


**Figure 10.5. Trends of soil water depletion and diameter increment per tree for average dominant loblolly pine during one growing season, thinned plots and unthinned plots. (Redrawn from Zahner and Whitmore, 1960; after Zahner, 1968.)**

Drought years leave their record in the growth rings of trees, and the high correlation of ring width with summer water deficit is widely documented. Up to 90 percent of the variation in width of annual rings of conifers has been attributed to water stress in semiarid climates (Douglass, 1919; Fritts et al., 1965) and up to 80 percent in humid temperate climates (Zahner and Donnelly, 1967).

Plants resist drought by avoiding and tolerating drought stress (Levitt, 1972). Woody species primarily avoid drought stress and do so by morphological and anatomical adaptations that maintain favorable internal water potentials despite the low potential in the environment to which they are exposed. The major avoidance mechanisms are: (1) the ability of roots to extract large amounts of water from the soil, (2) high root-to-shoot ratio, (3) reduced leaf surface area (including rolling, folding, shedding of leaves, and maintaining a dense pyramidal crown), (4) stomatal control to reduce transpiration—stomata closed most of the day and the ability to close stomata very rapidly in response to stress, (5) thick cuticle to reduce cuticular transpiration, and (6) a high proportion of water-conducting tissue to nonconducting tissue—veins close together so more water can be delivered per unit area. Reduced transpiration may be of less importance in resisting drought stress than the ability of trees to absorb large quantities of water. The efficient water-absorbing species can keep stomata open for photosynthesis and can transpire and still maintain a favorable internal water potential. Avoidance by stomatal closure alone has the disadvantage that it leads to low photosynthesis and starvation. Drought tolerance is less common than drought avoidance in woody plants. Nevertheless, many species can withstand considerable dehydration before closing their stomata.

## PRECIPITATION AND FORESTS

Soil moisture depleted through evapotranspiration must be replenished by precipitation or ground water. Since ground water itself originates from precipitation, however, and since soil-moisture relationships are most critical on well-drained soils where the root systems of trees do not reach the water table, we may confine our attention to the replenishment of soil moisture through precipitation.

That the amount of water reaching the soil under a forest may differ considerably from the amount reaching a rain gauge in the open has already been discussed in Chapter 6. Considerable rainfall and snow may be intercepted by tree crowns and evaporated directly back to the atmosphere. On the other hand, considerable fog vapor in some forests may be intercepted, condensed, and dripped to the ground by vegetation. The combined effects of interception and stem flow result in the concentration of rainfall into small openings in the canopy and in the immediate area of tree boles (Voigt, 1960; Eschner, 1967).

Surface run-off of rain water represents water lost to the plant. The useful water for tree growth is that portion of precipitation that infiltrates the soil. The best infiltration is obtained with forest soils characterized by a good cover of litter and a low bulk density. Root and animal (worms, insects, mammals) tunnels are highly beneficial. A forest soil under hardwoods in southeastern Ohio, for example, was found to contain more than 1000 large vertical channels per hectare from decaying tap roots alone, an amount capable of handling most of the free water that reached the surface as precipitation.

In undisturbed forests, the infiltration rate is generally high. At Coweeta in the southern Appalachians, infiltration rates in such conditions generally are in excess of 250 mm of water per hour—far greater than observed maximum rainfall intensities. In sandy soils of Michigan, infiltration rates may be as high as 1000 mm per hour.

When the surface litter is removed from fine-textured soils, however, permitting the striking energy of raindrops to destroy surface soil structures, the infiltration of rainwater is markedly inhibited. Compacting of the surface soil by cattle in grazed woodlots or by humans in forest park areas is even more destructive of infiltration capacity.

## Forests and Water Yield

About 70 percent of the total precipitation falling in the continental United States is returned to the atmosphere through evapotranspiration, and forests contribute significantly to this loss. The management of forested watersheds for water is assuming an ever increasing role since inadequate water supplies are no longer restricted to arid and semiarid regions. In general, reduction of forest cover increases stream flow while establishment of forest cover on sparsely vegetated land decreases it. Hibbert (1967) reviewed 30 experiments from various parts of the world on the effects of forest removal on stream flow. He found the response highly variable and often unpredictable due to interactions of climatic, soil, and vegetative factors. Nevertheless, the average increase for the first year following cutting was about 2 mm per 1 percent of forest cover removed. No increase was found with removals below 14 percent. In the Appalachian Highlands of the United States, Douglass and Swank (1972) found an increase only if more than 12 percent of forest cover was removed. With 90 percent removed, an increase of 250 mm was reported (2.8 mm per each 1 percent cut).

Increases in water yield typically show a decline soon after treatment. The rate of decline is positively correlated to the rapidity of revegetation. Clearcutting the forest overstory does not save all the water which otherwise would be transpired or intercepted. Evaporation from the soil and litter and use of water by understory vegeta tion mean the average increase will only be about 40 percent of the water used by the forest (Pereira, 1967).

The manner of cutting may affect water yield. For example, by removing 50 percent of the conifer cover on a watershed in Colorado in clear-cut strips, Goodell (1958) recorded increases in stream flow of 270 and 220 mm in the 2 years following cutting. However, using a method of partial cutting removing 36 percent of the basal area of a mixed conifer stand in Arizona, large gaps were not created and stream flow was not increased significantly (Rich, 1959). Studies of snowpack accumulation in the central Sierra Nevada in California by Anderson (1967) suggest that snow accumulation would differ greatly under contrasting types of forest cutting—individual-tree selective cutting versus strip clearcutting. Assuming that both cuts remove 60 percent of the stand, the average percentage increase in accumulation over a 300-mm base for a dense forest would be 12 percent for the individual tree removal and 43 percent for the strip clearcutting (Anderson, 1970).

In mountainous areas of high snowfall, the snowpack is an important consideration in influencing water yield. By manipulating forest stands in various ways, snow-water storage may be increased and snow melt may be delayed (Anderson, 1970). Besides an expected increase in snow accumulation and water yield following forest removal, increases in stream flow may also occur following cutting, through rearrangement of a given amount of snow. For example, in an experimental watershed in the Rocky Mountains of Colorado, alternate clearcutting of strips of timber resulted in an average increase of 25 percent in stream flow, most of which occurred in the spring (Hoover and Leaf, 1967). Comparisons of snow storage before and after cutting indicated no change in total snow on the watershed. Thus the increased stream flow was not attributed to decreased interception but to differential distribution of the same total snow pack over the watershed. Compared to the area before treatment, much more snow was trapped in the openings than in the forested portion. The integrated result of the increased snow in the clear-cut openings and proportional decrease under forest was to increase the amount of stream flow from the same amount of snow pack.

Manipulation of forest cover is not only important in increasing water yield or decreasing peak discharges that may cause floods but in regulating the timing of water flow (Johnson, 1967; Dortignac, 1967) and decreasing erosion on denuded slopes (Margaropoulos, 1967; McClurkin, 1967). However, measures that increase water yield, cutting or fire, may adversely affect water quality by causing erosion and thereby increasing the amount of suspended sediment in stream water or by causing undesirable chemical changes. An undesirable circumstance is the occurrence of heavy rains following cutting and before vegetation has stabilized the soil. For example, the effect of complete clearcutting of all woody vegetation in a 180-ha watershed in north-central Arizona on water yield and other products was reported by Brown (1971). Slightly more water was produced the first summer than might have been expected without treatment, but a high sediment yield (60 tons per hectare) was produced, much of it from a single, heavy summer rain one month following cutting. Sediment yields in the following two years, both wetter than average, dropped to 0.2 ton per hectare. Strip clearcutting of ponderosa pine forests in the same area promises to give increased water, herbage, and wildlife yields and reduced losses in sediment compared to complete clearcutting. In general, forest land can be managed so that there is little or no increase in soil erosion. However, severe erosion may follow if revegetation is prevented by farming, grazing, intense burning, spraying with herbicides, or if the land is otherwise devegetated after the tree cover is removed.

Clearcutting generally increases stream flow significantly in small watersheds until revegetation occurs. However, in large basins such increases may be greatly overshadowed by water flowing from uncut watersheds. The effects of forest management on stream flow, snow pack, floods, sedimentation, and water supply have received considerable attention. They are properly the domain of watershed management and thus beyond the scope of this chapter. Important publications concerning forests and water use are those of Lull and Reinhart (1972), Leaf (1975a,b), Harr (1976), and Anderson et al. (1976).

Forest cutting may affect fish populations significantly. Although the indirect effects of timber harvest on fish populations may be minor compared to industrial pollution, the effects of certain practices have been deleterious and are of concern. Forest harvesting affects streams by increasing sedimentation, changing water chemistry, raising stream temperatures, increasing stream flow, and changing streamside vegetation. The literature review and annotated bibliography by Gibbons and Salo (1973) provide a good perspective on forest and fish relationships.

The total effect of watershed treatments on the forest, wildlife, and human consumers of water is only partially understood and is becoming better known through integrated studies of entire ecosystems (Chapter 18).

## Precipitation and the Distribution of Forests

Now that we have considered the place of precipitation in the soil-plant water cycle, we are in a position to summarize the relationships between precipitation and the distribution of trees. Precipitation is not a factor directly affecting tree distribution and growth, but rather is important because of its indirect effect in supplying moisture to the soil. It is thus discussed here under soil-water factors of the site rather than as a climatic factor.

For an area to support tree growth, enough water must be supplied to satisfy the minimum requirements of the trees for transpiration and photosynthesis. This amounts to about 125 to 200 mm of precipitation a year. For forests of moderate growth rates with tree crowns touching, the minimum supply approximates 400 mm a year. High growth rates require at least 500 to 700 mm a year. The precipitation must be sufficiently larger than these values to allow for interception, surface runoff, direct evaporation, and subsurface drainage.

In the far north, permafrost prevents drainage of water through the soil, no precipitation is lost during the long winter, and evaporation losses during the short summer are relatively slight. Under such conditions, good forest growth is possible with low precipitation. Vigorous forests grow in the vicinity of Fairbanks, Alaska where the annual precipitation is only 300 mm (of which 125 fall during June, July, and August, however). Forests may grow in these latitudes with annual precipitation values as low as 180 mm.

In the eastern United States, the presence of forests is nowhere limited by insufficient rainfall. In general, the soil is wet to or near the field capacity at the beginning of the growing season. Tree growth depletes the soil of moisture during the growing season and utilizes in addition water added by rain during this period. In the northern states, at least 600 mm of precipitation are needed for moderate forest growth, with 900 to 1000 mm being similarly required in the southern states.

Across the Great Plains, the 500-mm (20-inch) precipitation line roughly delimits the boundary east of which it is possible to grow trees without irrigation or access to ground water. This line, which more or less coincides with the 100th meridian, passes through the Dakotas, Nebraska, Kansas, Oklahoma, and Texas. Actually, less precipitation is needed in the northern states and more in the southern; a range from 380 mm in North Dakota to 640 mm in Texas would probably be more accurate. It is this boundary that limits the western expansion of shelter belts in the Great Plains.

In the western states, transitions from forest to nonforest vegetation occur in all states, and many correlations have been made between these boundaries and mean annual precipitation. Generalizing, approximately 380 mm seem to be required to sustain an open woodland type of vegetation (pinyon-juniper, oak woodland, chaparral, etc.), about 500 mm for an open ponderosa pine forest, and over 640 for a closed-canopy mixed coniferous forest. Since the higher rainfalls occur at higher elevations (Chapter 6), temperature influences are interrelated with precipitation influences in these relationships.

## SUGGESTED READINGS


Anderson, H. W. 1970. Storage and delivery of rainfall and snowmelt water as related to forest environments. *Proc. 3rd Forest Microclimate Symp.*, pp. 51–67. Canad. For. Serv., Calgary, Alberta.

Anderson, Henry W., Marvin D. Hoover, and Kenneth G. Reinhart. 1976. Forests and water: effects of forest management on floods, sedimentation, and water supply. USDA For. Serv. Gen. Tech. Report PSW-18. Pacific Southwest For. and Rge. Exp. Sta., Berkeley, Calif. 115 pp.

Brady, Nyle C. 1974. *The Nature and Properties of Soils*, 8th ed. (Chapter 7, Soil water: characteristics and behavior, pp. 164–199. The Macmillian Co., Toronto, Ontario. 639 pp.

Gindel, I. 1973. *A New Ecophysiological Approach to Forest-Water Relationships in Arid Climates.* W. Junk B. V., Publishers. The Hague. 142 pp.

Hibbert, Alden R. 1967. Forest treatment effects on water yield. *In* William E. Sopper and Howard W. Lull (eds.), *Forest Hydrology*. Pergamon Press, Inc., New York.

Kramer, Paul J. 1969. *Plant and Soil Water Relationships: A Modern Synthesis*. McGraw-Hill Book Co., Inc., New York. 482 pp.

Kramer, Paul J. 1974. Fifty years of progress in water relations research. *Plant Physiol*. 54:463–471.

Lull, Howard W., and Kenneth G. Reinhart. 1972. Forests and floods in the eastern United States. USDA For. Serv. Res. Paper NE-226. Northeastern For. Exp. Sta., Upper Darby, Pa. 94 pp.

Patric, J. H. 1976. Soil erosion in the eastern forest. *J. For.* 74:671–677.

Penman, H. L. 1963. Vegetation and hydrology. Bur. Soils, Harpenden. Tech. Commun. 53. 125 pp.

Rutter, A. J. 1968. Water consumption by forests. *In* T. T. Kozlowski (ed.), *Water Deficits and Plant Growth* II. Academic Press, New York.

Sopper, William E., and Howard W. Lull (eds.). 1967. *Forest Hydrology*. Pergamon Press, Inc., New York. 813 pp.

Zahner, R. 1968. Means and effects of manipulating soil water in managed forests. *In Forest Fertilization*. Tennessee Valley Authority, Muscle Shoals, Ala. 306 pp.

———. 1968. Water deficits and growth of trees. *In* T. T. Kozlowski (ed.), *Water Deficits and Plant Growth* II. Academic Press, Inc., New York.

# 11
# Fire

Fire has affected a substantial portion of the forests of the world at one time or another. In North America, virtually all of the upland forest in the South, the Lake States and adjacent central Canada, the West, and much of that of the Northeast, Appalachian Region, and Central States has been burned over more or less frequently. In Alaska and the Canadian North, fire has been a powerful natural factor affecting forests and wildlife. Lowlands, such as swamps (Cypert, 1973), bogs, marshes, prairies, and semitropical forests of high humidity also have been burned, and their vegetation markedly affected. In addition, fire has patterned the forests and savannas of the Mediterranean Region (Naveh, 1974), much of Africa (Komarek, 1972; Phillips, 1974), and of Europe, Australia, and Asia. The publications of Wright and Heinselman (1973), Kozlowski and Ahlgren (1974), and the papers of the Annual Tall Timbers Fire Ecology Conferences[1] provide an entry into the voluminous literature on fire.

Fire has always been a natural and extremely important environmental factor. It is a principal influence on species traits and life history as well as ecosystem characteristics and processes—carbon, nutrient, and water cycling, productivity, succession, and diversity. Specifically, fire plays several major roles in fire-dependent ecosystems around the world (Wright and Heinselman, 1973). It influences:

- The physical and chemical properties of the site.
- Dry-matter accumulation.
- Genetic adaptations of plant species.
- Species establishment, development, composition, and diversity; it thereby often determines community relations, especially succession.
- Wildlife habitat and wildlife populations.
- The presence and abundance of forest insects, parasites, and fungi.

In this chapter we consider fire as a physical site factor, examining its effects on forest species and forest site quality. The role of fire in forest communities is considered in Chapter 16.

[1]Tall Timbers Research Station, Tallahassee, Fla.

## FIRE AND THE FOREST TREE

### Causes and Kinds of Fires

Evidence of natural fires in the form of fossil charcoal, termed **fusain**, has been found in the Carboniferous coal deposits of 400 million years ago and in Tertiary deposits of brown coal (Harris, 1958; Komarek, 1973). Lightning was the prime cause of these fires before the advent of humans. Meteorites were an ignition source, and falling igneous rock from volcanic eruptions undoubtedly caused local fires then as it does today. Spontaneous combustion (Viosca, 1931) and sparks from falling quartzite rocks (Henniker-Gotley, 1936) are rare but documented possible causes of fires.

Today, lightning is estimated to cause about 50,000 wildland fires worldwide each year (Taylor, 1974b). This is less than 1 percent of the estimated 182 million cloud-to-ground discharges occurring annually in the forest and grasslands of the world. About 10,000 lightning-caused fires occur in the United States each year, and about 80 percent of these are in the Rocky Mountain and Pacific Coast states. Here a single lightning storm may start many small fires when fuel and climatic conditions are conducive to ignition by lightning.

Throughout much of the modern world, however, humans have been the most significant cause of fires. In pre-Colonial America, Indians set many forest fires. Europeans have followed suit. Because of their high intensity and frequency, such fires, often associated with logging and land clearing, have changed the character of forests and affected site quality.

Three kinds of fire, according to the level at which they burn, are ground fires, surface fires, and crown fires. The most common type, the **surface fire**, burns over the forest floor consuming litter and humus, killing herbaceous plants and shrubs, and typically scorching the bases and crowns of trees. The greater the fuel accumulated on the surface, the greater the mortality of shrubs and trees. In addition to the intensity of the fire, the amount of tree mortality depends on the species, the age of the tree, and rooting habit. Young pines may succumb to a surface fire, whereas older individuals of the same species survive due to thicker bark protecting the cambium from heat damage and the higher elevation of the crown above the flames. A shallow rooting habit, whether due to the inherent nature of the species or the site conditions (rock outcrops or swamp), increases the susceptibility to fire injury compared to that of the deep-rooting habit typical of many upland species, such as oaks and hickories.

Surface fires tend to kill young trees of all species (often, however, just the above-ground portion) and most of the trees of less fire-resistant species of all sizes. However, pole-size to mature trees of fire-resistant species survive light surface fires in varying proportions. Survival in a surface fire for most fire-resistant tree species is not typically dictated by damage to the stem cambium but by their susceptibility to root injury and to scorching of the crown by hot gases rising above the flames. For example, Van Wagner (1970) reported that a light surface fire will leave mature crowns of red pine undamaged, whereas a hot surface fire will kill a red pine stand just as surely as a crown fire 10 times as intense. Observations of fire-damaged red and white pines suggest that if more than 75 percent of the crown is killed, the tree will die. Mortality of pole-size ponderosa pine in northeastern Washington was directly related to crown injury; mortality was only 6 to 24 percent when less than 80 percent of the crown was burned (Lynch, 1959). Surface roots are highly susceptible to fire injury; in eastern white pine mortality was found related more closely to the percentage of surface roots killed or severely damaged than to crown scorch (McConkey and Gedney, 1951).

Fires sweeping the forest surface may generate **ground fires**, which burn in thick accumulations of organic matter, often peat, which overlies mineral soil. They burn below the surface; they are flameless, and they may kill most plants with roots growing in organic matter. Ground fires burn slowly and usually generate very high temperatures. In moist organic matter, heat from the fire dries out material adjacent to the burning zone and perpetuates a zone of combustible fuel. Ground fires tend to be persistent and serve as reignition sources for surface fires.

Surface fires fueled by accumulations of organic matter and whipped by winds may scorch and ignite crowns of trees, thus generating a **crown fire**. Traveling from one crown to another in dense even-aged stands, most trees are killed in its path. Where the forest is patchy and broken, consisting of small groups of trees such as in much of the dry-climate, ponderosa pine forests of the West, some groups may carry a crown fire, but an extensive and devastating crown fire is highly unlikely. Conifers are most susceptible to crown fires because of the high flammability of their foliage and the greater likelihood of their occurrence in pure stands than broad-leafed species. Sparks and burning debris may start new surface fires often far away from the site of the crown fire.

Fire is irregular in frequency, intensity, and burning pattern. These characteristics are primarily controlled by climate, fuel accumulation and flammability, and soil-site conditions, especially topography. Fire frequency was probably greatest in grasslands where burns every 2 to 3 years were common. In the dry ponderosa pine forests of the West, the average interval between fires can vary from 6 or 7 years to about 18 years (Weaver, 1974). Averages vary because pine forests of the drier low elevations burn more frequently than those of moist slopes and higher elevations.

Arno (1976) reported fire frequency over an elevational gradient (1150–2600 m) for three watersheds in the Bitterroot Mountains of western Montana for the period 1735 to 1900. Low elevation forests of ponderosa pine and Douglas-fir burned at about 9-year intervals; the minimum and maximum intervals between burns was from 2 to 20 years. Intervals between burns increased with increasing elevation; at high elevations the average interval was about 35 years (range 2–78 years).

Intense surface fires may damage the cambium and leave a scar. The fire scar record of individual trees indicates the actual fire frequency at a single place (Spurr, 1954). One ponderosa pine in Arno's study was scarred by 21 fires from 1659 to 1915, an average interval of 13 years (Figure 11.1). Nearby trees were scarred by additional fires, demonstrating that fires do not always burn hot enough to cause injury, particularly when they occur so frequently that only light accumulations of fuel have built up. Records of fire frequency, therefore, are typically conservative for most forests, with the likelihood that at least some light surface fires have gone undetected over the period investigated.

In the Sierra Nevada Mountains of California, a frequency of 9 years between fires was found at one locality in the giant sequoia—mixed conifer forest during the period 1705 to 1873 (Kilgore, 1973). This is comparable to the overall 8-year frequency of fire reported for Sierra Nevada forests (Wagener, 1961) and for the Southwestern ponderosa pine forests (Weaver, 1951). As is typical of many other western forests, the moister east and north slopes do not burn as readily as drier south and west slopes. However, when the mesic slopes burn, they may burn more intensely than those that burn more frequently. A similar relationship develops along elevational gradients from warm and dry lowlands to the cool and moist slopes of high altitudes.

In the conifer forest of northern Minnesota (Boundary Waters Canoe Area), detectable fires were relatively frequent (Heinselman, 1973). Fires burned somewhere in the area at 4-year intervals during the presettlement period 1727 to 1886. Settlement activities increased fire frequency, and the interval dropped to 2 years. Major fires, burning over large areas, occurred at longer intervals, 21 to 28 years. At nearby Itasca Park, the average interval between major burns for the period 1712 to 1918 was 10 years (Frissell, 1973). The effect of settlement is seen in the change in interval between fires

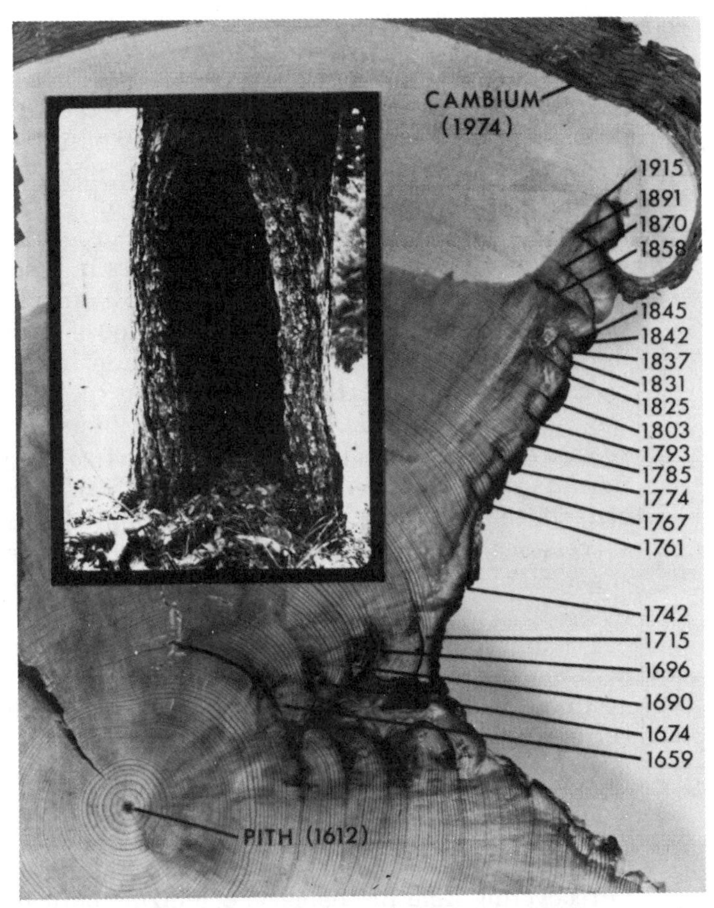


**Figure 11.1. Multiple fire-scar cross-section of ponderosa pine showing 21 fire scars from 1659 through 1915 (Arno, 1976. U.S. Forest Service photo.)**

from 16 years for the period 1712 to 1885 to 4 years from 1885 to 1918.

The emerging pattern from studies of fire history is one of cyclic occurrence of fire primarily regulated by fuel accumulation and flammability and the occurrence of droughts. Thus fires of different intensity occurred: from frequent and light surface fires that reduced fuel accumulations to rare but intense crown fires that regenerated entire stands. This range of fire conditions was instrumental in the evolution of the life span and characteristics of different species, and more generally the kind of vegetation present in each region (Chapter 16).

## Species Adaptations

Through the selective pressures of recurrent fires, certain adaptive characteristics of forest trees and shrubs were elicited, particularly in response to the frequency and intensity of fire. Many individual adaptations are typically cited: for example, the thick insulative bark of Douglas-fir and red pine. However, few adaptations probably can be exclusively related to fire; selective forces of various site factors also influence the characteristics of species in fire-dependent communities. The characteristics probably most directly elicited by fire are the closed-cone condition **(cone serotiny)** of certain pine species, thick fire-resistant bark, sprouting ability of many species, rapid juvenile growth, early flowering of some species, and the "grass stage" in longleaf pine. In addition, the life span of many forest species has been determined in part by the frequency of devastating fires. As a result, longer-lived species, such as Douglas-fir and eastern white pine, are able to compete with long-lived, late-successional associates (western hemlock and western red cedar; sugar maple and American beech) on mesic sites that burn infrequently.

The traits of forest species that enable them to compete effectively in fire-dependent communities may be grouped in four general headings to emphasize major life history features related to fire and site. Thus species exhibit traits that act (1) to **prevent** fire damage, (2) to **recover** following fire damage, (3) to **colonize** sites after fires, and (4) to **promote** fire in their habitat. Especially fire-resistant or susceptible species typically exhibit the extreme expression of these traits. The adaptive mechanisms that act in preventing fire damage, in recovering from fire damage, colonizing burned-over areas, and promoting fire occurrence are described in the following list.

### Preventing Fire Damage

- Thick insulated bark (many pines and oaks, western larch, giant sequoia, redwood).
- The "grass stage" in longleaf pine (see Chapter 16, p. 430).
- Deep rooting; tap root in young plants (upland oaks and hickories).
- Rapid juvenile growth—crown grows above the surface fire zone and heat-resistant bark is formed (pines).
- Basal crook—dormant buds on the lowermost stem are protected from fire by a crook of the stem that brings the buds in contact with mineral soil (pitch, shortleaf, pond, and other hard pines).
- Branch habit and self-pruning ability—rapid self-pruning of branches decreases likelihood of crown fire, whereas low or

drooping branching habit and poor self-pruning ability increase the likelihood of crown fire (larches, pines, and Douglas-fir self-prune well under stand conditions, whereas true firs, hemlocks, and spruces retain their branches and often exhibit drooping branches).

- Stand habit—open-grown stands decrease the probability of crown fire and also afford less fuel (western larch, ponderosa pine, longleaf pine).
- Fire-resistant live foliage (hardwoods much less flammable than conifers, for example, maples, basswoods, beeches; among conifers larches have less flammable foliage than pines, Douglas-fir, and true firs).
- Rapid foliage decomposition—retards fuel accumulation and reduces the opportunity for fire ignition and spread (sugar maple, basswood).

The differential resistance of tree species to fire damage and mortality was shown by Flint (1930). He determined the relative resistance of the major tree species of the northern Rocky Mountains using characteristics of bark thickness of old trees, rooting habit, resin in old bark, branch and stand habit, relative flammability of foliage, and abundance of lichens on stems. Very resistant were western larch, ponderosa pine, and Douglas-fir, whereas western hemlock and subalpine fir were low to very low in resistance. Although species vary greatly in resistance, severe fires largely erase differences in resistance (Wellner, 1970).

## Recovering from Fire Damage

- Sprouting—new shoots arise from various portions of the plant following fire damage:

  Root collar or stump (oaks, paper birch, black cherry, chaparral species, redwood). Sprouting is rare in conifers but in hard pines (shortleaf, pitch, longleaf, Virginia, Monterey, etc.) sprouts may occasionally arise from dormant buds formed in the axils of primary needles at the base of the stem (Stone and Stone, 1954).

  Root (trembling and bigtooth aspen, rock elm, sweet gum, sassafras).

  Rhizomes (many herbaceous plants and some shrubs).

  Branches (layering)—branches are pressed into contact with the soil by snow or accumulating litter and take root (firs, black spruce, chaparral shrubs).

  Bole—dormant buds along the bole initiate new shoots after the crown is killed (pitch pine, big-cone Douglas-fir, some eucalyptus species).

- Tap root or well-developed deep root system—undamaged root system provides food reserves for rapid regeneration of new shoots (upland oaks and hickories).

**Colonizing Burned-Over Areas**

- Early seed production—enables a species to reproduce itself sexually on a site where there are short intervals between fires (jack pine, lodgepole pine, pitch pine).
- Light, wind-borne seeds—enable species growing at some distance from a burn to disperse seeds to that site. For example, in presettlement times trembling aspen tended to be concentrated along watercourses and in swampy sites except where burns occurred in upland pine and hardwood forests. Its production of many, light, cotton-tufted seeds gave it access to sites not otherwise available for colonization.
- Serotinous cones—closed cones containing viable seeds and persisting on branches are typical of many trees of various pine species (jack pine, lodgepole pine, pitch pine, and certain southern and western pines). Black spruce exhibits semiserotinous cones; the cone scales open upon drying and close when wet. Seed dispersal is therefore periodic, sometimes occurring over a 2-year period.
- No dormancy—certain species in fire-dominated habitats have seeds that do not enter true dormancy and thus germinate readily at any favorable time following postfire rains (jack pine, lodgepole pine, longleaf pine, pitch pine).
- Heat-induced germination—hard-coated seeds of certain species of *Arctostaphylos*, *Ceanothus*, and *Rhus* tend to lie dormant in the soil (in redstem ceanothus up to 150 years (Mutch, 1976)). Germination is favored by fire, which cracks the seed coats and generates the heat needed to stimulate germination.

**Promoting Fire Occurrence.** Traits that increase the likelihood of fire:

- Flammable foliage and bark—needles of pines and many conifers are highly flammable, decompose slowly, and form a ready fuel source for surface fires (Mutch, 1970). The bark of certain species, such as paper birch and some eucalyptus species, is flammable.
- Retention of foliage—promotes crown fires (firs, northern white and western red cedar, juvenile oaks).
- Short stature—brings foliage close to the ground where a surface fire may spread to the crown (young or slow growing trees with flammable foliage, jack pine).

Some communities are strongly affected by fire, and characteristics of their sites (climate, topography, and soil) increase the likelihood of fire. Seasonally hot, dry sites favor fire and support species that not only have fire adaptations but are physically adapted to live in such a severe environment. Thus most of the above traits evolved not only in response to fire, but in response to various selection pressures of the plant's total environment. Development of tap roots of many young pines, oaks, and hickories prevents fire damage and provides resources for revegetation following a destructive fire (Whitford, 1976). However, the tap root is probably of greater importance as an adaptation to moisture stress, which occurs regularly every year, whereas fire is more irregular in occurrence. Also, seed size is probably not as strongly adapted to fire as to the soil moisture conditions of the seedbed. Species native to dry sites typically have larger seeds than those adapted to mesic or wet-mesic sites.

Nevertheless, where fire plays a dominant role many life history characteristics appear strongly adapted to fire. This is the case with strongly fire-dependent species such as jack pine and lodgepole pine of northern and western North America, longleaf pine of the American South, and ponderosa pine of the West. Jack pine and lodgepole pine exhibit many of the fire adaptations cited above and cone serotiny in particular. Closed cones may persist on the tree 10 to 25 years and still bear viable seeds, even after being overgrown by wood. In lodgepole pine, millions of seeds per ha may be stored in serotinous cones until a fire releases them.

Both jack and lodgepole pines may bear cones that open readily (nonserotinous) or remain closed under normal climatic conditions. The cone scales are bonded by resin, which melts from fire-generated heat above 45°C; the scales open as they dry and the seeds are disseminated. Cones may be exposed to temperatures of 900°C for 30 seconds, and seeds still have high viability. The range of closed cones per tree in jack pine is from 0 to 100 percent; the average for the species is about 78 percent (Schoenike, 1976). Open-cone trees are found in the more southerly portion of the species' range, whereas closed-cone trees predominate in the northern and western range (Figure 11.2). In the southern portion of its range, jack pine occurs in mixed stands with oaks and other pines, and fires are of the lighter, surface type than in the northern forest. In the northern forest, jack pine is regenerated by crown fires, which open the cones and prepare the seedbed, thereby perpetuating the predominance of the serotinous habit.

In pitch pine, a local portion of its distribution on the coastal plain of New Jersey, the pine barrens, has nearly 100 percent closed-cone trees. The frequency of serotiny decreases markedly away from this pocket (Ledig and Fryer, 1972). Frequent and severe fires on the

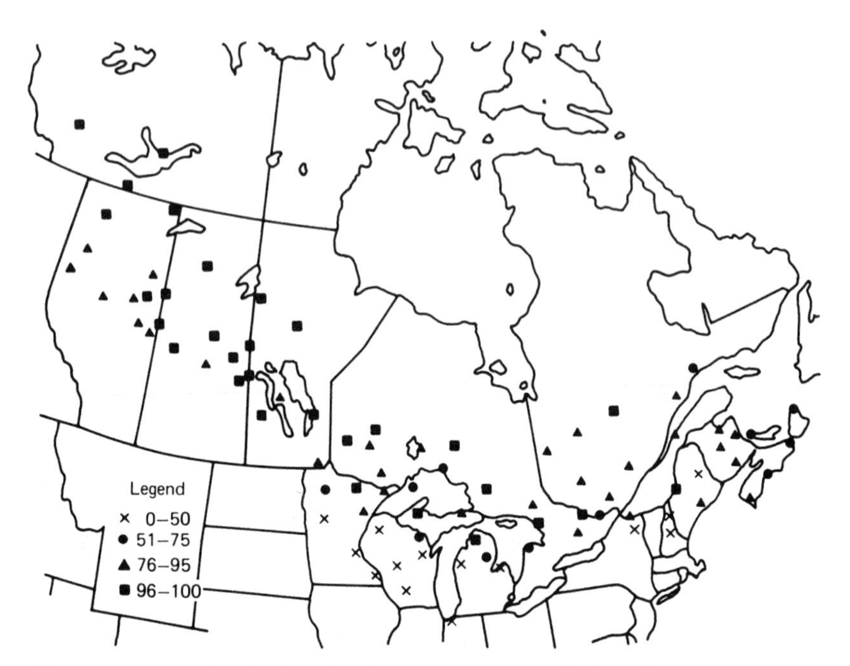


**Figure 11.2. Variation in the percentage of closed cones per tree in jack pine (After Schoenike, 1976. Figure reprinted courtesy of Roland E. Schoenike, Agricultural Experiment Station, University of Minnesota.)**

barrens are probably responsible for maintaining this pocket of high cone serotiny.

Fire also acts to sanitize stands of jack and lodgepole pines, both of which are parasitized by dwarf mistletoe (*Arceuthobium americanum* Nutt. ex Engelm.). Fire acts as the primary natural control agent where the entire stand, and hence the infection source, is destroyed by a severe crown fire (Alexander and Hawksworth, 1975). The newly regenerated stand is relatively free of infection because pine seedlings reestablish on the burned area much faster than the parasite.

## FIRE AND THE SITE

In early days, the tendency of forestry writers was to consider fire as a destructive agent with few or no beneficial aspects. The development of interest in prescribed or controlled burning as a tool in silviculture, fire hazard reduction, and fire management, however, has caused a reevaluation of the effect of fire on the site. This reevaluation has led to the realization that the effect of fire on forest land and its productivity is complex and may often be entirely beneficial. We may distinguish between the indirect effects of fire on site quality through its effects on vegetation, litter accumulation, and soil organisms, and its direct effects on soil properties and microclimate.

### Indirect Effects

The indirect effects of fire depend upon the changes in the vegetation. These are discussed in detail in Chapter 15. Since an intense fire will kill most or all of the plant life above the soil surface, the succeeding vegetation tends to be made up of light-seeded species that can move in from outside the burned area, species with perennial root systems capable of sending up new sprouts, and species with dormant seeds stimulated by heat. Many legumes fall in these categories, and the abundance of these and other nitrogen-fixing plants is often increased by burning. In such a case, although previously accumulated nitrogen is volatilized, there may shortly be a net increase in available nitrogen, and the overall site quality may be temporarily improved. However, in many parts of the world, recurrent fires favor the development of a shrubby vegetation composed of sprouting species with characteristically tough foliage, low in nutritive value, and slow to decompose. Heather in northern Europe, blueberry species and bracken in many countries, junipers around the Northern Hemisphere, scrub oaks in an equally widespread region, and the many species of chaparral (the broad-sclerophyll scrub or brushland type of vegetation) of California and the Southwest—all are plants that become dominant after heavy and repeated fires. The heaths, the blueberry barrens, the juniper and oak woodlands, and the chaparral are fire types. In them the soil will usually deteriorate under the influence of woody and impoverished litter until the soil can no longer support the original forest vegetation. The generalization is dangerous, for among the junipers, oaks, and chaparral species are some soil-building species, but the overall detrimental effect of most of these brushland fire types on the site cannot be denied.

To reclaim heathland back into forest, it is frequently necessary to break the hardpan with a deep soil plow and to mulch and fertilize the planted tree seedlings. Even then, only trees such as Scots pine, black pine, and lodgepole pine that make little demand on the soil can be expected to thrive, and these will grow slowly for many years.

Fire regulates dry matter accumulation, thereby controlling the severity of burning. This affects the density and composition of forest vegetation, which influence site quality. Abnormally long intervals between fires, such as those caused by prolonged fire exclusion, usually lead to high concentrations of organic matter. The intense and destructive fires, which follow sooner or later, may preclude or delay the reestablishment of normal vegetation on the site or change the kind of vegetation present. The indirect effects of change of vegetation or lack of vegetation, however, are much less than the direct effects of severe burning, which are described below.

Soil flora and fauna significantly influence site quality by decomposing organic matter, fixing nitrogen, and providing aeration. The effects of burning on soil organisms are highly variable depending on the intensity of the fire, depth below the surface, time elapsed following burning, and the nature of the soil and vegetation of the site (I. Ahlgren, 1974). Changes in populations of bacteria, actinomycetes, and fungi are most evident in the upper 2 to 5 cm. Decreases are typically observed immediately following intense fires. No postfire decreases are normally found on light burns, in soil below the surface 5 to 8 cm, and at 6 months to 1 year after burning. Increases in microorganisms are typically found after fire, often multifold, apparently due to the sudden availability of mineral nutrients, increases in soil pH, and other soil chemical changes associated with burning. Reinoculation may occur quickly from windblown spores or other debris and by invasion from subsurface layers. Moisture is very important; moist soil and rainfall following fire favor reinoculation and increase microbial populations.

Little is known about the response of soil fauna to fire. Populations of earthworms, beetles, spiders, mites, collembola, centipedes, and millipedes are typically reduced by burning but increase thereafter (I. Ahlgren, 1974). Earthworms tend to be affected more by postfire loss of soil moisture than by the heat generated in burning. Ants are less affected than other fauna because their habits enable them to survive in lower soil layers. Furthermore, they are adapted to xeric conditions of postfire topsoil. Although many ants are often destroyed by fire, their social organization and rapid colonizing ability enable them to reestablish populations rapidly after burning. In general, soil fauna are more severely affected in forest sites than in grasslands. The effect of fire is greater in the forest because forest fires may be hotter than grassland fires due to a greater accumulation of fuel. In addition there is a change from cool and moist forest floor to xeric postfire conditions, there are greater temperature fluctuations, and the fauna may lack food.

## Direct Effects

The direct effects of fire on site quality arise from two principal sources: the burning of organic matter above and on the mineral soil and the heating of the surface layers of the soil. The burning of organic matter results in the release of carbon dioxide, nitrogenous gases, and ash to the atmosphere and the deposit of the minerals in the form of ash. The wood and litter ash is more soluble than the organic matter from which it was formed. Thus the effect of fire is to increase the amount of available minerals, at least temporarily, to lessen the soil acidity and increase base saturation, to decrease the supply of total nitrogen, and to change the moisture and temperature conditions of the site.

Two examples are indicative of many studies on fire effects. In the ponderosa pine region of Arizona, burning was found to increase the soluble nutrients as a result of the ashing of the surface layer of unincorporated organic matter (Fuller et al., 1955). This caused an increase in the pH, available phosphorus, exchangeable bases, and total soluble salts, and a decrease in organic matter and nitrogen to a depth of 20 to 30 cm. Microbiological activity, particularly of bacteria, increased as a result of burning. On the negative side, the surface was compacted by rains following the removal of litter, resulting in a decrease in the rate of water penetration. In another study in eastern Washington, (Tarrant, 1956a), however, burning was not found to be detrimental and perhaps even slightly beneficial in its effects on permeability and associated physical properties of the soil.

In the Douglas-fir region, Tarrant (1956b) has investigated the effects of slash burning on physical soil properties. Light burning increased the percolation rate of water within the surface 8 cm of soil, but severe burning confined to less than 5 percent of intentionally burned, logged-over sites did seriously impede water drainage about 70 percent.

Unfortunately, the same characteristics of small particle size and high solubility that render the ash minerals readily available to plants also render the ash susceptible to leaching and erosion by rainwater. If the ash is washed down into the soil so that the roots can absorb the nutrient ions dissolved from it, site quality is usually improved, at least temporarily. If, on the other hand, the ash is leached down below the tree roots (or is washed off the surface) then site quality is lowered. In general, the former may occur on level soils of sandy to loamy texture while the latter is apt to happen on very coarse sands or heavy soils, particularly those with considerable slope.

The loss of total nitrogen through volatilization is widely recognized and is related to the intensity of the fire. Knight (1966), working in the coastal Douglas-fir region, found no nitrogen loss in soils heated to 200°C, a 25 percent loss at 300°C, and a 64 percent loss at 700°C. Nitrogen loss is also proportional to the amount of dry matter of fuel consumed, and considerable nitrogen may be lost during intense fires. For example, the severe Entiat fire in a second-growth, mixed-conifer stand in north-central Washington resulted in a loss of about 97 percent nitrogen originally in the forest floor and a loss of two-thirds of the nitrogen of the $A_1$ horizon of mineral soil (Grier, 1975). Although replacement of the nitrogen by precipitation alone would require about 900 years, nitrogen reaccumulation will result much sooner due to the combined action of symbionts associated with snowbrush *(Ceanothus velutinus)*, increased activity of soil organisms, and weathering of soil parent materials. Some losses of calcium, magnesium, potassium, and sodium also occurred during the fire, primarily by ash convection and volatilization. Mineralized cations of these nutrients, however, were rapidly leached into the mineral soil, and large amounts of them were retained by the soil.

Much of the nitrogen lost through burning of litter, humus, and vegetation is not in a form available to the plant. For example, the humus layer in coniferous forests, especially in boreal regions, contains a large amount of nitrogen, but only a minute part of it is in a form that the vegetation can use (Viro, 1974). To be made available for uptake it must first be mineralized, that is, it must be converted into ammonia or nitrate nitrogen (Chapter 9, Figure 9.6). The ability of the succeeding vegetation and soil bacteria to replace the available nitrogen lost in burning is an important factor determining the effect of fire on site quality. The higher pH due to release of mineral bases in the soluble ash can provide a more favorable soil environment for the free-living, nitrogen-fixing bacteria and thus results in an immediate increase in available nitrogen. Although the loss of nitrogen is widely cited as a deleterious effect of fire, the significance of the loss for the new regeneration and the overall nutrition of the ecosystem is not well known. For boreal forests dominated by Norway spruce, Viro (1974) concludes: "burning unquestionably results in great losses of total nitrogen from the site, but simultaneously results in an increase in mineralized nitrogen. The former is practically unimportant, the latter of great consequence."

The actual heating of the mineral soil is of relatively less importance than the action of fire on the organic matter. The heat of the fire does not penetrate far into the soil. Even under a hot fire in logging slash, temperatures seldom exceed 90°C at 3 cm down in the mineral soil. Light surface fires only heat the top centimeter of the mineral soil to near the boiling point.

In this heated zone, soil aggregates may be broken down, first by the heat and later by the direct striking action of raindrops, resulting in loss of soil structure and in lowered infiltration capacity of the surface soil. In extreme cases—which are rare—clay soils may be baked hard, and soil organisms will be destroyed. Burning of logging debris can provide such an extreme example. Slash burning is typically practiced following clearcutting to reduce the fire hazard, particularly of coniferous debris. Slash is either scattered and "broadcast" burned or piled and burned. The latter technique concentrates the fuel and the effects of burning over 20 to 30 percent or less of the area; similarly, the nutrients in the debris are also concentrated. In many cases, burning heavy accumulations of slash in piles causes intensely hot fires and alters the physical structure of the soil. The underlying soil may be baked and its structure markedly altered. For example, burning piles of coniferous slash in northwestern Montana resulted in lower tree densities and markedly slower growth of conifers on burned slash pile sites compared to adjacent unburned areas (Vogl and Ryder, 1969). The growth depression was attributed to impaired physical soil properties, especially decreased water infiltration. In general, broadcast burning is the more desirable technique because it approximates naturally occurring wildfire in spreading burning effects over a greater surface area (DeByle, 1976).

Although the effects of direct heating on the mineral soil are many and varied, in general their sum total does not alter the site quality to any marked extent for any substantial period of time. Except in the rare burn that creates extreme heat within the mineral soil, the effects of fire on site quality are best interpreted in the light of its effect upon the soil through the destruction of organic matter.

## Detrimental and Beneficial Effects of Fire

There are some situations where fire is obviously catastrophic. These include cases where the soil is composed almost entirely of organic matter and those where the destruction of organic matter exposes highly erodible soils to heavy rain.

The burning of a peat bog after drainage or a series of dry seasons literally results in the complete destruction of the soil and the return to swamp conditions. Thousands of years of peat accumulation are necessary to replace the lost organic soil, and such areas can be virtually eliminated from our productive sites. In the United States, many bogs in the Atlantic Coastal Plain and in the recently glaciated parts of the Northeast and the Lake States have been destroyed as forest sites by fire.

Similarly, the burning of humus lying directly on top of rock will eliminate the soil and destroy the site. In glaciated portions of Canada and the northeastern United States, and in many mountain regions, thin accumulation of humus provides the only nutrition for forest trees. Fire in such cases often burns down to bedrock with disastrous results.

Still another bad situation occurs where highly erodible soil of steep slopes is exposed by fire burning the organic protection. Fires can cause serious erosion in the northern Rocky Mountains where erosive soils occur on steep slopes; the fires tend to be catastrophic, and multiple burns frequently occur. The classic example is in southern California, where chaparral species and the organic matter from them protect the granitic or heavy soils lying at approximately the angle of repose in steep mountains. Even in the absence of fire, erosion is high. In the San Gabriel Mountains near Los Angeles, studies of debris movement on steep slopes covered by old chaparral revealed that each year an average of 8000 kg of debris per hectare moved down the slopes to the stream channels (Anderson et al., 1959; Krammes, 1965). Chaparral is highly susceptible to burning, especially when it reaches about 30 years of age, because of the dense, highly flammable fuel accumulations, shrubby growth habit, and seasonal dry, windy periods (Biswell, 1974). Following wildfire, debris movement increased dramatically, and on south-facing slopes reached 10 times that of the already high prefire rate (Figure 11.3). In this instance, most of the movement was during the dry period. However, if, after a fire, the heavy rainstorms characteristic of this semiarid region strike before revegetation of the burn, whole slopes may wash downhill. Such massive erosion not only lowers the growing potential of the soil but frequently wreaks havoc on the valleys below.

The burning of litter and organic matter in the soil may be significant in causing reduced infiltration, increased surface runoff, and erosion in many areas of the western United States where water-repellent soils have been reported (DeBano et al., 1967; Meeuwig, 1971; Dyrness, 1976). For a number of years California scientists were puzzled by the sight of "dusty tracks in the mud" in freshly burned watersheds after fall and winter rains. Now it is known that a variety of soils can become resistant to wetting. These are soils in which the particles repel water; droplets do not readily penetrate and infiltrate but "ball up" and remain on the soil surface for variable periods of time. This phenomenon is widespread in sandy soils throughout much of the wildland areas of western North America supporting chaparral or coniferous vegetation, as well as many other parts of the world (Foggin and DeBano, 1971).

Fire plays an important role in the formation of water-repellent soils. In unburned areas, litter decomposition produces nonwettable or hydrophobic organic molecules that coat surface soil particles, creating a weak water-repellent layer in the upper soil profile between the litter layer and the mineral soil (Figure 11.4; DeBano, 1969). During a fire, litter is consumed and the hydrophobic sub-

**Figure 11.3. Dry-creep erosion may be severe immediately after intense wildfire in old-growth chaparral on slopes above the angle of repose. (Photo courtesy of the Pacific Southwest Forest and Range Experiment Station, U.S. Forest Service.)**

stances are volatilized and diffuse downward into the soil and condense on cooler soil particles (Figure 11.4). After summer and fall wildfires, water repellency is high and rain falling on the soil surface infiltrates readily until impeded by a nonwettable layer. After the wetting front encounters the repellent layer, infiltration is slowed, surface runoff begins, and erosion may readily occur.

Just as there are cases where fire is catastrophic in its effects on site, so there are cases where it is clearly beneficial. Such is the case for sites in the far north where dampness and cold retard the decomposition of organic matter, giving rise to thick mats of highly acid raw humus. In Norway, Sweden, and Finland, considerable success has been achieved by burning such sites to raise the site quality by improving nutrition, moisture, and temperature conditions (Viro, 1974).

Postfire temperature conditions are often changed drastically by fire. Temperature extremes are typically greater on burned sites than on unburned sites. Average maximum soil surface temperatures on

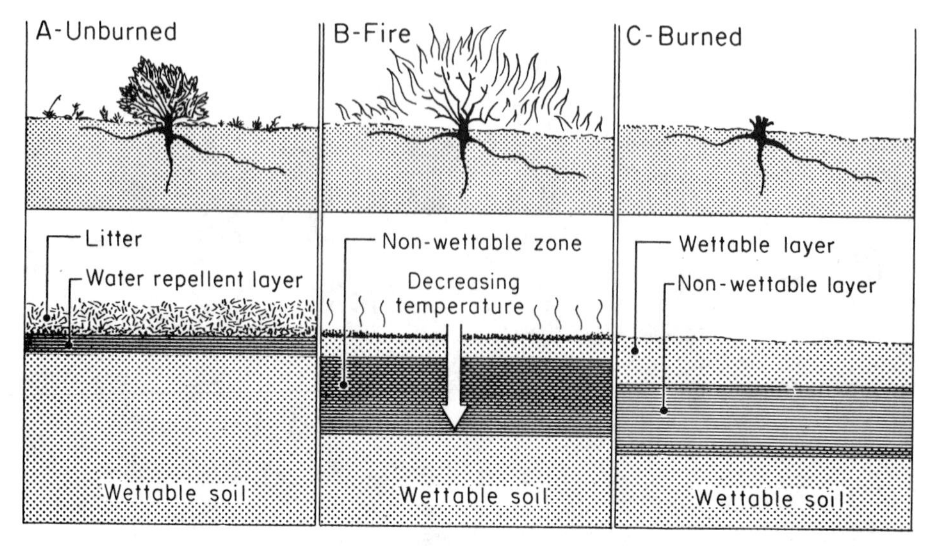


**Figure 11.4. Soil non-wettability before, during, and after fire. (A) Before fire, the non-wettable substances accumulate in the litter layer and mineral soil immediately beneath it. (B) Fire burns vegetation and litter layer, causing non-wettable substances to move downward along temperature gradients. (C) After a fire, non-wettable substances are located below and parallel to soil surface on the burned area. (After DeBano et al., 1967.)**

burned sites may be from 3° to 16°C higher than on comparable unburned areas (C. Ahlgren, 1974). Increased soil temperatures hasten spring development of roots and shoots on burned areas, speed decomposition, and promote the activity of soil organisms. Extremely high temperatures, due to the blackened soil surface and the presence of charcoal, may cause seedling mortality and delay forest development. On fire habitats, however, fire-dependent species are typically adapted to tolerate extreme conditions. In boreal climates, where thick layers of raw humus tend to develop under spruce and fir vegetation, site quality is markedly improved by burning because it improves the postfire thermal regime.

In most situations, however, the effect of fire on site quality is relatively less pronounced. Repeated burning—once haphazard and now more or less controlled—in the sand plains of the southern pine region has apparently had no major detrimental effect on site quality and, in fact, has been shown locally to be beneficial to the soil (Metz et al., 1961). In loess soils under even-aged shortleaf and loblolly pines on flat terrain in Arkansas, nine successive annual burns had little effect on the nutrient content or structure of the topsoil (Moehring et al., 1966). Studies of soil microorganisms after 20 years of annual prescribed burning on a very fine sandy loam in the coastal plain indicated that burning had no effect on the total number of fungi per gram of soil, although it did reduce their total number through a decrease in weight of the organic horizon (Jorgensen and Hodges, 1970). The number of bacteria in the organic layer was reduced by annual burning but not in mineral soil. For the organisms studied, there was little indication that prescribed burning adversely affected metabolic processes.

Controlled burning in a Douglas-fir—larch forest on sandy loam soils in northwestern Montana generally reduced accumulated fuels without nutrient loss, runoff, or erosion (Stark, 1977). The "biological life of the soil," the years that a particular soil is capable chemically of supporting trees, was quantified using a formula based on nutrient losses through burning, erosion, harvesting, and the like. The estimate of 55,000 years showed that burning could be conducted on this soil a very long time with no problems of soil fertility. The biological life of the soil was so long that major catastrophic climatic, erosional, or glacial events are likely to occur and change natural processes over such a long time span.

In the sand plains of the Lake States, organic matter provides the major source of colloids for soil nutrition. Burning the sand plains, therefore, may be undesirable and has been shown by Stoeckler (1948, 1960) to reduce site quality of trembling aspen, a species of relatively high nutrient requirements. In contrast, 10 years' experience with light prescribed burning on sandy soils of Minnesota indicates that site productivity will not be altered for red pine, a species with low nutrient requirements (Alban, 1977). There seems no reason to fear site deterioration in burning the more level pinelands of the Western states. When dealing with clay soils and steep slopes, however, fire may do no harm if gentle rains impound the ash and revegetation anchors the soil. And then, again, if this does not happen, the site may be greatly deteriorated.

## SUGGESTED READINGS

Ahlgren, I. F., and C. E. Ahlgren. 1960. Ecological effects of forest fires. *Bot. Rev.* 26:483–533.

Ahlgren, Isabel F. 1974. The effect of fire on soil organisms. *In* T. T. Kozlowski and C. E. Ahlgren (eds.), *Fire and Ecosystems.* Academic Press, Inc., New York.

Biswell, Harold H. 1974. Effects of fire on chaparral. *In* T. T. Kozlowski and C. E. Ahlgren (eds.), *Fire and Ecosystems.* Academic Press, Inc., New York.

Harvey, A. E., M. F. Jurgenson, and J. J. Larsen. 1976. Intensive fiber utilization and prescribed fire: effects on the microbial ecology of forests. USDA For. Ser. Gen. Tech. Rep. INT–28. Intermountain For. and Rge. Exp. Sta., Ogden, Utah. 46 pp.

Komarek, E. V. 1973. Ancient fires. *In Proc. Tall Timbers Fire Ecology Conference,* No. 12, 1972:219–240. Tall Timbers Res. Sta., Tallahassee, Fla.

Lotan, James E. 1976. Cone serotiny—fire relationships in lodgepole pine. *In Proc. Montana Tall Timbers Fire Ecology Conference and Fire and Land Management Symposium,* No. 14, 1974:267–278. Tall Timbers Res. Sta., Tallahassee, Fla.

Mutch, Robert W. 1970. Wildland fires and ecosystems—a hypothesis. *Ecology* 51:1046–1051.

USDA Forest Service. 1971. Prescribed burning symposium proceedings. Southeastern For. Exp. Sta., Asheville, N.C. 160 pp.

Viro, P. J. 1974. Effects of forest fires on soil. *In* T. T. Kozlowski and C. E. Ahlgren (eds.), *Fire and Ecosystems.* Academic Press, Inc., New York.

Vogl, Richard J. 1968. Fire adaptations of some southern California plants. *In Proc. California Tall Timbers Fire Ecology Conference,* 7:79–110. Tall Timbers Res. Sta., Tallahassee, Fla.

Wright, H. E., Jr., and M. L. Heinselman. 1973. The ecological role of fire in natural conifer forests of western and northern North America: introduction. *Quaternary Res.* 3:319–328.

# Part III

# The Ecosystem: Site, Community, and Ecosystem Analysis

# 12

# Site

In the preceding chapters, the forest tree and the individual environmental factors that affect it have been considered separately, yet within an ecosystem framework. In the succeeding chapters, we recognize the ecosystem (biogeocoenosis) as the overall focus while examining the site (the collective of physical and biotic factors, Chapter 12), the forest community and its animal (Chapter 13) and plant (Chapters 14–17) components, and the integrated analysis of whole ecosystems (Chapter 18). It is clear that communities and individuals are interrelated with one another and their environment through the cycling of water and nutrients, the circulation, transformation, and accumulation of matter, and the multiplicity of regulatory mechanisms, which limit the numbers of plants and animals and influence their physiology and behavior. The term ecosystem is the most concise formulation of this concept of an interacting system comprising living organisms together with their nonliving habitat.

The consideration of the forest ecosystem in Part III as a dynamic entity, changing in time and in space, is illustrated in Part IV by reference to the historical development and distribution of the present-day forests themselves, primarily those of North America. This final portion, dealing with historical forest geography, provides an understanding of the forest as it actually is.

Forest scientists are primarily concerned with the forest tree segment of the ecosystem. They wish to evaluate the suitability of various tree species or other genetic entities for a given site, to rate the competitive ability of alternative species that are capable of growing in that locality, and to estimate the growth potential—particularly the productivity of wood substance—of the forest communities that occupy the site.

The forest ecosystem may be divided into the forest trees and associated plants and animals (the **biome** or **biocoenosis**), the sites they occupy (defined by position in space), and the environmental conditions associated with these sites. The term **site** (habitat) usually includes both the position in space and the associated environment. The forest site quality thus is defined as the sum total of all of the factors affecting the capacity to produce forests or other vegetation: climatic factors, soil (edaphic) factors, and biological factors.

The forest scientist is faced with the problem of integrating all the various site factors, to produce an estimate of the forest site quality. Statistically, the site factors are treated as independent variables and some measure of forest growth as the dependent variable. Yet, in fact, all are part of the same interacting ecosystem. The site factors are not only interdependent, but are also dependent in part upon the forest, which is itself a major site-forming factor. Because of these interactions, the simple regression technique of estimating site quality from an evaluation of a few important site factors, important as it is in practical forest ecology, can only be approximate. Only by considering the forest and the site together as a complex interrelated ecosystem can the true dynamic nature of both be fully understood.

Nevertheless, the estimation of forest productivity is of the utmost importance in both forest ecology and in silvicultural management. This productivity, or actual site quality, may be measured directly for a few forests where accurate long-term records of stand development and growth have been maintained. Generally however, it can only be estimated indirectly by one or more of these alternatives:

A. Vegetation of the forest
   1. Trees (site index)
   2. Ground vegetation (indicator species and species groups)
   3. Overstory and understory vegetation in combination
B. Factors of the physical environment
   1. Climate
   2. Soil and topography
C. Multiple factor or combined methods (using some or all of the foregoing factors and forest land-use history)

These are the alternatives discussed in the present chapter. Significant reviews of forest site quality have been published by Coile (1952), Rennie (1962), Ralston (1964), Jones (1969), and Carmean (1970a; 1977). A comprehensive review of site quality evaluation, its history, methods, and applications was prepared by Carmean (1975).

## DIRECT MEASUREMENT OF FOREST PRODUCTIVITY

Actual forest productivity is generally measured in terms of the gross volume of bole wood per acre or hectare per year over the normal rotation. This gross mean annual increment (m.a.i.) may be computed from long-term permanent sample plot data. For instance, on pumice soil sites on the North Island of New Zealand, Douglas-fir has been computed to yield a gross m.a.i. of 439 $ft^3$ $acre^{-1}$ $yr^{-1}$, or 31 $m^3$ $ha^{-1}$ $yr^{-1}$ (Spurr, 1963). Yields of 36 $m^3$ $ha^{-1}$ $yr^{-1}$ (516 $ft^3$ $acre^{-1}$ $yr^{-1}$) may be expected from Monterey pine on similar sites (Spurr, 1962). In the United States, gross mean annual increments range upward to perhaps 15 $m^3$ $ha^{-1}$ $yr^{-1}$ (210 $ft^3$ $acre^{-1}$ $yr^{-1}$) on the best sites. Average productivity in the temperate forests of North America and Europe is approximately 5 $m^3$ $ha^{-1}$ $yr^{-1}$ (70 $ft^3$ $acre^{-1}$ $yr^{-1}$). The growth is presented in gross values—the total amount of wood put on by all trees within a given unit of time without deduction for natural mortality removal by humans or decrease in wood volumes by rot. By consistently using such gross values, comparable increment measurements can be obtained.

Unfortunately, actual gross productivity data like these are scarce. Furthermore actual yield is conditioned not only by site factors but also by genetic factors (of species and race), age or rotation, by the biotic history of the stand, and by stand density. Nevertheless, actual growth represents the proven productivity of a site and therefore may be taken as the closest available approximation of potential productivity.

*Theoretically, a stand of a given species of a given age on a given site will produce the same amount of wood a year at various densities of stocking as long as the site is fully occupied.* As long as the trees in the unthinned stands retain good crown development and vigor, they will fully occupy the site. Conversely, if the trees in thinned stands or even in open stands fully occupy the soil to the extent of being able to fully utilize available soil moisture, they will normally fully occupy the site even if excess crown space is available in the stand. These being the cases, thinned and unthinned stands, otherwise comparable, should give the same total increment. To make such a comparison correct, however, gross growth should be computed by adding back in mortality within a growth period, whether from man's cutting activity or from natural causes.

In recent years, forest ecologists have been attempting to estimate forest productivity in terms of all components of the forest ecosystem rather than the growth of the tree boles alone. Researchers around the world have been sampling not only the stems but also the branches, the leaves, the organic matter in the forest floor, and even the animals inhabiting the forest, to provide a more exact appraisal of the entire forest ecosystem; their studies are considered in Chapter 18. For practical forest site evaluation purposes, however, the wood content of the boles or main stems of the forest trees remains the best-known and most useful measure of forest productivity.

## TREE HEIGHT AS A MEASURE OF SITE

The height of free-grown trees[1] of a given species and of a given age is more closely related to the capacity of a given site to produce wood of that species than any other one measure. Furthermore, height of free-grown trees is less influenced by stand density than other measures of tree dimensions and may thus be used as an index of site quality in even-aged stands of varying density and silvicultural history.

The height of the dominant portion of a forest stand at a specified standard age is commonly termed **site index** even though tree height is but one of many indices of site quality used in forest ecological and silvicultural investigations (Figure 12.1).

In the United States, site index has long been defined as the average height that the dominant, or dominant and codominant, portion of the even-aged stand will have at a specified age. This standard age is generally 50 years in the eastern United States, and 100 years for the longer-lived species of the West Coast. Occasionally, other standard ages are specified for a particular species or region, for example, 35 years for pulpwood rotations in the South.

Still, measuring both dominant and codominant trees for height is not always satisfactory. First, these tree classes are subjective, and opinions may differ widely as to what constitutes dominant and codominant trees. Second, many of the codominant trees will drop out of the main canopy as the stand ages and, therefore, perhaps should not be measured. Third, thinning and other cutting operations may artificially change the average height of the dominant and codominant trees without, of course, changing the actual site quality. Finally, it is often very difficult to see the tops of codominant trees in tall and dense timber, and it is therefore difficult to measure their heights accurately.

For these reasons there is a tendency to restrict more carefully the trees that should be measured for site determination. A preferable practice is to measure the height of an objectively determined sample of the larger trees in the stand. In the British Commonwealth, the concept of **top height** is widely used. This refers to the arithmetic mean height of the 250 largest-diameter trees per hectare. Since, however, it is seldom feasible to measure so many trees per hectare for height, and since heavy thinning practices frequently reduce the stand to fewer than 250 trees per hectare, the present tendency is to restrict top height to a smaller sample of the very largest trees. Thus a mean height based on the largest 100 trees per hectare (sometimes

[1]Trees that have grown from the time of their establishment without an overstory canopy above them to suppress them or slow their development.

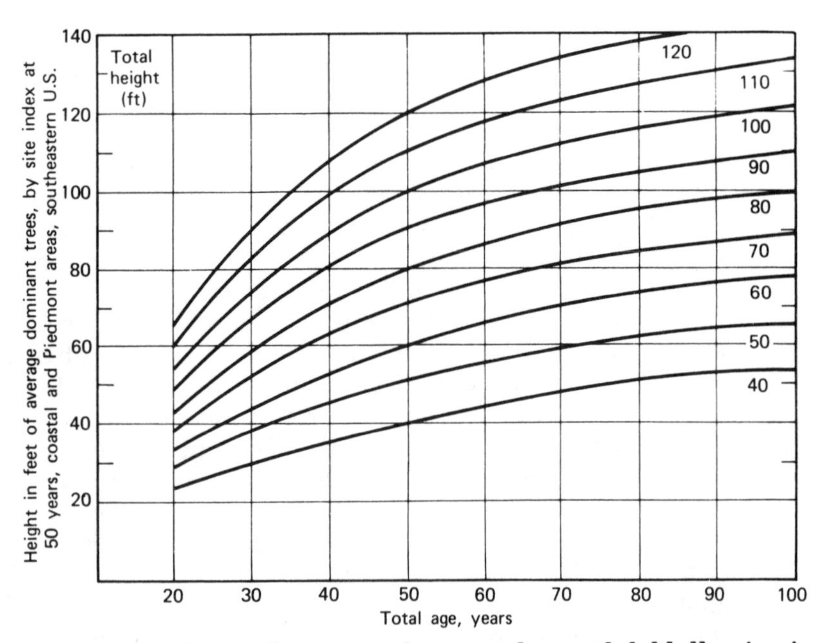


**Figure 12.1. Site-index curves for second-growth loblolly pine in the coastal plain and Piedmont areas of the southeastern United States. Example: If the total height of dominants in a second-growth loblolly pine stand averaged 55 feet (*y* axis) at age 60 (*x* axis), the site index is 50, indicating a relatively poor site. If the corresponding height and age values are 85 feet at 40 years in another stand, the site index is 95, indicating a relatively good site. (After Hampf, 1965. Source: U. S. Dept. Agr. Misc. Publ. 50, 1929; revised by Coile and Schumacher, 1953.)**

defined as mean **predominant height**) is used. Top heights based upon the 60 largest-diameter trees per hectare, the 8 to 12 largest trees, or even upon the 1 largest tree per hectare are also used.

Although height is perhaps the single tree measure best related to the site productivity of a given species, it does not necessarily follow that it is completely unrelated to other factors, nor that a perfect correlation may be obtained between stand top height and site productivity. In particular, stand density, particularly extremes of stand density, may influence height growth. Under such circumstances, site-index curves should be developed separately for different stand density classes, as has been done for ponderosa pine in the Inland Empire (Lynch, 1958) and lodgepole pine in the Rocky Mountains (Alexander, 1966; Alexander et al., 1967).

The genetic factors of a tree or population may well control height growth to a great extent on a regional basis. The many plantings of trees of different provenances testify eloquently to this (Chapter 2). Within a given region, however, the strength of genetic control (heritability) of height growth tends to be low and is demonstrated by the strong control of height growth by site factors such as topography, soil moisture, and soil texture. Nevertheless, in a given stand, significant differences in individual genotypes may occur, primarily due to genetic factors. Naturally occurring aspen clones on sandy soils in northern lower Michigan have been found to differ greatly in height on the same site (Zahner and Crawford, 1965). Some clones are more than twice as high as adjacent clones of the same age. Such variations are more likely in species developing natural clones than in species where each stem is a different genotype. When stems of different genotypes compete in a stand (as in pines and maples), height growth is of survival value and slow-growing stems tend to be eliminated. In clones, however, competition is primarily between stems of the same genetic constitution, and slow-growing clones are more likely to survive.

Finally, the condition of the site at the time the stand is established as well as competition from other vegetation in early years may affect height growth markedly. Both naturally reseeding and planted pines will usually show different growth trends and amounts on old fields as contrasted with cutover sites.

## Site-Index Curves

The usual method of determining site index on the basis of tree height depends upon the use of a height-over-age growth curve to estimate the height at a standard age. Most such curves for American species in the past have been developed from a series of regression curves, based upon a single guiding curve and harmonized to have the same form and trend (Figure 12.1).

The weaknesses of this approach are by now well known (Spurr, 1952b). First and foremost, the technique is sound only if the average site quality is the same for each age class. If, however, as is often the case, younger stands are found on generally better sites (perhaps because of early logging on these sites) while the remaining old growth stands are concentrated on the poorer sites, the average curve will be warped upwards at younger ages and downward at older ages. This seems to be the situation for the standard Douglas-fir site-index curves (Spurr, 1956c) when checked by the growth of permanent sample plots. The reverse situation can also occur.

A second major weakness of the conventional technique is the assumption that the shape of the height-growth curve is the same for all sites. Although this generalization gives good results in many instances, it does not hold for all soil conditions. For instance, if the depth of a soil is limited either by physical or physiological reasons, growth of a tree may be normal up to the point that the depth of the soil becomes a limiting factor (curve *B*; Figure 12.2). On another soil, the same species may grow slowly until the roots reach an underlying enriched horizon or a deep-lying water supply, after which growth will be accelerated (curve C). The shape of these two growth curves may differ markedly from the normal growth curve on a normal soil (curve *A*).

A corollary of this problem is the assumption in the standard technique that site differences are apparent at early ages. The process of harmonizing site-index curves assumes that, if a site produces a higher tree at age 50 or 100, that tree will be higher at all preceding and all subsequent ages. The assumption is in contrast to the fact that many plantations and even-aged natural stands on marginal sites may grow normally in youth and only in middle life exhibit sharply decreased growth. Planted black walnut trees on seven contrasting sites in southern Illinois (Figure 12.3) show rapid

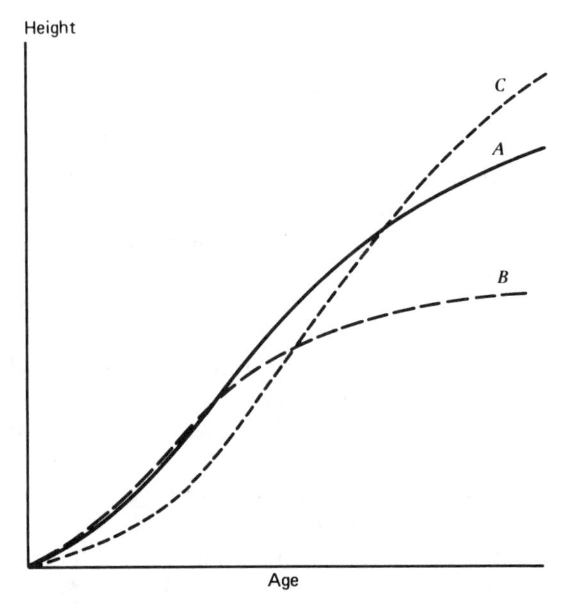


**Figure 12.2. Theoretical effect of soil profile on height growth. *(A)* Normal height growth in homogeneous soil. *(B)* Height growth on good but shallow soil. *(C)* Height growth on soil poor at surface but with rich horizon beneath.**

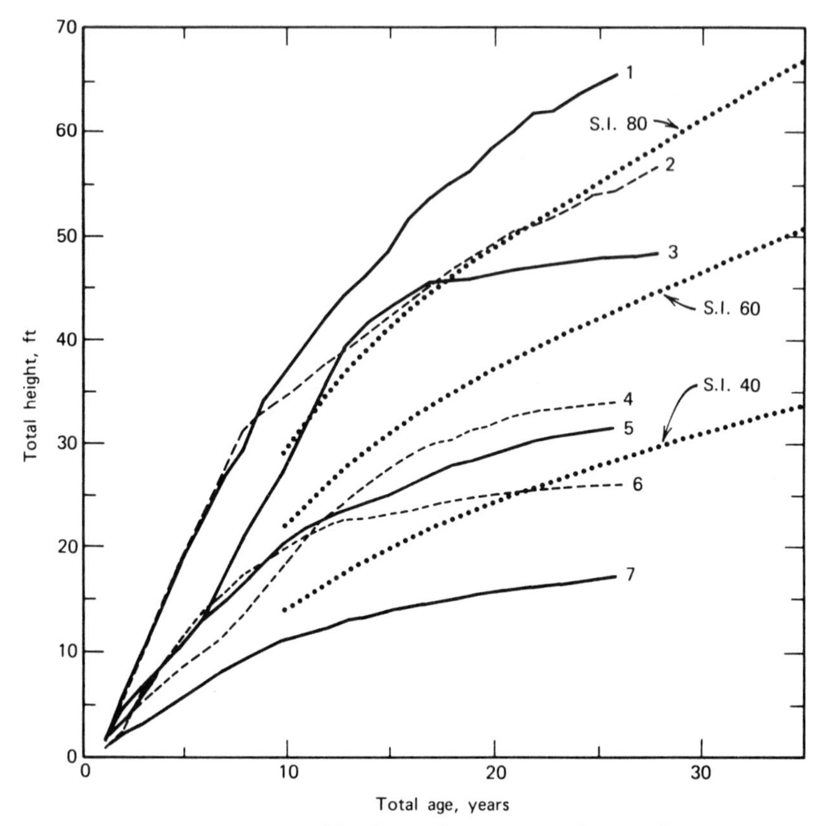


**Figure 12.3.** **Dominant black walnut trees from plantations in southern Illinois have marked polymorphic patterns of height growth. The height-age curves are averages from 36 sectioned trees growing on 7 contrasting-site plots. (Plot 1—deep, well-drained alluvial silt loam; plot 3—similar to 1 but with restricted internal drainage; plots 4–6—bottomland silt loam underlain at 0.5 to 1 m by gravelly subsoil; plot 7—bottomland silt loam underlain at 0.3 m by a gravelly, cherty subsoil.) Also shown are the regional harmonized site-index curves for black walnut (Kellogg, 1939). (After Carmean, 1970b. Reprinted with permission from *Tree Growth and Forest Soils,* © 1970 by Oregon State University Press.)**

early growth, even on the poorer sites, but may slow abruptly after 10 years (Carmean, 1970b). The polymorphic patterns are closely related to soil conditions. Trees on plot 1 are growing on a deep well-drained alluvial silt loam, while those on plots 4 to 7 are growing on a bottomland silt loam soil underlain at 1 m or less by a gravelly subsoil. A similar example is that of red pine plantations in the Saginaw Forest of the University of Michigan that were planted on glacial till soils of much heavier texture than those on which the species is generally found. Marked reduction in growth was noticed in the permanent sample plot data after 30 years. Concurrently, symptoms of malnutrition appeared resembling the littleleaf disease of shortleaf and loblolly pine in the southeastern states. Stem analysis revealed that a marked reduction in growth had occurred for 12 years before external symptoms had become evident. If, as seems to be the case, this reduction is due to commonly occurring soil conditions in the area, it should be taken into account in the site-index curves for red pine growing on those soils. Curves based on the premise of harmonization with a standard average curve cannot show such a plant-soil relationship.

Height-growth patterns are known not only to vary in different parts of the range of a species but also in local areas of contrasting soil and topography. Height-growth patterns of oak, for example, not only vary between different soil texture groups but also vary with aspect and slope within soil groups. Therefore, it is not surprising that polymorphic site-index curves have been repeatedly demonstrated to characterize the variable height-growth patterns of forest trees better than the simple monomorphic pattern portrayed by regional harmonized curves (Carmean, 1970b, 1975). Polymorphic site-index curves have been prepared for many species, including red pine (Bull, 1931; Van Eck and Whiteside, 1963; DeMent and Stone, 1968), eastern white pine (Beck, 1971), loblolly pine (Zahner, 1962; Trousdell et al., 1974), ponderosa pine (Daubenmire, 1961), Douglas-fir (Carmean, 1956), western larch (Roe, 1967b), black spruce (Jameson, 1965), yellow-poplar (Beck, 1962), upland oaks (Carmean, 1965, 1972), and black walnut (Carmean, 1966). Unfortunately, such curves often are not presented individually according to the site type or soil conditions responsible for the characteristic growth pattern. Examples of such polymorphic curves for various site-types (site units) for Norway spruce and other species of southwestern Germany (see pp. 324 to 328) are available (Günther, 1955; Moosmayer, 1957; Werner, 1962; Hasenmaier, 1964).

Considering the weaknesses of the standard site-index curve techniques, height-growth curves should be based upon actual measured growth of trees on specific soils or site-types and not upon the harmonized method of averaging together height and age values from plots for the entire or regional range of sites upon which the species is found. Such a conclusion is not new. Most present-day European growth curves are based upon actual measured growth rather than upon temporary plots.

In studying height aspects of forest growth for correlation with soil characteristics, then, our basic objective should be to work with actual recorded tree growth and to collect our data in such a way that separate growth curves can be evolved for different soil-site conditions should evidence of differing curve shapes become apparent. There are three general sources of growth information that meet these requirements. The first and the best are the records of long-term permanent sample plots. Where enough plots have been established and measured for a considerable period of years, growth curves should be evolved from the data these plots provide.

In the absence of sufficient permanent plot data, recourse may be had to stem analysis. By sectioning trees from top to bottom, their course of growth can be reconstructed. This technique, too, has its pitfalls, but they are less serious and more easily overcome than those of the temporary plot techniques. The appropriate regression methods for estimating site index based on stem analysis are discussed by Curtis et al. (1974a,b).

A third source of information is the internode distance on the boles of trees which put out distinct annual whorls. On Douglas-fir and many of the pines, the whorl pattern shows clearly the height of the tree each year in the past back to an early age. Careful measurements of such trees can be used to produce accurate height-growth curves (Bull, 1931; King, 1966; Beck, 1971). Growth curves can be produced by any of the above three techniques—or better yet by the combination of these techniques.

## Comparisons Between Species

Just as trees of differing genetic character within a species may be expected to show variations in height response to a given forest site, different species growing on the same site will show the same variations but to a much greater degree. The height-over-age site-index curve may be quite different for two different species on the same site. Nevertheless, the site index as predicted from the measurement of height of one species may be used to give an estimation of the site index of the other. Carmean (1975) lists 16 such studies, including examples from the southern pines, eastern hardwoods, and western conifers.

In a number of American studies, site indices of different species have been correlated statistically to permit the estimation of site index of one species on the basis of a knowledge of that of another. Most of these studies are limited in that they are based upon harmonized or generalized site-index curves rather than upon specific growth curves developed separately for the site in question. Nevertheless, these studies are useful in practical forestry provided that their limitations are understood. For example, in several studies in the southern United States, site indices for various species and species-groups have been correlated. A site-sensitive species such as yellow-poplar proved to have the highest site index on the best sites and the lowest on the poorest sites (Olson and Della-Bianca, 1959). In the southern Appalachians, site index of the oak group (excluding white oak) could be estimated with a standard error of 1.6 m from a knowledge of the site index of other species (white oak, white pine, shortleaf and pitch pine grouped, and yellow-poplar; Doolittle, 1958).

Much more information is possible when the site-index comparisons are based upon stem analysis of paired trees, such as a study of white pine and red maple site index by Foster (1959). Here it was found that red maple height growth gave a better indication of potential white pine height growth than did a composite of 11 site factors based on soil and topographic characteristics. White pine site index could best be predicted by an equation based upon the independent variables of red maple height, red maple age, and aspect. On northerly aspects, white pine grew relatively better than red maple with the opposite being true on southerly aspects.

## Height Growth for a Portion of the Life Span

Since the total height at a given age of a tree is an expression of all past growing conditions, such a measure may be influenced by conditions that prevailed for a few years—such as grass competition in the seedling state, absence of mycorrhizal infection in newly planted stock, insect attack, and drought. It is therefore sometimes preferable to estimate height-growth potential in terms of the measured growth over a period of a relatively few years in the history of the stand. This approach is particularly feasible for the white pines and other species which put on a single well-defined whorl of lateral branches each growing season.

The best indication of current growing conditions is, of course, current height increment, and this can usually be discerned readily from an inspection of the terminal part of a tree. For taller trees, of course, this is a difficult and time-consuming task.

For the middle part of the height-over-age growth curve, height growth is relatively constant, and the average annual distance between whorls may be assumed to hold true for a future short period.

Once a stand approaches maturity, however, height growth will diminish with the passage of time, and either present height or present age should be added as a second variable in the prediction of future heights.

The use of mean annual height increment over the middle period of height growth as a measure of site quality is old and well established. This method has been given the name of the **growth-intercept** method and has received renewed emphasis. For loblolly, shortleaf, and slash pines (but not for longleaf), the 5-year intercept above breast height has proved better correlated with site quality than total height in a test of trees from plantations about 20 years old (Wakeley and Marrero, 1958). The technique has been adapted to red pine, Douglas-fir, southern pines and other species on which the annual whorls are apparent (Carmean, 1975).

Of the various methods for determining site quality, site index is the one most often employed in North American forests, Site index can provide a convenient and reliable estimate of forest productivity. It is also a useful guide in selecting the most productive tree species for specific sites. However, site index is not an appropriate guide for determining silvicultural practices for regeneration and care of stands. Therefore, site-quality studies and site classification using vegetation and physical site factors have attracted universal attention. They not only provide a basis for estimating productivity,.but they also provide the ecological basis for determining silvicultural practices for various management goals. In the following sections, we will discuss these other approaches used to estimate site quality—not only as a means of understanding timber production potential, but also in understanding the larger ecological significance of site quality in managing forests for wildlife, recreation, and wilderness as well as timber.

## VEGETATION AS AN INDICATOR OF SITE QUALITY

The presence, relative abundance, and relative size of the various species in the forest reflect the nature of the forest ecosystem of which they are a part and thus may serve as indicators of site quality. The correlation may or may not be apparent because the vegetation also reflects the effects of happenstance: plant competition; past events in the history of the vegetation such as drought, fire, and insect outbreaks; and many other factors in the ecological complex giving rise to the plant community. Nevertheless, site characteristics are sufficiently reflected in the vegetation to make the use of the latter a successful index of site quality in many instances. The plants themselves are used as the measure of site: they are **phytometers**.

Tree species are useful indicators. They are long-lived, relatively unaffected by stand density, and easily identified in all seasons of the year. Some species have such a narrow ecological amplitude that their occurrence is indicative of a particular site. Demanding hardwoods such as black walnut, white ash, and yellow-poplar reach their best development only on moist, well-drained, protected sites rich in soil nutrients and characterized by a well-developed forest floor. Most trees, however, have a wide ecological amplitude: they may occur and prosper on a wide variety of sites. Their presence is thus of little indicator value. Their relative abundance and their relative size, however, may be. In the same eastern hardwood forests, the greater the proportion of red oak to black and white oak in the forest, the better will be the soil moisture conditions and the general site quality. The sizes of free-grown dominant oaks of these three species in even-aged stands of uniform density may be used as an index of site quality.

Understory plant species in the forest—although they are more apt to be influenced by stand density, past history, and the composition of the forest than the tree species—have in many cases a more restricted ecological tolerance and may therefore be more useful as plant indicators. This is particularly true in the circumpolar boreal forest, where the dominant tree species—the spruces, firs, pines, birches, and aspens—are few in number and widespread in their distribution on various sites, thus having relatively poor indicator value. Under such circumstances, site classification schemes based upon indicator species in the understory have been markedly successful. They are most easily applied in regions where variations in altitude and precipitation are great and where humans have not markedly altered the original vegetation.

The classic example is Cajander's system of site types for Finland (1926), designed to segregate minor quality classes in a forest characterized by spruce, pine, and birch in both pure and mixed stands. Height-growth curves of pine on four such site types are presented in Figure 12.4. These may be taken as indicative of the system. The *Cladina* type is characterized by a lichen understory (*Cladonia alpestria* in particular) and occurs on dry sandy heaths of the poorest site quality. On the *Calluna* type, mosses and lichens are generally present, but heather (*Calluna vulgaris*) is typically the predominant species. *Juniperus communis* is also fairly common in the understory, while birches and spruce commonly occur with the dominant pine in the overstory. The *Vaccinnium* type is typified by the lingonberry, *Vaccinium vitis-idaea*, and occurs on moderately dry sandy ground and on glacial ridges. Mosses and lichens are of less importance and dwarf-shrubs of greater importance than in the preceding types. Finally, the *Myrtillis* type, named for the predominant

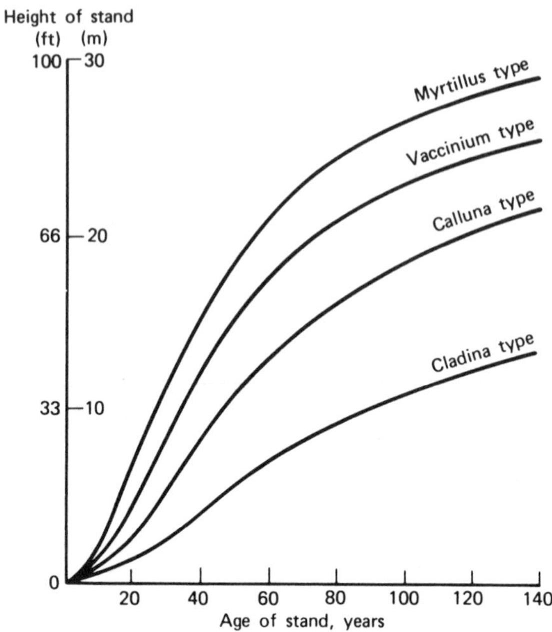


**Figure 12.4. Height-over-age curves of Scots pine on different site types in Finland. Note that the site type of each curve is specified.**

*Vaccinium myrtillis,* is developed on a richer soil supporting a climax vegetation of spruce but frequently converted to pine by fires, felling, and silvicultural control. Lichens are unimportant and herbs more important and richer in numbers than in the other types cited.

At a more complex level, all of the resources of vegetation description may be used in the segregation of forest site classes. It should be remembered, though, that site variation frequently takes the form of a gradient rather than of distinct and mutually exclusive site classes. The latter are found only when a distinct break in site factors occurs, such as a break between a sandstone-derived residual soil and a limestone-derived residual soil. Otherwise, site changes tend to be gradual, and the continuum may better be described in terms of an ecological gradient. Thus Rowe (1956), working with the mixed boreal forest in Manitoba and Saskatchewan, describes the changes in vegetation associated with a range of sites varying from a very dry forest to a wet forest. Similarly, Whittaker (1954) categorizes the vegetation of the Great Smoky Mountains in terms of the interaction of ecological gradients of elevation and of moisture conditions, the latter being related to topographic site. This latter paper summarizes the thinking of ecologists of plant indicators being an expression of the complex of the forest ecosystem rather than representing a specific effect of site quality.

The understory plant-indicator concept has been adapted successfully to the spruce-fir forests of North America, in eastern Canada, the Adirondacks, Maine, the Great Smokies, and central Canada. Studies by Linteau (1955), Rowe (1956), Crandall (1958), Damman (1964), Mueller-Dombois (1964), and Grandtner (1966) refine and summarize competent earlier work along similar lines. In modern practice, however, even in Scandinavia understory plant indicators are not used exclusively (Fennoscandian Forestry Union, 1962). Increasingly, the overstory dominants and physical factors of the environment are used in conjunction with ground vegetation to classify and map sites and estimate site quality.

In the more complex and more disturbed forests to the south or at lower elevations than the spruce-fir forest, however, similar indicator site types based upon a relatively few understory species have proved less applicable. Here the vegetation approach to site classification has involved an evaluation of the total vegetation, including not only the species present in all layers of the forest, but also the abundance, size, and vigor of all elements of the flora.

## Species Groups and Indicator Spectra

It is possible under certain conditions to characterize a site in terms of a very few species. However, the key indicator plants may or may not be present in a given locality because of chance, past forest history, or present competitive conditions. It is, for instance, subjectively difficult to classify a site as an *Oxalis-Cornus* site type if it contains no *Oxalis* and no *Cornus* because of past forest history or because of the flukes of local plant distribution. Yet the sum total of the vegetation and the growth characteristics of the forest may well fit the area into this category.

The disadvantages of the foregoing approach may be overcome by using many species either arrayed singly along an environmental gradient (indicator spectrum) or placed in groups, the species of each group having similar environmental requirements. A model of the indicator-spectrum approach was developed for spruce and fir in the northeastern United States (Table 12.1). The spectrum was modified by Marinus Westveld from a preliminary list of the senior author and is presented here to indicate the method rather than a final listing. The indicator spectrum is simply a list of plants, including trees and shrubs as well as herbs and other vegetation, classified according to the sites they indicate. Plants denoting dry, infertile sites are placed at the top of the list and those characteristic of moist and fertile sites are placed at the bottom. In the field, the indicator plants are checked as being present, common, or abundant. The center of the distribution curve produced by these checks denotes the site quality. In the sample, the site quality is clearly C although *Clintonia* and *Oxalis*, two characteristic plants of that site, are absent.

**Table 12.1 Indicator Plant Spectrum (Northeastern Spruce and Fir)**

| GENUS OR SPECIES | SITE | PRESENT | COMMON | ABUNDANT |
|---|---|---|---|---|
| *Myrica* | | | | |
| *Vaccinium* | | | | |
| *Gaultheria* | | | | |
| *Hylocomium* | A | | | |
| *Hypnum* | | | | |
| *Chiogenes* | | | | |
| *Pteridium* | | | | |
| *Coptis* | | | | |
| *Bazzania* | B | | | |
| *Corylus* | | | | |
| *Maianthemum* | | X | | |
| *Cornus* | | X | | |
| *Aralia* | | | X | |
| *Clintonia* | C | | | |
| *Oxalis* | | | | |
| *Dryopteris* | | | | X |
| *Acer saccharum* | | | X | |
| *Asplenium* | | | | |
| *Smilacina* | | | | |
| *Mitchella* | D | | | |
| *Viola* | | | | |
| *Oakesia* | | | | |

In one of the few trials of indicator plants in the southern United States, Hodgkins (1961, 1970) achieved limited success in predicting the site index of longleaf pine using a modified spectrum approach. The plant indicators failed to predict site index closely in deep soils with droughty surfaces because they apparently do not utilize the deeper layers of soil tapped by pine roots.

A refinement of the spectrum technique is to group plants of similar ecological requirements together and use the groups to distinguish different ecosystems. Besides the example presented immediately below, the use of ecological species groups is described in conjunction with the multifactor site classification system used in Germany and presented later in the chapter.

In undisturbed coniferous forests of the Northern Rocky Mountains, Daubenmire (1952) and Daubenmire and Daubenmire (1968) used groups of understory species, termed **subordinate unions**, in combination with late-successional overstory species (**dominant unions**) to distinguish forest associations. The collective area of a given forest association, the **habitat type** (literally, the type of climax vegetation on a particular habitat or site), indicates similar environmental and biotic conditions, hence an ecosystem. The understory unions are composed of from 1 to 24 species, for example, the *Physocarpus malvaceus* union of 6 species, the *Pachistima myrsinites* union of 24 species, and the *Xerophyllum tenax* union of 2 species. Different habitat types are distinguished by specific combinations of overstory and understory unions. In some cases the overstory union is the major determinant of the habitat type, whereas in other situations the understory union is definitive. The *Pinus ponderosa–Physocarpus* and *Pseudotsuga–Physocarpus* habitat types are distinguished by their respective overstory dominants. However, three habitat types with the same overstory, *Abies lasiocarpa-Pachistima, Abies lasiocarpa–Xerophyllum*, and *Abies lasiocarpa–Menziesia* occur on different sites and are distinguished by their understory unions. The polymorphic site-index curves of ponderosa pine in seven habitat types indicated that ecosystems delineated by vegetation are of substantially different site quality (Daubenmire, 1961).

**Operational Site Classification by Habitat Types.** A habitat-type approach to site classification has been applied throughout forested lands of Montana (Pfister et al., 1977; Arno and Pfister, 1977). It has gained widespread acceptance elsewhere in the western United States (Thilenius, 1972; Wirsing and Alexander, 1975; Hoffman and Alexander, 1976; Reed, 1976; Pfister, 1976). The rationale of the system is that late-successional (climax) overstory and understory vegetation integrate and express the environmental complex of climate, physiography, and soil. The classification is based upon the above-mentioned approach of Daubenmire in identifying late-successional overstory dominants that occur along an elevational gradient from grassland to alpine tundra (Daubenmire, 1966; Chapter 17, pp. 483 to 486). Figure 12.5 shows the sequence of these late-successional trees that are typical of the northern Rocky Mountains of northwestern Montana. Within each of these tree-species ***series*** there is one or more characteristic understory plant unions which together with the tree overstory define the ***habitat type***. A

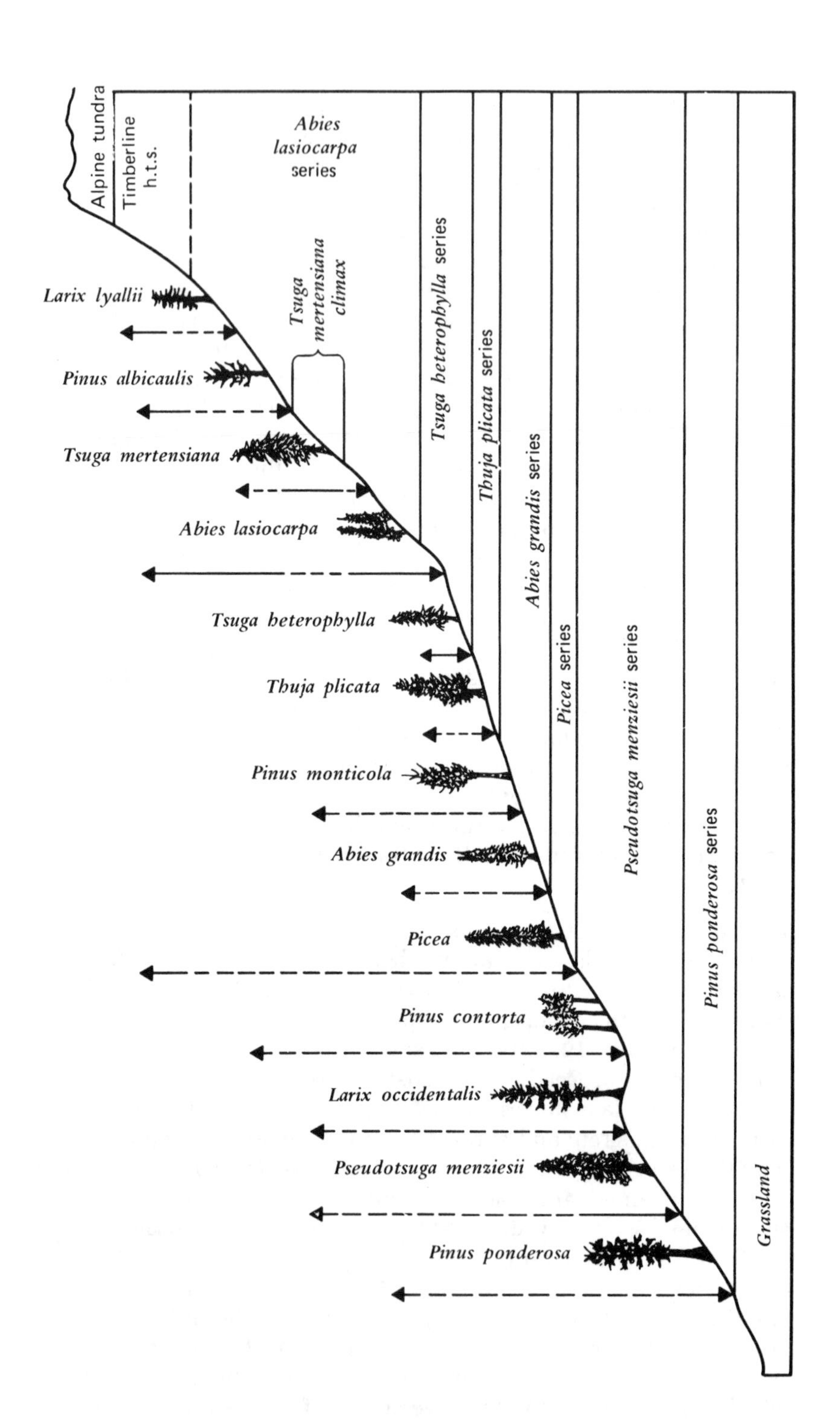
Alpine tundra
Timberline h.t.s.
Abies lasiocarpa series
Tsuga mertensiana climax
Tsuga heterophylla series
Thuja plicata series
Abies grandis series
Picea series
Pseudotsuga menziesii series
Pinus ponderosa series
Grassland
Larix lyallii
Pinus albicaulis
Tsuga mertensiana
Abies lasiocarpa
Tsuga heterophylla
Thuja plicata
Pinus monticola
Abies grandis
Picea
Pinus contorta
Larix occidentalis
Pseudotsuga menziesii
Pinus ponderosa

third level, or ***phase***, is sometimes necessary to recognize subdivisions of a habitat type. The habitat types are based on the potential climax overstory and understory vegetation, not on the currently existing vegetation such as early successional species that might follow logging or fire.

Sixty-four habitat types characterize forests of Montana, although only 12 to 15 would be expected to occur in any one locality. In the mountains of Montana that are relatively undisturbed by humans, the types occur in a generally consistent and predictable local pattern related to topography. Mapping of habitat types has helped refine the classification and has become an important management tool (Deitschman, 1973; Stage and Alley, 1973; Daubenmire, 1973). The classification and maps are used widely in wildland management (Layser, 1974; Pfister, 1976): in timber, wildlife, range, and watershed management, in recreational use studies (Helgath, 1975), in assessing the role and impact of fire, insects, and disease (Arno, 1976), and in natural area preservation (Schmidt and Dufour, 1975). The case for the use of vegetation in assessing forest productivity is presented by Daubenmire (1976). Although still applicable, the method would be increasingly difficult to apply as topographic differences diminish and as forests become more disturbed. Reliance on soil and drainage conditions would then become increasingly appropriate.

## ENVIRONMENTAL FACTORS AS A MEASURE OF SITE

Many situations exist where the forest vegetation, whether the trees or the understory plants, are less useful in providing an index of forest productivity. These include agricultural or other non-forested lands to be reforested; areas recently subjected to fire, logging, heavy grazing, or other disturbances; and areas to be converted from one forest type to another. In these and similar cases, site productivity must often be estimated from an analysis of the physical environment rather than from the vegetation.

Physical factors may be used singly or in combination as an index

**Figure 12.5. Distribution of forest trees in the Rocky Mountains of northwestern Montana. Arrows show the relative elevational range of each species; solid portion of the arrow indicates where a species is the potential climax dominant (late-successional) and the dashed portion shows where it is seral (early successional). (After Pfister et al., 1977.)**

of site quality. Included are all the climatic and edaphic factors discussed in the previous chapters. To be useful as an index, however, any factor should be capable of simple and inexpensive measurement and should furthermore be highly correlated to forest productivity. Many site factors are disqualified for index use because of lack of information concerning them or because of the difficulty or expense in obtaining this information. Others are not used because of their lack of sensitivity as a measure of site quality. In general, in accordance with the modern concept of the law of the minimum (Chapter 4), the most useful environmental factors for site-index purposes are those which are in short supply with regard to the demand by forest trees, so that small changes in the supply of the factor will result in measurable changes in the growth of trees. It follows that different environmental factors will prove useful in different forest situations. This is indeed the case.

Climatic site factors are generally useful in providing a rough index of productivity among adjacent forest regions or among altitudinal zones within a geographic region. They are more closely associated with (and largely responsible for) genetic differences over a species' geographic and altitudinal range than are soil factors. In particular, temperature and precipitation data may be used to compare forest growth in various geographic regions or altitudinal zones, assuming similar soil conditions, or at least assuming soil conditions that in themselves are closely related to the climate.

Within a given climatic region, growth will vary greatly, depending upon soil and topographic conditions. Soil factors, therefore, are apt to be particularly useful in local studies of forest site quality.

Since the sum total of the climatic, topographic, soil, and biotic factors define the site portion of the forest ecosystem, the more factors used as an index of forest productivity, the better will be the correlation. However, increasing the number of factors usually means increasing the cost and the time needed to complete the analysis. Thus accuracy and reliability must be weighed against the cost. In efforts to construct environmental site classifications applicable to a wide area, multiple factors should be used wisely: gross climatic factors to delineate regional or altitudinal differences and physiographic and soil factors to distinguish local site differences.

The emphasis of the forest ecologist on environmental factors should not blind the ecologist to the realization that growth depends upon the genetic composition of the forest as well as upon the environment in which the forest grows. What constitutes a good site for one species may prove a poor site for another. To cite but one of many possible examples, Monterey pine in New Zealand and Australia makes phenomenal growth on sites that previously supported very slow-grown indigenous forests. Even within a species, genetic variations may be highly important. One genotype of aspen, growing in a clonal colony, may make twice the growth of another in an adjacent clone on exactly the same site. We should never forget that growth is a function of the effect of the environment on the genotype, and that site factors alone can never account for all the variation in growth rates found between different species or even between races or different sexually regenerated individuals within a single species.

## Climatic Factors

Forest climate is obviously related to tree growth because the crowns and boles live in the air and are affected by it. Macroclimatic factors have been appropriately used to distinguish differences among major forest regions (Krajina, 1959; Schlenker, 1960; Rowe, 1972; Findlay, 1976). The applications are limited though, because long term climatic data from forest sites are either nonexistent or difficult to obtain. Data interpretation for the various interrelated variables is problematical, and furthermore, vegetation itself provides a useful and more easily measured integrator of the complex of climatic factors. In addition, most site quality evaluations for wildland management decisions take place within a forest region where average climate may not vary widely. However, local climate may vary significantly from place to place within the region.

Despite the obvious importance of local climate, however, it has seldom been used in site evaluation. For one reason, other variables such as land-use history, forest management history, and forest soils have obvious as well as masking effects on forest growth, and it has proved very difficult to obtain good correlation between local climate and growth. For another reason, local climate is often related to soils, or to local topography, and a site classification based upon soils and topography will also carry with it an implied classification with respect to local climate. Thus the same factors that make a soil very well drained are apt to insure good drainage of the cold air and to indicate a site with lower maximum and higher minimum temperatures than would be indicated by a regional climatic average. Similarly, a very poorly drained soil is apt to result from a topography that inhibits the drainage of cold air as well as of soil water. Thus the very poorly drained sites are apt to be characterized by temperature extremes and a short growing season.

Some general correlations between climate and forests have already been indicated. That both precipitation and the temperature regime influence the distribution and growth of the forests in a broad sense is illustrated by the rough correlation that exists between climatic classifications such as those of Merriam, Köppen, and

Thornthwaite and the occurrence of forest types. Similarly, forest species are strongly genetically adapted to the day length associated with the latitude and elevation of their habitats. Any planting of seedlings significantly south or north of their native site must take day length and elevational changes into account (Chapter 2).

Many efforts have been made to relate rainfall to forest growth. Good relationships have been obtained in many cases since soil moisture is frequently a limiting factor in forest growth, at least during the hottest part of the growing season. For example, in the longleaf pine region of the deep south from Mississippi to Texas, the January–June rainfall, ranging from 600 to 900 mm, has proved more important than any soil factor in predicting site quality (McClurkin, 1953). In the mountains of the American West, the effect of rainfall is complicated by elevation. Precipitation normally increases with elevation, but temperature similarly decreases: site quality may sometimes be increased and sometimes decreased by the combined effects of increasing precipitation and decreasing warmth.

Climatic summaries based upon a comparison of precipitation and evapotranspiration curves (Figure 7.3; Chapter 7), synthesizing as they do the combined effects of precipitation and temperature upon the water balance, have potential use in studying variation in site quality. Sites with large annual water deficits can be expected to produce less forest growth than sites with small annual water deficits. A related and important approach emphasizing the physiological responses of a plant to limiting local climatic factors has been developed by Waring et al. (1972). Plant response indices related to moisture and temperature (as they affect transpiration and growth) were successfully related to maximum tree height as a measure of site quality.

### Soil and Topographic Factors—Soil–Site Studies

The problem of relating soil and topography to site quality has attracted many investigators. Of principal concern has been the determination of site quality for areas that have highly variable soil, topography, and stand conditions and for areas that are either unstocked, stocked with unwanted species, or stocked with trees unsuited for site-index measurements. Such methods have received great emphasis. Carmean (1975, 1977) discusses the applications and limitations of this method and cites over 170 soil-site studies for the United States and Canada; Burger (1972) cites many additional ones for Canada.

Depending upon the nature of the specific site, many individual soil and topographic factors may serve as useful indices of forest productivity in that they may be correlated with the site index of the desired forest species. Specifically, soil-site studies involve measuring or scoring many soil and site variables, termed independent variables (e.g., soil depth, texture, and drainage class; slope position; aspect) and relating these through multiple regression analyses to tree height or site index. By combining these and other soil and topographic factors, many useful formulas have been evolved by which site index can be estimated approximately (standard errors of the estimate range from 1.5 to 3.0 m for trees 18 to 30 m high at 50 years). For example, Zahner (1958a), restricting his regressions to soil groups within a limited geographical region, related the site index of two southern pines to the thickness of the surface soil, the percentage of clay of the subsoil, the percentage of sand of the subsoil, and the slope percentage.

The equations derived from soil-site studies are used for developing site-prediction tables and graphs for estimating site index in the field. In successful soil-site studies, the combination of independent variables may explain 65 to 85 percent of the variation in tree height or site index observed in the field plots (Carmean, 1975). Although many variables may be used to develop precise equations, some of the variables may be difficult or tedious to measure in the field. Thus somewhat less precise equations are often developed using the variables that are most easily identified and used in the field.

Before examining soil-site studies further, however, two warnings are in order. First, the problem is one of correlation and not necessarily of cause and effect. Too many investigators have read into regressions based upon soil and topographic factors causal relationships which are not justified. Such correlations may merely reflect a causal relationship attributable to another and unmeasured soil characteristic. Second, frequently the dependent variable is site index read from harmonized site-index curves. As previously pointed out, such values are suspect by the very nature of the method used to construct the site-index curves. In any event, they are derived values read from a curve rather than actual values of forest productivity.

Generalizing from the many efforts to relate soil properties to forest site, the growth potential of forest trees is chiefly affected by the amount of soil occupied by tree roots and by the availability of soil moisture and nutrients in this limited space. Of prime importance, then, is the **effective depth** of the soil or surface soil depth—the depth of the portion of the soil that is either occupied or capable of being occupied by the roots of the tree for which the site index is desired. This effective depth may obviously be limited by the occurrence of bedrock near the surface. The position of the water table during the growing season likewise sharply limits root penetration. Less obviously but equally significantly, a coarse, dry stratum may prove an effective barrier to root penetration just as may a highly compact and impervious stratum such as a highly developed hardpan.

Consequently, many measures of effective soil depth have proved significant in correlating soil factors with site quality. Coile (1952) not only accomplished much of the basic work in this area but summarized earlier studies. The soil factors most frequently found important are the depth of the A horizon above a compact subsoil, the depth to the least permeable layer (usually the $B_2$ horizon), the depth to mottling (indicative of the mean depth to restricted drainage), and thickness of the soil mantle over bedrock. All these measures quantify the effective rooting depth of trees. They are important when soils are shallow but are relatively unimportant for deep soils where downward root development is unimpeded.

Next in importance are soil profile characteristics that affect soil moisture, soil drainage, and soil aeration. The physical nature of the profile—soil texture and structure of the least permeable horizon (again usually the $B_2$)—is of principal importance here.

Actually, however, the topographic position of the site is often closely related to microclimate and to the physical properties of the soil that govern soil-moisture and aeration relationships. Moreover, topographic site can be quickly recognized and evaluated, using aerial photographs and topographic maps, without the necessity of soil measurement. Many useful site relationships have been evolved based upon the relative topographic elevation, aspect, slope position, and degree of slope. Studies of oak site quality in the Appalachian Mountains and the Appalachian Plateau (Trimble and Weitzman, 1956; Doolittle, 1957; Carmean, 1967) have found that relative position between ridge top and cove, aspect, and degree of slope are all closely related to site quality. Carmean (1967) found that equations based solely on topographic features explained more than 75 percent of the variation in total height of black oak in southeastern Ohio. These close relations between topography and site occur because topography is closely related to important soil features such as A horizon depth, subsoil texture, stone content, and organic matter content. The relationships of aspect, slope steepness, and site index are illustrated in Figure 12.6.

The general relationship of site quality to aspect for mixed upland oak forests in the Appalachian Mountains resembles a cosine curve (Figure 12.7; Lloyd and Lemmon, 1970). In hilly terrain, the importance of topography and its close relationship with microclimate must be stressed. Northeast aspects and lower slopes usually have cool, moist microclimates and thus are the better sites; southwest aspects and upper slopes and ridges have dry and warm microclimates and hence are usually the poorer sites.

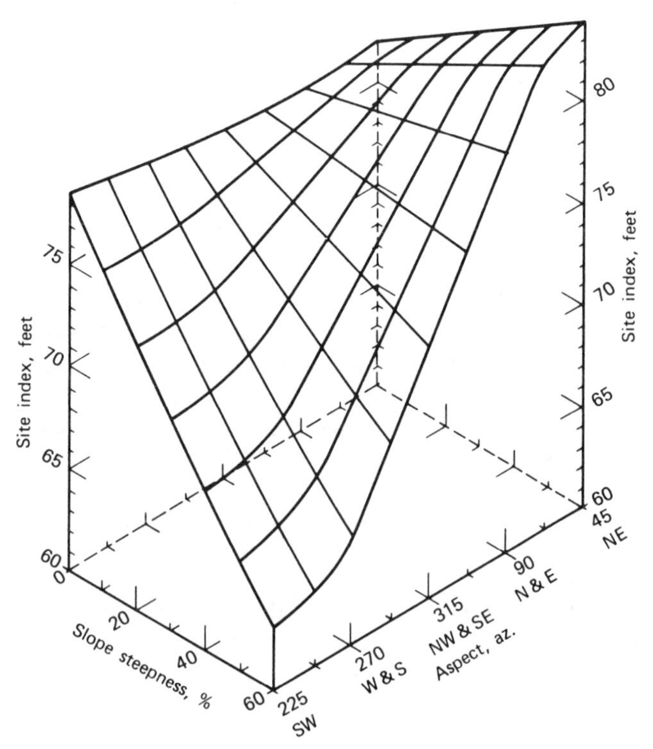


**Figure 12.6. Relation between aspect, slope steepness, and site index for black oak growing on medium-textured, well-drained soils. Site index increases from southwest-facing slopes to northeast-facing slopes. These increases are very pronounced for steep slopes, but site increases related to aspect are relatively minor on gentle slopes. For southwest-facing slopes, site decreases drastically with increased slope steepness whereas site increases slightly on northeast slopes as slopes become steeper. (After Carmean, 1967. Reproduced from *Soil Science Society of America Proceedings,* Vol. 31, p. 808, 1967 by permission of the Soil Science Society of America.)**

The effective depth of the soil, however, remains highly significant, along with topographic position. In the Douglas-fir region, Lemmon (1955) listed effective soil depth, aspect, and position on the slope as the most useful factors in delineating site quality. Minor variations in topography may be highly important in very flat locations, reflecting the effective depth of the soil over poorly aerated lower horizons. For instance, Beaufait (1956), working with south-

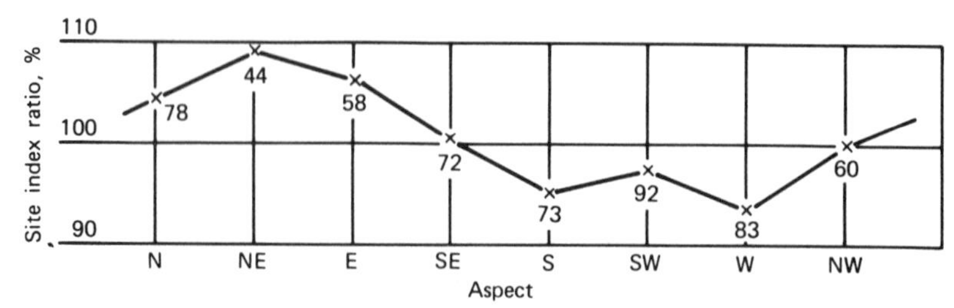


**Figure 12.7. Productivity curve with site-index ratio over aspect; based on 560 soil-site index plots on 27 soil series in mixed upland oak forests of the Appalachian Mountains. Site-index ratio is the ratio of the plot site index to the average site index of all plots of that soil series. (After Lloyd and Lemmon, 1970. Reprinted with permission from *Tree Growth and Forest Soils*, © 1970 by Oregon State University Press.)**

ern river bottomland oak sites, found that topographic features varying only centimeters in elevation were correlated with marked differences in the silt and clay content of the soil, with flooding conditions, and with soil aeration, and thus were highly related to site quality.

Predicting site index using soil factors has worked well in many upland areas for hardwoods and pine (Ralston, 1964; Carmean, 1975, 1977). However, it has proved discouraging for lowland hardwoods on alluvial soils of the mid-south except for very small and uniform areas (Broadfoot, 1969). Working primarily on alluvial soils of the Mississippi River over a six-state area, Broadfoot developed multiple regression equations for soil-factor relationships for each of seven hardwood species. None of the equations predicted site index with sufficient precision for investment planning. He concluded that over broad areas and complex land patterns the relationship between soil characters and height growth seems to defy quantification.

Although, under most American forest conditions, physical soil factors affecting the soil moisture regime are of the greatest indicator value in site studies, deficiency situations exist where the nutrient levels in the soil markedly affect site quality. Experimentally, good results have been obtained in such circumstances by relating the available nutrients, measured from chemical analysis of the soil, of the humus, or from foliar analysis of tree material, to site quality. Practically, such determinations are time-consuming and expensive, although approximations of soil nutrition may be made on the basis of a thorough knowledge of soil geology.

Soil survey mapping, using soil series and phases (Chapter 8), provides a taxonomic classification of forest soils but has generally proved unsatisfactory for the precise estimation of tree site quality. Typically, the variation in forest productivity, as estimated by site index, within a given soil-taxonomic unit is too great to be acceptable. On soils of the Rustin series, loblolly pine site index ranged from 59 to 105 (Covell and McClurkin, 1967). In California, ponderosa pine site index ranged from 89 to 182 on the Shaver series; average annual precipitation in areas of Shaver soils ranged from 300 to 1400 mm (Zinke, 1961). Excessive site variation within soil-taxonomic units also has been reported for numerous species in eastern hardwood forests (Carmean, 1970a; 1975).

It is now widely recognized (Coile, 1960; Rowe, 1962; Jones, 1969; Carmean, 1970a; 1975) that soil series alone are too heterogeneous to serve as a basis of site evaluation. They can prove satisfactory if they are refined to incorporate specific soil and topographic factors that are closely related to forest productivity (Richards and Stone, 1964; Carmean, 1967; 1970a). Carmean (1967) did this effectively in predicting site quality for black oak by combining existing soil series into two "woodland suitability" groups and subdividing them into phases based on topographic features known to be closely related to oak site quality. Because of their flexibility to provide phases or other subdivisions and their systematic and widespread application, soil surveys closely coordinated with soil-site research have a promising future in forest site evaluation (Byrd et al., 1965; Bartelli and DeMent, 1970; Lemmon, 1970). For example, in Washington and Oregon the Weyerhaeuser Company has prepared detailed soil-landform maps of their tree farm system and find them extremely useful in many management decisions, from the location of logging roads and fire-fighting access to timber production and recreation planning (Steinbrenner, 1975).

## MULTIPLE-FACTOR METHODS OF SITE CLASSIFICATION

In the previous sections we have considered simple approaches to the evaluation of site quality, such as site index, and indicators such as soils and vegetation. However, these represent only individual elements of the ecosystem whereas site quality is the sum total of factors affecting the capacity of land to produce forests. The more factors taken into account, the better is the estimate of site productivity and for understanding the site potential for managing forested lands.

Intensive methods involving multiple factors have been employed with success for over 50 years in Europe, and similar but more extensive methods have proliferated widely in Canada.

### The Baden-Württemberg System

Baden-Württemberg, a state in southwestern Germany of approximately 36,000 square kilometers (about the size of Connecticut, Massachusetts, and Rhode Island), has an enormous diversity of climatic, geological, and soil conditions. A mosaic of vegetation patterns exists, partly due to this variable environment and partly as a result of a long history of disturbance by humans. To cope with the complexity of these problems a multiple-factor system, integrating geography, geology, climatology, soil science, plant geography and sociology, pollen analysis, and forest history, was developed to classify and map forest sites. Following classification and mapping, the growth, site index, and productivity in stem volume of the major tree species are determined for the major site units. Then silvicultural and management recommendations are made for each species and species mixture for each site. A model of this ecosystem approach is illustrated in Figure 12.8. Together with a similar system in eastern Germany, it is the most intensively developed and applied system of its kind. Although the system was developed for practical resource management purposes, it is one approach for describing and studying the structure, productivity, and processes of an ecosystem. It provides a framework of landscape ecosystems (Rowe, 1969) that has applications not only in forestry but also in the physical, biological, and social sciences.

The multifactor approach evolved from the pioneering work of G. A. Krauss who, beginning about 1926 in Tharandt, initiated a team approach to study complex site relationships. The method consists of a synthesis of important site factors at regional and local ecosystem levels (Krauss, 1936). The state is initially subdivided into major landscapes or **growth areas**, and each of these is then divided into minor landscapes termed **growth districts**. This feature acts to limit sweeping generalizations and practical prescriptions that were often made for a species over wide areas having vastly different environments, biota, and histories. On the local level a series of **site units**, each site unit identified by a particular combination of physical and biotic site factors, is distinguished for each growth district.

Parts of Baden-Württemberg have been settled for over 2500 years, and the presettlement forest of beech and oak has been replaced by a monoculture, often several generations old, of Norway spruce. The system was initially employed to determine whether major site differences could still be distinguished in these areas, all of which appeared as one vast forest of spruce. Despite the homogeneity of the spruce stands and the changed ground flora induced by spruce, sites of markedly different productivity and silvicultural potential were identified (Arbeitsgemeinschaft, 1964). The concept and scope of the

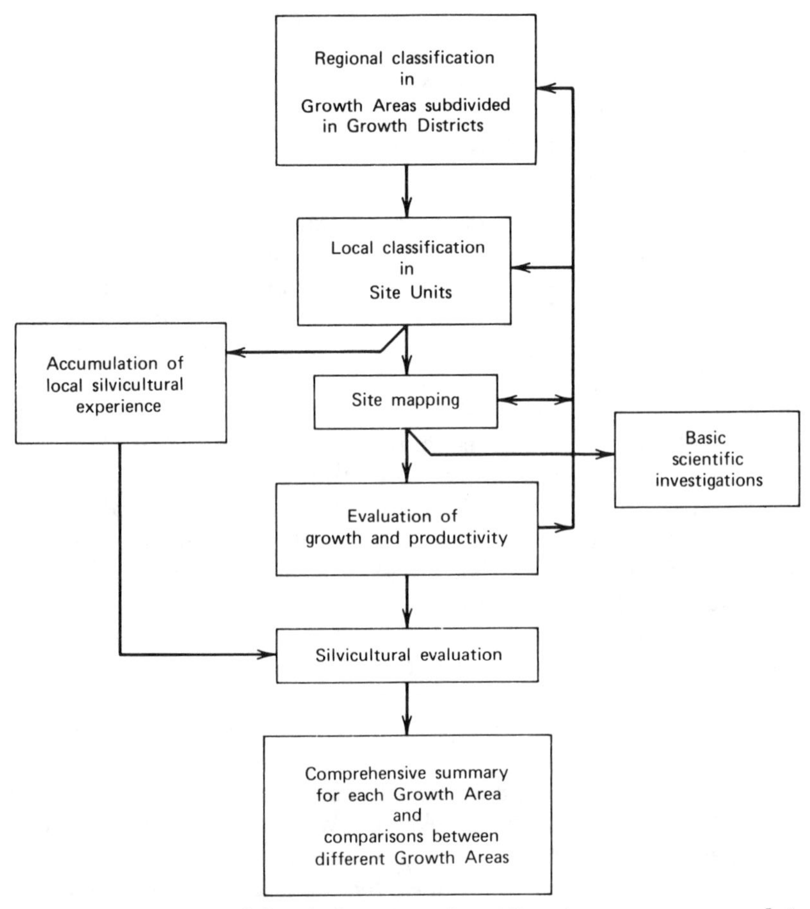


**Figure 12.8. Model of the site-classification system used in Baden-Württemberg, Germany.**

system as well as a brief discussion of methods, following the sequence shown in Figure 12.8, are presented below.

**Regional Classification.** Seven broad forest landscapes, the growth areas, are distinguished by major differences in climate, geology, and soil (e.g., the Upper Rhine River Basin, the Black Forest, and Swabian Alb). The growth areas are not homogeneous and are subdivided into growth districts based on finer distinctions, in macroclimate especially, but also in parent material, soil, and vegetation. Each district is characterized by one or more dominant, indigenous forest types whose composition is determined largely by climate, e.g., lowland mixed-oak forest, montane beech-fir forest, boreal-montane fir-spruce-pine forest, etc. The natural forest vegetation is of major importance in determining the limits of growth districts, and since many of the stands have been disturbed by humans, great attention is placed on pollen analysis, forest history, and studies of the ground vegetation. Through these studies the indigenous vegetation of the period begining about 1500 B.C. and ending in the Middle Ages is reconstructed.

**Local Classification and Mapping.** Each growth district is subdivided into site units, the number and character of which depend upon the conditions of the respective district. A site unit includes individual sites (the area occupied by a group of trees) which, though not identical, have similar silvicultural potential, growth rates, and productivity for the major tree species. The site unit is delineated in the field by local differences in topography, soil factors such as soil texture, structure, acidity, depth, and moisture-holding capacity, microclimate, and overstory and understory vegetation.

Each site unit is characterized by a local overstory type and in addition is floristically delineated through the use of **ecological species groups** (Table 12.2; Sebald, 1964; Schlenker, 1964; Dieterich, 1970). Each group is composed of several plant species, which, because of similar environmental requirements or tolerances, indicate certain site-factor complexes, for example, soil-moisture or soil-acidity gradients. Some species groups have a wide ecological amplitude, like the *Milium effusum* group, whereas others have a narrow amplitude, like the *Aruncus silvester* group (Table 12.2).

A representative sample of the 24 ecological species groups and 30 site units of the Upper Neckar Growth District (Sebald, 1964) illustrates the use of species groups to differentiate site units (Table 12.2). A site unit is characterized by the presence or absence of groups or the relative abundance of the species in the respective groups. The gradual trend of differences when site units are arranged along two moisture gradients (from moderately fresh to moderately dry, units 1 to 10, and from fresh to wet and imperfectly drained, units 11 to 15) is clearly seen. Units at opposite ends of the respective gradients are easily distinguished by the species groups. However, adjacent site units may be similar in their species groups and would be differentiated in the field by soil and topographic features. For example, site units 1 and 2 have similar vegetation, but 1 is a podzolized loamy sand on level terrain whereas 2 is a podzolized sand on a moderately steep slope of south or southwest aspect.

In field mapping, ecological species groups are used simultaneously with topography and soil characters to delineate site unit boundaries. The indicator value of each group is reliable only within the rooting zone of the species in the group. As seen in Table 12.2,

**Table 12.2 Representative Ecological Species Groups and Site Units of the Upper Neckar Growth District[a]**

| ECOLOGICAL SPECIES GROUP | SITE UNIT: MOD. FRESH → MOD. DRY | | | | | | | | | | FRESH→WET | | | | |
|---|---|---|---|---|---|---|---|---|---|---|---|---|---|---|---|
| | 1 | 2 | 3 | 4 | 5 | 6 | 7 | 8 | 9 | 10 | 11 | 12 | 13 | 14 | 15 |
| *Vaccinium myrtillus* group | ● | ● | ● | ● | · | | | | | | ● | · | | | |
| *Leucobryum glaucum* group | • | • | · | | | | | | | | | | | | |
| *Bazzania trilobata* group | • | · | | | | | | | | | · | | | | |
| *Deschampsia flexuosa* group | ● | • | ● | ● | • | · | · | | | | ● | • | • | | |
| *Pirola secunda* group | | | | · | · | · | • | · | · | · | | · | | | |
| *Milium effusum* group | · | · | • | • | ● | ● | • | · | · | • | · | ● | ● | ● | • |
| *Elymus europaeus* group | | | | | | • | • | | | · | | | | • | |
| *Aruncus silvester* group | | | | | | | | | | | | | | · | ● |
| *Ajuga reptans* group | | | | · | · | • | · | | | | · | • | ● | • | • |
| *Stachys silvaticus* group | | | | | · | · | | | | | | · | · | • | ● |
| *Chrysanthemum corymb.* group | | | | | | | · | • | · | · | | | | | |
| *Carex glauca* group | | | | | | | • | ● | ● | • | | | | | |
| *Molinia coerulea* group | | | | | | | | | | | · | | · | | |

Source: After Sebald, 1964.

[a] Spaces between lines indicate major differences between species groups (13 of Sebald's 24 groups are shown). Two groups of site units are ordered along gradients from moderately fresh to moderately dry (units 1 to 10) and from fresh to wet and imperfectly drained (units 11 to 15).

Key: ● species of the group abundant
• species of the group moderately abundant
· species of the group rare

certain site units are well defined by vegetation and could be mapped by vegetation alone. However, the combined technique, using soils and topography as well, is always faster and more reliable.

Mapping proceeds only after a reconnaissance survey of the forest to be mapped and determination of a tentative list of site units. Mapping is conducted systematically over 100 percent of each major forest of a district, and the end result is a detailed map of the site units of each forest to a scale of 1:10,000. In addition, a detailed report is prepared describing the major site features and the site units; for each site unit recommendations for choice of species, the risk of windthrow or fungus attack, rotation age, and other silvicultural and managerial features are presented.

**Growth and Productivity.** Growth rate, site index, and productivity are determined for the major species in a district or related group of districts upon completion of mapping. Using permanent sample plots and stem analyses, height-over-age curves are constructed for the major species of the most important site units. In a given site unit, substantial differences are often found between species. For a given species, polymorphic curves are prepared for the major site units within a growth district. This then is an example of how actual growth data on specific soil-site types have been used to resolve the major objections to site index listed earlier in the chapter.

The productivity in stem volume for the major species on each site unit can be determined, and the productivity of sites within a district or between districts can be compared (Moosmayer, 1955, 1957; Werner, 1962). Through such studies the site classification and mapping phases can be critically evaluated, and site units of similar yield classed together into "productivity groups." For example, Werner (1962) found more than threefold differences in stem volume of Norway spruce for the 13 major site units in the central Swabian Alb (Table 12.3). Obviously, the recognition of such differences in forest growth has great significance in forest management and land evaluation. Thus an intensive knowledge of soils, vegetation, climate, geology, and forest history are not academic exercises but can be integrated to give a meaningful and eminently practical result.

**Table 12.3 Productivity of Norway Spruce in the Central Swabian Alb**

| PRODUCTIVITY GROUP | | SITE UNIT | AVERAGE MEAN ANNUAL INCREMENT[a] m³ | ft³ |
|---|---|---|---|---|
| I | decreasing soil moisture ↓ | 1 | 16.3 | 228 |
| II | | 2 | 14.9 | 209 |
| | | 3 | 14.7 | 206 |
| III | | 4 | 13.9 | 195 |
| | | 5 | 13.7 | 192 |
| IV | | 6 | 12.9 | 181 |
| | | 7 | 12.8 | 179 |
| | | 8 | 12.4 | 174 |
| V | | 9 | 10.5 | 147 |
| | | 10 | 10.2 | 143 |
| VI | | 11 | 9.0 | 126 |
| VII | | 12 | 6.1 | 85 |
| | | 13 | 5.1 | 71 |

[a] Volume in cubic meters per hectare per year at age 100.
Source: Werner, 1962.

**Silvicultural Evaluation.** The value of site classification lies not only in the ability to predict productivity but also in the silvicultural handling and the management of forested land. Differences in site units, that is, local ecosystems, directly affect decisions as to the choice of species, establishment techniques, thinning regimes, and the risks of windthrow, soil degradation, disease, and insect attack.

The importance of distinguishing site units and knowledge of their characteristics may be illustrated by three examples. The culture of Norway spruce, the principal timber species in Baden-Württemberg, is undesirable on some site units because of high susceptibility to heart-rot fungi *(Fomes annosus)*; on other units there is a high risk of windthrow or soil compaction. Douglas-fir is of increasing importance and is faster growing than Norway spruce on many site units. However, on certain site units with high concentrations of calcium in the topsoil its establishment is virtually impossible. On sites having heavy clay soils, European silver fir is recommended for planting in mixtures with spruce and beech due to its intensive root development in these soils.

### Applications of Multifactor Methods in Europe and America.

Site classification systems similar to that in Baden-Württemberg have been used in other West German states, in Austria, and for many years in East Germany (Wagenknecht et al., 1956; Fiedler and Hunger, 1967; Eberhardt et al., 1967). In Baden-Württemberg, the intensive site classification provides a framework for management of forested lands and is also used as a basis for research in forest genetics, growth and yield, silviculture, soils, and pathology. It also has applications in land use planning, establishing natural areas, watershed management, and in environmental education.

As desirable as a system of this type may seem, it is usually regarded as too intensive, expensive, and impractical for American conditions. However, the model is sound, and the level of intensity can be chosen to meet the level of management, extensive or intensive. The delineation of major and minor landscapes is certainly possible, has been in application in Canada for many years (Halliday, 1937; Rowe, 1959, 1972), and has been initiated in the form of physiographic provinces in the Pacific Northwest (Franklin, 1965; Franklin and Dyrness, 1969; 1973), and in the northeastern (Committee on Site Classification, 1961) and the southeastern United States (Hodgkins, 1965). Within provinces multiple-factor classifications have appeared (Driscoll, 1964; Corliss and Dyrness, 1965; Franklin, 1966). For example, in the central Oregon juniper zone, vegetation, soil, and topographic features were used simultaneously to define nine ecosystems, which are useful in range inventory, in evaluation of range conditions, and in designing range rehabilitation programs (Driscoll, 1964). A map of "ecoregions" of the United States, based upon climate, physiography, and vegetation, has been published (Bailey, 1976), and a land system inventory stressing geology, landform, and soils is being developed and applied on National Forest lands (Wertz and Arnold, 1972; 1975; Wendt et al., 1975). North American programs have been influenced by the strong geomorphologic survey approach to land inventory as applied in Australia (Christian and Stewart, 1968; Stewart, 1968).

Multiple-factor methods employing soil, topography, and vegetation have been used extensively in California, where soil-vegetation surveys of public land have been under way since 1947 (Calif. Div. For., 1969). Maps providing general information on soils and vegetation types for forest and range management planning have been prepared for over five million hectares (Wieslander and Storie, 1952, 1953; Bradshaw, 1965). A site-index rating of the major species, where it can be determined from the existing stands, is indicated on the map units.

## Forest Land Classification in Canada

The most active use of multiple-factor methods outside Europe has been in Canada, where, coupled with airphoto techniques, they promise site inventory and site-quality evaluation at any desired management level. Site classification in Canada, as in Europe, is characterized by many different systems and complex terminologies, but it is strikingly different due to its necessarily extensive scale. Because of the need to describe and classify Canada's immense land resources for multiple-use management, a combined approach of physiography and vegetation, relying heavily on aerial photointerpretation, is being widely applied.

Much of the early attention in Canada was given to ground vegetation due to the influence of Cajander's method and that of the Zürich-Montpellier school of plant sociology (Lemieux, 1965; Burger, 1972). Studies of vegetation-soil relationships and those emphasizing physiography have evolved into the combined use of these features in many of the different systems used in Canadian provinces. Because of the diversity of site classification efforts, a significant development has been the formation of a nationwide group which has sought common ground among Canadian workers. It has worked to develop a truly ecosystematic classification system involving both biological and physical features of the land. Following pilot projects throughout Canada and conceptual inputs from previous studies, guidelines for biophysical land classification were published (Lacate, 1969). Emphasis was placed primarily on methods for reconaissance surveys whereby information for large inaccessible regions can be gathered and areas where more detail may be needed in the future can be identified. The Ecological (Biophysical) Land Classification program is now operational. Surveys have been completed in many provinces (Jurdant et al., 1975), and one of these is described briefly below. The continuing objective of this nationwide group (Thie and Ironside, 1977) is to further a coordinated approach of ecosystem classification among Canadian site workers and yet to permit regional variation.

Despite the diversity and dynamic state of Canadian methods, two examples must suffice to illustrate the design and practical features of Canadian methods. An overview of Canadian systems is available in the publications of Burger (1972), Jurdant et al., (1975), and Thie and Ironside (1977).

**Hills' Method in Ontario.** The "total site" classification system developed by Hills and his associates in Ontario (Hills, 1952; Hills and Pierpoint, 1960; Burger, 1972; Hills, 1977) since about 1940 is the most comprehensive and extensive system developed in a Canadian province. The method allows extensive application of airphoto interpretation in the classification, mapping, and evaluation of large, often inaccessible land areas. As in Baden-Württemberg, Hills has stressed the all-inclusive or holistic concept of site, defining it as the integrated complex of land and forest features within a prescribed area. Physiographic features are the basic frame of reference because "they remain most easily recognizable in a world of constant change." Thus **physiographic site types** and **forest types** (characterized by both overstory and ground vegetation) are combined to form the **total site types** (Figure 12.9).

To provide a framework for detailed classification, Ontario is subdivided into 13 site regions based on vegetation-physiography relationships reflecting major differences in climate. To gain additional homogeneity, a site region may be subdivided into site districts on the basis of relief and type of bedrock or parent materials. Details of classification are described by Hills (1952), Hills and Pierpoint (1960), and Burger (1972). Examples of field application of the system are available in the works of Pierpoint (1962) and Zoltai (1965).

The present and potential production of tree species on the various physiographic sites are estimated, using ratings of I (very high) to V (very low). In addition, a capability rating of site types (A = excellent to G = extremely poor) is obtained by integrating the potential productivity and a rating of degree of effort required to achieve this production (Hills and Pierpoint, 1960). Such evaluations are much less precise than determinations of absolute productivity, as accomplished in the Baden-Württemberg system. However,

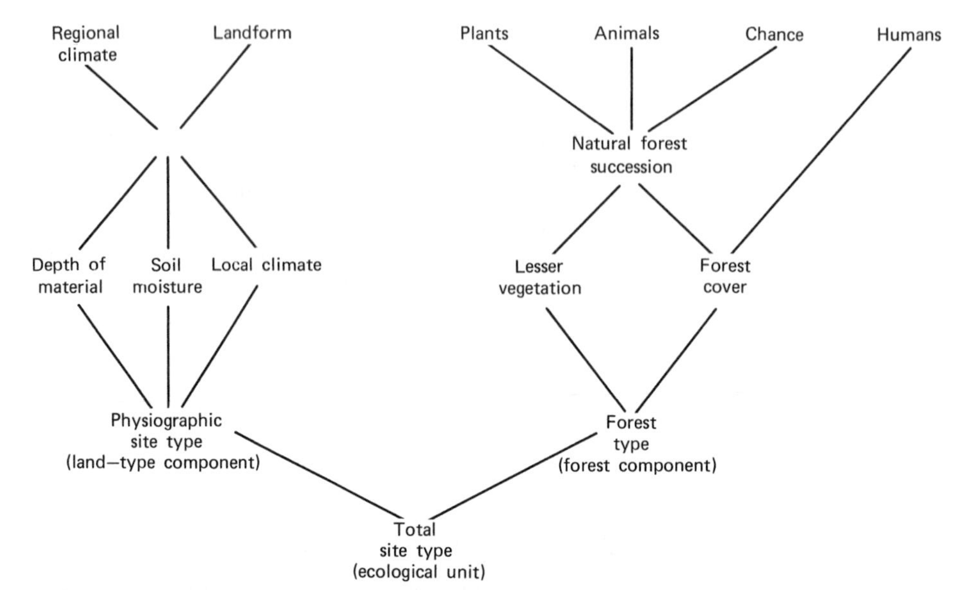


**Figure 12.9. Model of the classification of total site types by Hills' method. (After Hills and Pierpoint, 1960.)**

they provide an important first step in extensive management by ranking sites in timber productivity and capability for a number of species, or alternatively, providing wildlife or recreational capability ratings.

**Ecological (Biophysical) Land Classification.** The major objective of this system is to describe and characterize the biological and physical features of the land and organize the knowledge into a useful framework for land management (Jurdant et al., 1975). It is essentially a system to inventory ecosystems for land-use planning purposes. Because the inventory of ecosystems must be made at different scales and intensities to serve different objectives, a hierarchical system of five levels is used: Land Region, Land District, Land System, Land Type, and Land Phase (Figure 12.10):

- Land Region (mapping scale 1:1,000,000 to 1:3,000,000): characterized by a distinctive regional climate as expressed by vegetation.
- Land District (scale 1:250,000 to 1:1,000,000): characterized by a distinctive pattern of relief, geology, geomorphology, and associated vegetation.

- Land System (scale 1:100,000 to 1:250,000): characterized by a recurring pattern of landforms, soils, and succession of vegetation (chronosequences). This is the working level of most Canadian biophysical surveys.
- Land Type (scale 1:10,000 to 1:60,000): characterized by a fairly homogeneous combination of soil and chronosequence of vegetation. This is the basic ecological unit and the one on which the biological productivity and other interpretive ratings are made.
- Land Phase: a subdivision of the Land Type based on the stage of vegetational succession found on the area at the time of the survey.

Because of the small-scale survey nature of the work, landform is the dominant feature of the system. Landform classification and mapping provide the framework within which climatic, vegetational, soils, and hydrologic data are described and classified. The rationale is that although all site factors influence each other, landform affects the other factors more than it is affected by them.

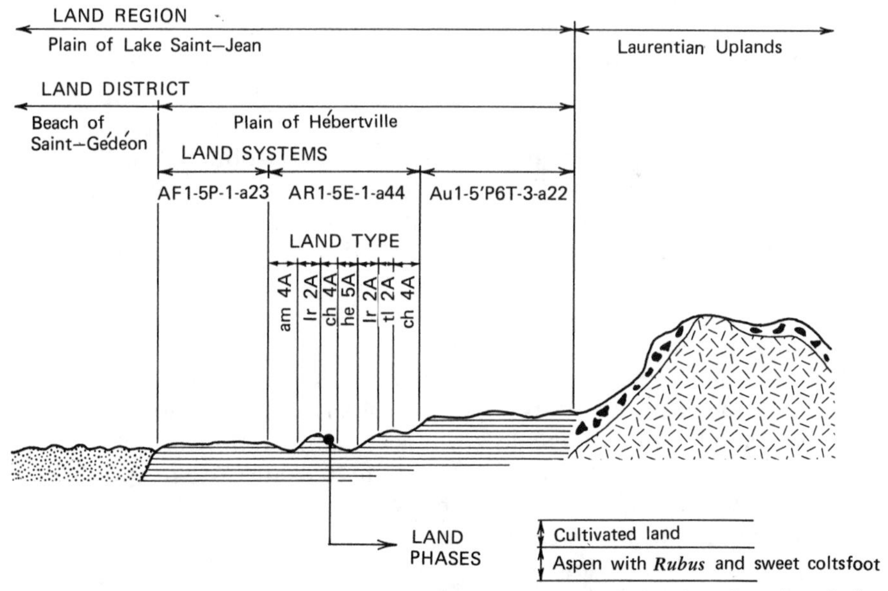


**Figure 12.10. An example from Quebec of the five levels of classification used in the Canadian ecological (biophysical) land classification. (After Jurdant et al., 1977. Reproduced by permission of the Minister of Supply and Services Canada.)**

A key feature of the system is in developing the basic classification from an a priori integration of knowledge of geomorphology, soils, and vegetation that is collected by an interdisciplinary team of specialists working together in the field and office. Such an approach is preferred to an a posteriori integration obtained by superimposing overlay maps of geomorphology, soils, and vegetation. These are not only more expensive but create complications when developed independently by each discipline because unit boundaries rarely conform to one another.

Following a pilot reconaissance survey in the Saguenay-Lac-Saint-Jean region of Quebec (Jurdant et al., 1972) to test and refine methods, a survey was conducted in the James Bay region of western Quebec encompassing an area of 350,000 km². The purpose of the survey was to obtain information and interpretations relating to agriculture, forestry, recreation, wildlife, water, and engineering prior to the installation of major hydroelectric plants on several rivers. Such development would require construction of new access routes, large dams, settlement locations, and transmission facilities as well as the preparation of extensive impoundment areas.

The Land System (scale 1:250,000) is the primary mapping unit. These units have recurring patterns of landform (Figure 12.10), soils, drainage, and vegetation such that they are discernible environments with distinct boundaries. They are initially delineated by interdisciplinary teams that examine small-scale aerial photographs for recurring patterns of topography, depth of unconsolidated surficial materials, and landform. Extensive field work is then conducted to sample and describe the ecosystems, check mapping boundaries, and determine the Land Types. The details of the methods and the results of this major undertaking are effectively presented and illustrated by Jurdant et al. (1977).

In summary, site classification and forest land mapping from airphotos using physiographic and vegetation types can provide meaningful units for intensive or extensive forest resource management and more broadly for integrated resource planning (Rowe, 1971; Duffy, 1975). The European methods illustrate that site classification and productivity evaluation can be developed to meet the most intensive management practices even in areas of highly variable soil, topography, and climate, and for sites which have been substantially disturbed by man. With the increasing refinement of satellite imagery, airphoto techniques, and methods of soil and vegetation classification, systematic site classification and mapping are becoming increasingly important in multiple-use land management and planning in the United States and Canada.

## SUGGESTED READINGS

Bernier, B., and C. H. Winget (eds.). 1975. *Forest soils and forest land management*. Proc. Fourth North Am. For. Soils Conf. Laval Univ. Press, Quebec. 675 pp.

Burger, D. 1972. Forest site classification in Canada, *Mitt. Vereins forstl. Standortsk. Forstpflz.* 21:20–36.

Carmean, Willard H. 1975. Forest site quality evaluation in the United States. *Adv. Agronomy* 27:209–269.

———. 1977. Site classification for northern forest species. *In* Proc. Symposium on Intensive Culture of Northern Forest Types, pp. 205–239. USDA For. Serv. Gen. Tech. Report NE-29. Northeastern For. Exp. Sta., Upper Darby, Pa.

Coile, T. S. 1952. Soil and the growth of forests. *Adv. Agronomy* 4:330–398.

Damman, A. W. H. 1964. Some forest types of central Newfoundland and their relation to environmental factors. For. Sci. Monogr. 8. 62 pp.

Daubenmire, R. 1976. The use of vegetation in assessing the productivity of forest lands. *Bot. Rev.* 42:115–143.

———, and Jean B. Daubenmire. 1968. Forest vegetation of eastern Washington and northern Idaho. Washington Agric. Exp. Sta., Tech. Bull. 60. 104 pp.

Hills, G. A. 1952. The classification and evaluation of site for forestry. Ontario Dept. Lands and For., Res. Rept. 24. 41 pp.

———, and G. Pierpoint. 1960. Forest site evaluation in Ontario. Ontario Dept. Lands and For., Res. Rept. 42. 64 pp.

Jurdant, M., D. S. Lacate, S. C. Zoltai, G. G. Runka, and R. Wells. 1975. Bio-physical land classification in Canada. *In* B. Bernier and C. H. Winget (eds.), *Forest Soils and Forest Land Management*. Laval Univ. Press, Quebec. 675pp.

Pfister, Robert D., Bernard L. Kovalchik, Stephen F. Arno, and Richard C. Presby. 1977. Forest habitat types of Montana. USDA For. Serv. Gen. Tech. Report INT-34. Intermountain For. and Rge. Exp. Sta., Ogden, Utah. 174 pp.

Ralston, Charles W. 1964. Evaluation of forest site productivity. *Int. Rev. For. Res.* 1:171–201.

Rennie, P. J. 1962. Methods of assessing forest site capacity. *Trans. 7th Inter. Soc. Soil Sci.*, Comm. IV and V, pp. 3–18.

Rowe, J. S. 1962. Soil, site and land classification. *For. Chron.* 38:420–432.

———. 1971. Why classify forest land? *For. Chron.* 47:144–148.

———. 1972. *Forest Regions of Canada*. Can. For. Serv., Dept. Env. Publ. No. 1300, Ottawa. 172 pp. + map.

Thie, J., and G. Ironside. 1977. Ecological (biophysical) land classification in Canada. Ecological land classification series, No. 1, Lands Directorate, Environment Canada, Ottawa. 296 pp.

# 13
# Animals

A host of animals of all sizes forms an indispensable part of the forest ecosystem. As biotic factors they markedly influence forest patterns and processes; the role of soil animals was described in Chapter 9. Animals, in turn, are strongly affected both by the physical environment and by the plant communities with which they associate.

Plants provide shelter for animals. The foods produced by green plants are the foundation of plant-animal relationships of the ecosystem; they initiate all food chains. Each food chain consists of a food plant, the animals that eat it (**browsers** and **grazers** or **phytophages**), the animal predators and parasites that feed on phytophages, and the scavengers that eat animal remains and excrement. The plant-animal cycle is completed by decomposers (both animals and plants) that degrade and mineralize plant litter and animal residues (Chapter 18, Figure 18.1).

In this brief treatment, we emphasize the reciprocal adaptations of plants and animals; the role of animals in regulating forest composition, development, and productivity; wildlife habitat and fire; and the effects of large animals on the quality of the site. Often we give most attention to the destructive attributes of forest animals. However, we cannot overemphasize the great but little-appreciated contributions of animals in the evolution of plants and their contributions to ecosystem processes. A detailed treatment of animals as components of Russian and European forest ecosystems is presented by Sukachev and Dylis (1964). Janzen (1966, 1969, 1970) has been especially instrumental in stimulating interest and providing an understanding of animal-plant interactions in tropical forest ecosystems. A greater emphasis on this phase of forest ecology is needed in temperate communities.

## ANIMALS AND THEIR ROLES IN THE FOREST ECOSYSTEM

Both invertebrate and vertebrate animals affect ecosystem processes (such as mineral and water cycling) and the regeneration and succession of forest trees by their major activities of dispersing pollen and seeds, feeding on live plant tissues (grazing), decomposing dead organic matter, and burrowing in the soil (Mattson, 1977). These activities affect many aspects of forest ecosystems.

## Macroevolution and Reciprocal Adaptations

Insects, birds, and mammals were apparently instrumental in the evolution of flowering plants (angiosperms) and their dominance in many ecosystems (Regal, 1977). Gymnosperms, including conifers, had achieved great abundance and diversity, with similarities to modern genera by the Jurassic period, 130 to 185 million years ago. Originating later, at least 125 million years ago in the Cretaceous period, angiosperms became diverse and spread rapidly in mesic lowlands by 110 million years ago, soon achieving worldwide dominance (Chapter 19). The precise reasons for this significant change are still unknown. It seems likely, however, that both animals and physical factors, including worldwide warming of climate (Raven, 1977), were involved.

Pollination and seed dispersal by animals may largely explain why flowering plants rapidly differentiated and replaced wind pollinated and wind-dispersed gymnosperms and ferns which had previously been dominant (Regal, 1977). Insect pollination facilitated an outcrossing mode of reproduction among individuals that were widely separated in the forest. Whereas wind is an efficient pollinator in stands of many individuals of the same species, it becomes less efficient as individuals of a species become widely separated in the forest. Possibly, the rise of seed- and seedling-eating animals (insects, birds, and mammals) during the Cretaceous Period generated selection pressures favoring individuals that were spaced at some distance from the parent tree. Thus they were not subject to high concentrations of seed and seedling eaters around the parent (Janzen, 1970). Cross-pollination generated the genetic variation and flexibility that could have enabled individuals to take advantage of new habitats at some distance from the parent tree. Birds were probably the major dispersal agents, increasing the incidence of long-distance dispersal and creating new patterns of local dispersal.

Wind pollination is still dominant in conifers and in most tree angiosperms of temperate and boreal regions (Whitehead, 1969). In these regions, climatic and soil extremes and fires restrict the diversity and abundance of animals. These factors also restrict niche diversity so that woody plant diversity is low compared to equable tropical forests. However, some insect-pollinated trees, notably basswood, cherries, and willows, occur in the northern forest, and various species (magnolias, yellow-poplar, sourwood) occur in the mesophytic forests of the Southern Appalachian Mountains.

A profound coevolutionary event occurred in the Tertiary period as ungulates (hoofed mammals) evolved digestive systems to deal with fibrous vegetation. The perissodactyl (horses, tapirs, and rhinos) and artiodactyls (deer, bovids, camels, and hippos) arose

GREEN MOUNTAIN
COLLEGE
Poultney, Vermont 05764

Temperatures are usually higher on burned sites (black surface) vs green surface

Temp may be 3-16° higher than on unburned.

Increased soil temp hasten spring development of roots and shoots in burned area

Speed decomposition

Promotes activity of soil organisms

independently in the early Tertiary, and both developed fermentation systems to digest cellulose (Janis, 1976). The artiodactyls are the dominant large herbivorous mammals of grasslands, savannas, and temperate forests (deer, elk, moose, caribou, antelope, bison, and domesticated cattle, sheep, and goats), and have a marked influence today on species composition, productivity, and site quality. Members of this group (ruminants) evolved a compartmentalized stomach or rumen, which functions as a storage chamber for plant material during digestion and as a fermentation chamber where a diverse bacterial and protozoan community digests cellulose and other plant substances. Members of this group are termed ruminants because they ruminate, or chew the cud. In addition to being able to accommodate large amounts of fibrous vegetation, the rumen system is also efficient in making nitrogen available and has detoxifying bacteria to cope with plant toxins. The microorganisms of the rumen adapt to the kinds of food ingested. For example, if a toxic food is given and the amount slowly increased, there is selection for strains of bacteria capable of degrading the toxin (Freeland and Janzen, 1974). Deer accustomed to Douglas-fir foliage can consume up to 50 percent of it in their diet, whereas floras of deer rumen not experienced with Douglas-fir are severely inhibited by it (Oh et al., 1967).

Although many herbivores feed primarily on fruits and seeds, ungulates eat the leaves, shoots, and woody stems. In the tropical forest environment of the early Tertiary Period, the ancestral ruminants selected a largely cellulose-free diet of fruits, seeds, and young growth all the year round. However, as the climate of North America and Eurasia became cooler and drier (Dix, 1964) plant growth became more seasonal, and animals were increasingly forced to cope with a diet that was more fibrous during parts of the year. Thus an efficient fermentation system gradually evolved in response to gradual changes in vegetation.

## Plant Defense Adaptations

Many examples of mutual adaptations of woody species and animals are observed. We accept many as logical deductions without rigorous demonstration of cause-and-effect relations. The animal-plant adaptations, however, must also be consistent with other selective pressures of the physical environment, including fire. A recurrent feature is the various defense mechanisms woody plants evolve under the selection pressure of herbivores and seed eaters. Plants are subject to attack at all stages in their life cycle, and all parts of a tree are subject to attack.

The kinds of defenses employed by woody plants include texture and composition of the plant surface, presence of specialized organs or tissues such as thorns or resin ducts, the absence of nutrients required by the pest, the presence of hormonelike substances that affect the development of insects, unsuitable pH or osmotic pressure, or accumulations of secondary compounds such as alkaloids, terpenes, phenolics, and cyanogenic glycosides (Levin, 1976). In all higher plants, an enormous number and diversity of secondary compounds are known; thousands—4000 alkaloids alone—have been described. Detailed reviews of the chemical defenses of plants and how animals detoxify or cope with these plant products are given by Levin (1976) and Swain (1977). Freeland and Janzen (1974) have considered the strategies of herbivorous mammals in relation to plant secondary compounds.

Many tree species, including black locust, honey locust, osage orange, hawthorns, junipers, and some hard pines, as well as many woody shrubs and vines (roses, smilaxes, blackberries, raspberries) have prickles, spines, thorns, or sharp needle-leaves that deter browsing. In tree species, these structures are particularly concentrated in the juvenile stage when foliage and stem feeding by rodents, rabbits, and other herbivores is most likely.

In conifers, and in particular the Pinaceae, the physical (rate and duration of flow, quantity produced, viscosity, exudation pressure) and chemical properties of oleoresin exuded from resin ducts in needles, shoots, and bark significantly deter foliage feeders and bark beetles that mine and construct egg galleries in the inner bark tissues. At the same time, however, insects use resin vapors to find their hosts. They are often attracted to damaged trees where the resin vapor is highly concentrated. Vigorous, standing trees are typically more resistant than lightning damaged or freshly cut trees (Hanover, 1975). For example, western larch trees are virtually immune to bark beetle attack while standing but are immediately attacked after being felled (Furniss, 1972).

In young and vigorously growing pines, oleoresin pressure is high and beetles entering the bark contact the resin directly. They are either physically repelled ("pitched out") or physiologically impaired by chemical properties of the resin, thus preventing them from breeding and reproducing. Resin also deters establishment of the European pine shoot moth on Scots pine and black pine (Harris, 1960). In ponderosa pine, terpenes myrcene and limonene kill western pine beetles feeding on its needles or bark (Smith, 1966). Bark beetles exhibit various degrees of tolerance to the toxicity of resins. In general, beetles as a group are often host specific and are more tolerant to resins of their own host species than those of other pines.

The severity of bark beetle attack is greatest when the resin defenses are low. This may be caused by seasonal variation in resin level, by the natural aging process, or by major stresses on the plant, such as extreme competition, drought, pollution, logging damage, disease, and defoliation.

In certain hardwood trees, resins may also deter feeding of insects as in cottonwoods (Curtis and Lersten, 1974). Young leaves of creosote bush, a desert shrub, have two to three times the resin content of older leaves and are accordingly more resistant to defoliation by various insects (Rhoades, 1976). Other chemicals are also important. Juglone, produced by shagbark hickory, deters feeding by some bark beetles but not others. Tannins, when in high concentration, can deter insect feeding on oak leaves. However, during leaf and shoot development in the spring, when tannins are absent or scarce, oaks may be infested with insects. The late and rapid flushing of preformed leaves and shoots of the oak minimizes the time insects may feed and reproduce using these tissues.

This rapid leaf-flushing trait of trees of temperate forests has apparently elicited reciprocal adaptations of insects and their hosts with respect to time of flushing. For example, flushing time may differ by as much as three weeks between different trembling aspen clones (Barnes, 1969). Populations of tortricid caterpillars infest predominantly the leaves of the early flushing clones (Witter and Waisanen, 1978). A similar relationship was also reported for larvae of tortrix moths on oaks in Russia (Sukachev and Dylis, 1964) and Europe (Stern and Roche, 1974).

Mattson and Addy (1975) and White (1978) have emphasized how food quality affects insect feeders; moisture stress and other stresses may increase the concentrations of sugar and other usable substances in the foliage or create a more favorable balance of nutrients. Plant-animal biochemical systems, having evolved for millions of years, are closely interdependent, and the balance between their systems is usually finely tuned. Plant foliage may be only marginally adequate nutritionally for its usual consumers (Southwood, 1973). Thus insects respond readily to slight changes to plant chemistry which make the food more nutritional, less toxic, or allow easier access to the plant tissues, as in the case of bark beetles.

Many adaptations in the reproductive process may either promote or deter animal feeding: yearly periodicity and seasonal timing of seed production, abundance and placement of fruits and seeds, secondary compounds in fruits and seeds, and fruit shape, color, and palatability. Many tropical and temperate trees and shrubs have evolved attractive, palatable, fleshy fruits that favor widespread animal dispersal.

A general defense mechanism against seed eaters in nearly all tree species of temperate forests is periodicity of major seed crops. Red pine is a good example; large seed crops occur at intervals of 3 to 7 years. Insects, birds, and mammals may consume all seeds in years of low seed production. However, in years of excellent seed crops the food supply is more than enough to satiate seed-eating animals and seedlings may establish providing other factors are favorable. Such periodicity prevents long-term build-up of consumer populations.

Although many plants rely on the chemical or timing defenses described above, the swollen-thorn acacias (Leguminosae) of Central America have substituted the abilities of obligate acacia-ants (*Pseudomyrmex* spp.) for their defense (Janzen, 1966). The ants take shelter and nest in large stipular thorns of the swollen-thorn acacias. They feed on nectaries and leaflet tips that are produced nearly year round. The obligate acacia-ants protect the swollen-thorn acacias from grazing insects and neighboring plants, especially vines. The ants patrol the surface of the acacia day and night with 25 percent of the ant colony normally outside of the thorns. Without ants the acacias are a favored host of many insects and would soon be killed. Thus the swollen-thorn acacias are unable to survive without protection from the obligate acacia-ants that in turn receive shelter and food.

The adaptations cited above are not elicited exclusively in response to animal pressures. However, selection pressures by animals may be especially strong in tropical regions wherever climatic factors do not severely limit animal diversity and abundance. Other selection pressures, such as temperature, moisture, fire, and seedbed habitat, may be equally or more important in temperate forests.

Not only do individual species have their own defense mechanisms, but groups of plants, termed **guilds**, may be functionally interdependent in providing a defense against herbivores. Botanically, the term guild has been used to characterize a group of plants in some way dependent on other plants, such as climbing vines, epiphytes, and parasites. However, the term is also more broadly used to describe ecologically unified or functional groups of organisms. For example, the presence of toxic or noxious plants may deter animals from feeding on associated species. The several functions of guild members are described by Atsatt and O'Dowd (1976). They cite Monteith (1960), who found that the amount of parasitism of the larch sawfly by *Bessa harveyi* was mainly influenced by the odors from shrubs beneath eastern larch and by the proximity of white spruce and other tree species. The guild concept suggests that plants grown in monocultures may suffer more herbivore damage than the same species grown in association with their native flora.

## Pollination

Since pollination is a critical process in the life cycle of plant species, the role of animals as important pollinators of woody plants cannot be overstated. Wind pollination is dominant in temperate forests, and coniferous trees are exclusively wind pollinated. Animal pollination is widespread among woody species in tropical regions but much less common in temperate forests. Pollination is accomplished by insects (bees, wasps, flies, beetles, butterflies, and moths), birds (in the New World especially hummingbirds), and bats. Only rarely do other mammals other than bats pollinate trees and shrubs; a small primate, the bush baby, is known to pollinate the African baobab tree (Coe and Isaac, 1965). The primary attractants of animal pollinators are nectar; fragrance; flower color, shape, and size; and in the case of birds insects visiting the blossoms. A comprehensive account of animal-plant interactions in pollination ecology is given by Faegri and van der Pijl (1971).

In temperate forests, dominantly or wholly animal-pollinated groups include all the Ericaceae and species of *Prunus, Acer, Aesculus, Robinia, Tilia, Magnolia, Liriodendron, Catalpa, Salix*, and *Rhamnus*. In the case of yellow-poplar, insect pollination is inefficient (Boyce and Kaeiser, 1961), and only about 10 percent of the seeds may be viable. Nevertheless, enough germinable seeds are produced per tree that natural regeneration of disturbed areas is usually abundant.

## Seed Dispersal

Animals are significant dispersal agents of seeds and thus instrumental in maintaining and spreading woody plant populations. Seed dispersal also promotes population differentiation and gene flow in populations. Animal agents of seed dispersal are not only important for many angiosperms, especially in equable climates, but may be dominant for some gymnosperms as well. A comprehensive treatment of dispersal ecology is presented by van der Pijl (1972), and a monumental early reference is available in the work of Ridley (1930). The anatomical mechanisms of seed dispersal are described by Fahn and Werker (1972), and dispersal in relation to seed predation is reviewed by Janzen (1971).

Vertebrate animals (fish, reptiles, birds, and mammals) are the primary dispersal agents of woody plants; insects transport fungi and moss spores, and earthworms are thought to disperse seeds of some orchids. Regardless of the vertebrate dispersal agent, three key features must characterize the animal-plant dispersal system to guarantee establishment and persistence of the plant species. First, animals must be attracted to the fruit or seed by smell, sight, or taste. Second, the attractant must be timed to coincide with the maturity of the seed; premature ingestion destroys the developing seed. Finally, enough viable seeds must "escape" the dispersal agent and be deposited in a site where the chances of successful establishment are high. Some escape mechanisms include burying of seeds by vertebrates, regurgitation of seeds by some vertebrates, hard, smooth, seed coatings that assure undamaged passage through digestive tracts, and darkly or inconspicuously colored fruits coupled with brightly colored accessory parts. In the latter case, the animal is attracted to the fruit by a red or orange aril, peduncle, or bract, but if the dark-colored fruit is dropped, it is not readily found (Janzen, 1969).

**Fish and Reptiles.** Fish eat pulpy seeds of various woody species growing along tropical rivers, but the fate of these seeds is unknown. Although the seeds of many tropical trees are dispersed by water, the fish may provide upstream transport that is not normally possible.

Reptiles, turtles and tortoises, alligators, and lizards have a keen sense of smell and eat fruits after they have dropped off trees or when borne close to the ground. The fruits of *Celtis iguana* are eaten by climbing iguanas. Most modern reptiles, however, are not vegetarians. Fruits eaten by reptiles typically have a smell, they may be colored, and they are often borne near the ground or dropped at maturity.

**Birds.** Birds are primary dispersal agents, and many adaptive devices have appeared. The two main dispersal methods are disgorging fruits or seeds that they carry in the mouth and excreting seeds contained in fruits that have been eaten. Rarely do birds carry tree fruits on the outside of their body. However, fruits of the dwarf mistletoe are carried on the bodies of birds.

Birds destroy many seeds when they eat and digest them. However, many wood pigeons, thrushes, nutcrackers, crows, and waxwings can disgorge fruits and seeds they are carrying in their beak. More important are the birds that store part of their food but neglect to eat it. The wingless-seeded soft pines of semiarid (pinyon pines) and subalpine (stone pines) environments are dependent on dispersal by jays, woodpeckers, and nutcrackers. Siberian nutcrackers store stone pine nuts in caches of 6 to 12 under mosses and lichens up to several kilometers from the parent tree (Sukachev and Dylis, 1964). They store from 4000 to 34,000 seeds per hectare and eat only about half. Seeds of many woody angiosperms (oaks, beeches, walnuts, chestnuts, hazels) are also spread in this manner. Jays and rooks have been observed to bury acorns and hazelnuts (Chettleburgh, 1952). Russian jays hide oak acorns and hundreds of oak seedlings may be established in this way. The California woodpecker is noted for imbedding thousands of acorns, almonds, and hickory nuts in the bark of standing trees (Figure 13.1). Squirrels and other rodents may then carry them to a germination site, completing the dispersal.

**Figure 13.1. The California woodpecker *(Melanerpes formicivorus)* has imbedded many acorns of the California black oak (right) in this Jeffrey pine tree. (Near Julian, California; Photos by Terry Bowyer.)**

Birds may disperse large numbers of seeds, sometimes for long distances. For example, Schuster (1950) reported that jays in Germany dispersed approximately 4600 acorns per bird in a season. Many birds carried the acorns 4 km. Through such long-distance transport, birds could significantly speed the migration of tree species following glaciation, especially pines and oaks.

The amount of fruit a tree produces may be adapted to the effectiveness of its dispersal agents. A tree producing a limited number of accessible fruits should rely on an efficient and dependable obligate fruit-feeding dispersal agent (McKey, 1975). Howe (1977) describes such a mutualistic relationship of the Costa Rican rain forest tree *Casearia corymbosa* whose seeds are reliably dispersed by one bird species, although 21 other avian visitors feed on its fruit. The masked tityra (*Tityra semifasciata*) is an effective seed dispersal agent because it is a common and regular visitor throughout the fruiting season, has high feeding rates, removes seeds from the vicinity of the parent tree before processing them, and regurgitates viable seeds. In contrast, two parrot species remain in the tree feeding on arils bearing the seeds, that then drop to the ground under the tree. Here seedling mortality is virtually complete.

Dispersal by birds is efficient for plants having fruit adapted to fruit-eating birds that excrete the seeds undamaged and sometimes prepared for germination. In the Temperate Zone, species of *Cornus*, *Prunus*, *Ribes*, and *Sorbus* ripen fruits in the late summer and fall. Fruits of other species (in *Cotoneaster*, *Ilex*, *Ligustrum*, *Rosa*, *Viburnum*, and *Vitis*) last through the winter to be consumed in preferential sequence.

In temperate forests, canopy position may be correlated with mode of dispersal. Most shrubs and small trees of the lower canopy have fleshy fruits for bird and mammal dispersal, whereas the fruits of upper canopy trees are mainly wind or water dispersed. However, some large overstory trees also have fleshy fruits or seeds for bird dispersal: black cherry, magnolias, hackberries, and gum species (*Nyssa* spp.).

*Juniperus* is an exceptional conifer genus in producing berrylike cones that are eaten and the seeds widely dispersed by birds. In contrast, seeds of the closely related genus *Cupressus* are encased in round, heavy, woody cones that travel only as far as they roll down slope or rodents carry them up slope. This may help to explain the fact that junipers are widely distributed in North America, whereas species of *Cupressus* are limited to a few isolated groves (Grant, 1963).

**Mammals.** Mammals, including rodents, ungulates, bats, and humans, are also important dispersal agents. Tropical forests are rich in species having fruits adapted for mammal dispersal the year round; many fruits are eaten by both birds and mammals. In temperate forests, rodents are the primary agents.

Some animals, especially rodents, destroy the seeds of oaks, pines, and various other groups. However, many seeds are cached and forgotten so that germination and establishment are accomplished. Acorns and hickory nuts are dispersed up and down slope by squirrels for distances up to 50 m from the parent tree. Many hickory species rely on gaps in the oak forest or the forest edge for their eventual development into the overstory. Wind and gravity provide negligible dispersal away from the parent tree except on steep slopes where fruits might roll down hill.

Squirrels prefer acorns of the white oak group to those of the black oak group (Short, 1976). Tannins in the acorns of both groups (three to four times higher in the black oak group) deter rodent feeding so that if other more palatable foods are available, acorns may be cached rather than eaten on the spot. Acorns of the white oak group germinate soon after falling and diminish in palatability after sprouting (Smith and Follmer, 1972). Acorns of the black oak group lie dormant over winter and germinate the following spring. They are available but of low palatability during this time. Compared to other foods available to squirrels (legume and other seeds, dried and fleshy fruits), acorns have relatively low levels of protein and phosphorus (Short, 1976). They supply adequate energy but do not satisfy the metabolic requirements of squirrels for nitrogen and probably phosphorus. Therefore, a finely balanced adaptation system between squirrels and oaks has evolved chemically, physiologically, and morphologically accounting at least in part for the persistence of both groups.

The fruit characteristics of plants are therefore adapted closely to the behavior of their dispersal mammals. Mammals differ from birds in that they have a keener sense of smell, possess teeth, masticate much better; they are also typically larger, rarely lead an arboreal life, and are mostly color-blind night feeders (van der Pijl, 1972). The corresponding characteristics of fruits eaten by mammals are a favorable smell; a hard skin; a more evident protection of the seed proper against mechanical destruction (often a stone-like covering of the seed as in all drupes), often assisted or replaced by the presence of toxic or bitter substances in the seed; nonessentiality of color; and large size in many cases, typically causing them to drop to the ground.

Ungulates, including tropical and savanna ruminants and elephants, consume a wide variety of vegetation. African ruminants of savannas (springbok, gemsbok, eland) consume considerable amounts of tree fruits, particularly legumes. Many acacias provide leathery, nutritious pods containing extremely hard and smooth seeds, which evade or resist strong molars.

Bats are important dispersal agents of woody species, primarily in tropical Asia and Africa (van der Pijl, 1957). Fruit bats are nocturnal, color blind, and have a keen sense of smell. Bats eat fruits that are of drab color, have a musky odor, and are often large and exposed outside the foliage. Bats typically consume the juice after intense chewing of the fruit, and the remnants are regurgitated. They transport seeds within about 200 m of the fruit source.

Monkeys and apes are mostly destructive, eating everything edible, ripe or unripe, with apparently a limited dispersal role. Humans, however, are significant dispersal agents.

Animals not only disperse seeds, but small mammals disperse the spores of mycorrhizal fungi that form fruiting bodies below ground (Maser et al., 1978). Although chipmunks and deer mice eat tree seeds, they may have an even more important positive role in the coniferous ecosystem by dispersing the mycorrhizal inoculum that is necessary to the establishment and vigorous growth of conifers.

## Germination and Establishment

Animals influence germination of temperate tree species by caching seeds in the forest floor where dormancy requirements (if any) may be satisfied. In addition, many seeds are adapted to pass through the digestive tracts of birds and mammals unharmed (Krefting and Roe, 1949). In fact, digestive juices may weaken the seed coat, thus favoring the absorption of water and increasing eventual germination. Sukachev and Dylis (1964) cite Russian work indicating that from 72 to 92 percent of the seeds having tough coverings retain their germinability, and the germinability of some seeds is increased after passing through digestive tracts.

The digging activities of various animals turn up mineral soil that provides a seedbed suitable for establishment. Sukachev and Dylis (1964) give an example in which 2 to 3 percent of the surface area is covered with molehills where the germination of oak and maple seeds is twice as high as in soil undisturbed by moles. Rooting by wild boars may also remove thick moss or other vegetation that prevents seeds from germinating and establishing. However, rooting animals may destroy young seedlings and cause widespread damage. Rooting by wild boars in Russia may cause the replacement of hardwoods by Norway spruce. They may enhance the establishment of European white birch by consuming oak and beech seeds and leaving a seedbed suitable for birch. Rooting of feral pigs in the southern United States once caused great mortality to pine seedlings, especially longleaf pine. Grazing by domestic animals can preclude establishment of all tree seedlings and may severely affect site quality.

## Decomposition, Mineral Cycling, and Soil Improvement

The significant role of soil animals in organic matter decomposition and mineral cycling has already been cited in Chapter 9. Soil animals aid in the breakdown of organic matter by:

1. Physically disintegrating tissues and increasing the surface area available for bacterial and fungal action.
2. Selectively decomposing material such as sugar, cellulose, and even lignin.
3. Transforming plant residues into humic materials.
4. Mixing decomposed organic matter into the upper layer of soil.
5. Forming complex aggregates between organic matter and the mineral fractions of soil (Edwards, 1974).

Soil-dwelling animals, microfauna (invisible or nearly so to the naked eye) and mesofauna (larger invertebrates), are primarily saprophytes. Microfauna include protozoa, various worms (nematodes and enchytracids), mites, primitive wingless insects, and larvae of some winged insects. Mesofauna comprise earthworms, molluscs (snails and slugs), myriopods, and some insects (including larvae and adults). Detailed treatments of the biology of organic matter decomposition by specific groups of soil animals are presented by Dickinson and Pugh (1974).

The physical environment, forest composition, and the amount and kind of humus (mull or mor) strongly affect the number and diversity of micro- and mesofauna. The sequential development of vegetation on a site largely determines the corresponding sequence of soil invertebrates. Together these changes constitute the continuous succession of biota of an ecosystem.

Invertebrates directly decompose organic matter and catalyze the action of microflora; microfauna excrete substances that create conditions that stimulate bacteria and other microflora. Invertebrates also aid in horizontal and vertical distribution of organic matter for more intense activity by microorganisms. In turn, microorganisms initially prepare the food for soil-dwelling invertebrates.

Higher and more stable humidity caused by animal activity (increasing water-holding capacity, decreasing evaporation, reducing topography of the forest floor) has a profound effect in increasing microbial activity. For example, microbial decomposition of leaves buried in the humid, mucus-rich opening of an earthworm's burrow probably greatly exceeds the decomposition by enzymes of the earthworm's gut (Satchell, 1974).

In tropical and subtropical regions, termites, including many mound-building species, dominate the soil fauna. Termite mounds, built of soil and carton (a substance consisting primarily of organic excreta) are highly resistant to microbial decomposition (Satchell, 1974). According to Lee and Wood (1971), much of the nutrient content of organic matter that would be mineralized by microbes in temperate climates is, in the tropics, included in termite mounds and withheld from circulation.

Severe insect defoliation of the overstory usually results in more light, warmth, and moisture reaching the forest floor. Insect defoliation may regulate nutrient cycling by increasing the rate of litter fall, influencing the rate of nutrient leaching from foliage, stimulating the redistribution of nutrients within plants, increasing light penetration through foliage canopy, and stimulating the activity of decomposer organisms (Mattson and Addy, 1975).

The physical and chemical properties of soils are improved by soil-dwelling invertebrates and burrowing mammals. Organic matter is mixed with mineral soil, and aeration, moisture-holding capacity, and nutrient content are improved. Soil structure is aided by various soil fauna, including insects, myriopods, woodlice, earthworms, moles, and other rodents. Soil invertebrates ingest soil particles and mix them with finely ground and digested organic matter. This material when excreted forms small but durable structural aggregates. Soil animals, through their excreta and carcasses, may return considerable amounts of nutrients to the soil. For an ash-oak forest in England, Satchell (1967) estimated that earthworms annually returned about 100 kg of nitrogen per hectare.

Soil animals move through the profile and leave their feces at different depths; earthworms carry fragments of litter with them. They deepen the humus zone and bring about an even distribution of organic matter, thus improving aeration, porosity, and moisture–holding capacity. Earthworms also help neutralize acid soil by secretions of calcium carbonate from calciferous glands.

Soil moisture is profoundly influenced. First, the action of soil-dwelling invertebrates and burrowing vertebrates increases the percolation of water in the soil. Second, the organic matter distributed by soil animals increases the water-holding capacity of the soil. Third, severe defoliation by canopy insects limits interception, evaporation, and throughfall and thereby increases the incident precipitation available to the understory and forest floor. Finally, the drainage system of lands adjacent to streams and small lakes may be significantly changed by dam building activities of beavers.

## Damage in Forest Stands

Animals eat plants, so it is not surprising that much of the literature reports animal damage to forests. Animals browse, chew, gnaw, pierce, strip, debark, girdle, fell, and trample woody species. Nearly every animal group inhabiting forest land has been cited for some kind of damage (Crouch, 1976). Nevertheless, the incalculable positive influence of animals cannot be overlooked.

In the presettlement forest, animals were natural thinning agents whose populations fluctuated in relation to food supply, climate, and other animal predators and parasites. Major concern over animal damage arose in North America in the early part of the twentieth century when humans began regenerating forest lands. This typically followed deforestation, which in many cases had disrupted natural plant-animal relationships of the ecosystem. Modern forest management, which is increasingly devoted to artificial regeneration, must often contend therefore with severe human-caused problems such as high populations of herbivores and lack of predators. For example, deer browsing is probably the most common mammal injury in Europe and North America. In many areas of eastern North America, deer populations exploded following heavy cutting in the nineteenth or early twentieth century. Figure 13.2 illustrates the

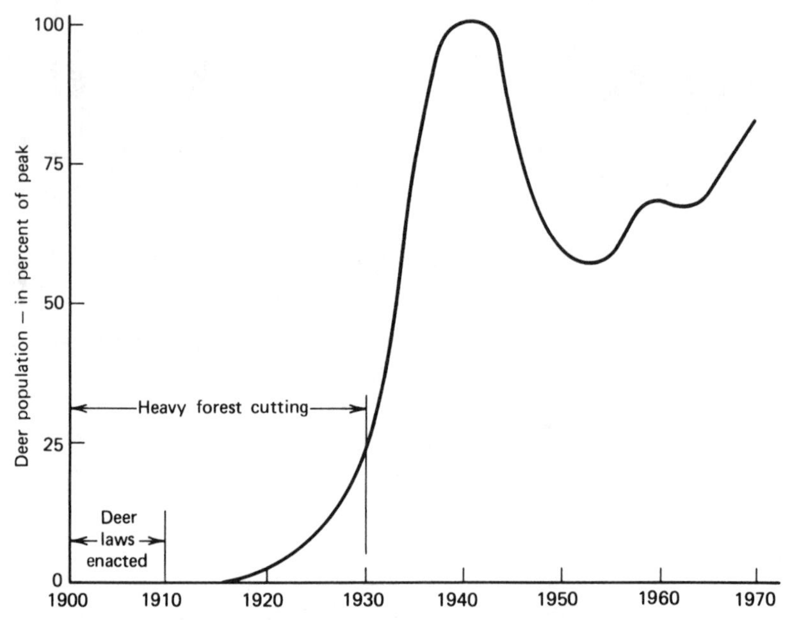


**Figure 13.2. Deer populations in Pennsylvania from 1900 to 1970. (After Marquis, 1975.)**

rapid rise of deer populations in Pennsylvania. Under such conditions, stands become parklike, a distinct browse line is maintained, and no regeneration is possible. The lush regeneration within areas from which deer have been excluded (**exclosures**) contrast vividly with severely browsed areas outside the exclosure (Figure 13.3). In Central Europe, high deer populations for centuries have severely limited forest regeneration and stand development; most plantations must be fenced from deer for several years after establishment.

**Figure 13.3. Deer browsing has completely eliminated tree regeneration outside this fenced deer exclosure, Allegheny Plateau, Pennsylvania. (Marquis, 1974; U.S. Forest Service Photo.)**

From seed and seedling to mature tree, no forest tree is exempt from animal damage. Seeds and cones are destroyed by various insects, often causing major damage. Squirrels cut conifer cones and cache them by the hundreds, storing them for 2 years or more. In hard pines, cones of 3 years (the primordium, the immature cone, and the mature cone) can be lost by indiscriminate cutting of pine shoots by animals in autumn. What seed reaches maturity may be widely consumed by birds and small mammals (mice, chipmunks, voles, shrews, and squirrels). In a western Oregon clearcut, birds and mammals caused a 63 percent loss of Douglas-fir seeds, whereas little loss was sustained by the smaller-seeded western hemlock (16 percent) and western red cedar (negligible) (Gashwiler, 1967). In a 2-year study of the fate of seeds of the relatively large-seeded ponderosa pine, only 4 percent of all seeds reaching maturity were available for germination (Schmidt and Shearer, 1971):

| FATE OF SEEDS (per 100 reaching maturity) | NUMBER OF SEEDS | PERCENT |
|---|---|---|
| Seeds used by animals before dispersal | 66 | 66 |
| Seeds dispersed | 34 | |
| Seeds used by animals after dispersal | 30 | 88 |
| Seeds available for germination | 4 | |

In excellent seed years, however, millions of seeds may be produced with the result that more than enough survive even heavy losses to germinate and live.

After establishment, seedlings of all woody species are subject to clipping and bark removal by hares (Dimock, 1970; Crouch, 1976) and stem and root girdling by voles, mice, shrews, mountain beaver, and pocket gophers (Crouch, 1976), as well as trampling and browsing by deer and other ungulates. In nurseries and plantations in particular, seedlings are subject to root girdling by white grubs (larvae of scarabacid beetles) causing death or markedly reduced growth (Sutton and Stone, 1974). Sapling and pole-sized trees are subject to browsing by deer (Graham et al., 1963; Crouch, 1969; Marquis, 1974, 1975), moose (Snyder and Janke, 1976), elk (Gaffney, 1941), girdling by beavers (Crawford et al., 1976), and porcupines (Krefting et al., 1962), and defoliation by a variety of insects. They are even subject to damage by sapsuckers that make distinctive drill holes in the bark, killing or damaging a wide variety of orchard, shade, and forest trees (Rushmore, 1969; Erdmann and Oberg, 1974). Mature and overmature trees are subject to severe attack by a variety of insect feeders. This is a primary consideration of the following section.

## Productivity and Regeneration

The typical insect consumption of from 5 to 30 percent of annual foliage usually does not impair tree growth or total annual plant production (Franklin, 1970; Mattson and Addy, 1975). Under epidemic infestation conditions, however, insect grazers may consume 100 percent of the foliage. Insect outbreaks are most likely in overmature stands that have passed their efficiency in production. Thus over the long run, insects, together with other agents, recycle aging temperate forests, which are then replaced by fast growing, productive young stands (Sukachev and Dylis, 1964; Mattson and Addy, 1975). Spruce budworm outbreaks in overmature balsam fir stands are a good example (Mattson and Addy, 1975). To this may be added the pine-bark beetle cycles. Insects usually do not unilaterally cause the regeneration of senescent or stressed forests. They typically act in conjunction with fire (Chapter 11), climatic stress (White, 1969, 1974), or windthrow.

Thus in summary, a complete life cycle of reciprocal relations exists between forest trees and animals—beginning with pollination, seed dispersal, and establishment, through stand development and thinning, and finally to regeneration of mature trees and stands. Throughout this cycle, organic matter is continuously being decomposed and mineralized.

## WILDLIFE HABITAT AND FIRE

All temperate forests and grasslands are influenced more or less by fire, thus markedly affecting the evolution and behavior of wildlife. Scientists who have reviewed the significance of fire in each part of North America have stressed the importance of fire in providing a mosaic of habitats necessary for diverse wildlife species (Wright and Heinselman, 1973). The effects of fire on birds and mammals have been reviewed by Bendell (1974).

### Adaptation to Fire Habitats

Within temperate forests, no clear difference in animal species diversity has been found between the more combustible conifer forests and the less combustible hardwood forests (Bendell, 1974). Due to the patchiness of the forests after burning and the close proximity or mixture of hardwoods and conifers, the distinctness of habitats is blurred and animal species are not typically exclusive to one type or the other. For example, relatively few bird species specialize in conifer trees for breeding, feeding, and living (Udvardy, 1969).

The habitat conditions created by fire include: an abundance of vegetative growth on or near the ground; growth of shrubs and trees in open stands with thick branches and twigs, large fruits and seeds that may be retained on the plant; and slowly rotting litter (Knight and Loucks, 1969; Bendell, 1974). These features tend to favor animals that browse and graze and are of relatively large size. Functional adaptations of animals to habitats dominated by fire are many (Handley, 1969; Komarek, 1962): the ability to fly or run quickly and for long distances, the ability to burrow and live underground, effective camouflage and ability to press flat to avoid detection in open areas, and the ability to store food.

Many animals have evolved to exploit periodically burned grasslands and forests, and in turn some may promote fire and otherwise help perpetuate their habitat. Squirrels typically feed at the base of trees, and the dry cone scales accumulated there provide fine, resinous fuel to ignite a fire following a lightning discharge (Rowe, 1970).

Similarly, the placement of nests of grass and fine woody material by birds and mammals on or inside large trees may enhance their flammability (Bendell, 1974). Animals also disperse seeds of favored forage species in new burns (Ahlgren, 1960; West, 1968).

## Fire and Kinds and Abundance of Animals

Fire habitats are remarkably stable in the kinds of birds and mammals found before and after burning. Bendell (1974), in summarizing the findings of many authors (Table 13.1), found that most birds and small mammals stayed on the area after fire. Over 80 percent of the birds and small mammals present following fire were present before fire. Only a few species disappeared, and only a few new species moved in. Most bird populations either showed no change or increased. The greatest increase was shown by ground-foraging birds; the majority of tree dwellers did not change. Populations of small mammals showed little change in grassland and shrub zones and in the forest. In addition, Bendell found that there was little change in abundance and density of animals following fire compared to prefire conditions. The major factors causing the persistence of birds and mammals on fire habitats are their adaptations to fire itself, their ability to tolerate the wide fluctuation of prefire to postfire conditions, and the fact that due to the erratic burning pattern some prefire habitat is typically left interspersed with varying degrees of burned habitat.

**Table 13.1 Change in Species of Breeding Birds and Mammals After Burning**

| FORAGING ZONE | BEFORE BURN | AFTER BURN | GAINED[b] %[c] | LOST[b] %[c] | NO CHANGE[b] %[c] |
|---|---|---|---|---|---|
| | No. of species of birds [a] | | | | |
| Grassland and shrub | 48 | 62 | 38 (18) | 8 (4) | 92 (44) |
| Tree trunk | 25 | 26 | 20 (5) | 16 (4) | 84 (21) |
| Tree | 63 | 58 | 10 (6) | 17 (11) | 82 (52) |
| Totals | 136 | 146 | 21 (29) | 14 (19) | 86 (117) |
| | No. of species of mammals [a] | | | | |
| Grassland and shrub | 42 | 45 | 17 (7) | 10 (4) | 90 (38) |
| Forest | 16 | 14 | 13 (2) | 25 (4) | 75 (12) |
| Totals | 58 | 59 | 16 (9) | 14 (8) | 86 (50) |

[a] Sources: Modified from Bendell, 1974, p. 105.
[b] Numbers of species are in parentheses.
[c] Based on number of species before burn.

The above findings substantiate many observations that animals of fire habitats have a high tolerance for surviving fire; burning does not cause much immediate loss of life (Bendell, 1974). Some animals are killed but most apparently avoid or escape fire. Few are killed by direct heat; most mortality is apparently due to suffocation. The stability of species composition and abundance indicates that wildlife of fire-dominated habitats are broadly adaptable. Wildlife can tolerate wide fluctuations of the physical environment as a result of fire in addition to the fire itself.

## Factors Affecting Animal Response to Fire

Food supply and the pattern and structure of the vegetation are probably the most important factors controlling the kinds and abundance of animals in fire habitats. The quantity of food of the right kind is probably more important than food quality. Protein content of browse plants is apparently maintained at a relatively constant level throughout fire cycles, whereas mineral nutrients may vary more widely.

The mosaic of forest, shrub, and open land created by the marked variation in the intensity, frequency, shape, and extent of fire affects not only the physical environment (wind, temperature, snow cover, light) but also interspecific competition, predators, parasites, and diseases, all of which influence the response of animals to burning. Birds and mammals typically benefit from reduced incidence of parasites for several years after fire. The size of the burn, amount of forest edge, and interspersion of openings with different kinds of cover types positively influence the response of specific animals such as moose (Buckley, 1959) and ruffed grouse (Gullion, 1972). In several instances, the effects of logging on wildlife populations could not be distinguished from that of fire. This suggests that the environmental conditions and diversity of new cover created is more important than how it is achieved (Hagar, 1960; Redfield et al., 1970).

The pattern of cover greatly influences predator-prey relationships. In an open burn, small mammals may be exposed to new predators. On the other hand, ruffed grouse may be preyed upon by birds concealed in thick clumps of conifers (Rusch and Keith, 1971; Gullion, 1972) and by mammal predators hiding in slash and debris on the ground. Therefore, burning is of value in removing both kinds of cover. The ruffed grouse is strongly dependent on periodic fire which removes conifers and produces a mix of cover types: dense clonal stands of trembling or bigtooth aspen less than 10 years old for breeding and broods, 10- to 25-year old aspen clones for food (male flower buds borne on thick, stubby branches where birds can rest), nesting, and wintering (Gullion, 1972). The relationship of forest structure and composition to ruffed grouse life history and survival is illustrated in Figure 13.4. Breeding densities and longevity of males are greatest in young aspen stands. Stands of aspen and conifers over 50 years of age are not highly productive habitat for grouse.

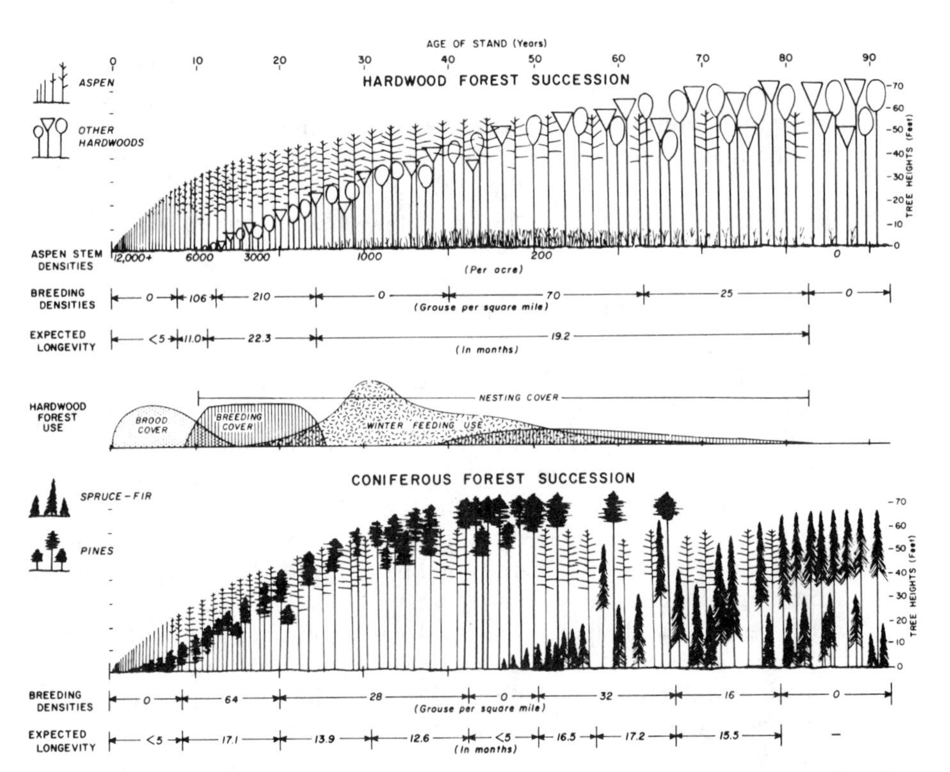


**Figure 13.4. The effect of forest composition, structure, density, and succession on the reproduction and survival of ruffed grouse. (After Gullion, 1972. Figure supplied courtesy of G. W. Gullion, Agricultural Experiment Station, University of Minnesota.)**

Like the ruffed grouse of the northern forest, the bobwhite quail of southern pinelands is favored by burning. Stoddard (1931) pioneered the use of prescribed fire for improvement of quail habitat in pine forests against the prevailing ideas of the time. Studies by Ellis et al. (1969) confirm Stoddard's early work indicating that both fire and cultivation of patches of crops markedly increase quail populations. Many other game birds are favored by fire: wild turkey, ring-necked pheasant, various grouse species (sharp tailed, prairie chicken, ruffed grouse, blue grouse, willow ptarmigan), and some waterfowl. Exclusion of wildfire in the sandy, scrub oak plains of the Northeast was probably instrumental in the decline and extinction of the heath hen (Thompson and Smith, 1970).

Large mammals that respond favorably following fire include moose, white- and black-tailed deer, elk, cougar, coyote, black bear, beaver, and hare (Bendell, 1974). On the other hand, various animals of mature forests may be displaced or eliminated by burning: caribou (mountain, woodland, and barren-ground), marten, grizzly bear, wolverine, and fisher. In summary, the response of animals to fire or mechanical disturbance is highly variable and related to both the genetic quality and behavior of the animals in addition to the complex of plant and environmental changes that we have stressed.

## IMPACT OF LARGE ANIMALS ON THE SITE

The activities of humans, deer, cattle, sheep, and other large animals result in substantial changes in the forest site. The hoofed grazing and browsing animals are the most effective, but logging equipment and other human implements are locally of the greatest importance.

The impact of herbivores on the forest site quality is both indirect and direct. Grazing and browsing animals change the vegetation through their selective feeding habits and the differential ability of different plants to survive and prosper. The changed vegetation in turn results in changed litter and soil biotic activity and thus in changed site conditions. The change may be for the better or worse, but is most apt to be the latter as the less palatable species are generally those with lower nutrient content, and the woody species are generally those with lower nutrient content and more woody structure. The litter of such plants will generally decompose but slowly and will inhibit soil biotic activity. Thus changes in vegetation result in changes in soil organic matter, soil organisms, soil chemistry, and therefore soil productivity.

Heavy, single-species grazing pressure can cause entire changes in the structure of the plant community. For example, in the nineteenth century following white settlement of the intermountain American West, grazing pressure from livestock, particularly cattle, depleted the bunch-grass vegetation (Wagner, 1969). As a result of the reduced competition from grasses, woody species such as big sage (*Artemisia tridentata*), junipers (*Juniperus osteosperma* and *J. scopulorum*), bitterbrush (*Purshia tridentata*), and serviceberry (*Amelanchier alnifolia*) increased greatly and turned grassland into brushland. (The reverse is reported in Africa where elephants and other browsing animals may turn woodlands and brush types into grasslands [Wagner, 1969].) Then in the twentieth century widespread increases of the mule deer placed heavy pressure on the shrubs, causing them to disappear slowly. In many areas the vegetation is returning to the original bunchgrass type.

The direct effects of grazing on site result largely from the action of animal hoofs in compacting the surface soil and in breaking up the ground cover. Both are detrimental under most circumstances and often highly so. The pounding of animal hoofs results in a breaking down of the soil aggregates which give a crumb structure to the surface soil. The pore space of the surface soil is greatly reduced, often to the point of seriously reducing the supply of air in the soil, and of preventing the infiltration of rain water at a rate sufficient to prevent surface runoff. As a result, heavily grazed soils are apt to be poorly aerated, have a lessened water absorbing capacity, and become subject to sheet erosion. Biotic activity is minimized in highly compacted grazed soils. Site quality is often lowered substantially.

In a study of paired grazed and ungrazed woodlots in southern Wisconsin, Steinbrenner (1951) found that, in general, the highly compacted soils of the heavily grazed woodlots had a lower initial moisture content in the spring and dried out faster in the summer and late fall, evidently because of the lowered soil permeability and increased run-off. The grazed soils were so compacted that water permeabilities averaged about one-tenth those of comparable ungrazed woodlots. The organic matter content and the available potassium were significantly higher in the surface soils of the grazed woodlots, apparently due to the manuring action of livestock.

Grazing has greatly affected the forests and the site throughout most of the farm woodlands of the East, Central States, and South, and throughout most of the open woodland and ponderosa pine forests of the American West. Cattle do much damage because of their sharp hoofs, heavy weight, and tendency to congregate around water, salt, or bedding areas. Sheep may be harmful if run in excessive numbers, but, when herded from place to place so as to prevent overgrazing, usually leave the site in fairly good condition. Deer are generally not present in numbers sufficient to compact the soil excessively but have markedly changed the forest composition through the differential browsing of seedlings in many localities in the Appalachians, Lake States, western pine forests, and elsewhere. Where deer have been introduced into forests composed predominantly of palatable species previously unbrowsed, they may virtually eliminate all vegetation within reach. This has occurred in many indigenous forest areas in New Zealand. Other grazing and browsing animals such as elk, caribou, and bison have affected the site in local situations. ***Around the world, grazing by livestock has probably been more important than any other factor in reducing the productive capacity of uncultivated land.***

A compatible relationship of plants, soil, and animals may be achieved by matching plant diversity and the diversity of grazing animals. For example, Holsworth (1960) reported that the major game populations of Elk Island National Park, Alberta, Canada, were bison, elk, and moose. The three animals showed distinctly different habitat and food preferences such that their complementary feeding habits tended to maintain the vegetation, and thus the soil, in a form usable by all species. And in the East African savanna the specialization of feeding habits is strikingly seen. There herds of wild animals, composed mainly of nine herbivores and one carnivore with a biomass of 82,200 to 117,500 kg per square kilometer yearly, coexist, with no evidence of overgrazing or serious soil deterioration (Talbot et al., 1965). This biomass is six times that of cattle, sheep, and goats supported in the same area under native herding with moderate to severe overgrazing and three times the biomass under European-type cattle ranching with slight to moderate overgrazing. According to Wagner (1969), the variety of herbivores is an important influence in maintaining the vegetative diversity and consequently the natural condition of the soil as well.

At the same time, the destructiveness of soil productivity by human activity should not be minimized. The effects of agricultural practices, both good and bad, on site quality are obviously great but are outside the scope of the present work. It is equally obvious that the great changes in forest vegetation brought about by logging, burning, and other human activities have resulted in great changes in soil productivity.

Confining ourselves to direct effects, human activity is responsible for considerable site deterioration through soil compaction. Trucks, tractors, and other heavy equipment used in logging result in substantial soil compaction (Moehring and Rawls, 1970; Dickerson, 1976). In the Douglas-fir region in Washington, skid roads and other affected areas may occupy a substantial percentage of the logging area. Such compaction on skid roads has been shown to reduce the soil permeability 92 percent and the microscopic pore space by 53 percent, thus increasing the bulk density by 35 percent (Steinbrenner and Gessel, 1955). Similarly in the Atlantic coastal plain, soil compaction on skid roads was found to reduce soil infiltration rate and pore space by 84 and 34 percent, respectively, and to increase bulk density 33 percent (Hatchell et al., 1970). Forty years may be required for infiltration to recover on severely compacted logging roads (Perry, 1964).

Compaction from logging traffic is much more pronounced on wet than dry soils and more severe on clayey than on sandy soils (Steinbrenner, 1955; Moehring and Rawls, 1970). Severe skidding traffic on three or four sides of trees in wet weather significantly reduced growth of trees up to 60 percent of that of trees in an undisturbed stand (Moehring 1970). Although this intensity is not normally encountered except at landings and along major skid trails, wet weather logging can cause soil compaction that may markedly reduce growth rates of established seedlings and significantly reduce seedling establishment on skid trails (Youngberg, 1959; Perry, 1964; Hatchell et al., 1970). Hatchell and Ralston (1971) estimated that 18 years may be required for severely disturbed soils to attain normal tree densities.

The human foot itself is an effective compacting agent. On the Mall in Washington, D. C., for example, heavily trampled soils were found to have a bulk density and particle density similar to that of asphalt and concrete (Patterson, 1976). The problem of soil compaction in forest parks and other recreational areas within the forest has been recognized for over 50 years (Meinecke, 1928) and is reaching serious proportions. Death of large and famous trees has been attributed to compaction, and decreased growth rate is frequently apparent. Because of compaction, it has been necessary to fence out tourists from the immediate neighborhood of famous trees, and to move public camp grounds out of old-growth areas as in the redwood and bigtree localities in California. The increasingly intensive use of the forests for camping and other recreation is giving added importance to the dangers of site deterioration directly from humans and their vehicles.

## SUGGESTED READINGS


Bendell, J. F. 1974. Effects of fire on birds and mammals. *In* T. T. Kozlowski and C. E. Ahlgren (eds.), *Fire and Ecosystems*. Academic Press, Inc., New York.

Dickinson, C. H., and G. J. F. Pugh. 1974. *Biology of Plant Litter Decomposition*. Vol. 1, Part I. *Types of Litter*. Vol. 2, Part II. *The Organisms*, Part III. *The Environment*. Academic Press, Inc., New York. 775 pp.

Hanover, James W. 1975. Physiology of tree resistance to insects. *Ann. Rev. Entomol.* 20:75–95.

Janis, Christine. 1976. The evolution strategy of the Equidae and the origins of rumen and cecal digestion. *Evolution* 30:757–774.

Janzen, Daniel H. 1970. Herbivores and the number of tree species in tropical forests. *Am. Nat.* 104:501–528.

———. 1971. Seed predation by animals. *Ann. Rev. Ecol. Syst.* 2:465–492.

Mattson, W. J. (ed.). 1977. *The Role of Arthropods in Forest Ecosystems*. Springer-Verlag, New York. 104 pp.
———, and Norton D. Addy. 1975. Phytophagous insects as regulators of forest primary production. *Science* 190:515–522.
Pijl, L. van der. 1972. *Principles of Dispersal in Higher Plants*. Springer-Verlag, New York. 162 pp.
Regal, Philip J. 1977. Ecology and evolution of flowering plant dominance. *Science* 196:622–629.
Sukachev, V. N., and N. V. Dylis. 1964. *Fundamentals of Forest Biogeocoenology*. (Chapter IV, Animal life as a component of a forest biogeocoenose, pp. 253–354) Oliver and Boyd, Ltd., Edinburgh. 672 pp.

# 14
# Competition and Survival

## SYNECOLOGY

The individual organism, whether it is a forest tree or something else, is the product of its genetic constitution as affected by the environment. The study of the individual organism in relation to its environment falls within the scope of **autecology**.

The forest, however, is a complex of organisms, both plant and animal, mutually occupying a complex of environments. These organisms are in competition for the light, air, water, warmth, and nutrients necessary for life. Each, in its turn, creates part of the environment affecting the others. The study of the community and the interaction of the organisms which compose it falls within the province of **synecology**. Synecology, thus, is broader than autecology, and a more integrative phase of ecology, dealing with living communities rather than with individuals (Weaver and Clements, 1938; Oosting, 1948; Braun-Blanquet, 1964; Daubenmire, 1968a).

When two forest trees occupy the same site in close proximity to one another, they inevitably come into competition for the same necessities of life. The presence of one will affect the life of the other as each is a part of the habitat in which the other lives. A consideration of competition for survival in this chapter arises naturally out of autecology and is basic to an understanding of the community. The forest is never static, changing in its composition, structure, and general character continually; the dynamics of these changes over time is treated in Chapters 15 and 16, the former dealing with forest succession under disturbed conditions, and the latter with the effects of disturbances of various sorts. Not only does the forest vary continuously in time, but it varies continuously in space. The effect of variations in the forest site on the forest community is considered in Chapter 17.

In contrast to autecology, which is largely factual and scientific, synecology, because of its very complexity, has tended to be subjective and philosophical. Our state of knowledge of the innumerable interrelationships between all the millions of organisms in any forest community and between these organisms and the physical factors of the site is limited and we have no choice but to generalize and theorize without too many facts at our disposal. Ecologists traditionally have tended to study the forest community separately from the forest environment, and with good reason. Through this separation, however, descriptive and classificatory studies of the community have flourished while studies of community processes and the interaction of communities and environment were neglected. The modern trend toward the ecosystem approach, the systematic study and modelling of community processes that are inseparable from environmental factors (Chapter 18), is shifting our thinking toward an understanding of the whole ecosystem, rather than just its component parts.

## THE FOREST COMMUNITY

Trees occur with other plants and with animals in natural groupings that are more or less repeated from place to place over a period of years whenever similar conditions recur. An astute observer familiar with a forested region can usually identify the more distinct forest communities and can infer with reasonable probability the interplay of factors that have brought the vegetation into the conditions one observes. Consequently there exists a mass of valid descriptive data of specific forest areas that have been studied and written up by foresters and ecologists.

Our problem arises when one begins to generalize. The great complexity of ecological problems—involving the interaction of many organisms and many environmental factors over a long span of time, with much of the past history of the interaction virtually unknown—has resulted in a great gap between the description of given areas and the evolution of the laws and principles which are the ultimate goal of the scientist. Different individuals and groups, influenced by stimulating teachers and the problems in their particular geographical area, have assayed various paths to this goal and have developed languages of their own to express their thinking and progress. Nevertheless, an essential similarity of purpose exists between the different schools of ecological thought.

A community is a united body of individuals. It follows therefore, that a forest community is one dominated by trees—a body or group of individual trees of one or more species growing in a specific area and in association and mutual interaction with one another and with a complex of other plants and of animals.

Rarely is a forest community unique. A mixture of Norway spruce and Asiatic spruce growing on abandoned farmland in the Harvard Forest in central Massachusetts, for example, is probably not duplicated or even approximated anywhere else in the world.

More commonly, however, the community is similar to other communities composed of the same or similar species and growing on the same or similar soil as a result of the same or similar sequence of events. Such a grouping of similar communities constitutes a vegetation or forest type. In forest communities, such a type is formed by jack pine, occurring in pure or nearly pure stands after fires on sand plains in the Lake States and eastern Canada. No two jack pine communities are alike in all respects; yet the vegetation complex occurring on these soils following fire is so characteristic that it can be immediately recognized and named by almost anyone. Although any community might be considered unique if described in great detail, strong similarities exist among many communities. The groupings that are therefore possible are valuable for scientific study and particularly for the management of forests.

The nature of a given forest community is governed by the interaction of three groups of factors: (1) the site, or habitat, available for plant growth; (2) the plants and animals available to colonize and occupy that site or habitat; and (3) the changes in the site and the biota over a period of time as influenced by changing seasons, climates, soils, vegetation, and animals—in other words, the history of that habitat.

### The Community and the Phenotype

As we have defined it, the forest or plant community is the product of the interaction of the living organisms and the site or habitat over a period of time. The biologist will at once recognize an analogy to the modern conception of the phenotype, the body of the individual plant or animal as we see it. The individual organism is the product of the interaction of its genetic make-up and its environment over its life. When we confine our attention to one individual, the problem becomes relatively concrete. We are confronted with a single set of genes and only those environmental conditions that have existed from the time that the gene combination was put together up to the present. At the community level, the same principles apply but in exponential proportions. Our community is affected by all environmental conditions that have existed since the oldest plant invaded the habitat. Since conditions at that time were undoubtedly affected by the vegetation that existed then, we are concerned with even the earlier history of the site and the communities that occupied it in the past.

The community is infinitely more complex than the phenotype but the analogy is valid. Both are the product of the interaction of genetic and environmental factors over a period of time.

Some ecologists have carried the analogy of the community and the individual phenotype or organism to the point of classifying communities as individual organisms, even to the point of giving them binomial Latin scientific names. In such a holistic philosophy, there is a danger in forgetting that the individual community is only analogous to the individual plant or animal and not equivalent to it. Actually, the grouping of plants and animals on a given site into a community is a broader, more general, and less precise assemblage than the grouping of genes in a single body that exists in a single habitat. At the community level, we are dealing with an infinite and constantly changing number of gene combinations in a large and constantly changing number of organic bodies in a large and constantly changing number of microenvironments. *A given community exists only at a single point in space and time.* We should consider the community for what it is: a rather indefinite and constantly changing grouping of plants and animals on a given site; a broader, less intelligible, and less definable category of classification than the individual phenotype.

## CHANGE IN THE ECOSYSTEM

The forest community consists of an assemblage of plants and animals living in an environment of air, soil, and water. Each of these organisms is interrelated either directly or indirectly with virtually every organism in the community. The health and welfare of the organisms are dependent upon the factors of the environment surrounding them; and the environment surrounding them is itself conditioned to a considerable degree by the biotic community itself. In other words, the plants, the animals, and the environment—including the air, the soil, and the water—constitute a complex ecological system in which each factor and each individual is conditioned by, and in itself conditions, the other factors constituting the complex. There is perhaps nothing really new in this concept of an ecological system, but in recent years, when we envisioned it more simply as an **ecosystem**, we have made considerable progress in understanding the complex interrelationships that exist. In contrast, in former years by trying to simplify particular elements of the ecosystem into simple cause-and-effect relationships, we too often drew misleading and inaccurate generalizations.

The **forest ecosystem**, then, is the complex of trees, shrubs, herbs, bacteria, fungi, protozoa, arthropods, other invertebrates of all sizes, sorts, and description, vertebrates, oxygen, carbon dioxide, water, minerals, and dead organic matter that in its totality constitutes a forest. Such a complex never does and never can reach any balance or permanence. It is constantly changing both in time and in space.

The changes of the forest ecosystem in time take many patterns. First, there are diurnal changes. The balance of the forest community at midnight—when the plants are taking up oxygen and giving off carbon dioxide, when some animals are dormant and others are active, when temperature is lowered, and when moisture and humidity are relatively high—is quite different from the forest ecosystem at midday when the reverse of all these processes is going on.

The forest ecosystem changes seasonally around the year. As the cycle of activity of each of the organisms changes, and as the climate changes, the balance of the ecosystem itself changes. The temperate forest is not the same biotic community in midwinter that it is in midsummer, or even in one week that it was during the previous week.

Nor should we ignore long-term climatic changes. Whether climatic fluctuation is cyclic or not is immaterial; the fact is that it does change from one set of years to another. The relative warmth of the 1940s is in sharp contrast to the relative cold of the early 1800s. The warm xerothermic times of 4000 to 8000 years ago are an even greater contrast to the late Pleistocene of 16,000 to 18,000 years ago when the glaciers reached their last maximum. There is no such thing as a constant climate, and there can be no such thing as a constant ecosystem if that does not exist.

Finally, the plants and animals that constitute the ecosystem biota never remain the same for more than a given instant of time. There is a continual introduction of new species and the elimination of old—rarely when we consider the large and well-established trees, mammals, and birds, but frequently when we concern ourselves with the fungi, the bacteria, and the protozoa that far outnumber the larger organisms in the forest and approach them in overall importance. Even if we could prevent new organisms from constantly migrating into the forest and old organisms from constantly being eliminated from it, the organisms that remain there do not stay the same. Evolution is continual. The loblolly pine of today is not quite the loblolly pine of 100 years ago and certainly not that of 10,000 years ago. The bark beetle is not the bark beetle of the previous decade. The blister rust of this year is not the blister rust of last year. Through mutation, through natural selection, and through new population distributions of the combinations of genes, the organisms constituting the forest ecosystem biota are continually in a state of flux.

Equally important the ecosystem changes constantly in space. At any given instant of time the forest ecosystems high in the mountains will be different from those in the lowlands. Even on a level plain the ecosystem will vary from north to south and from east to west. The extremes are obvious but changes over short distances may be equally important. A distance of 100 m may not include any noticeable change in the distribution, size, and the vigor of the more visible components of the forest. The difference is there nonetheless. When one looks at the microorganisms in the forest floor which play such an important part in ecosystem processes, changes in even 1 meter may be real, measurable, and important in affecting the total ecosystem.

In short, the specific forest ecosystem exists only at a given instant of time and a given instant in space. Regardless of appearances, the ecosystem is never the same on succeeding days, succeeding years, or succeeding centuries, up the slope 1000 m, north 10 km, or east or west 100 km. Change characterizes the forest ecosystem continuously. *Stability is only relative and is only superficial.*

If we adopt the biocentric concept of the ecosystem, we may draw some ecologic generalizations that modify substantially some long-standing and accepted ecological principles. Once we accept the fact that the ecosystem exists only at a given instant in time and at a given point in space, we can take a fresh view of ecological theory.

First, it follows immediately that there is no meaning from a biocentric viewpoint to the contrasting concepts of native and introduced species. Characterizing a plant or animal as being exotic or endemic characterizes it only from the standpoint of our relationship to it. Actually, all plants and all animals are introduced or exotic from a biocentric standpoint except at the very point in space where the particular gene combination was first put together. Whether the subsequent migratory pattern of that organism took place independently of humans or with the help of humans is important to us but the forest ecosystem may be permanently altered in either case.

Examples are many. It is of interest to us to know whether we carried the coconut to a given tropical island or not. To the coconut it is of little importance as to whether it floated by itself in an ocean current or was lodged in the hull of a native dugout canoe, which floated on the ocean current. To the maple thriving in a given spot, it is immaterial whether its seed flew there on its own wings or whether it was aided and abetted by the wings of an airplane. A wild cherry is unaffected by concern as to whether its seed was deposited by a sea gull who spotted a target below or by a human recreationist who brought it thither in a paper bag. The forest insect is just as much a member of the California ecosystem if it flew in as an endemic insect from Oregon or was brought in as an exotic insect from Aragon. In short, from the viewpoint of the forest, there is no distinction between native and introduced species. However, whether native or introduced, the species' persistence depends upon its adaptation to compete in the forest community. The ecosystem consists of all of the plants and animals that are there at a given point of space and at a given instant of time. All were migrants there: some recently, some from the dim geologic past, some carried in by wind, some by animals, some by water, some by humans. Once they are there and adapted to the environmental conditions, they are members of the local ecosystem from that time on until they or their descendants are eliminated. The true endemic, perhaps, is really that plant or animal that prospers in the local environment, that is competitive in the local ecosystem, and that can maintain, reproduce and establish itself where it lives.

Human-caused disturbances may differ substantially in kind, severity, and effect from those resulting from climatic or other biological agents. Our unwitting introduction of the chestnut blight fungus, the Dutch elm disease, and domestic livestock to many forested areas have had devastating effects on enormous areas of forest. Human activities in setting fires and in reducing the spread of fires through the construction of roads have greatly affected the extent and severity of forest fires. Logging creates substantial changes in the forest dissimilar to those occurring in uncut forests. The impact of humans is great, frequently devastating, and must be taken into account. The kind, severity, and periodicity of human disturbances are of primary concern to the wildland managers. The forest ecosystem in which humans play a major role differs in its development from the forest ecosystem in which the activities of humans are minimal. It is hardly correct to argue, however, that one is more "natural" than the other.

Second, it becomes clear in our philosophic approach that there is no such thing as a climax community in the sense of a permanent, stable condition. It makes no difference whether humans have interfered or not. Change is perpetual and will go on at a rapid or at a gradual rate depending upon the rate of changes in the climate, the soils, the landforms, the fire history, and the biotic composition of the ecosystem. No matter how old the trees are, no matter how long fire has been kept out, the populations of the insects, the fungi, the plants, the animals all will change, and the ecosystem itself will as a consequence change.

Third, since the ecosystem exists only at a given instant in time and in a given instant of space, it follows that natural succession will never recreate an old pattern but will instead constantly create new patterns. The vegetation on Krakatoa is developing successionally, but it is already quite apparent that it will never equate itself with that of undisturbed islands nearby. After each retreat of the glaciers in the Lake States in the Pleistocene the forests moved back and reestablished themselves, but always in a new pattern with some new species present, some old absent, and the balance in many cases quite different. With widespread farm abandonment in New England the forests have reestablished themselves and have moved on into a new successional pattern with different balances of composition and different stand structures. In short, forest succession will lead to the development of a mature and long-lived forest community but one in which change under the surface is still going on and one which will never repeat exactly the pattern of a previous forest developed under a previous forest succession.

## COMPETITION

Changes in the structure and composition of the forest result from the constant demand of each individual tree for more space and from the eventual death of even the most dominant individuals. The increasing size of the main story trees results in competition for growing space, with a few individuals gaining space, a few more holding their own, and an increasing majority losing space and eventually succumbing. The death of the dominants, due to lightning, wind, insects, diseases, and even to old age, releases from the main canopy a portion of the site for occupancy by a growing and developing understory.

Competition between the trees of the same species does not affect the composition of the forest type and therefore has no effect on forest succession. Competition between individuals of different species, however, results in a natural succession from one forest composition to another. The rate of species replacement slows as succession proceeds, and ultimately a group of species having complementary ecological roles characterizes the mature late successional forest. Competition takes place in even- and uneven-aged stands and in the main canopy, or overstory, of the forest and in the understory.

### Competition, Stand Structure, and Density

In species that regenerate en masse at about the same time, as following fire or flooding, an **even-aged** stand structure is formed. All pines, Douglas-fir, black spruce, cottonwoods, eucalypts, and many other species form natural even-aged stands. As the name implies, the individuals are nearly the same age, and the stems are often of a similar height and diameter. In contrast, various long-lived hardwoods and conifers form stands in which great diversity exists in tree age and size. Such **uneven-aged** stands are typical of northern hardwood forests (comprising sugar maple, beech, yellow birch, basswood, and hemlock, among others) and of some spruce-fir forests of Central Europe. Here the trees of a given species on a single hectare may range in age 300 years or more. In many forests, a mosaic of small even-aged groups, when considered over a large area, forms the uneven-aged forest. This was probably the characteristic structure of presettlement upland oak and ponderosa pine forests.

In even-aged stands, competition for light, moisture, and nutrients depends largely on the number of stems per unit area. In time, crowns of the trees come together and crown class differentiation becomes pronounced (Figure 14.1). Due to a combination of environmental and genetic factors, some trees develop rapidly and exhibit large, well-formed crowns. Other trees grow more slowly and their crowns become more or less restricted; in time some trees are gradually overtopped and suppressed. Root competition is often severe in the even-aged stand, although it is much less observable than crown competition. In even-aged pine stands, for example, the root system of a tree may compete with several hundred trees, whereas its crown competes only with a few adjacent trees.

A simple classification, modified from the crown classes of Kraft (Dengler, 1944), illustrates the results of intense competition in even-aged stands (Figure 14.1). The silviculturist uses the crown classes as a basis for judging the vigor of the stand and for conducting thinnings and other cultural operations. The major crown classes are:

- Dominant: trees with crowns extending above the general level of the canopy and receiving full light from above and partly from the sides; larger than the average trees in the stand; crowns well developed but possibly somewhat crowded on the sides.
- Codominant: trees with crowns forming the general level of the canopy or somewhat below; receiving full light from above but only moderate amounts from the sides; usually with medium-sized crowns, and more or less crowded on the sides.
- Intermediate: trees shorter than the preceding classes but with crowns extending into the canopy formed by the dominants and codominants; receiving some direct light from above but little from the sides; usually with small crowns, considerably crowded on the sides.
- Suppressed: trees with their crowns entirely below the general canopy level; receiving no direct light from above or from the sides.

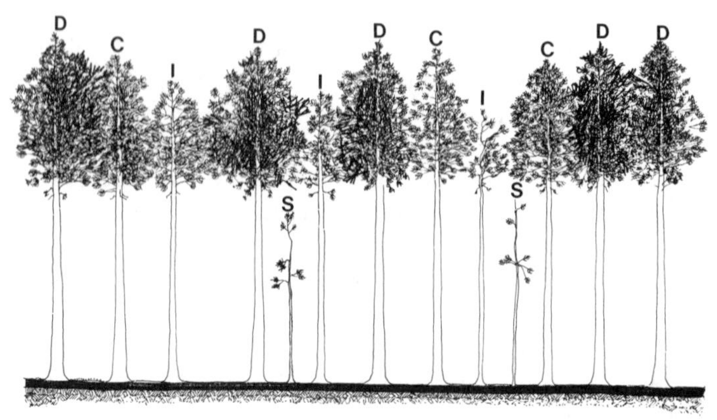


**Figure 14.1. Diagram of crown classes in an even-aged stand. D = dominant, C = codominant, I = intermediate, S = suppressed. (Modified from Kraft, see Dengler, 1944; Drawing by M. F. Orlando.)**

To study the competition within a species population in an even-aged stand or among different species in an uneven-aged stand, **life-table analyses** provide a convenient method of describing the mortality schedule of a population (Krebs, 1972). The life-table techniques that have proved a powerful tool in interpreting changes in animal populations are being applied to trees. The application has been limited because of the difficulty of determining the age of trees and because of the long life span of individuals. Nevertheless, studies such as that by Leak (1975), in an old-growth hardwood forest in New Hampshire, illustrate the general shape of the age-distribution curve over time as well as differences among several species (Figure 14.2). Sugar maple and beech exhibit the sharply declining populations of species that establish many seedlings periodically in the shaded forest floor. Many survive and eventually some replace the overstory veterans. In contrast, red spruce exhibits a lack of young seedlings, and over time red spruce may diminish in abundance. Without disturbance it is apparently not competitive in the younger age classes with beech and maple on this site. Such curves, developed at one point in time, do not tell us how the populations actually change over time. However, if the major past disturbances of the forest are known, the general trend of the populations over time can be determined. This population dynamics approach will become increasingly useful in modeling changes in tree populations through time, whereby the impact of natural and human disturbances may be monitored.

**Vertical Structure.** One result of competition among plant species of the forest is the development of a vertical structure of the vegeta-

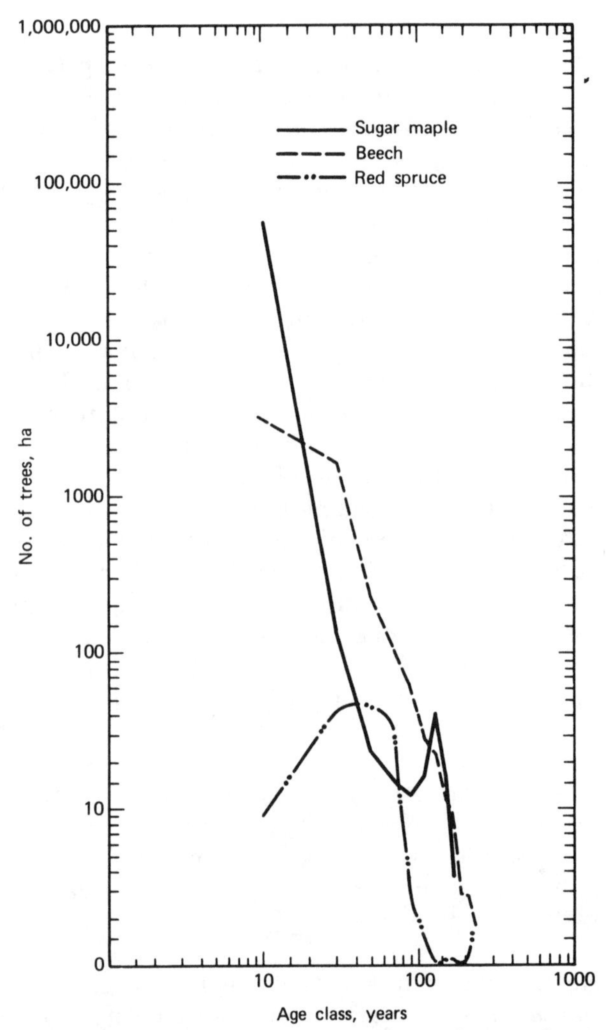


**Figure 14.2.** **Numbers of trees per hectare (log scale) in an old-growth northern hardwood forest over midpoints of 20-yr age classes (log scale) for sugar maple, beech, and red spruce. (After Leak, 1975. © 1975 by the Ecological Society of America.)**

tion. Forest stands exhibit several major vertical layers, characterized by trees, shrubs, herbs, and thallophytes (mosses, lichens, and liverworts). The multistoried forest has an upper layer of overstory trees, one or more subdominant layers (more or less distinct) composed of younger stems of the overstory trees and mature trees of other species that do not reach the overstory, layers of high and low shrubs, layers of high and low herbs, and a ground layer of mosses, lichens, and liverworts. The species of each layer are modified or genetically adapted to make the best use of the space, light, and microclimatic resources of their respective vertical position.

In the tree layer alone, considerable variation occurs depending on species composition, stand age, and site conditions. Generally, the more favorable the site conditions (especially moisture) the greater the number of layers. In North America, the complex structure of the southern Appalachian Mountain cove forests is noteworthy. In addition to a species-rich overstory, there may be several understory tree layers that contain all the species of the overstory and various subdominant species such as slippery elm, red mulberry, redbud, umbrella magnolia, Fraser magnolia, flowering dogwood, blue-beech, and pawpaw (Braun, 1950). Below these layers is often a dense layer of the tall shrub (or small tree), rosebay rhododendron. In the tropical rain forest, even a more pronounced vertical structure may exist, including huge emergent trees whose crowns are fully exposed to sunlight because they extend 5 to 10 m above the general canopy level. In such multistoried forests, each layer has a different microclimate and usually its distinct assemblage of insects and other animals. In contrast, many species, particularly pioneer species that colonize areas following major disturbances such as fire and flooding (virtually all pines, Douglas-fir, larches, aspens, some spruces and firs, among others), typically exhibit initially only a single tree layer. In time, such stands are usually invaded by other tree species, and a more complex vertical structure gradually develops. Although vertical structure has received much less attention than species composition in forest community analyses, a given forest ecosystem can usually be distinguished by its characteristic structure as well as by its species composition.

The overstory canopy intercepts much of the incoming irradiance, and given an unbroken overstory layer, species of the lower layers are physiologically and structurally adapted to use the continuously decreasing amount of light available as one approaches the forest floor. However, since the forest overstory is never completely unbroken over large stretches, various subdominant trees, shrubs, and herbs that require relatively high light levels to survive utilize the well-lighted microsites and extend their crowns into openings. In addition, overstory trees differ in the density of their crowns so that considerably more light may penetrate one overstory than another. This is one factor that may cause differences in the vertical structure of different forest communities. For example, the shrub layer is particularly well developed in oak forests of eastern North America compared to beech-maple forests. The oak crowns are less dense and the canopy is more open than that of beech-maple forests. The increased light, together with more frequent fires in the oak forests, favor the growth, sprouting, and clonal spread of shrubs in the understory.

**Stand Density.** The number or mass of trees occupying a site has important implications for the trees themselves, the site, and for the silviculturist responsible for controlling forest reproduction, growth, and composition. Density is typically measured in terms of numbers of trees or basal area[1] per unit area.

Except at very low densities, mortality is caused by competition among trees and between trees and associated vegetation for light, water, nutrients, and other site resources. Mortality is greatest in the seedling stage when the number of seedlings per hectare is highest. Curtis (1959) reported a mortality of 99 percent of sugar maple seedlings over a 2-year period. Hett (1971) followed up this work and found that the mortality rate is relatively independent of age in the early years but declines as the seedlings mature. As plants die the remaining individuals become larger; the smaller plants are continually eliminated from the population. Therefore in stands of plant species, there is a strong relationship between plant size and the density of the stand: larger size and weight of biomass of the stems are associated with fewer stems per unit area. The relationship is represented by the formula: $\log \bar{w} = a + b \log p$, where $\bar{w}$ = the mean weight of surviving plants, $a$ = a constant, $b$ = the slope of the regression line, and $p$ = density. Furthermore, for pure stands of annual herbaceous species, Japanese workers found that in those stands undergoing self-thinning, a single line with the slope of $-1.5$ represented the maximum plant size—density relationship regardless of age (Drew and Flewelling, 1977; White and Harper, 1970). This relationship, termed the $-3/2$ power law of self-thinning, describes the maximum size of a species in relation to stand density. It is reported to be valid for stands of any species, of any age on any site. For a variety of species in pure and mixed stands in north-temperate forests of the western United States the mean tree aboveground biomass was plotted over density (Long, 1980). The slope of the regression line was $-1.48$, suggesting a similar relationship for all species considered. Although the $-3/2$ power law has not been tested for most North American tree species there is as yet no evidence to doubt its reliability for stem biomass.

The relationship of stand density, tree growth, and site quality has been studied extensively, and the major findings are important for the ecologist and silviculturist. If very few trees occupy a site they

[1]Basal area is the cross-sectional area of a tree at 1.3 m above the ground. It may be expressed on a per-tree basis or summed for all trees giving an amount per unit area (e.g., $40 m^2 ha^{-1}$).

generally form large, branchy crowns, exhibit slow growth in height, and are typically undesirable in form and quality for commercial use. In such cases, the trees are typically not fully utilizing the water, nutrients, light, and other site resources, and their productivity is low. Because full sunlight reaches most of the ground area, grasses and shrubs may come to dominate the site, compete with the trees for resources, and affect soil development in a much different way than if the area were dominated fully by trees.

For the silviculturist, control of stand density is extremely important. Diameter growth is strongly influenced by density; diameter decreases with increasing density. Thus if large trees are desired, density must be controlled by thinning to an appropriate number of trees (or amount of basal area) per hectare to maximize diameter growth yet still fully utilize site resources. However, for some uses wood quality is adversely affected by low density (trees become branchy and the wood knotty; growth rings may be too large) so that a balance must be reached for a given species on a given site through appropriate density control depending on the end product desired.

Within a wide range of densities, tree height of free-growing trees is unaffected by stand density. This relationship, together with the fact that different site qualities strongly affect height growth, enables tree height to be used as an effective indicator of site quality (see Chapter 12). Finally, as brought out in Chapter 12, as long as the site is fully occupied (trees making their full use of available resources), the species will produce the same amount of wood per year at various densities. Whether there are many small trees or fewer large trees, a similar wood volume is produced. However, as site quality improves (higher fertility and more favorable water supply) a greater number of trees can be grown per unit area and a greater volume of wood can be obtained than on the poor site. Understanding the site quality is therefore of great importance in knowing how to regulate stand density. Detailed considerations of stand density from the standpoint of growth and yield and silvicultural management of forests are available in mensurational and silvicultural texts (Spurr, 1952b; Smith, 1962; Husch et al., 1972).

## Competition and Overstory Composition

The trees in the main canopy of the forest tend to increase yearly in height, bole size, length of each growing branch, and number of leaves. If the tree is to remain alive and vigorous, it must grow. To grow, it must increase its growing space—its utilization of the site.

In any fully stocked stand, then, it follows that competition between growing individuals in the overstory will result in the elimination of some trees, particularly of those species genetically less suited for survival under the particular environmental conditions that may exist. The result is a gradual change in the composition of the overstory and thus in the forest community. Examples are numerous.

In central New England, a mixture of hardwoods forms a single-storied stand following the cutting or windthrow of old-field eastern white pine. These include long-lived "timber species" such as red oak, red maple, and white ash, as well as short-lived, pioneer species such as gray birch, pin cherry, and trembling aspen. The birches, cherries, and aspens (there are several species of each) seed into the former white pine site in tremendous numbers and dominate the sapling hardwood stand. In terms of the dominant tree species, the young stand may be termed a birch—cherry—aspen type.

Mortality due to the tent caterpillar, the gypsy moth, and simple inability to compete and survive, however, continually reduces the numbers of these species so that the pole-size stand becomes dominated with time by the more persistent oaks, maples, and the ash. Eventually, a red oak–red maple–white ash type will evolve characteristically with only an occasional black cherry, black birch, and a few of the other more persistent, earlier-dominant group.

Even within the developing longer-lived hardwood type, competition will favor one species against another. On the better hardwood sites in central Massachusetts referred to above, red oak continually expands its crown at the expense of neighboring white ash, with the result that the percentage of the stand basal area occupied by the former species increases with time while that of the latter species decreases.

The internal changes in a developing temperate forest are illustrated in Table 14.1, which summarizes the growth over 45 years of a middle-aged, mixed hardwood stand growing on an island in St. Mary's River between Michigan and Ontario. During this period, the short-lived aspens have practically disappeared from the stand, while the slightly longer-lived birches have shown a steady decline both in number of stems and basal area. Red maple has also declined greatly in numbers, but the growth of the residual trees has been such that the basal area of this species has remained more or less the same. Sugar maple has also suffered mortality, but the growth of the residual trees has been great, with the result that the species has risen from 38 to 48 percent of the stand basal area over the 45-year period. The few red oaks have survived well and are making up a substantial proportion of the stand basal area. The better competitive ability of the sugar maple, red maple, and red oak at the expense of the aspen and birches has brought about major changes in stand

**Table 14.1 Changing Composition After 45 Years in a Hardwood Stand on Sugar Island, St. Mary's River, Michigan**

| | 1933 | 1938 | 1944 | 1949 | 1955 | 1960 | 1971 | 1978 |
|---|---|---|---|---|---|---|---|---|
| | Number of trees per hectare | | | | | | | |
| Sugar maple | 796 | 778 | 722 | 692 | 657 | 531 | 479 | 452 |
| Red maple | 309 | 292 | 200 | 178 | 170 | 156 | 136 | 128 |
| Red oak | 44 | 44 | 42 | 44 | 42 | 37 | 37 | 35 |
| Paper birch | 86 | 57 | 42 | 30 | 20 | 22 | 7 | 7 |
| Yellow birch | 40 | 22 | 17 | 15 | 15 | 15 | 10 | 10 |
| Bigtooth aspen | 49 | 27 | 7 | 7 | 5 | 7 | 7 | 7 |
| All trees[a] | 1354 | 1243 | 1048 | 983 | 919 | 781 | 677 | 655 |
| | Basal area in square meters per hectare | | | | | | | |
| Sugar maple | 9.0 | 10.1 | 11.3 | 12.1 | 12.8 | 13.3 | 14.1 | 14.0 |
| Red maple | 6.8 | 7.5 | 7.0 | 7.0 | 7.6 | 7.8 | 7.7 | 7.7 |
| Red oak | 2.0 | 2.3 | 2.8 | 3.3 | 3.4 | 3.9 | 5.0 | 5.0 |
| Paper birch | 2.8 | 2.2 | 1.7 | 1.4 | 1.2 | 1.7 | 0.5 | 0.6 |
| Yellow birch | 1.0 | 1.8 | 0.7 | 0.6 | 0.7 | 0.8 | 0.6 | 0.6 |
| Bigtooth aspen | 1.4 | 0.9 | 0.4 | 0.5 | 0.4 | 0.6 | 0.8 | 0.8 |
| All trees[a] | 23.4 | 24.0 | 24.3 | 25.4 | 26.5 | 28.7 | 29.1 | 29.2 |

[a] Includes some individuals of species not listed above.

composition. The stand basal area has increased slowly and appears to be leveling off in total and for each species.

While permanent sample plot data record the changing composition of the overstory with time, they do not chronicle the cause of the death of the trees that disappear in the intervals between measurements. We generally assume that most of these trees simply are suppressed, that is, their leaf area and feeding root area are reduced to the point that life can no longer be sustained and the tree dies.

In actual fact, death may be due to many causes. Even in the case of so-called natural mortality, the tree probably dies during some period of extreme stress, such as in a late summer, hot, dry spell or in a severe unseasonable frost, or in an extremely cold winter. In other cases, the tree becomes weakened to the point that it falls prey to some insect or disease. In many cases, several factors are interrelated and together account for death.

For example, black cherry seedlings are common in young hardwood stands in central Massachusetts. Yet, mature trees are seldom found in older stands on similar sites. Close observation of tagged cherry seedlings on sample plots revealed that the eastern tent caterpillar, *Malacasoma americana* F., typically defoliated the faster-growing dominant cherries year after year. As a result, these trees put on little annual increment in height, soon become overtopped by competing oak, maple, and other species, and then die because of inability to survive under shaded conditions as an understory tree. Over a 4-year period, the dominant 2 m high black cherry seedlings grew only 13 cm in an untreated plot as compared to 30 cm in adjacent plots where the egg clusters of the tent caterpillar were removed annually.

Again, Monterey pine in New Zealand and Australia is subject to mortality which has occasionally reached serious proportions. Present evidence indicates that epidemic mortality is primarily due to unusually dry years in which the trees are under severe moisture stress and readily succumb to attack by *Sirex noctilio* (F.), a wood wasp that oviposits in the bole cambium. If an unusually wet year precedes the dry period, *Phytophthora* fungi in the soil may cause substantial mortality to the feeding rootlets of the trees, predisposing the trees to even greater losses. Mortality in this instance may be due to a combination of wet and dry years, a fungus, and an insect.

## Competition in the Understory

Whereas competition in the overstory involves relatively few individuals and is spread out over many years, competition near the forest floor involves not only trees but many shrubs and herbaceous plants, often in very large numbers and changing in relationships year by year.

Under a more or less fully stocked forest stand, the supplies of light and soil moisture are limited. Light is available at a sufficient intensity for a net gain in plant weight through photosynthesis chiefly in the sunflecks that move over the ground as the angle of incidence of the sunlight is changed by the rotation of the earth and as wind moves the foliage of the overstory. Moisture on well-drained sites may be largely restricted to current precipitation which reaches the forest floor partially through direct fall and partially through stem flow and drip from the overstory trees.

Many plants are adapted to spending their entire life cycle in the understory. Included are some herbs that carry on a major portion of their annual life activities before the trees reach full leaf. Included also are other herbs, many shrubs, and small trees that are tolerant of understory environmental conditions. Some trees can both thrive in the understory and also occupy the overstory itself when the opportunity occurs. Such species will normally succeed to dominance of the forest community.

In assessing competition within the understory, the problem of the successful plant attracts little attention. It is merely carrying on its growth under existing environmental conditions without incident. As Decker (1959) points out, it is with the failing and dying plant, the plant intolerant of understory conditions, that we must be concerned. The unsuccessful competitor loses vigor and dies, but the causes of death may be various. Starvation resulting from inadequate light and consequent inadequate photosynthesis is certainly a contributory factor. Inadequate soil moisture is another. Insects and diseases, too, play their role, particularly when the plant has reached a weakened condition. The factors of the environment relating to understory survival are discussed in some detail in the following section.

## UNDERSTORY TOLERANCE

The problem of survival in the understory is basic to an understanding of forest succession, since those forest trees capable both of surviving as understory plants and responding to release to reach overstory size will inevitably form a major portion of the evolving forest community. A forest tree that can survive and prosper under a forest canopy is said to be **tolerant** while one that can thrive only in the main canopy or in the open is classified as **intolerant**.

This use of the term **tolerance** to refer to the relative capacity of a forest plant to survive and thrive in the understory is a restricted application of the general botanical meaning of the term, which deals with the general capacity of a plant to be genetically adapted and physiologically compatible with unfavorable conditions. Thus a salt-tolerant plant is one that is adapted to grow in soil with a high salt content, and a fume-tolerant plant is one that can grow in the presence of gases noxious to most other plants. In forestry, however, the unmodified use of the term tolerance refers to a plant's vigor in the forest understory due to its genetic and physiological adaptations to this environment. Understory tolerance is a more precise term.

In many cases, shade tolerance or intolerance may be the major factor accounting for a plant's performance in the understory. Although all successful understory species are adapted to shade, they may have other adaptations to the understory environment such as to moisture stress, microclimatic factors, browsing, and disease. Hence the term shade tolerance is usually accurate as far as it goes, but it may lead to oversimplified thinking about competition in the forest understory.

Relative tolerance can be recognized and rated in general terms with little difficulty. An understanding of the nature of tolerance, however, has been the subject of much conjecture and controversy among forest scientists. If the term is used only in the sense of the ability of a plant to survive and prosper in the understory, then its nature can be investigated in terms of the ecology and physiology of understory plants and the understory environment (see following discussion). Often many other characteristics of tolerants and intolerants are cited, and with their inclusion a broader understanding of the nature of tolerant and intolerant species is necessary.

## Recognition of Tolerance

Since, by definition, a tolerant tree is one that grows and thrives under a forest canopy, it may be recognized by various values that relate to its vigor under such conditions. In contrast, an intolerant tree is frequently characterized by opposite extremes of the same traits. In addition, other characteristics have been cited as criteria or indicators for the determination of tolerance (Toumey, 1947; Baker, 1950):

**Ocurrence in the Understory.** Tolerant trees live for many years as understory plants under dense forest canopies. Many trees may germinate and survive a few years under such conditions, but only tolerant trees will persist and continue to grow for decades.

**Response to Release.** The vigor of tolerant trees is demonstrated by the fact that, following removal of the overstory, they have the ability of responding by initiating immediate and rapid growth. A tolerant tree may survive for years in the understory while putting on as many as 12 to 24 rings per cm of radius, and yet begin growing rapidly soon after release.

**Crown and Bole Development.** The lower branches on tolerant trees will be foliated longer and to a greater extent than those on intolerant trees. Consequently, tolerant trees will have deeper and denser crowns. It follows that tolerant trees will prune naturally at a slower rate, will maintain a greater number of leaf layers, and will maintain healthy and vigorous leaves deeper into the crown. The boles of tolerant trees tend to be more tapered than those of intolerant trees because of the greater depth of crown (Larson, 1963b).

Shoot growth and branching pattern of a species are adapted to provide light appropriate to the light requirements of its foliage. For example, the relatively fast branch growth and spreading form of American elm, sycamore, and silver maple allow a well-lighted crown that is required for leaves of these species. In contrast, sugar maple and beech leaves tolerate considerable shade and thrive even in the interior of a crown characterized by a more compact form and slower branch growth than that of the more shade-intolerant species.

**Stand Structure.** Tolerant trees persist over long periods of time in natural mixed stands, tend to be successful in mixture with other species of equal size, and consequently form denser stands with more stems per hectare than do comparable intolerant trees.

**Growth and Reproductive Characteristics.** Tolerant trees frequently grow faster, particularly in height, than intolerant trees in the understory. In contrast, intolerants will normally outstrip tolerants under comparable conditions in the open. Tolerant trees typically mature later, flower later and more irregularly, and live longer than intolerant trees.

Although the understory tolerance of a species is defined on the basis of its ability to survive and prosper in the understory, it is not this attribute that causes the many other distinctive and important differences between tolerant and intolerant species. Populations of the respective types have evolved under different selection pressures of environment and plant competition and hence belong to two complex and markedly different adaptation systems. Tolerant species establish themselves, grow in, and have become adapted to conditions markedly different from that of intolerants. Although it is often instructive to compare the two extremes, we are dealing with species of varying degrees of intermediacy along a cline from very intolerant to extremely tolerant.

Intolerant species are typically pioneers that may colonize a wide variety of sites. They are successful because of two major types of adaptations. First, their capacity for rapid establishment on disturbed areas, fast growth in the open, early seed production, and widespread seed dispersal have enabled them to perpetuate themselves wherever fire, windthrow, flooding, cultivation, or other disturbances have eliminated or reduced the existing vegetation. In some climatic regions, disturbance alone may account for the perpetuation of intolerant species since they are replaced in time on most sites by more tolerant species.

However, intolerants have a second strategy—adaptation to extreme site conditions. They are typically adapted to some type of xeric (warm or cold) or infertile site as well as to the climatically extreme, initial conditions of the disturbed site. Thus they may form relatively permanent communities on extreme sites where more tolerant species are at a competitive disadvantage. For example, willows in annually flooded bottomlands, jack pine on nutrient–poor sands of the Lake States, Table Mountain pine on the driest and least fertile soils of the Appalachian Mountains, and sand-live oak on sands of the southern Coastal Plain are all examples of intolerants occupying extreme sites of various kinds. Although fire contributes to the maintenance of Table Mountain pine, without fire and human-caused disturbance the species would exist, but only on extremely dry and nutrient–poor rock outcrops and steep, shale slopes (Zobel, 1969). From such sanctuaries, especially in glaciated terrain with a mosaic of landforms and soil types, pioneer species are in a position to colonize readily adjacent mesic sites following disturbance.

In contrast, tolerant species occupy and are adapted to more moist, sheltered, and fertile sites (mesic conditions). They replace intolerant species and perpetuate themselves through adaptations favoring survival and growth in a shaded understory, establishment in undisturbed litter or duff layers, and a long life span.

## Scoring Tolerance of Forest Trees

A relative tolerance rating for species in a given forest region and on a given site class may be obtained by arbitrarily scoring the various criteria of tolerance. Graham (1954) evolved a technique of rating tolerance of forest trees in the Upper Peninsula of Michigan. Graham's system is based on scoring a sample of trees. at least 10 dominant or codominant individuals, on (1) crown density, (2) ratio of leaf-bearing length to total length of branches, and (3) ratio of crown length to total height of trees growing in an unbroken forest. This basic score is modified by nonquantitative observations representing the opinion of the observer.

By thus scoring each species in a stand or locality they may be arranged in order of relative tolerance. The following listing gives the ratings so obtained for trees growing in Iron County, Michigan:

| TOLERANT | SCORE | LOW MIDTOLERANT | SCORE |
|---|---|---|---|
| Hemlock | 10.0 | Black cherry | 2.4 |
| Balsam fir | 9.8 | Black ash | 2.4 |
| Sugar maple | 9.7 | Red pine | 2.4 |
| Basswood | 8.2 | | |
| **HIGH MIDTOLERANT** | **SCORE** | **INTOLERANT** | **SCORE** |
| White spruce | 6.8 | Jack pine | 1.8 |
| Black spruce | 6.4 | Paper birch | 1.0 |
| Yellow birch | 6.3 | Tamarack | 0.8 |
| Red maple | 5.9 | Aspens | 0.7 |
| White cedar | 5.0 | | |
| White pine | 4.4 | | |

## Tolerance Ratings of Species

Tolerance is not constant for a given species: it varies with genetically different individuals and races, with different regional climates, with different local sites, with different plant associates, with different vegetative conditions (seedlings vs. sprouts), and especially with age. A given species may be more tolerant in one part of its range than in another, on one site in comparison to another, and in one forest type than in another.

For example, eastern white pine is more tolerant in Minnesota, where it is considered to be midtolerant, than in New England, where it is relatively intolerant. In central New England, it is more tolerant on dry sandy soils than on moister sandy loams—at least it occurs more consistently under the less dense stands on the former sites. Like most species, it is more tolerant as a seedling than it is as it grows older. Sprouts with well developed root systems, such as occur in oaks, hickories, and yellow-poplar, appear more tolerant than seedlings of the same species because the sprouts are better able to absorb water and nutrients.

Tolerance, then, is not only a relative matter, but the relative ranking of a species with regard to tolerance will depend in part upon its region, site, age, and associates. Tolerance ratings must be interpreted with great care and with these points in mind.

At the same time, the approximate tolerance rating of the more important and characteristic forest tree species should be known and understood by the practicing silviculturist as a general frame of reference. In addition, tolerance is also related to many important attributes of tree species. Thus such a frame of reference, an estimate of the relative understory tolerance of selected species, assuming representative site conditions for each species, is presented in Table 14.2. Obviously the practicing ecologist must learn to understand

**Table 14.2 Relative Understory Tolerance of Selected North American Forest Trees [a]**

| EASTERN NORTH AMERICA | | WESTERN NORTH AMERICA |
|---|---|---|
| GYMNOSPERMS | ANGIOSPERMS | |
| | Very tolerant | |
| Eastern hemlock | Flowering dogwood | Western hemlock |
| Balsam fir | Hophornbeam | Pacific silver fir |
| Red spruce | Sugar maple | Pacific yew |
| | American beech | |

**Table 14.2** *(continued)*

| EASTERN NORTH AMERICA | | WESTERN NORTH AMERICA |
|---|---|---|
| GYMNOSPERMS | ANGIOSPERMS | |
| | Tolerant | |
| White spruce | Basswood | Spruces |
| Black spruce | Red maple | Western red cedar |
| | | White fir |
| | | Grand fir |
| | | Alpine fir |
| | | Redwood |
| | Intermediate | |
| Eastern white pine | Yellow birch | Western white pine |
| Slash pine | Silver maple | Sugar pine |
| | Most oaks | Douglas-fir |
| | Hickories | Noble fir |
| | White ash | |
| | Elms | |
| | Intolerant | |
| Red pine | Black cherry | Ponderosa pine |
| Shortleaf pine | Yellow-poplar | Junipers |
| Loblolly pine | Sweet gum | Red alder |
| Eastern red cedar | Sycamore | Madrone |
| | Black walnut | |
| | Scarlet oak | |
| | Sassafras | |
| | Very intolerant | |
| Jack pine | Paper birch | Lodgepole pine |
| Longleaf pine | Aspens | Whitebark pine |
| Virginia pine | Black locust | Digger pine |
| Tamarack | Eastern cottonwood | Western larch |
| | Pin cherry | Cottonwoods |

[a] Based on representative site conditions for the respective species.

Survival in the understory is related to light irradiance, moisture stress, and other factors. As a general guide to the light irradiance component, we estimate the range of *minimum* percentage of full sunlight for a species to survive in the understory at each of the five arbitrary levels of tolerance:

*Very tolerant* species may occur when light irradiance is as low as 1 to 3 percent of full sunlight; *Tolerant* species typically require 3 to 10 percent of full sunlight; *Intermediate* species 10 to 30 percent; *Intolerant* species, 30 to 60 percent; *Very intolerant* species, at least 60 percent. For example, an intolerant species competing in the understory is unlikely to survive with less than about 30 percent of full sunlight (unless other compensating factors are favorable).

the relative tolerance and competitive ability of trees under field conditions and not from textbook tables. Nevertheless, a general starting point and an awareness of the nature of tolerance are appropriate.

## Nature of Understory Tolerance

Although the relative tolerance of a given species growing in a given community on a given site can be recognized with some degree of accuracy, the explanation of the nature of tolerance is much more difficult. This problem has intrigued silviculturists for many years and has been the subject for much controversy and semantic debates.

The nature of tolerance may be examined from the broad species adaptation level, and as we have seen, the so-called tolerant and intolerant species belong to two complex and markedly different adaptation systems. Usually, however, the specific trait of survival in the understory is investigated by studying the environmental factors and the physiological processes involved.

**Environmental Factors Relating to Understory Tolerance.** The most obvious ecological feature of the understory environment is low light irradiance. In fact, many ecologists to this day associate the capacity of a plant to survive in the understory solely with the capacity of a plant to survive under low light irradiances, equating the general concept of understory tolerance with the specific concept of shade tolerance.

As data have accumulated from studies of the effect of light irradiance on tree growth under controlled or semicontrolled forest conditions, however, it has become evident that the light irradiances under most forest canopies are more than sufficient to permit most forest trees to carry on photosynthesis at rates substantially higher than required to balance respiration losses (i.e., the **light compensation point**). Under moderate covers, such as those created by pine and oak forests, ample light reaches the forest floor to provide energy for photosynthesis by many forest tree seedlings. Even under dense covers, such as those formed by spruce-fir and tropical rain forest species, light flecks sweeping through the forest permit occasional plants to survive and grow in the understory. Only in those forests where the relative illumination on the forest floor is less than about 2 percent is light obviously a single limiting factor in understory survival.

Light, though, is not the only environmental factor greatly modified by the forest canopy. Under a dense forest canopy, almost all the factors of the climate and soil differ from those characterizing similar open sites. Soil moisture is foremost among the affected environmental factors.

The importance of soil moisture in regulating understory occurrence and growth was demonstrated spectacularly in Germany in 1904 by Fricke, who cut the roots of competing understory trees by trenching around small, poorly developed Scots pine seedlings growing under a stand of the same species. These seedlings responded with vigorous growth, indicating that they had been inhibited principally by a shortage of soil moisture created by the competing roots of the overstory trees rather than by low light irradiances. Trenched plots have been used with similar results under eastern white pine in New Hampshire (Toumey and Kienholz, 1931), loblolly pine in North Carolina (Korstian and Coile, 1938), and others. Generally speaking, trenching a small plot of a few square meters or so in size under a pine stand so as to remove root competition by severing entering roots will result in a great increase in available soil moisture and the consequent appearance of luxurious vegetation.

Usually, either trenching or watering understory plants will substantially increase their height growth. However, tying overstory tops back to allow greater amounts of light to reach their leaves will have little effect.

It would be a mistake, however, to attribute tolerance solely to soil moisture just as it would to attribute it to light alone. Continued study of Toumey's trenched plot experiments (Lutz, 1945) indicated that the more intolerant species do poorly and eventually die even when root competition is kept low, so that the trenched plots gradually become dominated by tolerant plants. After 21 years, hemlock still survived and grew while white pine, which initially was the most abundant tree, had died out completely. Obviously, both soil moisture and light are involved in understory survival, and there may be other factors such as carbon dioxide content of the air as well.

A series of studies of the ecological nature of tolerance carried out with loblolly pine and associated hardwoods in North Carolina has done much to explain the cause of death of loblolly pine and other intolerant seedlings under loblolly pine overstories. In contrast to tolerant seedlings, which are able to survive and even grow under pine canopies, the relatively intolerant loblolly pine seedling seems to photosynthesize more than enough to counterbalance respiration losses but not enough to permit its root system to expand and reach the deeper soil strata. Over the years, therefore, the loblolly pine seedling develops a somewhat etiolated top without a compensatory root system of sufficient extent and depth. Sooner or later, these seedlings will die during a period of unusually severe moisture stress under hot, dry midsummer conditions. In contrast, tolerant hardwood species under similar conditions will develop root systems big and deep enough to permit them to survive these droughts.

Ability to survive under the moderate light intensities and severe soil-moisture shortages characteristic of pine forests, then, appears to be dependent upon a plant's carrying on sufficient photosynthesis to develop a sufficient root system to survive midsummer drought in soils kept at low moisture levels by competing roots of overstory trees. Under other situations, either light or root competition may be relatively more important. Under dense Sitka spruce and western hemlock in the Pacific Northwest rain forest, light is at extremely low levels whereas the site is almost always wet or at least damp. Here, light is obviously the more important factor. Under open oak woodland types or under ponderosa pine in the drier parts of its range, light under the forest is well above any critical levels while moisture is always in short supply. Here, soil moisture is obviously the more important factor. Always, however, it is the interaction of light, moisture, and possibly other environmental factors as well which together determine understory survival and growth. It is the understory site in toto, viewed as an integrated whole, that determines the ability of a plant to survive in the understory and not any single site factor taken by itself.

**Physiological Factors Relating to Tolerance.** That some understory plants can develop structures necessary for survival under the site conditions of the understory while others cannot indicates a basic genetic difference between species which is exhibited in their physiological response to these environmental conditions. The surviving plants must exhibit some superiority over failing plants such as (1) maintaining a greater photosynthesizing leaf area in the understory, (2) photosynthesizing more efficiently per unit leaf area in the understory, (3) maintaining lower rates of respiration per unit leaf area in the understory, (4) controlling water loss more efficiently, (5) converting a greater portion of their photosynthate into growth, or (6) absorbing water more efficiently.

Present evidence from controlled laboratory experiments of shade tolerance indicates that plant characteristics leading to failure of a genotype or species in one environment may be an indirect consequence of adaptations necessary for survival in another (Grime 1965a). Thus the adaptation of photosynthetic and respiration mechanisms of intolerant species for full productivity under full sunlight is achieved at the cost of lowered efficiency under shade conditions. Although the rates of photosynthesis have been found closely associated with performance in shade (Logan and Krotkov, 1969; Logan, 1970), the differences in rates of respiration between tolerant and intolerant species may be the most important determinants of success or failure in forest shade, where the plant may spend many more hours below than above the light compensation point (Grime, 1965b; Loach, 1967). Intolerant species have high rates of photosynthesis, but they are offset by high rates of respiration and rapid conversion of photosynthate into growth. These species are highly productive in open environments but less adapted to shaded conditions.

Tolerant species, according to Went (1957), are more competitive in shaded environments through selection for low respiration rates; they also tend to have lower photosynthetic rates and hence grow slowly in all environments. In addition, shade-tolerant species such as beech and sugar maple not only have the ability to open their stomates in dim light but to open them rapidly to take advantage of light flecks over short periods for photosynthesis, even though their absolute rates of photosyntheis may be low (Woods and Turner, 1971; Davies and Kozlowski, 1974).

A comparison of rates of photosynthesis and respiration for sun and shade leaves of several tolerant and intolerant species illustrates these relationships (Table 14.3; Loach, 1967). Striking differences occur between intolerants and tolerants at $P_{max}$ (the maximum rate

**Table 14.3 Summary of the Major Differences in the Rates of Photosynthesis and Respiration of the Sun (100% Daylight) and Shade (17% Daylight) Leaves of Shade-Tolerant and -Intolerant Species**

| | TOLERANT SPECIES | | INTOLERANT SPECIES | |
|---|---|---|---|---|
| | Sun | Shade | Sun | Shade |
| $P_{max}$ (mg $CO_2$/dm$^2$ hr) | 7.0 | 6.4 | 18.0 | 12.2 |
| $P^{250}$ (mg $CO_2$/dm$^2$ hr) | 0.9 | 2.0 | 0.3 | 1.4 |
| $R$ (mg $CO_2$/dm$^2$ hr) | 1.2 | 1.0 | 2.5 | 3.2 |

Source: Loach, 1967.

$P_{max}$ is the rate of photosynthesis at saturating light intensity and $P^{250}$ the rate of photosynthesis at 2700 lux (250 ft-c). $R$ is the rate of respiration in darkness.

of photosynthesis attained by increasing illumination), indicating that intolerants can make more efficient use of strong light. Intolerant species, however, suffer the greatest proportional reduction in photosynthesis when grown in shade ($P^{250}$). Significantly, the respiration rates of tolerant species are consistently less than those of intolerants. Thus selection for a high rate of photosynthesis at high light irradiances and high growth rate in full sunlight may inevitably limit the plant in shade. It is clear that there are genetic differences between species in rates of photosynthesis and respiration and probably the other factors as well. It will remain, however, for careful physiological studies and modelling of the growth system of tolerants and intolerants, particularly under natural conditions where both moisture and light may be limiting, to determine the relative importance of photosynthesis, respiration, and other factors to survival and growth in the understory.

### Examples of Tolerance in Forest Stands

A series of photographs, Figures 14.3 to 14.6, serves to illustrate various aspects of tolerance in four forest communities. Figure 14.3 depicts a young second-growth stand of intolerant yellow-poplar in a cove site of the Appalachian Mountains. Sufficient light and moisture reach the understory to favor development of a conspicuous and diverse community of midtolerant and tolerant species, such as oaks, maples, and beech, as well as a rich shrub and herbaceous flora.

The stand of midtolerant oaks of the Missouri Ozark Region (Figure 14.4) illustrates the dominant oak overstory and an open, partially shaded understory of oaks and associated vegetation. The shade cast by the overstory and the low soil moisture during the growing season combine to curtail growth of the understory. This is in contrast to the luxuriant understory of the yellow-poplar stand in the mesic cove site of Figure 14.3.

The dense shade from the overstory of sugar maples in a northern Michigan stand (Figure 14.5) favors the very tolerant sugar maple seedlings over all other species. In deep shade, they survive but grow slowly; in small openings where more light is available (background of Figure 14.5) they respond with accelerated growth and, barring disturbance, will perpetuate the dominance of sugar maple.

The western white pines of the northern Idaho stand in Figure 14.6 originally colonized the site following a fire. Now overmature, they form a dense stand and their crowns intercept most of the incoming light. However, they will be gradually replaced by the very tolerant species growing in the understory, western hemlock, grand fir, and western red cedar, in the absence of disturbances such as fire, windthrow, and logging.

## OVERSTORY MORTALITY

The question of tolerance is associated primarily with understory survival. The trees in the overstory, too, have a more or less limited life span. There are many reasons for overstory mortality, including

**Figure 14.3. Young second-growth stand of yellow-poplar in a cove of the Appalachian Mountains, North Carolina. (U.S. Forest Service photo.)**

competition, senescence, and death caused by external factors such as insects, diseases, wind, and lightning. Examples have already been given.

### Competition Among Overstory Trees

We have already seen that the surviving trees must inevitably expand and take up more growing space with the passage of years. Other trees will become suppressed and will either drop from the overstory through failure to maintain growth, or will die. Frequently they will first drop from the overstory and then die after they have become overtopped.

The causes of death of suppressed overstory trees under severe

**Figure 14.4. Mature, uncut oak stand in the Ozark Mountains of southeastern Missouri. (U.S. Forest Service photo.)**

competition are akin to those of intolerant smaller trees in the understory. Most obviously, the crowns of these trees recede because of lack of growth of the branch terminals coupled with the gradual lignification of the inner portions of the branch and associated loss of foliage. The friction of other tree crowns whipped back and forth in strong winds plays an important part in mechanically reducing crown size of those trees that have ceased to grow vigorously. Presumably, cessation of crown growth and reduction of crown size denote a comparable cessation of root growth and reduction of root extent.

In any event, such trees will eventually die, as they are less able than the dominant trees of the stand to survive periods of unfavor-

**Figure 14.5. Sugar maple stand in the Upper Peninsula of Michigan. Unbrowsed sugar maple seedling shown on left and browsed seedling on right. (U.S. Forest Service photo.)**

able environmental conditions. Mortality will normally be concentrated in periods of extreme heat and drought, extreme cold, or other critical periods.

### Senescence

As a tree grows larger, the distance between the feeding roots at the end of the root system and the active leaves at the top of the crown increases. Soil moisture and nutrients must move a greater and greater distance to reach the foliage, and food substances must move a greater and greater distance down to reach the roots. As it becomes larger the tree becomes less and less efficient. Eventually, a given individual of a given species reaches a size at which it is barely able to maintain life without further growth, and further height increment becomes impossible. Thus any tree has a maximum height which it can reach on a given site. Considerable variation will exist in this regard between various species. Even different individuals of

**Figure 14.6. The overmature western white pines will eventually be replaced, barring disturbance, by very tolerant conifers that have become established in the understory; northern Idaho. (U.S. Forest Service photo.)**

the same species will vary in their genetic capacity in maximum height.

As a tree approaches its maximum size, the ability of its leaves to supply needed foods to the bole will decrease, particularly during periods of unfavorable growing conditions. Cells laid down by the bole cambium, therefore, tend to become fewer and somewhat smaller with time, thus increasing the inefficiency of the total organism.

The increasing inefficiency of a tree as it approaches maximum size is coincident with the declining growth curve of old age. During this period of senescence, the tree gradually becomes less and less able to withstand climatic extremes. Sooner or later it will be weakened to the point that it will succumb to some insect attack, fungal attack, or other external enemy, particularly following some extreme dry spell, wet spell, hot spell, or cold spell.

While it is obvious that trees become less efficient organisms with increasing size, it is less clear as to whether or not they also become senescent in the sense of actual aging and loss of vigor of the meristematic tissues and the living protoplasm of the cells that make up the functional tissues of the tree. Different physiological phases, such as the juvenile and adult stages described in Chapter 3, do exist. However, the deterioration of meristematic tissue itself as a primary cause of death has not been demonstrated.

### External Factors

In addition to competition and senescence, insects, fungi, and climatic factors also kill overstory trees. These external causes sometimes attack healthy, vigorous trees, and sometimes merely complete the process initiated by weakening of the tree as a result of competition and senescence.

In many parts of the world, lightning plays a prominent role in killing large dominant trees. Among other climatic factors, snow and ice damage is particularly important, notably in dense, unthinned, even-aged stands. Trees of intermediate crown classes, the so-called "whips," are most apt to be bowed and broken down by snow and ice; but once a hole in the forest canopy has been opened, it frequently will be enlarged year after year by the bending and breaking of the exposed edge trees under the combined influence of snow, ice, and wind. Holes in Pacific Northwest stands of Douglas-fir and Norway spruce in Europe are continually being created and enlarged by this gradual process.

Winds, too, play a major part in removing the mature forest overstory. In the New England states, major hurricanes in 1635, 1815, and 1938, as well as many lesser storms, have blown down many thousands of hectares of the taller forests. Tornadoes, thunderstorm fronts, and other storms have been destructive to forests throughout much of the country.

Although many insects and fungi do attack healthy, mature trees, the greatest number prey upon trees in a weakened condition. Oviposition in the cambium or phloem by insects is more frequent and more apt to be successful on trees under moisture stress; low bole moisture content is a characteristic of a weakening and dying tree. Fungal attack is more frequent and more apt to be successful in trees whose bark and wood are opened or cracked by fire, ice, or wind so that both air and moisture are available to the attacking organisms.

## TRANSITION OF TREES FROM UNDERSTORY TO OVERSTORY

Forest succession under undisturbed conditions implies the development of understory trees into overstory. Several avenues are available for such movement.

## Growth Up Through the Overstory

If the main canopy is not too dense, tolerant species may sometimes grow directly into it, becoming a part of the overstory or even forming a superior canopy over the former overstory.

In central New England, old fields are frequently restocked with gray birch and white pine, the former growing much faster in height and forming the initial overstory. The birch, however, maintains its vigor for only a few decades. As it declines, its utilization of the site diminishes, allowing the white pine to grow up through, suppressing it, and forming a pure white pine stand.

Similarly, white pine can grow up through a declining jack pine type in the Lake States; and balsam fir and black spruce can penetrate a decadent aspen canopy on the better sites in the same region (Figure 14.7).

## Response to Release

Second, tolerant forest trees capable of occupying the overstory can persist in the understory for many years until death of one or more overstory trees creates a hole or opening which they can fill. Eastern hemlock in New England (Marshall, 1927) is notable for its capacity of living as an understory tree for decades and even centuries and yet responding almost immediately to release with greatly accelerated growth. Yet the same species on the drier and hotter (in summer) sites of the northern Lake States will frequently deteriorate and die when released rather than respond with vigorous growth (Secrest et al., 1941).

Although many understory trees can respond to release and assume overstory position through rapid growth, many others can not. Some tolerant understory tree species, such as flowering dogwood, hophornbeam, blue beech, and holly, simply do not have the growth potential to accelerate in growth upon release, or the genetic adaptation to grow as overstory trees. They are understory trees by nature and remain that way. Others persist in the understory but in such a debilitated condition that they have lost the capacity to respond, even though a seedling of the same species could germinate and grow successfully in the same forest clearing.

## Colonization of Openings—Gap Phase Replacement

A third avenue of access to the main canopy of an established forest is followed by species of intermediate tolerance which can colonize an opening quickly, and then proceed to outgrow competition to reach the overstory. The term **gap phase replacement** was given to this process by Watt (1947) who studied English beech forests. Bray

**Figure 14.7. Shade-tolerant balsam fir and spruces penetrating and becoming part of the overstory together with the intolerant trembling aspen; northern Wisconsin. (U.S. Forest Service photo.)**

(1956) described gap phase species in Lake States hardwood forests. Yellow birch in the northern hardwood forest of the eastern United States–Canadian border and white ash a few hundred miles to the south are such gap phase species. White pine in the East and Douglas-fir in the Pacific Northwest are also of this group. Many gap phase species have seed which is chiefly wind-disseminated and have the capacity of germinating on small patches of exposed mineral soil such as that created by the uptorn roots of a windthrown tree. Juvenile growth under conditions of partial shade and partial root competition is rapid so that the seedlings, or at least a portion of them, can outgrow the advance growth of more tolerant species already established in the openings; these may be, however, in more or less of a state of shock and therefore unable to respond quickly to the release.

## Reaching the Overstory as Vines

Rarely in the North Temperate Zone but frequently in the Tropics and even in the South Temperate Zone, some trees can penetrate even a dense crown canopy as woody lianas or vines. Upon reaching the top of the overstory, they expand and strangle their supporting tree, eventually forming a woody and self-supporting bole of their own. The strangler figs (*Ficus*) in the tropics and the northern rata (*Metrosideros robusta*) in New Zealand are examples of such vines that transform themselves into overstory trees. In the Appalachian Mountains, grape vines pose serious problems to newly developing stands on the better sites (Trimble, 1977). They may destroy or damage as much as 75 percent of the new tree stems. Likewise, Japanese Honeysuckle (*Lonicera japonica*) is a serious problem for tree regeneration on moist sites in the southern Piedmont and Coastal Plain of the southeastern United States. It has spread northward into the midwestern states and occurs sporadically as far north as Connecticut, Pennsylvania, and the lower peninsula of Michigan.

## SUGGESTED READINGS

Daubenmire, Rexford. 1968. *Plant Communities*. (Introduction, pp. 3–35.) Harper & Row, Inc., New York. 300 pp.

Grime, J. P. 1965. Shade tolerance in flowering plants. *Nature* 208:161–163.

Korstian, C. F., and T. S. Coile. 1938. Plant competition in forest stands. Duke Univ. School For. Bull. 3. 125 pp.

Loach, K. 1967. Shade tolerance in tree seedlings. 1. Leaf photosynthesis and respiration in plants raised under artificial shade. *New Phytol.* 66:607–621.

# 15
# *Forest Succession*

The bases of dynamic changes in the forest by which one community succeeds another have been detailed in the previous chapter. In synecology, **succession** refers to the replacement of the biota of an area by one of a different nature. Animals as well as plants are involved. Changes in the fauna, however, more often follow than lead the changes in the vegetation described in the present chapter. The development of the biota, beginning with unoccupied sites and proceeding in the absence of a catastrophic disturbance, is termed **primary succession** and is discussed in the present chapter. Succession which is subsequent to a disturbance that disrupts rather than destroys an existing biotic community is termed **secondary.** Primary succession is sometimes termed **autogenic** in that the displacement of one group of species by another results from the development within the ecosystem itself, being a part of the concomitant development of the vegetation, soil, and microclimate of the site. In contrast, secondary succession is **allogenic** in that it is induced by external forces which change the ecosystem (i.e., by forest destruction). Since disturbances are normal to the life of the forest, and since some disturbances such as changes in the microclimate and changes in the soil are brought about in part by changes in the vegetation, the distinction between primary and secondary succession is arbitrary rather than real. It is followed here simply as a matter of convenience in organizing material on dynamic changes in the forest composition and structure.

Succession is a continuing but not necessarily unidirectional process marked by myriads of changes in the vegetation, the fauna, the soil, and the microclimate of an area with the passage of time. These changes occur together, mutually affecting one another, with seldom any simple cause-and-effect relationships becoming evident.

## EVOLUTION OF THE CONCEPT OF FOREST SUCCESSION

The dynamic nature of the forest has been recognized at least from the time of the earliest observers who put their thoughts into writing. The formalization of the study of forest succession as a scientific discipline, however, has taken place in the last century.

## Historical Antecedents

The origin of the concept of forest succession (Spurr, 1952a) can be traced back through the writings of eighteenth-century foresters in Europe and read, by implication at least, in the words of early Roman natural historians.

In America, Jeremy Belknap early recognized the transitory nature of forest types. Writing in his history of New Hampshire, which was published in 1792, he observed:

> *There are evident signs of a change in the growth on the same soil, in the course of time; for which no causes can be assigned. In some places the old standing trees, and the fallen decayed trees, appear to be the same, whilst the most thriving trees are of a different kind. For instance, the old growth in some places is red oak, or white ash; whilst the other trees are beech and maple, without any young oak or ash among them. It is probable that the growth is thus changed in many places; . . .*

The term "succession" was used in a letter by John Adlum included in a memoir of the Philosophical Society for Promoting Agriculture as part of material published by Richard Peters in 1806 on "Departure of the southern pine timber, a proof of the tendency in nature to a change of products on the same soil." In northwestern Pennsylvania, the occurrence of old red and white oak in northern hardwood stands suggested the succession to sugar maple, beech, and yellow birch in that area. In the mid-Atlantic states, oak and hickory were observed to follow the clearcutting of pitch pine, while white pine was noted as appearing "spontaneously" on old fields.

In Europe, beginning with Hundeshagen in 1830, observed changes in forest composition were the subject of specific articles by professional foresters and botanists. Hundeshagen pointed out instances of spruce replacing beech and other hardwoods in Switzerland and Germany, and of spruce and other species taking the place of birch, aspen, and Scots pine.

The first detailed North American report of composition changes was apparently that of Dawson in 1847, dealing primarily with the Maritime Provinces of eastern Canada. He recognized the effects of windthrow and fire in the forests found by the original European settlers, and distinguished between successional trends in small clearings, following cutting, following a single fire, following repeated fires, and as a result of agricultural use of the land.

As early as 1863, Henry David Thoreau recognized that pine stands on upland soils in central New England were succeeded after logging by even-aged hardwood stands which today constitute the principal forest type of the region. He named this trend **forest succession.** A few years later, Douglas, in articles published in 1875 and 1888, discussed at some length the concepts of forest succession and pioneer species, and presented an explanation of how it is that short-lived, light-seeded pioneer species formed the first forest types on burned-over pine land.

The concept of forest succession, then, dates back well into the beginnings of forestry and ecological science. It evolved slowly, but was well established by the beginning of the twentieth century, when Cowles, Clements, and other American ecologists systematized its study.

### Formal Ecological Theory

A general theory of plant succession, and indeed the foundations of plant ecology as a study of community dynamics, were initiated by Henry C. Cowles (1899) with an analysis of the succession on sand dunes of Lake Michigan, beginning with uncolonized sand and ending with a mature forest.

It remained for a contemporary, Frederic E. Clements, to fabricate an elaborate philosophical structure of plant succession (1916, 1949) which attempted to formalize all eventualities of plant community change. Specific examples of forest succession were early documented by William S. Cooper, with his studies of Isle Royale in Michigan (1913) and the colonization by plants following glacial retreat in Glacier Bay, Alaska (1923, et seq.). A detailed analysis of plant succession has been given by Daubenmire (1968a).

Clements, in particular, evolved an elaborate nomenclature to describe plant succession, a system which has both facilitated and greatly complicated the efforts of his successors. Some of his terms have taken a permanent place in the vocabulary of ecologists, others have persisted but with broadened and changed meanings, while still others have been finding less and less general usage. In the belief that good general English usage is preferable to a formal, precisely defined vocabulary understandable only to the initiated, only the most common and most widely understood ecological terms are introduced.

## THE STAGES OF SUCCESSION

The simplest approach to understanding plant succession is to postulate an unvegetated substrate and then to deduce the successive plant communities that will occupy this site under the assumptions that (1) the regional climate will remain unchanged, and (2) catastrophic disturbances such as windstorm, fire, or epidemic will not occur. In view of the hundreds of years involved in most forest successions, these assumptions are completely unrealistic. Their adoption, however, does provide for an understanding of the development of vegetation on many areas as an orderly, successional sequence depending upon the character of the original physical habitat and the climate.

The recognition of stages, too, is a matter of convenience rather than of their actual occurrence. Actually, the plant community on a given site is continuously changing as new species invade the site and existing species either reproduce or disappear through failure to reproduce. The community is a continuum in time as it is also in space. Nevertheless, the arbitrary classification of this continuum into stages characterized by the dominance or presence of certain species or certain life forms of plants is a convenience worth maintaining.

Initial unvegetated sites range from pure mineral material (rock, soil, or detritus) to water, with mixtures of soil and water (i.e., moist, well-drained mineral soil) the most favorable for plant colonization and growth. Thus a continuous range in site exists. Nevertheless, it is convenient to select points along this range at which to postulate vegetational development. Primary plant succession beginning with dry rock material (either as rock or as mineral soil) is termed a **xerarch succession**; that beginning with water is termed a **hydrarch succession**; while that beginning with moist but aerated soil materials is a **mesarch succession.**

The major stages of primary succession are generally consistent and worthy of detailed study. However, one should remember that many types of primary successions exist and that both the specific successions and the vegetational stages within each are arbitrarily chosen. In Table 15.1, series of 10 typical stages are given for each representative type of primary succession, following the general scheme of Graham (1955). Some stages may be omitted under conditions where the next successional life form (as tree, shrub, herb, liana, etc.) is capable of directly colonizing an earlier vegetational type. These series of stages are termed **seres** (Clementsian terminology) or more recently **chronosequences** (Major, 1951). Another useful delineation of stages in succession is that of Dansereau (1957), who recognizes four: (1) pioneer stage, (2) consolidation stage, (3) subclimax stage, and (4) climax stage.

In relating the possible stages to real world succession, several misconceptions should be dealt with immediately. First, succession does not necessarily begin with stage 1 and proceed through each successive stage in a unidirectional sequence. Second, stages do not typically proceed separately one after another in relay fashion; considerable overlap usually occurs. Third, there is no set time period for each stage to begin and end; one stage, depending on the site, may occupy the site for a long time and seemingly terminate vegetational development until site conditions change. As with taxonomic classes, a false sense of uniformity and rigidity is often conveyed by classes, despite their convenience.

**Table 15.1 Stages in Primary Succession**

| STAGE | XERARCH | MESARCH | HYDRARCH |
|---|---|---|---|
| 1 | Dry rock or soil | Moist rock or soil | Water |
| 2 | Crustose lichens | (usually omitted) | Submerged water plants |
| 3 | Foliose lichens and mosses | (usually omitted) | Floating or partly floating plants |
| 4 | Mosses and annuals | Mostly annuals | Emergents |
| 5 | Perennial forbs and grasses | Perennial forbs and grasses | Sedges, sphagnum and mat plants |
| 6 | Mixed herbaceous | Mixed herbaceous | Mixed herbaceous |
| 7 | Shrubs | Shrubs | Shrubs |
| 8 | Intolerant trees | Intolerant trees | Intolerant trees |
| 9 | Midtolerant trees | Midtolerant trees | Midtolerant trees |
| 10 | Tolerant trees | Tolerant trees | Tolerant trees |

The actual composition of the different stages will be dependent upon those species that have access to the site in question either by virtue of their proximity or by the capacity of their seed to reach the site by various avenues of dissemination. The actual plant communities on any given site, of course, depend upon the available plants as well as upon the site; they will change from place to place and even in the same place from time to time.

It will be noted that the stages detailed in Table 15.1 are not mutually exclusive. The various life forms and developmental stages may be characteristic of more than one stage, and, indeed, many persist through many stages. Some mosses, for instance, may invade a site early in the succession and persist through to the later vegetational stages characterized by tolerant trees. Tree seedlings often establish themselves at an early stage of vegetational development along with annuals and grasses. This is particularly apparent in secondary succession when tree seed sources surround areas burned by fire or cleared by agriculture (Chapter 16). Thus all plants of each stage do not necessarily appear and die out abruptly at an appointed time; rather they may overlap in various sequences according to the physiological traits of each species. For example, many clonal shrubs and alder and rhododendron thickets persist for 20 to 40 years or more, preventing the establishment of tree seedlings within them by chemical inhibition, shading, or limiting soil moisture (Niering and Goodwin, 1974; Damman, 1975).

In general, species diversity tends to be low in the early and very late stages of forest development. It typically peaks in midsuccession when early, mid, and late successional trees may all be present. The severity of the physical and biotic factors of a habitat and the diversity of habitat conditions control the species diversity at any point in this process. Species diversity has been considered in detail by Loucks (1970), Peet (1974), and Whittaker (1975).

It may be inferred from Table 15.1 that the stages of the different primary successions become more and more similar as the succession develops, inasmuch as the last five stages are characterized by the same general terms. While this is true, and while one school of thought holds that, given indeterminate time, all successions in the same general climate will eventually lead to a vegetational community of the same composition and structure, in actual vegetation patterns this does not occur. Although the latest successional stage will usually be composed of tolerant trees in a climatic region characterized by forests, the identity and relative abundance of the different species will vary with the different sites within the region.

Finally, it must be emphasized that complete, stage-by-stage primary successions rarely, if ever, occur in nature. Disturbances disrupt the gradual internal changes of the ecosystem and may set back, accelerate, or permanently change the course of vegetational development. On many sites a more or less predictable sequence of vegetational stages over time, a chronosequence, can be expected. Often disturbances only temporarily alter the chronosequence. However, disturbances such as erosion, deposition, and fire may permanently change the habitat thereby initiating a new successional sequence. For example, on well-drained sand soils, nutrients are retained primarily in living vegetation and in the accumulated organic matter on the soil surface. A severe fire that destroys the stand and the organic matter of the forest floor may actually change soil conditions so drastically that a new chronosequence is initiated (Damman, 1975). Even a vegetational stage itself can so permanently change the habitat that there is a concomitant change in succession. In Newfoundland, for example, Damman (1975) reported a relatively predictable succession following fire on most sites. However, once *Kalmia* heath became firmly established after fire, it initiated soil changes leading to thin, iron-pan formation, water logging, and peat bog formation. This prevented the return of forest vegetation and created a new chronosequence. Although vegetational development does not always follow the established or expected pattern, changes in site conditions may strongly indicate the course of the new pattern. These examples emphasize that careful attention should be given to all components of the ecosystem, not just the vegetation.

## PRIMARY SUCCESSION

Sequences of vegetational development are initiated by disturbances that expose substrates that are essentially devoid of plant growth at the beginning. Primary successions may begin with water or mineral soil under a wide variety of climates. Mineral soil may be exposed in many ways: through glacial retreat, volcanic ash deposition, avalanches and landslides, spoils bank formation following strip-mining, extremely hot forest fires, sand dune formation, emergence of coastal strands, and so on. Since the specific succession will vary not only with the type of site exposed, but also with the climate of the locality and the variety and abundance of plants accessible to the site for colonization, it is manifestly impossible to detail all the major types of primary forest succession. In the present section, therefore, attention will be focused on a few sample primary successions illustrative of the stages of vegetation development following different types of site exposure in various geographical regions and under differing climatic conditions.

### Bog Succession in Eastern Canada

The succession beginning in shallow freshwater lakes of the spruce-fir boreal forest of eastern Canada and adjacent sections of the northeastern United States has attracted much attention. The lakes are of relatively recent origin. They were mostly formed following the retreat of the last continental ice sheet from 6,000 or so to 10,000 years ago, so that succession is actively proceeding at the present time. The various stages in succession are obvious as concentric bands or zones of vegetation spanning the distance from open water in the middle of many lakes to mature forest at the borders growing on peat deposits that occupy what was once open water. Finally, the area has long been accessible to plant ecologists from the heavily settled areas immediately to the south. As a result, many ecological studies have been concerned with bog forest succession in boreal North America (Rigg, 1940, 1951; Dansereau and Segadas-Vianna, 1952). In northern lower Michigan (Gates, 1942), for example, the most common typical sere is from open water through aquatic associations to the mat-forming sedge, *Carex lasiocarpa*, followed by

*Chamaedaphne calyculata*, which invades the floating *Carex* sedge mat. Eventually, the *Chamaedaphne* is replaced by high bog shrubs (*Nemopanthus*, willow, alder, birch), and these eventually give way to swamp conifers such as tamarack, black spruce, and (under aerated seepage conditions) northern white cedar. In one of the bogs that Gates studied, changes over a 55-year period revealed that vegetation had advanced into the bog pool, the bog forest had become well established and then died, and the *Chamaedaphyne* stage was reestablishing itself (Schwintzer and Williams, 1974). A natural rise in the water table due to natural weather cycles probably caused the tree mortality.

We should not infer that all bogs in the boreal forest are formed by the filling up of water bodies. Under cool and wet climatic conditions, bogs are also formed by the swamping out of previously well-drained forests. Either natural succession to a dense spruce forest, or secondary succession to heath (*Calluna* spp.), may bring about a type of vegetation whose litter forms an acid raw-humus mat sufficiently unfavorable to litter-destroying organisms that it will accumulate, retain more and more moisture, and eventually be transformed into upland peat, supporting a *Sphagnum* ground cover. As the upland peat builds up under cool, wet climates, it eventually develops a characteristic bog flora, so that bogs actually develop, even on steep slopes. This process is important in Scandinavia, Finland (Huikari, 1956), western Scotland and Ireland, and the wet Pacific Coast of southeastern Alaska and British Columbia.

Heinselman (1970) describes a similar process of peatland evolution in northern Minnesota. No fixed or unidirectional succession was evident. Instead, a general swamping of the landscape, rise of water tables, deterioration of tree growth, and formation of diverse habitats have occurred. The plant communities are seen to change predictably—provided the full complexity of interacting factors is known.

### Mangrove Succession in the Tropics

Coastal swamps of mangrove characterize shallow salt water bodies throughout the tropics. In these, plant succession is evidenced by zones of vegetation extending from the open sea to the interior high forest, just as in the case of the boreal bog succession.

The term mangrove refers to tropical maritime trees and shrubs, especially of the genus *Rhizophora*, but also including other plants similar in appearance and in ecological preference for coastal mudlands. With extensive aerial root systems, mangroves are important soil builders. They become established in shallow water, where they obstruct currents, speed up the rate of deposition, and bind the soil with their roots and incorporated humus. A review of the ecology of mangroves is presented by Lugo and Snedaker (1974).

Mangrove swamps commonly show zonation of the dominant species more or less parallel to the shoreline, with each successive interior zone being characterized by less flooding and characteristic mangrove species (Richards, 1952). Davis (1940) considered these zones to represent successive stages in a hydrarch succession originating with salt water mud flats and ending with tropical high forest. In Florida for example, continually submerged soil is first invaded by the red mangrove *(Rhizophora mangle)*, the viviparous seedlings of which float in the sea and become established on shoals and sandbanks. With time, a mature *Rhizophora* forest develops to a height of 10 meters or more, resulting in substantial soil anchoring and accumulation. With better drainage conditions, *Avicennia* replaces *Rhizophora*, extending even to relatively dry sites. Further inland, in a zone seldom reached by tides, *Conocarpus* and other semimangrove species characterize the community.

In general, as the mangrove swamp extends seaward, the interior portions become denser and populated with a greater variety of species. Impedance of water movement from the sea by the mass of roots and accumulating debris, coupled with transpiration pumping of the water, permits fresh water to move seaward into the swamp, thus reducing the salinity of the water. With the gradual invasion of freshwater plants into the freshening site, a freshwater swamp forest eventually develops in which mangroves are replaced by a variety of tropical swamp species. Under climatic conditions where the water table can be lowered by the high rate of transpiration possible in the tropics, the site can even be invaded in time by high forest species. Clearcutting of the forest or other destruction to the forest, however, will eliminate the transpiration pump, raise the water level of the site, and bring about a return to freshwater swamp conditions.

It should be pointed out that measurable rates of succession along the maritime strand can be obtained only under conditions where the coastline is actually advancing into shallow seas and the coastal marshes are actually filling in. In many, if not most, situations, these changes are not occurring and the different zones of vegetation lying parallel to the shore represent past succession that has ceased, so that the different types are each more or less permanent until the physiography of the site is changed again.

## Following Glacial Retreat in Southeastern Alaska

The exposure of fresh deposits of moraines and outwash following retreat of glaciers provides one of the clearest and best studied examples of primary plant succession. In Glacier Bay, Alaska, studies of the development of pioneer plants on permanently marked plots established by Cooper in 1916 on surfaces left free by the ice on known dates as early as 1879 have made available much information about primary forest succession over a 75-year period in this particular locale (Lawrence, 1958). The stages run from pioneers through the establishment of alder thickets to a spruce-fir forest, followed by forest deterioration leading to muskeg and pit-pond development.

The pioneer plants, small and slow growing, invade as seeds or spores blown in by wind or carried in the digestive tracts of birds and mammals. Dryas forms a prostrate mat and is the most abundant pioneer plant. The second stage is marked by the invasion of Sitka alder to form a thicket. The alder, and probably the dryas as well, fix atmospheric nitrogen. Nitrogen accumulation in the soil makes possible the establishment and growth of more demanding tree species. In the next stage, the alder is mature—about 60 years after ice recession. First, black cottonwood and later Sitka spruce and western hemlock infiltrate the alder thicket. Maximum forest development is reached in the fourth stage with a dense spruce-hemlock forest. About two centuries are required for the development of this stage after ice recession.

With the invasion by sphagnum mosses of the ground cover of this spruce-hemlock forest, water retention is greatly increased. As aeration of the forest soil is impaired, the older trees gradually die, sphagnum mosses succeed one another in more and more luxuriant development, and the more level sites eventually are transformed into muskeg (a bog characterized by an abundance of sphagnum moss and tussocks). Forests maintain themselves, however, on the steeper slopes where lateral soil drainage is effective. Pacific Coast lodgepole pine alone is capable of surviving and growing in the developing muskegs. In the last stage of muskeg development on level and slightly sloping ground, pit-ponds develop, creating surfaces partially of water and partially of muskeg.

### Bare Rock Succession

The classic xerarch succession beginning with bare rock surfaces is frequently cited in ecological texts. An example is provided by Oosting and Anderson (1939) for granitic rock in the Piedmont of the southeastern United States, in which the more or less level rock surfaces are invaded by a mat-forming moss *(Grimmia laevigata)* upon which a lichen *(Cladonia leporina)* becomes established. As the mat thickens, herbs come in with the eventual dominance by *Andropogon* spp. of bunch grasses. Shrubs, such as sumac *(Rhus copallina)*, form the next successional stage, followed over the years by the development of an oak-hickory forest.

In detailing plant successions, we should remember that animals are equally involved in the succession of ecosystems. Not only do the species and relative abundance of the species change continually as the plant community changes, but the changes in the fauna play a part in causing changes in the flora as well as responding to such changes. An example of the interaction between animals and plants is provided by a detailed study of the contribution of rock ants to the afforestation of rocks in south Finland (Oinonen, 1956). This ant, *Lasius flavus*, reaches optimum development on rocks covered with lichens and mosses. The structure and location of its nests are favorable for the natural establishment of Scots pine and Norway spruce seedlings, whose roots can usually be retained by the nests for the first 5 to 10 years. The root aphids associated with the pines provide food for the ants, which in turn create new sites for pine seedling establishment.

### Following Severe Fire

Fire normally initiates secondary succession. When the fire is hot enough to destroy all higher plants, however, including their root stocks so that no sprouting occurs, all that is left is an ash-covered mineral soil, and the resulting invasion may accurately be described as primary succession. Succession following fire is discussed in Chapter 16.

### Dune Sands Along Lake Michigan

It was in the colonization of the dune sands along the southern margin of Lake Michigan that Cowles made his pioneering study of plant succession. The shifting sand dunes are first anchored by various dune grasses *(Ammophila, Calamouilfa, Andropogon)* and succession leads eventually to pine (jack and white) and black oak (Olson, 1958). The succession from barren dune sand to black oak forest requires about 1000 years after stabilization, during which time the litter of the developing vegetation continually improves the soil. The changes and improvements slow down with time, however, so that there seems to be little prospect for continued change toward a more mesophytic forest or a better soil after this time. Fire history on the drier sites plays an important role in holding back succession, and the development of the more mesic basswood–red oak–sugar maple communities is restricted to the moister and more fertile sites on lower lee slopes and in dune pockets that are protected from drying and burning. Here the mesic community usually develops without passing through stages like those leading to the black oak–blueberry communities.

### Mining Spoils

The waste of mining creates exposed mineral deposits that are suitable in many cases for colonization by vegetation. Abandoned rock quarries, slag from coal mines (pit heaps) in England, the dredgings of gold dredges in California and Alaska, and the spoils banks left by strip mining for coal from Pennsylvania to Illinois, and in Washington, Montana, and British Columbia—all are sites on which primary succession takes place (Schramm, 1966).

Frequently, the unweathered minerals in such deposits are so acid (as in the case of much coal overburden) or so basic (as in the case of limestone and chalk quarries) that they are unsuitable for plant growth until after years or even centuries of leaching and weathering. If barren 3 or 4 years after exposure under climatic conditions favorable for plant growth, these sites are apt to remain barren for at least several decades, although grasses, shrubs, and some trees tolerant of exposure and unweathered soil conditions may eventually become established. Eventually, however, vegetation on the spoils may develop into communities similar to those on nearly undisturbed lands. On century-old iron-ore spoils in northern West Virginia, Tryon and Markus (1953) found no indication of significant differences in forest tree, shrub, or herb composition between spoils and undisturbed soils in a forest now characterized by red oak, chestnut oaks, red maple, and other central hardwoods.

Considerable effort has been put into the task of afforesting the spoils banks left by strip mining for coal in the eastern United States, but the success of the plantations has been relatively limited. For all types of surface mining throughout the United States the reclamation success with vegetation has been poor; as of the 1960s 71 percent of the disturbed lands had inadequate cover or were incapable of supporting vegetation (U.S. Department of Interior, 1967). State reclamation laws now have become more stringent, and a federal Surface Mining Control and Reclamation Act was passed in 1977. Stockpiling and replacing top soil and returning land to the approximate original contour may now be required. In some eastern states, the newer laws specify planting of low-growing grasses and legumes rather than trees to achieve rapid plant cover. Thus there is less emphasis on tree planting than formerly, and those trees that are planted must compete with a herbaceous cover. An annotated bibliography (Czapowskyj, 1976) and two books on the ecology and reclamation of devastated lands are available (Hutnik and Davis, 1973; Schaller and Sutton, 1978).

### On Volcanic Ash

Fresh mineral exposures are created by vulcanism. Lava flows and volcanic ash both provide substrates for plant growth, although lava may take centuries and even thousands of years to weather sufficiently to support vegetation (cf. the Mesa basalt flow in Oregon and adjacent states). Ash deposits, however, are generally quickly colonized. For example, following the 1886 eruption of Mt. Tarawera in New Zealand, new communities scarcely different from the old have evolved in the subsequent 70-year period (Nicholls, 1959). More generally, succession following older and more destructive ash showers in the North Island of New Zealand follows a well-defined pattern (McKelvey, 1953; 1963). A scrub community—probably *Leptospernum* spp.—apparently pioneered on the skeletal pumice soils, providing a nurse for the podocarps, the seeds of which were bird-disseminated into the devastated zone. Later, hardwoods invaded the podocarp forests, first as an understory, and eventually forming the overstory, with the more tolerant hardwoods gradually replacing the less tolerant species.

Within the tropics, the revegetation of Krakatoa, a volcanic island in Indonesia where all the vegetation was destroyed by a spectacular eruption in 1883, has been summarized by Richards (1952); succession on the Soufrière of St. Vincent in the West Indies has been recorded by Beard (1945); and that on recent volcanoes in Papua New Guinea was discussed by Taylor (1957). The vegetation in all these tropical situations trends toward the original undisturbed forest, but the progress is slow. On Krakatoa, the terrain was a barren desert the year after the eruption of 1883 and was vegetated chiefly by ferns 3 years after, while the interior of the island was clothed by a dense growth of grasses 14 years after the eruption. Woodland zones were well developed after 23 years. By 1919, after 36 years, much of the savanna had been converted into woodland by trees spreading upwards and outwards from the ravines, and by 1931, after 48 years, a secondary forest had developed similar to that elsewhere in the Malayan region. Many species, however, migrate very slowly into devastated areas. In Papua New Guinea, even after 80 years, the number of species present on volcanic soils is only a very small proportion of those in nearby undisturbed forests (25 vs. 500).

## NATURAL SUCCESSION WITHIN THE FOREST

In managing forested lands, the stages of succession from the exposure of the site to the appearance of a closed forest are of relatively less importance than the stages of succession from one forest type to another. Forest succession, in the limited sense, begins with the establishment of the pioneer forest trees and proceeds with their replacement by successor species which profit by the changing environment.

In general, the pioneer trees are intolerant, with midtolerant species characterizing the second tree stage, and tolerant species the late-successional forest types (Spurr and Cline, 1942). The correlation between tolerance and successional appearance, however, is not perfect. Some relatively tolerant species have the capacity of invading forest sites relatively early in the succession; while other tolerant trees, either because of a relatively short life span (as many of the true firs) or because of an inability to reach the overstory and survive in overstory environmental conditions (for instance, dogwood) may never form a major part of late-successional forest canopy.

In the complex hardwood deciduous forest of the eastern United States, sugar maple, beech, and basswood continually invade midtolerant forests characterized by various oaks, ashes, and elms. The same relationships between genera occur in the deciduous forests of central Europe and far-eastern Asia.

Douglas-fir of the Pacific northwestern United States is a midtolerant. With time, pure Douglas-fir stands are gradually changed to communities dominated by the more tolerant western red cedar, western hemlock, or the true firs, depending upon the locality and site. A similar situation occurs in Tasmania, where the midtolerant eucalyptus forest is slowly replaced by mixed, tolerant, rain-forest species in the absence of disturbances.

Under certain conditions, however, intolerant species maintain themselves essentially without replacement by more tolerant species. Ponderosa pine in many parts of the Southwest, trembling aspen in parts of the Great Basin, and jack pine and black oak of the sand plains and dunes in the Lake States are examples. Thus we cannot conclude that tolerant species always characterize late-successional forests. Many of these and other examples are presented in greater detail in the next chapter that deals with the effect of disturbances upon forest succession.

## THE CONCEPT OF CLIMAX

The process of plant and forest succession is incontrovertible and has been recognized from the earliest days of natural history study. The question of what the last stage of succession is, if any, however, has been the subject of much discussion and debate, a controversy confused by semantic difficulties in that the participants have frequently used the same terms but with different shades of meaning.

Under specified site conditions—a series of stages of plant succession can be deduced which begins with the colonization of that site and proceeds to a stage typically characterized by tolerant and long-lived organisms that constitute a more or less balanced and relatively permanent community. In each case, the tacit assumption is that site is changed only through the internal interactions of the community and its environment, and not through any external factor such as fire, wind, human activity, or regional climatic change.

Under such an assumption—that the site is constant except for the changes brought about by the plant succession itself—the last successional stage is termed the **climax**. It is theoretically a final, mature, stable, self-maintaining, and self-reproducing state of vegetational development that culminates plant succession on any given site.

Clements, who more than any other developed the theory and nomenclature of plant succession, believed that, given indefinite time without disturbance to the community or site, the plant communities in a given climatic region would approach the same composition and structure. In his so-called **monoclimax** theory, climate was the dominant community-forming factor, while the other factors (soil, topographic relief, and biota—time being another dimension and not a factor in the same sense), although important, were in some fashion of secondary importance. With such a philosophical frame of reference, one can theorize that, if disturbances such as fire, extensive windthrow, cutting of trees, and externally caused climatic change can be eliminated for thousands and even millions of years, the swamps will fill up, the hills will be eroded away, and the whole landscape will be a peneplane clothed with a uniform plant and animal community. The rationale of Clements' monoclimax represents an application to vegetation of the peneplane theory of Davis.

However, such conditions seldom if ever exist in nature. Furthermore, there seems to be no fundamental reason why climate should be a more important community-controlling factor than soil, topographic relief, and biota. If not, then it is equally possible to theorize a different climax community for each soil type, each topographic position, and indeed for each assemblage of plants and animals that through historical accident find themselves living and growing together. This is the **polyclimax** theory, a theory that holds that for any combination of organisms and environment, biotic succession will take place toward a climax but that the specific nature of the climax will vary with the specific environmental and biotic conditions. The polyclimax theory has its roots in the contributions of Nichols

(1923), who argued for a different **physiographic climax** on each site, differing more or less from the regional **climatic climax**; and of Gleason (1926), who maintained that plant communities were not individuals in themselves but more or less chance aggregations of the individuals which happened to have access to a particular site.

For example, in Cowles' classic studies of plant succession on the sand dunes of lower Lake Michigan, it was at first thought that succession on all sand dune sites would eventually reach the same mixed mesophytic hardwood forest stage. Yet, in a later study of the same area, Olson (1958) concluded that the drier and more exposed sites would never support such a community but would be more or less permanently clothed with a black oak–blueberry type because of the dryness and low fertility of the soil itself. Furthermore, he felt that although vegetational changes would continue with time, they would become slower and slower and never reach complete stability:

> *Vegetation, soil and other properties of the ecosystem usually change rapidly at first and more slowly later on. If they approach some limit asymptotically or fluctuate around it, this limit should describe the climax community on mature soil. . . . The limit itself may vary with time and place. Ideally, it describes a gradational "climax pattern" of communities or ecosystems in any region—generally not a uniform "climatic climax."*

Over the years, the concept of climax has become broader and more all-inclusive in line with the polyclimax approach. Tansley (1949) was instrumental in broadening the concept of climax by adopting a "dynamic" viewpoint in recognizing that

> *natural and semi-natural vegetation is constantly changing, that certain uniformities in the direction, methods, and causes of change can be detected, and that positions of relative equilibrium are reached in which the conditions and composition of the vegetation remain approximately constant for a longer or shorter time.*

The climax as a "position of relative stability" is a succinct and acceptable formulation of the climax concept. Furthermore, Tansley recognized that "these 'positions of equilibrium' are seldom if ever really 'stable'" and that they contained many elements of instability—as we have seen in the foregoing examples. This is similar to the viewpoint of many ecologists of the climax as a steady state in which opposing forces of change and stability are evident but tend to balance one another. Dramatic changes may occur at specific points in the ecosystem, but over the entire area there may be no net change in species composition.

As a result of the broadening concept of climax, many of the terms of Clements have fallen into disuse, while others have been used more and more in broader and less formal senses.[1] However, the polyclimax theory is also beset with its own terminology of climaxes: climatic, edaphic, topographic, topoedaphic, fire, zootic, salt, etc. (Oosting, 1948; Daubenmire, 1968a). Many of these are the same as or are merely more specific distinctions of Clementsian climaxes. For example, in polyclimax terminology the monoclimax of Clements becomes the climatic climax; Clements' subclimax may be either an edaphic or topographic climax; a disclimax becomes a fire or a zootic climax, etc. Many of the polyclimax terms will undoubtedly also fall into disuse. The major contribution of polyclimax theory, nevertheless, is in recognizing that factors other than climate may be determining in controlling community dynamics.

We may visualize a geographic area not only as comprising a potential mosaic of different polyclimax communities but a continuum of climaxes (Whittaker, 1953, 1975). This so-called "climax pattern" is a logical extension of the polyclimax approach and emphasizes that climax communities may change gradually or abruptly as the site factors change along environmental gradients. Phillips (1931, 1934–35), Tansley (1935), Cain (1939), Whittaker (1953), Daubenmire (1968a), and Langford and Buell (1969) summarized differing viewpoints in the evolution of climax theory.

## THE INSTABILITY OF THE FOREST

The so-called climax community, then, is a relatively stable, steady state, culminating a sequence of vegetational development on a particular site. If site conditions are changed by allogenic forces, a different pattern of biotic succession and a different steady state will follow. As discussed below, this steady state is not final nor is it

[1]Among the kinds of climax recognized and named by Clements are: subclimax, essentially equivalent to the physiographic climax of Nichols, being a more or less permanent but "imperfect" stage of development in which the vegetation is held indefinitely either by natural or artificial factors other than climate—such as grazing, burning, or cutting; disclimax, a replacement of the "true" climax, chiefly as a consequence of disturbance by man or domesticated animals; postclimax, the next more mesophytic climax to the local climax (i.e., the climax of a wetter and cooler climatic zone); and preclimax, the next more xerophytic climax to the local climax (i.e., the climax of a warmer and drier zone). All of these types are grouped as proclimaxes.

stable in the sense of being unchanging or constant in composition and structure. Secondary successions are ever present (see following chapter for discussion). Changes in both climatic and soil conditions vary from place to place within the forest, and from time to time, depending upon the evolution of the forest community itself as well as upon regional or worldwide climatic change. Species such as the American chestnut and American elm are eliminated from the forest community by disease while others such as Scots pine, the gypsy moth, and the chestnut blight fungus in northeastern United States are introduced and become part of the forest ecosystem.

More and more, it has become apparent that the forest is never stable, but remains a dynamic community in the later successional stages just as it was in the earlier stages of plant succession. Just as in a forest composed of pioneer intolerant tree species, a forest composed of tolerant tree species is constantly changing in composition and structure, and in associated fauna and flora as well. The rate of change may be less, it may approach an asymptote or fluctuate around it, but change itself is still characteristic of the community.

Climax, rather than referring to an ultimate and finite end stage of plant successions as it did in earlier ecological thought, now becomes a term somewhat loosely applied to a more or less stable and long-lived community that develops late in the course of vegetational development on a specific site. It is a steady state or position of relative stability. In this modern sense, however, the climax is no longer considered final since changes are taking place within the biotic community all the time even if at a slower and less obvious pace.

The geographic scale of reference must also be kept in mind in considering the climax condition of an ecosystem. A small gap in a mature mesic forest of 400 $m^2$ area, caused by windthrow of a large tree, may be in an early successional stage undergoing rapid secondary succession. In the surrounding ecosystem of several hectares, there may be other such gaps or patches, whereas the remainder of the forest ecosystem is in a very slowly changing late successional status in composition and structure. Although a small sample plot in the opening might be classed as early successional, the forest as a whole may be considered relatively stable. In the latter case the gaps and the seemingly "stable" surrounding community are considered together—forming the relatively stable steady state in the ecosystem as a whole. Thus it is important to specify geographic scale when considering the relative stability of an ecosystem.

Many examples may be cited pointing to the instability of all forest communities, even of the so-called climax types. The concept of a stable and enduring climax, developed in temperate vegetational zones, not only does not hold for these zones, but has even less reality in the arctic regions or in the tropics.

As Hewetson (1956) describes the tropical forest of India after many years of its study:

> *I would describe the Tropical Forest as a continuum in which the parts are in unstable equilibrium. All the species can survive but some species are more closely adapted to the sum total of environmental factors and in average climatic conditions are more likely to be successful and to form the greater part of the growing stock. The average may, however, be deflected by exceptional events such as tornados or droughts or land clearance or fires. The effect of these exceptional events may be seen 200 or 300 years later, and the growing stock we see before us today can only be understood in the light of conditions in the past. The interplay of the individual species extends from the trees in the top canopy down to the herb layer. The density and the composition of the lower strata may control what trees will succeed the present overwood as much as the potentiality for reproduction of the dominant species. It is quite possible the equilibrium may be maintained for a period in one forest and the same trees succeed their ancestors. Other forests may be in a condition of complete change with the present growing stock being replaced by different species. Between these two extremes many variations are possible.*

Raup (1957) has given examples of the basic instability of the site. A consideration of these results leads to the conclusion that repeated major disturbances by factors largely external to the vegetation should not be considered as unusual but as a part of the normal itself. As a result, actual succession should be considered as consisting of fragments of the theoretical complete succession, with the climax becoming a purely theoretical speculation. The grasslands of North America, as well as the forests, have had much the same history of oft-repeated disturbances (Malin, 1956).

The continual changes in the forest, which result in its common instability, are of many types. These include (Yaroshenko, 1946): (1) seasonal change, (2) annual changes due to year-to-year climatic variation, (3) short-term succession changes as discussed in this and the following chapter, and (4) long-term "historical" changes as discussed in Part IV. All result in a constantly changing forest ecosystem.

## Boreal North American Forest

In the preceding portion of the chapter the primary succession following glacial retreat in southeastern Alaska is detailed, and it is indicated that a mature forest composed of Sitka spruce and western hemlock is eventually followed on many level and less sloping sites by muskeg formation leading even to the swamping out of much of the muskeg in pit-pond formation. The successional trends are clear, but, if one is attempting to apply the traditional concept of climax to the succession, the question may be asked as to whether the conifer forest or the muskeg is the climax (Zach, 1950). In the Clementsian sense, the forest community qualifies as being mesophytic and composed of tolerant and long-lived species. Yet the fact remains that, given time, sphagnum invasion of the conifer forests on the more level sites will bring about the accumulation of upland peat, the deterioration of the conifers, and the development of a muskeg-pond type which could therefore be considered as the climax. This is a case of succession under one set of environmental conditions leading to conifer forest. Site changes (increasing moisture) then lead to another vegetational sequence proceeding in dynamic association with site conditions.

Similar situations occur elsewhere in the boreal forest of North America, although the species involved may differ. In the interior of Alaska, the effect of the development of a spruce forest canopy, according to Benninghoff (1952), is to decrease the amount of insolation reaching the ground, with the consequence that the ground becomes frozen throughout the year closer to the surface. Frozen ground acts as an impermeable stratum in the soil to perch the water table, with the result that the surface will sooner or later swamp out, leading again to muskeg and open-pond development.

The boreal vegetation, by and large, does not reach a stable and long-lived climax, but rather remains in a state of instability as a result of the complex interactions between vegetation, soil-water relationships, frost action, and permanently frozen ground. There is a basic instability in the environment and the vegetation (Churchill and Hanson, 1958). The concept of climax has little meaning in its original Clementsian sense.

## European Spruce–Fir–Beech

Within the Temperate Zone, the classic example of the instability of even the climax forest is provided by Norway spruce–silver fir–beech forests of central Europe, particularly in Switzerland and adjacent mountain areas in France, Germany, and Czechoslovakia. These species are all tolerant, long-lived, and capable of forming a many-aged, many-storied mixture which can be managed by single-tree selection methods of silviculture. By all conventional standards of the climax, this community qualified. Yet, on any given spot within this "climax" forest, the composition as well as the structure is unstable, with spruce replacing fir, fir replacing spruce, and similar changes taking place involving beech. This phenomenon of "alternation of species" has long been studied by European silviculturists (Nagel, 1950; Simak, 1951; Schaeffer and Moreau, 1958).

### Lake States Tolerant Hardwoods

The climax forest of the northern Lake States is composed of tolerant, long-lived species capable of forming a long-enduring, mixed, uneven-aged community. Prominent among the components are sugar maple, basswood, beech (in the eastern portion of the region), and hemlock (Graham, 1941). In general, however, the forest never reaches an equilibrium but, because of the continual action of local windthrow, fire, drought, insect attack, and fungus infestation, consists of a mosaic of patches each of which is constantly changing in composition and structure (Stearns, 1949). Midtolerant species such as yellow birch and white pine are able to maintain themselves almost indefinitely in competition with the more tolerant species through exploitation of small gaps in the forest canopy. It is clear that the forests are instable, undergoing continual change.

## SUGGESTED READINGS


Beard, J. S. 1944. Climax vegetation in tropical America. *Ecology* 25:127–158.

Cain, Stanley A. 1939. The climax and its complexities. *Amer. Midl. Natl.* 21:146–181.

Clements, Frederic E. 1949. *Dynamics of Vegetation: Selections from the Writings of Frederic E. Clements, Ph.D.* The H. W. Wilson Co., New York. 296 pp.

Cooper, W. S. 1913. The climax forest of Isle Royale, Lake Superior, and its development. *Bot. Gaz.* 55:1–44, 115–140, 189–235.

Daubenmire, Rexford. 1968. *Plant Communities.* (Plant succession, pp. 99–246.) Harper & Row, Inc., New York. 300 pp.

Drury, W. H., and I. C. T. Nisbet. 1973. Succession. *J. Arnold Arbor.* 54:331–368.

Egler, Frank E. 1977. *The Nature of Vegetation.* (pp. 138–176). Privately printed by Frank Egler. Aton Forest, Norfolk, Conn.

Hewetson, C. E. 1956. A discussion on the "climax" concept in relation to the tropical rain and deciduous forest. *Emp. For. Rev.* 35:274–291.

Horn, Henry S. 1976. Succession. *In* Robert M. May (ed.), *Theoretical Ecology*. W. B. Saunders Co., Philadelphia.
Lawrence, Donald B. 1958. Glaciers and vegetation in southeastern Alaska. *Amer. Sci.* 46:81–122.
Major, Jack. 1951. A functional, factorial approach to plant ecology. *Ecology* 32:392–412.
Olson, Jerry S. 1958. Rates of succession and soil changes on southern Lake Michigan sand dunes. *Bot. Gaz.* 119:125–170.
Tansley, A. G. 1935. The use and abuse of vegetational concepts and terms. *Ecology* 16:284–307.
Whittaker, R. H. 1953. A consideration of climax theory: the climax as a population and pattern. *Ecol. Monogr.* 23:41–78.

# 16
# *Disturbance Effects*

Although primary plant successions are the most obvious manifestations of succession and consequently have attracted the most study by plant ecologists, it is with secondary successions that the silviculturist and forest ecologist are mostly concerned. Once established, the forest is seldom completely destroyed, and the areas of new soil or site being formed within a forested region are negligible compared to the areas of existing forest. Primary succession in forested regions is the exception, therefore, rather than the rule. Throughout any forest region, disturbances of one sort or another are constantly altering the course of forest succession, and initiating secondary succession.

Disturbances to the forest can be grouped into three classes; first, disturbances altering the structure of the forest; second, disturbances altering the species composition of the forest; and third, disturbances altering the long-term climate in which the forest grows. The first class includes fire, windthrow, logging, and land-clearing activities. The second involves the introduction of new plants or animals into the forest ecosystem or the elimination of plants or animals from that system. The third is concerned with climatic changes over a period of years as well as climatic extremes which affect the relative vigor and competitive ability of the species making up the forest.

## FOREST DESTRUCTION

The most obvious disturbances to the existing forest are those which partially or completely destroy the forest structure by killing and overthrowing either the trees in the overstory or the trees and other plants in the understory. The prominent factors causing forest destruction and initiating secondary forest succession are: fire, windthrow, logging, and land clearing.

### Fire

***Fire is the dominant fact of forest history.*** The great majority of the forests of the world—excepting only the perpetually wet rain forest, such as that of southeastern Alaska, the coast of northwestern

Europe, and the wettest belts of the tropics—have been burned over at more or less frequent intervals for many thousands of years. Even under present-day conditions, marked by a great awareness of forest fires, the separation of forest tracts by intervening tracts of farmland and settlement, and the crossing of the forest by many roads and trails, fire continues to be a major disturbing factor in much of the North American forest. However, from about 1900 to 1940 organized fire protection activities were mounted and have become increasingly effective in reducing the size of fires.

The condition was quite different up to the present century. Primitive peoples throughout the world, and most civilized peoples as well, had until recently no compunction about burning the forest, and no desire or intent to put out existing fires, whether lit by man or lightning. In fact, throughout the world, fires have been set deliberately for thousands of years to clear the underbrush, improve grazing, drive game, combat insects, without thought, or just for the hell of it. As more and more historical research is carried on into the ecological history of fire, the more it is realized that frequent burning has been the rule for the vast majority of the forests of the world as far back as we have any evidence (Chapter 11).

Within forest regions, fire has been primarily responsible for heathlands and moors of western Europe and the British Isles, for many of the savannas within the tropical forest belts, for upland meadows within the forests of the American mountains, and in general for the persistence of grassland areas on upland sites within forest regions. Around the world, the dominance of pine and oak forests of virtually all species and in virtually all regions is due predominantly to fire. So is the vast acreage of Douglas-fir in the Rocky Mountains and Pacific Northwest and of eucalyptus in Australia. Even the vast areas of spruce in the boreal forest of North America and Eurasia are structured to a great extent by past fires (Bloomberg, 1950; Sirén, 1955; Viereck, 1973).

The foregoing statements are sweeping and perhaps overstated. Nonetheless, they reflect the feeling of many silviculturists and forest ecologists who, wherever they have studied and worked, have come increasingly to realize the great importance of forest fires and secondary succession following forest fires in framing the local forest in its composition and structure (Cooper, 1961).

From the viewpoint of the established and growing forest, fire is clearly a disturbance that disrupts and drastically alters the development of the existing stand. However, from the viewpoint of the fire-dependent community and species, fire is a natural factor whose effects have long been incorporated in species' adaptations and ecosystem dynamics. It is not therefore a disturbance when viewed over the long term, cyclic history of a fire-dependent ecosystem. In such systems, **fire exclusion** is actually the disturbance about which we are becoming increasingly concerned. The importance of fire in the ecosystem is highlighted by examining the similar roles it plays in many different fire-dependent communities.

### Role of Fire in the Forest Ecosystem

Around the world fire has played important and similar roles in fire-dependent conifer and hardwood forests in presettlement times. Many ecosystems are regularly recycled by fire; life for many forest species literally begins and ends with fire. The following major functions and processes are regulated by fire:

- **Regeneration and Reproduction.** Catastrophic fires kill existing stands and set the process of regeneration in motion for the next forest. Fire, as a selective force, also elicits the following reproductive adaptations of forest trees.

  Asexual reproduction, primarily sprouting, occurs in all fire-dependent hardwoods and in some conifers (Chapter 11).

  Sexual reproduction by light, wind-blown seeds is favored by fire, and self-fertility in pines (red pine in particular) may have evolved as a result of fire. Although a small percent of self-fertilized seeds are viable in many pines, red pine is highly self-fertile. The resulting seeds are viable, and the seedlings show no growth depression due to selfing, unlike most other tree species (Fowler, 1965a,b). Thus a single isolated red pine that survives a severe fire may self-fertilize and perpetuate itself by establishing a colony of seedlings. A similar situation may occur in Monterey pine introduced into New Zealand (Bannister, 1965).
- **Seedbed Preparation and Dry-Matter Accumulation.** Fire reduces the amount of litter, sometimes bares mineral soil, and greatly enhances seedling establishment. Seeds are often partially buried in the ash layer, thus favoring germination. Fire regulates dry-matter accumulation, thus influencing the severity of burning.
- **Competition Reduction.** Severe fires eliminate much tree, shrub, and herbaceous competition, as well as woody debris, thereby favoring establishment of a new stand. Light surface fires thereafter reduce encroaching vegetation and reduce competition for soil moisture and nutrients. This is typical in many mixed conifer-hardwood stands in the South where understory hardwood competition is reduced by periodic burning. Similarly, in oak forests, periodic surface fires kill seedlings of shade-

tolerant species that continuously establish on the forest floor and grow into the understory.

- **Nutrition.** Throughout the life of the stand recurrent surface fires reduce litter and woody debris to basic components of water, $CO_2$, and minerals. Minerals that are otherwise unavailable to the forest vegetation become available (Chapter 11). This is particularly important in northern conifer forests and wherever site conditions are unfavorable for decomposition. Although nitrogen is lost in burning, surface soil conditions often favor nitrogen-fixing bacteria and soil organisms.
- **Thinning.** In many conifer communities, dense pure stands of seedlings, that become established after fire, are thinned by periodic surface fires. The larger, faster growing seedlings are favored, and competition for moisture and nutrients is reduced.
- **Sanitation.** Fire creates the dense, even-aged stand conditions that are conducive to disease and insect outbreaks. Such epidemics generate fuel concentrations leading to intense and widespread fires. Fire terminates the outbreaks (such as those of bark beetles and the spruce budworm), destroying living as well as dead trees. It creates conditions for the establishment of a new even-aged stand, which is resistant for a time to disease and insect attack. Sooner or later the maturing stand is again susceptible to epidemics, and thus a self-perpetuating cycle is established. Fires also eliminate plant parasites such as mistletoes on ponderosa pine, lodgepole pine, and black spruce. In addition, smoke from fires has an inhibitory effect on the germination of several rusts and fungi (Parmeter and Uhrenholdt, 1976).
- **Succession.** Depending on site conditions, fire tends to retain fire-dependent species on an area as long as fire frequency and intensity are balanced with the species' fire adaptations. Ponderosa pine, southern pines, jack pine, oaks, and many other species may be recycled generation after generation although they are not the climax species of their regions. However, if the fire frequency or intensity increases, a given species is likely to be replaced by a more fire-dependent species. For example, in the Rocky Mountains, Engelmann spruce is replaced by Douglas-fir, Douglas-fir is replaced by ponderosa pine, and ponderosa pine is replaced by grassland. On the other hand, if fire frequency decreases (as with fire exclusion), a less fire-dependent type would prevail in each elevational zone: Engelmann spruce would replace Douglas-fir, Douglas-fir would replace ponderosa pine, and ponderosa pine would replace grassland. Thus a mosaic of successional fire-dependent communities of different age classes

was typical of many North American upland presettlement forests because of the varying frequency and intensity of fire and its erratic burning pattern—all superimposed over a variety of soil-site conditions.

Very shade-tolerant species of both conifers and hardwoods tend to be susceptible to fire (sugar maple, red maple, beech, hemlocks, firs, northern white and western red cedars) and typically occupy protected or moist sites that are less susceptible to burning. In intervals between significant fires, they colonize adjacent areas, establishing and eventually replacing fire-dependant species in the absence of fires intense enough to kill them.

- **Wildlife.** Fire universally created a mosaic of habitats and niches for wildlife of all kinds (Chapter 13). Species diversity, as with plants, tends to increase following fire until crown closure occurs and then declines.

  Out of the many examples that may be cited to demonstrate these roles and the impact of fire, a selection representing the most important North American trees and regions, together with one from Australia, will illustrate the great importance of the secondary successions following fires.

**Pine in New England and the Lake States.** In the northeastern United States and southeastern Canada, the occurrence of the two- and three-needled pines (red pine, jack pine, and pitch pine) as well as of even-aged pure stands of white pine (but not of individual white pine in mixed forests) is largely controlled by the past occurrence of forest fires. In colonial days, many of the fires resulted from burning operations by white settlers in land-clearing operations, but the evidence is ample that fires were commonly set by Indians for many hundreds of years before the coming of the white man (Cline and Spurr, 1942; Day, 1953; Curtis, 1959; Little, 1974). Because of the heavy precipitation commonly associated with summer thunderstorms, lightning apparently has played only a minor role in causing fires in this region.

Ecological studies of a few relict old-growth pine stands have all shown that fire played an important part in their formation. In both northwestern Pennsylvania (Lutz, 1930b) and southwestern New Hampshire (Cline and Spurr, 1942), more or less pure even-aged stands of old-growth white pine have been shown to have originated from past forest fires, while nearby mixed types with occasional dominant white pine were relatively free from evidence of past burns. In northwestern Minnesota, the extensive even-aged stands of old-growth red pine clearly date from a series of forest fires, many of which antedate the advent of the white settler (Spurr, 1954; Frissell, 1973).

Jack pine in the Lake States and Canada, and pitch pine on sand soils near the mid-Atlantic coast (Little, 1974), are virtually completely fire-controlled. Jack pine, for instance, grows in nearly pure stands on dry sandy soils, forming a highly flammable vegetational type. At intervals of a few decades, the jack pine stands are burned during hot dry periods, with the fire characteristically crowning and killing all the vegetation above the surface of the earth. The next generation of trees arises from four sources: (1) in the case of jack pine, from seeds stored in many years' accumulation of serotinous cones in the tree crown (Roe, 1963)—the cones being held closed by resin deposits which are melted above 45°C by the heat of the crown fire; (2) in the case of red pine, by seeds from residual veteran seed trees with bark of sufficient thickness and clear bole of sufficient length to permit them to survive the fire without crowning out; (3) in the case of the hardwoods such as trembling and bigtooth aspen, red oak, and red maple, by sprouts arising either from roots or portions of the stem unkilled by the fire; and (4) in the case of pioneer hardwoods, by the dissemination of seeds into the area from afar either by wind, as for aspens and birches, or by birds, as for the cherries; (5) in the case of pin cherry and black cherry from seeds stored in the seed bank of the forest floor. The composition of the postfire stand will depend upon the relative supply of seedlings or sprouts by each of the above avenues.

Once a jack pine type has become established, however, it will persist over approximately the same tract of land as long as a fire occurs every few decades. If fire is excluded, however, jack pine will begin to deteriorate and die out after 50 to 60 years, leaving the stand to associated longer-lived red pine; to white pines, which come in over a period of years and gradually infiltrate the overstory; and on mesic sites to tolerant species such as sugar maple, balsam fir, and spruces, which come in slowly as understory plants and eventually form a late-successional stage.

**Western Pines and Aspen.** The remarks made concerning the northeastern United States pines can be applied with little modification to virtually all other pine species growing in the United States. In the interior of the western forest, lodgepole pine plays an analogous part to that taken by its close relative, jack pine, in the Lake States. It seeds in recent burns, largely from seeds stored in serotinous cones of trees killed by the fire, to form dense, even-aged, postfire pioneer stands (Figure 16.1). Given several hundreds of years free from forest fire, the lodgepole pine gradually will be replaced by tolerant Engelmann spruce and alpine fir, with white and black spruce also playing a part as late-successional species in Alberta, and Douglas-fir and other more tolerant western conifers becoming

**Figure 16.1** Lodgepole pine regeneration following fire in Oregon. (U.S. Forest Service photos.) *(a)* Lodgepole pine revegetates a large burn in central Oregon. Reseeding the area to grass after the burn delayed the establishment of pine. *(b)* Under ideal conditions of seedbed and seed supply, dense lodgepole pine thickets develop following fire. This 65-year-old stand of about 25,000 stems per hectare in eastern Oregon averages only 8 m high and 5 cm in diameter at breast height.

prominent toward the Pacific Northwest. In subalpine and high foothills in Alberta, Horton (1956) estimates that from 225 to 375 years' exclusion from fire is required for succession to take place from pine to spruce and fir. On the drier southern slopes, the succession takes much longer, if indeed it ever takes place.

At lower elevations and in warmer, drier portions of the western forest, pure stands of ponderosa pine are commonly a product of a long and complex fire history (Figure 16.2; Cooper, 1960; see also Chapter 20). In the cooler and moister portions of its range, pine comes in as a pioneer following fire and is gradually replaced by more tolerant conifers such as Douglas-fir, incense cedar, and white fir. In the warmer and drier portions of the ponderosa pine range, it may be a late-successional, fire climax species (see following discussions and Chapter 17).

Trembling aspen competes as a pioneer on the higher, cooler, and wetter sites, coming in as wind-disseminated seed in the northern Rocky Mountains and in western Canada. In the Great Basin and the central and southern Rocky Mountains, aspen root suckers predominate, and aspen is maintained as a late-successional type by fire in certain localities. In contrast to the eastern and northern parts of its range, aspen clones of the central and southern Rockies may become very large, up to 43 ha in extent and contain thousands of genetically identical stems (Kemperman and Barnes, 1976; Chapter 3, Figure 3.14). The presence of many large clones and a markedly different leaf morphology from the small clones on glaciated areas of northern and eastern North America suggest that individual clones may have been perpetuated by fire for many thousands of years (Barnes, 1975). However, many of these central and southern Rocky Mountain clones are now deteriorating, and, due to fire exclusion, they are not being regenerated (Schier, 1975).

**Southern Pines.** Nowhere is the dependence of the pine forest upon recurring fires more evident than in the southern pine belt of the southeastern United States. Here, earlier travelers wrote of the open character of the "piney woods" due to Indian burning. Ever since settlement, local farmers have periodically burned the woods to keep down the "rough" and bring about fresh postfire revegetation suitable for grazing by domestic stock. After many years of futile attempts at complete fire exclusion by foresters, the practice of "prescribed" or "controlled" burning has become well accepted in recent decades to reduce the hazard of a crown fire and to maintain the pine type without reversion to the hardwoods which otherwise would replace it through natural succession.

Of the four most common southern pines, longleaf pine is most clearly dependent upon recurring fires for its perpetuation (Chapman, 1932). Occurring primarily on seasonally dry sites, longleaf pine stands must be burned in the grass stage to control the brown

Figure 16.2. Fire and ponderosa pine. (U.S. Forest Service photos.) *(a)* A surface fire burns grass and litter of this stand in central Idaho. *(b)* Fires in presettlement times maintained open, grassy, parklike stands of ponderosa pine such as this one in western Montana.

spot disease (*Schirria acicola*) and repeatedly after a forest canopy is formed to prevent the establishment of understory hardwoods which invade the site and will gradually replace the pine in the absence of fire. Furthermore, the longleaf pine regenerates itself best in full light on mineral seedbeds created by forest fires which destroy all but scattered longleaf seed trees, the species being one of the most fire-resistant of all forest trees (Figure 16.3). Many of the seedlings in Figure 16.3 appear like dense bunches of grass, and this stage is known as the "grass stage." Longleaf pine seedlings, unlike those of all other North American pines, remain for many years (typically 6 years, but as many as 12) in the grass stage while a deep tap root develops. In the grass stage, the bud is protected from fire by the dense needles. If the needles are destroyed by fire they are replaced by a new set. Eventually rapid stem elongation begins, quickly elevating the crown above the level of periodically occurring surface fires.

The other common southern pines—loblolly pine, slash pine, and shortleaf pine—as well as sand pine, Virginia pine, pitch pine, and other species of more or less local occurrence—all are pioneer species that become established after destructive forest fires and give

**Figure 16.3. An excellent stand of natural longleaf pine seedlings in the Coastal Plain of South Carolina following a prescribed burn to prepare the seedbed. (U.S. Forest Service photo.)**

way to tolerant hardwood mixtures in the long-continued absence of forest fires (Little and Moore, 1949; Wahlenberg, 1949; Campbell, 1955). The clearly established dependence of these species upon fire has given rise to extensive research and practical use of fire as a silvicultural tool or of alternative chemical and mechanical treatments that likewise will hold the natural succession in the pioneer pine stage.

**Douglas-Fir in the Pacific Northwest.** In the Douglas-fir region of northern California, western Oregon and Washington, and southern British Columbia, the characteristic summer drought results in highly flammable conditions during the hottest period of the year with the result that extensive forest fires have not only been characteristic of forests of the present century but also of presettlement times. On the west side of the Cascades and over much of the Coastal Range, Douglas-fir is the pioneer species on burns, provided that adjacent Douglas-fir stands survive undamaged to furnish a source of seed. The commercially valuable pure Douglas-fir type, therefore, represents the first stage of a postfire secondary succession. More

**Figure 16.4. The dominant but overmature Douglas-fir (left) in this western Washington stand in the absence of fire will eventually be replaced by more tolerant firs and western hemlocks of the understory. (U.S. Forest Service photo.)**

**Figure 16.5. 1890–1970: Eighty years of fire exlcusion in the giant sequoia—mixed conifer forests of Yosemite National Park (Confederate Group, Mariposa Grove).**

**a) 1890: A parklike stand as the result of periodic fires. (Kilgore, 1972; Historical documentation by Mary and Bill Hood. Photo by George Reichel. Courtesy of Mrs. Dorothy Whitener.)**

tolerant conifers such as western hemlock, Sitka spruce, and western red cedar invade the Douglas-fir type to form an understory and eventually achieve dominance after 500 or more years with the decadence and death of the dominant Douglas-fir (Figure 16.4).

**Giant Sequoia.** The native giant sequoia—mixed conifer forests of the Sierra Nevada of California are fire-dependent (Kilgore, 1973). Fire plays a major role in sequoia forests by preparing a seedbed of soft, friable soil on which the seeds of sequoia fall and are lightly buried thus favoring germination (Hartesveldt and Harvey, 1967). Moisture is a limiting factor in seedling establishment (Rundel, 1972), and partially burned litter may contain more available mois-

**Figure 16.5b:** **1970: The parklike stand was invaded by thickets of white fir. By 1970 firs obscure all but the fire-scarred sequoia on the left. Such thickets provide ladder fuels that could support a crown fire fatal even to mature sequoias. (National Park Service photo by Dan Taylor.)**

ture than unburned litter (Stark, 1968). Fire also kills fungi and eliminates competition, thereby favoring sequoia establishment.

Periodic presettlement surface fires kept the forests open and parklike (Figure 16.5a; Biswell, 1961). This favored pioneer species and eliminated the small tolerant trees, particularly white fir, which continually invaded the understory. Upon settlement Indian burning was gradually eliminated, and fire suppression became increasingly efficient. Surface fuels built up, and white fir and other species grew up and often formed thickets under the sequoias (Figure 16.5b). A ladderlike vertical sequence of fuel was thus formed from ground level to low-hanging fir branches to the top of understory crowns, 10 to 30 m above the surface (Kilgore and Sando, 1975). Therefore, fire starting in surface fuels is likely to pass through understory crowns and torch out in sequoia crowns. Because sequoia, unlike redwood, does not sprout, once the crown is killed the tree will die. Prescribed burning at 5- to 8-year intervals is therefore required to reduce surface fuels, kill small understory trees and the lower crowns of larger understory trees, and prevent continued encroachment of a tolerant understory.

**Eastern United States Hardwood Forest.** Although the hardwood forests are less combustible than conifer forests, fire-controlled secondary successions seem to have been important factors in these as well. In the mixed hardwood forest complex of the eastern United States, many oak types on dry sites owe their existence to a past history of forest fires, frequently set by Indians for driving game (Cottam, 1949; Curtis, 1959; Brown, 1960). Throughout much of the vast oak region, total exclusion of fire usually results in the establishment of a tolerant hardwood understory characterized by beech, sugar maple, and basswood, with other species being important in addition, particularly in the southern oak types. It seems clear that the oak, under present climatic conditions, is a pioneer type following fire and maintained by recurrent fire and disturbances of diverse sorts; while beech-maple represents the late-successional stage reached after long exclusion of such disturbances.

**Eucalyptus in Australia and Tasmania.** The tall-timber *Eucalyptus* species of the mountains of southeastern Australia occupy an ecological position very similar to that of Douglas-fir in the Pacific Northwest of the United States. They form tall, dense stands of high quality following fire under climatic conditions where more tolerant species would succeed in the forest succession if disturbance could be avoided (Mount, 1969). The eucalyptus succession is exemplified in a study by Gilbert (1959) of forest succession in the *Eucalyptus regnans* forest of western Tasmania. This species, the giant mountain ash (Victoria) or swamp gum (Tasmania) that reaches heights well in excess of 100 meters, forms dense stands following the burning of older eucalyptus forest—which is highly flammable during summer hot spells because of the flaky bark and the leaf oils characteristic of the genus. Where fire has been excluded for many years because of swamps, cliffs, or other natural barriers to its spread, however, an understory develops in which *Nothofagus* (antarctic beech), *Atherosperma*, and the *Dicksonia* tree fern are common components. Under rare circumstances where fire has been excluded for a century or two, the relatively short-lived eucalyptus tends to disappear from the forest and a late-successional, temperate Southern Hemisphere rain forest evolves through natural succession of the understory trees to dominance. Although the acreage of this rain forest in southeastern Australia is small because of the commonness and frequency of fires, it is clear that the predominant eucalyptus forest is a pioneer community resulting from and perpetuated by frequent forest fires which prevent succession to the rain forest type.

## Fire History and Behavior

Detailed studies of fire history in fire-dependent ecosystems increase our awareness of fire as a dominant force over virtually all of the northern and western landscape. Three examples illustrate fire history, behavior, and effects.

**Northern Lake States.** Heinselman's (1973) classic study in the Boundary Waters Canoe Area in northern Minnesota documents fire occurrence and effects over nearly 400 years. Fire has been a major force in determining the composition and structure of the presettlement vegetation for nearly 10,000 years. By the time logging reached the area, about 1895, recurrent fires had kept nearly three-fourths of the region in recent burns and commercially immature timber. Thus little of the 215,000 ha tract, now reserved as a unit of the Wilderness Preservation System, was subjected to the timber cutting of the time—"high grading" the best pines. Effective fire control began about 1911.

A fire chronology was developed based on the age of existing stands, tree-ring counts from sections or cores from fire-scarred trees, and historical records. Nearly all of the forest burned one to several times in the period 1595–1972. Much of the burned area is accounted for by just a few major fires. In the settlement period, fires of 1875 and 1894 accounted for 80 percent of the total area burned; and in the presettlement period just 5 brief fire periods accounted for 84 percent of the total.

Heinselman found that the major fire periods coincided with prolonged summer droughts of subcontinental extent. Climate combined with fuel accumulations (associated with dry matter accumulations, spruce budworm outbreaks, blowdowns, etc.) to bring optimum burning conditions at rather long intervals. The fire of 1863–1864, caused in part by the major droughts of those years, burned 44 percent of the area (Figure 16.6). Fire in 1864 also burned over 400,000 ha in Wisconsin, and Clements (1910) working at Estes Park, Colorado, concluded that the burns of that year were the most extensive in his region.

Heinselman estimates the natural fire rotation, the average number of years required to burn and reproduce new forest generations over the entire area, to be about 100 years. The natural rotation for different communities would vary; aspen-birch forests probably

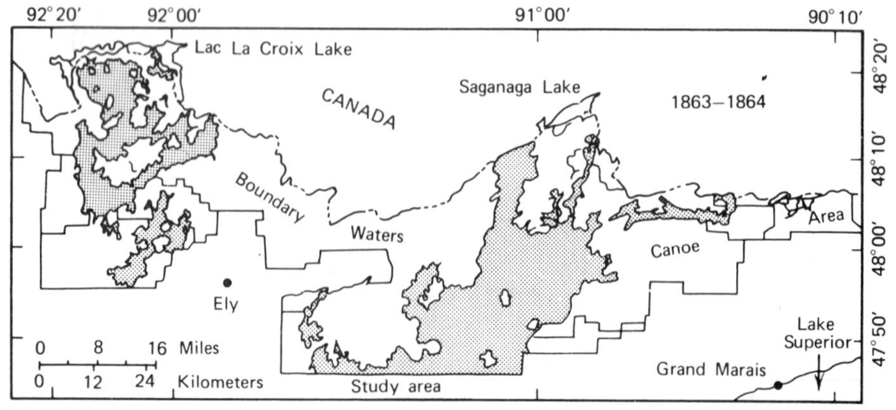


**Figure 16.6. Severe fires of 1863 and 1864 burned (shaded area) over 1800 km² (700 mi²) of the Boundary Waters Canoe Area, northern Minnesota. (After Heinselman, 1973.)**

burned at intervals of 50 years or less, and many red and white pine stands lived for 150 to 350 years before being regenerated. A typical sequence for red and white pine communities is occasional light surface fires that reduce the organic layer and understory competition, followed by a severe fire that kills the overstory and provides openings for regeneration. However, if a second major fire comes only a few years after a previous burn, many conifers could be eliminated, being too young to withstand the heat and needle scorch and too young to bear cones. Regeneration then comes from hardwood sprouters—trembling aspen, paper birch, red maple, and red oak.

The physiographic areas that burned most frequently were large upland ridges; long west-east ridges often burned entirely, acting as fire paths for eastward spread of fires. Jack pine, black spruce, trembling aspen, paper birch, and other hardwoods that sprout typically dominated these areas. Areas least frequently burned—swamps, valleys, ravines, islands, and the east, north, northeast, and southeast sides of lakes—support white pine, red pine, white spruce, northern white cedar, black ash, American elm, and balsam fir more abundantly than on the ridges.

Fires over the entire area produce a mosaic of stands of different composition and ages depending on the complex of site and fire factors. Thus fire generates diversity in stand and species composition. The entire biota are adapted to this stand and site mosaic. Moose, beaver, black bear, snowshoe hare, woodland caribou, small mammals and birds are all adapted and dependent on the mosaic of different habitat conditions created by fire. Fire exclusion is now restructuring the entire system, gradually eliminating the niches of many wildlife species as the forest composition changes. It will require monitored lightning fire and prescribed fires to restore the natural vegetation mosaic and wildlife habitats to this human-dominated wilderness.

**Boreal Forest and Taiga.** In the boreal forest and taiga of North America, as in the northern forest, a vegetation mosaic leading to plant and animal diversity is primarily the result of wildfires burning over diverse sites (Lutz, 1956; Slaughter et al., 1971; Rowe and Scotter, 1973; Viereck, 1973). Seven of the 10 major tree species are pioneers and have adaptations for rapid invasion of burned areas (Chapter 11). Only balsam fir and alpine fir are less well adapted for regenerating immediately after fire; their cones disintegrate at maturity and seeds are not retained in the crown. The understory and ground layers are characterized by clonal species whose underground stems or roots are undamaged by fire and regenerate rapidly. Fire maintains a patchy pattern of vegetation that assists in maintaining diverse wildlife populations.

North of the nearly continuous and fast-growing forest of the more southerly boreal forest is the taiga, a widespread area of open, slow-growing spruces interspersed with occasional well-developed forest stands and treeless bogs (Viereck, 1973). The pattern of the vegetation is closely related to fire history and permafrost. Permafrost is permanently frozen ground in which water is often incorporated as lenses of pure ice. Thick layers of moss in forest stands act as insulators during summer months, limiting the thaw of soils to depths of one meter or less. Permafrost is widespread and in many areas permafrost and vegetation are in delicate balance. Fire, by burning organic matter and warming the surface, can increase the annual depth of thaw of permafrost soils. This can significantly affect the presence, as well as the composition of the vegetation. Where ice lenses occur, increasing the depth of thaw following fire may cause subsidence of the areas over the ice, creating a polygonal mound-and-ditch pattern. This may occur in paper birch stands at least 40 to 50 years after fire. With succession to black spruce these sites may become stabilized, or small ponds may develop and alternate in a cycle with black spruce.

Fire control methods in permafrost areas, mainly the construction of firelines by large-tracked vehicles may have a greater effect on soil erosion and siltation than the fire itself and may cause more damage to the aquatic ecosystem than does the fire (Lotspeich and Mueller, 1971).

**Northern Rocky Mountains.** Fire has burned in every ecosystem and virtually every square meter of the coniferous forests and summer-dry mountainous forests of northern Idaho, western Montana, eastern Washington, and adjacent portions of Canada (Wellner, 1970; Habeck and Mutch, 1973; Arno, 1976). Prior to 1940 and the era of fire protection, fire was second only to precipitation as the major factor shaping the character of forests. Catastrophic fires, killing complete stands or nearly so, were common due to the extremely dry summer climate and the build up of fuels in the densely forested areas (Wellner, 1970). Forest insects, and at times diseases, contributed tremendous quantities of hazardous fuels by killing trees in dense stands over extensive areas. The series of 1910 fires alone burned between 1 and 2 million hectares.

The region is characterized by multiple burns, which usually remain without tree cover for long periods (Larsen, 1925). Fire breeds fire, and many single burns created such hazardous fuels that they reburned. Parts of the 1910 burn reburned in either 1917 or 1919, or both, and even again in subsequent years. Snags, (standing dead trees) were particularly important as a source of lightning ignition. Wind-blown burning material from tall snags could spot new fires far in advance of the burning front. Single burns in uncut forests reforested rapidly. Fire prevention since about 1940 has largely prevented multiple burns and increased the area of forest cover.

Fire was responsible for the widespread occurrence and even the existence of western larch, lodgepole pine, and western white pine. Fire maintains ponderosa pine throughout its range at the lower elevations and kills the ever-invading Douglas-fir and grand fir. Higher up, Douglas-fir is favored by fire over its more tolerant competitors, and higher still the distribution of Engelmann spruce was increased at the expense of alpine fir. In summary, fire has generated nearly continuous secondary successions, recycled forest vegetation on a regional scale, and maintained intolerant or midtolerant successional species as fire climaxes in their respective habitats.

**Fire Exclusion.** The foregoing sections illustrate the widespread occurrence of past fires and their intimate link with forest ecosystems. Seen in a time scale of centuries, fire was an integral part of forest evolution and existence rather than a disturbance. However, increasingly efficient fire suppression efforts of this century can, if continued indefinitely, provide a disturbance that will greatly change fire-dependent ecosystems in ways rarely or never before experienced.

Before settlement by Europeans fires were common, set by lightning and Indians. During settlement they were still a common sight, and unless they threatened human life, livestock, or buildings, they were little regarded except as a local nuisance. The rise in public opinion against fire, fueled by promiscuous burning and the large destructive fires of the late 1800s, led to the development of rigid fire exclusion policies. Fire was not considered a natural phenomenon but a disastrous threat to the forest and to human life and habitation. In the eyes of public opinion, fire killed trees, promoted erosion and floods, destroyed the habitat of animals, and often burned nests and killed the animals themselves.

Today, because of increasingly efficient fire control, due largely to sophisticated communications and rapid attack helicopters, it is increasingly possible to extinguish fires in an early stage of development. Although complete suppression is not possible, a significant reduction in the area burned is being achieved. Exclusion of fire leads to a marked buildup of fuel, decline in animal and plant species, and an increase in incidence of wind damage and insect and disease attack as the proportion of mature timber increases. These are all features experienced in presettlement forests in long intervals between major fires. However, prolonged buildup of fuel may eventually lead to fires more catastrophic and destructive to the site than typically occurred in the native forest. In addition, the mosaic of successional forest types is altered in that late-successional species increase in abundance, leading to a decline in wildlife habitat and in both plant and animal diversity (Chapter 13).

Reaction to fire exclusion came gradually. Research, especially in the South and West, led to the conclusion that fire was a natural factor—that prescribed burning could be beneficial to plants and animals. Therefore, in many parts of North America, attention is now turned to letting selected natural fires burn and to conducting prescribed burning. Fire control programs, formerly confined to fire prevention and suppression, are evolving into **fire management** programs.

Based on the work of pioneer fire scientists, such as Roland M. Harper, H. H. Chapman, Herbert L. Stoddard, Harold H. Biswell, and Harold Weaver, the U.S. Park Service and the U.S. Forest Service have instituted fire management programs with the goal that fires play a more natural role in forest ecosystems. Personnel of many national parks and monuments now let selected natural fires burn and conduct prescribed burning to maintain the natural diversity of plants and animals that evolved as a result of past fires. Fire management and prescribed burning have been considered from many standpoints and in relation to many different forest ecosystems (Slaughter et al., 1971; Wright and Heinselman, 1973; Kayll, 1974; Kilgore, 1975, 1976a, b, c; Mutch, 1976; Martin et al., 1977).

## Windthrow

Scattered windthrow of large, overmature trees is a prime cause of mortality in the old-growth forest and creates gaps in the main canopy into which advance growth from the understory or newly germinated seedlings may develop. Such gaps are frequently enlarged year after year by the attrition of surrounding trees under the weight of ice and snow and the pressure of high winds.

Occasional severe storms, particularly hurricanes and tornados, may destroy the timber on hundreds or thousands of hectares initiating secondary forest succession on a large scale. It is with such extensive windthrow that the present section is concerned.

The evidence of windthrow is preserved for many years in the forest by such signs as these: (1) hollows marking the holes made by uprooted trees; (2) overturned soil strata in a mound created by the upturned roots at the edge of the hollow; (This pit and mound microrelief provides habitats for the establishment of different species, most of which occurs on the mounds (Lyford and MacLean, 1966; Stone, 1975)); (3) ridges extending away from these mounds marking the remains of the fallen bole; (4) the frequent appearance of a row of younger trees along this ridge, particularly of species such as hemlock (both eastern and western) that characteristically seed in on rotten wood; and (5) evidence of release from suppression in the growth rings of the oldest adjacent trees, particularly of trees in the understory which were released by the windthrow at the time of the storm. The sharp release characteristic of hemlock in the eastern part of its range can easily be discerned in increment cores and used to date the time of windthrow. The dating of past windthrow by these techniques is a fascinating detective game and yields firm evidence of the prevalence of windthrow in the American forest for the past few centuries.

The great 1938 hurricane in New England created an awareness of the past occurrences of severe windstorms in this region at intervals ranging from a few to more than a hundred years. Other storms in the same region, including major hurricanes in 1635 and 1815, apparently inflicted similar damage to the forest. The study of natural regeneration following the 1938 hurricane provides an example of the effects of windthrow in initiating secondary forest succession (Spurr, 1956a).

In contrast to surface fires which may destroy the understory to a greater extent than the main forest canopy, windthrow damage is chiefly confined to the overstory. In forests where a tolerant understory is developing under a pioneer stage of succession, the effect of windthrow may merely be to end the pioneer stage and release the understory to form a new canopy. In southwestern New Hampshire, in an old-growth white pine stand blown down by the 1938 hurricane, the understory hemlock, beech, and red maple was simply released to form a low stand—but one of substantial age considering the great number of years these trees had existed in the understory prior to release. With the released hemlock and beech in the new stand, however, were mixed paper birch, yellow birch, and black cherry, all of which had seeded in on disturbed mineral soil upturned by the tearing out of the roots of the windthrown trees. In many cases, therefore, the secondary succession following windthrow will consist of a mixture of tolerant species from the released understory together with pioneers that have invaded the site.

## Logging

Logging is similar to windthrow in that the cutting of commercially valuable trees tends to remove the overstory and release the understory. Many understory hardwoods cut in brushing operations connected with logging will resprout to form vigorous and fast-growing stems competing for overstory space in the developing new stand.

As a general rule, the intensity and pattern of the cut will affect the competitive ability of the new crop. Light partial cuts will favor tolerant species, particularly those already established in the understory, and thus will tend to push forest succession forward rather than initiate an earlier stage in the process (Figure 16.7). Moderately heavy partial cuts will favor midtolerant species; as will, the cutting of small groups of trees to create gaps in the forest of approximately the same size that would be made by the death of an overmature dominant tree. Clearcutting, or the cutting of large gaps with a diameter of at least twice the height of the stand, will favor the invasion of pioneer species, particularly if mineral soil is exposed by the logging operation. Thus the silviculturist, by regulating the intensity and pattern of logging, can greatly influence subsequent forest composition and the rate of succession.

The effect of partial- versus clearcutting on forest succession is well illustrated by experience in the Douglas-fir region of the west side of the Cascades in Washington and Oregon. Much of the cutting in old-growth stands occurs in more or less even-aged stands composed predominantly of Douglas-fir and dating from past forest fires 150 to 500 years ago. Clearcutting in such stands at the time of, or immediately after, a seed year results in the reestablishment of even-aged Douglas-fir. Patches of Douglas-fir ranging from 1 to 15 ha in size will normally come back to Douglas-fir within a few years, with seed blowing in from surrounding uncut timber. On the lower and wetter sites in the coast range, red alder comes in prolifically as a pioneer species on clearcut sites. On any site, the new forest is composed of pioneer species which range from intolerant to midtolerant in their ability to compete in the understory in that region.

In contrast, partial cuttings in similar old growth Douglas-fir

**Figure 16.7. Partial cutting in this eastern white pine stand in Maine has favored advance regeneration of tolerant red spruce. (U.S. Forest Service photo.)**

forests result in a quite different forest structure and composition (Isaac, 1956). Following cutting of from 20 to 50 percent of the gross volume in a variety of Douglas-fir types, mortality by windfall in the overstory has proven high in the first 5 years after felling, and often in the second 5-year period. Felling and skidding also caused considerable damage to the residual stand. As the cutting was concentrated in the Douglas-fir, and since the residual Douglas-fir tended to be taller and slimmer than associated conifers of other species, the net result was to decrease the amount of Douglas-fir in the residual overstory and to increase the percentage of other species. In the openings created by the logging, few Douglas-fir seedlings appeared, but seedlings of the more tolerant species (western hemlock, grand fir, Pacific silver fir, and western red cedar) were well established in many stands. The end product of partial cutting in this type, then, is a deteriorating partial overstory with a vigorous understory composed of conifers more tolerant and less valuable commercially than Douglas-fir.

Another effect of logging on forest succession results from the differential removal of one species and the leaving of another—thus changing the composition of the forest. The effect of logging a favored species on forest composition is exemplified in the mixed-wood forests of Maine where logging has been more or less continuous for more than 100 years. Early logging was concentrated primarily on large white pines suitable for masts for wooden sailing ships, house construction, and floating down rivers from forest to the mill. The next stage of logging followed the building of sulfite and ground wood pulp mills and the large-scale cutting of red spruce to supply them. Since the red spruce grew in association with tolerant balsam fir, sugar maple, and beech, the logging of the spruce created a mixed wood of these species. However, the proportion of spruce was drastically reduced. Balsam fir then became more and more utilized as the supply of red spruce decreased. As clearcutting followed partial cutting, swamping out of many of the wetter flats occurred, and pioneer species such as aspen and paper birch invaded the drier sites. In the mixed-wood forests, removal of spruce and fir resulted in turning the site over to residual tolerant hardwoods. These have formed extensive tracts of culled old growth, which have been devastated in time by dieback of the yellow birch and sugar maple due to a combination of insects and disease attack, climatic conditions, and the low vigor of the residual old-growth trees. Thus successive waves of logging, each concentrating upon different species, have greatly modified the composition of the forests of central and northern Maine.

### Land Clearing

Forested regions are moist regions and therefore eminently suitable for the raising of agricultural crops. Since the great majority of the world's population lives in forest regions, it is inevitable that much of the world's forest has been cut and the land cleared for agriculture. In the tropics, the use of forest land for agriculture is often transitory, with worn-out fields being allowed to revert to forest again after a few years of cropping. Even in the Temperate Zone, much land used for farming in the past has been found unsuitable for continued cropping, or has been supplanted by bringing better lands into production, with the result that it has been planted to forest or allowed to revert naturally to forest. In recent years, the trend of farm land abandonment has been intensified by the concentration of crop production on the best farm lands through the use of improved strains of plants, and better fertilization and cropping techniques.

Secondary forest succession following land abandonment, therefore, is an important process, taking place over many hundreds of thousands of hectares in well-settled forest areas. The old-field succession in the eastern United States has been studied in particular detail, while the old-field succession in the tropical rain forest is of great long-term importance in the economy of tropical countries.

## Old-Field Succession in the Eastern United States

Agricultural use of land was at its most extensive development in the New England states about 1815 to 1830, and began to decline with the opening of the West and its ready access via the Erie Canal (1815) and the trans-Appalachian railroads. In the Atlantic-facing Piedmont of the southeastern states, agricultural use of lands reached its peak at the beginning of the Civil War. Throughout the entire eastern seaboard, most upland sites were cleared and were farmed until the 1815–1860 period, when the industrialization of the Northeast, coupled with the opening of the farmlands of the Midwest, initiated a long decline in agriculture acreage, a decline that is still in existence.

Secondary succession of forest on the abandoned upland fields and pastures has thus involved great areas, and has been instrumental in reforesting much of the eastern landscape. Since Thoreau's essay on the topic in 1860, many ecological studies have described the major stages of forest succession on these sites.

In most cases, fields were abandoned as grass-bearing hayfields or pastures that had ceased to yield sufficient hay to justify annual mowing. In such cases, conifers form the initial old-field tree invaders. The old-field conifer in northern New England is red spruce, with white pine coming in on these sites in central New England and New York State, red cedar in southern New England and the mid-Atlantic states, Virginia pine in the upper South, and loblolly pine and shortleaf pine throughout most of the rest of the South. Only when fields have been abandoned as fallow cultivated croplands do hardwoods—such as gray birch in the Northeast and sweet gum in the Southeast—predominate in the pioneer stage of forest succession.

The greater ability of conifers to invade and establish themselves on old grasslands seems to be due to several factors: (1) the large numbers of relatively heavy wind-disseminated seed, which can work down through the sod to make contact with the soil; (2) the presence of enough stored food in the seed to develop a seedling sufficiently large to compete with the grass; (3) the drought resistance of conifer seedlings, which permits them to survive summer droughts occasioned by root competition with grass; (4) the probable presence of inhibiting substances associated with the grasses which retard the growth of invading hardwoods; and (5) the ability of pines and their ectomycorrhizae to obtain nitrogen from the soil.

In general, old-field succession is characterized by increasing plant diversity, reaching a maximum in the forest stage when mixtures of early and late successional species are present. Thereafter, diversity declines as the shorter-lived early successional species are replaced by more tolerant, late successional species.

The conifer forest is not usually pure but is mixed with varying proportions of hardwood pioneers which become established mostly on small bare patches, brush patches, or other non-grassy patches in the old field. In the North, gray and paper birch, pin and black cherry, and bigtooth and trembling aspen are the principal hardwood pioneers although some white ash, red oak, and other midtolerants will come in with the first wave of tree invaders. In the South, sweet gum, red maple, and many other hardwoods come in under these conditions.

Once the conifer forest is well established, it is itself invaded with hardwoods which form a more or less abundant understory by the time the overstory is 20 to 40 years of age. In central New England, white ash, red oak, sugar maple, red maple, and black birch are the most common component. In the Piedmont of North Carolina, sweet gum, black gum, dogwood, and sourwood are commonly present. The overstory conifers seldom reproduce themselves under their own canopy except in the most open stands on the driest sites. As the years pass, some of the understory plants persist, others die back only to sprout again from the persisting root system. Others die back completely and are replaced in time with seedlings of more tolerant species such as hemlock, sugar maple, beech, and basswood.

The even-aged, old-field conifer stands are shorter-lived than mixed stands in which the same species is dominant because of over-growing which results in tall, slender trees and in subsequent wind friction that whips the crowns back and forth past one another, abrading and reducing the crowns. The conifers tend to become overmature on these sites and in these stand structures before a century is up (in 50 to 60 years for red cedar to 70 to 80 years for loblolly pine and a little older for white pine). By the end of the second century, in the absence of any further disturbance, replacement by a tolerant community, consisting mostly of hardwoods, is complete. The late-successional stage is usually a beech–sugar maple–basswood type in northern New England, a hemlock–red oak–red maple type in central New England and New York, and an oak-hickory complex to the south. In the deep South, beech, magnolias, and a great variety of tolerant hardwoods constitute the late-successional community (Delcourt and Delcourt, 1977).

The old-field white pine succession in central New England has produced much white pine of great commercial importance. It has been detailed in many studies at the Harvard Forest (Spurr, 1956b) and is pictorially depicted in the Harvard Forest models, which present in three-dimensional dioramas the history of land use in central Massachusetts (Figures 16.8–16.13).

The old-field red cedar succession has been studied by Lutz (1928) in Connecticut and Bard (1952) in New Jersey. The commercially important loblolly pine old-field succession has been the subject of many studies (Billings, 1938; McQuilken, 1940; Oosting, 1942; Barrett and Downs, 1943; Bormann, 1953), particularly in the Piedmont of North Carolina.

## Abandoned Farmland Succession in the Tropical Rain Forest

Throughout the Tropics, the pattern of agriculture in the rain forest zones is one of shifting cultivation. Many names and many different tree species are involved, but the general story is much the same (Richards, 1952, 1973; Bartlett, 1956). The native people fell the trees (except for an occasional large tree here and there), burn the brush, and plant their crops. One or more crops (such as hill rice, cassava, maize, yams, or bananas) may be grown before leaching of the humus and compaction of the upper soil horizon of the oxisols reduce soil fertility and structure to the point where further cultivation is useless. Fertilizers other than wood ash from the burned forest are not used. In the higher rainfall areas, it may take only 6 months to leach a cleared forest soil to an unproductive state.

With abandonment, a short-lived herb stage of the secondary succession is usually quickly followed by a dense stand of fast-growing, intolerant trees which form an even-aged stand composed of very few species as compared to the primary tropical rain forest. Balsa and various cecropias are characteristic of this pioneer forest community in tropical America, and various *Musanga* spp. commonly occur in tropical Africa. The trees in this stage are fast-growing, frequently have low-density wood, and commonly occur with many lianas, forming a tangled forest which is much lower, simpler in structure, and evener in crown canopy than the original forest. The "low bush" or "fallow" may be replaced rapidly by many species of a transitional stage, followed by slow invasion of shade tolerant species that eventually form the primary rain forest. The last successional phase involves a long period of time, but the time scale is but little known.

The secondary succession following land clearing in the tropics is very similar, in the principal stages and the ecological characteristics of the trees composing each, to the forest succession following land clearing in the eastern United States. The chief difference is that this succession in the Temperate Zone is characterized by a

Figure 16.8. The Harvard Forest models. A reconstruction of the mixed precolonial forest, with hemlock, tolerant hardwoods, and occasional white pine. (Model 1, courtesy of Harvard Forest, Harvard University.)

Figure 16.9. The same view in central Massachusetts in 1740 shortly after settlement. (Model 2, courtesy of Harvard Forest, Harvard University.)

Figure 16.10. The same view at height of farming development, 1830. (Model 3, courtesy of Harvard Forest, Harvard University.)

Figure 16.11. Farm abandonment and the seeding in of old-field white pine, 1850. (Model 4, courtesy of Harvard Forest, Harvard University.)

Figure 16.12. Harvesting the old-field pine in 1909, which seeded in after farm abandonment in the same view. (Model 5, courtesy of Harvard Forest, Harvard University.)

Figure 16.13. The young stand matures into the second-growth hardwoods characteristic of central New England of 1930. (Model 7, courtesy of Harvard Forest, Harvard University.)

coniferous pioneer stage, while that in the tropical rain forest is of necessity composed entirely of angiosperms.

## COMPOSITION CHANGES

Although secondary successions are considered to originate primarily from disturbances such as the factors of fire, windthrow, logging, and land clearing, the addition or subtraction of a species, whether plant or animal, will inevitably change the succession of the forest. Since these changes in the flora or fauna of the forest may result in considerable disturbance to the existing community, they may well be considered as initiating secondary successions in a very real sense.

### Elimination of Species

When a species is eliminated or greatly reduced in abundance, its place in the ecosystem must be taken by other species. An outstanding example of this is the virtual elimination of the chestnut in the eastern hardwood forest by the blight caused by *Endothia parasitica*. This disease, introduced from Asia about 1904, killed most of the mature chestnut in New England within 20 years, and had completed its work in the southern end of the commercial range of the species in the southern Appalachians by the 1940s. Seldom if ever before in historical times has a major forest tree been so nearly eradicated. Although the chestnut still sprouts profusely from root systems after a half century, few trees survive long enough to reach the main canopy of a mature forest stand.

Succession following the elimination of the chestnut has often resulted in the simple replacement of that species by its former associates (Korstian and Stickel, 1927; Keever, 1953; Woods, 1953; Nelson, 1955; Woods and Shanks, 1959; Day and Monk, 1974). Chief among the succeeding trees in the southern Appalachians are the oaks (especially chestnut oak and red oak), along with various hickories and, on the better sites, yellow-poplar. Thus the eradication of the chestnut in this region has resulted in the replacement of the former oak-chestnut type with an oak-hickory or oak type. Similar changes, but involving some different species, previously occurred in the middle Atlantic states and southern New England (Good, 1968).

In the Allegheny Mountains of western Pennsylvania, logging and fire following the death of chestnut created open sites for invasion by early and midsuccessional species, black cherry, black birch, black oak, sassafras, and sour gum together with species present on the site such as sugar maple, red maple, white ash, and beech (Mackey and Sivec, 1973). Without fire or other disturbances in the former chestnut forest, succession gradually proceeds to more tolerant and mesophytic species such as hemlock, sugar maple, and beech.

More recently the Dutch elm disease, caused by the fungus *Ceratocystis ulmi* (introduced into Ohio from Europe about 1930), has in about 40 years, together with another disease, phloem necrosis, virtually eliminated mature American elm trees from mesic and lowland forests of the eastern United States. Studies of succession following the elimination of overstory elms in midwest forests indicate that elms are replaced by a number of different species depending on regional site conditions. In Illinois woodlands, sugar maple is the species most likely to increase in dominance except where soils are too poorly drained (Boggess, 1964; Boggess and Bailey, 1964; Boggess and Geis, 1966). In addition to sugar maple, hackberry and ashes, particularly white ash, also show gains, and their seedlings are abundant in areas of heavy elm mortality. In southeastern Iowa, seedling and sapling regeneration under dead elms and adjacent living canopies were studied in four different forest types (McBride, 1973). Hackberry was most frequent in two of the types and box elder most frequent in the other two types. Young American elms were present in all types but relatively frequent in only one.

In southeastern Michigan, American elm was formerly a climax dominant in deciduous swamp forests together with red maple, yellow birch, and black ash. Investigations in three swamps, 5 to 10 years following death of the elms, revealed that saplings and seedlings of these three overstory dominants were abundant together with smaller amounts of blue beech, American elm, basswood, and several other species (Barnes, 1976). American elm has not been eliminated from deciduous swamps but makes up about 10 to 15 percent of the understory. Old fields and other open upland areas are much more important habitats for regeneration of elm than swamp communities. Indications are that unlike the American chestnut, American elm will be perpetuated for generations by seeds from young elm trees; however, the average life span will be drastically reduced.

In the spruce-fir forest of eastern Canada and adjacent northeastern United States, periodic epidemics of the endemic spruce budworm have resulted in the killing of overmature and mature balsam fir over large areas together with lesser amounts of killing of the black and white spruces. The budworm apparently has played a major role in the mixed softwood forests in holding down the proportion of the very tolerant balsam fir as compared to that of the somewhat less tolerant spruces.

In the animal portion of the forest ecosystem, the virtual elimination of many predators—particularly the wolf, bear, cougar, and lynx—from most of the American forest has played a role in increasing the number of deer, rabbits, and other herbivores. This in turn results in greater browsing of the understory, including tree regeneration.

### Addition of Species

The species occupying a given site are not necessarily those best adapted to compete and grow on that site, but merely the best of those that have access to that site at the time of its availability. Invasion of the site by better competitors often results in substantial changes in forest succession. It makes no difference whether this invasion occurs as a result of natural immigration or through introduction by humans. The result is ecologically the same.

The chestnut blight fungus already cited in the previous section is an example of an accidental introduction that has greatly modified the forest. Other organisms may be trees, other higher plants, fungi, bacteria, and animals of all levels.

Of forest trees, many have become naturalized in new geographical areas in historical times to become a part of existing plant communities so as to modify successional trends. Beech has moved northward through England in the last 2000 years in response to postglacial revegetational trends. The European silver poplar was introduced widely into North America and in Michigan has hybridized with bigtooth and trembling aspen, forming vigorous clones (Spies, 1978). Coconut and *Casuarina* spp. are representative of many littoral species that have been widely disseminated throughout the Tropics, partly through human and partly through other forms of transport.

Important deliberate forest-tree introductions include American Douglas-fir, white pine, red oak, and black locust in Europe; Scots pine, Norway spruce, and European larch in the northeastern United States; and Monterey pine, patula pine, and slash pine in temperate zones of the Southern Hemisphere. These and many others have become vigorous and dominant species in the local flora and must be considered as natural parts of the present and future plant communities and succession.

**Introductions to New Zealand.** A spectacular example of the rapid successional changes in both the flora and fauna of a region that can occur when better-adapted plants and animals are introduced is provided by the natural history of New Zealand. In this south temperate land, only Southern Hemisphere conifers (primarily podocarps) and hardwoods (primarily *Nothofagus* spp.) had access to the land prior to the coming of the Maoris a thousand years or so ago and of the white men in the past century or two. Birds—many of them without functional wings—constituted virtually all the higher animal life. The climate, however, is temperate, moist, and mild, ideal for many Northern Hemisphere plants and animals. Following extensive introduction by humans in the last hundred years, the introduced flora and fauna are rapidly and effectively replacing their native counterparts (Figures 16.14, 16.15). Monterey pine, lodgepole pine, and Douglas-fir among the trees, gorse, blue grass, and ragwort among the smaller plants, and the European red deer, sheep, rainbow trout, Australian possum, and Himalayan thar among animals are but a few examples of vigorous organisms better suited to the site than those isolated there by the accidents of geological history. The forest ecosystems and the successional trends within them will inevitably be more and more influenced by these new and vigorous plants and animals.

**Increased Animal Use.** It is not necessary that a species be either added or eliminated in order greatly to affect forest succession. Change in the abundance of a species may have a considerable effect in itself. This is illustrated by the great changes brought about by the increase in number of herbivores in the forest through reduction of predators, burning of the woods to increase browse, and other human activities. The herbivores may be domestic or wild—ecologically there is little difference in the end result.

Among domestic stock, goats are by far the most destructive of forest regeneration, followed by pigs, sheep, and cattle, in approximately that order. Long-continued overgrazing by any of these will result in the elimination of palatable species from the ground up to the browse line, compaction of the forest soil, and eventual conversion of the forest to an open scrub of unpalatable species or to grassland (Chapter 13). This has been the history of much of the forest in the Mediterranean regions of Europe and Africa, virtually all of Asia Minor and the countries to its east, and of large areas in the U.S.S.R., China, India, and elsewhere. Jarosenko (1956) has detailed this history for Transcaucasia in a Russian book. Much of the scrub and grassland in the drier and warmer forest regions of the world owes its origin to the long-continued overgrazing by goats, sheep, and cattle. Many of these sites could support high forest if this were controlled.

This condition is not unique to the Old World, but is equally important in the New, particularly in Latin America. In the United

**Figure 16.14. The original old-growth forest of New Zealand dominated by the podocarp *Rimu*, with characteristic understory of hardwoods and tree ferns. Under impact of logging and grazing, this forest is being replaced by North American conifers, both planted and naturally seeded. (New Zealand Forest Service photo by J. H. Johns, A.R.P.S.)**

States, the open character of much of the ponderosa pine forest, to cite but one example, and the holding of the succession at the pine stage instead of allowing it to move forward to the tolerant conifer stage, are attributable to heavy grazing by sheep, cattle, deer, and elk, coupled with periodic burning in the past (Arnold, 1950). In Mountain Meadows Valley, Utah (Cottam and Stewart, 1940), heavy grazing by cattle and sheep since 1864 has resulted not only in heavy erosion, but also in the replacement of grasses, rushes, and sedges by sagebrush, rabbitbrush, and juniper. With protection from heavy

**Figure 16.15.** ***Pinus radiata*** **(Monterey pine) forest in New Zealand, typical of the developing present-day forest there. These 47-year-old trees form a taller and higher-volume forest than the old-growth** ***Rimu*****, several hundred years of age. (New Zealand Forest Service photo by J. H. Johns, A.R.P.S.)**

grazing, however, the range grasses can still successfully compete on the site.

In the eastern hardwood forests, deer may by themselves retard the forest succession to as great an extent as the combination of domestic and wild animals in the ponderosa pine. Overpopulation resulting from elimination of predators and restricted and inadequate hunting has resulted in the development of parklike forests virtually free from undergrowth in parts of the Appalachians, the Lake States, and in western Europe. Graham et al. (1963) have detailed the effect of large deer populations in preventing regeneration in the aspen forests of lower Michigan, and the necessity of managing the deer herd concurrently with silvicultural treatment if yields both from the forest products and from the deer herd are to be maximized. The establishment of a tolerant understory is prevented, natural succession cannot proceed, and an open grassy woodland inevitably develops if the deer population is not substantially reduced for at least a period of years long enough to permit regeneration to grow up past the browse line (Habeck, 1960).

It is not only the larger herbivores that hold back forest succession; smaller mammals may play an equally important role. Squirrels in the eastern oak forest, deer mice in the Douglas-fir and Sierra forests, pocket gophers in the California red fir forests (Tevis, 1956), and many others may prevent the establishment of tolerant understory trees simply by consuming virtually all the available seed or by girdling the succulent seedlings that do manage to germinate.

## CLIMATIC CHANGE

A third source of disturbance initiating secondary succession in the forest is climatic change. The change may be short and extreme such as a severe drought or cold spell that kills part of the existing forest, thus initiating a secondary succession with a different complement of potential competitors for space than existed previously; or it may be a long-term climatic change of lesser intensity, but one which changes the relative competitive ability of the competing species. The first type is obvious and well understood. The second is being increasingly realized as playing an important role in changing forest succession.

Evidence for long-term climatic change is presented in Chapter 19 where its effect upon the distribution and composition of the forest is discussed in some detail. There is ample evidence that the climate of today is substantially different in many parts of the world from that of 150 years ago, when many existing forest stands were formed. It follows that forest succession on the same site may have been quite different under differing climates from what it is now. Many examples can be cited. In the Lake States, for example, tree species near the southern edge of their natural range such as white pine, trembling aspen, and black spruce are much poorer competitors in the present warmer climate of the region, and thus form a substantially smaller proportion of newly forming forest communities. Islands of the more northern white spruce and balsam fir in the maple-basswood forest of northwestern Minnesota have been declining and dying out. In Wisconsin, the decrease of beech abundance near its western border and the consequent increase of sugar maple (Ward, 1956) may well be attributable to the warmer present-day climate of the area. Similarly, changes in water levels in swamps resulting from recent increased precipitation result in slower growth and general deterioration of tamarack on these sites in the same region (Isaak et al., 1959).

The effects of climatic changes upon forest succession are usually not clearly self-evident and are frequently confounded with other factors. For instance, in the ponderosa pine forests of the Southwest, it is difficult to separate out the relative effects of climatic changes, overgrazing, and past burning—all three are and have been important in determining the secondary forest succession. Nevertheless, major and continued climatic change is by now a well-established fact and must constantly be kept in mind in any study of forest succession. The Clementsian assumption of a long-continued stable climate simply does not hold for much of the forested regions of the world.

## SUGGESTED READINGS


Auclair, Allan N., and Grant Cottam. 1971. Dynamics of black cherry (*Prunus serotina* Erhr.) in southern Wisconsin oak forests. *Ecol. Monogr.* 41:153–177.

Cooper, Charles F. 1960. Changes in vegetation, structure, and growth of southwestern pine forests since white settlement. *Ecol. Monogr.* 30:129–164.

———. 1961. The ecology of fire. *Sci. American* 204 (4):150–160.

Curtis, John T. 1959. *The Vegetation of Wisconsin*. (Pages 456–472.) Univ. Wisconsin Press, Madison. 657 pp.

Heinselman, Miron L. 1973. Fire in the virgin forests of the boundary waters canoe area, Minnesota. *Quaternary Res.* 3:329–382.

Isaac, Leo A. 1956. Place of partial cutting in old-growth stands of the Douglas-fir region. U.S. For. Serv., Pac. Northwest For. Rge. Exp. Sta., Res. Paper 16. 48 pp.

Kilgore, Bruce M. 1973. The ecological role of fire in Sierran conifer forests: its application to national park management. *Quaternary Res.* 3:496–513.

Mount, A. B. 1969. Eucalypt ecology as related to fire. *In Proc. Tall Timbers Fire Ecology Conference*, 1969, Tall Timbers Res. Sta., Tallahassee, Fla.

Rowe, J. S., and G. W. Scotter. 1973. Fire in the boreal forest. *Quaternary Res.* 3:444–464.

Spurr, Stephen H. 1956. Natural restocking of forests following the 1938 hurricane in central New England. *Ecology* 37:443–451.

Wellner, C. A. 1970. Fire history in the northern Rocky Mountains. *In Proc. Intermountain Fire Research Council and Symposium—The Role of Fire in the Intermountain West*, pp. 42–64. Univ. Montana, School of Forestry, Missoula, Mont.

# 17
# Spatial Variation in the Forest

The composition and structure of the forest differ not only with time but also in space. Both the site (including both climate and soil) and the geographical distribution of forest organisms help to determine the nature of a given forest stand. A forest on one site will obviously differ from that on a different adjacent site; a forest in one geographical area will also differ from a forest in another because of differing animal and plant populations in the two areas, differing histories, and, simply, different chance happenings.

The spatial variation within a forest may include abrupt changes from one type of forest to another as well as gradual changes in which the character of the forest alters with varying dominance or abundance of the component species, the appearance of new species, and the dropping out of others. That is, forest communities or types may change in composition either drastically or gradually over space. The concept of the forest community (association) is basic to an understanding of spatial variation in the forest.

## CONCEPT OF THE FOREST COMMUNITY

It is common knowledge that certain patterns of forest composition characterize extensive areas of forest. A belt of spruce and fir forest extends around much of the boreal zone and has a characteristic physiognomy, composition, and structure, which makes it immediately recognizable as a "spruce-fir" community, association, or forest type. Furthermore, many of the smaller plants and many of the animals found in one spruce-fir forest will be found occupying similar niches in other spruce-fir forests. True, the individual species of spruce, fir, other plants, and animals will vary from continent to continent and within the continent. Even within a forest characterized by the same species of spruce and fir, the races of these species will vary from place to place within the forest. Certain plants and animals will be found in one part of the community and not in others. Nevertheless, it is clear that a common denominator exists, so that when we speak of the boreal or montane spruce-fir forest, we convey an immediately recognizable concept and picture to others. It is a group of interrelated plants and animals that occur together more frequently than can be ascribed to chance. We identify a forest community with sufficient precision to be meaningful. We are speaking of a specific forest community or forest type.

Unfortunately, at this point, we become involved inevitably in nomenclature and in the semantic confusion that has beset the problem of identifying and naming plant communities. Only a few of the many terms that have been proposed and deposed, however, have become an important part of our common vocabulary of community ecology.

Community (a unified body of individuals) is a general term of convenience used to designate sociological units of any degree of extent and complexity (Cain and Castro, 1959). Traditionally, ecologists have established a hierarchy of communities and assigned special names to distinguish them. Thus a **formation** or **biome** is the largest and most comprehensive kind of plant community. In Europe, the formation is defined as a community of a given physiognomic form that recurs on similar sites (i.e., tropical rain forests, temperate evergreen forests, mangrove forests). In the United States, however, Clements originally defined the formation in both geographical and climatic terms (i.e., boreal or spruce-fir forest). This broader definition makes it possible for the Clementsian formation to include several physiognomic types within the same unit.

Each formation is composed of various other distinctive communities termed **associations**. The term is more broadly defined in current United States practice than in European systems. Thus the Deciduous Forest Formation is composed of many different associations—beech-maple, oak-hickory, aspen-birch—and not necessarily only climax associations. Küchler (1964) has published a non-hierarchical classification of the vegetation units of the United States; Rowe (1972) has described the forest regions of Canada, and a world-wide classification of plant formations has been published by Mueller-Dombois and Ellenberg (1974).

In the Clementsian system and European schools of phytosociology, finer subdivisions of the association are recognized and named. The subdivisions of association or community most widely used by American ecologists are "layer," "union," or "synusia"—terms that are more or less synonymous (but which are defined differently by different ecologists). Thus the spruce-fir formation of northeastern United States and eastern Canada includes the red spruce—balsam fir association (community), which may in turn be subdivided into a tree layer, a shrub layer, and a ground-cover layer.

An historical perspective given by Egler (1968) may be helpful at this point. Clementsian ecology was a very tidy and orderly science—far more orderly than nature itself. The analogy of the community with the organism was one of its key beliefs. At that time individuals of a species were thought to be more or less uniform; thus it was easy to believe in uniformity of plant communities.

About the same time a special kind of community (corresponding to the species in plant taxonomy), the "association," was defined in a technical sense by ecologists of the Zürich-Montpellier school of phytosociology and adopted at the International Botanical Congress of 1910:

> *An association is a plant community of definite floristic composition, presenting a uniform physiognomy, and growing in uniform habitat conditions.*

Again, Europeans and Americans followed markedly different courses in the use of the term association (Mueller-Dombois and Ellenberg, 1974). Partly because the term association was strongly identified with its rigid European definition and usage and partly because in Clementsian usage it related only to "climax" communities, the term association went into disfavor among many American ecologists. Today many ecologists read "community" in its place. A hierarchical system with the narrow definition of association as its core is preferred by many European ecologists involved in describing and classifying vegetation. Other ecologists not so involved also may use the term association but often in a much broader sense without organismal overtones. Although in this sense the term community appears synonymous with association, there is still much ambiguity in usage. The term community is not restrictive by itself, probably accounting for much of its modern popularity. Its particular meaning in a given study depends on the context in which it is used and the modifiers applied to it—oak-pine community, climax community, etc.

Clements and ecologists who follow his rules of nomenclature have restricted the use of the term association to relate only to climax communities. This usage, in company with the concept of the climax as a permanent and stable community, is losing ground in favor of the more general and flexible uses of the term.

The community, of course, consists of plants and animals living in a physical environment. Together they form an **ecosystem** or **biogeocoenosis.** The plants and animals in an ecosystem may be termed a **biome**, or **biocoenosis**. Plant ecologists, however, tend to characterize communities in terms of their plant composition **(phytocoenosis)**, only implying that certain animal patterns are usually associated with certain plant patterns.

The term ***forest type*** refers to a forest community defined only by composition of the overstory. Since the community or association is or should be defined by the sum total of the ecosystem, its naming usually takes into account characteristic ground-cover plants as well, or, alternatively, its characteristic site. For instance, the ponderosa pine type in northern Idaho and eastern Washington has been subdivided into six associations according to whether the undergrowth is characterized by grasses on stony, coarse-textured soils (*Agropyron spicatum, Festuca idahoensis, Stipa comata*) or shrubs on heavier-textured, more fertile soils (*Purshia tridentata, Symphoricarpos albus*, and *Physocarpus malvaceus*) (Daubenmire and Daubenmire, 1968). A gradient exists from the associations on driest areas bordering grassland or semidesert containing the grasses (*Pinus ponderosa–Festuca, Pinus ponderosa–Agropyron, Pinus ponderosa–Stipa*) to dry areas containing the shrubs (*Pinus ponderosa–Symphoricarpos, Pinus ponderosa–Physocarpus*) and grading into a series of late-successional associations of Douglas-fir on moister sites.

Conceptions of the plant association by ecologists vary from organismic views in which the association is conceived more or less mystically as some sort of superorganism, to individualistic views which treat each of the species comprising the association as occupying its range, and sites within its range, independently of one another. There seems little justification in fact for either extreme viewpoint. The degree of species interrelationship varies greatly, ranging from situations where two species have similar ecological requirements and are therefore apt to be found on the same site (such as spruce and fir; or beech and maple), to cases of obvious symbiosis (such as in mycorrhizae and lichens).

Communities are not composed of successive, mutually exclusive sets of species. It has been shown time and time again that individual species have different physiological and genetic tolerances and may exist in several different communities. A given species may be highly competitive in one community and hence predominate there. Although it may also exist in adjacent communities having different site conditions, other species may be more competitive in time and space and predominate in this different environment. Most forest species of a given climatic region probably have their optimal development under similar site conditions. Hence it is competition in space and time, coupled with adaptation of species to particular site conditions, that elicits differentiation of what we may recognize arbitrarily as distinct communities. For example, in a large part of central Europe many forest species reach their optimal development under similar environmental conditions of moisture and acidity (Figure 17.1; Ellenberg, 1963). However, because of the great competitive ability of European beech in space and time, the other species are competitive only in restricted portions of their potential range.

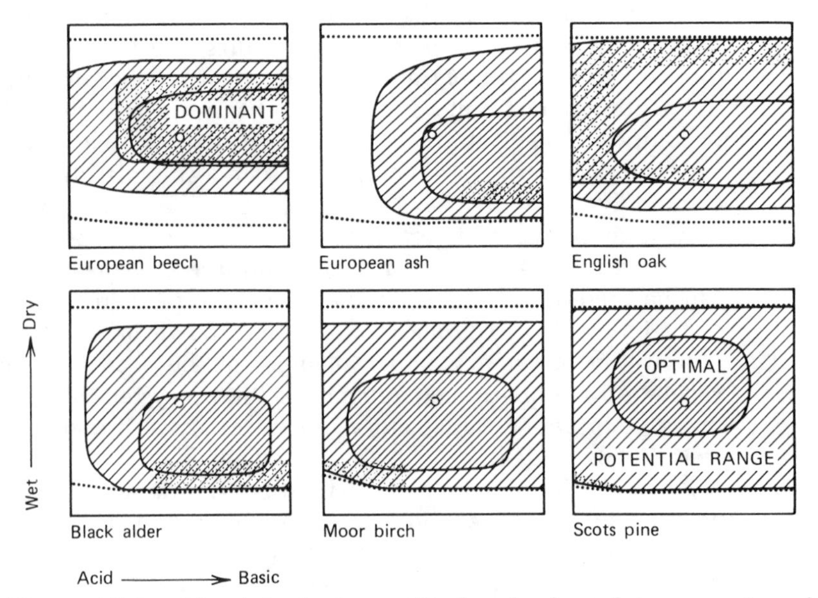


**Figure 17.1. Physiological amplitude of selected tree species of Central Europe of the submontane zone. Ordinate = moisture level from very wet to very dry. Abscissa = soil reaction from very acid to basic. Above dotted line at top is too dry for forest. Below dotted line at bottom is too wet for forest. Widely spread diagonal lines = potential range of species; closely spaced diagonal lines = optimal range; dotted area = portion of range in which through natural competition a given species attains a dominant position. (Modified from Ellenberg, 1963.)**

Although plants may have similar ecological requirements, they occur together in recognizable communities not because they react in the same way, but because they compete successfully in various ways due to differences in their genetic makeup and physiological responses. Different tree, shrub, and herb species may exist in a given arid, cold, or wet environment, and seemingly have similar requirements, but they occupy different **niches**, (Chapter 2). That is, they occupy different sites or microsites, or they utilize a given site differently through their physiological responses and life history traits such as time of germination, amount and timing of shoot and root growth, and adaptive mechanisms for obtaining water and retaining it. Thus species are co-adapted to one another and their physical and temporal environment; they are niche differentiated from one another. Actually, we know very little about the multiple dimensions of the niches of forest species and the interactions of these organisms with one another and with their environment.

The forest community, then, cannot be precisely defined. It varies in composition and structure. It has no genetic or other basic biological unity. Nevertheless, some plants and animals do "associate" commonly in that they have similar ecological requirements and that they themselves tend to create environmental conditions suitable to the existence of one another. In general terms, a forest community can be identified by the occurrence of one or more characteristic plant indicators. Accurate identification, however, must involve a description of a much larger segment of the ecosystem. Because vegetation is only one component of the ecosystem, it alone is an incomplete and unsatisfactory object for scientific study (Rowe, 1961, 1966, 1969).

A word about nomenclature might be in order at this point. Any nomenclature involves abbreviation in the sense that a few words are used to define a relatively complex structure. Thus the association described as *Pinus ponderosa—Agropyron* above is in reality a complex and variable community. The two plants named, one a tree and the other a grass, both occur commonly in other communities, and many other plants occur in the community being described.

Many plant sociologists go further and give the association a binomial Latin name modeled on taxonomic nomenclature rules. In the above example, the name would be *Pinetum agropyronum*, or, more fully, *Pinetum ponderosae agropyronum*. Giving binomial names to plant communities implies belief in the organismic conception of the plant community and is a particular fetish of European and European-influenced plant sociologists.

## The Zürich-Montpellier School of Phytosociology

The most important and influential European school of phytosociology is the Zürich-Montpellier school; its principal exponent is J. Braun-Blanquet (Poore, 1955; Becking, 1957; Whittaker, 1962; Braun-Blanquet, 1964; Shimwell, 1971; Mueller-Dombois and Ellenberg, 1974). In this system the association is seen as an organism and is comparable to the species of plant taxonomy. The association is the basic unit of a hierarchical system of vegetation classification which has become the primary goal of the school identified with

Braun-Blanquet. The association has a type specimen complete with author, date, and description analogous to that of a plant species: for example, "*Abieti-Fagetum* Oberdorfer 38"; and the "*Galio-Carpinetum* (Buck-Feuct 37) Oberdorfer 57 em. Th. Müll. 66." As we have indicated earlier, however, the association is not an organism and there is no single genetic control determining the nature of the association. Thus there is little justification for the naming of communities as if they were species.

In the Braun-Blanquet method, the association, an abstraction, is determined by the comparison of a number of lists of species made in selected sites in the field. It is not defined by site factors, but entirely by floristic composition. The field plots are subjectively selected for uniformity, which cannot be defined with precision (Poore, 1955). At each of the plots a complete vegetation list is made of species from all layers. One of the major advantages of the method is the completeness of the species list; many North American studies appear incomplete by comparison.

The plant lists from the plots are grouped into association tables, from which the associations are determined (Ellenberg, 1956; Poore, 1955; Mueller-Dombois and Ellenberg, 1974). Although associations are the cornerstone of the system, it is recognized that they intergrade extensively (Becking, 1957). However, erection of seemingly discrete classes, due primarily to the nonrandom method of plot selection, has conferred a false sense of uniformity within associations and discontinuity between them.

Many English and American ecologists have either been alienated by the system or have adopted an attitude of critical disinterest. This has been due to the questionable modelling on the plant taxonomic system, the seemingly rigid classification, and subjective methods. One result has been that many ecologists have hesitated to use the terms "association" and "phytosociology" and renounced for a time the use of classification, in part, at least, because of the close relationship of these terms and methods with this school. Criticism from without and within has led to a proliferation of modifications in the approach until today a single and dominant school cannot be clearly recognized. Shimwell (1971) and Mueller-Dombois and Ellenberg (1974) provide excellent descriptions of the Zürich-Montpellier School and Braun-Blanquet system, their advantages, and comparisons with other European and North American schools.

Despite its shortcomings, this approach has made and continues to make important contributions to the description and classification of vegetation. A major advantage for those versed in the method is the rapidity with which major associations can be described. The approach is widely applied in continental Europe in one variety or another and used by many plant sociologists throughout the world.

Detailed considerations of this and related European phytosociological methods are given by Kershaw (1964); Küchler (1967); Shimwell (1971); and Mueller-Dombois and Ellenberg (1974).

## SPATIAL CONTINUITY OF THE FOREST COMMUNITY

Accepting the concept of the forest community as a group of interrelated plants and animals that occur together more frequently than can be ascribed to chance, the question then arises as to how the forest varies spatially with changing conditions of site and geographical location.

One can readily find cases where one community is separated from an entirely different community by a sharp boundary. A field is distinct from an adjacent woods; and a pine plantation from an adjacent natural hardwood stand; and a tamarack bog from an adjacent upland conifer type. One can just as readily, though, find cases where the nature of the community changes gradually. A redwood type grades into a Douglas-fir-type upslope; a slash pine type near the border of a cypress-gum bottom grades into a loblolly pine type; and a good white oak type into a poor white oak type.

In nature one observes all degrees of change between communities—from abrupt to imperceptible. Yet a major controversy developed around the abstract conception of the distinctness or continuity of communities. The concept of continuous change in vegetation, the continuum concept, was conceived and developed in reaction to the organismal concept of the plant association as a discrete unit of vegetation and its use as the basis for a seemingly rigid classification of vegetation. Unfortunately, the controversy has usually been framed in terms of two opposing viewpoints: the association-unit (community-type) approach versus the continuum approach. However, this is a serious oversimplification of viewpoints and concepts. Having polarized ecologists for a time, the controversy today is largely academic. All parties agree that floras[1] are continuous and that abrupt as well as gradual changes exist between communities on the ground. Furthermore most ecologists agree that the community is not as well integrated as an organism nor are species making up the community completely independent of one another. What is not well known is the rate of

[1]The ***flora*** is the total list of plant species present, irrespective of the abundance of each species. ***Vegetation*** is the collective plant life of an area, the combinations of species present and their relative abundance. The vegetation of an area may be described in various ways (structure, abundance, density, etc.) without listing the species present.

change between communities and the degree of integration and mutual interdependence of species of a community.

A significant viewpoint in light of the controversy has been the approach of English workers who have seen neither the concept of discrete classes nor the continuum as a useful basis for investigation (Lambert and Dale, 1964). They have developed objective, numerical methods and have applied them in studying vegetation and associated habitat conditions in a variety of temperate and tropical ecosystems.

The continuum concept will be discussed in the next section. Thereafter, discrete and merging forest communities of the real world will be considered giving specific examples of the continuum and community-type approaches.

## The Continuum Concept

The concept of a vegetational continuum was introduced in the United States by John T. Curtis and co-workers at the University of Wisconsin (Curtis and McIntosh, 1951; Bray and Curtis, 1957; Curtis, 1959). The continuum was defined as:

> *An adjectival noun referring to the situation where the stands of a community or large vegetational unit are not segregated into discrete units but rather form a continuously varying series.*

The continuum meant that all communities of a vegetation type, for example, forest or grassland, could be ordered or ordinated[2] in an *abstract* series whose species composition changed gradually—typically along one or more environmental gradients. Thus although distinct, adjacent communities were encountered in the field (Curtis, 1959), they could be fitted into their respective position in the abstract continuum. Thus according to Curtis (1959):

> *It is not possible to erect a classification scheme which will place the plant communities of any large portion of the earth's surface into a series of discrete pigeonholes, each with recognizable and describable characteristics and boundary limitations. The plant communities, although composed of plant species, are not capable of being taxonomically classified, as are the species themselves.*

[2]An ordination is an arrangement of species, communities, or environments in sequence along axes with their respective properties determining their position.

The strong position against a taxonomy of vegetation is not surprising since the continuum concept was directed primarily to a different facet of vegetation study—the relationship of species and communities to one another and either indirectly or directly to environmental factors. A significant continuum approach, direct gradient analysis (Whittaker, 1956, 1967b, 1975), has been applied extensively in mountainous terrain. In this method, vegetation is directly related to environmental gradients, whereas in Curtis' approach vegetational composition could be used only to infer gradients such as moisture or temperature. The major contributions of the continuum approach have been the demonstration of considerable variation of species composition within stands and communities, the impetus it has given for the use of more objective techniques in vegetation analysis, and the demonstration of the individuality of species occurrence and dominance along environmental gradients. Various publications document the history and details of the controversy over the philosophy and methods of the continuum and community-type approaches (Whittaker, 1962, 1975; Daubenmire, 1966; McIntosh, 1967; Dansereau, 1968; Langford and Buell, 1969; Mueller-Dombois and Ellenberg, 1974). Two contrasting viewpoints on how species are distributed along environmental gradients are presented by Whittaker (1975, pp. 112–120) and Mueller-Dombois and Ellenberg (1974, pp. 331–333).

There is no question even by continuum critics that vegetation or floras are continua (Daubenmire, 1966; Daubenmire and Daubenmire, 1968). This is true for a number of reasons: (1) environmental conditions change in space at varying rates affecting the establishment and, through competition, the composition of the vegetation; (2) genecological variation of tree species is typically clinal, and in addition a given genotype often has a wide environmental tolerance; (3) historical and chance events reinforce points 1 and 2 so that no communities are exactly alike; and (4) a continuum of successional changes in time is superimposed upon changes in composition in space.

Continuum ecologists were opposed to classification, but according to McIntosh (1967) agree that "if classification is urged simply as a desirable convenience for mapping or information storage (Daubenmire, 1966) for practical ends and specific purposes, the classification being arbitrary and directed to these specific ends (Rowe, 1960), there is no contest." Much of the controversy falls within this realm.

Concerning classification and ordination, Lambert and Dale (1964) summarize the alternatives:

*. . . There seems to be a common misconception that classification is only properly applicable to "discontinuous" data, while ordination techniques are more appropriate to continuous systems. In contrast, it cannot be too strongly emphasized that there is no* a priori *reason why the use of either method should be restricted in this way: continuous systems can be efficiently classified if classification is desired, while "discontinuous" (i.e., markedly heterogeneous) systems can be ordinated if ordination is thought more useful for the immediate purpose in hand. Moreover, there is in principle no reason why classification and ordination techniques should be mutually exclusive: classified units can be ordinated, and ordinated units classified. Which method to adopt at a given stage of the investigation is entirely a matter for the user, irrespective of any subjective concept of the "real" nature of vegetation.*

For the silviculturist the two views are not incompatible, and often the character of nature itself may indicate the appropriate methods (Lieth, 1968). Unfortunately, however, throughout much of the world, forests and sometimes their sites have been drastically altered by humans. This is the case particularly in Central and Southern Europe, the Mediterranian areas, and also in the eastern half of North America. Because historical records are incomplete at best, we have little appreciation for the multiple cuttings (often extensive clearcutting), slash fires, and agricultural practices (cropping and grazing) that may have occurred in a given area (Chapter 20). Although their effects may be appreciated, ecologists and silviculturists tend more often than not to overlook the historical record, probably because it is unknown or not readily apparent. Nevertheless, soil-site changes, alteration of seed availability, and other deviations from presettlement conditions not only influence the composition of present-day stands but tend to exaggerate the continuity of existing communities in space. The vegetational continuum is the product of a continuum in space (species and communities influenced by site and biotic factors) *and* a continuum in time, that is, biotic succession. A major concern about continuum methods is that the continuum often generated is an artifact of sampling vegetation as it exists—usually in an unnaturally disturbed state. When vegetational samples in different successional stages are included, the result necessarily may be a more gradual change in the series of communities than if late-successional stands only were sampled. In the studies of Curtis and associates in Wisconsin, stands of varying successional stages were used and the resulting continuum is, to an unknown degree, partly temporal and partly spatial. This aspect of continuum methods has important theoretical and practical implications.

Studying vegetation as it exists, regardless of successional stage, will tend to give one picture of gradually changing communities. Sampling only late-successional stands throughout the same spectrum of sites might show a relatively abrupt rate of change in some places and gradual rates in others. Thus both Daubenmire (1966) and Langford and Buell (1969) have argued that if climax stands were used one might find more or less discrete associations connected by ecotones (transition zones between communities). The crux of the problem according to Daubenmire (1966) is:

> *There is no denying that vegetation presents a continuous variable by virtue of ecotones; the argument hinges on the existence or absence of plateau-like areas exhibiting minor gradients separated by areas of steeper gradients, with the plateau-like areas being of sufficient similarity to warrant being designated as a class.*

In studies of forests of the northern Rocky Mountains that are essentially undisturbed by man, Daubenmire presented evidence to substantiate the reality of these "plateau-like areas" (Daubenmire and Daubenmire, 1968). He found the associations, determined subjectively, extended across hundreds of kilometers without losing their identity. Although these types intergrade, he believes the discontinuities become important enough to "compel the segmentation of the landscape into categories that are objective in that different workers commonly recognize the same discontinuities independently." He credits part of the distinctness to the use of both overstory and understory vegetation, a combination that has not been a major determinant of continua in most investigations by continuum workers.

From a practical standpoint the forest land manager is concerned with the site potential. Neither an existing forest-type classification nor a continuum based on a mixture of successional stages may be useful in decision making. We agree with Rowe's (1962, 1969) recommendation in forest mapping that the landscape pattern first be broken into geomorphological and climatic parts, each relatively constant as to surface materials, which in turn can be subdivided into relatively homogeneous forest-land patches or ecosystems. Then late-successional vegetation for each ecosystem may be determined and used together with environmental factors in determining site quality (as in the Baden-Württemberg system).

In summary, it is possible, then, to approach the problem of the spatial relationships of vegetation either from the viewpoint of communities as separate units of vegetation which may be identified as a type and delineated on maps; or to consider the vegetation as a plastic community which changes gradually from place to place, without any specific point of maximum change. Both approaches, the community-type approach and the continuum approach, have widespread application in forest ecosystem studies. For the silviculturist or forest manager most forest operations (harvest cutting, thinning, fertilization, planting, etc.) are conducted most efficiently on an area basis. They must not only recognize changing conditions but more importantly determine similar units along the continuum and be able to draw boundaries for the prescribed treatments. Thus in practice community-types are typically identified.

## Numerical Analysis of Vegetation

English workers, predominantly Williams, Lambert, and associates, have developed an alternative system of vegetational analysis (Lambert and Dale, 1964). More objective than any previous approach, there is much to recommend it from ecological and mathematical standpoints. They found neither the approach using discrete communities nor that of a vegetational continuum particularly useful in extracting maximum ecological information from given areas. Making no assumptions except that vegetation was heterogeneous, they developed and adapted various classification techniques and used computer methods to describe and analyze vegetation-site relationships.

The English group has used both classification and ordination methods in studying species-site relationships in complex rain-forest communities (Webb et al., 1967a, b). In further studies, Williams and colleagues used numerical methods to analyze local forest-environment patterns (Williams et al., 1969), secondary succession (Williams et al., 1969), and the floristic-versus-physiognomic classification of complex rain forests (Webb et al., 1970). In one study (Webb et al., 1967a) ordination was found the single most informative method, but the larger part of the information was recovered by classification, which is simpler and faster. Thus they recommended classification first, followed by ordination only if classification proves unprofitable. Other workers agree; for example, Greig-Smith et al. (1967) indicate that classification is more satisfactory at high levels of variation in composition of vegetation and ordination more satisfactory at lower levels. Thus a combination of the two approaches will be more informative than either alone.

## DISCRETE FOREST COMMUNITIES

Wherever one type of forest community abuts on another distinctly different type of forest community, it will be found that this abrupt change is related to an abrupt change in site conditions or to a completely different vegetational history of the two communities. The existence of discrete communities, therefore, is evidence of the existence of discrete differences in growing conditions, either now or in the past.

The boundary between two communities is usually a belt rather than a sharp line. It is a belt or zone, though, which may vary widely in width. In the forest-grassland transition, there will always be an outer belt of forest which will be modified by the adjacent open areas, and an inner belt of grassland which will be modified by the adjacent forest. As mentioned above, the transition zone between two communities is termed an **ecotone**. It usually embodies some of the ecological features of the two communities, but has a characteristic ecological structure of its own.

Many persisting, abrupt site differences can give rise to sharp forest-type boundaries. Among these are: (1) a sharp boundary between two geological formations, giving rise to a sharp boundary between two soil types that differ markedly in the mineral sites they provide to vegetation (Chapter 8); (2) a sharp boundary in soil drainage or soil-moisture conditions, such as between a poorly drained swamp and a well-drained upland; (3) a sharp boundary in topographic position affecting local climate, such as a knife-edged ridge separating a north from a south slope or an air dam impeding cold air drainage so that frost pocket conditions exist below the level of the dammed air; and (4) a sharp boundary in the structure of the vegetation affecting the local climate and soil conditions, such as a forest edge facing grassland or a shrub community impinging upon open rock surfaces. Among the historical accidents that may give rise to sharp boundaries between plant communities are fires, tornados and other windstorms, salt spray from the sea, fumes from smelters, logging, and agricultural development of land. Many examples could be cited, but these are, in general, obvious. Abrupt changes in vegetation associated with changes in the parent geological material are dealt with in Chapter 8. Such changes may be found at the contact between sandstone and limestone outcrops, serpentine and adjacent soils, and organic and mineral soils. Relationships between vegetation and soils around the world are summarized by Eyre (1962).

The ecotones between forest and adjacent low-vegetation types have long attracted the attention of ecologists, particularly the forest-grassland ecotone and the alpine timberline.

## Forest-Grassland Ecotone

Abrupt changes between forest and grassland in the tropical and temperate zones may or may not be associated with abrupt changes in site. Once the forest edge has been established, such as by fire or land clearing, site conditions—both climatic and edaphic—within the forest may differ so substantially from those in the grassland as to perpetuate the forest border.

This is not to say that abrupt site changes cannot be found in many instances. In moist climates, forests are frequently found on well-drained upland soils, with grasslands becoming dominant on poorly drained sites. In the American tropics, all types of climate in the lowlands are adequate to support woody growth of some kind, but savannas occur upon ill-drained country of little relief such as an old alluvial plain or reduced upland (Beard, 1953). In semiarid regions, forests may be confined to coarse-textured or rocky soils while grassland occurs on the finer-textured soil types. For instance, in the Black Forest of Colorado, arborescent communities occur under the same climatic conditions as grasslands, but on sites where coarse-textured soils result in more rapid infiltration of the limited precipitation, reduced runoff, and lower wilting percentages in the soils (Livingston, 1949).

Often, however, the grassland originates in forested country as a result of a fire which destroys the forest and creates an environment at the ground more suitable for the development of grasses than for the reestablishment of the forest. Once established, the grassland persists because of the inability of the adjacent forest trees to invade the site—whether due to the recurrent incidence of fire (Wells, 1965; Rowe, 1966); to the failure of the tree seeds to penetrate the sod and reach a medium suitable for germination; to biochemical antagonisms between the grasses and the tree seedlings; to excessive root competition for soil moisture provided by the grasses; to the absence of mycorrhizae (Langford and Buell, 1969); or to the damage of direct insolation to the seedlings on the exposed open sites. The alpine meadows of the western American mountains, the fingers of prairie extending up into the Black Hills of South Dakota, and the extension of the Prairie Peninsula east into Michigan, Indiana, and Ohio, all are examples of fire-caused grasslands that have persisted for hundreds of years under climates suitable for tree growth.

This does not imply that the forest-grassland border is ever static. Invasion of one type by the other does occur. In the present century, grasslands within forested regions are being invaded by forest throughout the world, partly because of the improvements in modern fire-suppression techniques and, in some cases, partly because of a change in climate. In northwestern Minnesota, to cite one case, the maple-basswood forest is invading adjacent prairie (Buell and Cantlon, 1951; Buell and Facey, 1960).

### Alpine Timberlines

The upper edge of forest in mountain ranges provides another spectacular forest edge. Many ecologists, ignoring the biotic aspects of the plants forming the timberline and their history, have attempted to define timberlines purely in terms of the site, and have tried to evolve rules defining the height of the timberline for a given latitude, aspect, and other physical aspects of the timberline position.

As with all ecological phenomena, the timberline is a result of the interaction of the trees and the site over a long period of time. The position of the timberline may differ greatly with the species available to form it. In New Zealand, where only *Nothofagus* spp. form the temperate tree flora at high altitudes in the mountains of the South Island, the timberline is much lower than would be the case if North American conifers had been in the local plant population. In fact, it seems that lodgepole pine—both planted trees and seedlings from seed blown in from planted trees—will eventually form a new timberline type 300 or more meters higher than the present *Nothofagus* type.

Naturally enough, the timberline tends to rise toward the Equator, but the greatest height at which trees grow occurs in warm-temperate belts (about 30° latitude in the Northern Hemisphere and 25° in the Southern) rather than in the deep tropics. Generalized data on American timberlines are summarized by Daubenmire (1954), who has brought together theories concerning the distribution of American timberlines.

Many causes have been ascribed to timberline formation. Above the **forest line** (the upper edge of continuous forest), trees grow up to the **tree line**, the altitude of the highest stunted tree. Within this zone of stunted and recumbent trees, wind, snow blown by the wind, snow pack, and other factors produce an exposed and rigorous climatic zone near the ground through which trees cannot grow. Other causal factors of timberline formation that have been cited include excessive light, carbon dioxide deficiency, desiccation during temperature inversions in the winter, precipitation deficiency, solufluction, and light deficiencies in certain mountain regions commonly clouded over. After reviewing the evidence, Daubenmire concluded that one of the principal factors determining timberline location (as contrasted with climatic factors that cause dwarfing and recumbent growth) is heat deficiency during the growing season at high altitudes, which prevents trees from surviving and growing.

Wardle (1968) found that summer temperature data bore out this conclusion in a study of timberlines in Colorado which are among the highest in the world despite desiccating wind and low winter temperatures. He also found a strong correlation between the distribution and growth form of Engelmann spruce and exposure to wind. Wind also kept sites blown free of snow so that there was a lack of protection for seedlings during the winter and an absence of melt-water in the spring to moisten the rocky, coarse-grained soils. In the Austrian Alps, Aulitzky (1967) reported that only the highest tree line was governed by the 10°C line of July temperature. In most situations, growth of trees was lower than this level due to other unfavorable factors, wind, snow depth, and snow duration.

## MERGING FOREST COMMUNITIES

While distinct plant communities, separated by transitional belts or ecotones, reflect abrupt changes in site or land history, gradual changes in site or vegetational history result in similar gradual changes in the composition and structure of the forest. Gradual changes are characteristic of forests of a generally similar history over a geographical stretch of many kilometers (such as north to south or east to west). They also occur over gradual changes in altitude, aspect, soil moisture, soil fertility, and local climate.

Continua characterize the composition and structure within forests in the absence of an abrupt change in site or vegetational history. The great deciduous hardwood forest of the eastern United States stretching from the Gulf of Mexico to Canada is, in its broader aspects, a great continuum within which distinctive communities may be recognized. Such also is the conifer forest of western United States and Canada, although here the intrusion of large areas of desert and grassland, the presence of abrupt superficial geological changes, and the lasting effects of past severe forest fires all tend to emphasize the discreteness of forest communities rather than the gradual sequence of change that is superimposed over these local influences.

### Eastern Deciduous Forest

The forest of the eastern United States—characterized by deciduous hardwood species, but containing evergreen hardwood species in the deep south and conifers such as white pine, hemlock, and red cedar in much of it—constitutes a single great vegetational complex that shows many gradual changes from place to place and from site to site in any given place. Related to similar hardwood forests in

Mexico, it reaches its greatest complexity and size of individual trees, however, in the southern Appalachian region.

Within this great forest complex, ecological studies in the southern Appalachians, central New England, and Wisconsin may be taken as illustrations of the principle of gradual change resulting in continua, or merging forest communities, and of how in each case specific types of communities of the continuum have been distinguished. It should be emphasized that in all these studies tree species have received either major or exclusive emphasis.

**Southern Appalachians.** The old eroded slopes, valleys, and ridges of the southern Appalachian Mountains and the Cumberland Plateau to the west in Kentucky and Tennessee provide ecological gradients on which forest composition is similarly graded. Detailed studies by Braun in the Cumberland Mountains (1942) and Whittaker in the Great Smoky Mountains (1956) illustrate many of these changes.

Extensive surveys of uncut, old-growth, mixed deciduous forest in the Cumberland Mountains have resulted in a series of papers by Braun, which present the general picture of an "undifferentiated" complex marked by gradients in composition occurring with changes in elevation, aspect, and soil-moisture relationships. For instance, around the head of a west-facing stream valley in the Log Mountains, sugar maple, basswood, and buckeye constituted the major portion of the forest canopy on the north-facing slopes, and diminished in abundance toward the west, being absent on the south. Similarly, yellow-poplar, chestnut, and red oak varied in numbers with the aspect, reaching their maximum abundance on the west slopes; while chestnut oak and white oak were most abundant on the south slopes. In the upper mountain slopes, hemlock–beech–yellow birch forest in the ravines was found to grade gradually into chestnut oak–chestnut–hickory forest on the upper slopes. No sharp line dividing two quite different forest types could be drawn. Rather, the transition from one to another is indicated by declining numbers of one species and expanding numbers of another.

In the Great Smoky Mountains, Cain. in a series of papers, analyzed the vegetational communities in considerable detail, followed by Whittaker, who studied gradients in composition with altitude and with "moisture gradients" within altitudinal belts. This latter term refers to the complex gradient from valley bottoms to dry slopes without any assumption as to its causation. In such a gradient, the most numerous trees vary from mesic species in the valley bottoms to xeric species on the driest and most exposed portions of the slope. For instance, between 750 and 1050 m in elevation, hemlock was most numerous in the bottoms, with silver bell, red maple, chestnut, chestnut oak, scarlet oak, pitch pine, and Table Mountain pine each entering the transect and becoming more numerous toward the drier end of the gradient. Similarly, altitudinal gradients were constructed for a given site moisture. On the mesic sites, hemlock and red maple were most numerous at low elevations (600 to 900 m) with silver bell, yellow birch, sugar maple, and basswood reaching maximum abundance in the 900- to 1200-m zone, and buckeye, mountain maple, and beech being most common at higher elevations.

Some tree species show multinodal distributions indicative of the existence of two or more populations, some of which, like red maple and yellow birch, overlap, and some, like beech and white oak, are separated by an altitudinal zone of 300 m or more.

Combining the two gradients, Whittaker synthesized a general vegetation pattern of the Great Smokies (Figure 17.2). He found the

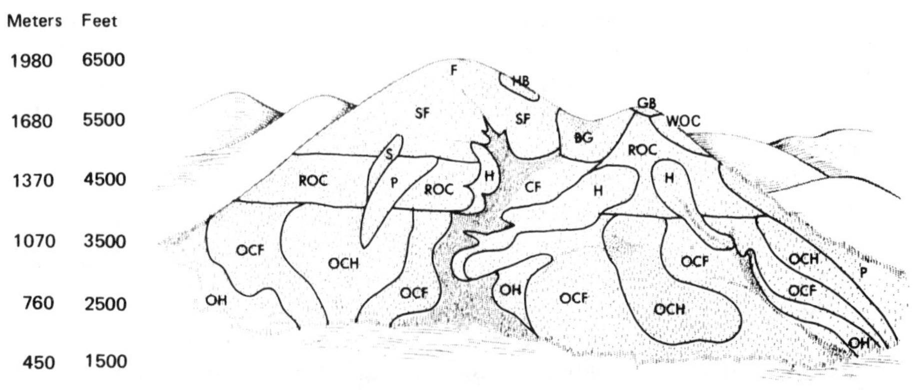


Redrawn, by permission, from R. H. Whittaker. 1956. Ecological Monographs 26: 1–80.

**Figure 17.2. Topographic distribution of vegetation types on an idealized west-facing mountain and valley in the Great Smoky Mountains. Vegetation types: BG, beech gap; CF, cove forest; F, Fraser fir forest; GB, grassy bald; H, hemlock forest; HB, heath bald; OCF, chestnut oak–chestnut forest; OCH, chestnut oak–chestnut heath; OH, oak-hickory forest; P, pine forest and pine heath; ROC, red oak–chestnut forest; S, spruce forest; SF, spruce-fir forest; WOC, white oak–chestnut forest. (After Kormondy, 1969, based on diagram by Whittaker, 1956. Reprinted from "Vegetation of the Great Smoky Mountains" (*Ecological Monographs* 26:1-80) by Robert H. Whittaker by permission of Duke University Press. © 1956 by the Ecological Society of America.)**

forest pattern one of continuous gradation of stands along environmental gradients. Certain relatively discontinuous types were also found. As described by Whittaker (1962):

> *The whole pattern was conceived to be a complex continuum of populations, with the relatively discontinuous types confined to "extreme" environments and forming a minor part of the whole. Allowing for discontinuities produced by disturbance and environmental discontinuity, the vegetation pattern could be regarded as a complex mixture of continuity and relative discontinuity.*

It is noteworthy that, while he disavowed the concept of the association, he did recognize the existence of abrupt discontinuities as between cove forests and beech stands on south-facing slopes, between grassy balds and forests, and between heath balds and spruce-fir forests. Also, he treated the boreal forest characterized by red spruce and Fraser fir as being a separate entity from the eastern forest, thus accepting the principle of the existence of two major forest formations in the mountains. The synthesized vegetational pattern portrays the acceptance of the convenience of community-type designations whether or not the investigator accepts their existence as discrete units.

Whereas Whittaker emphasized a modified continuum approach, Hack and Goodlett (1960), working in similar Appalachian hardwood forests but in the upper Shenandoah Valley region, found that species assemblages were generally coincident with topographic units and often change abruptly with changes in the form of the slope. Pine-oak forests were found to be generally restricted to noses, ridges, and other convex slopes; yellow birch–sugar maple–basswood forests were found to be generally restricted to hollows and other concave surfaces; while oak forests were found generally on straight slopes. The strong coincidences between the local distribution of species and groups of species are modified by: (1) the size of the valleys, a minimum area being essential for the establishment of the northern hardwood type in a hollow; (2) the orientation of the valleys and side slopes, north and east slopes being commonly forested with northern hardwoods, and south and west slopes being characteristically oak; and (3) the nature and attitude of the bedrock, formations favoring moisture accumulation and retention being more favorable for the more mesophytic species.

**New England.** In central New England the different successional stages of the old-field white pine succession occupy much of the forested landscape. It has occurred on soils of glacial origin ranging from poorly to excessively drained. As a result, the composition of the forest varies in response to two major ecological gradients: one gradient, in time, covering the development of old-field succession, and one, in space, covering the range of site conditions from wet to dry. In this region of moderate elevations and rolling topography, neither absolute elevation above sea level nor aspect is as important in affecting the composition of the forest as the two gradients named.

In the Harvard Forest in central Massachusetts, Spurr (1956b) classified all existing stands according to both relative position on the successional-development gradient (pioneer, transitional, or late-successional) and relative position on the soil-moisture gradient (from very well to very poorly drained). The frequency of occurrences of the individual tree species was found to vary consistently with the two gradients. Table 17.1 summarizes the change in the transitional type associated with changes in the moisture gradient. Two species were found to be practically omnipresent. Red oak and red maple were prominent in all successional stages, and one or the other was prominent on all sites. Both exhibited marked relationship to soil moisture, red oak being most frequent on the well-drained and red maple on the most poorly drained sites. Other species proved more specific in their site associations. White oak was most frequent on very well-drained, paper birch on well-drained, and white ash on imperfectly drained sites.

**Table 17.1 Occurrence of Species as Major Components in Harvard Forest Transitional Middle-Aged Stands**

| SITE | RED OAK | RED MAPLE | PAPER BIRCH | WHITE OAK | WHITE ASH |
|---|---|---|---|---|---|
| | | | *Frequency (percent)* | | |
| Very well drained | 67 | 0 | 0 | 33 | 0 |
| Well drained | 95 | 61 | 14 | 13 | 3 |
| Imperfectly drained | 81 | 81 | 9 | 6 | 19 |
| Poorly drained | 42 | 100 | 8 | 0 | 8 |
| Very poorly drained | 20 | 100 | 0 | 0 | 0 |

Everything considered, all the forest communities found in the Harvard Forest seem to represent a continuous gradational series correlated with successional stage and soil moisture (Table 17.2). Thus the old-field white pine and the pioneer hardwood types are pioneer communities which vary in composition according to the site where they occur. Among the transitional types, the white pine–hardwoods type on the driest sites grades into the transitional-hardwoods type on the intermediate sites, which in turn grades into the swamp-hardwoods type on the wettest sites. Among the late-successional types, the white pine–hemlock–hardwoods type on the driest sites grades into the softwood-swamp type of the wettest sites. A similar gradational series is found on each site as between the different successional stages. In general, the pioneer hardwood communities (but not pioneer white pine) are less than 30 years old, the transitional communities from about 30 to 60 years old, while the late-successional communities are older.

**Wisconsin.** In Wisconsin and adjacent areas of the Great Lakes region, J. T. Curtis and his students developed the continuum concept of variation in the composition of the forest (Curtis and McIntosh, 1951; Curtis, 1959; Maycock and Curtis, 1960). Curtis' classic book, *The Vegetation of Wisconsin*, is a definitive treatment of the vegetation in relation to environment, glacial history, and humans, as well as the presentation of his concept of the vegetational continuum.

The state was classified into two floristic provinces (southwest and northeast halves), and within each intergrading vegetation types (forests, savannas, and grasslands) were recognized on the basis of physiognomy. Finer subdivisions within vegetation types were based on floristic composition. The techniques of sampling and stand analysis used are open to question, but the results have clearly indicated the existence of vegetational continua, in both provinces. Each continuum could be arbitrarily divided into segments emphasizing the relationship between the community and a major environmental gradient, primarily soil moisture. Thus in the southwestern mixed hardwood forest, five segments of the continuum were recognized as communities: dry, dry-mesic, mesic, wet-mesic, and wet southern hardwoods. In all, 21 communities were identified.

In the Wisconsin studies, the forest stands chosen for sampling have typically been those of natural origin at least 6 ha in size, on upland landforms not subject to inundation, and free from recent disturbances within the lifetime of the stand as far as could be determined. From sampling at a number of points within each stand, the **relative frequency** (percentage of samples in which a species occurs), **relative density** (number of individuals of a given species per unit area as a percentage of the total of all species), and **relative dominance** (basal area represented by individuals of a given species as a percentage of the total basal area) of each tree species were determined. The value obtained by arbitrarily adding together the three separate measures of a given species, without exploring the interrelationships between them, is termed the **importance value**. It has the merits of quickness and simplicity and the demerits of un-

**Table 17.2 Tree Species Characterizing Communities in the Harvard Forest**

| SITE | PIONEER COMMUNITIES | TRANSITIONAL COMMUNITIES | LATE-SUCCESSIONAL COMMUNITIES |
|---|---|---|---|
| Very well drained | WHITE PINE<br>GRAY BIRCH<br>Red maple<br>Paper birch<br>Red oak<br>Black cherry | RED OAK<br>WHITE PINE<br>*White oak*<br>Red maple | HEMLOCK<br>WHITE PINE<br>RED OAK<br>Red maple<br>White oak |
| Well drained | WHITE PINE (old fields)<br>RED MAPLE<br>RED OAK<br>*Gray birch*<br>*Black cherry*<br>White ash<br>Paper birch | RED OAK<br>RED MAPLE<br>*Paper birch*<br>*White oak*<br>Black birch | HEMLOCK<br>RED OAK<br>RED MAPLE<br>*White pine*<br>*Yellow birch*<br>Paper birch<br>Black birch |
| Imperfectly drained | Insufficient data (Red maple, white ash, and birches predominate.) | RED OAK<br>RED MAPLE<br>*White ash*<br>*Yellow birch*<br>Paper birch | HEMLOCK<br>RED MAPLE<br>*Red oak*<br>Yellow birch |
| Poorly drained | Insufficient data (Red maple and birches predominate.) | RED MAPLE<br>*Red oak*<br>*Yellow birch* | HEMLOCK<br>RED MAPLE<br>*Yellow birch* |
| Very poorly drained | Insufficient data | RED MAPLE<br>*Red oak*<br>*Black birch*<br>*Elm* | HEMLOCK<br>RED MAPLE<br>*Spruce*<br>White pine<br>Tamarack<br>Yellow birch<br>Black gum |

Note: Relative importance is indicated by CAPITALS, followed by *italics*, followed by regular type.

proven assumptions, subjectivity, and the absence of analysis. Lambert and Dale (1964), in reviewing the use of statistics in phytosociology, state that there is little to be said for such mixed quantitative values and view the importance value as a sum of nonadditive numbers. Langford and Buell (1969) cited instances where use of composite indices such as importance values may tend to obscure differences between communities. In these cases artificial populations were generated having marked differences in relative density and relative dominance compared to a natural population yet having exactly the same importance values.

In early studies, the importance value was multiplied by an arbitrary climax adaptation number varying from 1 to 10 (essentially the same as the tolerance rating discussed in Chapter 14) and when summed for all species gave a **continuum index**. This index was then used to position stands in a linear continuum (Curtis, 1959). Later attempts were made to segregate out more complex patterns by the ordination of stand data on three interrelated axes (Bray and Curtis, 1957; Maycock and Curtis, 1960).

In the 1951 study of the upland (i.e., on well-drained soils) forests of the prairie-forest floristic province of southwestern Wisconsin, Curtis and McIntosh found that black oak, white oak, red oak, and sugar maple were the most important species. Relatively undisturbed forest stands in that region could be arranged in accordance with arbitrary and synthetic continuum index values, characterizing a gradient which runs from communities containing bur oak, aspens, and black oak at one end to those including hop hornbeam and sugar maple at the other. Since ordination was in terms of vegetation composition, the gradients could not be used to relate vegetation as the dependent variable directly to environmental or successional gradients as independent variables. It seems evident, however, that the vegetational continuum described is in part a successional continuum from pioneer to climax species and in part a response to a site continuum. In sampling stands in Wisconsin or the western Great Lakes region, inclusion of stands of different stages of succession apparently cannot be avoided.

In an extensive study of the boreal conifer-hardwood forests of the northern Great Lakes region, Maycock and Curtis (1960) did establish a moisture gradient broken into 5 arbitrary classes from dry to mesic to wet. Without taking into consideration the successional development of the stands, the occurrences of the various species were found to vary significantly with this moisture gradient. White pine and white spruce are of major importance on dry sites; sugar maple, red maple, yellow birch, and hemlock on mesic sites; and balsam fir and northern white cedar on wet sites.

In these "indirect gradient analyses," the vegetation is used to infer environmental or successional gradients (without directly measuring them) associated with the continuum of communities. For example, a given stand can be appropriately placed (by its species composition) between the hypothetical extremes of the continuum—pure bur oak stands at one extreme and pure sugar maple at the other. The controlling site factors of the given stand (for example, low vs. high soil moisture) or its successional status (from early to late) may not be known but can be inferred from its relative place in the continuum. However, whether the stand composition was primarily due to a site factor (such as moisture) or to succession (or both) cannot be directly determined. A critical evaluation of the older continuum methods led Peet and Loucks (1977) to conduct "direct gradient analysis" of southern Wisconsin forests employing soil moisture and nutrient analyses in addition to vegetational studies. They were able to characterize the forest composition along both a moisture-nutrient gradient and a successional gradient (from intolerant to tolerant). The direct gradient analysis method (measuring factors along an environmental gradient and relating them to species composition or productivity) has been employed primarily in mountainous regions where strong gradients are expressed. The work of Waring and Major (1964) and Zobel et al. (1976) provide good examples. Reviews of this method are given by Whittaker and Gauch (1973), Mueller-Dombois and Ellenberg (1974), and Whittaker (1973, 1975).

## Western Coniferous Forest

The discontinuities created by geographic separation, abrupt topographic changes, geological unconformities, and forest fires have resulted in many distinct vegetational zones within the forests of the mountains of the American West. Indeed, it is here that much of the important early work on the zonation of vegetation, particularly with reference to altitude, has been done (Merriam, 1890; Shreve, 1915, 1922; Daubenmire, 1943).

Nevertheless, recent ecological studies in the western coniferous forests have demonstrated the existence of vegetational gradients with altitude, soil-moisture conditions, parent materials, climate, and other ecological factors which create a gradual sequence of changes in forest composition and structure as well as some relatively abrupt transitions from one community to another. These may be illustrated with examples from the Inland Empire in northern Idaho and the mountains of Arizona.

**Northern Rocky Mountains.** The vegetation of relatively undisturbed stands of the conifer forest of eastern Washington and northern Idaho has been investigated intensively by Daubenmire (1952, 1966; Daubenmire and Daubenmire, 1968). Daubenmire recognized, in contrast to continuum methods, more or less distinct associations separated by relatively narrow ecotones. The major differences of approach are in data gathering and manipulation and Daubenmire's use of climax tree and understory dominants in defining his associations (Chapter 12).

In general, the approach was to identify late-successional community types subjectively in the field and then sample the vegetation, the environments, and the biotic features within these types. The first stage of classification was to determine the climax tree species—those that were demonstrating reproductive success in the face of intense competition. On this basis, eight late-successional forest types (all but one defined by a single tree species) were recognized along a macroclimatic gradient from warm and dry lowlands to cool and moist highlands (Figure 17.3). At a second level of classification, most of the 8 forest types were subdivided on the basis of shrubs and herbs dominant in the understory. Distinctions at this level reflect primarily soil and microclimatic variations within a macroclimatic zone. In total, 22 climax associations were distinguished. They can be identified even in the early stages of secondary succession following fires, due to the persistence of climax herbs and shrubs many of which sprout back after burning. This layer approximates stability long before it is achieved in the overstory.

Each kind of association indicates a different kind of ecosystem. This is indicated by correlations that can be shown between these vegetational units and specific edaphic and climatic conditions, differences in tree growth rate, and disease susceptibility. Thus the ecosystems recognized, termed **habitat types**, can be identified by and are named by the plant associations. For example, the *Pinus ponderosa–Symphoricarpos* association identifies an ecosystem termed the *Pinus ponderosa–Symphoricarpos albus* habitat type. The habitat types are similar in principle to the site units of the Baden-Württemberg system (Chapter 12) although they differ in scale and though vegetation is used almost exclusively in their identification.

Using continuum methods of sampling and importance values, which give all species equal weight regardless of successional status, Daubenmire (1966) demonstrated that a continuum could be shown for his data. In regard to his classification based on late-successional species he stated:

> *A basic aim of synecology is to predict the potentialities of disturbed areas from inspection of their current, usually disturbed, plant cover. To do this, any units of classification or ordination must emphasize trends rather than take a static view that emphasizes only current vegetation composition (Daubenmire and Daubenmire, 1968).*

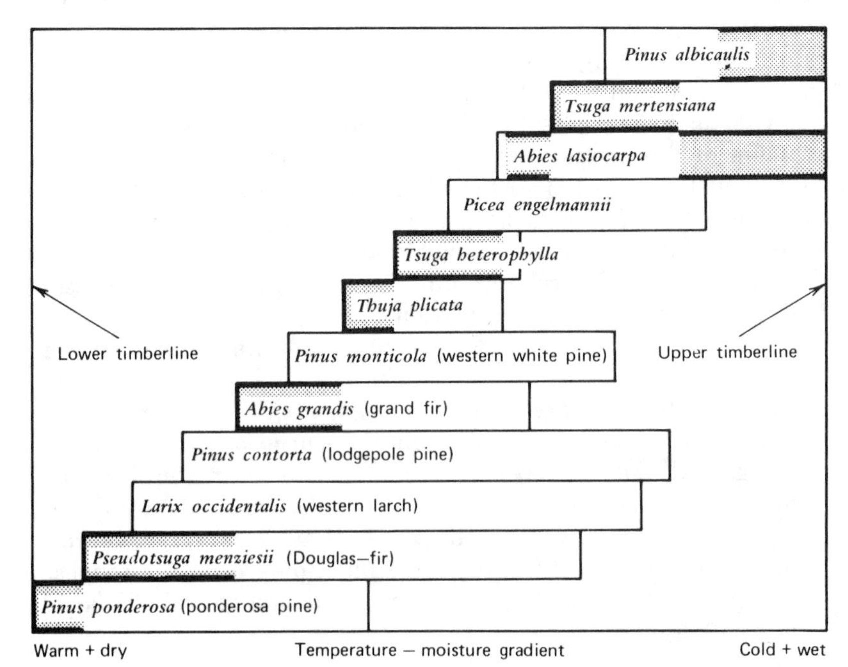


**Figure 17.3. Coniferous trees in the area centered on eastern Washington and northern Idaho, arranged vertically to show the usual order in which the species are encountered with increasing altitude. The horizontal bars designate upper and lower limits of the species relative to the climatic gradient. That portion of a species' altitudinal range in which it can maintain a self-reproducing population in the face of intense competition is indicated by the heavy lines. (After Daubenmire, 1966. Copyright 1966 by the American Association for the Advancement of Science.)**

Daubenmire's purpose is describing and classifying ecosystem units. This is clearly different from that of continuum workers who relate species and stands to environmental gradients. However, the respective approaches of Daubenmire and Whittaker are similar, not only in the mountainous terrain that facilitates their studies, but in their desire to identify ecosystem units. Whittaker emphasizes the environmental factors of the gradient in relation to the distribution of individual species, although he does recognize community types and discontinuities. Daubenmire emphasizes the distinctness of types based on late-successional species rather than the continuity of species along gradients. The conifers he studied had individualistic ranges so that contiguous communities tended to share many of the same species (Figure 17.3)—just as was demonstrated for species in the Wisconsin and Great Smokies studies. However, only in one segment of this range does a given species achieve climax status. This is well illustrated in three associations all dominated in the understory by the Pachistima union—*Abies grandis–Pachistima, Thuja–Pachistima; Tsuga–Pachistima*. The respective associations can be distinguished reliably only by the reproductive success of their characteristic climax tree species.

Regarding the continuity of the habitat types, Daubenmire stated:

> *As with other biologic classifications, we do not imply that intergrades do not exist. Not only is variation within types recognized, but a deliberate attempt has been made to document as much of this variation as possible. The ecotones of the eight primary divisions based on overstory are relatively sharp, but those of the subdivisions based on undergrowth sometimes intergrade to a considerable extent (Daubenmire and Daubenmire, 1968).*

Thus despite differences in various continuum approaches and this community-type method, different degrees of distinctness are typically found in natural environments. This is particularly noteworthy in the work of Daubenmire where by studying communities independently of secondary succession he has found varying degrees of continuity—neither the perfectly distinct association nor the completely continuous continuum.

**Arizona.** The mountains of Arizona, rising abruptly out of the cactus desert to an elevation of 3300 m and more, provide unusual opportunities for the study of vegetational change with altitude. It was here that, in 1890, C. Hart Merriam evolved his concept of life zones in the course of a survey of the biota of the San Francisco Mountains northwest of Flagstaff. Although he was concerned chiefly with the fauna of the region, he did take notes on the altitudinal distribution of the different tree species. For purposes of classification, he recognized in descending order an alpine zone, timberline zone, spruce zone, fir zone, pine zone, pinyon zone, and desert zone. His sketch (Figure 17.4) shows diagrammatically the effects of slope exposure on the altitudinal distribution of these zones.

Working out from the old Carnegie Desert Laboratory, Forest Shreve (1915) concerned himself with the distribution of vegetation on the south face of the Santa Catalina Mountains, lying northeast of

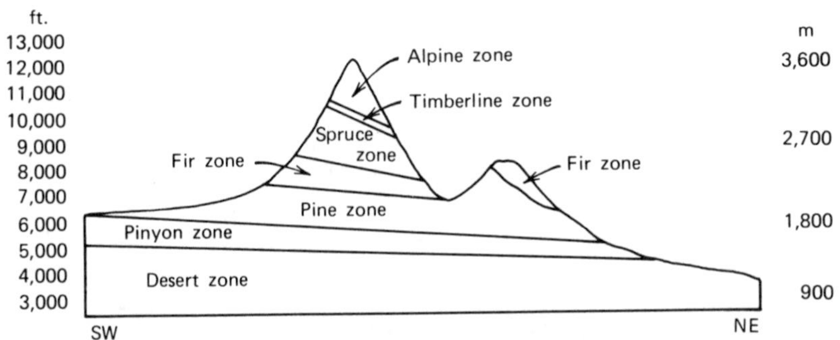


**Figure 17.4. Diagrammatic profile of San Francisco and O'Leary Peaks from southwest to northeast, showing the several life zones and effects of slope exposure. (After Merriam, 1890.)**

Tucson. Although he relied heavily upon the altitudinal distribution of the various plant species, he emphasized the effects of aspect and site on this distribution. Shreve was impressed with the independence of each species—the associated members of a plant community were not able to follow each other to a common geographic and habitat limit. He observed that the most closely associated species were not alike in their life requirements. The members of many diverse biological types or growth forms in a community found their soil moisture at different levels, procured it at different seasons, and lost it through dissimilar foliar organs, at the same time that they reacted differently to the same temperature conditions. Thus they did not live in the same climate but in different spatial or temporal sections of it.

In a comparative study of the vegetation and site conditions of north and south slopes of the Santa Catalina Mountains, Whittaker and Niering (1964, 1965, 1968a, b) have amplified Shreve's studies and further described the remarkable vegetational gradient from subalpine forest to Sonoran desert (Figure 17.5). Besides the vertical zonation of communities, one observes the effect of aspect on the altitudinal distribution of the communities. A more complex vegetational pattern existed on the north slope than on the south slope. This may be largely due to the mosaic of parent materials on the north slope (including both an acid granite-gneiss complex and basic limestone) in contrast to the dominant granite-gneiss complex of the south slope.

Species did not form distinct groups adapted and restricted to one parent material or another; they showed all degrees of relative restriction. This was also found in the Siskiyou Mountains of Oregon

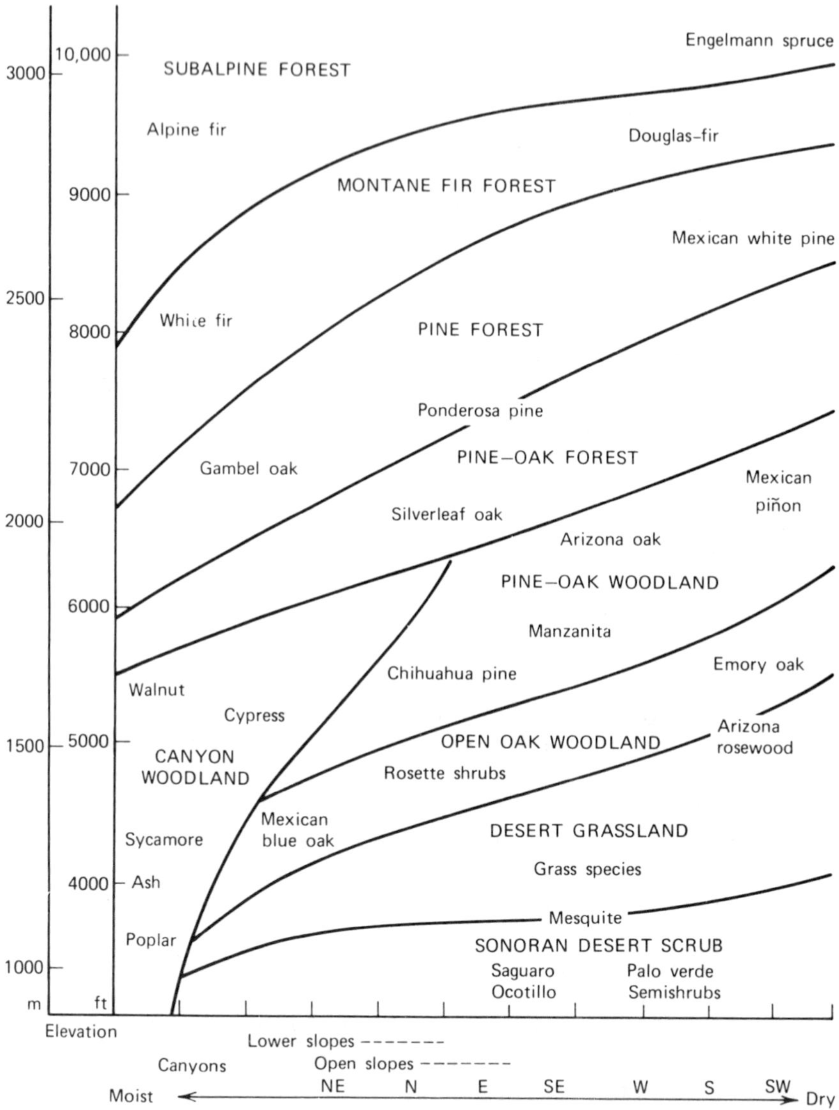


**Figure 17.5.** A diagram of the vegetation of the south slope of the Santa Catalina Mountains in southeastern Arizona. (The vegetation above 2,750 meters is for the nearby Pinaleño Mountains.) Major species are indicated by their centers of maximum importance. (After Whittaker and Niering, 1965. Reprinted with permission of Macmillan Publishing Co., Inc. from *Communities* and *Ecosystems* by Robert H. Whittaker, Copyright © 1975, Robert H. Whittaker.)

(Whittaker, 1960) and in the Coast Range of central California (McMillan, 1956). Rasmussen (1941) made similar findings in his studies on the Kaibab Plateau in the northwestern portion of Arizona. His work is noteworthy in that he found it impractical to recognize different forest zones except in a very broad sense. Rather, he emphasized that each species had a different altitudinal range and different abundance at each altitude, so that a series of intergrading forests occurred with the various species appearing, gaining abundance, and dropping out, more or less independently of one another. Similarly, the altitudinal distribution of forest trees on the mountains of southern Arizona is illustrated diagrammatically in Figure 17.6, based upon unpublished work by the senior author. The individualistic nature of the distribution pattern of each species is clearly shown, ranging from the high-altitude fir to low-altitude oaks. Although Douglas-fir is more abundant at higher altitudes than ponderosa pine and has a generally higher altitudinal range, there is such a broad range of overlap that both species may be expected to occur together in the absence of a disturbance, such as a fire, that might eliminate the fir. Similarly, the pinyons, oaks, and juniper share a low altitudinal range, but each species shows different affinities, so that the characteristic community might vary widely and more or less continuously in the absence of abrupt changes in aspect, soil moisture, parent material, or land history that might result in equally abrupt changes in the vegetation pattern.

The foregoing relationships provide evidence for the often cited principle of species individuality, originally set forth by Gleason (1926, 1939) as the "individualistic concept of the plant association." Rephrased by Whittaker and Niering (1968a), it states that "each species responds to the various environmental factors involved differently from any other species, according to its own genetic structure, range of physiological tolerances, and population dynamics including effect of competition and other relations to other species." Paradoxically, it is probably due to species interrelationships—interaction of species through competition and coevolution—that tree species show the individual distributions invariably found along gradients. Community studies typically have been directed toward describing communities and their species composition. The more complex questions of the degree of mutual interaction of species within a community and of community stability shift the emphasis of study from the community itself to the ecosystem. Such investigations involve the construction of models and testing of hypotheses of how ecosystems work and, ultimately, the kinds of genetic and non-genetic interrelationships of organisms to one another and to their physical environment. This important

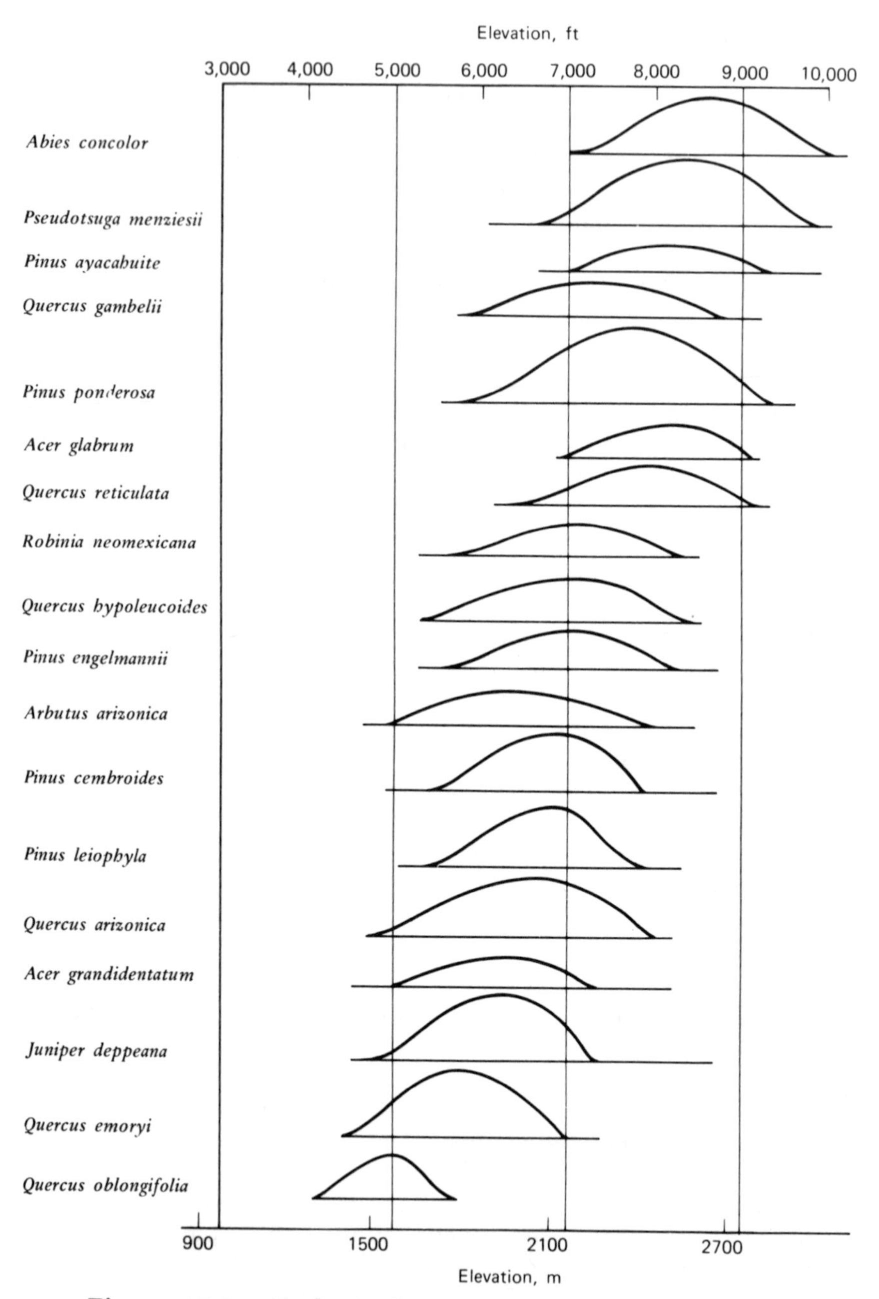


**Figure 17.6. Ecological range of mountain trees in southern Arizona.**

and difficult task of understanding ecosystems is undertaken in the following chapter.

## SUGGESTED READINGS


Curtis, John T. 1959. *The Vegetation of Wisconsin*. Univ. of Wisconsin Press, Madison. 657 pp.

Dansereau, Pierre (ed.). 1968. The continuum concept of vegetation: responses. *Bot. Rev.* 34:253–332.

Daubenmire, R. 1966. Vegetation: identification of typal communities. *Science* 151:291–298.

———. 1968. *Plant Communities.* (Chapter 4, Vegetation and ecosystem classification, pp. 249–273.) Harper & Row, Inc., New York. 300 pp.

———, and Jean B. Daubenmire. 1968. Forest vegetation of eastern Washington and northern Idaho. Washington Agric. Expt. Sta., Tech. Bull. 60. 104 pp.

Hack, John T., and John C. Goodlett. 1960. Geomorphology and forest ecology of a mountain region in the central Appalachians. U.S. Geol. Surv. Prof. Paper 347. 66 pp.

Lambert, J. M., and M. B. Dale. 1964. The use of statistics in phytosociology. *Adv. Ecol. Res.* 2:55–99.

Langford, Arthur N., and Murray F. Buell. 1969. Integration, identity and stability in the plant association. *Adv. Ecol. Res.* 6:83–135.

McIntosh, Robert P. 1967. The continuum concept of vegetation. *Bot. Rev.* 33:130–187.

Mueller-Dombois, Dieter, and Heinz Ellenberg. 1974. *Aims and Methods of Vegetation Ecology*. John Wiley & Sons, New York. 547 pp.

Rowe, J. S. 1961. The level-of-integration concept and ecology. *Ecology* 42:420–427.

———. 1969. Plant community as a landscape feature. *In* K. N. H. Greenidge (ed.), *Essays in Plant Geography and Ecology*. Nova Scotia Museum, Halifax, N.S.

Shimwell, David W. 1971. *The Description and Classification of Vegetation*. Univ. Washington Press, Seattle. 322 pp.

Spurr, Stephen H. 1956. Forest associations in the Harvard Forest. *Ecol. Monogr.* 26:245–262.

Waring, R. H., and J. Major. 1964. Some vegetation of the California coastal region in relation to gradients of moisture, nutrients, light, and temperature. *Ecol. Monogr.* 34:167–215.

Whittaker, R. H. 1956. Vegetation of the Great Smoky Mountains. *Ecol. Monogr.* 26:1–80.

———. 1975. *Communities and Ecosystems*. The Macmillan Co., 2nd ed., New York. 385 pp.

Zobel, Donald B., Arthur McKee, Glenn M. Hawk, and C. T. Dyrness. 1976. Relationships of environment to composition, structure, and diversity of forest communities of the central western Cascades of Oregon. *Ecol. Monogr.* 46:135–156.

# 18
# *Analysis of Forest Ecosystems*

Although the idea of the ecosystem is of great antiquity and is recognized universally among mankind (Major, 1969), it is only in the electronic and atomic age that we are able to deal with the ecosystem as a whole on a firm scientific basis. The goal of ecosystem analysis is to understand ecosystems and ecosystem processes. In forest systems, the goals may be directly related to a knowledge of the biological limits of productivity and the influence of such management practices as fertilization, timber harvest, grazing, irrigation, and drainage on the health, growth, and reproduction of the forest. Disturbances in the forest ecosystem and its breakdown through overgrazing, air pollution, and improper cutting have made us aware that we know too little about the functioning of the system and its interrelationship with adjacent terrestrial and aquatic systems. In the present chapter an overview of the approach to the analysis of ecological systems is given, energy flow and productivity of ecosystems are discussed, and examples of the ways in which forest ecosystems are being studied are presented.

## ECOSYSTEMS AND SYSTEMS ANALYSIS

An ecosystem is a special case of a general system—a collection of interacting entities or collection of parts, together with statements on the relationships between these parts (Dale, 1970). A system is termed *open* if there are inputs into and outputs from the system. An ecosystem is an open system in which at least one of the entities is living. In the system displayed in Figure 18.1, solar energy and precipitation are inputs into the system, and water, mineral substances, matter, and energy are outputs. In a managed forest system, timber, animals, water, aesthetics, and recreational values are the outputs that are utilized by humans.

Various kinds of ecological systems may be recognized and studied, depending on the level at which life is examined. The biosphere with its total environment is a gigantic ecological system. Within the biosphere are ecosystems of forests, grasslands, swamps, and cultivated fields, all of which fit the general scheme of Figure

18.1, that is, a biotic community together with its physical environment. At a lower rank each plant or animal individual, with its particular microenvironment, constitutes a biological system. Within the individual we can cite a further series of systems: organ–tissue–cell–organelle. An ecosystem can be broken into a hierarchy of subsystems based on processes. For example, subsystems of an ecosystem depicted by Figure 18.1 would include among others: the energy transfer system, the precipitation-evapotranspiration system, and the mineral cycling system.

The principal common denominator of terrestrial ecosystems is the food chain linking the physical environment with the three major living components, **producers, consumers,** and **decomposers** (Figure 18.1). Producers are usually green plants that produce carbohydrates and accumulate other organic substances. They are termed **autotrophic** (self-feeding) since the carbohydrates they produce satisfy their own nutritional requirements.

Herbivores, carnivores, and parasites are the consumers that feed on plants and are termed **heterotrophic** (other-feeding) because they derive their nutrition from other organisms. Decomposers (also heterotrophs), especially fungi and bacteria, live on dead organic material (termed **detritus**) and are essential in the process of mineral cycling, which many consider the heart of ecosystem dynamics. In contrast to nutrient flow, which is cyclic, energy transfer is unidirectional, and energy is lost irretrievably from a system in several ways.

Systems analysis is the method, employing conceptual and mathematical models, which allows us to handle the great complexity of ecosystems. It is simply the conscious application of scientific method to complex organizations in order that no important factor be overlooked. It may be described as an explicit statement of what goes on, and in what order, in a system. Systems analysis is similar to problem solving in that the same phases are characteristic of both (Dale, 1970): (1) identification of all significant components that are of interest, (2) definition of the relationship between the selected components, (3) specification of the mechanisms by which changes in the system (the distribution of properties across the components) take place, and (4) solution and validation of the model outputs by comparing them to real system outputs. This approach allows us to examine large segments of nature as integrated systems.

## DEVELOPING A SYSTEMS MODEL

Once the level and purposes of the analysis have been determined, the boundaries of the ecosystem must be established and the compo-

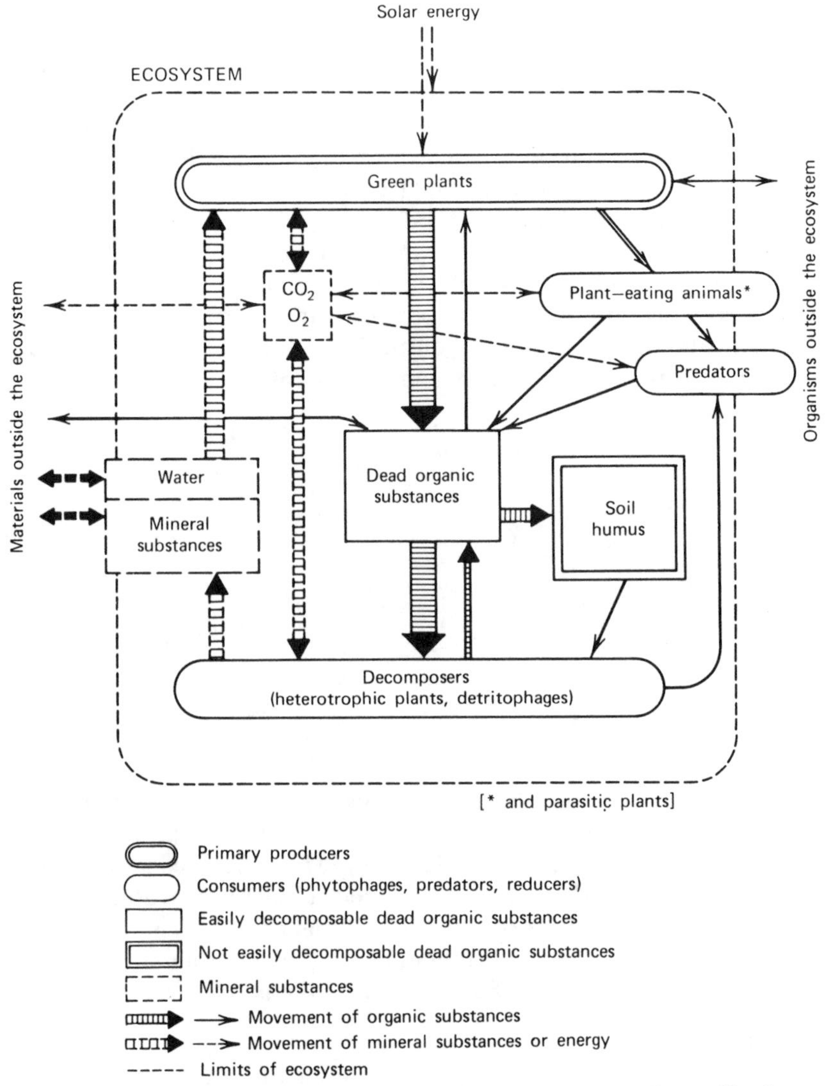


**Figure 18.1. General scheme of an ecosystem. (After Ellenberg, 1971b.)**

nents of interest identified. The boundaries are arbitrary except that the area must be large enough to contain all ecosystem components, processes, and their interactions. In addition, the boundary should be placed where inputs and outputs across it are most easily measured.

A good example of an ecosystem for studying the nutrient cycle-hydrological interaction is the small watershed approach at Hubbard

Brook, New Hampshire (Borman and Likens, 1969; Likens et al., 1977). The ecosystem is a forested watershed of uniform geology which is underlain by a tight bedrock. The inputs of water and minerals in precipitation could be measured. Since the watershed is watertight, all geological output (water, sediment, and dissolved minerals) can be measured as all drainage water flows over a notch in a weir at the mouth of the stream draining the watershed (Figure 18.2). The nutrient balance for a single element in the ecosystem is then: geological input + meteorological input + biological input − (geological output + biological output) = net loss or gain. Biological components would be animals and plants entering and leaving the system.

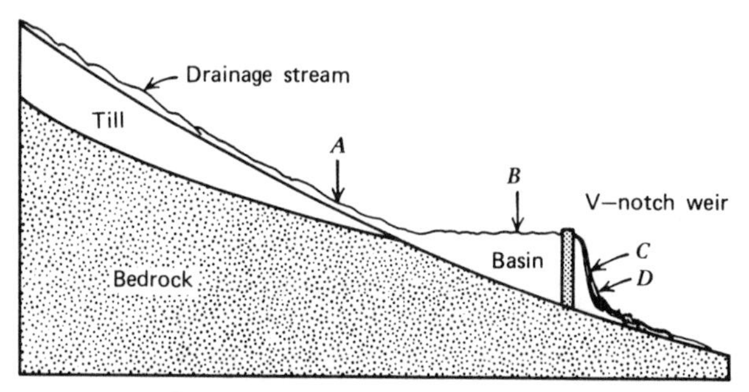


**Figure 18.2. Sampling design for measurement of output components in stream water of a terrestrial watershed ecosystem. *A*, water sample for dissolved substances; *B*, sediment load dropped in basin; *C*, water sample for millipore filtration; *D*, net sample. Total losses = dissolved substances (*A*) + particulate matter (*B* + *C* + *D*). (After Bormann and Likens, 1969.)**

Once the boundaries and components of interest are identified, the relationships between the components may be conceptually modeled and diagrammatically presented in a flow chart (Figure 18.3). Alternatively, a tabular format may be used to formalize a system; we will follow the one described by Smith (1970). First, assume the Hubbard Brook watershed has $n$ components, each of which can be described quantitatively by many criteria, such as amount of calcium, nitrogen, and phosphorus, water content, caloric content, and so on. The components could be the individual plant and animal species, or, more conveniently, trees, shrubs, herbivores, carnivores, fungi, and bacteria. For each variable (such as calcium) a set of tables can be used to describe the system. For example, Table

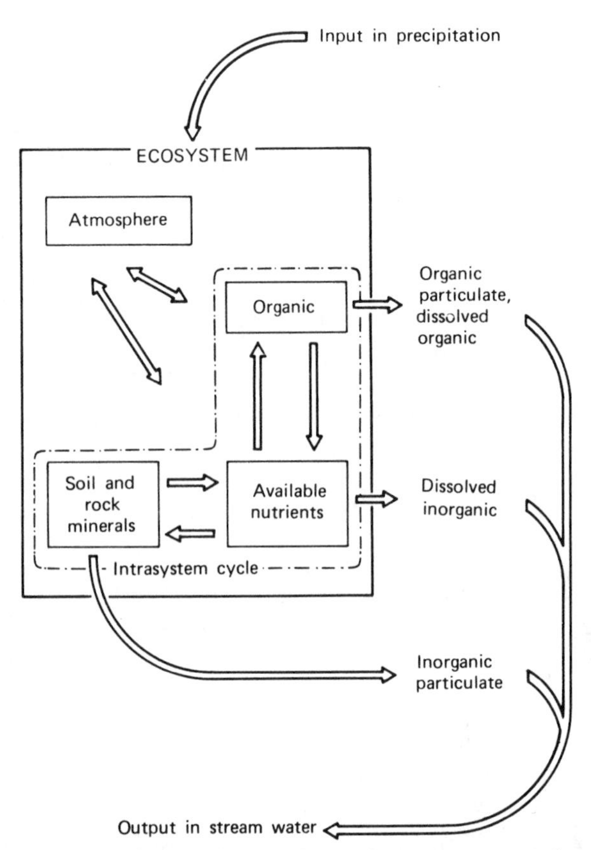


**Figure 18.3. Nutrient relationships for a terrestrial watershed ecosystem. Sites of accumulation, major pathways, and origins of chemical losses in stream water are shown. (After Bormann and Likens, 1969.)**

18.1 shows the amounts ($x_i$) of calcium in each component of the system; also shown are the inflow ($a_i$) and outflow ($z_i$) of calcium for each component of the system. The total amount of calcium in the system is the sum of the first row ($x_n$); the total inflow is the sum of the second row ($a_n$); and the total outflow is the sum of the third row ($z_n$). This table essentially summarizes the nutrient budget for calcium for each component and for the ecosystem as a whole. The budget for calcium, based on input and output data from a forest ecosystem at Hubbard Brook, is illustrated in Figure 18.4.

Calcium may be transferred from one component to another in the ecosystem in addition to inflow and outflow. This is illustrated in Figure 18.4 in the reciprocal transfer of exchangeable calcium between soil and trees and between components of the soil and the

**Table 18.1 The Amounts ($x_i$) of Calcium in Each of the $n$ Components of an Ecosystem, and the Rates at Which Calcium Is Entering ($a_i$) and Leaving ($z_i$) the System via Each Component**

| COMPONENT | 1 | 2 | 3 | • | • | • | $n$ |
|---|---|---|---|---|---|---|---|
| Amount | $x_1$ | $x_2$ | $x_3$ | • | • | • | $x_n$ |
| Inflow | $a_1$ | $a_2$ | $a_3$ | • | • | • | $a_n$ |
| Outflow | $z_1$ | $z_2$ | $z_3$ | • | • | • | $z_n$ |

Source: Smith, 1970.
Note: In any real system many of the rates will be virtually zero.

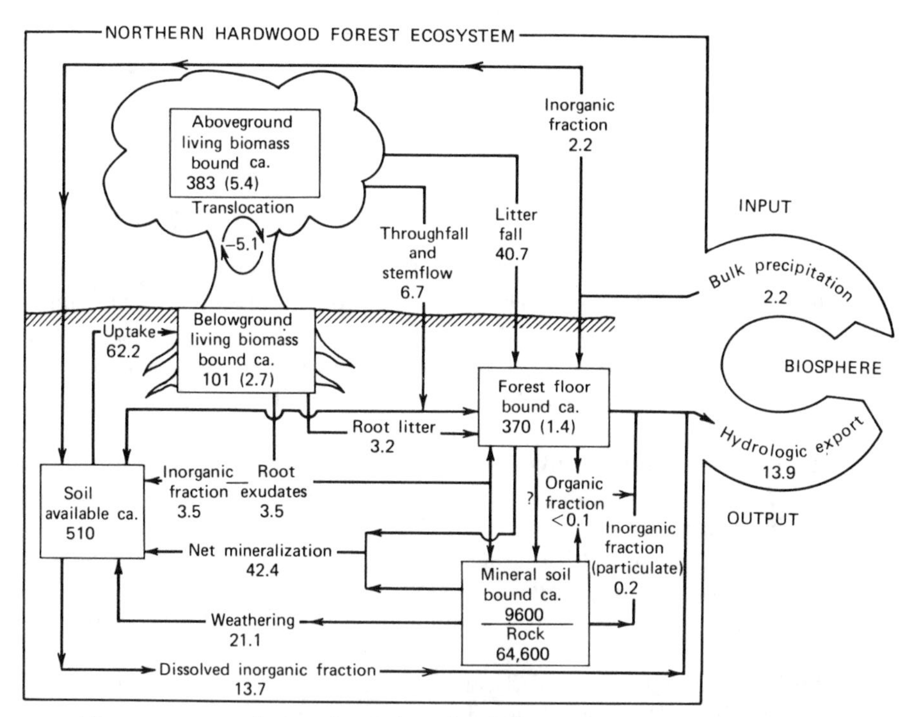


**Figure 18.4. Annual calcium budget of a forest ecosystem at Hubbard Brook. Standing crop values are in kg ha$^{-1}$ and calcium fluxes are in kg ha$^{-1}$ yr$^{-1}$. Values in parentheses are annual accretion rates. (After Likens et al., 1977.)**

minerals. Again a tabular presentation may be used to show transfer to and from each component (Table 18.2). Each row shows losses from a given component, and the row sum is the total rate of loss to other components. Each column shows gains to a particular component, and the column sum is the total rate of gain from all other

**Table 18.2 Transfer Rates of Calcium Among *n* Components of an Ecosystem**

| TRANSFER FROM EACH COMPONENT | TRANSFER TO EACH COMPONENT | | | | | | | |
|---|---|---|---|---|---|---|---|---|
| | 1 | 2 | 3 | 4 | • | • | • | $n$ |
| 1 | — | $y_{12}$ | $y_{13}$ | $y_{14}$ | • | • | • | $y_{1n}$ |
| 2 | $y_{21}$ | — | $y_{23}$ | $y_{24}$ | • | • | • | $y_{2n}$ |
| 3 | $y_{31}$ | $y_{32}$ | — | $y_{34}$ | • | • | • | $y_{3n}$ |
| 4 | $y_{41}$ | $y_{42}$ | $y_{43}$ | — | • | • | • | $y_{4n}$ |
| • | • | • | • | • | • | • | • | • |
| • | • | • | • | • | • | • | • | • |
| $n$ | $y_{n1}$ | $y_{n2}$ | $y_{n3}$ | $y_{n4}$ | • | • | • | — |

Source: Smith, 1970.
Note: In any real system many of the rates will be zero.

components. Many entries in such a table may be negligible or zero. An advantage of this format is that it can handle any degree of complexity within the system. Following Smith's (1970) description of the method, it is possible to write an equation for the rate of change of each variable (calcium, nitrogen, water, etc.) for each of the components with this information. Thus the equation for component 3 would be:

$$dx_3/dt = a_3 - z_3 + (y_{13} + y_{23} + \dots + y_{n3}) - (y_{31} + y_{32} + \dots + y_{3n})$$

If component 3 were leaf litter, the equation might represent the rate of change with time in the amount of calcium due to: litter moved inward across the ecosystem boundary, minus that moved outward, plus litter received from various plant components (other $y_{i3}$ terms may be zero), minus litter removed by consumers or decomposers, and minus litter transformed into humus (other $y_{3j}$ terms may be zero).

If a study of the Hubbard Brook ecosystem had been made using the tabular format, Tables 18.1 and 18.2 could be filled with numbers representing amounts and rates. One would learn where activity does and does not occur, an important gain in information for most ecosystems. The system could then be set into a computer and, beginning with all the $x_i$'s at their estimated levels, allowed to change through time according to the equations $dx_i/dt$.

Although we have learned many new things about the ecosystem, we will discover at this point that the data are inadequate to describe the functioning or dynamic nature of the ecosystem. Under computer simulation each component changes in the same direction at the same rate indefinitely. Because we are using average rates, the system we have modelled cannot respond to change with changes in rates. It is these responses that characterize the real world ecosystem which we wish to simulate. For example, if the growing season is exceedingly dry, the rates of calcium transfer from litter to soil and soil to plants may be markedly reduced and the output from the system substantially different from the average estimate. Thus the use of an average rate (a simple number), rather than a variable rate (a mathematical function) corresponding to changes in environmental factors, destroys the dynamic properties of the system.

Smith (1970) describes what is necessary as follows:

> *It turns out in fact that the new problem is an order of magnitude more complicated than the estimation of averages. In particular, the rates must be expressed as* functions *of the system, and not as simple numbers, if we wish to learn anything about the system. Each transfer rate,* $y_{ij}$, *and each outflow,* $z_i$, *is a set of functions, an equation, which relates the rate to all those direct causal factors that govern it. It is only then that we can predict how each rate will change if the system is changed. Furthermore, if all of these are combined in the computer, we can predict how the system will respond to change. This is the only way that we can discover how ecosystems operate.*

This development means that a great amount of experimental research that is relevant in the system, much of it physiological ecology, must be conducted. The expression of rates as sets of functions requires the estimation of many parameters other than the amounts and rates of calcium transfer. In simplified form, these can be arranged as shown in Table 18.3. In addition to $x_i$, $a_i$, and $y_i$, for each component, additional properties or attributes are usually needed. These $w_{ij}$'s may include the average size, age, and number of individuals in a component, distribution in space, growth habit, etc. Like the $y_{ij}$'s and the $z_i$'s they may be sets of functions. In addition, a set of additional inputs ($A_i$) to the system, shown at the bottom of Table 18.3, is necessary. These are ***external variables*** or ***driving variables***, such as climatic factors and season, that affect the system as a

**Table 18.3 Additional Descriptors Needed for the Specification of Functional Relations in an Ecosystem of *n* Components**

| COMPONENT | COMPONENT ATTRIBUTES (OPEN LIST) | | | | | | |
|---|---|---|---|---|---|---|---|
| 1 | $w_{11}$ | $w_{12}$ | $w_{13}$ | • | • | • | |
| 2 | $w_{21}$ | $w_{22}$ | • | • | • | | |
| 3 | $w_{31}$ | $w_{32}$ | $w_{33}$ | $w_{34}$ | • | • | • |
| • | | | | | | | |
| • | | | | | | | |
| • | | | | | | | |
| *n* | $w_{n1}$ | $w_{n2}$ | $w_{n3}$ | • | • | • | |

Source: Smith, 1970.
External Attributes (i.e., climate, weather; open list): $A_1$, $A_2$, $A_3$, $A_4$, etc.

whole. These and the inflows, $a_i$, are externally controlled variables that influence the system but are not affected by it. The rates, $y_{ij}$ and $z_i$, termed ***state variables***, measure system processes that may be functions of the external variables and any or all of the other amounts and rates that have been presented.

The immediate goal of system research workers is to find appropriate mathematical functions for the effect of external variables and for relationships among internal variables. The initial phase of ecosystem modelling is adequate once a satisfactory set of these functions is developed. The validity of the model can be assessed starting with an initial distribution of amounts, $x_i$, and a program through time of the input variables ($A_i$ and $a_i$) and then comparing the amounts of the components predicted by the computer with observations in the field. This is the validation phase of the systems procedure, and the primary interest lies in how well the model outputs mimic those of the real system. To obtain a high fidelity to a real system requires a process of successive approximation, with the model progressively changed until the desired fidelity is achieved. The effects on the system of a treatment such as fertilization may be simulated by increasing the inputs of mineral nutrients and following the computer program. The results may be compared with field trials and further adjustments made in the model as necessary. Once a satisfactory model has been constructed, it becomes a powerful tool of management for predicting treatment effects. At the same time it has extended biological and other knowledge of the system and thus provided an understanding of certain portions of an ecosystem. General reviews of modeling are provided by O'Neill (1975), and Wiegert (1975).

## PRODUCTIVITY

The greatest contribution of research on ecological systems is the elucidation of the functional processes of forests such as photosynthesis, respiration, and the cycling of water, carbon, and nutrients, and the bearing of these processes on forest productivity (Ovington, 1962). With progressively fewer exceptions most of the world's forests are modified by humans through management or mismanagement. Thus forest processes must be understood in detail to predict not only production but the other potential effects of manipulations. The first step in studying the processes as they relate to productivity is essentially an inventory of the levels of dry matter (g $m^{-2}$) in each component of the ecosystem. The most widely studied aspect of forest ecosystems in a systems context has been the flow of energy into photosynthate, its partition into dry-matter production of plants, and energy flow through the food chain of consumers and decomposers. Studies of production, although not currently applicable in site quality estimation, provide ways of characterizing and comparing forest communities and also gross estimates of the average productivity of major portions of the earth. A comprehensive account of the history, methods, and patterns of productivity of world ecosystems is available (Lieth and Whittaker, 1975; Reichle et al., 1975). The terminology of production and examples of productivity attributes of several forest communities are considered in this section.

A small fraction of the radiant energy reaching the earth's surface, about 2 percent, is converted by green plants into chemical energy in photosynthesis and is termed their **gross primary production.** This gross production may be utilized in the formation of plant tissues (biomass) or used in respiration. Thus gross primary production minus respiration of green plants (the primary producers) equals a biomass termed **net primary production.** The energy of net primary production may be partitioned in several ways. A portion is accumulated in the community in plant and animal tissues (biomass accumulation) and part is used in the respiration of animals and plants that feed on the green plants. These relationships are seen clearly when diagrammed:

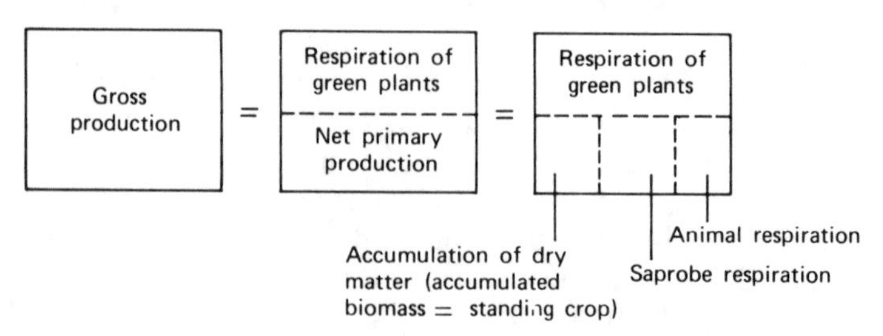

Productivity may be expressed either as an amount for a given period or an amount per unit time (rate), usually a year. The energy or dry matter in plants that is left after subtracting plant respiration per unit area and time is net primary productivity which may be expressed either as energy (cal $cm^{-2}$ $yr^{-1}$) or dry organic matter (g $m^{-2}$ $yr^{-1}$). The energy equivalents of dry-weight production are based on bomb calorimetric measurements for different tissues. Land plant tissues average about 4.25 kCal per dry gram. For example, the productivity of an oak-pine forest on Long Island, New York, has been estimated (Whittaker and Woodwell, 1969; Whittaker, 1975) and is partitioned as follows:

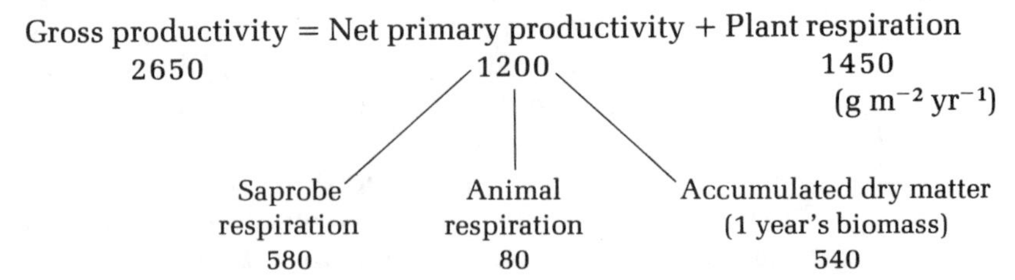


Most of the gross productivity goes to ecosystem metabolism, and less than half the net primary productivity and only 20 percent of the gross productivity remains in the standing crop of trees, understory plants, and animals.

The biomass, the weight of all organisms (usually expressed in dry-matter content), per unit area in an ecosystem at a given moment in time is termed the **standing crop.** In a forest community, standing crop is biomass accumulated to the time of measurement, whereas net primary productivity is but a fraction of this, the amount of dry matter produced in a given time period, usually a year. For example, the standing crop per hectare in a redwood forest 1500 years of age may be enormous compared to its present net productivity per year, which may be very low. An ecosystem supporting a fast-growing 15-year-old stand of eastern cottonwood, however, may have only a fraction of the biomass of the redwood ecosystem, but a net primary productivity many times that of the redwood system. The ratio of standing crop to net primary productivity, the biomass accumulation ratio, is typically low in young, fast-growing stands, where most energy is used for growth. It is higher in old stands, where most of the energy is used to maintain the high existing biomass.

Consumers (herbivores and carnivores) feeding directly or indirectly from primary producers create a secondary production (Figure 18.5). As energy flows or is dragged forcibly from one trophic level to the next, from primary producer to herbivore and from herbivore to carnivore, energy is lost and production decreases. Herbivores in the forest community utilize only a small part of the total gross

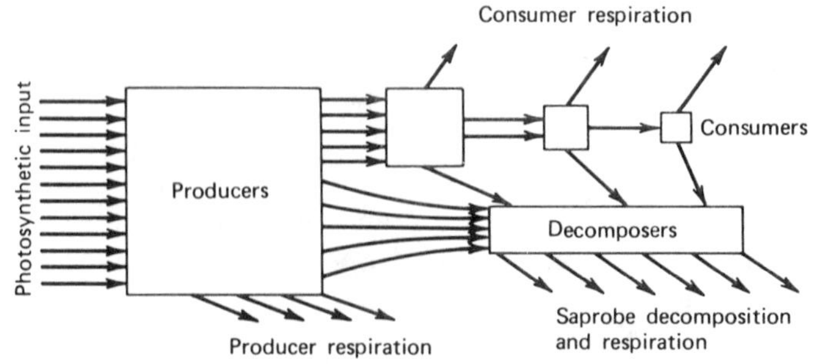


**Figure 18.5. Flow of energy in a natural community. In a steady-state community the intake of photosynthetic energy on the left, and the dissipation of energy back to environment toward the right, are in balance. The pool of energy of organic compounds within the community remains constant. (After Whittaker, 1975. Reprinted with permission of Macmillan Publishing Co., Inc. from *Communities and Ecosystems* by Robert H. Whittaker © 1975 by Robert H. Whittaker.)**

production of green plants. Part of it is utilized in respiration of primary producers, and much material is simply unavailable in roots, stem wood, bark, and branches. Even much of the available leaves, shoots, fruits, and seeds may be of a quality not acceptable to mammals, birds, and insects. Similarly, at the next trophic level secondary production by carnivores is substantially less than that of herbivores. Thus the energy entering the system is dissipated to the environment by respiration and biological activity and must be constantly replenished from the external radiation source.

Secondary production may be significant in a grassland or aquatic ecosystem where the herbivore-carnivore system (grazing chain) is especially well developed. However, in forest ecosystems, and especially plantations, litter decomposition and decomposers are primary sources of energy dissipation. Figure 18.6 illustrates the partition of energy in a young plantation of Scots pine. In an 18-year period most of the net production was stored in trees or in the litter or was decomposed by soil organisms. Instead of passing from herbivores to carnivores, much of the energy of net production was harvested by man.

Summaries of standing crop biomass and net primary productivity data (Art and Marks, 1971; Newbould, 1967; Ovington, 1962; Rodin and Bazilevich, 1968) show increasing productivity with increasing radiation (from arctic to tropical areas), decreasing elevation, and increasing moisture. Productivity estimates provide a very

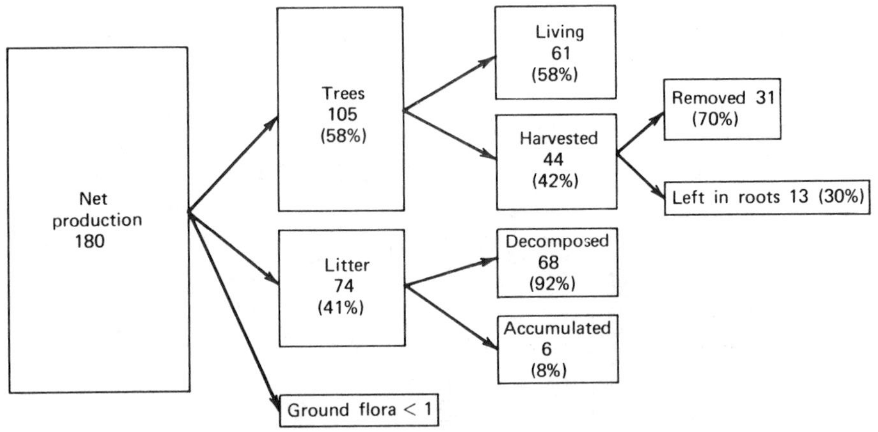


**Figure 18.6. Fate of energy in a Scots pine (*Pinus sylvestris*) plantation over 18 years, in $10^7$ kilocalories per hectare. (Based on data of Ovington, 1962; after Kormondy, 1969. With permission from Advances in Ecological Research 1:103-192. Copyright by Academic Press Inc. (London) Ltd.)**

general comparison of ecosystems of various kinds (Table 18.4). The estimates must be viewed with caution. For example, it would seem that temperate forests are about twice as productive as cultivated land in terms of net primary productivity (Table 18.4). However, this does not necessarily indicate that forest land is more fertile than cultivated land or that forests are more efficient than crop plants in producing organic matter. Actually, these productivities are not entirely comparable since that of agricultural crops is typically based on plants using only part of 1 growing season while annual production of forests is based on a growing stock that may have 30, 50, or more than 100 years of accumulated biomass. In Table 18.4, for example, the mean biomass accumulated in evergreen and deciduous temperate forests is 28 times that of cultivated land.

The standing crop is not a measure of potential site productivity, but rather a function of age. Stem biomass accumulates year by year and becomes an increasingly greater part of the total biomass of the community. The change of biomass with age in a natural stand of oaks and pitch pine (Figure 18.7) illustrates the increase in dry weight of trees and the declining biomass of the undergrowth shrubs as the trees become dominant.

Community or stand comparisons of biomass must also take stand density and past stand history into account, in addition to age. For example, markedly different estimates of biomass were found for

**Table 18.4 Net Primary Production and Plant Biomass for Major Ecosystems of the Earth's Surface**

| ECOSYSTEM TYPE | AREA[a] $10^6$ km$^2$ | NET PRIMARY PRODUCTIVITY, PER UNIT AREA[b] gm m$^{-2}$ yr$^{-1}$ | | WORLD NET PRIMARY PRODUCTION[c] $10^9$ t yr$^{-1}$ | BIOMASS PER UNIT AREA[d] kg m$^{-2}$ | | WORLD BIOMASS[c] $10^9$ t |
|---|---|---|---|---|---|---|---|
| | | NORMAL RANGE | MEAN | | NORMAL RANGE | MEAN | |
| Tropical rain forest | 17.0 | 1000–3500 | 2200 | 37.4 | 6–80 | 45 | 765 |
| Tropical seasonal forest | 7.5 | 1000–2500 | 1600 | 12.0 | 6–60 | 35 | 260 |
| Temperate evergreen forest | 5.0 | 600–2500 | 1300 | 6.5 | 6–200 | 35 | 175 |
| Temperate deciduous forest | 7.0 | 600–2500 | 1200 | 8.4 | 6–60 | 30 | 210 |
| Boreal forest | 12.0 | 400–2000 | 800 | 9.6 | 6–40 | 20 | 240 |
| Woodland and shrubland | 8.5 | 250–1200 | 700 | 6.0 | 2–20 | 6 | 50 |
| Savanna | 15.0 | 200–2000 | 900 | 13.5 | 0.2–15 | 4 | 60 |
| Temperate grassland | 9.0 | 200–1500 | 600 | 5.4 | 0.2–5 | 1.6 | 14 |
| Tundra and alpine | 8.0 | 10–400 | 140 | 1.1 | 0.1–3 | 0.6 | 5 |
| Desert and semidesert scrub | 18.0 | 10–250 | 90 | 1.6 | 0.1–4 | 0.7 | 13 |
| Extreme desert, rock, sand, and ice | 24.0 | 0–10 | 3 | 0.07 | 0–0.2 | 0.02 | 0.5 |

| | | | | | | | |
|---|---|---|---|---|---|---|---|
| Cultivated land | 14.0 | 100–3500 | 650 | 9.1 | 0.4–12 | 1 | 14 |
| Swamp and marsh | 2.0 | 800–3500 | 2000 | 4.0 | 3–50 | 15 | 30 |
| Lake and stream | 2.0 | 100–1500 | 250 | 0.5 | 0–0.1 | 0.02 | 0.05 |
| Total continental | 149. | | 773 | 115 | | 12.3 | 1837 |
| Open ocean | 332.0 | 2–400 | 125 | 41.5 | 0–0.005 | 0.003 | 1.0 |
| Upwelling zones | 0.4 | 400–1000 | 500 | 0.2 | 0.005–0.1 | 0.02 | 0.008 |
| Continental shelf | 26.6 | 200–600 | 360 | 9.6 | 0.001–0.04 | 0.01 | 0.27 |
| Algal beds and reefs | 0.6 | 500–4000 | 2500 | 1.6 | 0.04–4 | 2 | 1.2 |
| Estuaries | 1.4 | 200–3500 | 1500 | 2.1 | 0.01–6 | 1 | 1.4 |
| Total marine | 361 | | 152 | 55.0 | | 0.01 | 3.9 |
| Full total | 510 | | 333 | 170 | | 3.6 | 1841 |

Source: Whittaker, 1975. Reprinted with permission of Macmillan Publishing Co., Inc. from *Communities and Ecosystems* by Robert H. Whittaker, Copyright © 1975 by Robert H. Whittaker.

[a] Square kilometers × 0.3861 = square miles.

[b] Grams per square meter × 0.01 = t $ha^{-1}$, × 10 = kg $ha^{-1}$, × 8.92 = lbs $acre^{-1}$.

[c] Metric tons ($10^6$gm) × 1.1023 = English short tons.

[d] Kilograms per square meter × 10 = t $ha^{-1}$, × 8922 = lbs $acre^{-1}$, × 4.461 = English short tons per acre.

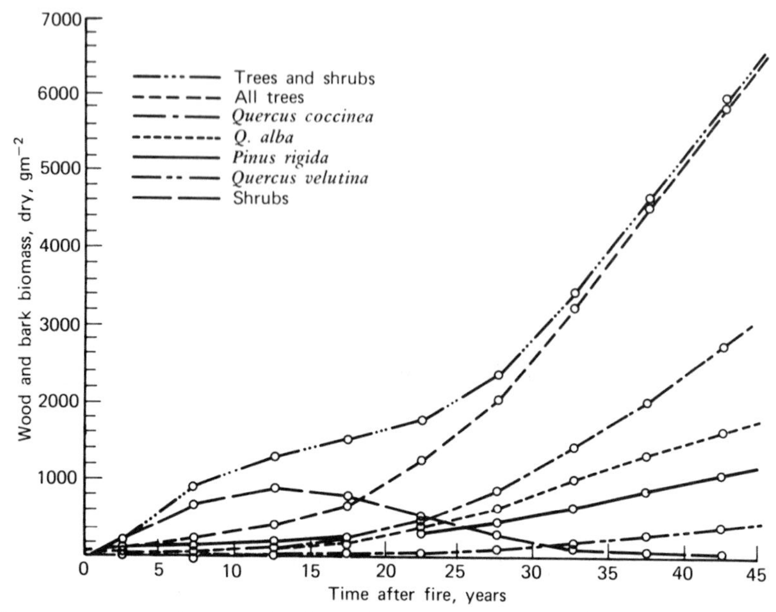


**Figure 18.7. Accumulation of wood and bark biomass in living trees above ground since a fire 45 years ago in the Brookhaven Forest, Long Island, New York. (After Whittaker and Woodwell, 1969.)**

three 100-year-old stands of lodgepole pine of different densities and past treatment in western Alberta, Canada (Table 18.5; Johnstone, 1971). The site conditions were similar for all stands, and the ground vegetation was negligible in the three stands. Stand 2 had been thinned at age 70, but the dry weight of the material removed is unknown. A major problem in using biomass to compare potential productivity of natural forest communities over large portions of North America is that their densities vary greatly and their past histories, often including harvesting, burning, grazing, or windthrow, are unknown.

The measurement of biomass and net annual productivity of the plant component of the ecosystem is an enormous task. In a plantation of trees of the same age, a sample of trees and other vegetation may be harvested and the dry weight per unit area of roots, stems, branches, and leaves determined to give the biomass of the existing community. In a 70-year-old stand, the total biomass divided by 70, plus corrections for annual production of leaves, flowers, and fruits as well as root, stem, and branch mortality, would give an estimate of average net productivity per year. In natural forest communities that

**Table 18.5 Stand Attributes and Standing Crop of Biomass for Three Stands of 100-year-old Lodgepole Pine**

| | STAND 1 | STAND 2 [a] | STAND 3 |
|---|---|---|---|
| Number of living stems per hectare | 2520 | 717 | 12,256 |
| Mean diameter (cm) | 16.3 | 24.9 | 5.6 |
| Mean height (m) | 16.8 | 20.1 | 5.8 |
| Stem volume ($m^3$ $ha^{-1}$) | 445 | 357 | 182 |
| Dry weight (of needles, branches, stem, and roots) (kg $ha^{-1}$) | 271,900 | 221,680 | 111,750 |

Source: Johnstone, 1971.
[a] Had been thinned.

have many species of different ages, complex sampling and analysis procedures are required (Newbould, 1967; Whittaker, 1966; Whittaker and Woodwell, 1968; Whittaker and Marks, 1975), and they are periodically improved.

Net primary productivity changes with age, probably not unlike the curve of mean annual increment of stem wood (Bruce and Schumacher, 1950), and hence, as with the standing crop, comparisons of community net productivity must consider age. The change in net productivity in a Scots pine plantation in England is illustrated in Figure 18.8 (Ovington, 1962). The mean productivity of trees increases to a maximum at about 35 years and decreases slightly thereafter. The current productivity of trees rises rapidly until about age 20, is maintained for several years, and declines rapidly after age 35. The understory vegetation makes its greatest contribution when the plantation is young and its least when tree dominance is greatest (dense crown cover and greatest root development). At later stages, particularly after age 35, thinning creates openings in the canopy that are not entirely closed so that additional light and moisture are available for undergrowth.

The productivity of a young oak-pine community on a poor site in central Long Island, New York, has been studied intensively by Whittaker and Woodwell (1968, 1969), and a comparison with a mixed hardwood cove forest in Tennessee and an old-growth Douglas-fir forest in Oregon (Table 18.6) illustrates several ways in which different ecosystems may be characterized. Small white, scarlet, and black oaks, along with scattered pitch pines, dominate the Long Island forest. They grow on well-drained podzolic soils derived from outwash sands and gravels. The community is floristically simple, spatially homogeneous, and of small stature. In contrast, the cove forest is rich floristically and spatially heterogeneous, with mature trees of large dimensions. The youth and small stature

**Figure 18.8. Annual net production of organic matter by the trees and ecosystem (trees + understory vegetation) in plantations of Scots pine. (After Ovington, 1962. With permission from Advances in Ecological Research 1:103-192. Copyright by Academic Press Inc. (London) Ltd.)**

of the oak-pine forest is evident from the low standing crop of biomass, high proportion of biomass in branch, bark, and leaves, and the low biomass accumulation ratio. The old-growth Douglas-fir forest is characterized by a higher biomass than the other forests, a high biomass accumulation ratio, and a high respiration (as expected for the mature forest).

The biomass of the root system of the oak-pine forest is about 2½ times as great as that of most mature forests (Whittaker and Woodwell, 1969) and exceeds that of the cove forest by a similar amount. The oaks and many shrubs of the oak-pine forest are adapted to fire and the shoots of the present stand arose from root systems which survived a fire about 1918. The roots have undoubtedly grown through more than one generation of shoots and are massive in comparison to the shoots they currently support (Table 18.6).

The oak-pine forest is atypical in the low percentage of stem wood of the total biomass compared with most of the forest communities for which estimates have been published. In general, the proportion of stem wood is particularly high (usually over 50 percent of above-

**Table 18.6 Net Production and Biomass in Temperate Zone Forests**

| | OAK–PINE FOREST[a] | | DECIDUOUS COVE FOREST[b] | | OLD–GROWTH DOUGLAS-FIR[c] | |
|---|---|---|---|---|---|---|
| | NET PRODUCTION | BIOMASS | NET PRODUCTION | BIOMASS | NET PRODUCTION | BIOMASS |
| Totals, net production ($gm\ m^{-2}\ yr^{-1}$) and biomass ($kg\ m^{-2}$), dry matter, for trees | 1,060 | 9.7 | 1,300 | 58.5 | 980 | 86.4 |
| Totals for undergrowth | 134 | 0.46 | 90 | 0.135 | 107 | 0.7 |
| Percentages of totals for trees in | | | | | | |
| Stem wood | 14.0 | 36.1 | 33.3 | 69.3 | 21.9 | 66.7 |
| Stem bark | 2.5 | 8.4 | 3.7 | 6.3 | 2.7 | 8.1 |
| Branch wood and bark | 23.3 | 16.9 | 13.1 | 10.3 | 21.2 | 6.1 |
| Leaves | 33.1 | 4.2 | 29.1 | .6 | 18.5 | 1.4 |
| Fruits and flowers | 2.1 | .2 | 1.8 | .03 | 4.8 | 0.1 |
| Roots | 25.0 | 34.2 | 19.0 | 13.5 | 30.9 | 17.7 |
| Biomass accumulation ratio | 8.5 | | 43.5 | | 80.1 | |
| Total respiration/gross productivity | 0.80 | | 1.0 | | 0.98 | |
| Age of canopy trees, years | 40–45 | | 150–400 | | 400–450 | |
| Mean tree height, m | 7.6 | | 34.0 | | 70.0 | |

Source: *a* and *b* modified from Whittaker (1975); Reprinted with permission of Macmillan Publishing Co., Inc. from *Communities and Ecosystems*, by Robert H. Whittaker, Copyright © 1975 by Robert H. Whittaker. *c* from Grier and Logan (1977).

[a] Young, oak-pine forest at Brookhaven, Long Island, New York.

[b] Climax deciduous cove forest, Great Smoky Mountains National Park, Tennessee.

[c] Old-growth Douglas-fir, H. J. Andrews Experimental Forest, west-central Cascade Mountains, Oregon.

ground biomass) in closed stands of conifers and for pioneer conifers and hardwoods in particular. Even the intolerant pitch pine of the oak-pine forest, due to its open-grown, branchy form, had only 19 percent of stem wood in the total above-ground biomass. Oaks and more tolerant species such as beech and maple have a greater proportion of branch, bark, and leaf dry matter than intolerants, due to their inherent branchiness and slow natural pruning. For example, the stem wood percentage of the total tree biomass of aspen and birch species in Japan was 47 to 64 percent while that for beech (*Fagus crenata*) was 34 percent (Satoo, 1970).

Although studies of productivity may be employed to compare different ecosystems, using the dimension of dry-matter production, their primary purpose lies in providing basic information for understanding dry-matter, carbon, and nutrient cycling in ecosystems, that is, in function rather than description. Moving beyond the current inventory stage of productivity estimates, the dynamics of biomass (carbon) transfer can be elucidated (see below) and the effect of external factors, such as drought, flooding, or fertilization, on biomass transfer in various systems can be evaluated. This will not only provide information on how ecosystems respond to treatments, but insights into the adaptation and integration of communities with their environments.

Productivity studies have generated increasing interest in the utilization of the total dry-matter content of stands as opposed to the traditional use of wood volume in tree boles (Keays, 1971; Young, 1964, 1971). The possibility of obtaining 40 to 50 percent more dry matter from a given stand is attractive to forest managers even if the quality yield (from leaves and bark) is lower. Total utilization might also open up for management many marginally productive or so-called unmerchantable stands, like the low-grade oak-pine stands on Long Island. The potential of complete-tree utilization is high, but the ramifications are many and complex (Keays, 1971). In particular, the massive biological problems of the effects of complete-tree utilization on the nutrient cycle, soil erosion, subsequent growth, natural or artificial regeneration, and the aesthetic-recreational problems are currently under investigation.

## EXAMPLES OF ECOSYSTEM ANALYSIS

Analysis of forest ecosystems calls for investigators of many disciplines to work together in constant communication, exchanging ideas and data, and studying that which is relevant in the system. Although many individual investigators may study the separate components of a system, unless there is overall planning, integration, and synthesis, we have merely a series of more or less related studies. Each may be very useful in its own right, but collectively they do not add up to an understanding of system dynamics. For this reason scientists in many parts of the world, recognizing the ever-increasing demands of people on their environment, initiated ambitious programs of ecosystem analysis under the auspices of the International Biological Program (IBP) (Blair, 1977). Although this program was discontinued in the early 1970s, various projects continue to pursue its major goal of broadening our understanding of productivity in nature. This goal not only demands an inventory of the productivity of the world's ecosystems, but more significantly it implies an understanding of the dynamics and functioning of processes of these systems. If successful these studies can serve as an effective guide to prediction of the changes that are likely when ecosystems are subject to stresses and manipulations. In addition, systems-oriented studies allow us to perceive and resolve environmental problems in a new way. As stated by Reichle (1971):

> *Ecologists cannot continue to respond to each new environmental crisis by simplistic "cause and effect" studies of isolated ecosystem components. The totality of environmental systems must be recognized and an understanding developed of the interactions and interdependencies of systems components. Only in this way can the effects of perturbation upon individual components be interpreted in the total context of the system.*

Research in some European IBP programs began in the mid-1960s, and the German Solling Project (Ellenberg, 1971a) was described and contrasted to one American program in the previous edition of this book (Spurr and Barnes, 1973). In North America, programs of wide scope were organized to study six major biomes: the deciduous forest (eastern United States), the coniferous forest (parts of the western United States), the grassland, the desert, the tundra, and the tropical biomes. The purpose of this section is to illustrate ecosystem analysis by examining the scope and approach of the Deciduous Forest Biome and highlights of research in the Coniferous Forest Biome and the Hubbard Brook Ecosystem project.

## Deciduous Forest Biome

Scientists of the program have sought to understand the dynamics and functioning of selected ecosystems in the eastern third of the United States. The forests of the biome region, 142 million hectares, are predominantly native deciduous forests but also include substantial areas of pine and other conifers. Almost all of the forests have been subjected to disturbance by humans. Approximately two-thirds of the population of the United States resides in the region, and it is the major area of human impact through pollution.

The biome program has been directed at all aspects of complexity of the ecosystem—biotic and abiotic, structure and function, interaction, and synthesis; the explicit goals were to:

1. Develop and provide a synthesis of knowledge of ecosystem processes.
2. Derive from this knowledge a scientific basis for resource management, including the basis for long-term utilization of land and water resources in ways that maintain or improve environmental quality.

The objectives are oriented toward ecosystem processes. A major emphasis is placed on developing models of productivity and related ecosystem processes of naturally forested and human-modified landscapes on a regional as well as a local basis (O'Neill, 1975).

Scientists also seek to understand the influence of terrestrial ecosystems on the biological productivity of aquatic ecosystems and the quality of water in these systems. One challenge has therefore been the coupling of the terrestrial system and its trophic dynamics through the hydrologic system with the aquatic system. Figure 18.9 illustrates the trophic-level approach of the biome program and particularly the linkage of the terrestrial and aquatic systems through the "Land-Water Interactions."

Research on terrestrial and aquatic systems was conducted at diverse sites in the biome. Deciduous forests received primary consideration at the Oak Ridge site in eastern Tennessee (see following discussions). In addition, biomewide studies on a larger geographic scale ("Biome and Regional Analysis" of Figure 18.9) were conducted. A main feature of this work was investigations of forest succession operating at the level of entire landscapes. The simulation of forest succession on a regional basis may be used to develop regional forest management or land-use plans. To do this it is necessary to characterize the mosaic of forest communities and determine the natural rate of change from one community to another. Thus regional models of succession, in the absence of human disturbances and major natural disturbances, were developed for forest types in the western Great Lakes (Shugart et al., 1973) and the northern Georgia Piedmont (Johnson and Sharpe, 1976).

For the latter region a simulation of the changes in the area oc-

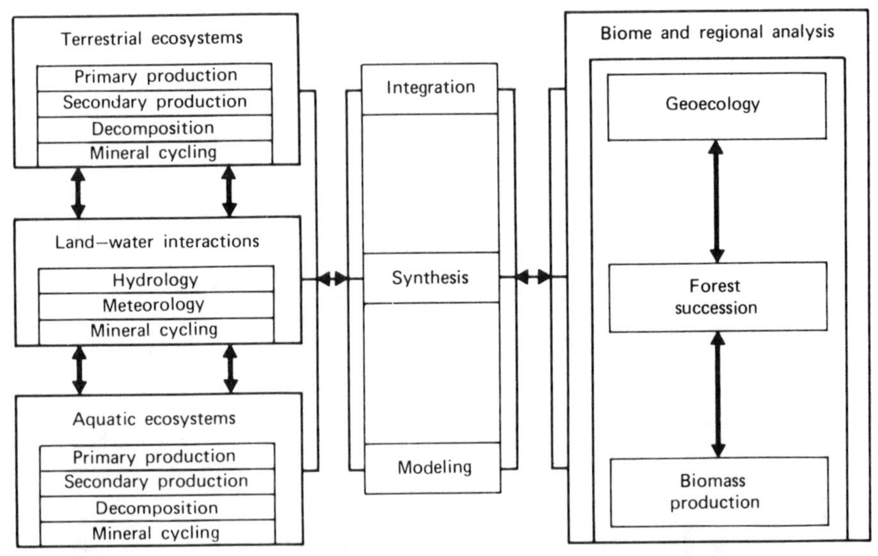


**Figure 18.9. Diagram of the interrelated parts of the deciduous forest biome program, a modified trophic-level approach. (Courtesy of the Environmental Sciences Division, Oak Ridge National Laboratory.)**

cupied by each forest type over a 30-year period was made. This took into consideration current rates of change (including natural succession rates and human disturbances through urbanization, reforestation, etc.) and hypothetical changes in the rates of land reversion to forest and harvesting (Figure 18.10). For purposes of the simulation, it was assumed that the rate of old-field abandonment and reversion to forest would decrease linearly to zero over the 30-year period; timber harvesting was assumed to increase at the rate of 1 percent per year. Under current rates of forest type change, the areal extent of loblolly pine is estimated to increase, and that of shortleaf pine is seen to decrease (Figure 18.10). However, assuming decreasing reversions to forest and increasing harvesting, the trend of increasing loblolly area is reversed with time, and further areal decline of shortleaf pine occurs. Only through markedly increasing management intensity (reforestation, prescribed burning, stand treatments, etc.) could the area of loblolly pine, an important commercial species, be maintained under these assumptions. The significance of such an example lies in the use of the simulation approach to natural succession in human-dominated ecosystems over a large region to determine present land-use patterns and to predict the consequences of proposed land-use policies.

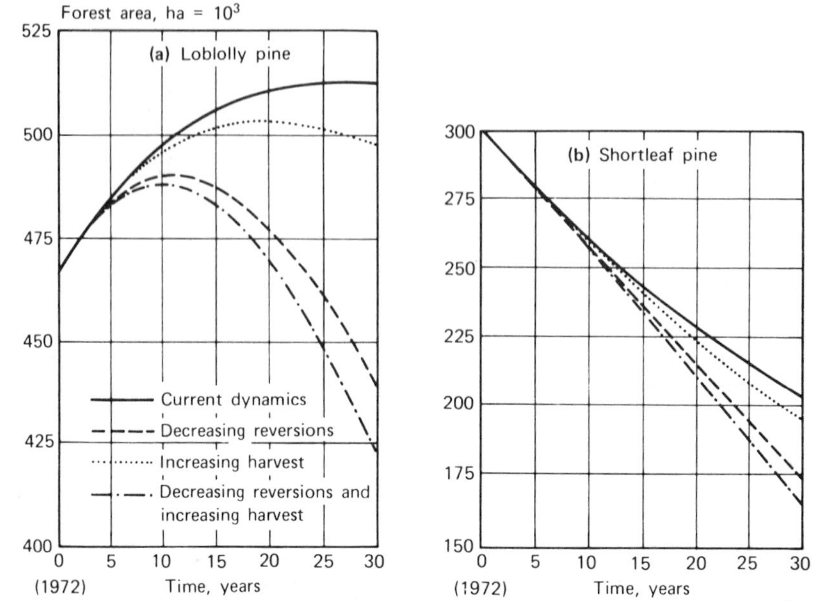


**Figure 18.10. Simulation of change in area of loblolly and shortleaf pine in the northern Piedmont of Georgia for a 30-year period based on current rates of change (current dynamics) and on hypothetical changes in the rates of timber harvesting and land reversion. *(a)* Loblolly pine. *(b)* Shortleaf pine. (After Johnson and Sharpe, 1976.)**

**Oak Ridge Site.** Through an example of the kinds of studies conducted at the Oak Ridge site we can gain an overview of the approach at a key site. The objectives were to develop mathematical models at three levels: I—the individual forest and aquatic ecosystems, II—the watershed ecosystem, and ultimately, synthesis at level III—the regional drainage basin. A model of individual forest and aquatic ecosystems (level I) and their coupling in a small-scale watershed of a regional drainage basin system (levels II and III) is shown in Figure 18.11. Research at the Oak Ridge site emphasized models of processes of a forested ecosystem and the manner in which nutrients and water affect forest growth and production. The ability to predict watershed responses enables the watershed to serve as the validation site for the coupling of the detailed forest and aquatic ecosystem models developed in level II. This is illustrated in the lower left portion of Figure 18.11 with the coupling of the aquatic system to a terrestrial system of one of the "geographic volumes."

The processes were studied through intensive investigations at

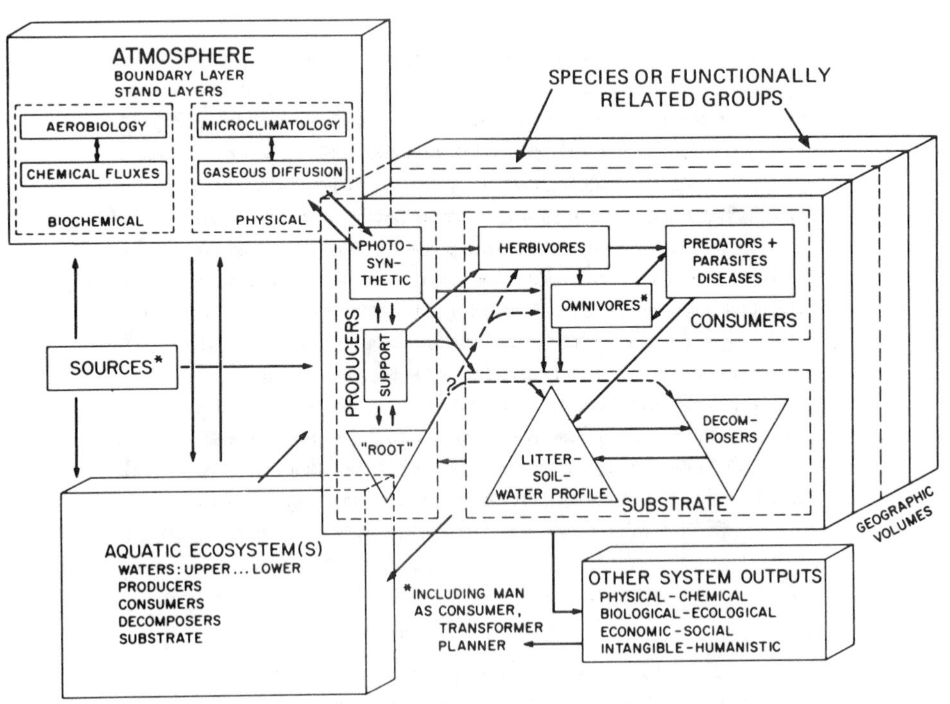


**Figure 18.11. Land ecosystem module. Model of levels I, II, and III of system complexity, Oak Ridge Site. The processes and components of terrestrial and aquatic ecosystems (level I, upper left and center) are shown coupled (lower left) in a watershed which is part of a regional drainage basin system (level II). The coupling of aquatic and terrestrial ecosystems in a number of watersheds, or "geographic volumes," constitutes the regional drainage basin system (level III). (Courtesy of the Environmental Sciences Division, Oak Ridge National Laboratory.)**

one research site, an ecosystem supporting a yellow-poplar community. The initial model of the terrestrial forest system considered the standing crop of dry matter or its carbon equivalent and the flow of the basic ecosystem components, water and nutrient elements. The generalized conceptual model for these key processes is shown in Figure 18.12. Research determines the levels of each constituent in the system and transfer rates of these constituents between compartments. From the measurements, functions which describe transfers can be postulated. A qualitative model that describes the tempo-

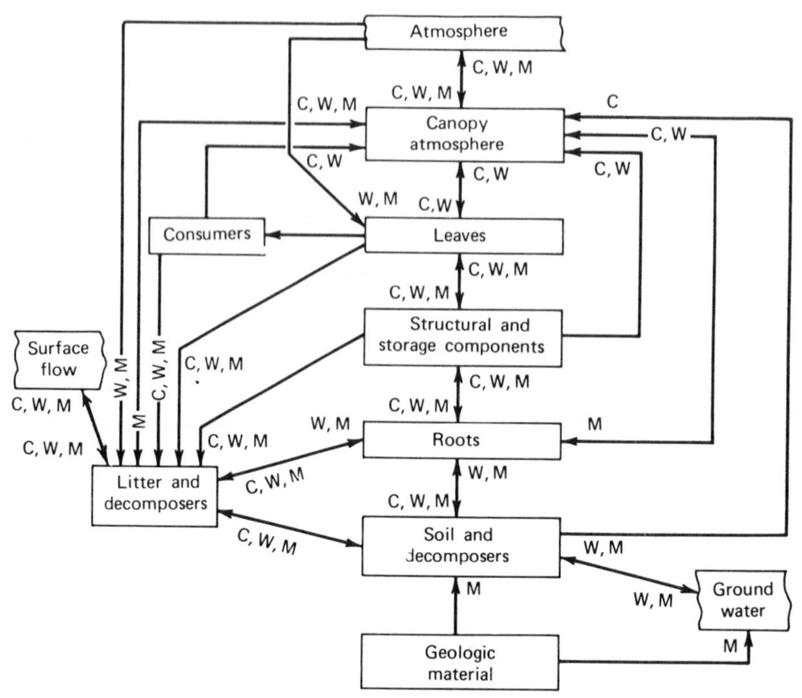


**Figure 18.12. Basic compartment model for carbon (C), water (W), and minerals (M), in a forest ecosystem. (Courtesy of the Environmental Sciences Division, Oak Ridge National Laboratory.)**

ral behavior of the system is the result, and the effects of changes or perturbations to the system can be predicted.

An example of a conceptual model of organic matter (Figure 18.13) in the yellow-poplar ecosystem illustrates a set of material pools or compartments with their respective level of organic matter, and the annual inputs to, losses from, or transfers between them. Through the use of these data and the diagrammatic model, a mathematical model of organic matter transformation was developed that can vary with season or stage of stand development (Sollins et al., 1976). The transfer values show total transfer during one year but are not meant to imply even distribution throughout the year. Litterfall is concentrated in about a 3-week period, and almost every flow is greater during the growing season. In this system, the rapid decomposition rate and substantial release of carbon through soil respiration are apparent (Figure 18.13). As in most forest systems, very little transfer occurs through herbivores (foliage feeders); large amounts of carbon are stored, particularly in the stem, litter, and soil.

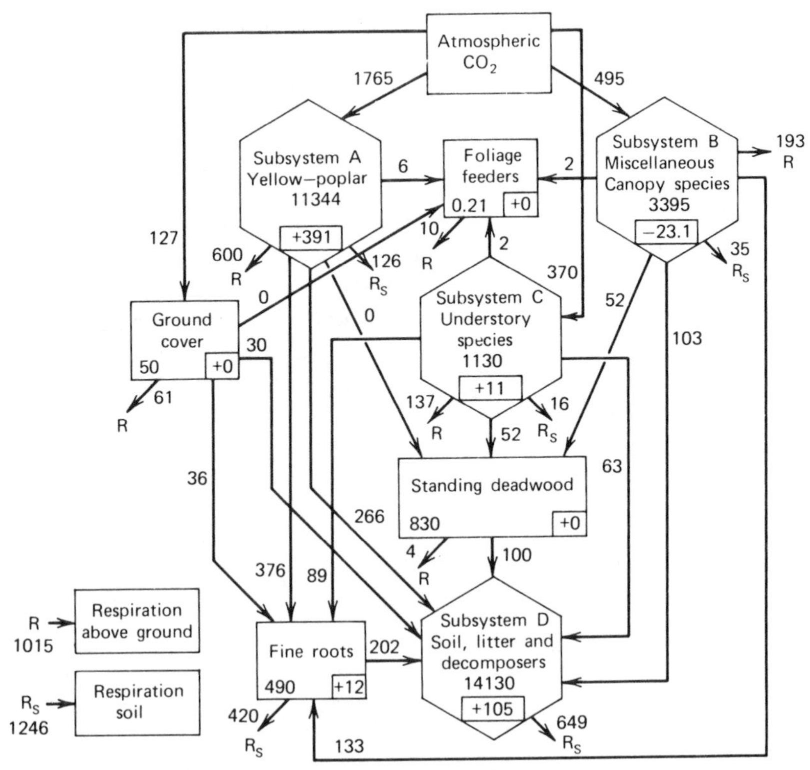


**Figure 18.13. Organic matter budget for a yellow-poplar-dominated mesic forest ecosystem at Oak Ridge, Tennessee. All compartment values are grams dry weight per square meter; all transfer and increment values are grams per square meter per year. Compartment increment values are shown in boxes within the compartment. *R* and *Rs* refer to aboveground and belowground respiration, respectively. (After Sollins et al., 1976.)**

The organic matter of the yellow-poplar ecosystem depicted here can then be converted to carbon and the carbon cycle diagrammatically presented (Figure 18.14; Harris et al., 1975). The chemical energy of organic-based carbon molecules is the "fuel" that runs biological systems (Reichle, 1975). It is produced in photosynthesis and necessary in all physiological processes in plants and animals. Carbon metabolism is a common denominator among ecosystems and allows the comparison of ecosystems of markedly different composition, such as forest, grassland, and tundra (Table 18.7; Reichle, 1975). From Table 18.7 it is apparent that although forests have a high gross primary production, they also have a very high respiration. Ecosystem respiration is 97 percent of gross primary production. Thus ecosystem productivity (0.03 g $m^{-2}$ $yr^{-1}$) is low compared to that of the oak-pine xeric forest and the prairie. Due primarily to high respiration, the overall residence time of carbon in the ecosystem is only 10 years. Although forests are a reservoir of carbon, there is considerable interchange with the atmosphere. Effective production ratios (NPP/GPP, Table 18.7) indicate that the prairie and tundra, dominantly annual communities with minimal perennial woody biomass, have the highest production relative to total carbon fixed. The ecosystem comparisons of Table 18.7 illustrate the level of ecosystem understanding that is being generated by studies of entire ecosystems and the kinds of ecosystem comparisons that are possible.

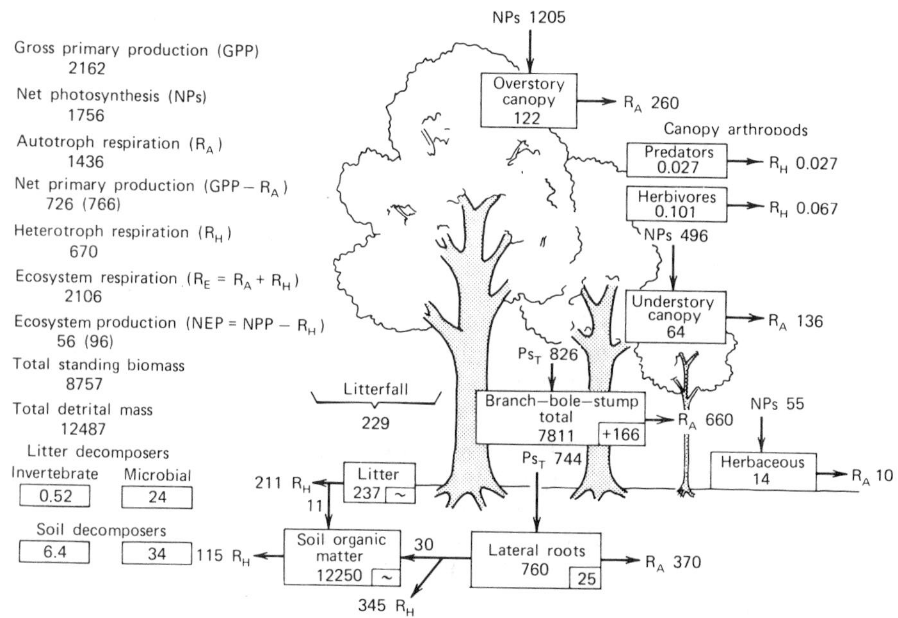


**Figure 18.14. Diagram of conceptual model of organic matter/carbon storage and flow in a temperate forest ecosystem. Ps = net photosynthesis; $R_A$ = autotrophic respiration; $R_H$ = heterotrophic respiration; L = losses due to litterfall or root mortality. Litter and soil decomposers include both microbial and invertebrate organisms. Standing deadwood is included in the branch-bole-stump compartment. (After Harris et al., 1975. Courtesy of the Environmental Sciences Division, Oak Ridge National Laboratory.)**

**Table 18.7 Comparative Metabolic Traits of Four Contrasting Ecosystems. (All Values in Grams of Carbon per Square Meter per year)**

| PROPERTY | MESIC FOREST[a] | XERIC FOREST[b] | PRAIRIE | TUNDRA |
|---|---|---|---|---|
| Gross Primary Production (GPP) | 2,162 | 1,320 | 635 | 240 |
| Autotrophic Respiration (RA) | 1,436 | 680 | 215 | 120 |
| Net Primary Production (NPP) | 726 | 600 | 420 | 120 |
| Heterotrophic Respiration (RH) | 670 | 370 | 271 | 108 |
| Net Ecosystem Production (NEP) | 56 | 270 | 149 | 12 |
| Ecosystem Respiration (RE) | 2,106 | 1,050 | 486 | 228 |
| | | | | |
| Production efficiency [RA/GPP] | 0.66 | 0.52 | 0.34 | 0.50 |
| Effective production [NPP/GPP] | 0.34 | 0.45 | 0.66 | 0.50 |
| Maintenance efficiency [RA/NPP] | 1.98 | 1.13 | 0.51 | 1.00 |
| Respiration allocation [RH/RA] | 0.47 | 0.54 | 1.26 | 0.90 |
| Ecosystem productivity [NEP/GPP] | 0.03 | 0.20 | 0.23 | 0.05 |

Source: Modified from Reichle, 1975. Reprinted, with permission, from April 1975 *BioScience* (Vol. 25, No. 4) published by the American Institute of Biological Sciences.

[a] Early successional deciduous yellow-poplar forest on alluvial soil (After Harris et al., 1975.)

[b] Oak-pine forest on sandy soil (After Woodwell and Botkin, 1970.)

Particular attention at the Oak Ridge site has been directed to belowground processes. Studies of aboveground and belowground biomass (Harris et al., 1973, 1977) together with simulation models of forest biomass dynamics, indicate that some temperate forests are characterized by a large amount of annual root biomass accumulation and rapid turnover (largely due to mortality of fine roots). The importance of the unseen belowground components in ecosystem functioning is that approximately 45 percent of the annual food production is utilized in belowground biomass production and in root respiration (Harris et al., 1975). The cyclic renewal of large and small roots and their associated mycorrhizae provide a reservoir of organic matter and essential nutrients. For example, in an oak-hickory ecosystem near Oak Ridge, root mortality was by far the most important mechanism for the recycling of nitrogen from vegetation to the soil (Henderson and Harris, 1975). Such studies emphasize that litterfall is not necessarily the most important source of nitrogen in cycling.

The coupling of the terrestrial processes with the aquatic ecosystem through the hydrologic cycle (Figure 18.11) has been studied at the Walker Branch Watershed. Walker Branch drains a 97-ha watershed having two subwatersheds dominated primarily by oak-hickory forests. Three phases of research have been emphasized (Harris, 1977):

1. Establishing the necessary baseline data on hydrology, determining chemical inputs and export for both macroelements (Henderson et al., 1977) and trace elements (Van Hook et al., 1977), and determining biomass pools and fluxes (Harris et al., 1973).
2. Manipulative studies of specific processes: for example, the effect of sulfate inputs to the forest floor on soil fertility and the effects of irrigating the forest ecosystem with high nitrate wastes have been investigated (Harris et al., 1976).
3. Mathematical modeling of soil-plant-water relationships (Huff et al., 1977).

In summary, the Oak Ridge site program has been one of intensive characterization of ecosystem processes, integration of process-level studies through development of mathematical models at the ecosystem level, and the study of land-water interactions of these processes.

### Coniferous Forest Biome

The program of the Coniferous Forest Biome is an interdisciplinary research effort dedicated to the understanding of the structure and functioning of coniferous forests and associated aquatic ecosystems. Coniferous forests of the Pacific Northwest have been the principal area of research. Intensive research sites are located in old-growth Douglas-fir forests of the H. J. Andrews Experimental Forest of the U.S. Forest Service in the Oregon Cascade Range and in the second-growth forests of the Cedar River–Lake Washington watershed of the Washington Cascade Range. The research program and specific objectives have been considered by Franklin (1972), Gessel (1972), Waring (1974), and Edmonds (1974). In addition to numerous scientific papers, a synthesis volume provides a detailed overview of the major results of biome research (Edmonds, 1980). A few highlights of this productive group are presented, illustrating several unique features of Douglas-fir and associated stream ecosystems.

**Forest Composition and Structure.** The study of terrestrial and aquatic processes is based on a solid understanding of the forest communities, their composition, structure, and dynamics. The extensive research of Jerry F. Franklin, C. T. Dyrness, and associates culminated in a superbly documented and illustrated treatment of the major vegetation units of Oregon and Washington (Franklin and Dyrness, 1973). In addition, the forest communities were effectively related to major environmental factors and their distribution plotted along gradients of moisture and temperature (Waring et al., 1972;

Dyrness et al., 1974; Zobel et al., 1974, 1976). This classification of the vegetation and its relation to environment provides a framework for making the detailed studies of ecosystem processes more widely applicable and thus provides the basis for intelligent wildland management.

A unique feature of the Pacific Northwest is the dominance of massive long-lived, evergreen conifers, often growing on steep mountain terrain (Waring and Franklin, 1979). The dense forests are dominated almost totally by conifers whose biomass accumulations far exceed those in almost every other temperate forest region (Chapter 7). An exciting feature of the program, therefore, is the new ecological insights to the old-growth forest ecosystems of the Douglas-fir region that provide a basis for their management. This management of decadence is a new challenge to forest ecologists and silviculturists. Several features of the old-growth ecosystem are described as follows.

Innovative studies in the crowns of these towering old-growth conifers on the H. J. Andrews Experimental Forest have provided new insights to ecosystem structure and functioning. Mountain climbing techniques were employed to climb and sample individual Douglas-fir trees up to 80 m tall and over 450 years old (Denison et al., 1972, Denison, 1973; Pike et al., 1977). Each tree is itself a landscape with variations in topography, climate, vegetation, and an associated fauna (Figure 18.15). Six zones supporting epiphytic lichens and bryophytes were recognized, including a basal zone, lower trunk dry side and moist side, branch axis and branchlets, and the upper trunk (Pike et al., 1975). Striking differences were found between the lower moist and dry sides of the trunk; bryophytes covered most of the moist side, whereas on the dry side crustose and scalelike lichens were present. The upper trunk had the most species of epiphytes and the greatest biomass of lichens per unit of bark. The biomass of epiphytes can only be reliably estimated by climbing because epiphytes are dislodged and branches fragmented if a tree is felled. The dry weight of epiphytes on branchlets and branch axes of one tree was 15 kg compared with 3 kg on the trunk and 198 kg of needles. The ecological contribution of epiphytes is more similar to the crown than the relatively inert biomass of the tree trunk since an epiphyte biomass is physiologically active when wet. In general, the epiphytic biomass (excluding crustose lichens) ranges between 10 and 20 percent of the foliage of the host tree. Nitrogen plays a key role in coniferous forests, and lichens, particularly *Lobaria oregana*, fix atmospheric nitrogen at an average rate of about 8 to 10 kg $ha^{-1}$ $yr^{-1}$ (see Chapter 9). In contrast, epiphytes play a negligible role in young (37-year old) second-growth stands (Grier et al., 1974).

Figure 18.15. Old-growth Douglas-fir, H. J. Andrews Experimental Forest, Blue River, Oregon. *(a)* Lower trunk showing sharp distinction between moist and dry sides. Tree is on a north-facing slope and leans toward the north, which is to the right in the picture. *(b)* View of two axes in the canopy showing mosses and accumulation of trapped needles. *(c)*

The large canopy also acts as a site for decomposition since much of the needle, twig, and bark litter becomes lodged in the crown and creates a perched soil habitat (Figure 18.15b). Fungi and bacteria infect living needles and decompose needles and other detritus lodged in the crown. Also, over 900 taxa of invertebrates, including direct foliage feeders (mites, springtails, fly larvae, etc.) have been collected. They are associated with epiphytes and lodged litter. Thus the epiphytic biomass through the fixation of nitrogen and interception of water and nutrients together with the crown habitat for decomposer flora and invertebrates, contributes significantly to the functioning of the forest as a whole.

**Large Organic Debris.** Another feature, unique to the massive old-growth forests with their very high aboveground biomass, is the presence of great amounts of large organic debris on the forest floor and in streams. In a 450-year-old Douglas-fir stand in the Andrews forest, 13 percent of the forest floor was covered with logs.[1] This amounted to 172 tons of biomass per hectare. In contrast, the 300-year-old yellow birch—red spruce forest studied in New York by McFee and Stone (1966) (Chapter 9) only had 43 t $ha^{-1}$. In addition to the woody debris on the forest floor, the Douglas-fir forest had 109 t $ha^{-1}$ of standing dead trees over 4 m tall.

The large biomass of low-nutrient materials is probably important in the nutrient relations of old-growth stands at the time they are naturally recycled by wildfire. The large logs are not entirely consumed by fires, and although large amounts of nitrogen may be volatilized from the litter, nitrogen still resides in the remaining debris. Decomposition is slow; 120 to 140 years are required to reduce Douglas-fir logs to 50 percent of their original weight. Therefore, the logs constitute a long-term reservoir for nutrients that are slowly available to the soil and are more readily available to plants with roots in the litter layer. Besides the nitrogen present in the logs, more is apparently added by invading fungi and by nitrogen fixation. The characteristic succession of Douglas-fir to western hemlock in the absence of severe wildfire may be dependent on woody debris. For example, over 67 percent of the western hemlock trees over 5 cm

[1]Results courtesy of the Coniferous Forest Biome and Oregon State University, Corvallis.

**Moist side of the lower trunk as viewed from the ground. *(d)* Dry side of lower trunk of the same tree shown in *(c)*. (After Pike et al., 1975. Reprinted courtesy of *The Bryologist*. Photos by William C. Denison.)**

diameter (at 1.3 m) in one stand had rooted in logs. The logs also provide an important habitat for small mammals that spread mycorrhizal fungi in their runways in the debris.

An example of the linkage between the coniferous forest and aquatic ecosystems is the large organic debris in small and intermediate-sized streams. For example, 56 tons of organic debris were measured in a 3-m-wide belt along a small stream in the Andrews Experimental Forest (Froehlich, 1973). Such debris, in the form of tree tops, limbs, root wads, and whole trees, is a principal factor determining the biological and physical character of these streams (Swanson et al., 1976; Swanson and Lienkaemper, 1978;) (Figures 18.16 and 18.17). Forest streams and their biota developed through a long history of high concentrations of debris, alternating between accumulation from the surrounding forest and movement through the stream channels by flotation at high water, in torrents, and as dissolved and fine particulate material following breakdown by wood processing organisms. The debris may reside in the channel over 100 years and acts to retain organic and inorganic sediment. The logs and associated gravel bars form a stream profile of steps, long, low-gradient sections separated by relatively short falls or cascades. This pattern can account for up to 100 percent of the total stream drop (Swanson et al., 1976), and much of the resulting stream bed may have a gradient less than that of the valley bottom. Energy is dissipated in short, steep falls and results in less energy available for erosion of stream bed and banks, more sediment storage in the channel, slower routing of organic materials, and greater habitat diversity than in straight, even gradient channels (Swanson and Lienkaemper, 1978). Except in periodic flushing episodes, the debris in the channel tends to route the sediment through the stream in a slow trickle, reducing the rate of downstream movement. The woody debris, through its very presence and the formation of the stepped profile, creates relatively stable habitats for a variety of stream biota. Consumer organisms tend to be concentrated in the wood and the wood-created habitat, which in small undisturbed streams may exceed 50 percent of total stream area. An energy base is constantly available (fine organic material and large debris) even during periods when leaves or needles are not available.

In streams adjacent to logged-over areas, large and fine debris may be largely eliminated (Figure 18.18) and the stream significantly altered physically and biologically. For example, on Mack Creek, extremely high winter storm flows moved logs, slash, and roots downstream leaving only about 4 percent of the large debris, compared to the uncut area upstream, 7 years after logging (Froehlich et

**Figure 18.16. Large organic debris in Lookout Creek, H. J. Andrews Experimental Forest, Oregon. Note figure in left center of picture. (Swanson et al., 1976; U.S. Forest Service photo. Courtesy of F. J. Swanson.)**

al., 1972). The consequences of removing large debris from streams are well summarized by Swanson et al., (1976):

> *A stream which had previously flowed over a series of steps formed by debris will assume a more uniformly steep profile and experience other changes in channel geometry. There will be a resulting decrease in diversity of stream habitat as biologically productive, debris-related depositional pools are eliminated. Increased water velocity will also contribute to the accelerated transport of fine organic matter through the channel system, thereby decreasing the opportunity of stream organisms to process the material. Consequently, the removal of large debris from streams may reduce long-term biological productivity and increase the rate of sediment transfer from headwater streams to downstream areas.*

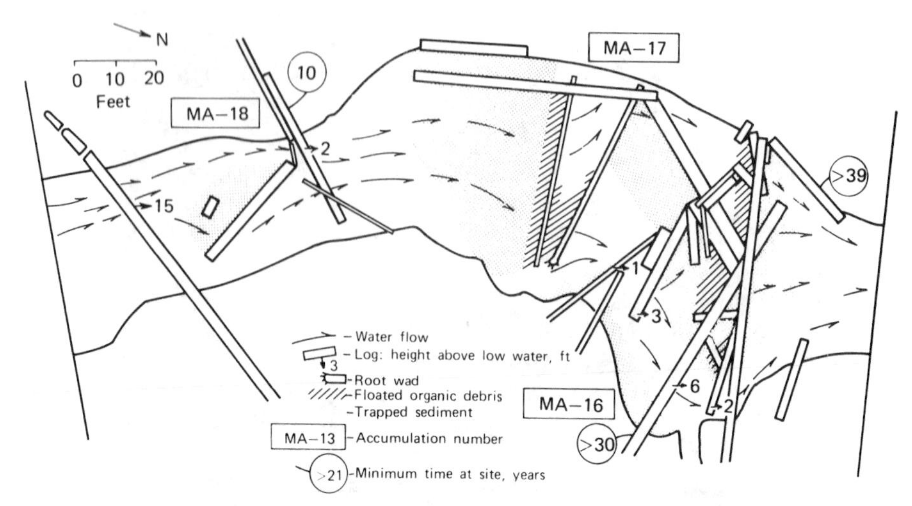


**Figure 18.17. Map of large organic debris and other material in a 60-m forest section of Mack Creek upstream from the section in Figure 18.18. H. J. Andrews Experimental Forest, Oregon. (After Swanson et al., 1976.)**

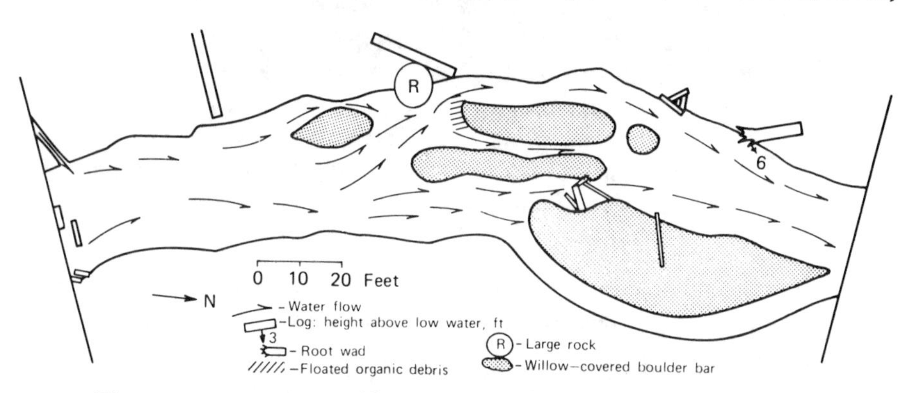


**Figure 18.18. Map of large organic debris and other material in a 60-m clearcut section of Mack Creek downstream from the section in Fig. 18.17. H. J. Andrews Experimental Forest, Oregon. (After Swanson et al., 1976.)**

It is apparent that the small coniferous forest stream is a part of an integrated forest-stream ecosystem. A consideration of the coniferous forest stream has been presented by Triska et al. (1980). The riparian zone serves as the interface between the old-growth forest and the stream. Its important functions have been reviewed by

Swanson, Sedell, and Trisca (1980). Erosional processes, described in the next section, also have important roles in the functioning of the forest-stream ecosystem.

**Erosion and Geomorphology.** Steep forest terrain combined with heavy winter precipitation, youthful soils, and massive trees make linkages between physical and biological processes and between terrestrial and aquatic environments more apparent than in gently sloping terrain of geologically older regions. Thus studies of erosion and forest geomorphology have been an important feature of biome research (Dyrness, 1967; Fredriksen, 1970; Swanson et al., 1980).

Past studies of erosion have identified the hillslope and stream channel processes which transfer organic matter and vegetation in old-growth stands (Figure 18.19). For example, subtle, slow mass movement processes of creep, slump, and earth flow operate simultaneously and sequentially to bring about rapid soil mass movements on hillslopes and debris torrents in stream channels (Swanson and Swanston, 1977). Tree-ring analyses suggest that such mass movements have spanned centuries. These mass movements are responsible for microenvironments affecting the composition, growth, and functioning of forests and streams. An analysis of erosion processes for a watershed of the Andrews Experimental Forest revealed that on a long-term basis the most episodic and infrequent processes (debris avalanches on hillslopes occurring approximately once every 370 years, and debris torrents in streams, occurring about once every 580 years) transferred more mineral matter than the persistent, pervasive, and continuous processes (creep, surface erosion, etc.). Through such studies the impact of erosion on other ecosystem processes can be evaluated and management recommendations given to minimize damage to the forest ecosystem.

**Nutrient Cycling.** The extensive research and contributions to our understanding of nutrient cycling in Pacific Northwest forests, especially at the Cedar River site, have been reviewed by Johnson et al. (1980). Nitrogen cycling and its role in the development of Douglas-fir forests probably has been the single most important contribution. Nitrogen is the most important and limiting nutrient in coniferous forests and has been found to be tightly conserved in system cycling. In old-growth forests, a small annual accumulation is observed (Fredriksen, 1972). Following clearcutting or fire, decaying fine roots and developing understory vegetation act as reservoirs for nitrogen, reducing its loss. Some understory species fix nitrogen, and rapidly developing riparian vegetation absorbs nitrogen in solution before it can be lost to the aquatic system.

In a young Douglas-fir stand, much of the required nitrogen is

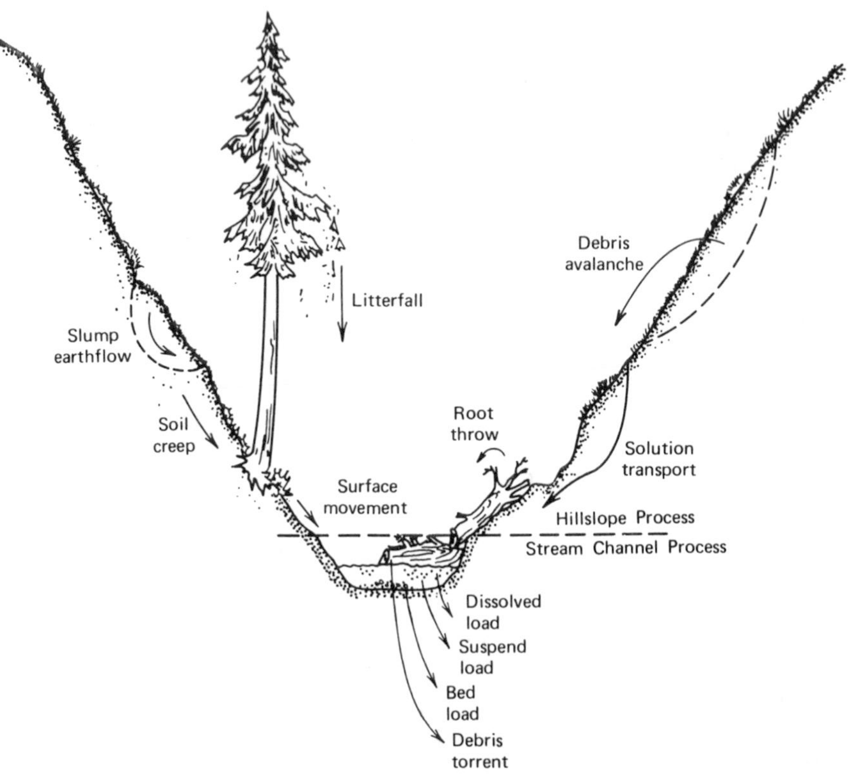


**Figure 18.19. Hillslope and stream channel processes that transfer organic and inorganic material in coniferous forest ecosystems. (Modified from Swanson et al., 1980. Courtesy of the Coniferous Forest Biome and Oregon State University. Drawing by M. F. Orlando.)**

taken up from the forest floor. As the stand ages and crowns develop, needle persistence increases from approximately 2 to 6 years. The nitrogen supply is thus accumulated in the needles, and this tends to offset the continuously decreasing uptake of nitrogen from the forest floor (where it may be tied up in unavailable forms). To meet growth needs, nitrogen in the tree crowns is internally translocated from older needles to developing new tissues. Nitrogen also plays a key role because it is taken up faster than other nutrient ions, and it tends to regulate their uptake. Such an understanding of nutrient dynamics in Douglas-fir forests, based on records in stands from 9 to 450 years old, is seen as a model for other ecosystems. These studies have emphasized the dynamic nature of nutrient cycles, that is, the transfers from one part of the system to another. However, it is clear that nutrient cycling must be studied in a systems context, not in isolation, because of the interacting effects of different temperature and moisture regimes, different soils and species, and different developmental stages of a given species.

### The Hubbard Brook Ecosystem Study

The best known and most comprehensive study of a system of watersheds is that of the Hubbard Brook Valley in the White Mountains of New Hampshire. The second-growth northern hardwood forests are typical of those covering a large area in New York and New England. The initial studies were of the hydrologic cycles of small watersheds. Research has expanded to include studies of nutrient cycles and budgets, and the structure, function, and dynamics of the forest communities. Various kinds of experimental treatments, including patch and strip clearcutting and denudation, have been employed to study system processes and vegetation responses following disturbance. In addition to numerous papers, two books provide an excellent overview of the Hubbard Brook study. These are devoted to the biogeochemistry of the forest (Likens et al., 1977) and the structure, function, and succession of the forest communities (Bormann and Likens, 1979).

In summary, these three major efforts to study entire forest ecosystems represent one of the most significant events in the history of ecological research. The study of whole ecosystems is in effect the creation of a whole new level of ecological science (Smith, 1970). Despite problems, this approach has sought out properties unique to whole ecosystems, and stressed how water, nutrients, and energy enter, move through, and leave forest ecosystems. It has given a new appreciation of the significant linkages between terrestrial and aquatic systems and has provided new insights to plant-animal interactions. In running counter to the traditional ways of scientific research, it has, however, examined the behavior of all ecosystem parts, revealing new linkages. Finally, this approach has caused mathematicians and biologists from many disciplines to cooperate in describing, modeling, and understanding the functioning of whole ecosystems.

## SUGGESTED READINGS

Dale, M. B. 1970. Systems analysis and ecology. *Ecology* 51:2–16.

Edmonds, R. L. (ed.). 1980. *Analysis of Coniferous Forest Ecosystems in the Western United States.* US/IBP Synthesis Series. Dowden, Hutchinson and Ross, Inc., Stroudsburg. Pa.

Kormondy, Edward J. 1969. *Concepts of Ecology*. (Chapter 1, The nature of ecosystems; Chapter 2, Energy flow in ecosystems; Chapter 3, Biogeochemical cycles and ecosystems.) Prentice-Hall, Inc., Englewood Cliffs, N.J. 209 pp.

Lieth, Helmut, and Robert H. Whittaker (eds.). 1975. *Primary Productivity of the Biosphere*. Ecological Studies 14. Springer-Verlag, New York. 339 pp.

Likens, Gene E., F. Herbert Bormann, Robert S. Pierce, John S. Eaton, and Noye M. Johnson. 1977. *Biogeochemistry of a Forested Ecosystem*. Springer-Verlag, New York. 146 pp.

Odum, Eugene P. 1969. The strategy of ecosystem development. *Science* 164:262–270.

Ovington, J. D. 1962. Quantitative ecology and the woodland ecosystem concept. *Adv. Ecol. Res.* 1:103–192.

Reichle, David. 1975. Advances in ecosystem analysis. *BioScience* 25:257–264.

Smith, Frederick E. 1970. Analysis of ecosystems. *In* David E. Reichle (ed.), *Analysis of Temperate Forest Ecosystems*. Springer-Verlag, New York. 304 pp.

Walters, Carl J. 1971. Systems ecology: the systems approach and mathematical models in ecology. *In* Eugene P. Odum, *Fundamentals of Ecology*. W. B. Saunders Co., Philadelphia. 574 pp.

Whittaker, Robert H. 1975. *Communities and Ecosystems*. Second edition (Chapter 5, Production.) The Macmillan Co., New York. 385 pp.

———, and G. M. Woodwell. 1969. Structure, production and diversity of the oak-pine forest at Brookhaven, New York. *J. Ecol.* 57:157–176.

# *Part* IV The Forest

# 19
# Historical Development of Forests

## PHYTOGEOGRAPHY

In this final part, we are concerned with providing a brief description of the present-day forests of the world with particular reference to the forests of temperate North America. Description, however, is not enough. An understanding must be provided also by summarizing briefly the historical development of present-day forests (this chapter), and, in the description of these forests, bringing in as much as possible of the climatic, edaphic, and dynamic relationships which go far toward explaining why a particular type of forest is growing on a particular site.

That phase of botany in which geographical problems of plant distribution are emphasized is termed **plant geography**, or **phytogeography**. Although much of the concern and knowledge is shared with plant ecology, it differs in that it has arisen from studies of the distribution of particular taxa of plants, rather than from studies of the environment. In other words, plant geography basically is floristically oriented whereas plant ecology is environmentally oriented, although both are concerned with the interrelationship of plants and their environment.

Plant geography is concerned with historical and present-day distributions of plant taxa, the location of their origins, studies of dispersal and migration, and in general with the evolution and present distribution of our flora. Only those phases pertinent to a brief survey of the present distribution of the world's forests can be summarized here. Excellent textbooks (Wulff, 1943; Cain, 1944, Good, 1974; Polunin, 1960; Walter, 1973; Cox et al., 1976; Daubenmire, 1978) are available for those wishing to delve more deeply.

## PALEOECOLOGY

In order to understand the present-day distribution of the forests of the United States and Canada—and indeed of the world—it is essential to know something of the evolution of modern forms of trees through geological times, the widespread changes in their distribution during Pleistocene glaciations, the changes in forest composition in postglacial times, and the effects of the development of civilizations on the distribution and character of forests.

The historical development of plants is known variously as **historical plant geography** if the emphasis is on geographical distribution of the various elements of the flora, and as **paleoecology** if environmental changes are of primary concern. Plant geographical knowledge with particular emphasis upon its historical development is covered by Cain (1944) while European literature and viewpoints are presented by Firbas (1949, 1952), Frenzel (1968), and Straka (1970). Recent advances in paleoecology are reported in the comprehensive treatment of the Quaternary of the United States (Wright and Frey, 1965) and in several symposium volumes (Hopkins, 1967a; Dort and Jones, 1970; Turekian, 1971; Birks and West, 1973).

## EVOLUTION OF MODERN TREE SPECIES

Aside from a few tree ferns and cycads which reach tree proportions, modern trees may be grouped as **gymnosperms** and **angiosperms**. The appearance of their prototypes and the gradual development of modern forms are recorded in fossils with which the study of paleobotany is concerned (Andrews, 1961; Beck, 1976). A few salient points may be summarized briefly.

The oldest lineage of modern trees is possessed by the conifers, which can be traced back to late Paleozoic. Some prototype conifers had leaves and branches very like those of modern species of *Araucaria*, a Southern Hemisphere genus that persists today with species such as Norfolk Island pine (*Araucaria excelsa*), hoop pine of Queensland (*A. cunninghamii*), Parana pine of Brazil (*A. angustifolia*), and the monkey-puzzle tree of Chile (*A. araucana*). These primitive conifers, however, constituted but a small part of the forests which formed the coal beds of Carboniferous (Mississippian and Pennsylvanian) times. Seed ferns (Pteridosperms), horse-tails (Calamites), club-mosses (Lepidodendrids), and an extinct line of primitive conifers (Cordaites) formed the dominant tree flora. These lowland, mild-climate plants were largely eliminated during the continental uplift and large-scale glaciation of the Permian times, leaving the land open to colonization by the developing modern arborescent groups.

In Mesozoic times, the conifers evolved into many forms and achieved great abundance. By Jurassic times, forms similar to modern *Libocedrus*, *Thuja*, *Sequoia*, and *Agathis* (the kauri of New Zealand and nearby lands) had appeared. Since identifications must be made frequently on impressions made by fragments of leaves and other vegetative parts, many are highly tentative. The general predominance of conifers in the fossil record of Mesozoic times is clear, however, as is the fact that most present-day genera of conifers are much less widely distributed than they apparently were in late Jurassic and early Cretaceous times (Li, 1953). The genus *Sequoia*, for example, is known from fossil records across Europe, central Asia, and North America as well as from Greenland, Spitsbergen, and the Canadian Arctic, but is restricted today to one species with local distributions in California. *Metasequoia*, also, was abundant at high northern latitudes during Cretaceous and early Tertiary times, but now exists only locally over a very restricted range in the interior of China (Chaney, 1949). *Cupressus* is still another once widespread coniferous genus, which now occurs only in widely scattered relict stands. The now widespread pines were less abundant during the Mesozoic. However, a Cretaceous pine described from the Cretaceous of Minnesota is similar to red pine (*Pinus resinosa*), which occupies the same region today (Chaney, 1954). In Australia and Tasmania, conifers similar to present-day podocarps of the area have been described from the Tertiary (Cookson and Pike, 1953).

One of the earliest and most primitive forms that has survived in a closely related form to the present day is the ginkgo (*Ginkgo biloba*), the sole surviving member of a once numerous group of gymnosperms. Species with leaves similar in form and in the structure of the epidermal layer to the present-day tree, which occurs naturally only in western China but which is widely introduced, are common in Triassic floras.

The angiosperms developed later than the conifers, but by the middle of the Cretaceous had become common and well represented in fossil floras. Pollen and macrofossil evidence indicates that the early angiosperms of the lower Cretaceous or possibly the late Jurassic showed little taxonomic diversity and were probably not closely similar to modern flowering plants. The first angiosperms apparently developed slowly on disturbed, semiarid areas, including loose talus slopes and stream banks, of the ancient landmass of Gondwanaland (Axelrod, 1970; Doyle and Hickey, 1976; Stebbins, 1976). Thereafter angiosperms evolved together with insects, birds, and mammals during periods of great environmental change, which probably elicited increasing diversity in plant forms (Chapter 13). Continents changed in position, configuration, size, and altitude through ocean-floor spreading, continental rafting (drift), and the fragmentation and joining of plates of the earth's crust. Mountains elevated as the plates collided (Himalayas and American Cordillera), and some desert lands became moist tropics; climates changed drastically. As summarized by Axelrod (1970):

*The great diversity of taxa in numerous families in the tropics and subtropics, as well as the evolution of unique floras of arid to semiarid regions and those of the temperate climates as well, seems directly related to the breakup of Gondwanaland following the medial Cretaceous and subsequent evolution in isolation.*

During the Tertiary, flowering plants became widely distributed and well differentiated throughout the world. In Eocene times, the warmest Tertiary epoch, tropical plants were intermingled with warm temperate types as far north as London, England, and the coastal plain of eastern United States. In the Eocene London coal beds, Nipa palm was one of the commonest species—it is now a salt-water swamp species of southeastern Asia. The vegetation around the north Pacific basin from Oregon to Alaska and southwest to Japan was tropical to subtropical in character. Then a major deterioration in climate took place in the Oligocene with broadleaved deciduous forests replacing subtropical forests. Except for minor fluctuations, the cooling trend continued in the Miocene and Pliocene and culminated in the Pleistocene ice age.

In Miocene times, a forest of temperate mesophytic species (ancestor of the modern Mixed Mesophytic Forest) replaced the warm temperate and subtropical types in much of the Northern Hemisphere. A continuous band of this broadleaved deciduous forest existed around the north Pacific from Oregon to Japan. The resemblance of forests in these areas is remarkable even on the species level. Furthermore, as late as 15 million years ago (middle Miocene) floristic continuity existed between eastern North America and eastern Asia. The Mixed Mesophytic Forest included several species that ranged across the broad land bridge in the Bering Sea area. This land bridge, open throughout much of Cenozoic times, aided in securing a broad interchange between Asiatic and American floras and faunas (Hopkins, 1967a).

By late Miocene, coniferous forests began to occupy large upland areas in Siberia and northern North America as the mesophytic hardwoods retreated southward (Wolfe and Leopold, 1967). A coniferous forest of spruce, fir, and hemlock for the first time extended from the uplands of Oregon northward through British Columbia into Alaska. The mesophytic forest of the northwest gradually became extinct with the coming of a cooler climate and the rise of the Cascades, which brought dry and even arid conditions to vast areas of the interior. A southern extension of the eastern American forest, represented by such trees as sweet gum, white pine, blue beech, American beech, the sugar maple group, and black gum in the cloud forest of Mexico and Central America (especially in the Sierra Madre Oriental), apparently originated about this time (Martin and Harrell, 1957). In Europe, the mesophytic flora was largely eliminated. Examples of genera that became extinct in Europe with progressive cooling are the following (Van der Hammen et al., 1971):

| At the End of | Extinct Genera of Europe |
| --- | --- |
| Miocene | *Castanopsis, Clethra, Libocedrus, Metasequoia* |
| Pliocene | *Aesculus, Diospyros, Elaeagnus, Liquidambar, Nyssa, Palmae, Pseudolarix, Rhus* (and probably *Sequoia* and *Taxodium*) |
| Pleistocene | *Carya, Castanea, Celtis, Juglans, Liriodendron, Magnolia, Ostrya, Tsuga* |

That many of these genera survived in North America and eastern Asia is ascribed to the north-south orientation of the mountain chains of these areas, permitting the temperate species to migrate southward to warmer areas and then expand northward in the interglacial intervals. In Europe, the Pleistocene continental glaciations, combined with east-west chains of the glaciated mountains (Alps, Pyrenees, Carpathian and Caucasian Mountains), are thought to account for the extinction of these elements in Europe and western Asia.

The extensive glaciation of the Pleistocene denuded much of the northern portions of the Northern Hemisphere. The distribution of trees during the maximum extent of the latest Pleistocene glaciation and the revegetation of the glaciated terrain are discussed in the following section.

## GLACIAL AND POSTGLACIAL FORESTS

One of the major scientific advances of recent decades has been the unraveling of much of the complex history of the vegetation and climates of the world from the time of the last great glacial advance to the present. By bringing together data from archeological sources, fossil studies and in particular pollen analysis, tree-ring analysis, and radio-carbon dating, it has become possible to spell out in considerable detail the overwhelming of the forests before the advancing Pleistocene ice, and their migration back onto the glaciated terrain as the ice melted. The irrefutable evidence that vegetation has been in an almost constant state of instability and adjustment due to an almost constantly changing climate over the past 10,000 years, and even over the past several hundred years, has done more than anything else to demonstrate the necessity for recognizing that present vegetation patterns are closely related to events in recent geological history.

In considering glacial and postglacial forest history, it is well to summarize first the techniques by which the various lines of evidence have been accumulated, then the geological history of ice advance and retreat, and then the evidence of changing forest composition over the last 10,000 years, which is one of the major proofs of rapid and continuing climatic changes extending up to and including the present.

## Lines of Evidence

The dating and ecological interpretations of the prehistoric past are made possible only by the combination of many lines of evidence. Archaeological stratigraphy, geomorphological evidence such as glacial varves and stream-terrace studies, dendrochronology or tree-ring studies, peat analysis, pollen analysis, carbon-14 and other types of radio-isotope dating, and many other research techniques all contribute bits of information and evidence which add together to give a picture of past vegetation and climate.

**Fossil Wood.** Since wood rots only under moist, aerobic conditions, wood may be preserved many years under anaerobic (i.e., without air) or under very dry conditions (without water). Wood may thus be preserved for many thousands of years in acid peat bogs, buried under glacial drift deeper than surface air can penetrate, or in the desert, particularly under protected cliffs. Charcoal, being carbonized wood, is chemically inactive under ordinary conditions of exposure and may be preserved for thousands of years in the soil, giving evidence of past fires. It can be frequently identified as to genus and dated as to age.

Important deposits of preserved wood are many. Buried under glacial till in the Lake States, the Two Creeks buried spruce forest dates the time of a late ice advance in northeastern Wisconsin, while spruces and larches buried under older tills in Indiana and elsewhere in the Central States give evidence both of the vegetation at the time of glacial advance and of the time of advance. Inundated forests in the Columbia River gorge date a prehistoric flooding of that river by a landslide in the Cascades; while similar inundated forests standing *in situ* below the tide level in Alaska record recent lowering of the land with consequent rising of the sea. Wood preserved in the pyramids of Egypt and in old Pueblo dwellings in the southwest American desert can be used as evidence of the vegetation at the time of construction and of the date of construction. Tree-ring analysis and carbon-14 dating are the chief methods of analysis of these and other fossil woods.

**Tree-Ring Analysis.** The science of dendrochronology owes much of its impetus and achievement to A. E. Douglass, his successor, Edmond Schulman, and the continuing work at the tree-ring laboratory at the University of Arizona (Schulman, 1956; Ferguson, 1968; Fritts, 1976). Through the comparison of tree-ring widths from old living trees growing on sites where they are sensitive to annual changes in limiting soil-moisture supply, and similar sections taken from prehistoric ruins, it has been possible to establish tree-ring chronologies extending over the past 2000 years for the American Southwest. From these, the dates of growth of timbers found in pueblo ruins and elsewhere usually can be established. For example, tree-ring materials from Mesa Verde National Park, Colorado, definitely indicate that a severe local drought was evident in the thirteenth century (Fritts et al., 1965). At this time the prehistoric cliff dwellings of Mesa Verde were apparently abandoned, and the severe drought from 1273 to 1285 was probably one of the factors in a chain of events which led to the disappearance of prehistoric Indian dwellers of Mesa Verde. Even earlier information, chronologies of 7000 years is provided by ring-width studies of 4000-year-old bristlecone pines, 2000-year-old limber pines, and 3200-year-old giant sequoias (Ferguson, 1968). From studies of relative width of rings, it is possible not only to date past severe droughts, but tree-growth response also can be used to reconstruct the occurrence of past climatic cycles (Fritts, 1976).

**Carbon-14 Dating.** The actual age of fossil wood and of much other old organic material can be dated within the past 70,000 or so years by measuring the ratio of radioactive carbon-14 to stable carbon-12 in the sample. Carbon-14 is produced by the action of cosmic rays on the earth's atmosphere, and recently synthesized wood will have approximately the same proportion of it as will the atmosphere. As time goes on, however, the amount of carbon-14 in wood will decline until, at approximately 5568 years, one half of the radioactive carbon-14 atoms will have decayed. By computing the amount of carbon-14 to carbon-12 as determined by careful analysis of radioactivity in a wood or other organic sample, and comparing this amount with the disintegration curve of carbon-14 based on a half-life of 5568 years, the approximate age of the sample can be computed. Spruce wood of the Two Creeks forest in northeastern Wisconsin has been dated many times by several different carbon-14 laboratories as being 11,850 $\pm$ 100 years old, thus dating the last major glacial advance of the Wisconsin period affecting the continental United

States. It is through carbon-14 dating that the absolute chronology of late-glacial and postglacial events is being established.

**Peat Analysis.** Acid peat bogs preserve plant remains in great abundance. Through studies of these fragmentary remains, it is possible to isolate fossil wood in the various levels and date them by carbon-14 dating. It is also possible to evaluate the relative abundance of various types of peat-forming bog plants and from these to derive some knowledge of past climatic fluctuations. As the bog builds up, different strata of peat indicate the kinds of vegetation present in the bog at different times. Peats may be derived from mosses (including sphagnums), sedges, and woody plants. Sphagnum peats are indicative of cool and humid times while alternating ligneous or woody peats in the same bog denote milder climatic conditions. Prominent ligneous peat horizons in European peatlands are found over large areas and seem to date from the same period of formation. Volcanic ash horizons, such as commonly occur in the Pacific Northwest, can frequently be dated. The Katmai ash from the 1912 eruption on the Alaska peninsula provides a datum from which the current rate of peat accumulation may be dated. It is from the study of trapped pollen rain, however, that the greatest information is obtained by means of peat bog analysis.

**Pollen Analysis.** Lake sediments are rich reservoirs of fossil pollen and plant remains. Lakes are common in glaciated areas, and the pollen of their sediments reflects the regional character of the vegetation. In addition, the cold, wet, anaerobic environment provided by acid peat bogs constitutes a perfect preservative for hard plant parts. None is preserved better than pollen, whose hard outer coats, or **exines**, are extremely resistant to decay. However, trees growing on the bog surface are likely to contribute disproportionately large amounts of pollen. Thus lake sediments are the chief source for pollen analytical studies. Many pollen grains can be identified as to genus and some as to species.

Pollen analysis, although perhaps the most useful single tool in the deciphering of past vegetational history, is fraught with many dangers of misinterpretation. The pollen rain falling on a lake, open bog, or other repository in which the pollen may be enclosed in sediments and preserved will contain an over-representation of species which produce pollen abundantly and whose pollens are disseminated by wind, and an underrepresentation of species that produce pollen sparsely and whose pollens are disseminated by insects or other means.

A major problem in interpretation of the fossil pollen record has been the nearly universal reliance on percentages of pollen of the various genera (or species) in a sample to determine the composition of past vegetation. Such compositional percentages tell us which types of pollen were represented better than others, but there is not necessarily a high correlation between the forest composition indicated by the percentages and the actual composition of the forests at that time. This problem has been resolved in part by the geographical study of modern pollen rain and a comparison of the pollen assemblages of modern forest types, whose composition is known, with fossil pollen assemblages.

Pollen grains, being produced in various quantities by different species and carried by the wind, are related in a complex way to the vegetation from which they came. Once this relationship is understood for modern forest communities, the fossil pollen record can be more accurately interpreted. To accomplish this task, the pollen content of sediment surfaces from lakes and peat bogs in various forest and tundra regions has been determined. Distinct pollen assemblages (Figures 19.1 and 19.2) characterize regions of distinct vegetation such as tundra, boreal conifer forest, mixed conifer-deciduous forest, and deciduous forest. The same pollen assemblages that are recognized over a climatic gradient in modern forest communities may be recognized in a vertical sequence in deposits of ancient pollen. For example, fossil pollen assemblages from southern New England and from the Great Lakes resemble charac-

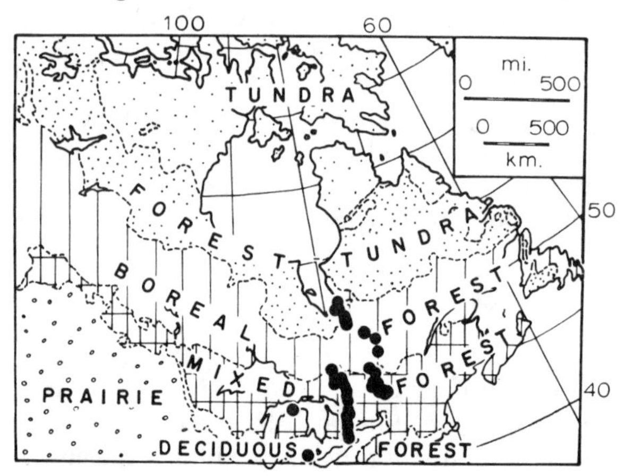


**Figure 19.1. Major vegetation regions in eastern and central Canada modified from Rowe (1959). The black dots represent sites where surface samples have been collected along a north-south transect. (After Davis, 1969. Reprinted by permission of *American Scientist*, Journal of Sigma Xi, The Scientific Research Society.)**

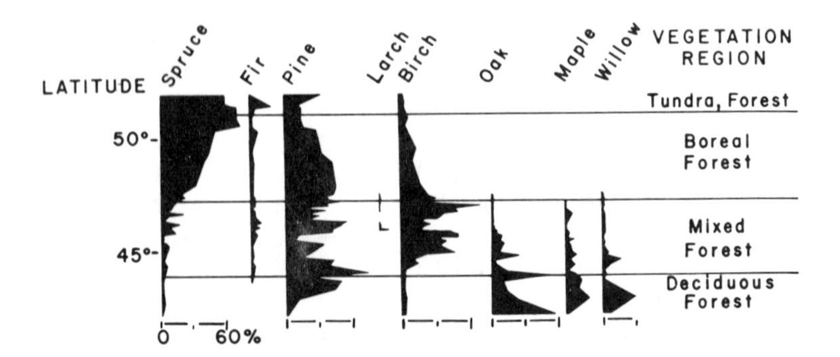


Figure 19.2. **Pollen assemblages in surface samples collected along the north-south transect through Canada shown in Figure 19.1. The percentage value for each type of pollen is shown on the abscissa. The latitude at which the sample was collected, and the forest region, are indicated on the ordinate. (After Davis, 1969. Reprinted by permission of *American Scientist*, Journal of Sigma Xi, The Scientific Research Society.)**

teristic but different modern vegetation types of the tundra and of forests of eastern, central, and west-central Canada (Davis, 1967). Moreover, sufficient pollen data are now available to permit the tracing of migration routes of trees through postglacial time from their glacial refugia to their present locations (Davis, 1976; Bernabo and Webb, 1977). Detailed reviews of the methods and progress of pollen analysis related to forest vegetation of the late Tertiary and Quaternary (Davis, 1963, 1969; Leopold, 1969; Wright, 1971) and an excellent book on palynology (Tschudy and Scott, 1969) are available.

## Pleistocene Glaciation

The advance of glaciers during the Pleistocene into the present temperate zones of the earth is, of course, the major event in late Cenozoic forest history. At its maximum, ice covered approximately 32 percent of the land area of the world as compared to the 10 percent covered at the present time.

There were perhaps as many as 20 major glaciations in the Pleistocene. They lasted 50,000 to 100,000 years each and were separated by relatively brief, warm interglacial periods of 10,000 to 20,000 years during which climates as mild and often milder than those of today existed (Davis, 1976). The last glaciation itself, the Wisconsin, insofar as the Great Lakes region is concerned, can be subdivided into several substages. The next to the last (the Woodfordian Substage in North America) apparently began to melt northward about 16,000 to 18,000 years before present (B.P.). During its general retreat the ice sheet reversed itself at least once, marking a minor readvance in the Lake Michigan basin about 11,850 years ago and continued to retreat shortly thereafter.

Obviously, forests were eliminated from the land under ice, and since these lands constitute some of the most important forest areas of the earth today, it is equally obvious that their present vegetation dates from the last retreat of the Pleistocene ice.

Not all the northern world, however, was under ice. A large area in the interior of Alaska and adjacent Yukon escaped glaciation, possibly because of low precipitation. This region apparently supported a tundra flora, probably because summers were too short and too cold to support forests (Hopkins, 1967a). Unglaciated areas on the islands off the Northwest Pacific coast such as the Queen Charlotte Islands of British Columbia likely served as refugia for Sitka spruce and western hemlock during glacial maxima. ***Nunatuks***—isolated mountain peaks surrounded by ice—were far less likely to support tree growth of any kind during the Pleistocene.

Considerable controversy existed as to the extent to which forests were displaced south of the glacial border. The difficulty arose because no comparable conditions exist today. Existing continental glaciers such as in Greenland, Baffin Land, and Antarctica are confined to high latitudes; while existing temperate zone glaciers are of limited extent and therefore exert relatively little influence on climate beyond their borders. One school held that the forests were displaced relatively little south of the glacial border, citing such instances as the west coast of New Zealand, where podocarps and tree ferns grow adjacent to glaciers coming from the Southern Alps. Others, citing the evidence of fossil spruce pollen from Georgia and eastern Texas, held out for a major southward displacement of forest types. Another point of controversy concerned the extent and persistence of tundra or other nonforest zone between the edge of the ice and the edge of the forest. Pollen evidence now supports the existence of treeless (tundra) vegetation along much of the southern margin of the retreating ice sheet.

Although many of the details have yet to be worked out, considerable literature is accumulating on these problems (Heusser, 1960; Davis, 1965, 1967, 1976; Wright, 1970, 1971; Whitehead, 1973). The truth seems to be that the forests were displaced considerably southward, but not to a distance comparable with the advance of the ice. The distribution of forests today is much farther north of their glacial distribution than was previously thought. Conifer forests dominated by spruce and pine (presumably jack pine), that are boreal species today, were present (perhaps mixed with deciduous forest stands in some localities) in much of the area south of the ice in main Wisconsin times. Furthermore, pollen and macrofossil records, indicate that a forest of jack pine and some black spruce, together with aquatic vegetation, occurred in the Piedmont region of northern Georgia 23,000 to 15,000 years ago (Watts, 1970). This vegetation, typical of modern forests 1100 km to the north, indicates a major climatic change (Wright, 1971). Boreal forest also occupied parts of the Middle West. However, in southern Illinois, the full-glacial vegetation apparently was not a true boreal conifer forest, but a mosaic of stands (composed of oak, spruce, and pine) and open areas (E. Grüger, 1972). Farther west in northeastern Kansas, pollen assemblages of 24,000 to about 15,000 years ago indicate a spruce forest with no pine and little oak (J. Grüger, 1973). Thus the full-glacial conifer forest differed considerably in composition from east to middle west in contrast to the relatively uniform boreal forest of spruce, fir, and pine across Canada today.

We have only sketchy and inconclusive evidence concerning the distribution and composition of temperate deciduous forests during the main Wisconsin glaciation (Whitehead, 1965, 1973; Wright, 1971). Oak pollen has been reported from the full glacial of northern Georgia (Watts, 1970), and it is likely that temperate deciduous trees existed on the more favorable sites in the southeast. Presumably, the subtropical and south temperate species were able to persist along the Gulf Coast and on the peninsula of Florida.

## Late-Glacial and Postglacial Forests and Climatic Changes

Late-glacial history begins with the fluctuating retreat of the last glacial maxima (16,000 to 18,000 B.P.) and ends with the final disappearance of the ice about 8,000 years ago. The general story of changes in forest and climate outlined below are well established even though subject to modification of detail. The various events can be traced more or less contemporaneously throughout the world although the mass of evidence concerns the changes in western Europe and eastern North America.

With the melting back of the Woodfordian ice in the Great Lakes region, the open land was first colonized by tundra-forming sedges and other plants or by boreal trees. A readvance of ice about 11,850 years ago extended as far south as the middle part of the Lake Michigan basin. It prolonged the spruce period in the forests of the surrounding area but had little effect on the revegetation patterns to the east (New England) and west (Minnesota and the Great Plains). This

minor readvance is now regarded by some as a local event and apparently was not associated with a major climatic change.

During the past 10,000 years or so, pollen studies have given a detailed picture of the appearance and disappearance of many forest genera and some other plants and at least an approximate idea of their relative abundance at any one time. In general, a warming trend is seen up to a time when the climate was somewhat warmer and drier than it is today. This has been followed by a general cooling up to the present time. However, much regional variation is seen in this overall pattern. The warmer-drier period in postglacial times seems to have occurred relatively early in the Southeast (10,000 to 6000 years B.P.), from 8000 to 4000 B.P. in the Middle West, where part of the deciduous forest changed to prairie and back again, but in New England the record is now considered "ambivalent" (Wright, 1971, 1976). These variations are not surprising in view of recent climatological evidence that a shift in air mass distribution does not have a uniform effect over continental North America.

The changes in pollen percentages with time may be considered a continuum, although abrupt as well as gradual changes have been observed. Although the cause of an abrupt change is open to interpretation (e.g., a climatic change), a first appearance by immigration of the species, or simply forest succession, the appearance in large numbers of some tree genera and the disappearance of others may be taken as demarcating zones in the postglacial chronology. At least three such arbitrary zones are frequently recognized: an initial cold period, a second warm period, and a final cooler period. Five zones are recognized in New England, seven in Great Britain, and 11 to 12 in northern Europe. It is obvious that different genera, or the same genera occurring in different relative numbers, will characterize different sites at the same time in the past as they do at present. These vegetation changes may be summarized for various parts of the world for which pollen chronologies have been prepared.

**Britain.** Early in the century, von Post recognized the three major postglacial stages of increasing warmth, maximum warmth, and decreasing warmth. In more recent studies, the details of vegetational changes are represented by pollen spectra. The postglacial history of Great Britain is representative (but not necessarily typical) of this work. Here, pollen sequences have been given absolute datings through carbon-14 determinations so that a well-defined history is available (Godwin, 1975). Pennington (1969) describes the history of British vegetation, and the effects of Neolithic and modern man on the forests are especially well documented.

**New England.** Throughout much of the Northeast, peat bogs and pond deposits consistently yield a maximum of spruce and fir pollen in the oldest and deepest stratum (A zone), followed by a pine maximum (B zone), and that by a hardwood zone (C zone) (Deevey, 1949). Each may be subdivided according to relative dominance of species, particularly of hardwood species in the top zone. These zones are not necessarily contemporaneous from one area to the next. Obviously the spruce zone will occur earlier and end earlier in the south than in the north under comparable conditions of ice retreat.

In southern New England, the period of increasing warmth following the retreat of Wisconsin ice was characterized by a long interval of tundra lasting until about 12,000 years ago, followed by a shorter period in which a transitional, open, spruce-hardwood woodland was succeeded by an open, spruce woodland (Davis, 1969). Then a dramatic change occurred about 9500 years ago as mixed deciduous-coniferous forest replaced the spruce woodland. The transition from spruce woodland to mixed deciduous-coniferous forest took place before a closed boreal forest could develop. Except for this missing community, the boreal forest, which today separates subarctic woodland and the mixed deciduous-coniferous forest, the march of vegetation in time closely resembles the arrangement of modern vegetation regions from north to south in Canada (Figure 19.3). The oldest pollen assemblages resemble those deposited in the modern pollen rain in tundra regions far from any forest. The spruce woodland of 10,500 years ago resembles modern open woodland north of the boreal forest in Quebec, and the deciduous-coniferous forests of 9500 B.P. are similar to modern forests of Canada and the northern Great Lakes region south of the present boreal forest. The resemblance of the late-glacial forest of New England to modern vegetation of central Canada indicates the climate was drier, cooler, and more continental than at present (Davis, 1967).

The period of maximum warmth (now termed Hypsithermal, but formerly termed xerothermic or climatic optimum), is not clearly marked in southern New England. On the other hand, in northern New England and southeastern Canada, the increase of hemlock and other conifers, beech, maple, and other cool-climate species following the warm period is usually attributed to a change of climate. Alternatively, it may be due to the activities of Indians and Europeans prior to the major forest clearings of the early nineteenth century. The effects of land clearing are well documented in the pollen record by the rise of nonarboreal pollen (grass, composites, plantain, ragweed, maize, and cereals) and pollen of certain tree species such as poplars, red maple, spruce, and fir.

**Western Great Lakes Region.** In much of the western Great Lakes region, tundra was apparently absent in late-glacial times, but it has

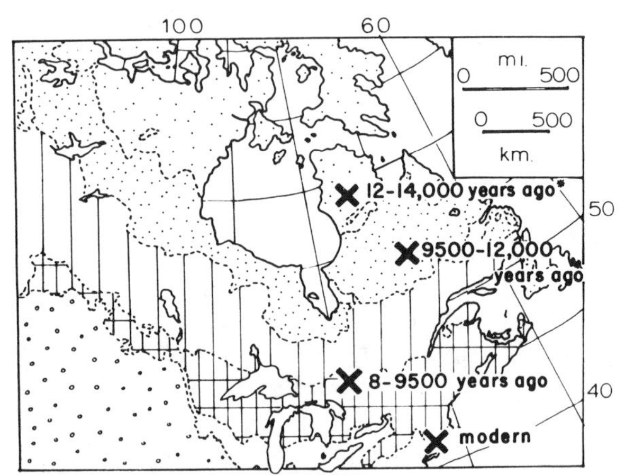


**Figure 19.3. Localities in Canada where surface pollen assemblages resemble fossil assemblages in southern New England. The sites are indicated by crosses. The age of the analogous fossil material is indicated for each locality. The location in the tundra indicated by an asterisk (*) is extrapolated from the resemblance of fossil material to assemblages from northernmost Quebec (Bartley, 1967). Surface samples are not yet available from the precise locality shown. (After Davis, 1969. Reprinted by permission of *American Scientist,* Journal of Sigma Xi, The Scientific Research Society.)**

been identified in northeastern Minnesota and Michigan. Farther south, spruce forest apparently extended to the edge of the ice and may have covered a zone of stagnant receding ice (Wright, 1971). Wright believes that some temperate deciduous trees, probably oak and ironwood, together with the boreal species birch and alder, were minor components of this late-glacial spruce forest; pine was generally absent from the area from Manitoba to northern Ohio.

Birch and alder were the first species to invade the spruce forest as a warming climate affected its outer portions, beginning about 12,000 B.P. in southern Minnesota (Wright, 1971). The rapid rise of pine pollen 10,000 years ago in eastern Minnesota marked a swift expansion of the jack pine–red pine type (jack pine and red pine are often considered together because of the similarity of their pollen) from the east. The westward expansion had been blocked by dense spruce forests in Ohio and by ice and glacial lakes in the Lake Michigan basin until about 11,000 B.P. Whereas pines migrated into the Great Lakes region from the east, deciduous elements moved in from the south. In Ohio, oak replaced spruce about 10,500 years ago

(Ogden, 1966), and elm, oak, and associated hardwoods entered southern Minnesota about 9500 B.P. and spread rapidly northward.

As the climate warmed and mean circulation of the westerlies increased, a wedge of air, dried by subsidence on crossing the Rockies, was driven eastward by the westerlies bringing with it dry conditions and an associated biota. In turn, oak savanna and true prairie expanded eastward. Fire was also a major associated factor in pushing back and maintaining the prairie-forest border. Reaching into Minnesota, Illinois, Indiana, southern Michigan, and Ohio, prairie openings apparently gradually replaced mixed pine-deciduous forests to the north and elsewhere deciduous forests and oak savannas. This ***Prairie Peninsula*** (Transeau, 1935) is evidenced by the widespread occurrence of prairie soils, remnants of oak savannas, and by the persistent fragments of the prairie until the initiation of farming by white man. The boundary dates of this period of maximum warmth and dryness are placed from 8000 to 4000 years ago; the peak of the warm period was apparently about 7000 B.P. The deterioration to a cooler, moister climate was slower than the rapid onset of the warm period had been. Pines and temperate hardwoods gradually replaced prairies, although prairie openings probably survived in east central Minnesota until 4000 years ago (Wright, 1971). The Prairie Peninsula posed a barrier to the northward migration of upland mesophytic species such as beech, hemlock, and yellow-poplar. These species migrated into the central and southern portion of Lower Michigan from the east, entering these areas before reaching the eastern edge of the prairie in Indiana (Benninghoff, 1963).

White pine, which may have had its refuge on the mid-Atlantic coast during full-glacial times, migrated more slowly than jack pine and red pine. It arrived in New England about 10,000 years ago and ultimately reached eastern Minnesota in quantity about 6800 B.P. (Figure 19.4b), just several hundred years after the prairies had arrived from the southwest (Wright, 1971; Davis, 1976). White pine was apparently halted here during the warm prairie period, but with the change toward cooler, wetter conditions, it renewed its westward migration. The uplands, although dominated by white pine, gained more mesic species, such as sugar maple and red maple as well as basswood, elm, and oak, that had survived the warm period. With the onset of cooler times, the jack pine-red pine type expanded out of its prairie-period refuge in northeasternmost Minnesota and reached the Itasca Park area about 2000 years ago. On lowland sites, macrofossil analysis indicated the cooler, wetter trend by a marked increase of remains of spruce, larch, and ericaceous bog vegetation. More recently, the arrival of man on the scene brought extensive cutting of the pine forests and their replacement in many areas with birch and aspen. Detailed accounts of vegetational history of the western Great Lakes are found in the works of Cushing (1965), McAndrews (1966), and Wright (1971; 1976).

The contrasting patterns of the migration of different trees, red and jack pines, white pine, beech, and hickory (*Carya* spp.), into the western Great Lakes region and New England are shown in Figure 19.4 (Davis, 1976). Migration of jack pine (which followed shortly after the spruces) into eastern and central United States and Canada took place from the southeast. In contrast, white pine followed somewhat later and migrated from areas on the mid-Atlantic coast. Beech moved northward from the south at a relatively rapid rate of 250 to 300 m per year. Hickory, apparently surviving glacial times in the southern Great Plains, moved north and east at a relatively slow rate.

**Pacific Northwest.** Postglacial history in the Pacific Northwest (from Oregon north to Alaska) is complicated by the fact that the Cordilleran ice radiated east and west from the high mountain ranges extending north and south rather than simply moving southward as in eastern North America; that maritime climatic conditions originating from the Pacific Ocean moderated at least some of the postglacial climatic fluctuations found elsewhere; and that refugia for various plants and animals, including Sitka spruce and western hemlock, along the Pacific Coast served to restock much of the glaciated terrain rather than simple migration from the south as in the case of the eastern part of the continent (Hansen, 1947, 1955; Heusser, 1960, 1965). An extensive refugium in unglaciated interior Alaska existed for forest trees such as spruce and birch, along with shrubs and herbs of the arctic tundra.

The Pacific Northwest has a late-glacial and postglacial history similar to that of eastern North America in that ice had begun to recede as early as 14,000 B.P., and that a late-glacial cold period was followed by a warmer and drier trend, which ultimately gave way to a cooler, more humid climate. Following retreat of the glaciers, a lodgepole pine parkland developed in the Pacific Northwest and northwestward as far as southeastern Alaska. As the climate ameliorated, alder, birch, and more tolerant coniferous species replaced pine. Douglas-fir became the dominant species of the Pacific Northwest during the warmer and drier times. Where rainfall was heavy, western hemlock replaced pine and mountain hemlock, and Sitka spruce became well represented from British Columbia to southeastern Alaska. In the Willamette Valley, the succession following pine was to Douglas-fir and Oregon oak, and east of the Cascades grassland achieved dominance on the dry plateaus. The approximate boundary dates of this period of maximum warmth and dryness are 8000 to 4000 B.P.

Reversal of the warming trend brought shifts to western hemlock

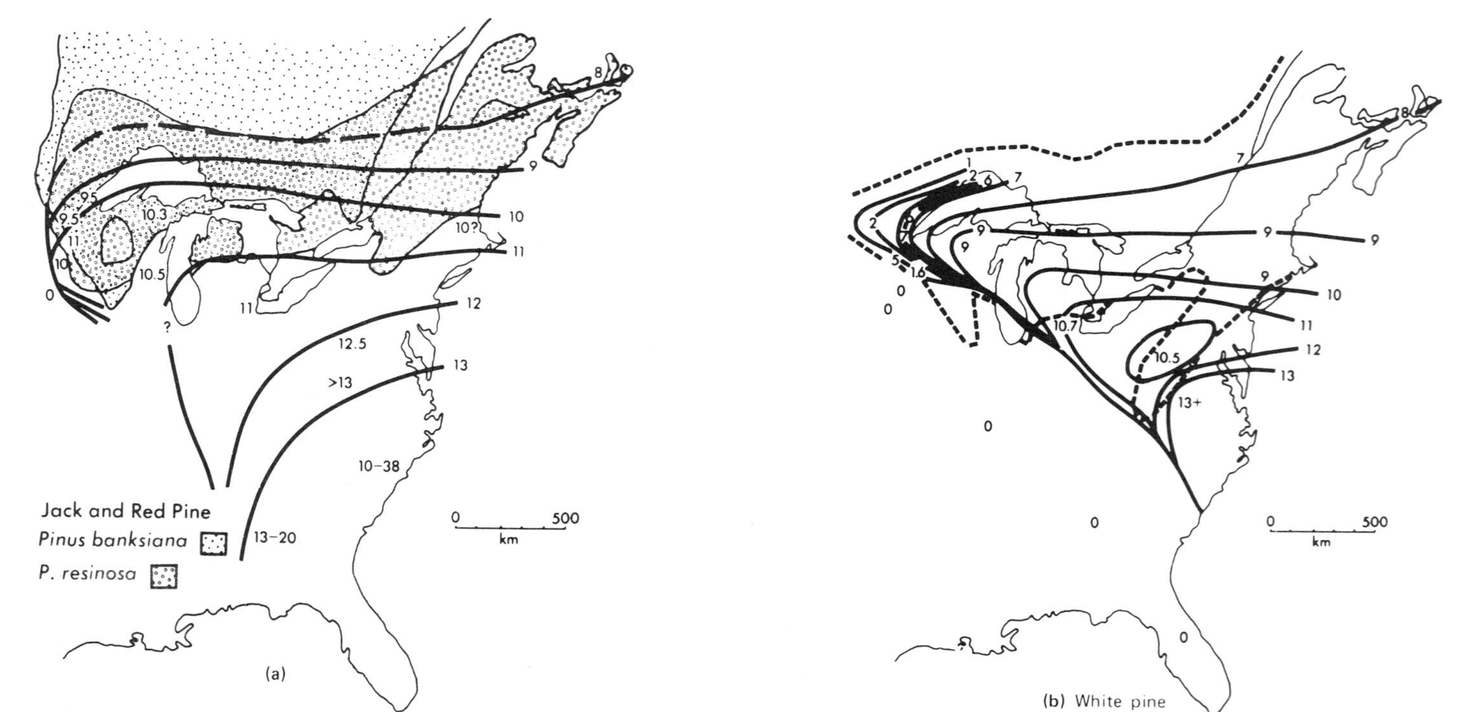


**Figure 19.4.** Migration routes of tree species in late-glacial and early Holocene (the present warm-climate interval, beginning 10,000 years ago). The small numbers on the map indicate the time of arrival at individual sites. Contours show the advancing frontier at 1000-year intervals. Dashed line surrounds the modern range in *b*, *c*, and *d*. *(a)* Jack and/or red pine migrated from the southeast; stippled area indicates the modern range for jack pine; open circles indicate the modern range for red pine. *(b)* White pine

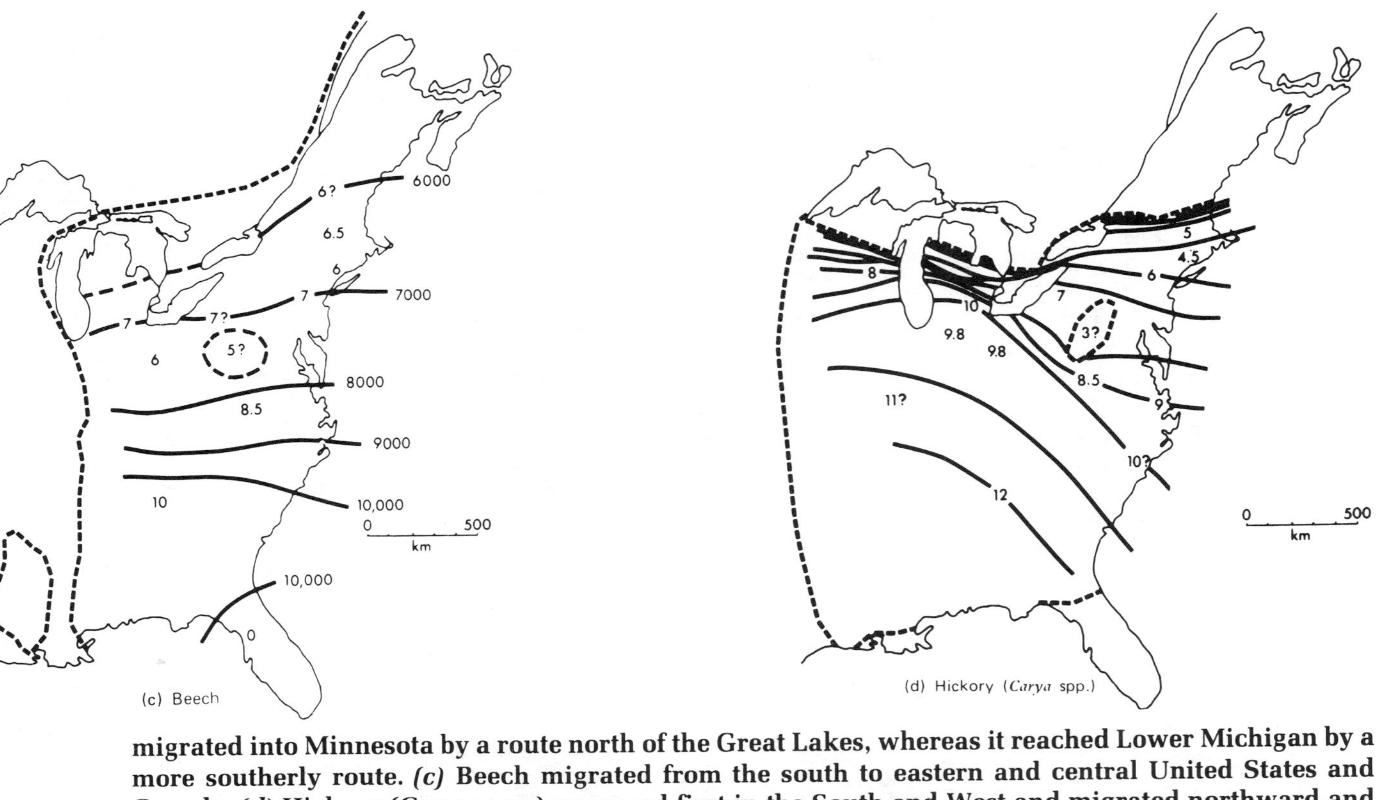


migrated into Minnesota by a route north of the Great Lakes, whereas it reached Lower Michigan by a more southerly route. *(c)* Beech migrated from the south to eastern and central United States and Canada. *(d)* Hickory *(Carya* spp.) appeared first in the South and West and migrated northward and eastward. (After Davis, 1976. Reprinted by permission of *American Scientist,* Journal of Sigma Xi, The Scientific Research Society.)

in northwest forests as far north as southeastern Alaska. Coastal areas too dry for hemlock supported Douglas-fir, as do these areas today. In eastern Washington, open ponderosa pine forests succeeded grasslands.

To the northern interior, records from lakes in Glacier National Park in northern Montana show that lodgepole pine and western white pine dominated the earliest forest; Engelmann spruce and alpine fir were also present. A trend to warmer and drier conditions is indicated (coincident with volcanic ashfall from Mount Mazama—now the remnant Crater Lake in Oregon) about 6600 B.P. by the rapid increase of pollen of Douglas-fir and ponderosa pine and a decrease in that of western white pine and lodgepole pine. Thereafter, a cooler and more humid climate ensued with western white pine, Engelmann spruce, and alpine fir becoming dominant. Lodgepole pine is also strongly represented in pollen spectra, and its abundance is attributed to a greater incidence of severe fires in more recent times (Hansen, 1948).

**Around the World.** The examples cited above from England, New England, the western Great Lakes region, and the Pacific Northwest are but a few of similar pollen sequences throughout the world. Elsewhere, the broad picture is much the same. The Pleistocene glaciers reached their last maximum extension about 18,000 years ago. The end of Pleistocene glaciation was marked by a clear, worldwide climatic change which was unidirectional in its major effects. For many thousands of years afterward, the climate in the ice-free areas ameliorated, and wave after wave of forest tree species migrated into the former glaciated sites. Not all postglacial changes were synchronous around the world since they were influenced greatly by regional air mass distributions. For example, the time of maximum warmth varied considerably from region to region in continental United States, depending on regional climatic patterns. From the warm period up to historic times, the climate has been becoming colder and in many places wetter. Many minor changes and stages are interpreted locally, but the general outline holds for the northern, southern, eastern, and western hemispheres.

## RECENT EVIDENCE OF CLIMATIC CHANGES

The continuing instability of vegetation as a result of continuing climatic change and continuing migrations of tree species becomes more and more apparent as we examine the more plentiful evidence of recent times (Dolf, 1960). Norsemen settled Greenland a thousand years ago, raising many head of cattle on what is now permanently frozen ground and traversing sea lanes blocked with floating ice but a century ago. Glaciers in Iceland were far less extensive from 900 to 1300 than they have been since. Glaciers in the Alps reached their maximum development during the period from about 1600 to 1850 and have since been receding, as indeed have most glaciers throughout all the rest of the world (Denton and Porter, 1970).

The evidence from the fluctuations of lakes, from tree rings, and from other sources point to the fact that the first millenium of the Christian era was relatively warm, but that a period of increasing cold set in about 1300 which culminated about 1800. During this latter period, mesophytic forest trees in many parts of the world were invading sites occupied by other species better suited to drier and warmer climates. European beech moved northward in Britain, American beech and sugar maple expanded their distribution in the Lake States, invading oak-hickory sites just as the oak and hickory were encroaching upon the eastern sections of the prairie. In New Zealand, evidence from existing old-growth stands is that *Nothofagus* was similarly invading podocarp forests (Holloway, 1954).

The best evidence to date is that these marked climatic fluctuations are not cyclic or periodic. Whether the recent long trend toward cooler climates or the even more recent but short trend toward warmer climates will prevail in the future cannot at present be predicted. It is clear, though, that our climate is unstable, that the balance between glaciers and climate is delicate, and that changes are to be expected both in future climates and in the forests as a result of the response of various species to changed environmental conditions.

## SUGGESTED READINGS


Axelrod, Daniel I. 1970. Mesozoic paleogeography and early angiosperm history. *Bot. Rev.* 36:277–319.

Davis, Margaret B. 1969. Palynology and environmental history during the Quaternary period. *Amer. Scientist* 57:317–332.

———. 1976. Pleistocene biogeography of temperate deciduous forests. *Geoscience and Man* 13:13–26.

Denton, George H., and Stephen C. Porter. 1970. Neoglaciation. *Sci. American* 222:100–110.

Ferguson, C. W. 1968. Bristlecone pine: science and esthetics. *Science* 159:839–846.

Fritts, Harold C. 1976. *Tree Rings and Climate.* Academic Press, Inc., New York. 567 pp.

Heusser, Calvin J. 1965. A Pleistocene phytogeographical sketch of the Pacific Northwest and Alaska. *In* H. E. Wright, Jr., and David G. Frey (eds.), *The Quaternary of the United States*. Princeton Univ. Press, Princeton, N.J.

Hopkins, David M. 1967. The Cenozoic history of Beringia—a synthesis. *In* David M. Hopkins (ed.), *The Bering Land Bridge*. Stanford Univ. Press, Stanford, Calif.

Pennington, Winifred. 1969. *The History of British Vegetation*. English Universities Press, London. 152 pp.

Regal, Philip J. 1977. Ecology and evolution of flowering plant dominance. *Science* 196:622–629.

Wolfe, Jack A. 1971. Tertiary climatic fluctuations and methods of analysis of Tertiary floras. *Palaeogeog., Palaeoclimatol., Palaeoecol.* 9:27–57.

———, and E. B. Leopold. 1967. Neogene and Early Quaternary vegetation of northwestern North America and northeastern Asia. *In* D. M. Hopkins (ed.), *The Bering Land Bridge*. Stanford Univ. Press, Stanford, Calif.

Wright, H. E., Jr. 1971. Late Quaternary vegetational history of North America. *In* Karl, K. Turekian (ed.), *The Late Cenozoic Glacial Ages*. Yale Univ. Press, New Haven.

———. 1976. The dynamic nature of Holocene vegetation. *Quaternary Res.* 6:581–596.

# 20
# The American Forest Since 1600

In the United States, written historical records of forest description and forest history date from the seventeenth century and settlement by the white man. Evidence from contemporary naturalists, surveyors who parceled off the wilderness, and from analysis of old-growth forest relicts have done much to give us an understanding of the pre-settlement forest in North America and the many changes that have taken place since settlement. A brief survey of the nature of the presettlement forest and of the effects on it of land clearing, logging, fire, and land abandonment is essential to an understanding of the present forests of temperate North America. A brief discussion of the effects of human activity on the European and near-east Asian forest is also included.

## THE PRESETTLEMENT FOREST

The question as to what constituted the original, or virgin, forest is an intriguing one. We have already seen that the forests in the glaciated region of the Northern Hemisphere have been in an almost constant state of instability and change over the past 10,000 years. Furthermore, most of this region has been inhabited by man since the retreat of the glaciers, and extensive burning by primitive people has been widespread throughout this period and throughout the world.

Actually, there is no such thing as an original forest or even a virgin forest. Our concern is often with the nature of the forest at the time of colonization of the land by white settlers. In eastern North America, we thus are interested in the composition and structure of the sixteenth-century forest. Farther west in the Central States, our interest is in the eighteenth-century forest; while in many of the more mountainous regions of the West we can still see the presettlement forest, simply because the land has never been settled and the forests have never been logged.

A further word about the concept of the virgin forest is in order. The terms "virgin forest" and Longfellow's "forest primeval" conjure up an image of great and old trees standing undisturbed and changeless for centuries. Specifically, any human disturbance is ruled out. They must be uncut and unharmed by man-set fires and the understory must be ungrazed by domestic stock. In other words, we conceive of the virgin forest as being simply an unharmed old-growth forest.

stop

Such stands simply do not exist. We have already seen that forests of any age are in a constant state of change, arising from the growth and senescence of the trees themselves, from consequent changes in the microclimate and edaphic site, from normal forest succession, and from regional climatic and geologic changes. Interferences with normal growth and development are common.

The forests that confronted the colonizing white settlers on the Atlantic seaboard of North America were in a state of constant change wrought by forest succession, climatic change, fire, wind, insects, fungi, browsing animals, and Indian activity. Nevertheless, since they are now vanished, it is worthwhile to reconstruct their composition and structure as an aid in understanding the present-day forests of the same region. It is less necessary to do this for the western American forest, where extensive old-growth stands still antedate in their history the presence of the white settler.

## Contemporary Observers

In a sense, it is surprising how little is recorded of the forests by the sea captains, explorers, first settlers, and travelling naturalists who saw and traveled in the presettlement forest. Their chronicles are almost inevitably concerned with their own comings and goings, with the Indians, and with human affairs generally. Nevertheless, here and there are preserved fragmentary comments that cast some light upon the early American forests. Early historical accounts have been put together in the greatest detail for the New England states, although the writings of such early eighteenth-century travellers as John Bartram in the American South and Alexander Mackenzie in the Canadian Northwest were notable for their description of the terrain and vegetation of the regions they traversed.

In most cases, comparisons of the present forest with early descriptions show that the forests of the two eras do not differ substantially, as in the case of Martha's Vineyard reported by Ogden (1962). A description of the coastal New England forests between 1629 and 1633 is given by William Wood in his *New England Prospect*, published in 1634. As quoted by Hawes (1923), he said in part:

> *The timber of the country grows straight and tall, some trees being twenty, some thirty foot high before they spread forth their branches; generally the trees be not very thicke, tho there*

*be many that will serve for mill posts, some being three foote and a half o're. And whereas it is generally conceived that the woods grow so thicke that there is no cleare ground than is hewed out by labour of man; it is nothing so; in many places divers acres being cleare, so that one may ride a hunting in most places of the land, if he will venture himself for being lost; there is no underwood saving in swamps, and low grounds that are wet in which the English get osiers and Hasles and such small wood as is for their use. Of these swamps some be ten, some twenty, some thirty miles long, being preserved by the wetness of the soile wherein they grow; for it being the custom of Indians to bourne the wood in November when the grass is withered and leaves dryed, it consumes all the underwood and rubbish, which otherwise would overgrow the country, making it impassable, and spoil their much affected hunting; so that by this means in those places where the Indians inhabit there is scarce a bush or bramble, or any combersome underwood to be seene in the more Champion ground.*

In Jeremy Belknap's history of New Hampshire (1792), he describes both the original forest and its clearing by the settlers:

*Another thing, worthy of observation, is the aged and majestic appearance of the trees, of which the most noble is the mast pine. This tree often grows to the height of one hundred and fifty, and sometimes two hundred feet. It is straight as an arrow, and has no branches but very near the top. It is from twenty to forty inches in diameter at its base, and appears like a stately pillar, adorned with a verdant capital, in form of a cone. Interspersed among these are the common forest trees, of various kinds, whose height is generally about sixty or eighty feet. In swamps, and near rivers, there is a thick growth of underwood, which renders travelling difficult. On high lands, it is not so troublesome; and on dry plains, it is quite inconsiderable.*

. . . . . . . . . . . . . . . . . . . . . .

*In the new and uncultivated parts, the soil is distinguished by the various kinds of woods which grow upon it, thus: white oak land is hard and stony, the undergrowth consisting of brakes and fern; this kind of soil will not bear grass till it has been ploughed and hoed; but it is good for Indian corn, and must be subdued by planting, before it can be converted into mowing or pasture. The same may be said of chestnut land.*

*Pitch pine land is dry and sandy; it will bear corn and rye with plowing; but is soon worn out, and need to lie fallow two or three years to recruite.*

*White pine land is also light and dry, but has a deeper soil, and is of course better; both these kinds of land bear brakes and fern; and wherever these grow in large quantities, it is an indication that ploughing is necessary to prepare the land for grass.*

*Spruce and hemlock, in the eastern parts of the state, denote a thin, cold soil, which, after much labor in the clearing, will indeed bear grass without ploughing, but the crops are small, and there is a natural tough sward commonly called a rug, which must either rot or be burned before any cultivation can be made. But in the western parts, the spruce and hemlock, with a mixture of birch, denote a moist soil, which is excellent for grass.*

. . . . . . . . . . . . . . . . . . . . . . .

*When the white pine and oyl-nut are found in the same land, it is commonly a deep moist loam, and is accounted very rich and profitable.*

*Beech and maple land is generally esteemed the most easy and advantageous for cultivation, as it is a warm, rich, loamy soil, which easily takes grass, corn and grain without ploughing; and not only bears good crops the first year, but turns immediately to mowing and pasture; the soil which is deepest, and of the darkest color, is esteemed the best.*

*Black and yellow birch, white ash, elm, and alder are indications of good soil, deep, rich and moist, which will admit grass and grain without ploughing.*

*Red oak and white birch are signs of strong land, and generally the strength of land is judged of by the largeness of the trees which it produces.*

. . . . . . . . . . . . . . . . . . . . . . .

*In the spring, the trees which have been felled the preceding year, are burned in the new plantations. If the season be dry, the flames spread in the woods, and a large extent of the forest is sometimes on fire at once. Fences and buildings are often destroyed by these raging conflagrations. The only effectual way to prevent the spreading of such a fire, is to kindle another at a distance, and to drive the flame along through the bushes, or dry grass, to meet the greater fire, that all the fuel may be consumed. In swamps, a fire has been known to penetrate several feet under the ground, and consume the roots of the trees.*

From such sources as the above as well as from the writings of Thomas Morton (1632), Peter Whitney (1793), and Timothy Dwight

(1795–1821), we may conclude that the precolonial forests of New England were made up of the same species that characterize the region today. Furthermore, specific kinds of sites were occupied by forest types generally similar to those occupying them today.

## Survey Records

Much more specific information is available from the notes of original land surveys where trees were commonly used for corners, bearing trees, and witness trees. Such records of the kind and size of trees at surveyed corners do not constitute a random sample of the forest, for the surveyor naturally chose trees having a durable wood, and of a size suitable for scribing and likely to be found again years later. Nevertheless, they do provide an important guide to the nature of the presettlement forests in areas where old-growth forests are virtually nonexistent today.

**Metes and Bounds Surveys.** Along the Eastern Seaboard, and throughout colonial times generally, American surveys were by metes and bounds, with the surveyor recording the compass bearing and distance of each line bounding a tract of land. The technique is well illustrated by the surveying notes of George Washington (Spurr, 1951) and a study of the original forest composition in northwestern Pennsylvania based on survey notes of 1814–1815 (Lutz, 1930a).

The job of the frontier surveyor was to lay out land for settlement as quickly and as inexpensively as possible. The surveying techniques of Washington were therefore of the rough-and-ready sort. He used a staff compass and chain. Magnetic bearings were recorded to the nearest degree and distances were measured to the nearest rod (termed "pole" by Washington). No corner stakes were set; rather, a convenient tree was chosen as a corner, blazed, and inscribed. It is this practice of the frontier surveyor of tying in to trees rather than stakes that provides the information concerning the forests of the time. A sample survey chosen at random from Washington's notes illustrates the type of survey and the nature of the information supplied:

*April 24th 1750*

*Plat drawn*

*Then Surveyed for Thomas Wiggans a certain tract of Waste Land Situate in Frederick County & on Potomack River about ½ mile aboe ye Mouth of great Cacapehon & bounded as followeth beg: at a white Oak a white Hickory & White Wood Tree just on ye Mount of Wiggan's Run & opposite to a nob of ye Mountains in Maryland & run thence S$^{o}$ 25 W$^{t}$ Two hund$^{d}$ &*

*twenty Eight Poles to a white hickory an Elm & Mulberry about 30 Pole from Cacapehon thence N*$^{o}$ *75 W*$^{t}$ *One hund*$^{d}$ *& forty Poles to a Chestnut Oak & white Oak thence N*$^{o}$ *25 E*$^{t}$ *Two hund*$^{d}$ *& Sixty Poles to a white Oak red Oak & Iron Wood on ye Riverside thence down ye several Meanders thereof S*$^{o}$ *67½ E*$^{t}$ *37 Po S*$^{o}$ *58½ E*$^{t}$ *74 Po S*$^{o}$ *55 E*$^{t}$ *to ye beg Con*$^{g}$ *210 Acres—*
*John Lonem*
*Isaac Dawson*
*William Wiggans*

There seems to be no reasonable doubt that Washington knew the trees of the middle Appalachian regions. His reputation as a surveyor would naturally depend in part upon the accuracy of his description of corners, and this in turn depended upon the correct naming of the corner trees. Furthermore, there is a wealth of internal evidence in the surveying notes confirming this knowledge. For example, he was very precise in his identification. He names separately seven oak species, including such a relatively uncommon and confusing species as swamp white oak. He readily distinguished the two *Juglans* from each other as he did also between the two maples. In all, he listed some 35 different tree species, all of which are known to occur in the area and on the sites and in the communities where he placed them.

The most commonly cited trees in Washington's surveying notes are listed in order of abundance in Table 20.1. These were listed from lot surveys in the valleys of the Shenandoah, the Cacapon, and the South Branch of the Potomac Rivers. White oak, red oak, and hickory were the characteristic species with considerable pine growing on the hills and mountain sides in the latter two areas.

All in all, the picture of the colonial forest of the central Appalachians as reconstructed from Washington's notes is a very believable one, and one that shows relatively little change in the forest of this area over the last 200 years.

Lutz (1930a) was able to compare data from an 1814–15 survey in Warren and McKean Counties, Pennsylvania, with information from the Heart's Content old-growth forest in the same part of northwestern Pennsylvania (Table 20.2). The similarity of forest composition, as regards dominant species, between the two sets of data is remarkable. Both indicated a beech-maple-hemlock forest with smaller amounts of chestnut and white pine.

**Rectangular Surveys.** As the inadequacies of metes and bounds surveys became appreciated, they were supplanted by various types of rectangular surveys. In the eastern United States, various unofficial types of rectangular grids were used; but in the nineteenth cen-

**Table 20.1. Trees Cited in George Washington's Surveying Notes in Northwestern Virginia and Northeastern West Virginia, 1751–1752**

| SPECIES | NUMBER |
|---|---|
| White oak | 366 |
| Red oak | 181 |
| Hickory | 142 |
| Pine (Virginia?) | 132 |
| Chestnut oak | 51 |
| Black oak | 37 |
| Black locust | 36 |
| Spanish oak (pin or scarlet?) | 28 |
| Others | 141 |
| Total | 1,114 |

tury, these were supplanted by the United States public lands survey in which most of the country was laid out into townships 6 miles square, these being subdivided into sections 1 mile square. Since witness trees were required to be blazed, measured, and recorded at each section and quarter-section corner, and since the surveyors were required to record the point at which their lines crossed rivers, prairies, swamps, peat beds, precipices, ravines, and tracks of fallen timber, it is possible to reconstruct the forest cover of the time of the original survey with some confidence and precision.

**Table 20.2. Percentages of the Most Common Forest Trees in the Old-Growth Forests of Northwestern Pennsylvania**

| | DALE SURVEY, 1814–1815 | HEART'S CONTENT OLD GROWTH, 1930 |
|---|---|---|
| Beech | 30.8 | 24.0 |
| Hemlock | 26.8 | 36.1 |
| Maple (red and sugar) | 13.0 | 10.9 |
| Birch (black and yellow) | 6.1 | 2.8 |
| White pine | 6.0 | 11.1 |
| Chestnut | 5.6 | 8.8 |
| Others | 11.7 | 6.3 |
| Total | 100.0 | 100.0 |

From Lutz (1930a).

There are many pitfalls inherent in interpretations of forest cover from early General Land Office Surveys (Bourdo, 1956). The procedure of the survey was not standardized until 1855, and even then practices varied according to the personnel involved. Some surveys are unreliable and some are strictly fictitious. Individual surveyors had their own preference in selecting trees for bearing and witness trees. In general, they were partial to medium-sized trees of species capable of attaining large size. Tree diameters were estimated by the surveyors, usually to the nearest 2-in class. Distances were also sometimes estimated. Nevertheless, if the modern investigator acquaints himself with the survey procedure of the area he studies and studies the work of the particular surveyors involved, he can interpret much about the nature of the presettlement forest.

Reconstructions of presettlement forest composition from public land survey records have been particularly fruitful in the American Midwest, where various investigators have constructed maps of presettlement vegetation for given districts, counties, and even states (Gordon, 1969). In northern Lower Michigan, for instance, Kilburn (1957) recorded the trees cited in the 1840 and 1855 surveys and found hemlock the most abundant species, although today, after complete clearcutting and extensive burning, the species is virtually absent in that area. The most extensive presettlement forest type was hemlock-beech-maple, with pine forests (red, white, and jack) being common on the sandier soils and swamp conifers (northern white cedar, tamarack, spruce, and balsam fir) in the bogs. Today, bigtooth and trembling aspen are the dominant tree species in the area with red oak and red maple the commonest hardwoods. Only the swamp conifers and the jack pine on the sand plains still occupy for the most part their presettlement sites.

**Old-Growth Remnants.** In areas long cutover, and consisting largely of second growth, occasional remnants of old-growth forests can be found. The detailed study of them provides valuable information on the forests at the time of settlement. Many of the more spectacular of these remnants are preserved as parks or as wilderness areas. Among those that have been studied in detail in the northeastern United States are the Heart's Content stand (Lutz, 1930b) and East Tionesta Creek forest (Hough, 1936) in northwestern Pennsylvania, the Pisgah forest in southwestern New Hampshire (Cline and Spurr, 1942), and the Itasca red pine stands in northwestern Minnesota (Spurr and Allison, 1956).

Similar studies have been made of small residual old-growth stands in western Europe. One example is the survey of the native Scots pine stands of Scotland by Steven and Carlisle (1959). The three dozen or so remnants were found to have a principal age class of from 140 to 190 years old and were lacking in the younger age classes. The low stocking of the overstory had had the consequence of giving rise to a rank growth of an understory so as to make natural regeneration difficult. Since most Scots pine do not survive more than 250 years in that region, positive action is needed to regenerate and protect the few residual woodlands.

### Second Growth Stands

As long as the forest itself is old, it is not necessary that it contain old trees in order to reconstruct the past forest history and cover of the tract. Many second-growth stands can yield valuable information on the presettlement forest.

The essential is that the land has never been completely cleared, ploughed, and farmed. If it has merely been logged from time to time, considerable evidence remains. Even in areas once largely farmed, careful surveys of land deeds indicate that certain areas have been described as woodlots back to the period of settlement (Raup and Carlson, 1941). Other evidence obtainable from the forest itself involves stump sprouts, root suckers, species that characteristically seed in on rotten wood, evidence of past windfalls, soil structure studies, and the occurrence of charcoal in the soil.

In the eastern hardwood forest of the United States, many species have a high coppicing ability throughout their life. A large percentage of the present oak forest, for instance, consists of stump sprouts that have arisen from previous stump sprouts, which in turn can be traced back in time to the period before settlement and land clearing. In such a case, it is obvious that the same species—and indeed the same genotype—occupied the same site several hundreds of years ago. The same is true for root-suckering species such as aspen, where the size of the clonal colony may indicate the length of time that a particular clone has grown on that particular spot.

In the case of species that characteristically seed in on rotten wood, the history of the forest can be traced well back. In the coastal Sitka spruce–western hemlock forests of British Columbia and southeastern Alaska, these species habitually become established on the elevated rotten wood of dead trees, the only sites in the forest dry enough for seedling establishment. Here, it is common to see a giant western hemlock growing 3 m from the ground on the stump of a prior western hemlock, which itself rises out of the rotten wood of still another gigantic tree of the same species. Assuming that each lived a life span of 400 years, a record of 1200 years is thus provided during which that particular spot always had a western hemlock growing on it.

The presence of old windfalls is indicated by a mound marking the former position of upturned roots and by a ridge marking the position of the former bole. Careful soil excavation of the mound will reveal upturned soil strata. Species such as eastern hemlock frequently line the site of the fallen bole where they seeded in on the rotten wood. Their age can be used to date the windfall.

Finally, charcoal can be found in the soil, old fire scars can be found on old-growth trees, and these can be used to date prehistoric forest fires. The charcoal can be dated by carbon-14 techniques and the fire scars by tree-ring analysis.

## HUMAN ACTIVITY

In the United States and Canada, the persistence of large areas of presettlement forest, particularly in the mountains of the West, is due primarily to the short span of time that the white settler in large numbers has been present on the scene. Undoubtedly, considerable changes in the forest cover had previously resulted from Indian fires. Evidence in Alaska is certainly indicative of this (Lutz, 1956) as is fragmentary evidence from the rest of North America. For instance, the use of thousands of ponderosa pine timbers in pueblos in the American Southwest located 90 or more kilometers from any present timber of the same species certainly points to widespread destruction of the ponderosa pine forest by Indian cutting and burning.

It is impossible to overestimate the effect of human activity on the forests of the world, whether by civilized or by prehistoric uncivilized man. The character of the vegetation of a large part of the world has been greatly altered by extensive burning, logging, land clearing for agriculture, and grazing by domestic stock. Although there is neither time nor space here to detail the historical evidence, illustrations from Europe and the United States are indicative of the story of the forests of the rest of the world.

### Europe and Adjacent Lands

Throughout the Mediterranean regions, in Asia Minor, throughout southern Europe, and to a lesser extent in northern Europe, the forests have been completely altered by hundreds of thousands of years' habitation by man. Everywhere, the evidence of fossil plant remains and fossil soil horizons indicates widespread destruction of the forest by repeated logging, burning, and continued grazing, particularly by goats. Climatic change is also involved, and it is impossible as yet to allocate responsibility for forest destruction between human activity and climatic change. Yet, there is no reason to doubt but that the barren nature of much of southern Europe—and of adjacent portions of Africa and Asia—is in large measure due to the destruction of forests in antiquity by human mismanagement (Darby, 1956).

Extensive forests once occurred in Greece, North Africa, Asia Minor, the northern or forest steppes of Russia, and elsewhere. In these regions today, the climate is favorable for tree growth once a forest is reestablished, but the eroded condition of the soils, the adverse microclimate of barren planting sites, and the difficulty of controlling browsing by goats and other domestic stock make reforestation extremely difficult if not impossible.

Forest deterioration in Europe, however, is not confined to the Mediterranean and dry eastern regions. In northern Europe, Neolithic activity about 5,000 years ago probably was associated with the abrupt fall in *Ulmus* pollen, marking the Elm Decline. Upon reaching northern Europe, Neolithic peoples encountered continuous forest and no large pastures. Therefore, leafy branches of forest trees, primarily elms, whose leaves are most nutritious for this purpose, were repeatedly gathered and fed to stalled domestic animals. This repeated pruning would reduce significantly the pollen production of elms. Furthermore, the development of polished stone axes led to deliberate clearing of all trees in limited areas by Neolithic agriculturists. Experiments in Denmark have indicated that three men equipped only with Neolithic axes could clear about 200 square meters of forest in 4 hours. Cereal crops, including Emmer wheat (*Triticum dicoccum*), were grown in woodland soil and the ashes resulting from slash burning. When soil fertility declined, in about 3 years, fields were abandoned and agriculture shifted to a new forested site. Abandoned lands reverted to scrub and through succession to forest again.

In England, many of the present moors were once deciduous forest. Pollen analysis of moorland in Yorkshire, for instance, establishes the presence of deciduous forest cover during Bronze Age human occupation (Dimbleby, 1952a). Moor formation is apparently due in large measure to repeated woods burnings by primitive people. In the wet climatic regions of Ireland, Scotland, and Scandinavia, upland peat bogs owe their origin to a combination of climatic cooling and prehistoric farming methods (Morrison, 1956). The prebog woodland was frequently pine, which degenerated into open woodland or scrub with a greater percentage of birch due to repeated burning, grazing, and the development of acid-humus forming ericaceous plants (*Calluna* heath and others). Under cold, wet climatic conditions of the past thousand years, raw humus profiles over strongly leached spodosols have gradually developed into blanket upland peats, which have eliminated the forests on many sites.

## United States

Extensive forest change, due to human activity of recent vintage in the eastern United States, is much easier to trace than in Europe, where higher levels of population have prevailed for a much longer period of time. Indians had a significant effect on forest composition, due primarily to the fires they set, both accidentally and intentionally. Curtis (1959) estimated that 50 percent of the land surface of Wisconsin was directly influenced by Indian fires. Although fire was by far the most important Indian influence on forest history, Indians also introduced various crop and tree species. An interesting example is the Kentucky coffeetree, whose large, hard seeds were used in a kind of dice game by various tribes. Apparently many seeds were lost in the vicinity of villages since reports in Wisconsin and from New York show the species has a very local distribution, in each case at or near an Indian village site.

Granting the undoubted effects of Indian burning on the forests, and in particular the effects of this practice on the distribution and abundance of the various pines, much has happened since extensive settlement by the white settler.

In central and southern New England, much of the land was cleared for crops, hay, and pasture by the middle of the nineteenth century. In central Massachusetts, the virgin forests were almost completely cut before the end of the eighteenth century and much of the area cleared for farming. For a generation or two in the first portion of the nineteenth century, the area was largely agricultural. To illustrate this with a specific example, the senior author prepared an agricultural land-use map of the Prospect Hill Block, a 400-ha forest in central Massachusetts. Areas which apparently have been always forested were mapped out on the basis of the Raup and Carlson study and of original field checks in which the ages of standing trees were determined and in which stumps and other evidence of former tree growth were noted. Nine percent of the tract fell into the continuously forested category. Less than 1 ha previously described is still covered with old-growth trees.

The remaining 91 percent has been cleared for agricultural use at one time or another, but not necessarily all at one time. Much of this has obviously never been cultivated repeatedly or thoroughly, but rather has been primarily for upland pasture. To estimate how much land has been cultivated, the stone walls were followed and studied (Figure 20.1). Stone walls made up of boulders taken from cultivated fields are typically larger and contain a greater variety of stone sizes and a greater number of stones than do simple stone walls that were erected merely to mark property lines or to fence in cattle and sheep. By correlating the nature of the stone wall with the character of the

**Figure 20.1. Stone walls and former tannery site with grindstone still in place under middle-aged hardwoods in central Massachusetts. (Courtesy of Harvard Forest, Harvard University.)**

ground surface on either side, it was possible to delineate those areas that had very likely been cultivated repeatedly or thoroughly. These areas covered 16 percent of the tract, the remaining 75 percent being mapped as having been cleared for upland pasture but not having been intensively cultivated.

Much the same history of extensive land clearing followed by large-scale farm abandonment, revegetation with old-field pines, and subsequent hardwood invasion holds for the southeastern United States. Here, the economic consequences of the Civil War coupled with soil deterioration resulting from erosion and lack of adequate fertilization have resulted in the abandonment of millions of hectares of tobacco, cotton, and other cropland.

In the Georgia Piedmont, the original forests seem to have varied from pure hardwoods on red lands through mixed pine and hardwoods on the gray, sandy lands to predominantly pine on the sandier and drier granitic lands (Nelson, 1957). Early settlers soon cleared the forest and developed an agricultural economy based on cotton which began to decline with the onset of the Civil War. Loblolly pine seeded in most commonly on abandoned farm lands, but the pine stands, if unburned or otherwise left untreated, develop dense hardwood understories. These may develop with time to convert the pine forests to hardwood forests, particularly on the moister and richer soils.

Extensive changes in the forest due to human activity in the eastern United States are not confined to areas formerly farmed. In lands that have always been forested, the effects of repeated cutting, burning, and heavy domestic grazing have been equally marked. Prior to the development of the coal resources of Pennsylvania and West Virginia, extensive areas of hardwood were clearcut at frequent intervals to provide fuel wood for domestic use, brick yards, iron works, railroad engines, and other purposes. Wooded hillsides in southern Connecticut, in northeastern Pennsylvania, and elsewhere adjacent to long-settled industrial areas have been cut and burned so much that tolerant and fire-susceptible species such as hemlock and beech have been virtually eliminated except in protected ravines.

In the Ozarks, extensive burning in the last century and before had resulted in the development of extensive tracts of open, parklike forest and grassland (Beilmann and Brenner, 1951). With better fire control in recent years, coupled with an increase in human population, the forests have been reestablishing themselves widely. Eastern red cedar aggressively invades old fields and grasslands. Fire-resistant oak woodland types of blackjack oak and post oak are being invaded by the more mesophytic red oak, white oak, and hickories.

In the southwestern United States, pre-settlement ponderosa pine forests were open, parklike stands of predominantly mature pine (Cooper, 1960). According to Beale (1858):

> *We came to a glorious forest of lofty pines, through which we have travelled ten miles. The country was beautifully undulating, and although we usually associate the idea of barrenness with the pine regions, it was not so in this instance; every foot being covered with the finest grass, and beautiful broad grassy vales extending in every direction. The forest was perfectly open and unencumbered with brush wood, so that the travelling was excellent.*

Surface fires at regular intervals of 3 to 10 years maintained the clumped pattern of all-aged stands in even-aged groups and reduced excess fuel. Fires set naturally by lightning and by Indians thinned out young pine reproduction and counteracted the tendency of trees to take on a random distribution. The most important changes brought by white man were the attempted exclusion of fire, which was accomplished by an intense fire prevention program, and the introduction of livestock that reduced the flammable grasses. With few fires, dense pine thickets developed and with them accompanying problems of slow growth and stagnation.

The significant problems associated with fire exclusion have led to increasing use of prescribed burning in the American South and West. Furthermore, a "let burn" policy for wildfires is being implemented by many federal and state agencies (see Chapter 16).

## SUGGESTED READINGS


Cooper, Charles F. 1960. Changes in vegetation, structure, and growth of southwestern pine forests since white settlement. *Ecol. Monographs* 30:129–164.

Curtis, John T. 1959. *The Vegetation of Wisconsin*. (Chapter 23, The effect of man on the vegetation, pp. 456–475.) Univ. Wisconsin Press, Madison. 657 pp.

Gordon, Robert B. 1969. The natural vegetation of Ohio in pioneer days. Bull. Ohio Biol. Surv., New Series, Vol. III, No. 2. Ohio State Univ., Columbus. 113 pp.

Iversen, J. 1949. The influence of prehistoric man on vegetation. Geol. Survey of Denmark. N. series 3, No. 6, 25 pp.

Lutz, H. J. 1930. The vegetation of Heart's Content, a virgin forest in northwestern Pennsylvania. *Ecology* 11:1–29.

Pennington, Winifred. 1969. *The History of British Vegetation*. (Chapter 7, The Neolithic revolution; man begins to destroy the forests, pp. 62–77; Chapter 9, Later changes in vegetation, and species introduced into the flora by man, pp. 88–99.) English Universities Press, London. 152 pp.

Spurr, Stephen H. 1951. George Washington, surveyor and ecological observer. *Ecology* 32:544–549.

———. 1966. Wilderness management. The Horace M. Albright Conservation Lectureship, Univ. California, School of Forestry, Berkeley, Calif. 16 pp.

Thomas, William L., Jr. (ed.). 1956. *Man's Role in Changing the Face of the Earth*. Univ. Chicago Press. 1193 pp.

# 21
# Forests of the World

The description of the forests of the world on either a physiognomic-structural or a floristic basis involves the adoption of some type of classification within a geographical framework. The choice is large, for there are approximately as many classifications of vegetation types as there are writers on the subject. The fact that no single approach has achieved anything like widespread recognition and adoption testifies to the arbitrariness and artificiality of most classifications. The fact is that each geographical area has its own peculiar history, that each species and each genus have their own particular distributions, and that there is and can be no compartmentalized classification scheme of vegetation types that can set up mutually exclusive divisions. Nevertheless, there are broad floristic provinces which have a general validity and usefulness in denoting geographical areas that are characterized by given groups of plants. Within these provinces, various communities occur which show less variation within the communities than between different communities. On this basis, the general forest patterns of the world are described in this chapter. A concise treatment of world vegetation is presented by Walter (1973).

Before entering upon a description of the forests of the world, however, it is well to note briefly those vegetational classification schemes which have achieved greatest recognition and usage.

## CLASSIFICATION OF VEGETATION

Within a geographical setting, vegetation may be described either on a physiognomic-structural or a floristic basis. Since the end result of vegetational classification frequently is a map of the vegetation of a district or region, geographical, climatic, physiognomic, and floristic considerations must eventually be integrated in order to prepare it. Much more than in the past, vegetational classification and maps are designed so that vegetation units designate similar site conditions. Thus there is a continuing evolution from strictly vegetation classification to the classification of all elements of the ecosystem, even though vegetation may be used to name each type of ecosystem recognized.

## Geographical-Climatic Classification

To even the most casual observer, the vegetation of the North Temperate Zone is characterized by common genera that set it apart from the South Temperate Zone. Similarly the tropical flora of the New World is quite different floristically from that of the Old World. Most plant geographers, therefore, recognize four plant kingdoms: (1) ***boreal and north polar*** (North Temperate Zone); (2) ***paleotropical*** (Old World Tropics); (3) ***neotropical*** (American Tropics); and (4) ***southern oceanic and subantarctic*** (South Temperate Zone).

These "kingdoms" are subdivided into geographical subkingdoms, or regions, which represent major subdivisions of continents on the basis of broad floristic differences. Here there is less consistency, but North America, for example, can be divided into: (1) arctic or tundra zone; (2) boreal conifer forest; (3) western temperate conifer forest; (4) eastern temperate pine-hardwood forest; (5) central prairie belt forest; (6) subtropical forest; and (7) tropical forest.

The regions can be further subdivided into provinces, the provinces can be further subdivided into sectors, and these into districts. The further the subdivision is carried, the more arbitrary and artificial it becomes and the less is the agreement between plant geographers on the classification of vegetation. The floristic province, however, characterized by a general coincidence of distributions of a group of species over a geographical range, has considerable validity and usefulness.

## Physiognomic-Structural Classification

The outward appearance of vegetation, that is, its physiognomy (closed forest, open woodland, grassland), and structural features (evergreen forest, deciduous forest, thorn forest) have been widely used to describe and map vegetation on a world scale. The classification by Mueller-Dombois and Ellenberg (1974) is a good example, and a modified portion of this classification is shown in Table 21.1. As seen in Table 21.1, the physiognomic-structural classification is closely related to environmental factors, especially temperature and precipitation (see Chapter 7, p. 186). For example, "Closed Forests" are not exclusively characteristic of any one geographic region, but are further subdivided on the basis of geographic differences of temperature and moisture. Webb (1959, 1968) has provided a useful physiognomic-structural classification of the forests of eastern Australia. The change in structural types along three environmental gradients is clearly depicted in Figure 21.1.

Physiognomic-structural features are not only suitable for classification at the broad reconnaissance level but have proved useful at

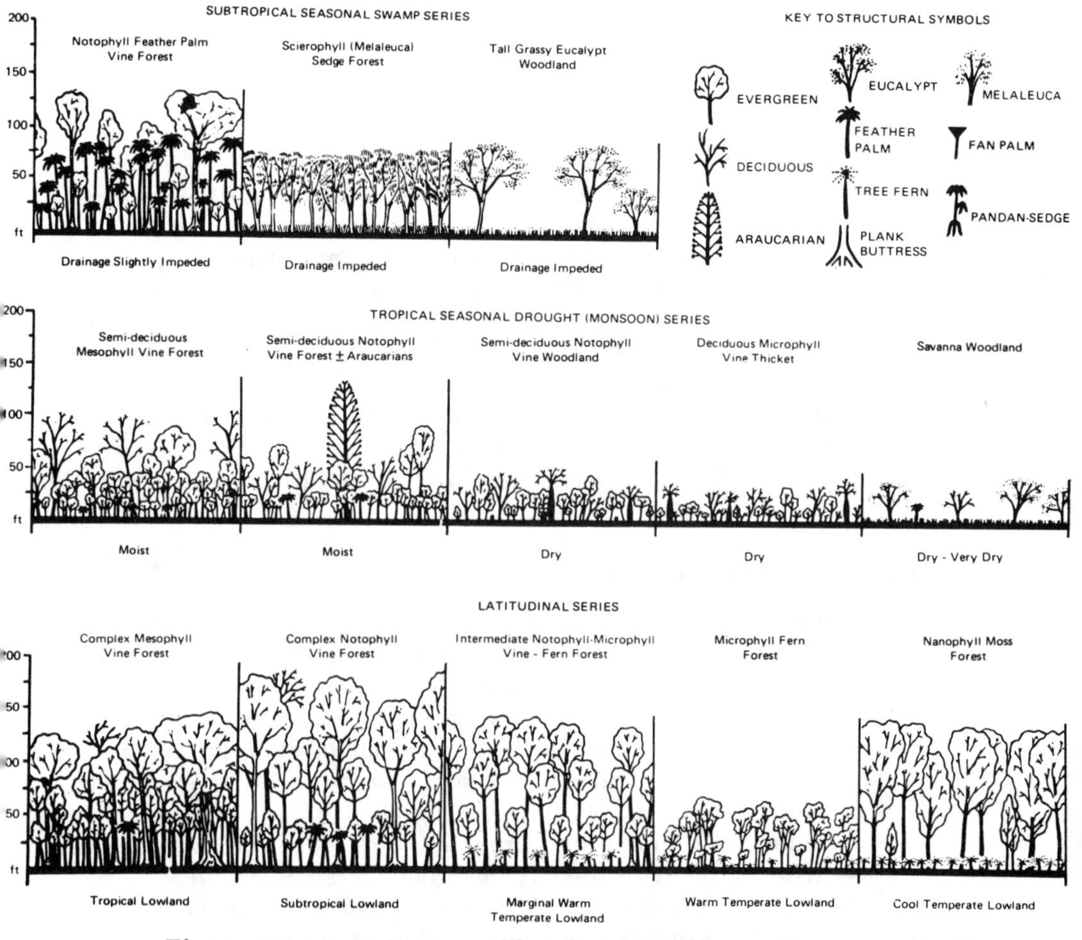


**Figure 21.1. Forest profile sketches illustrating structural changes along different environmental gradients for forests of eastern Australia. (After Webb, 1968. Reprinted from "Environmental relationships of the structural types of Australian rain forest vegetation" (Ecology 49:296–311) by L. J. Webb by permission of Duke University Press. Copyright 1968 by the Ecological Society of America.)**

lower levels as well, as in the tropical rain forests of eastern Australia (Webb et al., 1970). Here they were demonstrated to be as efficient as floristic classification in assessing the environmental conditions of an area. Structural classification is particularly useful in areas such as tropical rain forests where the flora is not well known and structural data may be rapidly collected, even by inexperienced personnel.

### Floristic Classifications

Plant sociologists have long attempted to develop systems for classifying plant communities along lines similar to those that have been evolved for plant individuals. Starting with the analogy between the plant species and the plant association, elaborate schemes of community classification and their binomial nomenclature have been proposed. In western Europe, where the greatest effort at systematization has occurred, the schools of Braun-Blanquet, Schmid, Du Rietz, Gaussen, and Aichinger (Küchler, 1967) are in conflict on many points. In northern Europe, classification of the forests is greatly influenced by Cajander's reliance upon the occurrence of characteristic ground-cover species, whereas in Russia, the work of Morosov, Sukachev, and others has led to quite different schemes of classifying the boreal forest.

### A Classification of Forests

Since it is necessary to espouse some system of classification in order to discuss logically the forests of the world, the following loose and approximate scheme will be followed:

As a basic major division, forests characteristic of frost-free regions (i.e., the Tropics) will be considered separately from those annually subject to frequent and severe frosts (i.e., the Temperate Zone and temperate belts within the Tropics).

Within the Tropics, forest formations are recognized on the basis of soil-water relationships arising from the amount and seasonal pattern of precipitation and upon drainage conditions in the soil.

Within the Temperate Zone, forest formations are recognized on the basis of dominant tree genera. Within these formations, forest types are recognized as characterized by dominant tree species. These in turn may be subdivided into communities on the basis of total floristic composition.

## TROPICAL FORESTS

In the frost-free tropical zone, many species of trees live and thrive, forming a great complex of vegetational formations and communities. The species are numbered in the hundreds and even in the thousands, and classification of forests on the basis of floristic composition is not only extremely difficult but results in the recognition of hundreds of taxa. Forests on similar sites, however, tend to have a similar physiognomy and structure regardless of their species composition. An ecological classification of vegetation based on site has proved much simpler and more useful than one based upon species. Such an approach will suffice for the present brief description of tropical forests.

A few tropical plants occur in habitats protected from frost north of the Tropic of Cancer and south of the Tropic of Capricorn. These are almost entirely confined to windward ocean sites where moderating maritime influences create a frost-free—but not necessarily a warm—climate. Palms are grown in Devon and Cornwall in southwestern England. Both the southern half of the Florida peninsula and the southern coast of California support frost-susceptible plants in large numbers and can be considered at least as subtropical if not actually tropical. Similar conditions occur in North Africa, northern New Zealand, and other areas of mild oceanic climate.

## Classification of Tropical Forests

Within the frost-free zone of the world, the physiognomy of vegetation varies primarily according to the amount and seasonal pattern of rainfall and secondarily in response to human activity, primarily land clearing, burning, and plant introduction. Many classifications of tropical forests have been proposed (Beard, 1955; Haden-Guest et al., 1956; Cain and Castro, 1959; Webb, 1968). All recognize a gradation ranging from swamp forests to tropical desert, a moisture cline which may be divided arbitrarily into few or many vegetational types. The following is perhaps the simplest possible classification:

- Swamp forest
  - Salt water swamp
  - Fresh water swamp
- Rain forest, climate wet the year around
  - Lowland rain forest, grading with increasing altitude into
  - Montane rain forest and cloud forest
- Monsoon forest, alternating wet and dry seasons
- Dry forest, no pronounced wet periods
  - Closed dry forest, grading with decreasing precipitation into
  - Savanna woodland

## Swamp Forest

Poorly drained and undrained sites in the Tropics support the most hydrophytic formations. These range from salt-water swamps and low beaches through brackish water swamps to fresh-water swamps. Swamp forests may be classified further according to the depth of the water, aeration of the water, the type of substrate, and the seasonal variation in the water level.

**Salt-Water Tropical Shorelines.** The vegetation closely associated with seawater is quite distinct from that associated with fresh water. The nature of the vegetation differs with geographical location, nearness to the shore, and soil—whether mud, sand, or coral.

Mangroves are evergreen trees and shrubs belonging to several unrelated families which form communities very similar in physiognomic appearance on tidal mud flats throughout frost-free climates of the world. Mangrove swamps grow to latitude 32° north as far as Bermuda, the Gulf of Aqaba in the Red Sea, and southern Japan; and in the Southern Hemisphere as far as south Natal and northern New Zealand. *Rhizophora* is the commonest and most widespread genus of mangroves. The species occurring in mangrove swamps of the Indian and western Pacific Oceans differ from those of America, the West Indies, and Africa, but the two formations are very similar in appearance and ecological relationships. Mangroves are of considerable local importance for fuel wood, tannin, and even for timber. The strongly developed zonation in mangrove swamps is indicative of primary succession (Chapter 15).

In brackish swamps inland, other species occur. Nipa palm, important for thatch, is widespread in occurrence in southeastern Asia and the western Pacific.

On sandy beaches, still other species are found. Species of *Barringtonia* (Indian and Pacific Oceans), *Pandanus* (Asia and Africa), and *Coccoloba* (New World) are particularly widespread genera of trees and shrubs. *Casuarina* is a characteristic littoral tree that has been planted even more widely throughout the tropics for its shelterbelt and fuel value. Most important of the shoreline trees, however, is the coconut, a species so widely spread by humans that its natural range is still in dispute.

**Fresh-Water Swamps.** Both the site and the vegetation of fresh-water swamps grade into upland tropical rain forest as the soil drainage improves. The swamp vegetation, because of limiting soil aeration, however, is more open and irregular in structure and consists of fewer species than nearby upland types. Various palms reach their optimum development and abundance in fresh-water swamps and are perhaps the most characteristic plants of these sites. Cabbage palm (*Sabal palmetto*) is such a tree in the fresh-water swamps of Florida. Others are *Euterpe* and *Mauritia* in South America, *Phoenix* in Africa, and sago palm (*Metroxylon*) in the western Pacific. Various *Cecropias* are important in developing forest successions in tropical America, where the largest late-successional fresh-water swamp species is the huge emergent, *Ceiba pentandra*.

**Table 21.1. A Portion of a Physiognomic-Structural Classification of Forest and Related Woody-Plant Ecosystems**

---

I. ***Closed Forests***
   A. Mainly Evergreen Forests
      1. Tropical rain forests
      2. Tropical and subtropical evergreen seasonal forests
      3. Tropical and subtropical semideciduous forests
      4. Subtropical seasonal rain forests
      5. Mangrove forests
      6. Temperate and subpolar evergreen rain forests
      7. Temperate evergreen seasonal broadleaved forests
      8. Winter-rain evergreen broadleaved sclerophyllous forests
      9. Temperate and subpolar evergreen coniferous forests
   B. Mainly Deciduous Forests
      1. Drought-deciduous forests (tropical, subtropical)
      2. Cold-deciduous forests with evergreens
      3. Cold-deciduous forests without evergreens
   C. Extremely Xeromorphic Forests
      1. Sclerophyllous-dominated forests
      2. Thorn forests
      3. Mainly succulent forests
II. ***Woodlands***
   A. Mainly Evergreen Woodlands
      1. Evergreen broadleaved woodlands
      2. Evergreen needle-leaved woodlands
   B. Mainly Deciduous Woodlands
      1. Drought-deciduous woodlands
      2. Cold-deciduous woodlands with evergreens
      3. Cold-deciduous woodlands
   C. Extremely Xeromorphic Woodlands
III. ***Scrub***

---

Source: Modified from Mueller-Dombois and Ellenberg (1974); from Mueller-Dombois and Ellenberg, *Aims and Methods of Vegetation Ecology* © 1974, John Wiley & Sons, Inc.

## Rain Forest

Under conditions of adequate precipitation, lush rain forests develop in the Tropics. The extent, however, is limited by climatic patterns, much of the Tropics being too dry. The rain forest is concentrated in three broad regions: (1) northern South and adjacent Central America, particularly in the Amazon Basin; (2) western Equatorial Africa from Sierra Leone to the Zaire (Congo) Basin; and

(3) the Indo-Malayan region, including the west coast of India, much of the Indo-Chinese Peninsula, Indonesia, Papua New Guinea, and the northeast coast of Australia. The vegetation differs greatly not only between the three main belts of tropical rain forest but in different portions of each belt. The number of species represented is not known. Individual regions such as the Congo and Borneo have been estimated to contain about 10,000 species of phanerogams (seed plants) each. In the Malay Peninsula alone, about 2,500 species of trees are known, while the great rain forest or ***hylaea*** of the Amazon contains at least that many species of large trees. Richards (1952, 1973) has brought together and interpreted our knowledge of the tropical rain forest. His book is an indispensable reference to those interested in the subject.

Fortunately, the general appearance and structure of the tropical rain forest is much the same everywhere. Its description in tropical Africa by Aubréville in *A World Geography of Forest Resources* (Haden-Guest et al., 1956) is typical and worth quoting:

> *The rain forest is very dense, with a tightly closed canopy. Three stories may be distinguished. The upper story is discontinuous, composed of a relatively few gigantic and usually isolated trees with mighty crowns rising 40 to 45 meters above the ground. At a height of about 25 to 30 meters a continuous middle story of crowns pressed one against another gives the forest, as seen from an airplane, a characteristically undulating and unbroken appearance and hides the trunks from view. The lowest story is made up of small trees and bushes whose crowns fill almost all of the remaining space.*
>
> *Not much light penetrates to the undergrowth through the higher levels of foliage, and the sun's rays seldom reach the ground. Nor can one, from the ground, see the crowns of the biggest trees, hidden as they are by those of the lower stories. Indeed, because of the dimness in the undergrowth and the clutter of small trees and lianas, one cannot see even the trunks of the largest trees, except from close at hand. Hardly ever does the rain forest offer those beautiful vistas of great columns that are presented by some of the mature forests of temperate lands. The ground, however, is bare, or garnished with a few sporadic herbaceous plants. It is easy enough to walk through the forest, but creepers frequently hinder rapid progress. The humidity is high, but the air seems fresh by contrast with that of sun-baked clearings unprotected by the forest screen. The secondary forests, on the other hand, are difficult to penetrate because of*

the density of the stems, which are often spiny, and the great quantity of lianas and herbaceous plants.

Lianas of all kinds also abound in the primary rain forest: filiform lianas, stretched like strings; huge woody lianas that curl around the tree trunks; rope-like lianas that hang to the ground from branches. Some of the lianas ramify out into the subshine in the crowns of larger trees, joining the upper stems and branches one to another, and making the forest canopy yet more dense. Epiphytes are equally abundant, gripping tree trunks and branches. Palms are infrequent except for those of the genus Raphia, which grows in stands in marshy bottoms, and the creeping rattan palms. The latter proliferate on the banks of streams and climb to the highest tree tops with the help of leaves provided with hooks along their rachises.

In the mountains and in humid ravines there are sometimes clumps of beautiful arborescent ferns. Here and there one finds the strangler fig. Springing from seed lodged in the crook of treetop branches, it grows into a small epiphytic bush; its roots reach the ground by spiraling down around the trunk and then grow into powerful tentacles that anastomose and stifle the supporting tree. After the latter's death, the fig may live on independently as a large tree. Some trees are cauliflorous: first their flowers and then their fruits—some of which may be huge—are produced directly from the tree trunk itself.

The crowns of the trees seldom have any one characteristic shape. Some are wide and strongly sculptured; others, where compressed by neighboring crowns, may be ovoid and astonishingly narrow. Yet more remarkable are the tabular crowns composed of the verticils of horizontal branches (e.g., of Terminalia superba and T. ivorensis). Other crowns—notably Mimosaceae—spread out like parasols.

Many tree trunks are remarkably free of branches to great heights. Standing almost perfectly upright, they look like narrow cylinders and give the impression, when they are seen from a distance, that they are inordinately tall and thin, although their actual diameters may be large. Indeed, some trees, free of branches up to a great height, have cylindrical trunks that seem technologically perfect (Entandrophragma utile, Mimusops heckelii, Terminalia superba). Others have trunks of irregular cross section or are fluted at the base. Trees with really large diameters (i.e., more than 2 or 3 meters) are exceptional. Eighty centimeters at the larger end of the utilizable portion of the trunk is about average diameter of the timber trees now being exploited.

*The most remarkable physiognomic characteristic of many of the great trees of these forests is the buttress structure at their bases. Usually triangular these buttresses serve as firm anchors for very tall trees that are otherwise attached to the soil by shallow root systems only. The buttresses rise several meters above the surface and sometimes stretch out 10 meters or more along the ground in the shape of winding flattened roots. Some of them are spectacular indeed. Their presence and in some cases their form characterize different species. Some species have no buttresses, but merely trunks thickened at the base, as in temperate-forest trees. The degree to which the buttresses are developed in individual trees depends on the depth of the soil. Sometimes parts of them are raised completely off the ground to form true aerial roots that let one see clear under the tree (*Tarrietia utilis*, many* Xylopia, Musanga smithii*, etc.). Other trees, yet more curious, are upheld by systems of adventitious curving and ramifying roots (which may be as thick as a man's arm), in such a way that each tree appears perched upon a network of intertwined roots. Among trees of this type are the mangroves (genus* Rhizophora*), found in marshy soil, and trees of genus* Uapaca*, which grow in dry soil.*

*Buttresses and aerial roots naturally make felling difficult. Cutting is commonly done above the buttress, and the woodsmen build a platform around the tree on which to work.*

*All types of bark are represented in the tropical rain forest: smooth, thin, thick, fissured, gnarled, prone to come off in sheets, and so forth. Some barks are conspicuously colored yellow or even bright red (*Distemonanthus benthamianus*, some of the* Copaifera *and* Xylopia*). The fragrance of certain barks is equally distinctive: the cedar smell of mahogany,* Guarea, Lovoa*; the garlic smell of* Scorodophloeus zenkeri*. Latex, appearing in a gashed bark, and dripping gum likewise give valuable clues to the forester, who can thereby recognize certain tree families (Moraceae, Sapotaceae, Apocynaceae, etc., with latex; Guttiferae and Hypericaceae, with dripping gum).*

The great number of tree species and the complexity of tropical rain forests have long intrigued biologists. Their very complexity and relative inaccessibility have limited detailed analytical studies of their nature. Nevertheless, the large number of species and the low density of adults of each species in the rain forest as compared to the temperate forest may be explained at least in part by several interacting factors: (1) favorable temperature, precipitation, and soil conditions permitting expression of many and varied mechanisms of plant growth and reproduction, that is, relatively free release and expression of genetic variability; (2) proportionately greater interspecific competition as individuals approach maturity (contrasted to greater intraspecific competition in temperate forests) leading to the evolution of mutually avoiding, hence ecologically complementary, species (Ashton, 1969); (3) environmental conditions favoring a rich fauna, particularly insect, bird, and bat pollinators which, being many and specialized, effect pollination among widely spaced individuals and promote speciation of plants (Chapter 13); and (4) the action of predators destroying seeds and seedlings, which tends to increase the distance between adult breeding trees of many species, leaving more space in the habitat for other species (Janzen, 1970). Summarizing his work in the species-rich southeast Asian mixed dipterocarp forests, Ashton (1969) stated that the complexity of the rain forest can be explained in terms of:

> *(i) The seasonal and geological stability of the climate which had led to selection for mutual avoidance, and through increased specialization, to increasingly narrow ecological amplitudes, leading to complex integrated ecosystems of high productive efficiency. As the complexity increases, the numbers of biotic niches into which evolution can take place increase but become increasingly narrow.*
>
> *(ii) Their great age.*

Commercially important rain forest trees include: mahogany (*Swietenia* spp.), rosewood (*Dalbergia* spp.), Spanish cedar (*Cedrela* spp.), balsa (*Ochroma* spp.), and rubber (*Hevea brasilensis*) in the American forest; African mahogany (*Khaya and Entandrophragma* spp.), okoume (Aucoumea klaineana), and ebony (*Diospyros* spp.), in Africa; and Philippine mahogany (*Shorea* and *Parashorea* spp.) and yang (*Dipterocarpus alatus*) among the dipterocarps of the southeast Asia—western Pacific area.

The tropical rain forest in any one geographical area represents a multidimensional continuum with gradual changes in composition and structure occurring with distance from the ocean, distance from rivers, increasing altitude, and changing geographical position. A given acre may contain 50 species; a hectare perhaps 100 species, while another 100 species may be found by extending the search over several hectares. Riparian forest communities are usually rather different from the forest communities away from streams, but even here the ecotone between communities is apt to be broad and indistinct.

With increasing altitude, the rain forest becomes shorter in stature, simpler in floristic composition, and increasingly characterized by luxuriant epiphytes, particularly mosses and lichens. At elevations above about 2000 m in the equatorial Tropics, this trend culminates in the ***montane rain forest***, also termed the ***cloud forest*** or ***mossy forest***. If dwarfed, it may be called ***elfin woodland***. As the names imply, the mountain climate around such forests is apt to be mostly cloudy with the air saturated with moisture and fog-drip providing daily precipitation. The climate is cool, constantly damp, and persistently misty. Whereas tree heights in the lowland rain forest frequently exceed 30 m (maximum 84 m for *Koompassia excelsa* in Borneo), heights in lower montane belts commonly are only 20 to 25 m, and in the high mossy forest may be only 6 m. The number of tree stories and the number of tree species become fewer with increasing elevation, but the number of individual trees increases.

## Monsoon Forest

Since evapotranspiration in the Tropics (outside of the cloud zone in the mountains) is extremely high, evergreen broadleaf forests are confined to the rain belts, where at least 100 mm of rain normally fall each month of the year and 2000 mm or more of rain fall during the year at least. In regions with 1000 to 2000 mm annual precipitation that are characterized by a dry period of a month or more, some of the dominant trees tend to lose their leaves, especially toward the end of this period. The term ***monsoon forest*** is given to the deciduous and semideciduous forests of southeastern Asia developed under a climate characterized by a very dry period of 2 to 6 or more months, broken by the very wet monsoon which comes in June or thereabouts depending upon the locality. Not too accurately, the term has been extended to refer to smaller areas of deciduous and semideciduous forests elsewhere.

Changes in the seasonal foliation of forests tend to occur gradually with climatic gradients in the Tropics. A marked deciduous period occurs in forests growing in areas with an equally marked dry season. A general lowering of annual precipitation without a strongly marked dry season, however, results in the forest being simpler in structure, smaller in size, and more xerophytic in character, but not necessarily deciduous.

The tropical deciduous forest is extremely important in southeast Asia, both for its extent and for the commercial importance of its tree species. Dipterocarps constitute the most important genus. In the wetter zone, the dipterocarps are tall and shed their leaves annually only after the new leaves are expanded. In drier zones, the dominants may have a leafless period while the understory shrubs and trees may remain evergreen. Teak (*Tectona grandis*) is commercially the most important of the monsoon forest trees. *Xylia xylocarpa*, a deciduous leguminous tree, however, is more abundant. *Terminalia* and *Shorea* are other common genera, sal (*Shorea robusta*) being a commercially important monsoon forest tree in India.

### Dry Forest

With less than about 1000 mm of rain, forests in the Tropics tend to be very xerophytic. Depending upon the amount and distribution of annual precipitation, such forests vary from low and rather simply structured closed forest to open woodland, thorn woodland, and open wooded savannas. These dry forests occupy most of tropical Africa south of the Sahara except for the equatorial rain forest, much of tropical Australia, and a good deal of South America both north and south of the Amazon Basin.

As with other tropical vegetation types, composition and structural changes are graded to climatic changes, there being no sharp cleavage between dry closed forest, dry open forest, and savanna forest. Fire is the dominant factor in these dry tropical types, having been widespread and frequent for many thousands of years. Vast areas of forest have been degraded by repeated burning into sparsely wooded savannas (Bartlett, 1956). On the driest sites, thorny shrubs and small trees with thin parasollike crowns replace the larger arborescents. *Acacia* is a particularly common dry tropical genus of wide distribution.

## FROST-HARDY FORESTS

As with the tropical forests of frost-free climates, frost-hardy forests subject to annual freezing temperatures are marked by discontinuity. The distribution of temperate genera and species is broken, however, not only by oceanic barriers, but by the barrier of the equatorial belt as well. Although high elevations with temperate or near-temperate climates occur in the tropics of all continents, these areas are not continuous. In many cases, however, closely related "***vicarious***" species occupy similar sites in widely separated areas. In still other places, vicarious species are missing through historical factors governing their distribution. Their place in the ecological scheme is taken by unrelated trees which can occupy the site in the absence of better competitors.

The other major factor governing the distribution of frost-hardy forests is the existence of a circumpolar boreal forest belt populated around the world with various closely related species of spruces, firs, larches, birches, and aspens. This belt is of relatively new origin, much of it growing on sites that have been glaciated up to within the last 10,000 years. As a result, it is inhabited by northward migrants of various populations that survived the Pleistocene in cool moist refugia. Many tree genera are represented in several of the temperate forest centers in the Northern Hemisphere, and others are similarly represented in temperate forest centers of the Southern Hemisphere. Since the species of each genus tend to occupy similar ecological habitats and niches wherever they occur, the homologous nature of the temperate forests of the world can perhaps best be approached by considering the world distribution and ecological habits of the most important tree taxa. These may, for convenience, be considered in two groups: boreal trees which comprise the circumboreal forest of the high latitudes and which also occur at high altitudes in the temperate zone; and temperate forest trees which form discontinuous distributions between the boreal forest and the Tropics.

## Boreal Forest Taxa

The principal boreal forest groups include the spruces, firs, larches, birches, and aspens (section Leuce of *Populus*). Of these, the spruces are the most widespread in distribution and the most characteristic of the boreal forest. The spruce-fir, larch, and birch-aspen forests everywhere have a similar physiognomy, similar ecological relationships, and respond to similar silvicultural treatment.

**Spruces.** The circumpolar forest is characterized by spruces, except for eastern Siberia, where larches are more numerous. The principal species in North America are white and black spruce, which have similar ranges from Alaska to Newfoundland. Norway spruce in western Europe and the closely related Siberian spruce (*Picea obovata*) in Russia and Siberia occupy similar sites in similar latitudinal belts from Norway to eastern Siberia (Schmidt-Vogt, 1977).

South of the boreal belt, about 25 species of spruce occur in isolated mountain ranges as far south as the 25th parallel in Mexico (*P. chihuahuana*) and the 23rd in Taiwan (*P. morrisonicola*) (Wright, 1955). All are similar in appearance and in ecological habit and show close taxonomic relationships between widely spaced species in the different continents. They are characterized by a shallow root system, often largely confined to the organic humus layers of the typical mor-type soils which develop under all spruces. The genus is particularly well adapted to grow in cold soils and even on permanently frozen ground and on glaciers covered with detritus, for it is this shallow top layer which warms up and dries out sufficiently under arctic climatic conditions to make tree growth possible. On lighter-textured and warmer soils, however, spruces, particularly Norway spruce and white spruce, develop moderately deep root systems. This suggests considerable plasticity in their rooting depth, depending on the local soil and humus environment. Being characterized by sharply conical crowns, the spruces are well suited to bear winter snows and to shed them when slight warming or wind occurs. Partly because of their shallow root systems, and partly because of their low nutrient requirements, spruces are tolerant of acid, undrained soil conditions and are able to survive and even grow in northern bogs and (as in the case of Sitka spruce) in cold, wet, rain forest or cloud forest conditions. It is not by accident, therefore, that spruce is the characteristic genus of the northern forest, and that its world distribution may be taken as approximately delineating both the circumpolar forest and its outliers in the high mountains of the Northern Hemisphere.

**Firs.** Being similar to the spruces in general form and appearance, and occupying much the same ecological niche, it is not surprising that the genus *Abies* has much the same distribution as the genus *Picea*. The fact that spruces and firs are frequently lumped together as characterizing the boreal forest, however, should not obscure the fact that the firs are less tolerant of poorly drained conditions, less tolerant of fire, and more tolerant of warmer and drier climates. While firs are common upland species in the boreal forest of North America, therefore, they do not occur to any extent in the boreal forests of Eurasia. Also, they occur farther south and in somewhat drier mountain ranges than the spruces.

The principal boreal forest firs are the balsam fir of eastern North America and the alpine fir of western North America. In Eurasia, fir is primarily a high mountain genus of the Temperate Zone, important species being European silver fir (*A. alba*) in the mountains of central Europe and as far south as the Pyrenees, where it is the major tree species, *A. nordmanniana* in the mountains of the Caucasus and the Black Sea region of Turkey, and silver fir (*A. pindrow*) in the Himalayas. The genus occurs in several high mountain ranges far south, with *A. guatemalensis* and *A. religiosa* growing in the mountains of Mexico and Guatemala, *A. pinapso* and *A. numidica* in the mountains of North Africa, *A. cilicia* in the mountains of Asia Minor, and *A. pindrow* in the mountains of northern Indo-China.

**Larches.** In contrast to *Abies*, which is more a boreal group of the New World than of the Old, *Larix* is the principal boreal genus in Siberia, and occurs in the American boreal forest only as tamarack, a species typically restricted to swamps by its relatively poor competitive ability. Being deciduous, the larches are better adapted to survive in cold, dry climates than the spruces and firs, which are better suited to cold, wet climates. Being highly intolerant as a group, however, the larches require an open site and a dominant crown position throughout life and cannot compete successfully with spruces, firs, and other trees under moister climatic conditions.

Larch is the principal forest genus in eastern Siberia, forming extensive stands from longitude 90 to 150 and almost as far south as the southern border of the U.S.S.R. In fact, larch (chiefly *L. sibirica*) is by far the most widespread tree of the U.S.S.R., occupying some 43 percent of the total forest area of the country, mostly in the cold, dry region described.

Elsewhere, with the exception of the boreal swamp-inhabiting tamarack, larch is primarily a mountain tree of cold, dry climate. Such are *L. occidentalis* of northern Idaho and western Montana, *L. decidua* in the mountains of central Europe, *L. griffithii* in the Himalayas, *L. gmelini* in Korea, and *L. leptolepis* in Japan. None of these species, however, extends as far to the south as species of spruces and firs in the same localities, nor do they occur as far down the slopes when they do occur in the same latitudes.

**Birches and Aspens.** Associated with spruces and firs, but growing into drier climates than the conifers, are the birches and aspens. Both are light-seeded pioneer trees which can colonize large burns, and which can regenerate from undamaged rootstocks after fire has killed their above-ground portions—the birches by stump sprouts and the aspens by root suckers. These are the fire species of the boreal forest *par excellence*, and their abundance is a direct measure of the severity and frequency of past forest fires.

Birches form a circumpolar population composed of separate migrations from various Pleistocene refugia which have merged and interbred to form a complex of closely related species. The *Betula papyrifera* complex in North America and the *Betula alba* complex in Europe (*B. pubescens* and *B. verrucosa*) form a single species-group of white birches which are similar in morphology and ecological habitat and closely related genetically. The birches and aspens extend farther east in Siberia than the spruces and larches, and are the principal forest trees of Kamchatka. Birches and aspens occur abundantly throughout the dry larch forests of eastern Siberia and commonly though less abundantly in the moist spruce-fir forests of Europe and western Siberia. In North America, birches and aspens are most abundant in the drier sections of the Canadian north and in the most heavily burned sections of the Lake States.

In the Arctic tundra, dwarf birches form a characteristic part of the vegetation, while south of the boreal forest, outliers of birch occur in major mountain ranges of the eastern United States, the Pyrenees, the Himalayas, and the Korean mountains. These temperate species vary from midtolerant to intolerant and in their site preferences from riparian to mountain-slope species. They perhaps reach their optimum development in the northeastern United States, where yellow and black birches form a minor but important part of the forest complex.

The aspens similarly form a circumboreal complex made of intergrading populations that have migrated north from various Pleistocene centers of refuge. *Populus tremula* in Europe and Asia, *P. davidiana* in eastern Siberia, and *P. tremuloides* in North America are the names given respectively to the Old and New World portions of this complex. In contrast to birches, which have only a limited distribution in the western United States, aspen extends southward at high elevations down the Rockies and Sierra Nevada to the mountains of northern Mexico. Some clonal populations of the central Rockies may be extremely old, and in morphology they resemble aspens of eastern Asia (Barnes, 1975). In the eastern United States, *P. grandidentata* has a range which overlaps the eastern portion of the *P. tremuloides* range, but the former species grows better on drier and warmer sites than the latter. In Eurasia, outliers of *P. tremula* occur in the Himalayas, China, Korea, and Japan. *Populus alba* is another poplar of section Leuce which is native to moister and riparian sites from central Europe through the Balkans as far west as Iran.

## Temperate Forest Taxa

A belt of frost-hardy forests occupies the temperate zones of the Northern Hemisphere between the boreal forest and the frost-free forest of the Tropics, and all of the Southern Hemisphere south of the Tropics (there being no austral equivalent of the boreal forest). It consists of many disjunct tree populations separated by oceanic and desert barriers of long geologic standing. In both the Northern and Southern Hemispheres, however, closely related species of the same genera (i.e., vicariads) frequently occupy similar ecological niches in the different regions. Thus a discussion of the temperate forest on the basis of the dominant forest taxa serves to indicate the large common denominator of these forest stands throughout the world. Pines, oaks, beeches, and eucalypts are the principal forest-forming trees, but other conifers and angiosperms must be included in the discussion.

**Pines.** The genus *Pinus* is the most important of all forest-tree genera. There are approximately 100 species and most of them are locally or nationally important in the timber or fiber economy. They occur from the boreal forest to south of the Equator in the equatorial mountains of Sumatra and Java (*Pinus merkusii*). An excellent monograph of the genus *Pinus* is available (Mirov, 1967).

Despite the taxonomic variety and the wide distribution of pines in the Northern Hemisphere, however, all have a great deal in common in their ecological place in the world's forest. Virtually all are characteristic of coarse dry soils, especially sands, gravels, and rock outcrops; and most owe their dominance to the frequent burning of such sites and their ability to regenerate abundantly in the ashes of the blackened site. The closed-cone pines are the most dependent on fire, but even the five-needled white pines, the most mesophytic of the group, are commonly found in pure, even-aged stands most often after destruction of the prior forest by fire. Their many adaptations to fire are cited in Chapter 11.

As a group, the pines have a deep root system, with the growing root tips requiring large soil interstices for penetration. Because of the deep root system, they do not grow well on frozen or poorly drained soils, and because of the poor penetrating ability of the roots, they grow best on deep coarse soils. They have a high ability to withstand hot, dry conditions, whether in the Tropics or in northern continental climates as in the interior of Canada and Siberia. They are widely used for lumber, fiber, naval stores (resin and turpentine), and edible nuts (pinyons and stone pines) throughout much of the Northern Hemisphere.

In the boreal forest, the vicariads, Pacific coast lodgepole, Rocky Mountain lodgepole, jack pine, and Scots pine form a circumpolar belt of forest across North America, Europe, and Asia. These species do not extend far into the northern zones of permanently frozen soils, but do extend quite far south along high mountain ranges, lodgepole occurring as far south as southern California and Scots pine growing in Spain, Italy, Greece, and Turkey.

Also north temperate for the most part are the five-needled white pines, of which closely related vicarious species occur in relatively cool, moist temperate zones around the world. The subsection Strobi include: sugar pine and western white pine in the western United States; eastern white pine in eastern North America and Chiapas; Mexican white pine (*P. ayacahuite*) in Mexico; Japanese white pine (*P. parviflora*); Formosan white pine (*P. formosana*); Macedonian white pine (*P. peuce*); and Himalayan white pine (*P. griffithii*) in the Himalayas. The other groups of white pines include the stone pines and the limber pines, both also widely distributed.

The hard pines, usually with two or three needles in a cluster, are more xerophytic than the white pines and generally grow in warmer drier sites (except for the boreal pines mentioned above which do, however, endure a hot, dry continental summer in much of their ranges). These pines have a similar physiognomic appearance and occupy similar sites whatever their name and wherever they grow.

In western North America, ponderosa pine is the principal hard pine, with lodgepole, and Jeffrey pines being also of major importance. The closed-cone relict pines of the Pacific Coast, particularly *P. radiata* and *P. muricata*, however, thrive extremely well when transferred to maritime climates elsewhere.

In eastern North America, red pine (*P. resinosa*) and the four principal southern pines (*P. taeda, P. elliottii, P. palustris,* and *P. echinata*) are the major hard pines, but, as in the west, many other species occur locally.

The Mexican and Central American pines, *P. caribaea, P. hondurensis*, and *P. patula*, have proven well suited for extensive planting in other continents. A considerable number of other species also occur, occupying chiefly gravelly and sandy soils at high elevations. *Pinus caribaea*, however, reaches the seacoast as far south as Honduras and Nicaragua.

The dry mountain ranges of southern Europe, the Mediterranean region in general, and Asia Minor shelter many species of pines, although their ranges and quality have been severely limited by centuries of grazing and burning. Perhaps the most important is *P. nigra*, which grows in moderate temperate climates from the Pyrenees east to Greece, Turkey, and Cyprus. Several races have been distinguished, the most important in world forestation being var. *corsicana* or Corsican pine and var. *austriaca* or Austrian pine. Other species of the region of widespread distribution are French maritime pine (*P. pinaster*), an important sandy coastal species growing from Portugal to Italy and reaching its optimum development along the Bay of Biscay in northern Spain and southwestern France; Aleppo pine (*P. halepensis*), which occurs from Spain on both sides of the Mediterranean into Jordan, Israel, and Turkey, enduring hot and dry climates; Calabrian pine (*P. brutea*) in the mountains of the Near East from Turkey to Iraq; and Italian stone pine (*P. pinea*), a nut or pinyon pine that provides edible seeds from Spain to Turkey. A purely tropical European pine is the Canary Island pine (*P. canariensis*), found only on those islands.

In Asia, aside from the great northern extent of Scots pine, temperate pines occur in the mountains of India and Pakistan, China, Korea, and Japan. In the Himalayan region chir pine (*P. roxburghii*) grows in the hot and dry lower elevations, with Himalayan white pine (*P. griffithii*) being the principal species of the higher zones. In China, *P. massoniana* and *P. tabulaeformis* are the principal pines;

while in Japan, Japanese black pine (*P. thunbergii*) and Japanese red pine (*P. densiflora*) are both important species. Two of the important Asiatic pines are tropical, growing widely in the southwest Pacific in the same regions as podocarps and other south-temperate genera. *Pinus merkusii* occurs in the mountains of Burma, Thailand, the Indo-Chinese peninsula, the Philippines, and one locality in Sumatra, where it extends south of the equator. *Pinus khasia* has a very similar range and frequently grows in conjunction with it but at higher elevations. The geography of each pine species is discussed by Mirov (1967).

**Oaks.** The members of the genus *Quercus* are the angiosperm equivalents of the pines, being a Northern Hemisphere, widely distributed group of deep-rooted, xerophytic trees that occupy dry sites from the southern edge of the boreal forest well into the Tropics. As with the pines, oak bark is thick and fire-resistant. When the bole is killed, the root system remains alive almost indefinitely, sending up generation after generation of coppice stump sprouts that perpetuate both the individual tree and the dominance of oaks in the forest. Throughout the North Temperate Zone and much of the high country of the north Tropics, oaks are the characteristic late-successional dominants in the drier forest zone. They also occur as mid-successional species on mesic sites and some occur as dominants in floodplains and swamps. Oaks do not extend as far north as the pines, and thrive on drier sites in southern latitudes. Otherwise, their ecological requirements are much the same as those of the pines. Fire-induced pine types in much of the world will be succeeded in the absence of repeated fire by oaks, which invade as understory plants and assume dominant position when the pines die or are overthrown by wind. In other words, the oaks are more tolerant of understory conditions than the pines as a group, and follow the pines in natural succession.

The oaks show a great deal of genetic variation. More than five hundred species have been recognized, but the merging of morphological characteristics between adjacent populations makes species identification difficult and nomenclature uncertain in many instances. With this wide diversity in morphology goes a similar wide tolerance of ecological conditions. Although primarily xerophytes, there are species of oaks that have become adapted to mesic and even hydric conditions. In the eastern United States, for instance, cherrybark oak (*Q. falcata.* var. *pagodaefolia*) is the most important of many bottomland oaks in the Mississippi Valley flood-plain; the red oaks (*Q. rubra* and *Q. falcata*) grow over a wide range of sites and regions but reach their best development under mesic conditions; white and black oaks (*Q. alba* and *Q. velutina*) similarly have a wide ecological amplitude but are most characteristic of drier sites; chestnut oak (Q. *prinus*) is dominant on very dry hillsides and mountain slopes; while a whole series of scrub oaks (such as Q. *ilicifolia*, Q. *laevis*, and Q. *marilandica*) grow on the very driest and most infertile sand plains, sites often too poor for good pine growth.

In the North Temperate Zone, the oaks are deciduous, but in the South Temperate Zone, oak species are frequently evergreen or have but a brief leafless season. Most oaks are ***sclerophylls,*** having a hard leaf with thick-walled structures suitable for preventing wilting of the leaves, even under protracted drought conditions. Mesophytic species of oaks, however, may have quite succulent leaves, particularly under shade-grown conditions. On the very driest sites, on the other hand, oaks tend not only to be highly sclerophyllous, but also to have narrow or deeply-cut leaves with small leaf surfaces.

The Temperate Zone has dry forest ranging from dry closed through dry open to dry scrub just as in the Tropic Zone. Oaks are frequently the characteristic species of these zones. Open oak woodlands are widespread in the South Temperate Zone from California east throughout the Southwest to Texas, throughout the Mediterranean region, Asia Minor, and southern Asia to New Guinea in the south of eastern Asia and Japan in the north. Scrub woodland stands, frequently with oak as a major species, occur similarly distributed on the drier sites. The ***chaparral*** ("little oak") of California, the ***scrub oak*** of the southeastern United States, and the ***maquis*** of the Mediterranean are examples. *Arbutus* and *Pistacia* are common associates of oak in the Mediterranean scrub.

Some of the major oak species around the world may be mentioned. In eastern North America, oaks are the most abundant and widely distributed tree genus, with red oak (Q. *rubra*), black oak (Q. *velutina*), and white oak (Q. *alba*) being perhaps the most important. Along the Gulf Coast, in the Southwest, and in Mexico, many of the oaks are evergreen or have only a brief leafless period. These live oaks include several species in both the Southeast (Q. *virginiana*) and the Southwest (Q. *agrifolia*, Q. *chrysolepis*, Q. *emoryi*, Q. *wislizenni*). As far south as Costa Rica, Q. *copeyensis* forms magnificent stands in the mountains of Central America, while other species occur as far south as Colombia and Ecuador.

White oak is the most important and widely distributed hardwood in Europe with two closely related species, Q. *robur* (= Q. *pedunculata*) and Q. *petraea* (= Q. *sessiliflora*) extending from Portugal and Ireland on the west to central Russia and the Caspian Sea on the east and from southern Scandinavia south to the Mediterranean (Schoenicher, 1933). The Mediterranean forests are dominated by other oak species including cork oak (Q. *suber*) in the west from southern France to Morocco; holm oak (*Q. ilex*) in the maquis of the entire Mediterranean; and *Q. macrolepis* or Vallonia oak, important for the tannin from its acorn cups in the eastern Mediterranean. In Asia Minor, *Q. aegylops*, is important; while *Q. aegylops*, *Q. castaneaefolia*, and *Q. infectoria* occur widely in Iraq, Iran, and surrounding regions. Farther east, the Himalayan oaks are numerous and range widely both in elevation and extent. To the north, oaks are important in China, Korea, and Japan; while to the south they grow in Assam, Indo-China, Indonesia, and the mountains of New Guinea. The Mongolian oak (*Q. mongolica*) is widely distributed throughout southeastern Siberia, Mongolia, Manchuria, and Korea.

**Beeches and Maples.** Several genera of angiosperms typically occupy the more mesic sites in the Temperate Zone of the Northern Hemisphere. Of these, the beeches and maples are particularly widespread and important. The species of these genera are commonly highly shade tolerant, as a result of which they frequently succeed oak in the absence of fire by seeding in to the understory and growing slowly many decades until they are released by the death or decadence of the oaks. In the moister and cooler sites, beech and maple occur in all stages of the forest succession and characterize the hardwood forest.

The principal beeches include *Fagus grandifolia*, which grows in eastern North America from Cape Breton Island and Ontario to Florida and east Texas on the Gulf of Mexico, with outlying relict colonies in Mexico; *Fagus sylvatica*, which occurs from England and southern Sweden south to northern Spain, the Italian mountains, and Greece; the closely allied *Fagus orientalis* from Greece through the high mountains of Asia Minor and the Caucasus to the Caspian forest of northern Iran; and five species in eastern Asia. These are all very similar in appearance and in ecological requirements, characterizing the moist, cool, temperate forest, and forming late-successional communities whether in pure stands or mixed with other tolerants in areas long unburned.

The maples have similar ecological preferences to the beeches, but being more variable and consisting of many more species (70 as against 8 or 9), have a wider distribution and a wider ecological amplitude. Some are large trees but others are primarily shrubs. In eastern North America, the sugar maple group (*Acer saccharum et al.*) has a very similar distribution to beech from Canada to Mexico and occurs in mixed beech-maple stands on the rich, cool, moist, undisturbed sites. The red maple group (*A. rubrum et al.*) has even a wider range and a wider ecological amplitude, growing on virtually all sites, and frequently forming pure stands in swales and swamps.

In western North America, bigleaf and vine maples (*A. mac-*

*rophyllum* and *A. circinatum*) are fire-susceptible tolerants that form a minor part of the conifer forest of the Pacific Northwest, especially in the moist ravines and draws; while in the Rocky Mountains, the bigtooth maple (*A. grandidentatum*) occurs in similar sites.

In Europe, the maples are less important than beech, but do occur widely as minor components of the forest. Sycamore maple (*A. pseudoplatanus*), Norway maple (*A. platinoides*), and field maple (*A. campestre*) occur throughout much of central Europe with Norway maple extending as far north as southern Norway and Finland, and field maple as far south as North Africa and Turkey.

Other maples are associated with the oak forests throughout Asia, forming minor portions of the stand, especially on the cooler moister sites in the Middle East, the Himalayas, China, southern Siberia, Korea, and Japan. They extend into the tropics and south of the equator in the mountains of Indonesia.

**Miscellaneous Hardwoods.** The distribution of the birches, aspens, oaks, beeches, and maples has already been discussed. Other hardwood or angiosperm tree taxa are also widely distributed in the North Temperate Zone. As with the more abundant genera, these are frequently represented with vicarious species in North America, Europe, southwestern Asia, the Himalayan region, and eastern Asia. The mixed hardwood complexes of the eastern United States, western Europe, the Himalayas, eastern Siberia, Manchuria, and Japan have much in common. Genera that occur with disjunct species pretty much around the Northern Hemisphere include the lindens and limes (*Tilia*), elms (*Ulmus*), ashes (*Fraxinus*), walnuts (*Juglans*), hornbeams (*Carpinus*), chestnuts (*Castanea*), planes or sycamores (*Platanus*), willows (*Salix*), and alders (*Alnus*). Still others are common to both the eastern American and east Asian forest, including *Liriodendron, Liquidambar, Magnolia,* and *Nyssa;* while only a few are confined to a single continent.

The extent of the southern temperate land zone is very limited and what there is is sparsely forested. Relatively few hardwoods, therefore, are abundant or widespread south of the Tropics. *Nothofagus*, the antarctic or southern beech, is the most important, with deciduous and evergreen species characterizing the cool wet forest of southern Chile and Argentina, and evergreen species the similar zones of New Zealand and southeastern Australia. In New Guinea, *Nothofagus* occurs with *Quercus*, this region forming a bridge between the north-temperate and south-temperate Fagaceae.

The other southern hemisphere hardwood genus of great importance is *Eucalyptus*, with several hundred species in Australia and the islands to its north. It occurs both in tropical and temperate climates, and in both xeric and mesic moisture zones. In fact, there is in Australia a eucalyptus community in every type of forest niche from cool, wet, rain forest through closed mesic forest to open xeric forest and sparsely wooded semidesert. The large, fast-growing eucalypts, however, grow in the higher rainfall belts under temperate climatic conditions. These include blackbut (*E. pilularis*) and flooded gum (*E. grandis*) in the east, alpine ash (*E. delegatensis*), mountain ash (*E. regnans*), and messmate (*E. obliqua*) in the southeast, and jarrah (*E. marginata*) and karri (*E. diversicolor*) in the far west. These trees reached heights of 60 to 98 m and are among the fastest-growing trees in the world.

**Miscellaneous Conifers.** The principal north-temperate conifers mentioned thus far—the spruces, firs, larches, and pines—belong to the pine family, the Pinaceae. There are others of the pine family of local importance. These include Douglas-fir of western North America, hemlock in western and eastern North America and western Asia, and the true cedars (*Cedrus libani*, Lebanon cedar, in the near east; and *C. deodar* in the Himalayas).

The members of the family Cupressaceae are also of great importance in the Northern Hemisphere, with some genera occurring locally in the southern. The junipers, in particular, occur around the Northern Hemisphere, occupying dry and infertile sites and forming open woodlands at the drier edge of the forest. Other members of the family of local importance include sha mu, or Chinese fir (*Cunninghamia lanceolata*), the most important timber tree of China; sugi (*Cryptomeria japonica*) in Japan; *Cupressus lusitanica* in the mountains from Mexico to Honduras; cypress pine (*Callitris cupressiformis*) in the dry interior of Queensland; southern bald cypress (*Taxodium distichum*) in the swamps of the American Southeast and redwood (*Sequoia sempervirens*) in coastal northern California.

In the Southern Hemisphere, however, Araucariaceae and Podocarpaceae are the two important gymnosperm families. The former includes *Araucaria* and *Agathis*, and the latter, *Phyllocladus*, *Podocarpus*, and *Dacridium*.

The araucarias are important timber trees in Australia, islands east and north of Australia, Chile, and Argentina. *Agathis*, including the kauri of New Zealand (*Agathis australis*), is a tropical genus of the southwestern Pacific with a number of commercially important species, frequently reaching gigantic size (Figure 21.2), extending from Borneo and the Philippines on the North to Queensland and northern New Zealand on the South. The kauris as a group are important not only for their lumber but for their gum, used in the manufacture of varnish.

The various podocarps are widely distributed in the Southern Hemisphere, occurring in Africa, southern Asia, and the southwest

**Figure 21.2. Giant kauri in Waipoua Forest, North Island, New Zealand. Diameters exceed 3 m with little stem taper. (New Zealand Forest Service photo by J. H. Johns, A.R.P.S.)**

Pacific, and the Americas from the West Indies to southern Chile. They constitute the native conifers of east and south Africa, and also occur in Indo-China, New Guinea, the Philippines, Australia, and New Zealand. In the latter country, rimu (*Dacrydium cupressinum*), totara (*Podocarpus totara*), and kahikatea (*P. dacrydioides*) are the principal native conifers. Among the American species is *P. coriaceous* in the uplands of the West Indies, Venezuela, and Colombia.

## SUGGESTED READINGS

Beard, J. S. 1955. The classification of tropical American vegetation-types. *Ecology* 36:89–100.

Burley, J., and B. T. Styles. 1976. *Tropical Trees, Variation, Breeding, and Conservation*. Linnean Soc. Symp. Series No. 2. Academic Press, Inc., New York. 243 pp.

Ellenberg, H., and D. Mueller-Dombois. 1970. Geographical index of world ecosystems. *In* David E. Reichle (ed.), *Analysis of Temperate Forest Ecosystems*. Springer-Verlag, New York.

Haden-Guest, Stephen, John K. Wright, and Eileen M. Teclaff (eds.). 1956. *A World Geography of Forest Resources*. Amer. Geog. Soc. Spec. Publ. 33. The Ronald Press Co., New York. 736 pp.

Küchler, A. W. 1967. *Vegetation Mapping*. (Part V. Application of vegetation maps, pp. 307–402; Vegetation classification, pp. 438–451.) The Ronald Press Co., New York. 472 pp.

Mirov, N. T. 1967. *The genus "Pinus."* (Chapters 1–3, pp. 1–320.) The Ronald Press Co., New York. 602 pp.

Richards, P. W. 1952. *The Tropical Rain Forest: An Ecological Study*. Cambridge Univ. Press, London. 450 pp.

———. 1973. The tropical rain forest. *Sci. American* 229:58–67.

Tomlinson, P. B., and Martin H. Zimmermann (eds.). 1978. *Tropical Trees as Living Systems*. Cambridge Univ. Press, New York. 675 pp.

Walter, Heinrich. 1973. *Vegetation of the Earth: In Relation to Climate and the Eco-Physiological Conditions*. Springer-Verlag, New York. 237 pp.

Webb, L. J. 1959. A physiognomic classification of Australian rain forests. *J. Ecol.* 47:551–570.

———. 1968. Environmental relationships of the structural types of Australian rain forest vegetation. *Ecology* 49:296–311.

# Scientific Names of Trees

| | |
|---|---|
| Acacia | *Acacia* spp. |
| Ailanthus or tree of heaven | *Ailanthus altissima* (Mill.) Swingle |
| Alder, European or black | *Alnus glutinosa* (L.) Gaertn. |
| Alder, red or Oregon | *Alnus rubra* Bong. |
| Alder, Sitka | *Alnus sinuata* (Reg.) Rydb. |
| Alder, speckled | *Alnus rugosa* (Du Roi) Spreng. |
| Ash, black | *Fraxinus nigra* Marsh. |
| Ash, European | *Fraxinus excelsior* L. |
| Ash, mountain (Australia) | *Eucalyptus regnans* F. Muell. |
| Ash, red or green | *Fraxinus pennsylvanica Marsh.* |
| Ash, white | *Fraxinus americana* L. |
| Aspen, bigtooth | *Populus grandidentata* Michx. |
| Aspen, quaking or trembling | *Populus tremuloides* Michx. |
| Balsa | *Ochroma pyramidale* (Cav.) Urban |
| Basswood | *Tilia americana* L. |
| Beech, American | *Fagus grandifolia* Ehrh. |
| Beech, Antarctic | *Nothofagus antarctica* (Forst.) Oerst. |
| Beech, blue | *Carpinus caroliniana* Walt. |
| Beech, European | *Fagus sylvatica* L. |
| Bigtree, or giant sequoia | *Sequoiadendron giganteum* (Lindl.) Buchholz |
| Birch, black or sweet | *Betula lenta* L. |
| Birch, bog, swamp, or low | *Betula pumila* L. |
| Birch, European | *Betula pendula* Roth. |
| Birch, gray | *Betula populifolia* Marsh. |
| Birch, moor | *Betula pubescens* Ehrh. |
| Birch, paper or white | *Betula papyrifera* Marsh. |
| Birch, river or red | *Betula nigra* L. |
| Birch, yellow | *Betula alleghaniensis* Britton |
| Buckeye, Ohio | *Aesculus glabra* Willd. |
| Buckeye, painted | *Aesculus sylvatica* Bartr. |
| Buckeye, yellow | *Aesculus octandra* Marsh. |
| Butternut | *Juglans cinerea* L. |
| Cedar, Alaska | *Chamaecyparis nootkatensis* (D. Don) Spach |
| Cedar, eastern red | *Juniperus virginiana* L. |
| Cedar, incense | *Libocedrus decurrens* Torr. |
| Cedar, northern white | *Thuja occidentalis* L. |
| Cedar, Port Orford | *Chamaecyparis lawsoniana* (A. Murr.) Parl. |
| Cedar, southern white | *Chamaecyparis thyoides* (L.) B. S. P. |

| | |
|---|---|
| Cedar, western red | *Thuja plicata* Donn |
| Cherry, black | *Prunus serotina* Ehrh. |
| Cherry, choke | *Prunus virginiana* L. |
| Cherry, pin | *Prunus pensylvanica* L. f. |
| Chestnut, American | *Castanea dentata* (Marsh.) Borkh. |
| Chinquapin, Ozark | *Castanea ozarkensis* Ashe |
| Coconut | *Cocos nucifèra* L. |
| Cottonwood, black | *Populus balsamifera* ssp. *trichocarpa* (Torr. & Gray) Brayshaw |
| Cottonwood, eastern | *Populus deltoides* Bartr. |
| Cottonwood, European or black poplar | *Populus nigra* L. |
| Cypress, Arizona | *Cupressus arizonica* Greene |
| Cypress, southern bald | *Taxodium distichum* (L.) Rich. |
| Dogwood, flowering | *Cornus florida* L. |
| Douglas-fir | *Pseudotsuga menziesii* (Mirb.) Franco |
| Elm, American | *Ulmus americana* L. |
| Elm, rock | *Ulmus thomasii* Sarg. |
| Elm, Siberian | *Ulmus pumila* L. |
| Elm, slippery | *Ulmus rubra* Mühl. |
| Elm, winged | *Ulmus alata* Michx. |
| Eucalyptus | *Eucalyptus* spp. |
| Fir, alpine or subalpine | *Abies lasiocarpa* (Hook.) Nutt. |
| Fir, balsam | *Abies balsamea* (L.) Mill. |
| Fir, Douglas- | *Pseudotsuga menziesii* (Mirb.) Franco |
| Fir, European silver | *Abies alba* Mill. |
| Fir, Fraser | *Abies fraseri* (Pursh) Poir. |
| Fir, grand or lowland white | *Abies grandis* (Dougl.) Lindl. |
| Fir, noble | *Abies procera* Rehd. |
| Fir, Pacific silver | *Abies amabilis* (Dougl.) Forbes |
| Fir, red | *Abies magnifica* A. Murr. |
| Fir, white | *Abies concolor* (Gord. & Glend.) Lindl. |
| Gum, black or sour | *Nyssa sylvatica* Marsh. |
| Gum, southern blue | *Eucalyptus globulus* Labill. |
| Gum, sweet | *Liquidambar stryaciflua* L. |
| Hackberry | *Celtis occidentalis* L. |
| Hawthorne | *Crataegus* spp. |
| Hazel | *Corylus cornuta* Marsh. |
| Hemlock, eastern | *Tsuga canadensis* (L.) Carr. |
| Hemlock, mountain | *Tsuga mertensiana* (Bong.) Carr. |
| Hemlock, western | *Tsuga heterophylla* (Raf.) Sarg. |
| Hickory, bitternut | *Carya cordiformis* (Wangenh.) K. Koch |
| Hickory, mockernut | *Carya tomentosa* Nutt. |
| Hickory, pignut | *Carya glabra* (Mill.) Sweet |
| Hickory, shagbark | *Carya ovata* (Mill.) K. Koch |
| Holly, American | *Ilex opaca* Ait. |
| Hophornbeam, or ironwood | *Ostrya virginiana* (Mill.) K. Koch |
| Ironwood, or hophornbeam | *Ostrya virginiana* (Mill.) K. Koch |

| | |
|---|---|
| Juniper, alligator | *Juniperus deppeana* Steud. |
| Juniper, common | *Juniperus communis* L. |
| Juniper, one-seed | *Juniperus monosperma* (Engelm.) Sarg. |
| Juniper, Rocky Mountain | *Juniperus scopulorum* Sarg. |
| Juniper, Utah | *Juniperus osteosperma* (Torr.) Little |
| Kentucky coffeetree | *Gymnocladus dioicus* (L.) K. Koch |
| Larch, European | *Larix decidua* Mill. |
| Larch, Japanese | *Larix leptolepis* (Sieb. & Zucc.) Gord. |
| Larch, western | *Larix occidentalis* Nutt. |
| Laurel, mountain | *Kalmia latifolia* L. |
| Locust, black | *Robinia pseudoacacia* L. |
| Locust, honey | *Gleditsia triacanthos* L. |
| Madrone, Pacific | *Arbutus menziesii* Pursh. |
| Magnolia, Fraser | *Magnolia fraseri* Walt. |
| Magnolia, sweetbay | *Magnolia virginiana* L. |
| Magnolia, umbrella | *Magnolia tripetala* L. |
| Mahogany, West Indies | *Swietenia mahagoni* Jacq. |
| Mangrove, black | *Avicennia nitida* Jacq. |
| Mangrove, red | *Rhizophora mangle* L. |
| Maple, bigleaf | *Acer macrophyllum* Pursh. |
| Maple, mountain | *Acer spicatum* Lam. |
| Maple, Norway | *Acer platanoides* L. |
| Maple, red | *Acer rubrum* L. |
| Maple, silver | *Acer saccharinum* L. |
| Maple, striped | *Acer pensylvanicum* L. |
| Maple, sugar | *Acer saccharum* Marsh. |
| Maple, vine | *Acer circinatum* Pursh. |
| Mesquite | *Prosopis juliflora* (Sw.) DC. |
| Mimosa | *Albizzia julibrissin* Durazz. |
| Mountain-ash, American | *Sorbus americana* Marsh. |
| Mulberry, red | *Morus rubra* L. |
| Oak, bear | *Quercus ilicifolia* Wangenh. |
| Oak, black | *Quercus velutina* Lam. |
| Oak, blackjack | *Quercus marilandica* Muenchh. |
| Oak, bur | *Quercus macrocarpa* Michx. |
| Oak, California black | *Quercus kelloggii* Newb. |
| Oak, California live | *Quercus agrifolia* Née |
| Oak, canyon live | *Quercus chrysolepis* Liebm. |
| Oak, cherrybark | *Quercus falcata* var. *pagodaefolia* Ell. |
| Oak, chestnut or rock chestnut | *Quercus prinus* L. |
| Oak, chinquapin or yellow | *Quercus muehlenbergii* Engelm. |
| Oak, Emory | *Quercus emoryi* Torr. |
| Oak, English or European | *Quercus robur* L. |
| Oak, interior live | *Quercus wislizenii* A. DC. |
| Oak, live | *Quercus virginiana* Mill. |
| Oak, Oregon | *Quercus garryana* Dougl. |
| Oak, pin | *Quercus palustris* Muenchh. |
| Oak, post | *Quercus stellata* Wangenh. |

| | |
|---|---|
| Oak, pubescent | *Quercus pubescens* Willd. |
| Oak, red | *Quercus rubra* L. |
| Oak, sand live | *Quercus geminata* Small. [= *Q. virginiana* var. *maritima* (Michx.) Sarg.] |
| Oak, scarlet | *Quercus coccinea* Muenchh. |
| Oak, sessile | *Quercus petraea* (Mattuschka) Lieblèin. |
| Oak, southern red | *Quercus falcata* Michx. |
| Oak, swamp white | *Quercus bicolor* Willd. |
| Oak, turkey | *Quercus laevis* Walt. |
| Oak, white | *Quercus alba* L. |
| Palm, cabbage | *Sabal palmetto* (Walt.) Lodd. |
| Palmetto, saw | *Serenoa repens* (Bartr.) Small |
| Pawpaw | *Asimina triloba* (L.) Dunal |
| Pine, Caribbean | *Pinus caribaea* Morelet |
| Pine, Corsican or Austrian | *Pinus nigra* Arnold |
| Pine, digger | *Pinus sabiniana* Dougl. |
| Pine, eastern white | *Pinus strobus* L. |
| Pine, erectcone or Calabrian | *Pinus brutia* Ten. |
| Pine, jack | *Pinus banksiana* Lamb. |
| Pine, Jeffrey | *Pinus jeffreyi* Grev. & Balf. |
| Pine, knobcone | *Pinus attenuata* Lemmon |
| Pine, limber | *Pinus flexilis* James |
| Pine, loblolly | *Pinus taeda* L. |
| Pine, lodgepole | *Pinus contorta* Dougl. |
| Pine, longleaf | *Pinus palustris* Mill. |
| Pine, maritime | *Pinus pinaster* Ait. |
| Pine, Mexican pinyon | *Pinus cembroides* Zucc. |
| Pine, Monterey or radiata | *Pinus radiata* D. Don |
| Pine, mugo or mountain | *Pinus mugo* Turra. [*P. montana* Mill.] |
| Pine, patula | *Pinus patula* Schl. & Cham. |
| Pine, pinyon | *Pinus edulis* Engelm. |
| Pine, pitch | *Pinus rigida* Mill. |
| Pine, ponderosa | *Pinus ponderosa* Laws. |
| Pine, red | *Pinus resinosa* Ait. |
| Pine, sand | *Pinus clausa* (Chapm.) Vasey |
| Pine, Scots | *Pinus sylvestris* L. |
| Pine, shortleaf | *Pinus echinata* Mill. |
| Pine, slash | *Pinus elliottii* Engelm. var. *elliottii* |
| Pine, sugar | *Pinus lambertiana* Dougl. |
| Pine, stone or Swiss stone | *Pinus cembra* L. |
| Pine, Table Mountain | *Pinus pungens* Lamb. |
| Pine, Virginia | *Pinus virginiana* Mill. |
| Pine, western white | *Pinus monticola* Dougl. |
| Pine, whitebark | *Pinus albicaulis* Englem. |
| Plum, Allegheny | *Prunus alleghaniensis* Porter |
| Poplar, balsam | *Populus balsamifera* L. |
| Poplar, Lombardy | *Populus nigra* cv. '*italica*' Muenchh. |
| Poplar, white | *Populus alba* L. |

| | |
|---|---|
| Redbud, eastern | *Cercis canadensis* L. |
| Redwood | *Sequoia sempervirens* (D. Don) Endl. |
| Rhododendron, rosebay | *Rhododendron maximum* L. |
| Sagebrush | *Artemisia tridentata* Nutt. |
| Sassafras | *Sassafras albidum* (Nutt.) Nees |
| Sequoia, giant or bigtree | *Sequoiadendron giganteum* (Lindl.) Buchholz |
| Silver bell or Carolina silverbell | *Halesia carolina* L. |
| Sourwood | *Oxydendrum arboreum* (L.) DC. |
| Spruce, black | *Picea mariana* (Mill.) B. S. P. |
| Spruce, Engelmann | *Picea engelmannii* Parry |
| Spruce, Norway or European | *Picea abies* (L.) Karst. |
| Spruce, red | *Picea rubens* Sarg. |
| Spruce, Sitka | *Picea sitchensis* (Bong.) Carr. |
| Spruce, white | *Picea glauca* (Moench) Voss |
| Sugarberry | *Celtis laevigata Willd.* |
| Sumac, smooth | *Rhus glabra* L. |
| Sycamore, American | *Platanus occidentalis* L. |
| Tamarack or eastern larch | *Larix laricina* (Du Roi) K. Koch |
| Tamarisk, five-stamen | *Tamarix pentandra* Pall. |
| Teak | *Tectona grandis* L. f. |
| Tree of heaven or ailanthus | *Ailanthus altissima* (Mill.) Swingle |
| Tupelo, swamp | *Nyssa aquatica* L. |
| Walnut, black | *Juglans nigra* L. |
| Willow, black | *Salix nigra* Marsh. |
| Yellow-poplar or tulip tree | *Liriodendron tulipifera* L. |
| Yew, Pacific | *Taxus brevifolia* Nutt. |

# Literature Cited


Aaltonen, V. T. 1948. *Boden und Wald, unter besonderer Berucksichtigung des nordeuropäischen Waldbaus*. Paul Parey, Berlin. 457 pp.

———. 1950. Die Blattanalyse als Bonitierungsgrundlage des Waldbodens. Commun. Inst. For. Fenn. 37(8). 41 pp.

Abbot, Charles Greeley. 1929. *The Sun*. Appleton-Century-Crofts, Inc., New York. 447 pp.

Adams, D. F., D. J. Mayhew, R. M. Gnagy, E. P. Richey, R. K. Kappe, and I. W. Allen. 1952. Atmospheric pollution in the ponderosa pine blight area. *Ind. Eng. Chem.* 44:1356–1365.

Adams, Michael S., and Orie L. Loucks. 1971. Summer air temperatures as a factor affecting net photosynthesis and distribution of eastern hemlock (*Tsuga canadensis* L. (Carriere)) in southwestern Wisconsin. *Am. Mid. Nat.* 85:1–10.

Ahlgren, Clifford E. 1960. Some effects of fire on reproduction and growth of vegetation in northeastern Minnesota. *Ecology* 41:431–445.

———. 1974. Effects of fires on temperate forests: north central United States. *In* T. T. Kozlowski and C. E. Ahlgren (eds.), *Fire and Ecosystems*. Academic Press, Inc., New York.

Ahlgren, Isabel F. 1974. The effect of fire on soil organisms. *In* T. T. Kozlowski and C. E. Ahlgren (eds.), *Fire and Ecosystems*. Academic Press, Inc., New York.

Aho, Paul E., Ramon J. Seidler, Harold J. Evans, and P. N. Raju. 1974. Distribution, enumeration, and identification of nitrogen–fixing bacteria associated with decay in living white fir trees. *Phytopathology* 64:1413–1420.

Alban, David H. 1977. Influence on soil properties of prescribed burning under mature red pine. USDA For. Serv. Res. Paper NC–139. North Central For. Exp. Sta., St. Paul, Minn. 8 pp.

Albertson, F. W., and J. E. Weaver. 1945. Injury and death or recovery of trees in prairie climate. *Ecol. Monogr.* 15:395–433.

Alexander, Martin E., and Frank G. Hawksworth. 1975. Wildland fires and dwarf mistletoes: a literature review of ecology and prescribed burning. USDA For. Serv. Gen. Tech. Report RM–14. Rocky Mt. For. and Rge. Exp. Sta., Ft. Collins, Colo. 12 pp.

Alexander, Robert R. 1964. Minimizing windfall around clear cuttings in spruce-fir forests. *For. Sci.* 10:130–142.

———. 1966. Site indexes for lodgepole pine, with corrections for stand density; instructions for field use. USDA For. Serv. Res. Paper RM–24. Rocky Mountain For. and Rge. Exp. Sta., Fort Collins, Colo. 7 pp.

———, and Jesse H. Buell. 1955. Determining the direction of destructive winds in a Rocky Mountain timber stand. *J. For.* 53:19–23.

———, David Tackle, and Walter G. Dahms. 1967. Site indexes for lodgepole pine, with corrections for stand density: methodology. USDA For. Serv. Res. Paper RM–29. Rocky Mountain For. and Rge. Exp. Sta., Fort Collins, Colo. 18 pp.

Allen, George S., and John N. Owens. 1972. The life history of Douglas-fir. Environment Canada, For. Serv., Ottawa. 139 pp.

Al-Naib, F. A., and E. L. Rice. 1971. Allelopathic effects of *Platanus occidentalis*. *Bull. Torrey Bot. Club* 98:75–82.

Anderson, Edgar. 1948. Hybridization of the habitat. *Evolution* 2:1–9.

———. 1949. *Introgressive Hybridization*. John Wiley & Sons, Inc., New York. 109 pp.

———. 1953. Introgressive hybridization. *Biol. Rev.* 28:280–307.

Anderson, H. W. 1967. Snow accumulation as related to meteorological, topographic, and forest variables in Central Sierra Nevada, California. Int. Ass. Sci. Hydrol. Publ. 76:215–224.

———. 1970. Storage and delivery of rainfall and snowmelt water as related to forest environments. *Proc. 3rd Forest Microclimate Symp.*, pp. 51–67. Canad. For. Serv., Calgary, Alberta.

———, Goeffery B. Coleman, and Paul J. Zinke. 1959. Summer slides and winter scours—dry-wet erosion in southern California mountains. USDA Pacific Southwest For. and Rge. Exp. Sta. Tech. Paper 36. 12 pp.

———, Marvin D. Hoover, and Kenneth G. Reinhart. 1976. Forests and water: effects of forest management on floods, sedimentation, and water supply. USDA For. Serv. Gen. Tech. Report PSW–18. Pacific Southwest For. and Rge. Exp. Sta., Berkeley, Calif. 115 pp.

Andersson, Enar. 1963. Seed stands and seed orchards in the breeding of conifers. World Consult. For. Gen. and For. Tree Imp. Proc. II FAO-FORGEN 63–8/1:1–18.

Andrews, Henry N. 1961. *Studies in Paleobotany*. John Wiley & Sons, Inc., New York. 487 pp.

Arbeitsgemeinschaft "Oberschwäbische Fichtenreviere." 1964. *Standort, Wald und Waldwirtschaft in Oberschwaben*. Verein forstl. Standortsk. Forstpflz., Stuttgart. 323 pp.

Armson, K. A. 1977. *Forest Soils: Properties and Processes*. Univ. Toronto Press, Toronto. 390 pp.

———, H. H. Krause, and G. F. Weetman. 1975. Fertilization response in the northern coniferous forest. *In* B. Bernier and C. H. Winget (eds.), *Forest Soils and Forest Land Management*. Laval Univ. Press, Quebec.

Arno, Stephen F. 1976. The historical role of fire on the Bitterroot National Forest. USDA For. Serv. Res. Paper INT–187. Intermountain For. and Rge. Exp. Sta., Ogden, Utah. 29 pp.

———, and Robert D. Pfister. 1977. Habitat types: an improved system for classifying Montana's forests. *Western Wildlands* 3:6–11.

Arnold, J. F. 1950. Changes in ponderosa pine bunchgrass ranges in northern Arizona resulting from pine regeneration and grazing. *J. For.* 48:118–126.

Art, H. W., and P. L. Marks. 1971. A summary table of biomass and net annual primary production in forest ecosystems of the world. *In Forest Biomass Studies*. Misc. Publ. 132, Life Sci. and Agr. Exp. Sta., Univ. Maine, Orono.

Ashton, P. S. 1969. Speciation among tropical forest trees: some deductions in light of recent evidence. *Biol. J. Linn. Soc.* 1:155–196.

Aston, J. L., and A. D. Bradshaw. 1966. Evolution in closely adjacent plant populations. II. *Agrostis stolonifera* in maritime habitats. *Heredity* 21:649–664.

Atsatt, Peter R., and Dennis J. O'Dowd. 1976. Plant defense guilds. *Science* 193:24–29.

Aulitzky, Herbert. 1967. Significance of small climatic differences for the proper afforestation of highlands in Austria. *In* William E. Sopper and Howard W. Lull (eds.), *Forest Hydrology*. Pergamon Press, Inc., New York.

Axelrod, Daniel I. 1970. Mesozoic paleogeography and early angiosperm history. *Bot. Rev.* 36:277–319.

Baes, C. F., Jr., H. E. Goeller, J. S. Olson, and R. M. Rotty. 1976. The global carbon dioxide problem. Oak Ridge National Lab. Report ORNL 5194, Oak Ridge, Tenn. 72 pp.

Bailey, Robert G. 1976. Ecoregions of the United States. USDA For. Serv., Ogden, Utah. Map.

Baker, Frederick S. 1929. Effect of excessively high temperatures on coniferous reproduction. *J. For.* 27:949–975.

———. 1944. Mountain climates of the western United States. *Ecol. Monogr.* 14:223–254.

———. 1945. Effects of shade upon coniferous seedlings grown in nutrient solutions. *J. For.* 43:428–435.

———. 1950. *Principles of Silviculture*. McGraw-Hill Book Co., Inc., New York. 414 pp.

Bannister, M. H. 1965. Variation in the breeding system of *Pinus radiata*. *In* H. G. Baker and G. Ledyard Stebbins (eds.), *The Genetics of Colonizing Species*. Academic Press, Inc., New York.

Barber, H. N., and W. D. Jackson. 1957. Natural selection in action in Eucalyptus. *Nature* 179:1267–1269.

Bard, G. E. 1946. The mineral nutrient content of the foliage of forest trees on three soil types of varying limestone content. *Proc. Soil. Sci. Soc. Amer.* 10:419–422.

———. 1952. Secondary succession on the Piedmont of New Jersey. *Ecol. Monogr.* 22:195–215.

Barnes, Burton V. 1967. The clonal growth habit of American aspens. *Ecology* 47:439–447.

———, 1969. Natural variation and delineation of clones of *Populus tremuloides* and *P. grandidentata* in northern Lower Michigan. *Silvae Genetica* 18:130–142.

———. 1975. Phenotypic variation of trembling aspen in western North America. *For. Sci.* 21:319–328.

———. 1976. Succession in deciduous swamp communities of southeastern

Michigan. formerly dominated by American elm. *Can. J. Bot.* 54:19–24.
———, Bruce P. Dancik, and Terry L. Sharik. 1974. Natural hybridization of yellow birch and paper birch. *For. Sci.* 20:215–221.
Barrett, L. I., and A. A. Downs. 1943. Hardwood invasion in pine forests of the Piedmont Plateau. *J. Agr. Res.* 67:111–128.
Bartelli, Lindo J., and James A. DeMent. 1970. Soil survey—a guide for forest management decisions in the southern Appalachians. *In* Chester T. Youngberg and Charles B. Davey (eds.), *Tree Growth and Forest Soils.* Oregon State Univ. Press, Corvallis, Ore.
Bartlett, H. H. 1956. Fire, primitive agriculture, and grazing in the tropics. *In* William L. Thomas (ed.), *Man's Role in Changing the Face of the Earth.* Univ. Chicago Press.
Bartley, D. D. 1967. Pollen analysis of surface samples of vegetation from arctic Quebec. *Pollen & Spores* 9:101–105.
Bassett, J. R. 1964. Diameter growth of loblolly pine trees as affected by soil moisture availability. USDA For. Serv. Res. Note SO–9, Southern For. Exp. Sta., New Orleans, La. 7 pp.
Bates, C. G., and Jacob Roeser, Jr. 1928. Light intensities required for growth of coniferous seedlings. *Amer. J. Bot.* 15:185–244.
Baumgartner, Albert. 1958. Nebel und Nebelniederschlag als Standortsfacktoren am grossen Falkenstein (Bayr. Wald). *Forstw. Centralbl.* 77:257–272.
———. 1967. Energetic bases for differential vaporization from forest and agricultural lands. *In* William E. Sopper and Howard W. Lull (eds.), *Forest Hydrology.* Pergamon Press, Inc., New York.
Beale, E. F. 1858. Wagon road from Fort Defiance to the Colorado River. 35 Cong. 1 Sess., Sen. Exec. Doc. 124.
Beard, J. S. 1945. The progress of plant succession on the Soufrière of St. Vincent. *J. Ecol.* 33:1–9.
———. 1953. The savanna vegetation of northern tropical America. *Ecol. Monogr.* 23:149–215.
———. 1955. The classification of tropical American vegetation-types. *Ecology* 36:89–100.
Beaufait, W. R. 1956. Influence of soil and topography on willow oak sites. U.S. For. Serv., Southern For. Exp. Sta. Occ. Paper 148. 12 pp.
Beck, Charles B. (ed.). 1976. *Origin and Early Evolution of Angiosperms.* Columbia Univ. Press, New York. 341 pp.
Beck, Donald E. 1962. Yellow-poplar site index curves. U.S. For. Serv., Southeastern For. Exp. Sta. Res. Note 180. 2 pp.
———. 1971. Polymorphic site index curves for white pine in the southern Appalachians. USDA For. Serv. Res. Note SE–80. Southeastern For. Exp. Sta., Asheville, N.C. 8 pp.
Becking, R. W. 1957. The Zürich-Montpellier school of phytosociology. *Bot. Rev.* 23:411–488.
Beilmann, A. P., and L. G. Brenner. 1951. The recent intrusion of forests in the Ozarks. *Ann. Mo. Bot. Gdn.* 38:261–282.
Bellinger, P. F. 1954. Studies of soil fauna with special reference to the Collembola. Conn. Agr. Exp. Sta. Bull. 583. 67 pp.

Bendell, J. F. 1974. Effects of fire on birds and mammals. *In* T. T. Kozlowski and C. E. Ahlgren (eds.), *Fire and Ecosystems.* Academic Press, Inc., New York.

Benninghoff, William S. 1952. Relationships between vegetation and frost in soils. Proc. Permafrost International Conference, pp. 9–13.

———. 1963. The Prairie Peninsula as a filter barrier to postglacial plant migration. *Proc. Indiana Acad. Sci.* 72:116–124.

Benson, L. B. 1962. *Plant Taxonomy: Methods and Principles.* The Ronald Press Co., New York. 494 pp.

Bernabo, J. Christopher, and Thompson Webb III. 1977. Changing patterns in the Holocene pollen record of northeastern North America: a mapped summary. *Quaternary Res.* 8:64–96.

Bernier, B., and C. H. Winget (eds.). 1975. *Forest Soils and Forest Land Management.* Proc. Fourth North American Forest Soils Conference. Laval Univ. Press, Quebec. 675 pp.

Berry, Charles R., and George H. Hepting. 1964. Injury to eastern white pine by unidentified atmospheric constituents. *For. Sci.* 10:2–13.

Bibelriether, H. 1964. Unterschiedliche Wurzelbildung bei Kiefern verschiedener Provenienz. *Forstwiss. Cbl.* 83:129–140.

Bigelow, R. S. 1965. Hybrid zones and reproductive isolation. *Evolution* 19:449–458.

Billings, W. D. 1938. The structure and development of old-field pine stands and certain associated physical properties of the soil. *Ecol. Monogr.* 8:437–499.

Bingham, R. T. and A. E. Squillace. 1955. Self-compatibility and effects of self-fertility in western white pine. *For. Sci.* 1:121–129.

———, and ———. 1957. Phenology and other features of the flowering of pines, with special reference to *Pinus monticola* Dougl. Intermountain For. and Rge. Exp. Sta. Res. Paper 53. 26 pp.

Birch, L. C., and D. P. Clark. 1953. Forest soil as an ecological community—with special reference to the fauna. *Quart. Rev. Biol.* 28(1):13–36.

Birks, H. J. B., and R. G. West. 1973. *Quaternary Plant Ecology:* The 14th Symposium of the British Ecological Society. Blackwell Sci. Publ., Oxford. 326 pp.

Biswell, Harold H. 1961. The big trees and fires. *Nat. Parks Mag.* 35:11–14.

———. 1974. Effects of fire on chaparral. *In* T. T. Kozlowski and C. E. Ahlgren (eds.), *Fire and Ecosystems.* Academic Press, Inc., New York.

Blair, W. Frank. 1977. *Big Biology, The US/IBP.* Vol. 7, US/IBP Synthesis series. Academic Press, Inc. New York. 272 pp.

Blais, J. R. 1952. The relationship of the spruce budworm (*Choristoneura fumiferana,* Clem.) to the flowering condition of balsam fir (*Abies balsamea* (L.) Mill.). *Can. J. Zool.* 30:1–29.

Bloomberg, W. J. 1950. Fire and spruce. *For. Chron.* 26(2):157–161.

Blow, F. E. 1955. Quantity and hydrologic characteristics of litter under upland oak forests in eastern Tennessee. *J. For.* 53:190–195.

Bode, Hans Robert. 1958. Beiträge zur Kenntnis allelopathischer Erscheinungen bei einigen Juglandaceen. *Planta* 51:440–480.

Boggess, W. R. 1956. Weekly diameter growth of shortleaf pine and white oak as related to soil moisture. *Proc. Soc. Am. Foresters*, 1956, pp. 83–89.

———. 1964. Trelease Woods, Champaign County, Illinois: woody vegetation and stand composition. *Trans. Ill. Acad. Sci.* 57:261–271.

———, and L. W. Bailey. 1964. Brownfield Woods, Illinois: woody vegetation and changes since 1925. *Am. Mid. Nat.* 71:392–401.

———, and J. W. Geis. 1966. The Funk Forest Natural Area, McLean County, Illinois: woody vegetation and ecological trends. *Trans. Ill. Acad. Sci.* 59:123–133.

Bolin, Bert. 1977. Changes of land biota and their importance for the carbon cycle. *Science* 196:613–615.

Bond, G. 1967. Fixation of nitrogen by higher plants other than legumes. *Ann. Rev. Plant Physiol.* 18:107–126.

Bonnemann, Alfred, and Ernst Röhrig. 1971. *Waldbau auf ökologischer Grundlage. I. Der Wald als Vegetationstyp und seine Bedeutung für den Menschen.* Verlag Paul Parey, Berlin. 229 pp.

———, and ———. 1972. *Waldbau auf ökologischer Grundlage. II. Holzartenwahl, Bestandesbegrundung und Bestandespflege.* Verlag Paul Parey, Berlin. 264 pp.

Books, David J., and Carl H. Tubbs. 1970. Relation of light to epicormic sprouting in sugar maple. USDA For. Serv. Res. Note NC–93. North Central For. Exp. Sta., St. Paul, Minn. 2 pp.

Borchert, J. R. 1950. The climate of the central North American grassland. *Ann. Assoc. Amer. Geog.* 40:1–39.

Bormann, F. H. 1953. Factors determining the role of loblolly pine and sweetgum in early old-field succession in the Piedmont of North Carolina. *Ecol. Monogr.* 23:339–358.

———. 1966. The structure, function, and ecological significance of root grafts in *Pinus strobus* L. *Ecol. Monogr.* 36:1–26.

———, and G. E. Likens. 1969. The watershed-ecosystem concept and studies of nutrient cycles. *In* George M. Van Dyne (ed.), *The Ecosystem Concept in Natural Resource Management.* Academic Press, Inc., New York.

———, and ———. 1979. *Pattern and Process in a Forested Ecosystem.* Springer-Verlag, New York. 256 pp.

Bourdo, Eric A., Jr. 1956. A review of the general land office survey and of its use in quantitative studies of former forests. *Ecology* 37:744–768.

Bowen, G. D. 1973. Mineral nutrition of ectomycorrhizae. *In* G. C. Marks and T. T. Kozlowski (eds.), *Ectomycorrhizae—Their Ecology and Physiology.* Academic Press, Inc., New York.

Bowen, Murray G. 1974. Selected metric (SI) units and conversion factors for Canadian forestry. For. Mgt. Inst. Can. For. Serv., Ottawa, Ontario.

Boyce, Stephen G., and Margaret Kaeiser. 1961. Why yellow-poplar seeds have low viability. USDA For. Serv. Central States For. Exp. Sta., Tech. Paper 186. 16 pp.

Boyle, James R. 1975. Nutrients in relation to intensive culture of forest

crops. *Iowa State J. Res.* 49:297–303.
———, John J. Phillips, and Alan R. Ek. 1973. Whole tree harvesting: nutrient budget evaluation. *J. For.* 71:760–762.
Bradshaw, A. D. 1965. Evolutionary significance of phenotypic plasticity in plants. *Adv. Genetics* 13:115–155.
Bradshaw, Kenneth E. 1965. Soil use and management in the national forests of California. *In* Chester T. Youngberg (ed.), *Forest-Soil Relationships in North America.* Oregon State Univ. Press, Corvallis, Ore.
Brady, Nyle C. 1974. *The Nature and Properties of Soils.* Macmillan Publishing Co., Inc. New York. 639 pp.
Braun, E. Lucy. 1942. Forests of the Cumberland Mountains. *Ecol. Monogr.* 12:413–447.
———. 1950. *Deciduous Forests of Eastern North America.* McGraw-Hill Book Co., Inc., New York. 596 pp.
Braun-Blanquet, J. 1964. *Pflanzensoziologie,* 3rd ed. Springer-Verlag, Vienna. 865 pp.
Bray, J. R. 1956. Gap phase replacement in a maple/basswood forest. *Ecology* 37(3):598–600.
———, and J. T. Curtis. 1957. An ordination of the upland forest communities of southern Wisconsin. *Ecol. Monog.* 22:217–234.
———, and Eville Gorham. 1964. Litter production in forests of the world. *Adv. Ecol. Res.* 2:101–157.
Brayshaw, T. C. 1965. Native poplars of southern Alberta and their hybrids. Canada, Dept. of Forestry Publ. No. 1109. 40 pp.
Brayton, R., and H. A. Mooney. 1966. Population variability of *Cercocarpus* in the White Mountains of California as related to habitat. *Evolution* 20:383–391.
Brittain, W. H., and W. F. Grant. 1967. Observations on Canadian birch collections at the Morgan Arboretum. V. B. *papyrifera* and *B. cordifolia* from eastern Canada. *Canad. Field-Naturalist* 81:251–262.
Broadfoot, W. M. 1969. Problems in relating soil to site index for southern hardwoods. *For. Sci.* 15:354–364.
Brooks, M. G. 1951. Effects of black walnut trees and their products on other vegetation. West Va., Agr. Exp. Sta. Bull. 347. 31 pp.
Brown, Claud L., Robert G. McAlpine, and Paul P. Kormanik. 1967. Apical dominance and form in woody plants: a reappraisal. *Amer. J. Bot.* 54:153–162.
Brown, Harry E. 1971. Evaluating watershed management alternatives. *J. Irrigation Drainage Div., Proc. Amer. Soc. Civil Engineers.* 97:93–108.
Brown, J. H., Jr. 1960. The role of fire in altering the species composition of forests in Rhode Island. *Ecology* 41:310–316.
Brown, J. M. B. 1955. Ecological investigations: shade and growth of oak seedlings. Rep. For. Res. Comm., London, 1953–54.
Buckley, John L. 1959. Effects of fire on Alaskan wildlife. *Proc. Soc. Amer. For.* 1958:123–126.
Buckman, Harry O., and Nyle C. Brady. 1969. *The Nature and Properties of Soils,* 7th ed. The Macmillan Co., Toronto, Ontario, Canada. 653 pp.

Buell, M. F., and J. E. Cantlon. 1951. A study of two forest stands in Minnesota with an interpretation of the prairie-forest margin. *Ecology* 32:294–316.

———, and V. Facey. 1960. Forest-prairie transition west of Itasca Park, Minnesota. *Bull. Torrey Bot. Cl.* 87(1):46–58.

Buffo, John, Leo J. Fritschen, and James L. Murphy. 1972. Direct solar radiation on various slopes from 0 to 60 degrees north latitude. USDA For. Serv. Res. Paper PNW–142. Pacific Northwest For. and Rge. Exp. Sta., Portland, Ore. 74 pp.

Bull, H. 1931. The use of polymorphic curves in determining site quality in young red pine plantations. *J. Agric. Res.* 43:1–28.

Burger, D. 1972. Forest site classification in Canada. *Mitt. Vereins forstl. Standortsk. Forstpflz.* 21:20–36.

Burgess, Alan, and D. F. Nicholas. 1961. The use of soil sections in studying the amount of fungal hyphae in soil. *Soil Sci.* 92:25–29.

Burgess, Robert L., and Robert V. O'Neill. 1975. Eastern deciduous forest biome progress report. Env. Sci. Div. Publ. 751, Oak Ridge Natl. Lab., Oak Ridge, Tenn. 252 pp.

Burns, George P. 1923. Studies in tolerance of New England forest trees IV. Minimum light requirements referred to a definite standard. Univ. Vermont Agr. Exp. Sta. Bull. 235.

Burns, G. Richard. 1942. Photosynthesis and absorption in blue radiation. *Amer. J. Bot.* 29:381–387.

Büsgen, M. and E. Münch. 1929. *The Structure and Life of Forest Trees,* 3d ed. Trans. by Thomas Thomson. Chapman & Hall, Ltd., London. 436 pp.

Buttrick, P. L. 1914. Notes on germination and reproduction of longleaf pine in southern Mississippi. *For. Quart.* 12:532–537.

Byers, H. R. 1953. Coast redwoods and fog drip. *Ecology* 34:192–193.

Byram, George M., and George M. Jemison. 1943. Solar radiation and forest fuel moisture. *J. Agr. Res.* 67:149–176.

Byrd, H. J., N. E. Sands, and Jack T. May. 1965. Forest management based on soil surveys in Georgia. *In* Chester T. Youngberg (ed.), *Forest-Soil Relationships in North America.* Oregon State Univ. Press, Corvallis, Ore.

Cain, Stanley A. 1939. The climax and its complexities. *Amer. Midl. Nat.* 21:146–181.

———. 1944. *Foundations of Plant Geography.* Harper & Row, Inc., New York. 556 pp.

———, and G. M. de Oliveira Castro. 1959. *Manual of Vegetation Analysis.* Harper & Row, Inc., New York. 325 pp.

Cajander, A. K. 1926. The theory of forest types. *Acta For. Fenn.* 29. 108 pp.

California Div. For. 1969. Soil-vegetation surveys in California. State of California, Dept. of Cons., Div. of Forestry, Sacramento. 31 pp.

Callaham, Robert Z. 1962. Geographic variability in growth of forest trees. *In* Theodore T. Kozlowski (ed.), *Tree Growth.* The Ronald Press Co., New York.

———, and A. R. Liddecoet. 1961. Altitudinal variation at 20 years in ponderosa and Jeffrey pines. *J. For.* 59:814–820.

Campbell, Robert K. 1976. Adaptational requirements of planting stock. *In*

*Global Forestry and the Western Role*, Perm. Assoc. Comm. Proc., 1975, West. For. and Conserv. Assoc., Portland, Ore.

Campbell, R. S. 1955. Vegetational changes and management in the cutover longleaf pine-slash pine area of the Gulf Coast. *Ecology* 36:29–34.

Canadian For. Serv. 1974. Proceedings of a workshop on forest fertilization in Canada. Can. For. Serv., Great Lakes For. Res. Centre For. Tech. Report 5. Sault Ste. Marie, Ontario. 127 pp.

Carlisle, A., A. Brown, and E. White. 1967. The nutrient content of tree stem flow and ground flora litter and leachates in a sessile oak (*Quercus petraea*) woodland. *J. Ecol.* 55:615–627.

Carlson, Clinton E., and George M. Blake. 1969. Hybridization of western and subalpine larch. Montana For. and Cons. Exp. Sta. Bull. 37. Univ. Montana, Missoula, Mont. 12 pp.

———, and Jerald E. Dewey. 1971. Environmental pollution by fluorides in Flathead National Forest and Glacier National Park. USDA For. Serv. Northern Region, Missoula, Mont. 57 pp.

———, Wayne E. Bousfield, and Mark D. McGregor. 1974. The relationship of an insect infestation on lodgepole pine to flourides emitted from a nearby aluminum plant in Montana. USDA For. Serv. Northern Region Insect Disease Report No. 74–14. Missoula, Mont. 21 pp.

Carmean, Willard H. 1956. Suggested modification of the standard Douglas-fir site curves for certain soils in southwestern Washington. *For. Sci.* 2:242–250.

———. 1965. Black oak site quality in relation to soil and topography in southeastern Ohio. *Proc. Soil Sci. Soc. Amer.* 29:308–312.

———. 1966. Soil and water requirements. *In* Black Walnut Culture. U.S. For. Serv., North Central For. Exp. Sta., St. Paul, Minn.

———. 1967. Soil refinements for predicting black oak site quality in southeastern Ohio. *Proc. Soil Sci. Soc. Amer.* 31:805–810.

———. 1970a. Site quality for eastern hardwoods. *In* The silviculture of oaks and associated species. USDA For. Serv. Res. Paper NE–144. Northeastern For. Exp. Sta., Upper Darby, Pa. 66 pp.

———. 1970b. Tree height-growth patterns in relation to soil and site. *In* Chester T. Youngberg and Charles B. Davey (eds.), *Tree Growth and Forest Soils.* Oregon State Univ. Press, Corvallis, Ore.

———. 1972. Site index curves for upland oaks in the central states. *For. Sci.* 18:109–120.

———. 1975. Forest site quality evaluation in the United States. *Adv. Agronomy* 27:209–269.

———. 1977. Site classification for northern forest species. *In Proc. Symposium on Intensive Culture of Northern Forest Types*, pp. 205–239. USDA For. Serv. Gen. Tech. Report NE–29. Northeastern For. Exp. Sta., Upper Darby, Pa.

———, F. Bryan Clark, Robert D. Williams, and Peter R. Hannah. 1976. Hardwoods planted in old fields favored by prior tree cover. USDA For. Serv. Res. Paper NC–134. North Central For. Exp. Sta., St. Paul, Minn. 16 pp.

Carter, G. S. 1934. Reports of the Cambridge expedition to British Guiana,

1933. Illumination in the rain forest at the ground level. *J. Linn. Soc., London. Zool.* 38:579–589.
Chaney, R. W. 1949. Redwoods—occidental and oriental. *Science* 110:551–552.
———. 1954. A new pine (*Pinus clementsii*) from the Cretaceous of Minnesota and its palaeoecological significance. *Ecology* 35:145–151.
Chapman, H. H. 1932. Is the longleaf pine a climax? *Ecology* 13:328–335.
Chapman, J. D. 1952. The climate of British Columbia. *Trans. 5th Brit. Columbia Natur. Resour. Conf.*, pp. 8–54.
Chettleburgh, M. R. 1952. Observations on the collection and burial of acorns by jays in Hainault. *Brit. Birds* 45:359–364.
Christian, C. S., and G. A. Stewart. 1968. Methodology of integrated surveys. *In Proc. Aerial Surveys and Integrated Studies Conf.*, pp. 233–280. UNESCO, Toulouse.
Christy, H. R. 1952. Vertical temperature gradients in a beech forest in central Ohio. *Ohio J. Sci.* 52(4):199–209.
Churchill, E. D., and H. C. Hanson. 1958. The concept of climax in arctic and alpine vegetation, *Bot. Rev.* 24(2/3):127–191.
Cieslar, Adolf. 1887. Über den Einfluss der Grosse der Fichtensamen auf die Entwicklung der Pflanzen nebst einigen Bemerkungen über schwedische Fichten und Weissfohrensamen. *Centbl. gesam. Forstw.* 13:149–153.
———. 1895. Über die Erblichkeit des Zuwachsvermögens bei den Waldbäumen. *Centralbl. gesam. Forstw.* 21:7–29.
———. 1899. Neues aus dem Gebiete der forstlichen Zuchtwahl. *Centbl. gesam. Forstw.* 25:99–117.
Clements, Frederic E. 1910. The life history of lodgepole burn forests. U.S. For. Serv. Bull. 79. 56 pp.
———. 1916. Plant succession: an analysis of the development of vegetation. Carneg. Inst. Wash. Publ. 242. 512 pp.
———. 1949. *Dynamics of Vegetation: Selections from the Writings of Frederic E. Clements, Ph.D.* The H. W. Wilson Co., New York. 296 pp.
Cline, A. C., and Stephen H. Spurr. 1942. The virgin upland forest of Central New England. Harv. For. Bull. 21. 51 pp.
Cobb, Fields W., Jr., and R. W. Stark. 1970. Decline and mortality of smog-injured ponderosa pine. *J. For.* 68:147–149.
Coe, H. J., and F. M. Isaac. 1965. Pollination of the baobab (*Adansonia digitata* L.) by the lesser bush baby (*Galego crassicaudatum* E. Geoffroy). *E. African Wildlife J.* 3:123–124.
Cogbill, Charles V., and Gene E. Likens. 1974. Acid precipitation in the northeastern United States. *Water Resources Res.* 10:1133–1137.
Coile, T. S. 1937. Distribution of forest tree roots in North Carolina Piedmont soils. *J. For.* 35:247–257.
———. 1952. Soil and the growth of forests. *Adv. Agronomy* 4:330–398.
———. 1960. Summary of soil-site evaluation. *In* P. Y. Burns (ed.), *Proc. Eighth Annual Forestry Symposium.* Louisiana State Univ. Press, Baton Rouge, La.
———, and F. X. Schumacher. 1953. Site index of young stands of loblolly and shortleaf pines in the Piedmont Plateau region. *J. For.* 51:432–435.

Cole, D. W., J. B. Crane, and C. C. Grier. 1975. The effect of forest management practices on water chemistry in a second-growth Douglas-fir ecosystem. *In* B. Bernier and C. H. Winget (eds.), *Forest Soils and Forest Land Management*. Laval Univ. Press, Quebec.

Committee on Site Classification, Northeastern Forest Soils Conference. 1961. Planting sites in the Northeast. Northeastern For. Exp. Sta. Paper 157. 24 pp.

Cookson, I. C. and K. M. Pike. 1953. The Tertiary occurrence and distribution of Podocarpus (section *Dacrycarpus*) in Australia and Tasmania. *Aust. J. Bot.* 1:71–82.

Cooper, Charles F. 1960. Changes in vegetation, structure, and growth of southwestern pine forests since white settlement. *Ecol. Monogr.* 30:129–164.

———. 1961. The ecology of fire. *Sci. American* 204(4):150–160.

———, and William C. Jolly. 1969. Ecological effects of weather modification: a problem analysis. Univ. of Michigan, School of Natural Resources, Ann Arbor. 160 pp.

Cooper, W. S. 1913. The climax forest of Isle Royale, Lake Superior, and its development. *Bot. Gaz.* 55:1–44, 115–140, 189–235.

———. 1923. The recent ecological history of Glacier Bay, Alaska. *Ecology* 4:93–128, 223–246, 355–365.

Corliss, J. F., and C. T. Dyrness. 1965. A detailed soil-vegetation survey of the Alsea area in the Oregon Coast Range. *In* Chester T. Youngberg (ed.), *Forest-Soil Relationships in North America*. Oregon State Univ. Press, Corvallis, Ore.

Cottam, G. 1949. The phytosociology of an oak woods in southwestern Wisconsin. *Ecology* 30:271–287.

Cottam, Walter P., and George Stewart. 1940. Plant succession as a result of grazing and meadow desiccation by erosion since settlement in 1862. *J. For.* 38:613–626.

Covell, R. R., and D. C. McClurkin. 1967. Site index of loblolly pine on Ruston soils in the southern Coastal Plain. *J. For.* 65:263–264.

Cowles, Henry C. 1899. The ecological relations of the vegetation on the sand dunes of Lake Michigan. *Bot. Gaz.* 27:95–116, 167–202, 281–308, 361–391.

Cox, Christopher B., Ian N. Healey, and Peter D. Moore. 1976. *Biogeography: An Ecological and Evolutionary Approach*. 2nd ed. Blackwell Sci. Publ., Oxford. 194 pp.

Crandall, Dorothy L. 1958. Ground vegetation patterns of the spruce-fir area of the Great Smoky Mountains National Park. *Ecol. Monogr.* 28:337–360.

Crawford, Hewlette S., R. G. Hooper, and R. F. Harlow. 1976. Woody plants selected by beavers in the Appalachian Ridge and Valley Province. USDA For. Serv. Res. Paper NE–346. Northeastern For. Exp. Sta., Upper Darby, Pa. 6 pp.

Croft, A. R., and L. V. Monninger. 1953. Evapotranspiration and other water losses on some aspen forest types in relation to water available for stream flow. *Trans. Amer. Geophys. Union* 34(4):563–574.

Crouch, Glenn L. 1969. Deer and reforestation in the Pacific Northwest. *In*

Hugh C. Black (ed.), *Proc. Wildlife and Reforestation in the Pacific Northwest*, pp. 63–66. School For., Ore. State Univ., Portland, Ore.

———. 1976. Wild animal damage to forests in the United States and Canada. *In Proc. XVI IUFRO World Congress, Div. II*, pp. 468–478. Oslo, Norway.

Curtis, J. D., and N. R. Lersten. 1974. Morphology, seasonal variation and function of resin glands on buds and leaves of *Populus deltoides* (Salicaceae). *Am. J. Bot.* 61:835–845.

Curtis, John T. 1959. *The Vegetation of Wisconsin*. Univ. Wisconsin Press, Madison. 657 pp.

———, and R. P. McIntosh. 1951. An upland forest continuum in the prairie-forest border region of Wisconsin. *Ecology* 32:476–496.

Curtis, Robert O., Donald J. DeMars, and Francis R. Herman. 1974a. Which dependent variable in site index—height-age regressions. *For. Sci.* 20:74–87.

———, Francis R. Herman, and Donald J. DeMars. 1974b. Height growth and site index for Douglas-fir in high-elevation forests of the Oregon-Washington Cascades. *For. Sci.* 20:307–315.

Cushing, Edward J. 1965. Problems in the Quaternary phytogeography of the Great Lakes Region. *In* H. E. Wright, Jr., and David G. Frey (eds.), *The Quaternary of the United States*. Princeton Univ. Press, Princeton, N.J.

Cypert, Eugene. 1973. Plant succession on burned areas in Okefenokee Swamp following the fires of 1954 and 1955. *In Proc. Annual Tall Timbers Fire Ecology Conference*, 12:199–217. Tall Timbers Res. Sta., Tallahassee, Fla.

Czapowskyj, Miroslaw M. 1976. Annotated bibliography on the ecology and reclamation of drastically disturbed areas. USDA For. Serv. Gen. Tech. Report NE–21. Northeastern For. Exp. Sta., Upper Darby, Pa. 98 pp.

Daft, M. J., and E. Hacskaylo. 1976. Arbuscular mycorrhizas in the anthracite and bituminous coal wastes of Pennsylvania. *J. Appl. Ecol.* 13:523.

———, and ———. 1977. Growth of endomycorrhizal and nonmycorrhizal red maple seedlings in sand and anthracite spoil. *For. Sci.* 23:207–216.

Dale, M. B. 1970. Systems analysis and ecology. *Ecology* 51:2–16.

Damman, A. W. H. 1964. Some forest types of central Newfoundland and their relation to environmental factors. *For. Sci. Monogr.* 8. 62 pp.

———. 1975. Permanent changes in the chronosequence of a boreal forest habitat induced by natural disturbances. *In* Wolfgang Schmidt (ed.), *Sukzessionsforschung*. Proc. Int. Symp. Int. Verein. Vegetationsk. J. Cramer, Vaduz.

Dancik, Bruce P., and Burton V. Barnes. 1972. Natural variation and hybridization of yellow birch and bog birch in southeastern Michigan. *Silvae Genetica* 21:1–9.

Daniel, Theodore, John A. Helms, and Frederick S. Baker. 1979. *Principles of Silviculture*. 2nd ed., McGraw-Hill Book Co., New York. 500 pp.

Dansereau, Pierre. 1957. *Biogeography, an Ecological Perspective*. The Ronald Press Co., New York. 394 pp.

——— (ed.). 1968. The continuum concept of vegetation: responses. *Bot. Rev.* 34:253–332.

———, and Fernando Segadas-Vianna. 1952. Ecological study of the peat bogs of eastern North America. I. Structure and evolution of vegetation. *Can. J. Bot.* 30:490–520.

Darby, H. C. 1956. The clearing of the woodland in Europe. *In* William L. Thomas, Jr. (ed.), *Man's Role in Changing the Face of the Earth.* Univ. Chicago Press, Chicago.

Daubenmire, R. F. 1943. Temperature gradients near the soil surface with reference to techniques of measurement in forest ecology. *J. For.* 41:601–603.

———. 1952. Forest vegetation of northern Idaho and adjacent Washington, and its bearing on concepts of vegetation classification. *Ecol. Monogr.* 22:301–330.

———. 1954. Alpine timberlines in the Americas and their interpretation. *Butler Univ. Bot. Stud.* 11:119–136.

———. 1956. Climate as a determinant of vegetation distribution in eastern Washington and northern Idaho. *Ecol. Monogr.* 26:131–154.

———. 1959. *Plants and Environment,* 2d ed. John Wiley & Sons, Inc., New York. 422 pp.

———. 1960. A seven year study of cone production as related to xylem layers in *Pinus ponderosa. Amer. Midl. Nat.* 64:187–193.

———. 1961. Vegetative indicators of rate of height growth in ponderosa pine. *For. Sci.* 7:24–32.

———. 1966. Vegetation: identification of typal communities. *Science* 151:291–298.

———. 1968a. *Plant Communities.* Harper & Row, Inc., New York. 300 pp.

———. 1968b. Some geographic variations in *Picea sitchensis* and their ecologic interpretation. *Can. J. Bot.* 46:787–798.

———. 1973. A comparison of approaches to the mapping of forest land for intensive management. *For. Chron.* 49:87–91.

———. 1976. The use of vegetation in assessing the productivity of forest lands. *Bot. Rev.* 42:115–143.

———. 1978. *Plant Geography.* Academic Press, Inc., New York. 338 pp.

———, and Jean B. Daubenmire. 1968. Forest vegetation of eastern Washington and northern Idaho. Washington Agric. Exp. Sta., Tech. Bull. 60. 104 pp.

Davies, W. J., and T. T. Kozlowski. 1974. Stomatal responses of five woody angiosperms to light intensity and humidity. *Can. J. Bot.* 52:1525–1534.

Davis, J. H. 1940. The ecology and geologic role of mangroves in Florida. Carneg. Inst. Wash. Publ. 517.

Davis, Margaret Bryan. 1958. Three pollen diagrams from central Massachusetts. *Amer. J. Sci.* 256:540–570.

———. 1963. On the theory of pollen analysis. *Amer. J. Science* 261:897–912.

———. 1965. Phytogeography and palynology of northeastern United States. *In* H. E. Wright, Jr., and David G. Frey (eds.), *The Quaternary of the United States.* Princeton Univ. Press, Princeton, N.J.

———. 1967. Late-glacial climate in northern United States: a comparison of New England and the Great Lakes region. *In* E. J. Cushing and H. E.

Wright, Jr. (eds.), *Quaternary Paleoecology*. Yale Univ. Press, New Haven.

———. 1969. Palynology and environmental history during the Quaternary period. *Amer. Scientist* 57:317–332.

———. 1976. Pleistocene biogeography of temperate deciduous forests. *Geosci. and Man* 13:13–26.

Davis, P. H., and V. H. Heywood. 1963. *Principles of Angiosperm Taxonomy*. D. Van Nostrand Co., Inc., New York. 558 pp.

Day, Frank P., Jr., and Carl D. Monk. 1974. Vegetation patterns on a southern Appalachian watershed. *Ecology* 55:1064–1074.

Day, G. M. 1953. The Indian as an ecological factor in the northeastern forest. *Ecology* 34:329–346.

DeBano, Leonard F. 1969. The relationship between heat treatment and water repellency in soils. *In* Leonard F. DeBano and John Letey (eds.), *Water-Repellent Soils*. Univ. Calif., Riverside.

———, Joseph F. Osborn, Jay S. Krammes, and John Letey, Jr. 1967. Soil wettability and wetting agents . . . our current knowledge of the problem. USDA For. Ser. Res. Paper PSW–43. Pacific Southwest For. and Rge. Exp. Sta., Berkeley, Calif. 13 pp.

DeBell, Dean S. 1971. Phytotoxic effects of cherrybark oak. *For. Sci.* 17:180–185.

DeByle, Norbert V. 1976. Soil fertility as affected by broadcast burning following clearcutting in Northern Rocky Mountain larch/fir forests. *In Proc. Annual Tall Timbers Fire Ecology Conference No. 14 and Intermountain Fire Research Council Fire and Land Management Symposium*. 14:447–464. Tall Timbers Res. Sta., Tallahassee, Fla.

Decker, J. P. 1947. The effect of air supply on apparent photosynthesis. *Plant Physiol.* 22(4):561–571.

———. 1959. A system for analysis of forest succession. *For. Sci.* 5:154–157.

Deevey, Edward S., Jr. 1949. Biogeography of the Pleistocene. *Bull. Geol. Soc. Amer.* 60:1315–1416.

Deitschman, G. H. 1973. Mapping of habitat types throughout a national forest. USDA For. Serv. Gen. Tech. Report INT–22. Intermountain For. and Rge. Exp. Sta., Ogden, Utah. 14 pp.

Delcourt, Hazel R., and Paul A. Delcourt. 1977. Presettlement magnolia-beech climax of the Gulf Coastal Plain: quantitative evidence from the Apalachicola River bluffs, north-central Florida. *Ecology* 58:1085–1093.

Delfs, J. 1967. Interception and streamflow in stands of Norway spruce and beech in West Germany. *In* William E. Sopper and Howard W. Lull (eds.), *Forest Hydrology*. Pergamon Press, Inc., New York.

Del Moral, Roger, and C. H. Muller. 1970. The allelopathic effects of *Eucalyptus camaldulensis*. *Amer. Midl. Nat.* 83:254–282.

———, and Rex G. Cates. 1971. Allelopathic potential of the dominant vegetation of western Washington. *Ecology* 52:1030–1037.

DeMent, J. A., and E. L. Stone. 1968. Influence of soil and site on red pine plantations in New York. II. Soil type and physical properties. Cornell Univ. Agric. Exp. Sta. Bull. 1020. 25 pp.

Dengler, Alfred. 1944. *Waldbau auf ökologische Grundlage*. 3d ed. Springer-Verlag, Berlin. 596 pp.
Denison, William C. 1973. Life in tall trees. *Sci. American* 228:74–80.
———, Diane M. Tracy, Frederick M. Rhoades, and Martha Sherwood. 1972. Direct, nondestructive measurement of biomass and structure in living, old-growth Douglas-fir. *In* J. F. Franklin, L. J. Dempster, and R. H. Waring (eds.), *Proc.—Research on Coniferous Forest Ecosystems—A Symposium*. USDA Pacific Northwest For. and Rge. Exp. Sta., Portland, Ore.
Denton, George H., and Stephen C. Porter. 1970. Neoglaciation. *Sci. American* 222:100–110.
Dickerson, B. P. 1976. Soil compaction after tree-length skidding in northern Mississippi. *Proc. Soil Sci. Soc. Amer.* 40:965–966.
Dickinson, C. H., and G. J. F. Pugh. 1974. *Biology of Plant Litter Decomposition*. Vol. 1, Part I. *Types of Litter*. Vol. 2, Part II. *The Organisms*, Part III. *The Environment*. Academic Press, Inc., New York. 775 pp.
Dieterich, Hermann. 1970. Die Bedeutung der Vegetationskunde für die forstliche Standortskunde. *Der Biologieunterricht* 6:48–60.
Dimbleby, G. W. 1952a. The historical status of moorland in north-east Yorkshire. *New Phytol.* 51:349–354.
———. 1952b. Soil regeneration on the north-east Yorkshire moors. *J. Ecol.* 40:331–341.
———, and J. M. Gill. 1955. The occurrence of podzols under deciduous woodland in the New Forest. *Forestry* 28:95–106.
Dimock, Edward J., II. 1970. Ten-year height growth of Douglas-fir damaged by hare and deer. *J. For.* 68:285–288.
Dix, Ralph L. 1964. A history of biotic and climatic changes within the North American grassland. *In* D. J. Crisp (ed.), *Grazing in Terrestrial and Marine Environments*. Symp. Brit. Ecol. Soc. Blackwell Sci. Publ. Ltd.,Oxford.
Dobzhansky, Theodosius. 1968. Adaptedness and fitness. *In* Richard C. Lewontin (ed.), *Population Biology and Evolution*. Syracuse Univ. Press, Syracuse, New York.
Dochinger, Leon S., and Carl E. Seliskar. 1965. Results from grafting chlorotic dwarf and healthy eastern white pine. *Phytopathology* 55:404–407.
———, and Carl E. Seliskar. 1970. Air pollution and the chlorotic dwarf disease of eastern white pine. *For. Sci.* 16:46–55.
———, and T. A. Seliga. 1976. Proceedings of the first international symposium on acid precipitation and the forest ecosystem. USDA For. Serv. Gen. Tech. Report NE–23. Northeastern For. Exp. Sta., Upper Darby, Pa. 1074 pp.
Dolf, Erling. 1960. Climatic changes of the past and present. *Amer. Sci.* 48:341–364.
Doolittle, W. T. 1957. Site index of scarlet and black oak in relation to southern Appalachian soil and topography. *For. Sci.* 3:114–124.
———. 1958. Site index comparisons for several forest species in the Southern Appalachians. *Proc. Soil Sci. Soc. Amer.* 22:455–458.
Dort, Wakefield, Jr., and J. Knox Jones, Jr. 1970. *Pleistocene and Recent Environments of the Central Great Plains*. Dept. Geol., Univ. Kansas

Special Publ. 3, Univ. Press of Kansas, Lawrence. 433 pp.
Dortignac, E. J. 1967. Forest water yield management opportunities. *In* William E. Sopper and Howard W. Lull (eds.), *Forest Hydrology*. Pergamon Press, Inc., New York.
Douglass, A. E. 1919. Climatic cycles and tree growth; a study of the annual rings of trees in relation to climate and solar activity. Carnegie Inst. Wash. Publ. 289.
Douglass, James E., and Wayne T. Swank. 1972. Streamflow and modification through management of eastern forests. USDA For. Serv. Res. Paper SE–94. Southeastern For. Exp. Sta., Asheville, N.C. 15 pp.
Downs, Robert Jack. 1962. Photocontrol of growth and dormancy in woody plants. *In* Theodore T. Kozlowski (ed.), *Tree Growth*. The Ronald Press Co., New York.
———, and H. A. Borthwick. 1956. Effects of photoperiod on growth of trees. *Bot. Gaz.* 117:310–326.
Doyle, James A., and Leo J. Hickey. 1976. Pollen and leaves from the Mid-Cretaceous Potomac Group and their bearing on early angiosperm evolution. *In* Charles B. Beck (ed.), *Origin and Early Evolution of Angiosperms*. Columbia Univ. Press, New York.
Drew, T. John, and James W. Flewelling. 1977. Some recent Japanese theories of yield-density relationships and their application to Monterey pine plantations. *For. Sci.* 23:517–534.
Driscoll, Richard S. 1964. Vegetation-soil units in the central Oregon juniper zone. USDA For. Serv. Res. Note PNW–19. Pacific Northwest For. and Rge. Exp. Sta., Portland, Ore. 60 pp.
Duchaufour, P. 1947. Le hêtre est-il une essence améliorante? *Rev. Eaux For.* 85(12):729–737.
Duffield, J. W., and E. B. Snyder. 1958. Benefits from hybridizing American forest trees. *J. For.* 56:809–815.
Duffy, P. J. B. 1975. Information requirements of land managers on large projects with potential for major environmental impact. *In* B. Bernier and C. H. Winget (eds.), *Forest Soils and Forest Land Management*. Laval Univ. Press, Quebec.
Duggar, Benjamin M. (ed.). 1936. *Biological Effects of Radiation*. 2 vols. McGraw-Hill Book Co., Inc., New York. 1343 pp.
Duvigneaud, P. 1968. Recherches sur l'écosystème forêt. La Chênaie-Frenaie a Coudrier du Bois de Wève. Aperçu sur la biomasse, la productivité et le cycle des éléments biogènes. *Bull. Soc. Roy. Botan. Belg.* 101:111–127.
———, and S. Denaeyer-De Smet. 1968. Biomass, productivity and mineral cycling in deciduous mixed forests in Belgium. *In* H. E. Young (ed.), *Symposium on Primary Productivity and Mineral Cycling in Natural Ecosystems*. Univ. Maine Press, Orono, Maine.
———, and ———. 1970. Biological cycling of minerals in temperate deciduous forests. *In* David E. Reichle (ed.), *Analysis of Temperate Forest Ecosystems*. Springer-Verlag, New York.
Dyrness, C. T. 1967. Mass soil movements in the H. J. Andrews Experimental Forest. USDA For. Serv. Res. Paper PNW–42. Pacific Northwest For. and Rge. Exp. Sta., Portland, Ore. 12 pp.

———. 1976. Effect of wildfire on soil wettability in the high Cascades of Oregon. USDA For. Serv. Res. Paper PNW–202. Pacific Northwest For. and Rge. Exp. Sta., Portland, Ore. 18 pp.

———, Jerry F. Franklin, and W. H. Moir. 1974. A preliminary classification of forest communities in the central portion of the Western Cascades in Oregon. Conif. For. Biome Bull. No. 4, 123 pp. Coniferous Forest Biome, US/IBP, Univ. Washington, Seattle, Wash.

Ebell, L. F., and R. L. Schmidt. 1964. Meteorological factors affecting conifer pollen dispersal on Vancouver Island. Canada, Dept. For. Publ. No. 1036. 28 pp.

Eberhardt, E., D. Kopp, and H. Passarge. 1967. Standorte und Vegetation des Kirchleerauer Waldes im Schweizerischen Mittelland. *In* Heinz Ellenberg (ed.), *Vegetations– und bodenkundliche Methoden der forstlichen Standortskartierung*. Geobotanische Institute, ETH, Zürich.

Edmonds, R. L. (ed.). 1974. An Initial Synthesis of Results in the Coniferous Forest Biome, 1970–1973. Conif. For. Biome Bull. No. 7, Coniferous Forest Biome, US/IBP, Univ. Washington, Seattle, Wash.

Edmonds, R. L. (ed.). 1980. *Analysis of Coniferous Forest Ecosystems in the Western United States*. US/IBP Synthesis Series. Dowden, Hutchinson and Ross, Inc., Stroudsburg, Pa.

Edwards, C. A. 1974. Macroarthropods. *In* C. H. Dickinson and G. J. F. Pugh (eds.), *Biology of Plant Litter Decomposition*. Academic Press, Inc., New York.

———, D. E. Reichle, and D. A. Crossley, Jr. 1970. The role of soil invertebrates in turnover of organic matter and nutrients. *In* David E. Reichle (ed.), *Analysis of Temperate Forest Ecosystems*. Springer-Verlag, New York.

Edwards, N. T., and W. F. Harris. 1977. Carbon cycling in a mixed deciduous forest floor. *Ecology* 58:431–437.

Egler, Frank E. 1968. The contumacious continuum. *In* Pierre Dansereau (ed.), The continuum concept of vegetation: responses. *Bot. Rev.* 34:253–332.

Ehrlich, Paul R., and Peter H. Raven. 1969. Differentiation of populations. *Science* 165:1228–1232.

Eidmann, F. E. 1959. Die Interception in Buchen- und Fichtenbeständen; Ergebnis mehrjähriger Untersuchungen im Rothaargebirge, Sauerland. *In Proc. Hannoveresch-Münden Symposium, Eau et Forêts*, pp. 8–25. Publ. Inter. Assoc. Sci. Hydrol. 48.

Eis, S. 1976. Association of western white pine cone crops with weather variables. *Can. J. For. Res.* 6:6–12.

———, E. H. Garman, and L. F. Ebell. 1965. Relation between cone production and diameter increment of Douglas-fir (*Pseudotsuga menziesii* (Mirb.) Franco), grand fir (*Abies grandis* (Dougl.) Lindl.), and western white pine (*Pinus monticola* Dougl.). *Can. J. Bot.* 43:1553–1559.

Ellenberg, Heinz. 1956. *Aufgaben und Methoden der Vegetationskunde*. Eugen Ulmer, Stuttgart. 136 pp.

———. 1963. *Vegetation Mitteleuropas mit den Alpen*. Eugen Ulmer, Stuttgart. 943 pp.

———. 1968. Wege der Geobotanik zum Verständnis der Pflanzendecke.

*Naturwissenschaften* 55:462–470.
———. 1971a. *Integrated Experimental Ecology*. Springer-Verlag, New York. 214 pp.
———. 1971b. Introductory survey. *In* H. Ellenberg (ed.), *Integrated Experimental Ecology*. Springer-Verlag, New York.
———, and D. Mueller-Dombois. 1966. Tentative physiolognomic-ecological classification of plant formations of the earth. *Ber. gebot. Forsch. Inst. Rübel* 37:21–55.
———, and ———. 1970. Geographical index of world ecosystems. *In* David E. Reichle (ed.), *Analysis of Temperate Forest Ecosystems*. Springer-Verlag, New York.
Ellis, Jack A., William R. Edwards, and Keith P. Thomas. 1969. Responses of bobwhites to management in Illinois. *J. Wildl. Manage.* 33:749–762.
Engler, Arnold. 1905. Einfluss der Provenienz des Samens auf die Eigenschaften der forstlichen Holzgewächse. *Mitt. schweiz. Centralanst. forstl. Versuchsw.* 8:81–236.
———. 1908. Tatsachen, Hypothesen und Irrtümer auf dem Gebiete der Samenprovenienz-Frage. *Forstwissenschaftliches Centralblatt* 30:295–314.
Erdmann, Gayne G., and Robert R. Oberg. 1974. Sapsucker feeding damages crown-released yellow birch trees. *J. For.* 72:760–763.
Esau, Katherine. 1977. *Anatomy of Seed Plants*. John Wiley & Sons, Inc. New York. 550 pp.
Eschner, Arthur R. 1967. Interception and soil moisture distribution. *In* William E. Sopper and Howard W. Lull (eds.), *Forest Hydrology*. Pergamon Press, Inc., New York.
Evans, G. G. 1956. An area survey method of investigating the distribution of light intensity in woodlands with particular reference to sunflecks. *J. Ecol.* 44:391–428.
Evers, Fritz Helmut. 1967. Kohlenstoffbezogene Nährelementverhältnisse (C/N, C/P, C/K, C/Ca) zur Charakterisierung der Ernährungssituation in Waldböden. *Mitt. Vereins Forstl. Standortk. Forstpflz.* 17:69–76.
———. 1968. Die Zusammenhänge zwischen Stickstoff-, Phosphor- und Kalium-Mengen (in kg/ha) und den C/N-, C/P- und C/K- Verhältnissen der Oberböden von Waldstandorten. *Mitt. Vereins forstl. Standortk. Forstpflz.* 18:59–71.
———. 1971. Untermauerung der Standortsgliederung durch Laboruntersuchungen. *Allgem. Forstzeitschr.* 26:443.
Eyre, S. R. 1962. *Vegetation and Soils. A World Picture*. Edward Arnold, Ltd., London. 324 pp.
Faegri, K., and L. van der Pijl. 1971. *The Principles of Pollination Ecology*. Pergamon Press, New York. 291 pp.
Fahn, Abraham, and Ella Werker. 1972. Anatomical mechanisms of seed dispersal. *In* T. T. Kozlowski (ed.), *Seed Biology*, Vol. 1. Academic Press, Inc., New York.
Farmer, Michele Marlene. 1977. Variation and identification of hybrid and backcross populations of *Populus grandidentata* and *P. tremuloides*. Master's Thesis. University of Michigan. 169 pp.

Federer, C. A., and C. B. Tanner. 1966. Spectral distribution of light in the forest. *Ecology* 47:555–560.

Fennoscandian Forestry Union. 1962. A symposium on forest land and classification of site in the Fennoscandian countries. (Translation of the paper, Skogsmark och Bonitering i de Nordiska Landerna. *Svenska Skogsvardsforeningens Tidskr.* 52:189–226.) U.S. Forest Service, Washington, D.C. 42 pp.

Ferguson, C. W. 1968. Bristlecone pine: science and esthetics. *Science* 159:839–846.

Fiedler, H. J., and W. Hunger. 1967. Grundzüge und Arbeitsgebiete der standortskundlichen Forschung am Institut für Bodenkunde und Standortslehre zu Tharandt aus der Zeit von 1945 bis 1965. *Mitt. Vereins forstl. Standortsk. Forstpflz.* 17:77–85.

Findlay, B. F. 1976. Recent developments in eco-climatic classifications. *In* J. Thie and G. Ironside (eds.), *Ecological (Biophysical) Land Classification in Canada.* Lands Directorate, Environment Canada, Ottawa.

Firbas, Franz. 1949. *Spät- und nacheiszeitliche Waldgeschichte Mitteleuropas nördlich der Alpen. I. Allgemeine Waldgeschichte.* Verlag Gustav Fischer, Jena. 480 pp.

———. 1952. *Spät- und nacheiszeitliche Waldgeschichte Mitteleuropas nördlich der Alpen. II. Waldgeschichte der einzelnen Landschaften.* Verlag Gustav Fischer, Jena. 256 pp.

Fisher, R. F., and E. L. Stone. 1969. Increased availability of nitrogen and phosphorus in the root zone of conifers. *Soil Sci. Soc. Amer. Proc.*, 33:955–961.

Flint, Howard R. 1930. Fire as a factor in the management of north Idaho national forests. *Northwest Sci.* 4:12–15.

Foggin, G. Thomas, III, and Leonard F. DeBano. 1971. Some geographic implications of water-repellent soils. *Prof. Geographer* 23:347–350.

Fons, Wallace L. 1940. Influence of forest cover on wind velocity. *J. For.* 38:481–486.

Fortescue, J. A. C., and G. G. Marten. 1970. Micronutrients: forest ecology and systems analysis. *In* David E. Reichle (ed.), *Analysis of Temperate Forest Ecosystems*. Springer-Verlag, New York.

Foster, R. W. 1959. Relation between site indexes of eastern white pine (*Pinus strobus*) and red maple (*Acer rubrum*). *For. Sci.* 5:279–291.

Foth, Henry D. 1978. *Fundamentals of Soil Science*. 6th ed. John Wiley & Sons, New York. 436 pp.

Fowells, H. A. 1948. The temperature profile in a forest. *J. For.* 46:897–899.

Fowler, D. P. 1965a. Effects of inbreeding in red pine, *Pinus resinosa* Ait. II. Pollination studies. *Silvae Genetica* 14:12–23.

———. 1965b. Effects of inbreeding in red pine, *Pinus resinosa* Ait. IV. Comparison with other Northeastern *Pinus* species. *Silvae Genetica* 14:76–81.

———, and C. Heimburger. 1969. Geographic variation in eastern white pine, 7-year results in Ontario. *Silvae Genetica* 18:123–129.

———, and R. E. Mullin. 1977. Upland-lowland ecotypes not well developed in black spruce in northern Ontario. *Can. J. For. Res.* 7:35–40.

Franklin, E. C. 1970. Survey of mutant forms and inbreeding depression in species of the family Pinaceae. USDA For. Serv. Res. Paper SE-61. Southeast For. Exp. Sta., Asheville, N. C. 21 pp.

Franklin, Jerry F. 1965. Tentative ecological provinces within the true fir-hemlock forest areas of the Pacific Northwest. USDA For. Serv. Res. Note PNW–22. Pacific Northwest For. and Rge. Exp. Sta., Portland, Ore. 31 pp.

———. 1966. Vegetation and soils in the subalpine forests of the southern Washington Cascade Range. Ph.D. dissertation, Washington State University, Dissertation Abstr. 27(6), No. 66–13, 558.

———. 1968. Cone production by upper-slope conifers. USDA For. Serv. Res. Paper PNW–60. Pacific Northwest For. and Rge. Exp. Sta., Portland, Ore. 21 pp.

———. 1972. Why a coniferous forest biome? *In* Jerry F. Franklin, L. J. Dempster, and Richard H. Waring (eds.), *Proc.—Research on Coniferous Forest Ecosystems—A Symposium*. USDA For. Serv., Pacific Northwest For. and Rge. Exp. Sta., Portland, Ore.

———, and C. T. Dyrness. 1969. Vegetation of Oregon and Washington. USDA For. Serv. Res. Note PNW–80. Pacific Northwest For. & Rge. Exp. Sta., Portland, Ore. 216 pp.

———, and ———. 1973. Natural vegetation of Oregon and Washington. USDA For. Serv. Gen. Tech. Report PNW–8. Pacific Northwest For. and Rge. Exp. Sta., Portland, Ore. 417 pp.

———, Kermit Cromack, Jr., William Denison, Arthur McKee, Chris Maser, James Sedell, Fred Swanson, and Glenn Juday. 1979. Ecological characteristics of old-growth forest ecosystems in the Douglas-fir region. USDA For. Serv. Gen. Tech. Report. In press. Pacific Northwest For. and Rge. Exp. Sta., Portland, Ore.

Franklin, Rudolph T. 1970. Insect influences on the forest canopy. *In* David E. Reichle (ed.), *Analysis of Temperate Forest Ecosystems*. Springer-Verlag, New York.

Fraser, D. A. 1956. Ecological studies of forest trees at Chalk River, Ontario, Canada. II. Ecological conditions and radial increment. *Ecology* 37:777–789.

———. 1962. Tree growth in relation to soil moisture. *In* T. T. Kozlowski (ed.), *Tree Growth*. The Ronald Press Co., New York.

Fredriksen, R. L. 1970. Erosion and sedimentation following road construction and timber harvest on unstable soils in three small western Oregon watersheds. USDA For. Serv. Res. Paper PNW–104. Pacific Northwest For. and Rge. Exp. Sta., Portland, Ore. 15 pp.

———. 1972. Nutrient budget of a Douglas-fir forest on an experimental watershed in western Oregon. *In* Jerry F. Franklin, L. J. Dempster, and Richard H. Waring (eds.), *Proc.—Research on Coniferous Forest Ecosystems—A Symposium*. USDA For. Serv. Pacific Northwest For. and Rge. Exp. Sta., Portland, Ore.

———, D. G. Moore, and L. A. Norris. 1975. The impact of timber harvest, fertilization, and herbicide treatment on streamwater quality in western

Oregon and Washington. *In* B. Bernier and C. H. Winget (eds.), *Forest Soils and Forest Land Management*. Laval Univ. Press, Quebec.

Freeland, W. J., and Daniel H. Janzen. 1974. Strategies in herbivory by mammals: the role of plant secondary compounds. *Am. Naturalist* 108:269–289.

Frenzel, B. 1968. *Grundzüge der pleistozänen Vegetationsgeschichte Nordeuropas*. Franz Steiner Verlag, Wiesbaden. 326 pp.

Frissell, Sidney S., Jr. 1973. The importance of fire as a natural ecological factor in Itasca State Park, Minnesota. *Quaternary Res.* 3:397–407.

Fritts, Harold C. 1971. Dendroclimatology and dendroecology. *Quaternary Res.* 1:419–449.

———. 1976. *Tree Rings and Climate*. Academic Press, Inc., New York. 567 pp.

———, D. G. Smith, and M. A. Stokes. 1965. The biological model for paleoclimatic interpretation of Mesa Verde tree-ring series. *Amer. Antiquity* 31:101–121.

Froehlich, H. A. 1973. Natural and man-caused slash in headwater streams. Loggers Handb., Vol. 33, 8 pp. Pac. Logging Congr., Portland, Ore.

———, D. McGreer, and J. R. Sedell. 1972. Natural debris within the stream environment. Coniferous For. Biome. U.S. Int. Biol. Program. Int. Rep. 96, Univ. Wash., Seattle. 10 pp.

Fuller, H. J. 1948. Carbon dioxide concentration of the atmosphere above Illinois forest and grassland. *Amer. Midl. Nat.* 39:247–249.

Fuller, W. H., Stanton Shannon, and P. S. Burgess. 1955. Effect of burning on certain forest soils of northern Arizona. *For. Sci.* 1:44–50.

Furniss, Malcolm. 1972. Observations on resistance and susceptibility to Douglas-fir beetles. *In* Program Abstracts, Second North American For. Biol. Workshop, Oregon State Univ., Corvallis. p. 24.

Gaffney, William S. 1941. The effects of winter elk browsing, south fork of the Flathead River, Montana. *J. Wildl. Manage.* 5:427–453.

Gaiser, R. N. 1952. Readily available water in forest soils. *Proc. Soil Sci. Soc. Amer.* 16:334–338.

———, and R. W. Merz. 1951. Stand density as a factor in estimating white oak site index. *J. For.* 49:572–574.

Galloway, James N., Gene E. Likens, and Eric S. Edgerton. 1976. Acid precipitation in the northeastern United States: pH and acidity. *Science* 194:722–724.

Garner, W. W. 1923. Further studies in photoperiodism in relation to hydrogen-ion concentration of the cell-sap and the carbohydrate content of the plant. *J. Agr. Res.* 23:871–920.

———, and H. A. Allard. 1920. Effect of the relative length of day and night and other factors of the environment on growth and reproduction in plants. *J. Agr. Res.* 18:553–606.

Gary, Howard L. 1972. Rime contributes to water balance in high-elevation aspen forests. *J. For.* 70:93–97.

Gashwiler, Jay S. 1967. Conifer seed survival in a western Oregon clearcut. *Ecology* 48:431–438.

Gates, David M. 1962. *Energy Exchange in the Biosphere*. Harper & Row, Inc., New York. 151 pp.

———. 1965. Heat transfer in plants. *Sci. Amer.* 213:76–84.

———. 1968. Energy exchange between organism and environment. *In* William P. Lowry (ed.), *Biometeorology*. Oregon State Univ. Press, Corvallis, Ore.

———. 1970. Physical and physiological properties of plants. *In Remote Sensing, with Special Reference to Agriculture and Forestry*. National Academy of Sciences, National Res. Council, Washington, D.C.

Gates, F. C. 1914. Winter as a factor in the xerophylly of certain evergreen ericads. *Bot. Gaz.* 57:445–489.

———. 1942. The bogs of northern lower Michigan. *Ecol. Monogr.* 12:213–254.

Geiger, R. 1950. *The Climate near the Ground*. Translation by M. N. Stewart and others of the 2nd German edition of *Das Klima der bodennahen Luftschicht*. Harvard Univ. Press, Cambridge, Mass. 482 pp.

Genys, John B. 1968. Intraspecific variation among 200 strains of two-year old Norway spruce. *In Proc. Eleventh Meeting, Comm. For. Tree Breeding in Canada*, pp. 195–203. MacDonald College, Quebec, Canada.

George, M. F., M. J. Burke, H. M. Pellet, and A. G. Johnson. 1974. Low temperature exotherms and woody plant distribution. *Hort. Sci.* 6:519–522.

Gessel, S. P. 1968. Progress and needs in tree nutrition research in the northwest. *In Forest Fertilization*. Tennessee Valley Authority, Muscle Shoals, Ala.

———. 1972. Organization and research program of the Western Coniferous Forest Biome. *In* Jerry F. Franklin, L. J. Dempster, and Richard H. Waring (eds.), *Proc.—Research on Coniferous Forest Ecosystems—A Symposium*. USDA For. Serv., Pacific Northwest For. and Rge. Exp. Sta., Portland, Ore.

Ghent, A. W. 1958. Studies of regeneration in forest stands devastated by the spruce budworm. II. Age, height, growth and related studies of balsam fir seedlings. *For. Sci.* 4:135–146.

Gholz, H. L., F. K. Fitz, and R. H. Waring. 1976. Leaf area difference associated with old-growth forest communities in the western Oregon Cascades. *Can. J. For. Res.* 6:49–57.

Gibbons, Dave R., and Ernest O. Salo. 1973. An annotated bibliography of the effects of logging on fish of the western United States and Canada. USDA For. Serv. Gen. Tech. Report PNW–10. Pacific Northwest For. and Rge. Exp. Sta., Portland, Ore. 145 pp.

Gilbert, J. M. 1959. Forest succession in the Florentine Valley, Tasmania. *Pap. Proc. Royal Soc. Tasmania* 93:129–151.

Gill, C. J. 1970. The flooding tolerance of woody species—a review. *For. Abstr.* 31:671–688.

Gilmour, J. S. L., and J. W. Gregor. 1939. Demes: a suggested new terminology. *Nature* 144:333–334.

Gindel, I. 1966. Attraction of atmospheric water by woody xerophytes in desert and semidesert conditions. *Emp. For. Rev.* 48:217–242.

———. 1973. *A New Ecophysiological Approach to Forest-Water Relationships in Arid Climates*. W. Junk B. V., Publishers, The Hague. 142 pp.
Gleason, H. A. 1926. The individualistic concept of the plant association. *Bull. Torrey Bot. Cl.* 53:7–26.
———. 1939. The individualistic concept of the plant association. *Amer. Midl. Nat.* 21:92–110.
Godwin, G. E. 1968. The influence of wind on forest management and planning. *In* R. W. V. Palmer (ed.), *Wind Effects on the Forest*. Suppl. to Forestry, Oxford Univ. Press. London.
Godwin, Harry. 1975. *The History of the British Flora: A Factual Basis for Phytogeography*, 2nd ed. Cambridge Univ. Press, Cambridge. 541 pp.
Good, Norma Frauendorf. 1968. A study of natural replacement of chestnut in six stands in the Highlands of New Jersey. *Bull. Torrey Bot. Club* 95:240–253.
Good, Ronald O. 1974. *The Geography of Flowering Plants*. 4th ed. Longman, London. 557 pp.
Goodell, B. C. 1958. A preliminary report on the first year's effects of timber harvesting on water yield from a Colorado watershed. USDA Rocky Mt. For. & Rge. Exp. Sta., Sta. Paper 36. 12 pp.
———. 1967. Watershed treatment effects on evapotranspiration. *In* William E. Sopper and Howard W. Lull (eds.), *Forest Hydrology*. Pergamon Press, Inc., New York.
Gordon, Alan. G. 1976. The taxonomy and genetics of *Picea rubens* and its relationship to *Picea mariana*. *Can. J. Bot.* 54:781–813.
———, and Eville Gorham. 1963. Ecological aspects of air pollution from an iron-sintering plant at Wawa, Ontario. *Can. J. Bot.* 41:1063–1078.
Gordon, Robert B. 1969. The natural vegetation of Ohio in pioneer days. Bull. Ohio Biol. Surv., New Series, Vol. III, No. 2. Ohio State Univ., Columbus. 113 pp.
Gorham, E. and A. G. Gordon. 1960a. Some effects of smelter pollution northeast of Falconbridge, Ontario. *Can. J. Bot.* 38:307–312.
———, and A. G. Gordon. 1960b. The influence of smelter fumes upon the chemical composition of lake waters near Sudbury, Ontario, and upon the surrounding vegetation. *Can. J. Bot.* 38:477–487.
Gosz, James R., Gene E. Likens, and F. Herbert Bormann. 1976. Organic matter and nutrient dynamics of the forest and forest floor in the Hubbard Brook forest. *Oecologia* 22:305–320.
Graber, R. E. 1970. Natural seed fall in white pine (*Pinus strobus* L.) stands of varying density. USDA For. Serv. Res. Note NE-119. Northeastern For. Exp. Sta., Upper Darby, Pa. 6 pp.
Graham, B. F., Jr., and F. H. Bormann. 1966. Natural root grafts. *Bot. Rev.* 32:255–292.
Graham, Samuel A. 1941. Climax forests of the upper peninsula of Michigan. *Ecology* 22:355–362.
———. 1954. Scoring tolerance of forest trees. *Michigan For.* Note 4. Univ. Michigan, Ann Arbor. 2 pp.
———. 1955. An ecological classification of vegetation types. *Michigan For.* Note 11. Univ. Michigan, Ann Arbor. 2 pp.

———, Robert P. Harrison, Jr., and Casey E. Westell, Jr. 1963. *Aspens: Phoenix trees of the Great Lakes Region*. Univ. Michigan Press, Ann Arbor. 272 pp.

Grandtner, M. M. 1966. La végétation forestière du Québec méridional. Laval Univ. Press, Quebec. 216 pp.

Grant, Verne. 1963. *The Origin of Adaptations*. Columbia Univ. Press, New York. 606 pp.

———. 1971. *Plant Speciation*. Columbia Univ. Press, New York.

———. 1977. *Organismic Evolution*. W. H. Freeman and Co., San Francisco. 418 pp.

Grasovsky, Amihud. 1929. Some aspects of light in the forest. Yale For. Bull. 23. 53 pp.

Gratkowski, H. J. 1956. Windthrow around staggered settings in old-growth Douglas-fir. *For. Sci*. 2:60–74.

Greenland, D. J., and J. M. L. Kowal. 1960. Nutrient content of the moist tropical forests of Ghana. *Plant & Soil* 12:154–174.

Gregor, J. W., and Patricia J. Watson. 1961. Ecotypic differentiation: observations and reflections. *Evolution* 15:166–173.

Greig-Smith, P., M. P. Austin, and T. C. Whitmore. 1967. The application of quantitative methods to vegetation survey. I. Association-analysis and principal component ordination of rain forest. *J. Ecol*. 55:483–503.

Grier, Charles C. 1975. Wildfire effects on nutrient distribution and leaching in a coniferous ecosystem. *Can. J. For. Res*. 5:599–607.

———, D. W. Cole, C. T. Dyrness, and R. L. Fredriksen. 1974. Nutrient cycling in 37– and 450-year-old Douglas-fir ecosystems. *In* R. H. Waring and R. L. Edmonds (eds.), *Integrated Research in the Coniferous Forest Biome*. Conif. For. Biome Bull. No. 5, Coniferous Forest Biome, US/IBP, Univ. Washington, Seattle, Wash.

———, and Robert S. Logan. 1977. Old-growth *Pseudotsuga menziesii* communities of a western Oregon watershed: biomass distribution and production budgets. *Ecol. Monogr*. 47:373–400.

———, and Steven W. Running. 1977. Leaf area of mature northwestern coniferous forests: relation to site water balance. *Ecology* 58:893–899.

Griffin, James R. 1965. Digger pine seedling response to serpentinite and nonserpentinite soil. *Ecology* 46:801–807.

Griffith, A. L., and R. S. Gupta. 1948. Soils in relation to teak with special reference to laterisation. Indian For. Bull. 141. 58 pp.

Griffith, B. G., E. W. Hartwell, and T. E. Shaw. 1930. The evolution of soils as affected by the old field white pine-mixed hardwood succession in central New England. Harv. For. Bull. 15. 82 pp.

Grime, J. P. 1965a. Comparative experiments as a key to the ecology of flowering plants. *Ecology* 46:513–515.

———. 1965b. Shade tolerance in flowering plants. *Nature* 208:161–163.

———. 1966. Shade avoidance and shade tolerance in flowering plants. *In* Richard Bainbridge, G. Clifford Evans, and Oliver Rackham (eds.), *Light as an Ecological Factor*. Blackwell Sci. Publ., Oxford, England.

———, and D. W. Jeffrey. 1965. Seedling establishment in vertical gradients of sunlight. *J. Ecol*. 53:621–642.

Gross, H. L. 1972. Crown deterioration and reduced growth associated with

excessive seed production by birch. *Can. J. Bot.* 50:2431–2437.
———, and A. A. Harnden. 1968. Dieback and abnormal growth of yellow birch induced by heavy fruiting. Canada Dept. For. and Rural Develop., For. Res. Lab. Info. Report 0–X–79, Sault Ste. Marie, Ontario. 12 pp.
Grüger, E. 1972. Pollen and seed studies of Wisconsinan vegetation in Illinois, U.S.A. *Geol. Soc. Am. Bull.* 83:2715–2734.
Grüger, J. 1973. Studies on the late Quaternary vegetation history of northeastern Kansas. *Geol. Soc. Am. Bull.* 84:239–250.
Guderian, R. 1977. *Air Pollution*. Springer-Verlag, New York. 127 pp.
Gullion, Gordon W. 1972. Improving your forested lands for ruffed grouse. Ruffed Grouse Soc. No. Am., Rochester, N.Y. 34 pp. (*Minn. Agr. Exp. Sta., Misc. J. Ser., Publ.* 1439).
Günther, M. 1955. Untersuchungen über das Ertragsvermögen der Hauptholzarten im Bereich verschiedener Standortseinheiten der württembergischen Neckarlandes. *Mitt. Vereins forstl. Standortsk.* 4:5–31.
Gustafson, Felix G. 1943. Influence of light upon tree growth. *J. For.* 41:212–213.
Gutschick, V. 1950. *Forstliche Standortskunde als Grundlage für den praktischen Waldbau*. Verlag M. & H. Schaper, Hannover. 259 pp.
Haase, Edward F. 1970. Environmental fluctuations on south-facing slopes in the Santa Catalina Mountains of Arizona. *Ecology* 51:959–974.
Habeck, J. R. 1958. White cedar ecotypes in Wisconsin. *Ecology* 39:457–463.
———. 1960. Winter deer activity in the white cedar swamps of northern Wisconsin. *Ecology* 41:327–333.
———, and Robert W. Mutch. 1973. Fire-dependent forests in the northern Rocky Mountains. *Quaternary Res.* 3:408–424.
Hack, John T., and John C. Goodlett. 1960. Geomorphology and forest ecology of a mountain region in the central Appalachians. U.S. Geol. Surv. Prof. Paper 347. 66 pp.
Hacskaylo, John, R. F. Finn, and J. P. Vimmerstedt. 1969. Deficiency symptoms of some forest trees. Ohio Agr. Res. and Dev. Cent. Res. Bull. 1015. 68 pp.
Haden-Guest, Stephen, John K. Wright, and Eileen M. Teclaff (eds.). 1956. *A World Geography of Forest Resources*. Amer. Geog. Soc. Spec. Publ. 33. The Ronald Press Co., New York. 736 pp.
Hagar, Donald C. 1960. The interrelationships of logging, birds, and timber regeneration in the Douglas-fir region of northwestern California. *Ecology* 41:116–125.
Hagem, O. 1947. The dry matter increase of coniferous seedlings in winter: investigations in oceanic climate. *Medd. Vestland. Forstl. Forsoksstа.* 8(1):317 pp.
Halliday, W. E. D. 1937. A forest classification for Canada. Can. Dept. Resources and Develop. Bull. 89. 56 pp.
Hamilton, E. L. 1954. Rainfall sampling on rugged terrain. U.S. Dept. Agr. Tech. Bull. 1096. 41 pp.
Hampf, Frederick E. (ed.). 1965. Site index curves for some forest species in the eastern United States. USDA For. Serv., Eastern Region. Upper Darby, Pa. 43 pp.
Handley, Charles O., Jr. 1969. Fire and mammals. *In Proc. Annual Tall*

*Timbers Fire Ecology Conference* 9:151–159. Tall Timbers Res. Sta., Tallahassee, Fla.

Hannah, Peter R. 1969. Stemwood production related to soils in Michigan red pine plantations. *For. Sci.* 15:320–326.

Hanover, James W. 1975. Physiology of tree resistance to insects. *Ann. Rev. Entomol.* 20:75–95.

———, and R. C. Wilkinson. 1970. Chemical evidence for introgressive hybridization in *Picea*. *Silvae Genetica* 19:17–22.

Hansen, H. P. 1947. Postglacial forest succession, climate, and chronology in the Pacific Northwest. *Trans. Amer. Philosoph. Soc.* 37(1). 130 pp.

———. 1948. Postglacial forests of the Glacier National Park region. *Ecology* 29:146–152.

———. 1955. Postglacial forests in south central and central British Columbia. *Amer. J. Sci.* 253:640–658.

Hanson, Herbert. 1962. *Dictionary of Ecology*. Peter Owen, London. 382 pp.

Hanson, Herbert C. 1917. Leaf-structure as related to environment. *Amer. J. Bot.* 4:533–560.

Hardin, James W. 1975. Hybridization and introgression in *Quercus alba*. *J. Arnold Arb.* 56:336–363.

Hare, F. Kenneth. 1950. Climate and zonal divisions of the boreal forest formation in eastern Canada. *Geogr. Rev.* 40:615–635.

Harley, J. L. 1939. The early growth of beech seedlings under natural and experimental conditions. *J. Ecol.* 27:384–400.

———. 1969. *The Biology of Mycorrhiza*. Leonard Hill, London. 334 pp.

———, *et al.* 1950–1955. The uptake of phosphate by excised mycorrhizal roots of the beech. *New Phytologist* 49:388–397; 51:56–64, 342–348; 52:124–132; 53:92–98, 240–252; 54:296–301.

Harper, J. L. 1965. Establishment, aggression, and cohabitation in weedy species. *In* H. G. Baker and G. L. Stebbins (eds.), *The Genetics of Colonizing Species*. Academic Press, Inc., New York.

———. 1977. *Population Biology of Plants*. Academic Press, Inc., New York. 892 pp.

———, and J. White. 1974. The demography of plants. *Ann. Rev. Ecol. Syst.* 5:419–463.

Harr, R. Dennis. 1976. Forest practices and streamflow in western Oregon. USDA For. Serv. Gen. Tech. Report PNW–49. Pacific Northwest For. and Rge. Exp. Sta., Portland, Ore. 18 pp.

Harris, P. 1960. Production of pine resin and its effect on survival of *Rhyacionia beolina*. *Can. J. Zool.* 38:121–130.

Harris, T. M. 1958. Forest fire in the mesozoic. *J. Ecology*. 46:447–453.

Harris, W. F. 1977. Walker Branch watershed: site description and research scope. *In* David L. Correll (ed.), *Watershed Research in Eastern North America: A Workshop to Compare Results*. Vol. 1. Smithsonian Inst., Edgewater, Md.

———, R. A. Goldstein, and G. S. Henderson. 1973. Analysis of forest biomass pools, annual primary production and turnover of biomass for a mixed deciduous forest watershed. *In* Harold Young (ed.), *Proc. Working Party on Forest Biomass of IUFRO*, pp. 41–64. Univ. Maine Press, Orono.

———, P. Sollins, N. T. Edwards, B. E. Dinger, and H. H. Shugart. 1975.

Analysis of carbon flow and productivity in a temperate deciduous forest ecosystem. *In* D. E. Reichle, J. F. Franklin, and D. W. Goodall (eds.), *Productivity of World Ecosystems*. Natl. Acad. Sci., Washington, D.C.

———, G. S. Henderson, and D. E. Todd. 1976. Disposal of industrial nitrate effluents by means of forest spray irrigation. *In Proc. National Conference on Disposal of Residues on Land,* pp. 166–173. U.S. Env. Prot. Agency, Wash., D.C.

———, R. S. Kinerson, Jr., and N. T. Edwards. 1977. Comparison of below-ground biomass of natural deciduous forests and loblolly pine plantations. *Pedobiol.* 17:369–381.

Hartesveldt, R. J., and H. T. Harvey. 1967. The fire ecology of sequoia regeneration. *In Proc. Annual Tall Timbers Fire Ecology Conf.*, 7:65–77. Tall Timbers Res. Sta., Tallahassee, Fla.

Hartmann, Franz. 1952. *Forstökologie.* Verlag Georg Fromme & Co., Vienna. 461 pp.

Hasenmaier, Erhardt. 1964. Versuch einer waldbaulichen Auswertung der Standortskartierung im Virngrund (Nordwürttemberg). *Mitt. Vereins forstl. Standortsk. Forstpflz.* 13:3–89.

Hatch, A. B. 1937. The physical basis of mycotrophy in the genus *Pinus. Black Rock For. Bull.* 6:1–168.

———, and C. W. Ralston. 1971. Natural recovery of surface soils disturbed in logging. *Tree Planters' Notes* 22:5–9.

Hatchell, G. E., C. W. Ralston, and R. R. Foil. 1970. Soil disturbances in logging. *J. For.* 68:772–778.

———, and C. W. Ralston. 1971. Natural recovery of surface soils disturbed in logging. *Tree Planters' Notes* 22:5–9.

Hawes, Austin F. 1923. New England forests in retrospect. *J. For.* 21:209–224.

Heath, G. W., M. K. Arnold, and C. A. Edwards. 1966. Studies in leaf litter breakdown. I. Breakdown rates among leaves of different species. *Pedobiol.* 6:1–12.

Heikurainen, Leo. 1964. Improvement of forest growth on poorly drained peat soils. *Int. Rev. For. Res.* 1:39–113.

Heiligmann, Randall, and G. Schneider. 1974. Effects of wind and soil moisture on black walnut seedlings. *For. Sci.* 20:331–335.

———, and ———. 1975. Black walnut seedling growth in wind protected microenvironments. *For. Sci.* 22:293–297.

Heinselman, M. L. 1970. Landscape evolution, peatland types, and the environment in the Lake Agassiz Peatlands Natural Area, Minnesota. *Ecol. Monogr.* 40:235–261.

———. 1973. Fire in the virgin forests of the boundary waters canoe area, Minnesota. *Quaternary Res.* 3:329–382.

Helgath, Sheila F. 1975. Trail deterioration in the Selway-Bitterroot Wilderness. USDA For. Serv. Res. Note INT–193. Intermountain For. and Rge. Exp. Sta., Ogden, Utah. 15 pp.

Hellmers, Henry. 1962. Temperature effect upon optimum tree growth. *In* TheodoreT. Kozlowski (ed.), *Tree Growth.* The Ronald Press Co., New York.

———. 1966a. Temperature action and interaction of temperature regimes in

the growth of red fir seedlings. *For. Sci.* 12:90–96.
———. 1966b. Growth response of redwood seedlings to thermoperiodism. *For. Sci.* 12:276–283.
———, and W. P. Sundahl. 1959. Response of *Sequoia sempervirens* (D. Don) Endl. and *Pseudotsuga menziesii* (Mirb. Franco) seedlings to temperature. *Nature* 184:1247–1248.
———, M. K. Genthe, and F. Ronco. 1970. Temperature affects growth and development of Engelmann spruce. *For. Sci.* 16:447–452.
Henderson, G. S., and W. F. Harris. 1975. An ecosystem approach to characterization of the nitrogen cycle in a deciduous forest watershed. *In* B. Bernier and C. H. Winget (eds.), *Forest Soils and Forest Land Management*. Proc. Fourth North Am. Forest Soils Conf. Laval Univ. Press, Quebec.
———, Arnold Hunley, and William Selvidge. 1977. Nutrient discharge from Walker Branch watershed. *In* David L. Correll (ed.), *Watershed Research in Eastern North America: A Workshop to Compare Results*, Vol. 1. Chesapeake Bay Center for Env. Studies, Smithsonian Inst., Edgewater. Md.
Henniker-Gotley, G. R. 1936. A forest fire caused by falling stones. *Indian Forester* 62:422–423.
Hermann, Richard K., and Denis P. Lavender. 1968. Early growth of Douglas-fir from various altitudes and aspects in southern Oregon. *Silvae Genetica* 17:143–151.
Heslop-Harrison, J. 1964. Forty years of genecology. *Adv. Ecol. Res.* 2:159–247.
———. 1967. *New Concepts in Flowering-Plant Taxonomy*. Harvard Univ. Press, Cambridge, Mass. 134 pp.
Hett, J. M. 1971. A dynamic analysis of age in sugar maple seedlings. *Ecology* 52:1071–1074.
Heusser, Calvin J. 1960. Late-Pleistocene environments of North Pacific North America. Amer. Geog. Soc. Spec. Publ. 35. 308 pp.
———. 1965. A Pleistocene phytogeographical sketch of the Pacific Northwest and Alaska. *In* H. E. Wright, Jr., and David G. Frey (eds.), *The Quaternary of the United States*. Princeton Univ. Press, Princeton, N.J.
———. 1969. Late-Pleistocene coniferous forests of the Northern Rocky Mountains. *In* Center for Natural Resources (ed.), Coniferous Forests of the Northern Rocky Mountains. Univ. of Montana Foundation, Missoula, Mont.
Hewetson, C. E. 1956. A discussion on the "climax" concept in relation to the tropical rain and deciduous forest. *Emp. For. Rev.* 35:274–291.
Hewlett, John D. 1967. Summary of forests and precipitation session. *In* William E. Sopper and Howard W. Lull (eds.), *Forest Hydrology*. Pergamon Press, Inc., New York.
———, and Wade L. Nutter. 1969. *An Outline of Forest Hydrology*. Univ. Georgia Press, Athens. 137 pp.
Hibbert, Alden R. 1967. Forest treatment effects on water yield. *In* William E. Sopper and Howard W. Lull (eds.), *Forest Hydrology*. Pergamon Press, Inc., New York.
Hills, G. A. 1952. The classification and evaluation of site for forestry. On-

tario Dept. Lands and For., Res. Rept. 24. 41 pp.
———. 1960. Regional site research. *For. Chron.* 36:401–423.
———. 1977. An integrated iterative holistic approach to ecosystem classification. *In* J. Thie and G. Ironside (eds.), *Ecological (Biophysical) Land Classification in Canada.* Ecological land classification series, No. 1, Lands Directorate, Env. Canada, Ottawa.
———, and G. Pierpoint. 1960. Forest site evaluation in Ontario. Ontario Dept. Lands and For., Res. Rept. 42. 64 pp.
Hilmon, J. B., and J. E. Douglass. 1968. Potential impact of forest fertilization on range, wildlife, and watershed management. *In Forest Fertilization.* Tennessee Valley Authority, Muscle Shoals, Ala.
Himelick, E. B., and Dan Neely. 1962. Root-grafting of city-planted American elms. *Plant Dis. Rep.* 46:86–87.
Hocking, D., and R. A. Blauel. 1977. Progressive heavy metal accumulation associated with forest decline near the nickel smelter at Thompson, Manitoba. Can. For. Serv. Inf. Report NOR–X–169, Northern For. Res. Cent., Edmonton, Alberta. 20 pp.
Hodges, J. D., and L. S. Pickard. 1971. Lightning in the ecology of the southern pine beetle, *Dendroctonus frontalis* (Coleoptera: Scolytidae). *Canad. Entomol.* 103:44–51.
Hodgkins, Earl J. 1961. Estimating site index for longleaf pine through quantitative evaluation of associated vegetation. *Proc. Soc. Am. For. 1960:* 28–32.
———. 1965. Southeastern forest habitat regions based on physiography. Auburn Univ., For. Dept. Series No. 2. 10 pp.
———. 1970. Productivity estimation by means of plant indicators in the longleaf pine forests of Alabama. *In* Chester T. Youngberg and Charles B. Davey (eds.), *Tree Growth and Forest Soils.* Oregon State Univ. Press, Corvallis, Ore.
Hoffman, George R., and Robert R. Alexander. 1976. Forest vegetation of the Bighorn Mountains, Wyoming: a habitat type classification. USDA For. Serv. Res. Paper RM–170. Rocky Mountain For. and Rge. Exp. Sta., Ft. Collins, Colo. 38 pp.
Holdridge, L. R. 1967. *Life Zone Ecology.* Tropical Sci. Center, San Jose, Costa Rica. 206 pp.
Holloway, John T. 1954. Forests and climates in the South Island of New Zealand. *Trans. Royal Soc. New Zealand.* 82(2):329–410.
Holmsgaard, Erik. 1955. Tree-ring analyses of Danish Forest trees. *Det. Forstl. Forsøgsvaesen i Danmark* 22:1–246.
———. 1962. Influence of weather on growth and reproduction of beech. *Commun. Inst. Forst. Fenni.* 55:1–5.
Holsworth, W. N. 1960. Interactions between moose, elk, and buffalo in Elk Island National Park, Alberta. M.S. thesis, British Columbia. 92 pp.
Hook, D. D., and Jack Stubbs. 1967. An observation of understory growth retardation under three species of oaks. USDA For. Serv. Res. Note SE–70. Southeast. For. Exp. Sta., Asheville, N.C. 7 pp.
Hoover, Marvin D. 1953. Interception of rainfall in a young loblolly pine plantation. U.S. For. Serv., Southeast. For. Exp. Sta. Paper 21. 13 pp.
———, and Charles F. Leaf. 1967. Process and significance of interception in

Colorado subalpine forest. *In* William E. Sopper and Howard W. Lull (eds.), *Forest Hydrology*. Pergamon Press, Inc., New York.

Hoover, W. H. 1937. The dependence of carbon dioxide assimilation in a higher plant on wave length of radiation. *Smithsn. Misc. Coll.*, Vol. 95. 13 pp.

Hopkins, David M. (ed.), 1967a. *The Bering Land Bridge*. Stanford Univ. Press, Stanford, Calif. 495 pp.

———. 1967b. The Cenozoic history of Beringia—a synthesis. *In* David M. Hopkins (ed.), *The Bering Land Bridge*. Stanford Univ. Press, Stanford, Calif.

Hori, T. (ed.). 1953. *Studies on Fogs in Relation to Fog-preventing Forest*. Foreign Books Dept. Tanne Trading Co., Ltd., Sapporo, Hokkaido. 399 pp.

Hornbeck, James W. 1977. Nutrients: a major consideration in intensive forest management. *In Proc. Symp. on Intensive Culture of Northern Forest Types*, pp. 241–250. USDA For. Serv. Gen. Tech. Report NE–29. Northeastern For. Exp. Sta., Upper Darby, Pa.

———, G. E. Likens, R. S. Pierce, and F. H. Bormann. 1975. Strip cutting as a means of protecting site and streamflow quality when clearcutting northern hardwoods. *In* B. Bernier and C. H. Winget (eds.), *Forest Soils and Forest Land Management*. Laval Univ. Press, Quebec.

Horsley, Stephen B. 1977. Allelopathic inhibition of black cherry by fern, grass, goldenrod, and aster. *Can. J. For. Res.* 7:205–216.

Horton, K. W. 1956. The ecology of lodgepole pine in Alberta and its role in forest succession. For. Br. Can. Tech. Note 45. 29 pp.

———. 1959. Characteristics of subalpine spruce in Alberta. Can. Dept. Northern Affairs and Nat. Res. For. Res. Div. Tech. Note No. 76. 20 pp.

Hosie, R. C. 1969. *Native Trees of Canada*, 7th ed. Can. For. Serv., Ottawa, Canada. 380 pp.

Hough, A. F. 1936. A climax forest community on East Tionesta Creek in Northwestern Pennsylvania. *Ecology* 17:1–28.

———. 1945. Frost pocket and other microclimates in forests of the northern Allegheny plateau. *Ecology* 26:235–250.

Howe, Henry F. 1977. Bird activity and seed dispersal of a tropical wet forest tree. *Ecology* 58:539–550.

Huberman, M. A. 1943. Sunscald of eastern white pine, *Pinus strobus* L. *Ecology* 24:456–471.

Huff, D. D., G. S. Henderson, C. L. Begovich, R. J. Luxmoore, and J. R. Jones. 1977. The application of analytic and mechanistic hydrologic models to the study of Walker Branch watershed. *In* David L. Correll (ed.), *Watershed Research in Eastern North America: A Workshop to Compare Results*, Vol. 1. Chesapeake Bay Center for Env. Studies, Smithsonian Inst., Edgewater, Md.

Huikari, O. 1956. Primäärisen soistumisen osuudesta Suomen soiden synnyssa. (The part played by primary bog formation in the origin of Finnish bogs.) Commun. Inst. For. Fenn. 46(6). 80 pp.

Hursch, C. R. 1948. Local climate in the Copper Basin of Tennessee as modified by the removal of vegetation. U.S. Dept. Agr. Circ. 774. 38 pp.

Husch, Bertram, Charles I. Miller, and Thomas W. Beers. 1972. *Forest Mensuration*. 2nd ed. The Ronald Press Co., New York. 410 pp.
Hutnik, R. J. and G. Davis (eds.). 1973. *Ecology and Reclamation of Devastated Land*. Vol. 1, 538 pp. Vol. 2, 504 pp. Gordon and Breach, Science Publ., Inc. New York.
Huxley, J. S. 1938. Clines, an auxiliary taxonomic principle. *Nature* 142:219–220.
———. 1939. Clines: an auxiliary method in taxonomy. *Bijdragen tot de Dierkunde* 27:491–520.
Ingestad, T. 1962. Macroelement nutrition of pine, spruce, and birch seedlings in nutrient solution. *Medd. Skogsforskninst. Stockh.* 51(7):1–131.
Irgens-Moller, H. 1968. Geographical variation in growth patterns of Douglas-fir. *Silvae Genetica* 17:106–110.
Isaac, Leo. A. 1938. Factors affecting establishment of Douglas-fir seedlings. U.S. Dept. Agr. Circ. 486. 45 pp.
———. 1946. Fog drip and rain interception in coastal forests. U.S. For. Serv., Pac. Northwest For. Rge. Exp. Sta. Paper 34. 16 pp.
———. 1956. Place of partial cutting in old-growth stands of the Douglas-fir region. U.S. For. Serv., Pac. Northwest For. Rge. Exp. Sta. Res. Paper 16. 48 pp.
Isaak, D., W. H. Marshall, and M. F. Buell. 1959. A record of reverse plant succession in a tamarack bog. *Ecology* 40:317–320.
Jackson, L. W. R. 1952. Radial growth of forest trees in the Georgia Piedmont. *Ecology* 33:336–341.
———. 1967. Effect of shade on leaf structure of deciduous tree species. *Ecology* 48:498–499.
———, and R. S. Harper. 1955. Relation of light intensity to basal area of short-leaf pine (*Pinus echinata*) stands in Georgia. *Ecology* 36:158–159.
Jackson, R. C. 1976. Evolution and systematic significance of polyploidy. *Ann. Rev. Ecol. Syst.* 7:209–234.
Jacobs, M. R. 1955. Growth habits of the eucalypts. Forestry and Timber Bur. (Canberra). 262 pp.
Jain, S. K., and A. D. Bradshaw. 1966. Evolution in closely adjacent plant populations. I. The evidence and its theoretical analysis. *Heredity* 22:407–441.
Jameson, J. S. 1965. Relation of jack pine height-growth to site in the Mixedwood Forest Section of Saskatchewan. *In* Chester T. Youngberg (ed.), *Forest-Soil Relationships in North America*. Oregon State Univ. Press, Corvallis, Ore.
Janis, Christine. 1976. The evolution strategy of the Equidae and the origins of rumen and cecal digestion. *Evolution* 30:757–774.
Janzen, Daniel H. 1966. Coevolution of mutualism between ants and acacias in Central America. *Evolution* 20:249–275.
———. 1969. Seed-eaters versus seed size, number, toxicity and dispersal. *Evolution* 23:1–27.
———. 1970. Herbivores and the number of tree species in tropical forests. *Am. Nat.* 104:501–528.
———. 1971. Seed predation by animals. *Ann. Rev. Ecol. Syst.* 2:465–492.

———. 1976. Why do bamboos wait so long to flower? *In* J. Burley and B. T. Styles (eds.), *Tropical Trees, Variation, Breeding and Conservation.* Linn. Soc. Symp. Series No. 2, Academic Press, Inc., New York.

Jarosenko, R. P. 1956. *Smeny rastitel'nogo pokrova Zakavkaz'ja. (Natural Succession in the Vegetation of Transcaucasia.)* Izdatel'stvo Akademii Nauk SSSR, Moscow and Leningrad. 242 pp.

Jensen, K. F., L. S. Dochinger, B. R. Roberts, and A. M. Townsend. 1976. Pollution responses. *In* J. P. Miksche (ed.), *Modern Methods in Forest Genetics.* Springer-Verlag, New York.

Jester, J. R., and P. J. Kramer. 1939. The effect of length of day in the height growth of certain forest tree seedlings. *J. For.* 37:796–803.

Johnson, D. W., D. W. Cole, C. S. Bledsoe, G. Carrol, K. Cromack, R. L. Edmonds, S. P. Gessel, C. C. Grier, B. N. Richards, J. Turner, and K. Vogt. 1980. Nutrient cycling in forests of the Pacific Northwest. *In* R. L. Edmonds (ed.), *Analysis of Coniferous Forest Ecosystems in the Western United States.* US/IBP Synthesis Series. Dowden, Hutchinson and Ross, Inc., Stroudsburg, Pa.

Johnson, Edward A. 1967. Effects of multiple use on peak flows and low flows. *In* William E. Sopper and Howard W. Lull (eds.), *Forest Hydrology.* Pergamon Press, Inc., New York.

———, and J. L. Kovner. 1956. Effect on streamflow of cutting a forest understory. *For. Sci.* 2:82–91.

Johnson, P. W. 1972. Factors affecting buttressing in *Triplochiton scleroxylon* K. Schum. *Ghana J. Agr. Sci.* 5:13–21.

Johnson, Philip L., and Wayne T. Swank. 1973. Studies of cation budgets in the Southern Appalachians on four experimental watersheds with contrasting vegetation. *Ecology* 54:70–80.

Johnson, W. Carter, and David M. Sharpe. 1976. An analysis of forest dynamics in the northern Georgia Piedmont. *For. Sci.* 22:307–322.

Johnstone, W. D. 1971. Total standing crop and tree component distributions in three stands of 100-year-old lodgepole pine. *In Forest Biomass Studies.* Misc. Publ. 132, Life Sci. and Agr. Exp. Sta., Univ. Maine, Orono.

Jones, John R. 1969. Review and comparison of site evaluation methods. USDA For. Serv. Res. Paper RM-51. Rocky Mountain For. and Rge. Exp. Sta., Fort Collins, Colo. 27 pp.

Jordan, Marilyn J. 1975. Effects of zinc smelter emissions and fire on a chestnut-oak woodland. *Ecology* 56:78–91.

Jorgensen, J. R., and C. S. Hodges, Jr. 1970. Microbial characteristics of a forest soil after twenty years of prescribed burning. *Mycologia* 62:721–726.

Jorgensen, Jacques R., Carol G. Wells, and Louis J. Metz. 1975. The nutrient cycle: key to continuous forest production. *J. For.* 73:400–403.

Jurdant, M., J. Beaubien, J. L. Bélair, J. C. Dionne, and V. Gerardin. 1972. *Carte écologique de la région du Saguenay/Lac-Saint-Jean. Notice explicative. Vol. 1: l'environnement et ses ressources; identification, analyse et évaluation.* Rapport d'Information Q-F-X-31, Centre de Re-

cherches forestières des Laurentides, Environnement Canada, Québec. 93 pp.
———. D. S. Lacate, S. C. Zoltai, G. G. Runka, and R. Wells. 1975. Biophysical land classification in Canada. *In* B. Bernier and C. H. Winget (eds.), *Forest Soils and Forest Land Management*. Laval Univ. Press, Quebec.
———, J. L. Bélair, V. Gerardin, and J. P. Ducruc. 1977. *L'inventaire du Capital-Nature*. Service des Études Écologiques Régionales, Direction Regionale des Terres, Pêches et Environnement Canada, Québec. 202 pp.
Kalela Erkki K. 1957. Über Veränderungen in den Wurzelverhältnissen der Kiefernbestände im Laufe der Vegetationsperiode. *Acta For. Fenn.* 65:1–41.
Katz, M. (ed.). 1939. Effect of sulfur dioxide on vegetation. Nat. Res. Counc. Canada. Publ. No. 815. 447 pp.
Kayll, A. J. 1974. Use of fire in land management. *In* T. T. Kozlowski and C. E. Ahlgren (eds.), *Fire and Ecosystems*. Academic Press, Inc., New York.
Keays, J. L. 1971. Complete-tree utilization, resumé of a literature review. *In Forest Biomass Studies*. Misc. Publ. 132, Life Sci. and Agr. Exp. Sta., Univ. Maine, Orono.
Keever, C. 1953. Present composition of some stands of the former oak, chestnut forest in the southern Blue Ridge Mountains. *Ecology* 34:44–54.
Kelley, A. P. 1950. *Mycotrophy in Plants: Lectures on the Biology of Mycorrhizae and Related Structures*. The Ronald Press Co., New York. 223 pp.
Kellogg, L. F. 1939. Site index curves for plantation black walnut in the Central States region. Central States For. Exp. Sta., Res. Note 35. 3 pp.
Kemperman, Jerry A., and Burton V. Barnes. 1976. Clone size in American aspens. *Can. J. Bot.* 54:2603–2607.
Kerfoot, O. 1968. Mist precipitation on vegetation. *For. Abstracts.* 29(1):8–20.
Kershaw, Kenneth A. 1964. *Quantitative and Dynamic Ecology*. American Elsevier Publishing Co., New York. 183 pp.
Khoshoo, T. N. 1959. Polyploidy in gymnosperms. *Evolution* 13:24–39.
Kienitz, H. 1879a. Vergleichende keimversuche mit Waldbaum-Samen aus klimatisch verschiedenen Orten Mitteleuropas. Bot. Unters. herausgegeben von N. J. C. Müller 2.
———. 1879b. Ueber Formen and Abarten heimischer Waldbäume. *Forstl. Zeitschr.* 1.
Kilburn, Paul D. 1957. Historical development and structure of the aspen, jack pine and oak vegetation types on sandy soils in northern Lower Michigan. Ph.D. thesis, University of Michigan, Ann Arbor. 267 pp.
Kilgore, Bruce M. 1972. Fire's role in a sequoia forest. *Naturalist* 23:26–37.
———. 1973. The ecological role of fire in Sierran conifer forests: its application to national park management. *Quaternary Res.* 3:496–513.
———. 1975. Restoring fire to national park wilderness. *Am. Forests* 81:16–19.
———. 1976a. Fire management in the national parks: an overview. *In Proc.*

*Annual Tall Timbers Fire Ecology Conference and Fire and Land Management Symposium*, pp. 45–57. Tall Timbers Res. Sta., Tallahassee, Fla.

———. 1976b. America's renewable resource potential—1975: the turning point. *Proc. Soc. Amer. For.* 1975: 178–188.

———. 1976c. From fire control to fire management: an ecological basis for policies. *In Trans. 41st North American Wildlife and Natural Resources Conference*, pp. 477–493. Wildlife Management Institute, Washington, D.C.

———, and Rodney W. Sando. 1975. Crown-fire potential in a sequoia forest after prescribed burning. *For. Sci.* 21:83–87.

Kincer, J. B. 1933. Is our climate changing? A study of long-time temperature trends. *Monthly Weather Rev.* 61:251–259.

King, J. E. 1966. Site index curves for Douglas-fir in the Pacific Northwest. Weyerhaeuser Forest. Paper 8. 49 pp.

Kittredge, Joseph. 1953. Influences of forests on snow in the ponderosa-sugar pine-fir zone of the central Sierra Nevada. *Hilgardia* 22:1–96.

Knight, Dennis H., and Orie L. Loucks. 1969. A quantitative analysis of Wisconsin forest vegetation on the basis of plant function and gross morphology. *Ecology* 50:219–234.

Knight, H. 1966. Loss of nitrogen from the forest floor by burning. *For. Chron.* 42:149–152.

Koch, Werner. 1969. Untersuchungen über die Wirkung von $CO_2$ auf die Photosynthese einiger Holzgewächse unter Laboratoriumsbedingungen. *Flora* 158B:402–428.

Kollmansperger, F. 1956. Lumbriciden in humiden und ariden Gebieten und ihre Bedeutung für die Fruchtbarkeit des Bodens. Rapp. VI Congr. Int. Sci. Sol. (Paris) C, 293–297.

Komarek, E. V. 1962. The use of fire: an historical background. *In Proc. Annual Tall Timbers Fire Ecology Conference.* 1:7–10. Tall Timbers Res. Sta., Tallahassee, Fla.

———. 1972. Lightning and fire ecology in Africa. *In Proc. Annual Tall Timbers Fire Ecology Conference* 11:473–511. Tall Timbers Res. Sta., Tallahassee, Fla.

———. 1973. Ancient fires. *In Proc. Annual Tall Timbers Fire Ecology Conference* 12:219–240. Tall Timbers Res. Sta., Tallahassee, Fla.

Kormanik, Paul P., and Claud L. Brown. 1969. Origin and development of epicormic branches in sweetgum. USDA For. Serv. Res. Paper SE–54. Southeast. For. Exp. Sta., Asheville, N.C. 17 pp.

Kormondy, Edward J. 1969. *Concepts of Ecology.* Prentice-Hall, Inc., Englewood Cliffs, N.J. 209 pp.

Korstian, C. F., and Paul W. Stickel. 1927. The natural replacement of blight-killed chestnut in the hardwood forests of the Northeast. *J. Agr. Res.* 34:631–648.

———, and T. S. Coile. 1938. Plant competition in forest stands. Duke Univ. School For. Bull. 3. 125 pp.

Koski, Veikko. 1970. A study of pollen dispersal as a mechanism of gene flow in conifers. *Comm. Inst. For. Fenn.* 70:1–78.

Köstler, J. N., E. Brückner, and H. Bibelriether. 1968. *Untersuchungen zur*

*Morphologie der Waldbäume in Mitteleuropa.* Verlag Paul Parey, Munich. 284 pp.

Kozlowski, Theodore T. 1949. Light and water in relation to growth and competition of Piedmont forest tree species. *Ecol. Monogr.* 19:207–231.

———. 1971. *Growth and development of trees. I. Seed Germination, Ontogeny, and Shoot Growth,* 443 pp. II. *Cambial Growth, Root Growth, and Reproductive Growth.* Academic Press, Inc., New York. 520 pp.

———, and Wayne H. Scholtes. 1948. Growth of roots and root hairs of pine and hardwood seedlings in the Piedmont. *J. For.* 46:750–754.

———, and C. E. Ahlgren (eds.). 1974. *Fire and Ecosystems.* Academic Press, Inc., New York. 542 pp.

Krajina, Vladimir J. 1959. Bioclimatic zones in British Columbia. Botanical Series No. 1, Univ. of British Columbia. 47 pp.

Kramer, Paul J. 1937. Effect of variation in length of day on growth and dormancy of trees. *Plant Physiol.* 11:127–137.

———. 1943. Amount and duration of growth of various species of tree seedlings. *Plant Physiol.* 18:239–251.

———. 1957. Some effects of various combinations of day and night temperatures and photoperiod on the growth of loblolly pine seedlings. *For. Sci.* 3:45–55.

———. 1969. *Plant and Soil Water Relationships: A Modern Synthesis.* McGraw-Hill Book Co., Inc., New York. 482 pp.

———. 1973. Forty years of research on plant water relations. *Proc. Am. Phil. Soc.* 117:381–387.

———. 1974. Fifty years of progress in water relations research. *Plant Physiol.* 54:463–471.

———, and J. P. Decker. 1944. Relation between light intensity and rate of photosynthesis of loblolly pine and certain hardwoods. *Plant Physiol.* 19:350–358.

———, and W. S. Clark. 1947. A comparison of photosynthesis in individual pine needles and entire seedlings at various light intensities. *Plant Physiol.* 22:51–57.

———, and T. T. Kozlowski. 1960. *Physiology of Trees.* McGraw-Hill Book Co., Inc., New York. 642 pp.

———, and ———. 1979. *Physiology of Woody Plants.* Academic Press, Inc., New York. 832 pp.

Krammes, Jay S. 1965. Seasonal debris movement from steep mountainside slopes in southern California. *USDA Misc. Publ.* 970:85–88.

Krauss, G. A. 1936. Aufgaben der Standortskunde. *Jahresber. deutschen Forstvereins Berlin* pp. 319–329.

Krebs, Charles J. 1972. *Ecology.* Harper & Row, Publishers, New York. 694 pp.

Krebs, R. D., and J. C. F. Tedrow. 1958. Genesis of red-yellow podzolic and related soils in New Jersey. *Soil Sci.* 85:28–37.

Krefting, Laurits W., and Eugene I. Roe. 1949. The role of some birds and mammals in seed germination. *Ecol. Monogr.* 19:269–286.

———, J. H. Stoeckeler, B. J. Bradle, and W. D. Fitzwater. 1962. Porcupine-timber relationships in the Lake States. *J. For.* 60:325–330.

Kriebel, Howard B. 1957. Patterns of genetic variation in sugar maple. Ohio Agr. Exp. Sta. Res. Bull. 791. 56 pp.

———. 1958. Geographic differentiation in seed dormancy and juvenile growth rate of Ontario sugar maple. Proc. 6th Meeting For. Tree Breeding Comm. Can., 1958(2):R7–11. Montreal.

Kruckeberg, Arthur R. 1967. Ecotypic response to ultramafic soils by some plant species of northwestern United States. *Brittonia* 19:133–151.

———. 1969. The implications of ecology for plant systematics. *Taxon* 18:92–120.

Kucera, C. L. 1959. Weathering characteristics of deciduous leaf litter. *Ecology* 40:485–487.

Küchler, A. W. 1964. The potential natural vegetation of the conterminous United States. *Am. Geogr. Soc. Spec. Publ.* No. 36. 154 pp.

———. 1967. *Vegetation Mapping*. The Ronald Press Co., New York. 472 pp.

Kurz, Herman, and Delzie Demaree. 1934. Cypress buttresses and knees in relation to water and air. *Ecology* 15:36–41.

Lacate, D. S. (ed.). 1969. Guidelines for bio-physical land classification. Can. Dept. Fish.Forest., Can. Forest Serv., Publ. 1264. 61 pp.

Ladefoged, K. 1952. The periodicity of wood formation. *Copenh. Biol. Skr.* 7(3):1–98.

Lambert, J. M., and W. T. Williams. 1962. Multivariate methods in plant ecology. IV. Nodal analysis. *J. Ecol.* 50:775–802.

———, and M. B. Dale. 1964. The use of statistics in phytosociology. *Adv. Ecol. Res.* 2:55–99.

Langford, Arthur N., and Murray F. Buell. 1969. Integration, identity and stability in the plant association. *Adv. Ecol. Res.* 6:83–135.

Langlet, Olof. 1959. A cline or not a cline—a question of Scots pine. *Silvae Genetica* 8:13–22.

———. 1963. Patterns and terms of intra-specific ecological variability. *Nature* 200:347–348.

———. 1971. Two hundred years' genecology. *Taxon* 20:653–721.

Langner, W. 1953. Eine Mendelspaltung bei Aurea-Formen von *Picea abies* (L.) Karst als Mittel zur Klärung der Befruchtungsverhältnisse im Walde. *Z. Forstg. Forstpflz.* 2:49–51.

Lanner, Ronald M. 1966. Needed: a new approach to the study of pollen dispersion. *Silvae Genetica* 15:50–52.

Larsen, J. A. 1925. Natural reproduction after forest fires in northern Idaho. *J. Agr. Res.* 31:1177–1197.

Larson, M. M. 1967. Effect of temperature on initial development of ponderosa pine seedlings from three sources. *For. Sci.* 13:286–294.

———. 1970. Root regeneration and early growth of red oak seedlings: influence of soil temperature. *For. Sci.* 16:442–446.

Larson, Philip R. 1963a. The indirect effect of drought on tracheid diameter in red pine. *For. Sci.* 9:52–62.

———. 1963b. Stem form development of forest trees. *For. Sci. Monogr.* 5. 42 pp.

———. 1964. Some indirect effects of environment on wood formation. *In* M. H. Zimmermann (ed.), *The Formation of Wood in Forest Trees*. Academic Press, Inc., New York.

Lassoie, James P. 1980. Physiological processes in Douglas-fir. *In* R. L. Edmonds (ed.), *Analysis of Coniferous Forest Ecosystems in the Western*

*United States*. US/IBP Synthesis Series. Dowden, Hutchinson and Ross, Inc., Stroudsburg, Pa.

Lavender, Denis P., and W. Scott Overton. 1972. Thermoperiods and soil temperatures as they affect growth and dormancy of Douglas-fir seedlings of different geographic origin. Ore. State Univ., For. Res. Lab. Res. Paper 13. 26 pp.

Lawrence, Donald B. 1958. Glaciers and vegetation in southeastern Alaska. *Amer. Sci.* 46:81–122.

Lawrence, William H., and J. H. Rediske. 1962. Fate of sown Douglas-fir seed. *For. Sci.* 8:210–218.

Layser, Earle F. 1974. Vegetative classification: its application to forestry in the northern Rocky Mountains. *J. For.* 73:354–357.

Leaf, Albert L. 1968. K, Mg, and S deficiencies in forest trees. *In Forest Fertilization*. Tennessee Valley Authority, Muscle Shoals, Ala.

———, R. E. Leonard, and N. A. Richards. 1975. Forest fertilization for non-wood production benefits in northeastern United States. *In* B. Bernier and C. H. Winget (eds.), *Forest Soils and Forest Land Management*. Laval Univ. Press, Quebec.

Leaf, Charles F. 1975a. Watershed management in the Rocky Mountain subalpine zone: the status of our knowledge. USDA For. Serv. Res. Paper RM–137. Rocky Mountain For. and Rge. Exp. Sta., Ft. Collins, Colo. 31 pp.

———. 1975b. Watershed management in the central and southern Rocky Mountains: a summary of the status of our knowledge by vegetation types. USDA For. Serv. Res. Paper RM–142. Rocky Mountain For. and Rge. Exp. Sta., Ft. Collins, Colo. 28 pp.

Leak, W. B. 1975. Age distribution in virgin red spruce and northern hardwoods. *Ecology* 56:1451–1454.

Leaphart, Charles D., and Albert R. Stage. 1971. Climate: a factor in the origin of the pole blight disease of *Pinus monticola* Dougl. *Ecology* 52:229–239.

Ledig, F. Thomas, and John H. Fryer. 1972. A pocket of variability in *Pinus rigida*. *Evolution* 26:259–266.

Lee, K. W., and T. G. Wood. 1971. Physical and chemical effects on soils of some Australian termites, and their pedological significance. *Pedobiol.* 11:376–409.

Leibundgut, Hans. 1970. *Der Wald, eine Lebensgemeinschaft*. Verlag Huber & Co., Stuttgart. 232 pp.

———, and Hans Heller. 1960. Photoperiodische Reaktion, Lichtbedarf und Austreiben von Jungpflanzen der Tanne (*Abies alba* Miller). *Beiheft Zeitschr. schweiz. Forstv.* 30:185–198.

Lemieux, G. J. 1965. Soil-vegetation relationships in the northern hardwoods of Quebec. *In* Chester T. Youngberg (ed.), *Forest-Soil Relationships in North America*. Oregon State Univ. Press, Corvallis, Ore.

Lemmon, Paul E. 1955. Factors affecting productivity of some lands in the Willamette Basin of Oregon for Douglas-fir timber. *J. For.* 53:323–330.

———. 1970. Grouping soils on the basis of woodland suitability. *In* Chester T. Youngberg and Charles B. Davey (eds.), *Tree Growth and Forest*

*Soils*. Oregon State Univ. Press, Corvallis, Ore.

Leopold, Estella B. 1969. Late Cenozoic palynology. *In* Robert H. Tschudy and Richard A. Scott (eds.), *Aspects of Palynology*. John Wiley & Sons, Inc., New York.

Lester, D. T. 1967. Variation in cone production of red pine in relation to weather. *Can. J. Bot.* 45:1683–1691.

Levin, Donald A. 1976. The chemical defenses of plants to pathogens and herbivores. *Ann. Rev. Ecol. Syst.* 7:121–159.

Levitt, J. 1972. *Responses of Plants to Environmental Stresses*. Academic Press, Inc., New York. 697 pp.

Leyton, L. 1952. The effect of pH and form of nitrogen on the growth of Sitka spruce seedlings. *Forestry* 25:32–40.

———. 1955. The influence of artificial shading of the ground vegetation on the nutrition and growth of Sitka spruce (*Picea sitchensis* Carr.) in a heathland plantation. *Forestry* 28:1–6.

———, and K. A. Armson. 1955. Mineral composition of the foliage in relation to the growth of Scots pine. *For. Sci.* 1:210–218.

Li, Hui-Lin. 1953. Present distribution and habitats of the conifers and taxads. *Evolution* 7:245–261.

Libby, W. J., R. F. Stettler, and F. W. Seitz. 1969. Forest genetics and forest-tree breeding. *Ann. Rev. Genetics* 3:469–494.

Lieth, Helmut 1968. Continuity and discontinuity in ecological gradients and plant communities. *In* Pierre Dansereau (ed.), The continuum concept of vegetation: responses. *Bot. Rev.* 34:253–332.

———, and Robert H. Whittaker (eds.). 1975. *Primary Productivity of the Biosphere*. Ecological studies 14. Springer-Verlag, New York. 339 pp.

Likens, Gene E. 1976. Acid precipitation. *Chemical and Engineering News* 54:29–44.

———, F. Herbert Bormann, N. M. Johnson, and R. S. Pierce. 1967. The calcium, magnesium, potassium, and sodium budgets for a small forested ecosystem. *Ecology* 48:772–785.

———, ———, ———, D. W. Fisher, and Robert S. Pierce. 1970. Effects of forest cutting and herbicide treatment on nutrient budgets in the Hubbard Brook watershed-ecosystem. *Ecol. Monogr.* 40:23–47.

———, and ———. 1974. Acid rain: a serious regional environmental problem. *Science* 184:1176–1179.

———, ———, Robert S. Pierce, John S. Eaton, and Noye M. Johnson. 1977. *Biogeochemistry of a Forested Ecosystem*. Springer-Verlag, New York. 146 pp.

———, ———, ———, and W. A. Reiners. 1978. Recovery of a deforested ecosystem. *Science* 199:492–496.

Linder, Sune, 1971. Photosynthetic action spectra of Scots pine needles of different ages from seedlings grown under different nursery conditions. *Physiol. Plant*. 25:58–63.

Linteau, A. 1955. Forest site classification of the northeastern coniferous section, boreal forest region, Quebec. For. Br. Can. Bull. 118. 78 pp.

List, Robert J. 1958. Smithsonian meteorological tables. Smithsonian Inst. Pub. 4014 (rev. ed. 6). 527 pp. (2nd. reprinting, 1963).

Little, Silas. 1974. Effects of fire on temperate forests: northeastern United States. *In* T. T. Kozlowski and C. E. Ahlgren (eds.), *Fire and Ecosystems*. Academic Press, Inc., New York.

———, and E. B. Moore. 1949. The ecological role of prescribed burns in the pine-oak forests of southern New Jersey. *Ecology* 30:223–233.

Livingston, R. B. 1949. An ecological study of the Black Forest, Colorado. *Ecol. Monogr.* 19:123–144.

Lloyd, William J., and Paul E. Lemmon. 1970. Rectifying azimuth (of aspect) in studies of soil-site index relationships. *In* Chester T. Youngberg and Charles B. Davey (eds.), *Tree Growth and Forest Soils*. Oregon State Univ. Press, Corvallis.

Loach, K. 1967. Shade tolerance in tree seedlings. I. Leaf photosynthesis and respiration in plants raised under artificial shade. *New Phytol.* 66:607–621.

Lodhi, M. A. K. 1976. Role of allelopathy as expressed by dominating trees in a lowland forest in controlling the productivity and pattern of herbaceous growth. *Amer. J. Bot.* 63:1–8.

———. 1978. Comparative inhibition of nitrifiers and nitrification in a forest community as a result of the allelopathic nature of various tree species. *Am. J. Bot.* 65:1135–1137.

———, and E. L. Rice. 1971. Allelopathic effects of *Celtis laevigata*. *Bull. Torrey Bot. Club* 98:83–89.

Logan, K. T. 1965. Growth of tree seedlings as affected by light intensity. I. White birch, yellow birch, sugar maple, and silver maple. Dept. For. Canada, Publ. 1121. 16 pp.

———. 1966a. Growth of tree seedlings as affected by light intensity. II. Red pine, white pine, jack pine and eastern larch. Dept. For. Canada, Publ. 1160. 19 pp.

———. 1966b. Growth of tree seedlings as affected by light intensity III. Basswood and white elm. Dept. For. Canada, Publ. 1176. 15 pp.

———. 1970. Adaptations of the photosynthetic apparatus of sun- and shade-grown yellow birch (*Betula alleghaniensis* Britt.). *Can. J. Bot.* 48:1681–1688.

———, and G. Krotkov. 1969. Adaptations of the photosynthetic mechanism of sugar maple (*Acer saccharum*) seedlings grown in various light intensities. *Physiol. Plant.* 22:104–116.

Long, J. N. 1980. Productivity of western coniferous forests. *In* R. L. Edmonds (ed.), *Analysis of Coniferous Forest Ecosystems in the Western United States*. US/IBP Synthesis Series. Dowden, Hutchinson and Ross, Inc., Stroudsburg, Pa.

Lorenz, Ralph W. 1939. High temperature tolerance of forest trees. Univ. Minn. Agr. Exp. Sta. Tech. Bull. 141. 25 pp.

Lotspeich, Frederick B., and Ernst W. Mueller. 1971. Effects of fire in the taiga on the environment. *In* C. W. Slaughter, Richard J. Barney, and G. M. Hansen (eds.), *Fire in the Northern Environment—A Symposium*. USDA For. Serv. Pacific Northwest For. and Rge. Exp. Sta., Portland, Ore.

Loucks, O. L. 1970. Evolution of diversity, efficiency, and community stabil-

ity. *Am. Zool.* 10:17–25.
Lowry, William P. 1969. *Weather and Life, An Introduction to Biometeorology*. Academic Press, Inc., New York. 305 pp.
Lugo, Ariel E., and Samuel C. Snedaker. 1974. The ecology of mangroves. *Ann. Rev. Ecol. Syst.* 5:39–64.
Lull, Howard W., and Kenneth G. Reinhart. 1972. Forests and floods in the eastern United States. USDA For. Serv. Res. Paper NE–226. Northeastern For. Exp. Sta., Upper Darby, Pa. 94 pp.
Lundegardh, Henrik. 1957. *Klima und Boden in ihrer Wirkung auf das Pflanzenleben*, 5th ed. (Also English translation, 1931, Edward Arnold, Ltd., London.) Gustav Fischer Verlag, Jena. 584 pp.
Lutz, H. J. 1928. Trends and silvicultural significance of upland forest successions in southern New England. Yale For. Sch. Bull. 22. 68 pp.
———. 1930a. Original forest composition in northwestern Pennsylvania as indicated by early land survey notes. *J. For.* 28:1098–1103.
———. 1930b. The vegetation of Heart's Content, a virgin forest in northwestern Pennsylvania. *Ecology* 11:1–29.
———. 1945. Vegetation on a trenched plot twenty-one years after establishment. *Ecology* 26:200–202.
———. 1956. Ecological effects of forest fires in the interior of Alaska. U.S. Dept. Agr. Tech. Bull. 1133. 121 pp.
———. 1959. Forest ecology, the biological basis of silviculture. Publ. Univ. Brit. Col. 8 pp.
———, and Robert F. Chandler, Jr. 1946. *Forest Soils*. John Wiley & Sons, Inc., New York. 514 pp.
Lyford, W. F., and D. W. MacLean. 1966. Mound and pit microrelief in relation to soil disturbance and tree distribution in New Brunswick, Canada. Harvard For. Paper No. 15. Harvard Univ., Harvard Forest, Petersham, Mass. 18 pp.
———, and B. F. Wilson. 1966. Controlled growth of forest tree roots: technique and application. Harvard Forest Paper No. 16. Harvard Univ. Petersham, Mass. 12 pp.
Lynch, Donald W. 1958. Effects of stocking on site measurement and yield of second-growth ponderosa pine in the Inland Empire. Intermountain For. and Rge. Exp. Sta. Res. Paper 56. 36 pp.
———. 1959. Effects of a wildfire on mortality and growth of young ponderosa pine trees. USDA Intermountain For. and Rge. Exp. Sta. Res. Note 66. 8 pp.
Lyr, Horst, and Günter Hoffmann. 1967. Growth rates and growth periodicity of tree roots. *Int. Rev. For. Res.* 2:181–236.
———, Hans Polster, and Hans-Joachim Fiedler. 1967. *Gehölzphysiologie*. Gustav Fischer Verlag, Jena. 444 pp.
McAndrews, J. H. 1966. Postglacial history of prairie, savanna and forest in northwestern Minnesota. *Torrey Bot. Club Mem.* 22(2). 72 pp.
McBride, Joe. 1973. Natural replacement of disease-killed elms. *Am. Midl. Nat.* 90:300–306.
McClurkin, D. C. 1953. Soil and climatic factors related to the growth of longleaf pine. U.S. For. Serv. Southern For. Exp. Sta. Occ. Paper 132. 12 pp.

———. 1967. Vegetation for erosion control in the southern coastal plain of the United States. *In* William E. Sopper and Howard W. Lull (eds.), *Forest Hydrology*. Pergamon Press, Inc., New York.

McColl, J. G., and D. W. Cole. 1968. A mechanism of cation transport in a forest soil. *Northwest Sci*. 42:134–140.

McConkey, Thomas W., and Donald R. Gedney. 1951. A guide for salvaging white pine injured by forest fires. USDA Northeastern For. Exp. Sta. Res. Note 11. 4 pp.

McCullough, Dale R., Charles E. Olson, Jr., and Leland M. Queal. 1969. Progress in large animal census by thermal mapping. *In* Philip L. Johnson (ed.), *Remote Sensing in Ecology*. Univ. Georgia Press, Athens, Ga.

McDermott, R. E. 1954. Seedling tolerance as a factor in bottomland timber succession. Mo. Agr. Exp. Sta. Res. Bull. 557. 11 pp.

McFee, W. W., and E. L. Stone. 1966. The persistence of decaying wood in the humus layers of northern forests. *Soil Sci. Soc. Amer. Proc*., 30:513–516.

McIntosh, Robert P. 1967. The continuum concept of vegetation. *Bot. Rev*. 33:130–187.

McKelvey, P. J. 1953. Forest colonization after recent volcanicity at West Taupo. *New Zealand J. For*. 6:435–448.

———. 1963. The synecology of the West Taupo indigenous forest. New Zealand For. Serv. Bull. 14. Wellington. 126 pp.

McKey, Doyle. 1973. The ecology of coevolved seed dispersal systems. *In* Lawrence E. Gilbert and Peter H. Raven (eds.), *Coevolution of Animals and Plants*. Univ. Texas Press, Austin.

McMillan, C. 1956. The edaphic restriction of *Cupressus* and *Pinus* in the coast ranges of central California. *Ecol. Monogr*. 26:177–212.

McNeil, Mary. 1964. Lateritic soils. *Am. Scientist* 211:96–102.

McPherson, James K., and Cornelius H. Muller. 1969. Allelopathic effects of *Adenostoma fasciculatum*, "chamise," in the California chaparral. *Ecol. Monogr*., 39:177–198.

McQuilken, W. E. 1940. The natural establishment of pine in abandoned fields in the Piedmont Plateau region. *Ecology* 21:135–147.

McVickar, J. S. 1949. Composition of white oak (*Quercus alba*) leaves in Illinois as influenced by soil type and soil composition. *Soil Sci*. 68:317–328.

Mackey, Halkard E., Jr., and Neal Sivec. 1973. The present composition of a former oak-chestnut forest in the Allegheny Mountains of western Pennsylvania. *Ecology* 54:915–919.

Madge, D. S. 1965. Leaf fall and litter disappearance in a tropical forest. *Pedobiol*. 5:273–288.

———. 1966. How leaf litter disappears. *New Sci*. 32(517):113–115.

Madgwick, H. A. I., and J. D. Ovington. 1959. The chemical composition of precipitation in adjacent forest and open plots. *Forestry* 32:14–22.

Maguire, William P. 1955. Radiation, surface temperatures, and seedling survival. *For. Sci*. 1:277–285.

Major, J. 1951. A functional, factorial approach to plant ecology. *Ecology* 32:392–412.

———. 1969. Historical development of the ecosystem concept. *In* George M. Van Dyne (ed.), *The Ecosystem Concept in Natural Resource Management*. Academic Press, Inc., New York.

Malin, James C. 1956. *The Grassland of North America. Prolegomena to Its History*. Publ. by the author, Lawrence, Kansas. 469 pp.

Margaropoulos, Panos. 1967. Woody revegetation as a pioneer action towards restoring of totally eroded slopes in mountainous watersheds. *In* William E. Sopper and Howard W. Lull (eds.), *Forest Hydrology*. Pergamon Press, Inc., New York.

Marks, G. C., and T. T. Kozlowski (eds.). 1973. *Ectomycorrhizae—Their Ecology and Physiology*. Academic Press, Inc., New York. 444 pp.

Marks, P. L., and F. H. Bormann. 1972. Revegetation following forest cutting: mechanisms for return to steady-state nutrient cycling. *Science* 176:914–915.

Marquis, David A. 1974. The impact of deer browsing on Allegheny hardwood regeneration. USDA For. Serv. Res. Paper NE–308. Northeastern For. Exp. Sta., Upper Darby, Pa. 8 pp.

———. 1975. The Allegheny hardwood forests of Pennsylvania. USDA For. Serv., Northeastern For. Exp. Sta. Gen. Tech. Report NE–15. Upper Darby, Pa. 32 pp.

Marr, J. W. 1948. Ecology of the forest-tundra ecotone on the east coast of Hudson Bay. *Ecol. Monogr.* 18:117–144.

———. 1977. The development and movement of tree islands near the upper limit of tree growth in the southern Rocky Mountains. *Ecology* 58:1159–1164.

Marshall, P. E., and T. T. Kozlowski. 1976. Importance of photosynthetic cotyledons for early growth of woody angiosperms. *Physiol. Plant.* 37:336–340.

———, and ———. 1977. Changes in structure and function of epigeous cotyledons of woody angiosperms during early seedling growth. *Can. J. Bot.* 55:208–215.

Marshall, Robert. 1927. The growth of hemlock before and after release from suppression. Harv. For. Bull. 11. 43 pp.

Marston, R. B. 1956. Air movement under an aspen forest and on an adjacent opening. *J. For.* 54:468–469.

Martin, Paul S., and Byron E. Harrell. 1957. The Pleistocene history of temperate biotas in Mexico and eastern United States. *Ecology* 38:468–480.

Martin, Robert E., Robert W. Cooper, A. Bigler Crow, James A. Cuming, Clinton B. Phillips. 1977. Report of task force on prescribed burning. *J. For.* 75:297–301.

Marx, D. H. 1969a. The influence of ecotrophic mycorrhizal fungi on the resistance of pine roots to pathogenic infections. I. Antagonism of mycorrhizal fungi to root pathogenic fungi and soil bacteria. *Phytopathology* 59:153–163.

———. 1969b. The influence of ectotrophic mycorrhizal fungi on the resistance of pine roots to pathogenic infections. II. Production, identification, and biological activity of antibiotics produced by *Leucopaxillus cerealis* var. *piceina*. *Phytopathology* 59:411–417.

———, and C. B. Davey. 1969a. The influence of ecotrophic mycorrhizal fungi on the resistance of pine roots to pathogenic infections. III. Resistance of aseptically formed mycorrhizae to infection by *Phytophthora cinnamomi*. *Phytopathology* 59:549–558.

———, and ———. 1969b. The influence of ecotrophic mycorrhizal fungi on the resistance of pine roots to pathogenic infections. IV. Resistance of naturally occurring mycorrhizae to infections by *Phytophthora cinnamomi*. *Phytopathology* 59:559–565.

———. 1970. The influence of ectotrophic mycorrhizal fungi on the resistance of pine roots to pathogenic infections. V. Resistance of mycorrhizae to infection by vegetative mycelium of *Phytophthora cinnamomi*. *Phytopathology* 60:1472–1473.

———. 1973. Mycorrhizae and feeder root diseases. *In* G. C. Marks and T. T. Kozlowski (eds.), *Ectomycorrhizae—Their Ecology and Physiology*. Academic Press, Inc., New York.

———. 1975. Mycorrhizae and establishment of trees on strip-mined land. *Ohio J. Sci.* 75:288–297.

———. 1976. Synthesis of ectomycorrhizae on loblolly pine seedlings with basidiospores of *Pisolithus tinctorius*. *For. Sci.* 22:13–20.

———. 1977. Tree host range and world distribution of the ectomycorrhizal fungus *Pisolithus tinctorius*. *Can. J. Microb.* 23:217–223.

———, William C. Bryan, and Charles E. Cordell. 1976. Growth and ectomycorrhizal development of pine seedlings in nursery soils infested with the fungal symbiont *Pisolithus tinctorius*. *For. Sci.* 22:91–100.

———, and Daniel J. Beattie. 1977. Mycorrhizae—promising aid to timber growers. *For. Farmer* 36:6–9.

———, William C. Bryan, and Charles E. Cordell. 1977. Survival and growth of pine seedlings with *Pisolithus* ectomycorrhizae after two years on reforestation sites in North Carolina and Florida. *For. Sci.* 23:363–373.

Maser, Chris, James M. Trappe, and Ronald A. Nussbaum. 1978. Fungal—small mammal interrelationships with emphasis on Oregon conifer forests. *Ecology* 59:799–809.

Mather, K. 1943. Polygenic inheritance and natural selection. *Biol. Rev.* 18:32–64.

Matthews, J. D. 1963. Factors affecting the production of seed by forest trees. *For. Abstr.* 24(1):i–xiii.

Mattoon, Wilbur R. 1915. The southern cypress. Bull. U.S. Dep. Agric. No. 272. 74 pp.

Mattson, W. J. (ed.). 1977. *The Role of Arthropods in Forest Ecosystems*. Springer-Verlag, New York. 104 pp.

———, and Norton D. Addy. 1975. Phytophagous insects as regulators of forest primary production. *Science* 190:515–522.

Maycock, P. F., and J. T. Curtis. 1960. The phytosociology of boreal conifer-hardwood forests of the Great Lakes region. *Ecol. Monogr.* 30:1–35.

Maze, Jack. 1968. Past hybridization between *Quercus macrocarpa* and *Quercus gambelii*. *Brittonia* 20:321–333.

Meeuwig, Richard O. 1971. Infiltration and water repellency in granitic soils. USDA For. Serv. Res. Paper INT–111. Intermountain For. and

Rge. Exp. Sta., Ogden, Utah. 20 pp.

Meinecke, E. P. 1928. A camp ground policy. Report, California Dept. Nat. Res., Parks Div., Sacramento, Calif. 16 pp.

Merriam, C. Hart. 1890. Results of a biological survey of the San Francisco Mountain region and the desert of the Little Colorado, Arizona. U.S. Dept. Agr. No. Am. Fauna 3:1–136.

———. 1898. Life zones and crop zones. U.S. Dept. Agr. Div. Biol. Surv. Bull. No. 10. 79 pp.

Mettler, Lawrence E., and Thomas G. Gregg. 1969. *Population Genetics and Evolution*. Prentice-Hall, Inc., Englewood Cliffs, N.J. 212 pp.

Metz, L. J., and J. E. Douglass. 1959. Soil moisture depletion under several Piedmont cover types. U.S. Dept. Agr. Tech. Bull. 1207. 23 pp.

———, Thomas Lotti, and Ralph A. Klawitter. 1961. Some effects of prescribed burning on coastal plain forest soil. U.S. For. Serv. Southeastern For. Exp. Sta., Sta. Paper 133. 10 pp.

Mikola, Peitsa. 1962. Temperature and tree growth near the northern timber line. *In* Theodore T. Kozlowski (ed.), *Tree Growth*. The Ronald Press Co., New York.

———. 1970. Mycorrhizal inoculation in afforestation. *Int. Rev. For. Res.* 3:123–196.

Miller, David H. 1956. The influence of pine forest on daytime temperature in the Sierra Nevada. *Geog. Rev.* 46:209–218.

Miller, Paul R. 1969. Air pollution and the forests of California. *Cal. Air Environment* 1:1–3.

———. 1973. Oxidant-induced community change in a mixed conifer forest. *Advan. Chem. Ser.* 122:101–117.

———, and Joe R. McBride. 1975. Effects of air pollutants on forests. *In* J. B. Mudd and T. T. Kozlowski (eds.), *Responses of Plants to Air Pollution*. Academic Press, Inc., New York.

Miller, Richard E., Denis P. Lavender, and Charles C. Grier. 1976. Nutrient cycling in the Douglas-fir type—silvicultural implications. *Proc. Soc. Amer. For.* 1975: 359–390.

Mirov, N. T. 1967. *The Genus "Pinus."* The Ronald Press Co., New York. 602 pp.

Mitchell, H. L., and R. F. Chandler, Jr. 1939. The nitrogen nutrition and growth of certain deciduous trees of northeastern United States. *Black Rock Forest Bull.* 11. 94 pp.

Mitscherlich, Gerhard. 1970. *Wald, Wachstum und Umwelt. I. Form und Wachstum von Baum und Bestand*. J. D. Sauerlander's Verlag, Frankfort. 142 pp.

———. 1971. *Wald, Wachstum und Umwelt. II. Waldklima and Wasserhaushalt*. J. D. Sauerlander's Verlag, Frankfort. 365 pp.

———, K.-G. Kern, and E. Künstle. 1963. Untersuchungen über den Kohlensauregehalt der Waldluft in Plenterwald and Fichtenreinbestand. *Allg. Forst- u. Jagdzeitschrift* 134:281–290.

Moehring, D. H. 1970. Forest soil improvement through cultivation. *J. For.* 68:328–331.

———, C. X. Grano, and J. R. Bassett. 1966. Properties of forested loess soils after repeated prescribed burns. USDA For. Serv. Res. Note SO–40. Southern For. Exp. Sta., New Orleans, La. 4 pp.

———, and Ike W. Rawls. 1970. Detrimental effects of wet weather logging. *J. For.* 68:166–167.

Mohr, E. C. J., and F. A. van Baren. 1954. *Tropical Soils: A Critical Study of Soil Genesis as Related to Climate, Rock and Vegetation*. N. V. Uitgeverij W. van Hoeve, The Hague and Bandung. 498 pp.

Mohr, H. 1969. Photomorphogenesis. *In* Malcolm B. Wilkins (ed.), *The Physiology of Plant Growth and Development*. McGraw-Hill Book Co., Inc., New York.

Molchanov, A. A. 1963. *The Hydrological Role of Forests*. Trans. from Russian, Israel Program for Sci. Transl., Ltd. Off. Tech. Serv., U. S. Dept. Comm., Washington, D. C. 407 pp.

Montieth, L. G. 1960. Influence of plants other than the food plants of their host on host-finding by Tachinid parasites. *Can. Entomol.* 92:641–652.

Moore, P. D., and D. J. Bellamy. 1974. *Peatlands*. Springer-Verlag, New York. 221 pp.

Moosmayer, H.-U. 1955. Die Wuchsleistungen der Fichte und Buche auf den wichtigsten Standortseinheiten im Forstamt Königsbronn (Nordostteil der Schwäbischen Alb). *Mitt. Vereins forstl. Standortsk.* 4:52–60.

———. 1957. Zur ertragskundlichen Auswertung der Standortsgliederung im Ostteil der Schwäbischen Alb. *Mitt. Vereins forstl. Standortsk. Forstpflz.* 7:3–41.

Morey, Philip R. 1973. *How Trees Grow*. Inst. of Biology, Studies in Biol., No. 39. Edward Arnold, Ltd. 59 pp.

Morris, R. F. 1951. The effects of flowering on the foliage production and growth of balsam fir. *For. Chron.* 27:40–57.

Morrison, I. K. 1974. Mineral nutrition of conifers with special reference to nutrient status interpretation: a review of literature. Dept. Env. Can. For. Serv. Publ. No. 1343. Great Lakes For. Res. Cent., Sault Ste. Marie, Ontario. 74 pp.

Morrison, M. E. S. 1956. Factors in the degeneration of the prehistoric woodland. *Irish Naturalists J. (Belfast)* 12:57–65.

Mosquin, Theodore. 1966. Reproductive specialization as a factor in the evolution of the Canadian flora. *In* Roy L. Taylor and R. A. Ludwig (eds.), *The Evolution of Canada's Flora*. Univ. Toronto Press. 137 pp.

Mount, A. B. 1969. Eucalypt ecology as related to fire. *In Proc. Annual Tall Timbers Fire Ecology Conference*. 9:75–108. Tall Timbers Res. Sta., Tallahassee, Fla.

Mudd, J. B., and T. T. Kozlowski. 1975. *Responses of Plants to Air Pollution*. Academic Press, Inc., New York. 383 pp.

Mueller-Dombois, Dieter. 1964. The forest habitat types of southeastern Manitoba and their application to forest management. *Can. J. Bot.* 42:1417–1444.

———, and Heinz Ellenberg. 1974. *Aims and Methods of Vegetation Ecology*. John Wiley & Sons, New York. 547 pp.

Muller, Cornelius H. 1966. The role of chemical inhibition (allelopathy) in vegetational composition. *Bull. Torrey Bot. Club* 93:332–351.

Müller, P. E. 1887. *Studien über die natürlichen Humusformen und deren Einwirkung auf Vegetation und Boden*. Julius Springer, Berlin. 324 pp.

Murphy, P. W. 1953. The biology of forest soils with special reference to the mesofauna or meiofauna. *J. Soil Sci*. 4:155–193.

Musselman, Robert C., Donald T. Lester, and Michael S. Adams. 1975. Localized ecotypes of *Thuja occidentalis* L. in Wisconsin. *Ecology* 56:647–655.

Mutch, Robert W. 1970. Wildland fires and ecosystems—a hypothesis. *Ecology* 51:1046–1051.

———. 1976. Fire management and land use planning today: tradition and change in the Forest Service. *Western Wildlands* 3:13–19.

Nagel, J. L. 1950. Changement d'essences. *Schweiz. Z. Forstw*. 101:95–104.

National Academy of Sciences. 1971. Biochemical interactions among plants. U.S. National Comm. for the Int. Biol. Program, Natl. Acad. Sci., Washington, D.C. 134 pp.

Naveh. Z. 1974. Effects of fire in the Mediterranean region. *In*. T. T. Kozlowski and C. E. Ahlgren (eds.), *Fire and Ecosystems*. Academic Press, Inc., New York.

Nelson, Thomas C. 1955. Chestnut replacement in the southern highlands. *Ecology* 36(2): 352–353.

———. 1957. The original forests of the Georgia Piedmont. *Ecology* 38:390–397.

Newbould, P. J. 1967. *Methods for Estimating the Primary Production of Forests*. IBP Handbook No. 2. Blackwell Sci. Publ., Oxford-Edinburgh. 62 pp.

Newnham, R. M. 1968. A classification of climate by principal component analysis and its relationship to tree species distribution. *For. Sci*. 14:254–264.

Nicholls, J. L. 1959. The volcanic eruptions of Mt. Tarawera and Lake Rotomahana and effects on surrounding forests. *N.Z.J. For*. 8(1):133–142.

Nichols, George E. 1923. A working basis for the ecological classification of plant communities. *Ecology* 4:11–23, 154–172.

Nienstaedt, Hans, and Abraham Teich. 1971. The genetics of white spruce. USDA For. Serv. Res. Paper WO–15. Washington D.C. 24 pp.

Niering, William A., and Richard H. Goodwin. 1974. Creation of relatively stable shrublands with herbicides: arresting "succession" on rights-of-way and pastureland. *Ecology* 55:784–795.

Nikles, D. G. 1970. Breeding for growth and yield. *Unasylva* 24(2–3):9–22.

Northeastern For. Exp. Sta. 1973. Forest fertilization symposium proceedings. USDA For. Serv. Gen. Tech. Report NE–3. Northeastern For. Exp. Sta., Upper Darby, Pa. 246 pp.

Nye, P. H. 1966. The effect of the nutrient intensity and buffering power of a soil, and the absorbing power, size and root hairs of a root, on nutrient absorption by diffusion. *Plant and Soil* 25:81–105.

Oberlander, G. T. 1956. Summer fog precipitation of the San Francisco peninsula. *Ecology* 37:851–852.

Odén, Svante. 1976. The acidity problem—an outline of concepts. *In Proc.*

*The First International Symposium on Acid Precipitation and the Forest Ecosystem*, pp. 1–36. USDA For. Serv. Gen. Tech. Report NE–23. Northeastern For. Exp. Sta., Upper Darby, Pa.

Odum, Eugene P. 1971. *Fundamentals of Ecology*, 3rd ed. W. B. Saunders Co., Philadelphia. 574 pp.

Ogden, J. G. 1962. Forest history of Martha's Vineyard, Massachusetts. I. Modern and precolonial forests. *Amer. Mid. Nat.* 66:417–430.

———. 1966. Forest history of Ohio. I. Radiocarbon dates and pollen stratigraphy of Silver Lake, Logan County, Ohio. *Ohio J. Sci.* 66:387–400.

Ogilvie, R. T., and E. von Rudloff. 1968. Chemosystematic studies in the genus *Picea (Pinaceae)*. IV. The introgression of white and Engelmann spruce as found along the Bow River. *Canad. J. Bot.* 46:901–908.

Oh, Hi Kon, T. Sakai, M. B. Jones, and W. M. Longhurst. 1967. Effect of various essential oils isolated from Douglas-fir needles upon sheep and deer rumen microbial activity. *Appl. Microbiol.* 15:777–784.

Oinonen, E. A. 1956. Kallioiden muurahaisista ja niiden osuudesta kallioiden metsitty-miseen Etela-Suomessa. (On rock ants, and their contribution to the afforestation of rocks in south Finland.) Acta Entomologica Fennica No. 12 (Helsinki). 212 pp.

Olson, D. F., and L. Della-Bianca. 1959. Site index comparisons for several tree species in the Virginia-Carolina Piedmont. U.S. For. Serv., Southeastern For. Exp. Sta. Paper 104. 9 pp.

Olson, Jerry S. 1958. Rates of succession and soil changes on southern Lake Michigan sand dunes. *Bot. Gaz.* 119:125–170.

———, Helen A. Pfuderer, and Yip Hoi Chan. 1978. Changes in the global carbon cycle and the biosphere. Oak Ridge National Lab. Report ORNL–EIS 109. Oak Ridge, Tenn.

O'Neill, Robert V. 1975. Modeling in the eastern deciduous forest biome. *In* B. C. Patton (ed.), *Systems Analysis and Simulation in Ecology*, Vol.III. Academic Press, Inc., New York.

Oosting, H. J. 1942. An ecological analysis of the plant communities of Piedmont, North Carolina. *Amer. Mid. Nat.* 28:1–26.

———. 1948. *The Study of Plant Communities: An Introduction to Plant Ecology*. W. H. Freeman & Co., San Francisco. 388 pp.

———, and L. E. Anderson. 1939. Plant succession on granite rock in eastern North Carolina. *Bot. Gaz.* 100:750–768.

Ovington, J. D. 1954. A comparison of rainfall in different woodlands. *Forestry* 27:41–53.

———. 1962. Quantitative ecology and the woodland ecosystem concept *Adv. Ecol. Res.* 1:103–192.

———. 1968. Some factors affecting nutrient distribution within ecosystems. *In* F. E. Eckardt (ed.), *Functioning of Terrestrial Ecosystems at the Primary Production Level*. UNESCO, Natural Resources Research V., Paris.

Owens, John N., Marje Molder, and Hilary Langer. 1977. Bud development in *Picea glauca*. I. Annual growth cycle of vegetative buds and shoot elongation as they relate to date and temperature sums. *Can. J. Bot.* 55:2728–2745.

Pallman, H., and E. Frei. 1943. Beitrag zur Kenntnis der Lokalklimat einiger kennzeichnender Waldgesellschaften des schweizerischen Nationalparkes. *Ergeb. wissens. Untersuch. schweiz. Nationalparkes* 1:437–464.

Parker, J. 1955. Annual trends in cold hardiness of ponderosa pine and grand fir. *Ecology* 36(3):377–380.

———. 1963. Cold resistance in woody plants. *Bot. Rev.* 29:123–201.

Parmeter, John R., Jr., and Bjarne Uhrenholdt. 1976. Effects of smoke on pathogens and other fungi. *In Proc. Tall Timbers Fire Ecology Conf. and Fire and Land Management Symp.* 14:299–304. Tall Timbers Res. Sta., Tallahassee, Fla.

Parris, G. K. 1967. "Needle curl" of loblolly pine. *Plant Dis. Rep.* 51:805–806.

Patric, James H., James E. Douglass, and John D. Hewlett. 1965. Soil water absorption by mountain and Piedmont forests. *Proc. Soil Sci. Soc.* 29:303–308.

———, and Peter E. Black. 1968. Potential evapotranspiration and climate in Alaska by Thornthwaite's classification. USDA For. Serv. Res. Paper PNW–71. Pacific Northwest For. and Rge. Exp. Sta., Portland, Ore. 28 pp.

———, and David W. Smith. 1975. Forest management and nutrient cycling in eastern hardwoods. USDA For. Serv. Res. Paper NE–324. Northeastern For. Exp. Sta., Upper Darby, Pa. 12 pp.

Patterson, James C. 1976. Soil compaction and its effects upon urban vegetation. *In* Frank S. Santamour, Jr., Henry D. Gerhold, and Silas Little (eds.), *Better Trees for Metropolitan Landscapes: Symposium Proceedings*. USDA For. Serv. Gen. Tech. Report NE–22. Northeastern For. Exp. Sta., Upper Darby, Pa.

Pauley, Scott S. 1958. Photoperiodism in relation to tree improvement. *In* Kenneth V. Thimann (ed.), *The Physiology of Forest Trees*. The Ronald Press Co., New York.

———, and T. O. Perry. 1954. Ecotypic variation of the photoperiodic response in *Populus*. *J. Arnold Arbor.* 35:167–188.

Pearson, G. A. 1914. A meteorological study of parks and timbered areas in the western yellow-pine forests of Arizona and New Mexico. *Monthly Weather Rev.* 41:1615–1629.

———. 1936. Some observations on the reaction of pine seedlings to shade. *Ecology* 17:270–276.

———. 1940. Shade effects in ponderosa pine. *J. For.* 38:778–780.

———. 1951. A comparison of the climate in four ponderosa pine regions. *J. For.* 49:256–258.

Peet, Robert K. 1974. The measurement of species diversity. *Ann. Rev. Ecol. Syst.* 5:285–307.

———, and Orie L. Loucks. 1977. A gradient analysis of southern Wisconsin forests. *Ecology* 58:485–499.

Penman, H. L. 1963. *Vegetation and Hydrology*. Bur. Soils, Harpenden. Tech. Commun. 53. 124 pp.

Pennington, Winifred. 1969. *The History of British Vegetation*. English Universities Press, London. 152 pp.

Pereira, H. C. 1967. Summary of forests and runoff session. *In* William E. Sopper and Howard W. Lull (eds.), *Forest Hydrology*. Pergamon Press, Inc., New York.

Perry, Thomas O. 1964. Soil compaction and loblolly pine growth. U.S. Dept. Agr. For. Serv. Tree Planters Notes 69:9.

———. 1971. Dormancy of trees in winter. *Science* 171:29–36.

———, and Chi Wu Wang. 1960. Genetic variation in the winter chilling requirement for date of dormancy break for *Acer rubrum*. *Ecology* 41:790–794.

———, Harold E. Sellers, and Charles O. Blanchard. 1969. Estimation of photosynthetically active radiation under a forest canopy with chlorophyll extracts and from basal area measurements. *Ecology* 50:39–44.

Pfister, Robert D. 1976. Land capability assessment by habitat types. *Proc. Soc. Amer. For.*, 1975: 312–325.

———, Bernard L. Kovalchik, Stephen F. Arno, and Richard C. Presby. 1977. Forest habitat types of Montana. USDA For. Serv. Gen. Tech. Report INT-34. Intermountain For. and Rge. Exp. Sta., Ogden, Utah. 174 pp.

Pharis, Richard P. 1974. Precocious flowering in conifers: the role of plant hormones. *In* F. Thomas Ledig (ed.), *Toward the Future Forest: Applying Physiology and Genetics to the Domestication of Trees*. Yale Univ. Sch. For. and Env. Studies, Bull. 85.

Phillips, J. 1931. The biotic community. *J. Ecol.* 19:1–24.

———. 1934–35. Succession, development, the climax, and the complex organism; an analysis of concepts. *J. Ecol.* 22:554–571; 23:210–246, 488–508.

Phillips, John. 1974. Effects of fire in forest and savanna ecosystems of sub-Saharan Africa. *In* T. T. Kozlowski and C. E. Ahlgren (eds.), *Fire and Ecosystems*. Academic Press, Inc., New York.

Phillips, W. S. 1963. Depth of roots in soil. *Ecology* 44:424.

Phipps, R. L. 1961. Analysis of five years dendrometer data obtained within three deciduous forest communities of Neotoma. Ohio Agr. Exp. Sta. Res. Circ. 105. 34 pp.

Pierpoint, Geoffrey. 1962. The sites of the kirkwood management unit. Ontario Dept. Lands and For., Res. Rept. 47. 91 pp.

Pijl, L. van der. 1957. The dispersal of plants by bats. *Acta bot. neerl.* 6:291–315.

———. 1972. *Principles of Dispersal in Higher Plants*. Springer-Verlag, New York. 162 pp.

Pike, Lawrence, H., William C. Denison, Diane M. Tracy, Martha A. Sherwood, and Frederick M. Rhoades. 1975. Floristic survey of epiphytic lichens and bryophytes growing on old-growth conifers in western Oregon. *Bryol.* 78:389–402.

———, Robert A. Rydell, and William C. Denison. 1977. A 400-year-old Douglas–fir tree and its epiphytes: biomass, surface area, and their distributions. *Can. J. For. Res.* 7:680–699.

Pisek, Arthur, Walter Larcher, Walter Moser, and Ida Pack. 1969. Kardinale Temperaturbereiche der Photosynthese und Grenztemperaturen des Lebens der Blätter verschiedener Spermatophyten. III. Temperaturabhängigkeit und optimaler Temperaturbereich der Netto-Photosynthese. *Flora* (Abt. B) 158:608–630.

Polunin, Nicholas. 1960. *Introduction to Plant Geography.* McGraw-Hill Book Co., Inc., New York. 640 pp.

Poore, M. E. D. 1955. The use of phytosociological methods in ecological investigations. *J. Ecol.* 43:226–269.

Poulson, Thomas L., and William B. White. 1969. The cave environment. *Science* 165:971–981.

Prescott, J. A., and R. L. Pendleton. 1952. Laterite and lateritic soils. Bur. Soil Sci. Tech. Commun. 47. 51 pp.

Pritchett, W. L., and W. H. Smith. 1975. Forest fertilization in the U.S. Southeast. *In* B. Bernier and C. H. Winget (eds.), *Forest Soils and Forest Land Management.* Laval Univ. Press, Quebec.

Pyatt, D. G. 1968. Forest management surveys in forests affected by winds. *In* R. W. V. Palmer (ed.), *Wind Effects on the Forest.* Suppl. to Forestry, Oxford Univ. Press, London.

Ralston, Charles W. 1964. Evaluation of forest site productivity. *Int. Rev. For. Res.* 1:171–201.

Randolph, L. F., Ira S. Nelson, and R. L. Plaisted. 1967. Negative evidence of introgression affecting the stability of Louisiana iris species. Cornell University, New York State College of Agriculture, Agricultural Exp. Sta. Memoir 398. 56 pp.

Rasmussen, D. I. 1941. Biotic communities of the Kaibab Plateau, Arizona. *Ecol. Monog.* 11:229–275.

Raunkiaer, C. 1937. *Plant Life Forms.* Transl. by H. Gilbert-Carter. Oxford Univ. Press, London. 104 pp.

Raup, Hugh M. 1957. Vegetational adjustment to the instability of the site. *Proc. Pap. 6th Tech. Meet. Int. Union Cons. Nat. & Nat. Res.* (Edinburgh), pp. 36–48.

———, and Reynold E. Carlson. 1941. The history of land use in the Harvard Forest. Harv. For. Bull. 20. 64 pp.

Raven, Peter H. 1977. A suggestion concerning the Cretaceous rise to dominance of the angiosperms. *Evolution* 31:451–452.

Read, R. A. 1952. Tree species occurrences as influenced by geology and soil on an Ozark north slope. *Ecology* 33:239–246.

Redfield, J. A., F. C. Zwickel, and J. F. Bendell. 1970. Effects of fire on numbers of blue grouse. *In Proc. Annual Tall Timbers Fire Ecology Conference.* 10:63–83. Tall Timbers Res. Sta., Tallahassee, Fla.

Reed, Robert M. 1976. Coniferous forest habitat types of the Wind River Mountains, Wyoming. *Am. Mid. Nat.* 95:159–173.

Reeves, Robert G. 1975. *Manual of Remote Sensing.* Am. Soc. Photogram., Falls Church, Va. 2 Vols. 2144 pp.

Regal, Philip J. 1977. Ecology and evolution of flowering plant dominance. *Science* 196:622–629.

Rehfeldt, G. E., A. R. Stage, and R. T. Bingham. 1971. Strobili development in western white pine: periodicity, prediction, and association with weather. *For. Sci.* 17:454–461.

Reichle, D. E. 1971. Oak Ridge National Laboratory site proposal for analysis of the structure and function of ecosystems in the Deciduous Forest Biome. Oak Ridge National Laboratory. Research proposal, FY 1972, Oak Ridge, Tenn. 162 pp.

———. 1975. Advances in ecosystem analysis. *BioScience* 25:257–264.

———, P. Duvigneaud, and J. S. Olson. 1971. International woodland synthesis symposium, a proposal. *In* T. Rosswall (ed.), Systems analysis in northern coniferous forests—IBP workshop. Bull. 14, Ecological Res. Comm., Swedish Natural Sci. Res. Council, Stockholm, Sweden.

———, J. F. Franklin, and D. W. Goodall (eds.). 1975. *Productivity of World Ecosystems.* National Acad. Sci., Washington, D.C. 166 pp.

Reifsnyder, William. 1955. Wind profiles in a small isolated forest stand. *For. Sci.* 1:289–297.

———. 1967. Forest meteorology: the forest energy balance. *Int. Rev. For. Res.* 2:127–179.

———, and Howard W. Lull. 1965. Radiant energy in relation to forests. USDA Agr. Tech. Bull. No. 1344. 111 pp.

Remington, Charles L. 1968. Suture-zones of hybrid interaction between recently joined biotas. *Evolutionary Biology* 2:321–428.

Renner, O. 1912. Versuche zur Mechanik der Wasserversorgung. *Ber. deut. Ges.* 30:576–580, 642–648.

Rennie, P. J. 1955. The uptake of nutrients by mature forest growth. *Plant & Soil* 7:49–95.

———. 1962. Methods of assessing forest site capacity. *Trans. 7th Inter. Soc. Soil Sci., Comm.* IV and V, pp. 3–18.

Reynolds, R. R. 1940. Lightning as a cause of timber mortality. U.S. For. Serv., Southern Forest Exp. Sta. Notes 31. 4 pp.

Rhoades, D. F. 1976. The anti-herbivore defenses of Larrea. *In* T. J. Mabry, J. H. Hunziker, and D. R. DiFeo (eds.), *The Biology and Chemistry of the Creosote Bush, A Desert Shrub.* Dowden, Hutchinson, and Ross, Stroudsburg, Pa.

Rice, Elroy L. 1974. *Allelopathy.* Academic Press, Inc., New York. 353 pp.

———. 1977. Some roles of allelopathic compounds in plant communities. *Biochem. Syst. and Ecol.* 5:201–206.

Rich, L. R. 1959. Watershed management research in the mixed conifer type. Rocky Mt. For. & Rge. Exp. Sta., Sta. Prog. Rept. 1959.

Richard, F. 1953. Über die Verwertbarkeit des Bodenwassers durch die Pflanze. *Mitt. schweiz. Anst. forstl. Versuchsw.* 29(1):17–37.

Richards, N. A., and E. L. Stone. 1964. The application of soil survey to planting site selection: an example from the Allegheny uplands of New York. *J. For.* 62:475–480.

Richards, P. W. 1952. *The Tropical Rain Forest: An Ecological Study.* Cambridge Univ. Press, London. 450 pp.

———. 1973. The tropical rain forest. *Sci. American* 229:58–67.

Richardson, S. D. 1956. Studies of root growth of *Acer saccharinum* L. IV: the effect of differential shoot and root temperature on root growth. *Proc. K. Ned. Akad. Wet.*, 59:428–438.

———. 1958. Bud dormancy and root development in *Acer saccharinum. In* (K. V. Thimann (ed.), *The Physiology of Forest Trees.* Ronald Press, Inc., New York.

Ridley, Henry N. 1930. *The Dispersal of Plants Throughout the World.* L. Reeve & Co., LTD. Ashford, Kent. 744 pp.

Rigg, G. B. 1940, 1951. The development of sphagnum bogs in North America. *Bot. Rev.* 6:666–693; 17:109–131.

Roche, L. 1969. Introgressive hybridization in the spruce species of British Columbia. *In* C. W. Yeatman (ed.), *Proc. Eleventh Meeting Com. For. Tree Breeding Canada*, Part 2. Dept. Fish. and For. Canada, Ottawa.

Rocky Mountain Forest Experiment Station. 1959. Research on Black Mesa. A progress report. Rocky Mountain For. Exp. Sta., Sta. Paper 41. 18 pp.

Rodin, L. E., and N. I. Bazilevich. 1967. *Production and Mineral Cycling in Terrestrial Vegetation.* Transl. ed. by G. E. Fogg. Oliver & Boyd, Edinburgh. 288 pp.

Roe, Arthur L. 1967a. Seed dispersal in a bumper spruce seed year. USDA For. Serv. Res. Paper INT-39. Intermountain For. and Rge. Exp. Sta., Ogden, Utah. 10 pp.

———. 1967b. Productivity indicators in western larch forests. USDA For. Serv. Res. Note INT–59. Intermountain For. and Rge. Exp. Sta., Ogden, Utah. 4 pp.

Roe, Eugene I. 1963. Seed stored in cones of some jack pine stands, northern Minnesota. USDA For. Serv. Res. Paper LS–1. Lake States For. Exp. Sta., St. Paul, Minn. 14 pp.

Röhrig, E. 1966. Die Wurzelentwicklung der Waldbäume in Abhängigkeit von den ökologischen Verhältnissen. *Forstarchiv* 37:217–229; 237–249.

Rohrman, F. A., and J. H. Ludwig. 1965. Sources of sulfur dioxide pollution. *Chem. Eng. Prog.* 61:59–63.

Romberger, J. A. 1963. Meristems, growth and development in woody plants. USDA Tech. Bull. No. 1293. 214 pp.

Rowe, J. S. 1956. Uses of undergrowth species in forestry. *Ecology* 37:461–473.

———. 1959. Forest regions of Canada. Can. Dept. Nor. Aff. and Nat. Res., Ottawa, Bull. 123. 71 pp.

———. 1960. Can we find a common platform for the different schools of forest type classification? *Silva Fennica* 105:82–88.

———. 1961. The level-of-integration concept and ecology. *Ecology* 42:420–427.

———. 1962. Soil, site and land classification. *For. Chron.* 38:420–432.

———. 1964. Environmental preconditioning, with special reference to forestry. *Ecology* 45:399–403.

———. 1966. Phytogeographic zonation: an ecological appreciation. *In* Roy L. Taylor and R. A. Ludwig (eds.), *The Evolution of Canada's Flora.* Univ. Toronto Press.

———. 1969. Plant community as a landscape feature. *In* K. N. H. Greenidge (ed.), *Essays in Plant Geography and Ecology.* Nova Scotia Museum, Halifax, N.S.

———. 1970. Spruce and fire in Northwest Canada and Alaska. *In Proc. Annual Tall Timbers Fire Ecology Conference* 10:245–254. Tall Timbers Res. Sta., Tallahassee, Fla.

———. 1971. Why classify forest land? *For. Chron.* 47:144–148.

———. 1972. *Forest Regions of Canada*. Can. For. Serv., Dept. Env. Publ. No. 1300, Ottawa. 172 pp. + map.

———. and G. W. Scotter. 1973. Fire in the boreal forest. *Quaternary Res.* 3:444–464.

Rowe, R. B., and T. M. Hendrix. 1951. Interception of rain and snow by second-growth ponderosa pine. *Trans. Amer. Geophys. Union* 32 (6):903–908.

Rubner, Konrad. 1960. *Die pflanzengeographischen Grundlagen des Waldbaues*. Neumann Verlag, Berlin. 620 pp.

———, and F. Reinhold. 1953. *Die pflanzengeographischen Grundlagen des Waldbaues*, 4th ed. Neumann Verlag, Radebeul and Berlin. 583 pp.

Rundel, Philip W. 1972. Habitat restriction in giant sequoia: the environmental control of grove boundaries. *Am. Midl. Nat.* 87:81–99.

Rusch, Donald H., and Lloyd B. Keith. 1971. Ruffed grouse—vegetation relationships in central Alberta. *J. Wildl. Manage.* 35:417–429.

Rushmore, Francis M. 1969. Sapsucker damage varies with tree species and seasons. USDA For. Serv. Res. Paper NE–136. Northeastern For. Exp. Sta., Upper Darby, Pa. 19 pp.

Russell, E. Walter. 1976. *Soil Conditions and Plant Growth*. 10th Ed. Longman Inc., New York. 849 pp.

Rutter, A. J. 1968. Water consumption by forests. *In* T. T. Kozlowski (ed.), *Water Deficits and Plant Growth II*. Academic Press, Inc., New York.

Sarvas, Risto. 1962. Investigations on the flowering and seed crop of *Pinus silvestris*. *Comm. Inst. For. Fenn.* 53:1–198.

———. 1967. Pollen dispersal within and between subpopulations; role of isolation and migration in microevolution of forest tree species. *Proc. XIV IUFRO Congress;* III (Sec. 22–AG 22/24):332–345.

Satchell, J. E. 1967. Lumbricidae. *In* Alan Burgis and Frank Raw (eds.), *Soil Biology*. Academic Press, Inc., New York.

———. 1974. Litter—interface of animate/inanimate matter. *In* C. H. Dickinson and G. J. F. Pugh (eds.), *Biology of Plant Litter Decomposition*. Academic Press, Inc., New York.

Satoo, Taisitiroo. 1970. A synthesis of studies by the harvest method: primary production relations in the temperate deciduous forests of Japan. *In* David E. Reichle (ed.), *Analysis of Temperate Forest Ecosystems*. Springer-Verlag, New York.

Satterlund, Donald R. 1972. *Wildland Watershed Management*. The Ronald Press Co., New York. 370 pp.

Schaeffer, R., and R. Moreau. 1958. L'alternance des essences. *Bull. Soc. For. Franche-Comté* 29:1–12, 76–84, 277–288.

Schaffalitzky De Muckadell, M. 1959. Investigations on aging of apical meristems in woody plants and its importance in silviculture. *Det Forstl. Forsgsv. i Danmark* 25:309–455.

———. 1962. Environmental factors in development stages of trees. *In* Theodore T. Kozlowski (ed.), *Tree Growth*. The Ronald Press Co., New York.

Schaller, F. W., and P. Sutton (eds.). 1978. *Reclamation of Drastically Dis-*

*turbed Lands.* Am. Soc. Agronomy. Madison, Wisconsin. 742 pp.

Schier, George A. 1975. Deterioration of aspen clones in the Middle Rocky Mountains. USDA For. Serv. Res. Paper INT–170. Intermountain For. and Rge. Exp. Sta., Ogden, Utah. 14 pp.

Schlenker, G. 1960. Zum Problem der Einordnung klimatischer Unterschiede in das System der Waldstandorte Baden-Württembergs. *Mitt. Vereins forstl. Standortsk. Forstpflz.* 9:3–15.

———. 1964. Entwicklung des in Südwestdeutschland angewandten Verfahrens der forstlichen Standortskunde. *In Standort, Wald und Waldwirschaft in Oberschwaben,* "Oberschwäbische Fichtenreviere." Stuttgart.

———. 1971. Die Auswirkungen der Laubholz- bzw. Fichtenbestockung auf den Regenwurmbesatz in Parabraunerden und Pseudogleyen des Neckarlandes. *Mitt. Vereins forstl. Standortsk. Forstpflz.* 20:60–66.

———. 1976. Einflusse des Standorts und der Bestandesverhaltnisse auf die Rotfäule (Kernfäule) der Fichte. *In* H. Frhr. v. Pechmann (ed.), *Forstwissenschaftliche Forschungen: Der Wurzelschwamm (Fomes annosus) und die Rotfäule der Fichte (Picea abies).* Beihefte z. Forstw. Centralb. Nr. 36. Verlag Paul Parey, Hamburg. 83 pp.

———, Ulrich Babel, and Hans Peter Blume. 1969. Untersuchungen über die Auswirkungen des Fichtenreinanbaus auf Parabraunerden und Pseudogleye des Neckarlandes. VII. Zusammenfassung und Folgerungen für die forstliche Standortsbewertung. *Mitt. Vereins forstl. Standortsk. Forstpflz.* 19:111–114.

Schlesinger, William H., and William A. Reiners. 1974. Deposition of water and cations on artificial foliar collectors in fir krummholz of New England mountains. *Ecology* 55:378–386.

Schmidt, H. 1970. Versuche über die Pollenverteilung in einem Kiefernbestand. Diss. Forstl. Fak. Univ. Göttingen, West Germany.

Schmidt, Wyman C., and Raymond C. Shearer. 1971. Ponderosa pine seed—for animals or trees? USDA For. Serv., Intermountain For. and Rge. Exp. Sta. Res. Paper INT–112. Ogden, Utah. 14 pp.

———, and W. P. Dufour. 1975. Building a natural area system for Montana. *Western Wildlands* 2:20–29.

Schmidt-Vogt, Helmut. 1977. *Die Fichte.* Verlag Paul Parey, Hamburg. 647 pp.

Schneider, G., Ghaus Khattak, and John N. Bright. 1970. Modifying sites for the establishment of black walnut. *In* Chester T. Youngberg and Charles B. Davey (eds.), *Tree Growth and Forest Soils.* Oregon State Univ. Press, Corvallis.

Schoenicher, W. 1933. *Deutsche Waldbäume und Waldtypen.* Gustav Fischer Verlag, Jena. 208 pp.

Schoenike, Roland R. 1976. Geographical variations in jack pine (*Pinus banksiana*). Univ. Minn. Agr. Exp. Sta. Tech. Bull. 304. 47 pp.

Schramm, J. R. 1966. Plant colonization studies on black wastes from anthracite mining in Pennsylvania. *Trans. Amer. Phil. Soc.,* Vol. 56, Part 1. 194 pp.

Schulman, Edmund. 1956. *Dendroclimatic Changes in Semiarid America.* Univ. Ariz. Press, Tucson, Arizona. 142 pp.

Schuster, L. 1950. Über den Sammeltrieb des Eichelhähers (*Garrulus*). *Vogelwelt* 71:9–17.

Schwintzer, Christa R., and Gary Williams. 1974. Vegetation changes in a small Michigan bog from 1917 to 1972. *Am. Mid. Nat.* 92:447–459.

Scott, D. R. M. 1955. Amount and chemical composition of the organic matter contributed by overstory and understory vegetation to forest soil. Yale Univ. Sch. For. Bull. 62. 73 pp.

Sebald, O. 1964. Ökologische Artengruppen für den Wuchsbezirk "Oberer Neckar." *Mitt. Vereins forstl. Standortsk. Forstpflz.* 14:60–63.

Secrest, H. C., J. J. MacAloney, and R. C. Lorenz. 1941. Causes of the decadence of hemlock at the Menominee Indian Reservation, Wisconsin. *J. For.* 39:3–12.

Selm, H. R. 1952. Carbon dioxide gradients in a beech forest in central Ohio. *Ohio J. Sci.* 52(4):187–198.

Senn, G. 1923. Über die Ursachen der Brettwurzelbildung bei der Pyramiden-Pappel. *Verh. naturf. Ges. Basel.* 35:405–435.

Shanks, R. E. 1956. Altitudinal and microclimatic relationships of soil temperature under natural vegetation. *Ecology* 37:1–7.

Sharik, Terry L., and Burton V. Barnes. 1976. Phenology of shoot growth among diverse populations of yellow birch (*Betula alleghaniensis*) and sweet birch (*B. lenta*). *Can. J. Bot.* 54:2122–2129.

Sharp, R. F. 1975. Nitrogen fixation in deteriorating wood: the incorporation of $^{15}N_2$ and the effects of environmental conditions on acetylene reduction. *Soil Biol. Biochem.* 7:9–14.

Shimwell, David W. 1971. *The Description and Classification of Vegetation.* Univ. Washington Press, Seattle. 322 pp.

Shirley, H. L. 1929. The influence of light intensity and light quality upon the growth of plants. *Amer. J. Bot.* 16:354–390.

———. 1932. Light intensity in relation to plant growth in a virgin Norway pine forest. *J. Agr. Res.* 44:227–244.

———. 1935, 1945a. Light as an ecological factor and its measurement. *Bot. Rev.* 1:355–381; 11:497–532.

———. 1945b. Reproduction of upland conifers in the Lake States as affected by root competition and light. *Amer. Mid. Nat.* 33:537–612.

Short, Henry L. 1976. Composition and squirrel use of acorns of black and white oak groups. *J. Wildl. Manage.* 40:479–483.

Shoulders, Eugene, and W. H. McKee, Jr. 1973. Pine nutrition in the west gulf coastal plain: a status report. USDA For. Serv. Gen. Tech. Report SO–2. Southern For. Exp. Sta., New Orleans. 26 pp.

Shreve, F. 1915. The vegetation of a desert mountain range as conditioned by climatic factors. Carnegie Inst. Wash., Publ. 217. 112 pp.

———. 1922. Conditions indirectly affecting vertical distribution on desert mountains. *Ecology* 3:269–274.

Shugart, H. H., T. R. Crow, and J. M. Hett. 1973. Forest succession models: a rationale and methodology for modeling forest succession over large regions. *For. Sci.* 19:203–212.

Siegelman, H. W. 1969. Phytochrome. *In* Malcolm B. Wilkins (ed.), *The Physiology of Plant Growth and Development.* McGraw-Hill Book Co., Inc., New York.

Simak, M. 1951. Untersuchungen über den natürlichen Baumartenwechsel in schweizerischen Plenterwäldern. *Mitt. schweiz. Anst. forstl. Versuchsw.* 27:406–468.

Sirén, G. 1955. The development of spruce forest on raw humus sites in northern Finland and its ecology. Acta For. Fenn. 62(4). 408 pp.

Slaughter, C. W., Richard J. Barnes, and G. M. Hansen (eds.). 1971. *Fire in the Northern Environment—A Symposium.* USDA For. Serv. Pacific Northwest For. and Rge. Exp. Sta., Portland, Ore. 275 pp.

Small, J. 1954. *Modern Aspects of pH, with Special Reference to Plants and Soils.* Bailliere, Tindall and Cox, London. 247 pp.

Smith, Christopher C. 1970. The coevolution of pine squirrels (*Tamiasciurus*) and conifers. *Ecol. Monogr.* 40:349–371.

———. 1973. The coevolution of plants and seed predators. *In* Lawrence E. Gilbert and Peter H. Raven (eds.), *Coevolution of Animals and Plants.* Univ. Texas Press, Austin.

Smith, C. C., and D. Follmer. 1972. Food preferences of squirrels. *Ecology* 53:82–91.

Smith, David M. 1951. The influence of seedbed conditions on the regeneration of eastern white pine. Conn. Agr. Exp. Sta. Bull. 545. 61 pp.

———. 1962. *The Practice of Silviculture,* 7th ed. John Wiley & Sons, Inc., New York. 578 pp.

Smith, Frederick E. 1970. Analysis of ecosystems. *In* David E. Reichle (ed.), *Analysis of Temperate Forest Ecosystems.* Springer-Verlag, New York.

Smith, R. H. 1966. Resin quality as a factor in the resistance of pines to bark beetles. *In* H. D. Gerhold, R. E. McDermott, E. J. Schreiner, and J. A. Winieski (eds.), *Breeding Pest-Resistant Trees.* Pergamon Press, New York.

Snyder, John D., and Robert A. Janke. 1976. Impact of moose browsing on boreal-type forests of Isle Royale National Park. *Am. Midl. Nat.* 95:79–92.

Soil Survey Staff. 1975. *Soil Taxonomy.* USDA, Agr. Handbook 436. Washington, D.C. 754 pp.

Solbrig, Otto T. 1970. *Principles and Methods of Plant Biosystematics.* The Macmillan Co., Collier-Macmillan Canada, Ltd., Toronto. 226 pp.

Sollins, P., W. F. Harris, and N. T. Edwards. 1976. Simulating the physiology of a temperate deciduous forest. *In* B. C. Patton (ed.), *Systems Analysis and Simulation in Ecology.* Vol. VI. Academic Press, Inc., New York.

Sondheimer, Ernest, and John B. Simeone (eds.). 1970. *Chemical ecology.* Academic Press, Inc., New York. 336 pp.

Sopper, William E., and Howard W. Lull (eds.). 1967. *Forest Hydrology.* Pergamon Press, Inc., New York. 813 pp.

Southwood, T. R. E. 1973. The insect/plant relationship—an evolutionary perspective. *In* H. F. van Emden (ed.), *Insect/Plant Relationships.* Sympos. Royal Ent. Soc. London No. 6. John Wiley & Sons, New York.

Spies, Thomas Allen. 1978. Occurrence, morphology, and reproductive biology of natural hybrids of *Populus alba* in southeastern Michigan. Master's thesis, University of Michigan, Ann Arbor. 125 pp.

Spurr, Stephen H. 1941. The influence of two *Juniperus* species on soil reaction. *Soil Sci.* 50:289–294.

———. 1945. A new definition of silviculture. *J. For.* 43:44.

———. 1951. George Washington, surveyor and ecological observer. *Ecology* 32:544–549.

———. 1952a. Origin of the concept of forest succession. *Ecology* 33:426–427.

———. 1952b. *Forest Inventory.* The Ronald Press Co., New York. 476 pp.

———. 1953. The vegetational significance of recent temperature changes along the Atlantic seaboard. *Amer. J. Sci.* 251:682–688.

———. 1954. The forests of Itasca in the nineteenth century as related to fire. *Ecology* 35:21–25.

———. 1956a. Natural restocking of forests following the 1938 hurricane in central New England. *Ecology* 37:443–451.

———. 1956b. Forest associations in the Harvard Forest. *Ecol. Monogr.* 26:245–262.

———. 1956c. Soils in relation to site index curves. *Proc. Soc. Amer. For.* 1955: 80–85.

———. 1957. Local climate in the Harvard Forest. *Ecology* 38:37–46.

———. 1960. *Photogrammetry and Photo-interpretation.* The Ronald Press Co., New York. 472 pp.

———. 1962. Growth and mortality of a 1925 planting of *Pinus radiata* on pumice. *N.Z.J. For.* 8:560–569.

———. 1963. Growth of Douglas-fir in New Zealand. New Zealand For. Serv. For. Res. Inst. Tech. Paper 43. 54 pp.

———, and A. C. Cline. 1942. Ecological forestry in central New England. *J. For.* 40:418–420.

———, and J. H. Allison. 1956. The growth of mature red pine in Minnesota. *J. For.* 54:446–451.

———, and James H. Zumberge. 1956. Late Pleistocene features of Cheboygan and Emmett Counties, Michigan. *Amer. J. Sci.* 254:96–109.

———, and Burton V. Barnes. 1973. *Forest Ecology.* The Ronald Press, Co., New York. 571 pp.

Squillace, A. E. 1966a. Racial variation in slash pine as affected by climatic factors. USDA For. Ser. Res. Paper SE–21. Southeast. For. Exp. Sta., Asheville, N.C. 10 pp.

———. 1966b. Geographic variation in slash pine. *For. Sci. Monogr.* 10. 56 pp.

———. 1967. Effectiveness of 400-foot isolation around a slash pine seed orchard. *J. For.* 61:281–283.

———. 1970. Genotype-environment interactions in forest trees. *In* Papers presented at the second meeting of the working group on quantitative genetics, Sect. 22, IUFRO, Aug. 18–19, 1969, Raleigh, N. C. Pub. by Southern For. Exp. Sta. 1970.

———, and R. T. Bingham. 1958. Localized ecotypic variation in western white pine. *For. Sci.* 4:20–34.

———, and Roy Silen. 1962. Racial variation in ponderosa pine. *For. Sci. Monogr.* 2. 27 pp.

Stage, Albert R., and Jack R. Alley. 1973. An inventory design using stand examinations for planning and programming timber management. USDA For. Serv. Res. Paper INT–126. Intermountain For. and Rge. Exp. Sta., Ogden, Utah. 17 pp.

Stålfelt, M. G. 1924. Tallens och Granens Kolsyreassimilation och dess Ekologiska Betingelser. *Medd. Statens Skogsförsöksan.* 21:181–258.

———. 1935. Einfluss der Durchforstung auf die Funktion der Nadeln und auf die Ausbildung der Baumkrone bei der Fichte. *Svenska Skogsvå–rdsför. Tidsk.* 33:149–176.

Stanhill, Gerald. 1970. The water flux in temperate forests: precipitation and evapotranspiration. *In* David E. Reichle (ed.), *Analysis of Temperate Forest Ecosystems.* Springer-Verlag, New York.

Stark, N. 1968. Seed ecology of *Sequoiadendron giganteum. Madroño* 19:267–277.

Stark, Nellie M. 1977. Fire and nutrient cycling in a Douglas-fir/larch forest. *Ecology* 58:16–30.

Stearns, F. W. 1949. Ninety-years change in a northern hardwood forest in Wisconsin. *Ecology* 30:350–358.

Stebbins, G. Ledyard. 1950. *Variation and Evolution in Plants.* Columbia Univ. Press, New York. 643 pp.

———. 1966. *Processes of Organic Evolution.* Prentice-Hall, Inc., Englewood Cliffs, N.J. 191 pp.

———. 1969. The significance of hybridization for plant taxonomy and evolution. *Taxon* 18:26–35.

———. 1970. Variation and evolution in plants: progress during the past twenty years. *In* Max K. Hecht and William C. Steere (eds.), *Essays in Evolution and Genetics.* Appleton-Century-Crofts, Inc., New York.

———. 1976. Seeds, seedlings, and the origin of angiosperms. *In* Charles B. Beck (ed.), *Origin and Early Evolution of Angiosperms.* Columbia Univ. Press, New York.

Steinbrenner, E. C. 1951. Effect of grazing on floristic composition and soil properties of farm woodlands in southern Wisconsin. *J. For.* 49:906–910.

———. 1955. The effect of repeated tractor trips on the physical properties of forest soils. *Northwest Sci.* 29:155–159.

———. 1968. Research in forest fertilization at Weyerhaeuser Company in the Pacific Northwest. *In Forest Fertilization.* Tennessee Valley Authority, Muscle Shoals, Ala.

———. 1975. Mapping forest soils on Weyerhaeuser lands in the Pacific Northwest. *In* B. Bernier and C. H. Winget (eds.), *Forest Soils and Forest Land Management.* Laval Univ. Press, Quebec.

———, and S. P. Gessel. 1955. The effect of tractor logging on physical properties of some forest soils of southwestern Washington. *Proc. Soil Sci. Soc. Amer.* 19:372–375.

Stern, Klaus, and Laurence Roche. 1974. *Genetics of Forest Ecosystems.* Springer-Verlag, New York. 330 pp.

Steven, H. M., and A. Carlisle. 1959. *The Native Pinewoods of Scotland.* Oliver and Boyd, Ltd., Edinburgh. 368 pp.

Stevens, Clark Leavitt. 1931. Root growth of white pine. Yale Univ. School For. Bull. 32. New Haven, Conn. 62 pp.
Stewart, G. A. (ed.). 1968. *Land Evaluation*. Macmillan Co., Melbourne. 392 pp.
Stewart, W. D. P. 1967. Nitrogen-fixing plants. *Science* 158:1426–1432.
Stoate, T. N. 1951. Nutrition of the pine. Austr. For. and Timb. Bur. Bull. 30. 61 pp.
Stoddard, Herbert L. 1931. *The Bobwhite Quail; Its Habits, Preservation and Increase*. Scribner's, New York. 500 pp.
Stoeckeler, Joseph H. 1948. The growth of quaking aspen as affected by soil properties and fire. *J. For.* 46:727–737.
———. 1960. Soil factors affecting the growth of quaking aspen forests in the Lake States. Univ. Minn. Agr. Exp. Sta. Tech. Bull. 233. 48 pp.
Stone, E. C. 1957. Dew as an ecological factor. *Ecology* 38:407–422.
———, A. Y. Shachori, and R. G. Stanley. 1956. Water absorption by needles of ponderosa pine seedlings and its internal redistribution. *Plant Physiol.* 31:120–126.
Stone, E. L. 1967. Potassium deficiency and response in young conifer forests in eastern North America. *In Proc. Colloquium on Forest Fertilization*, pp. 217–229. Int. Potash Inst., Berne, Switzerland.
———. 1968. Microelement nutrition of forest trees: a review. *In Forest Fertilization*. Tennessee Valley Authority, Muscle Shoals, Ala.
———. 1971. Effects of prescribed burning on long-term productivity of coastal plain soils. *In* Prescribed burning symposium proceedings. USDA For. Serv. Southeast. For. Exp. Sta., Asheville, N.C.
———. 1973. Regional objectives in forest fertilization: current and potential. *In Proc. Forest Fertilization Symposium*, pp. 10–18. USDA For. Serv. Gen. Tech. Report NE–3. Northeastern For. Exp. Sta., Upper Darby, Pa.
———. 1974. The communal root system of red pine: growth of girdled trees. *For. Sci.* 20:294–305.
———. 1975. Windthrow influences on spatial heterogeneity in a forest soil. *Mitt. Eidg. Anst. forstl. Versw.* 51:77–87.
———. 1977. Abrasion of tree roots by rock during wind stress. *For. Sci.* 23:333–336.
———, and C. McAuliffe. 1954. On the sources of soil phosphorous absorbed by mycorrhizal pines. *Science* 120:946–948.
———, and M. H. Stone. 1954. Root collar sprouts in pine. *J. For.* 52:487–491.
Straka, H. 1970. *Arealkunde/Floristisch-historische Geobotanik*. Verlag Eugen Ulmer, Stuttgart. 478 pp.
Stuiver, Minze. 1978. Atmospheric carbon dioxide and carbon reservoir changes. *Science* 199:253–258.
Sukachev, V. N., and N. V. Dylis. 1964. *Fundamentals of Forest Biogeocoenology*. Translated by J. M. MacLennan. Oliver and Boyd, Ltd., Edinburgh. 672 pp.
Sutton, R. F. 1969. Form and development of conifer root systems. Commonw. For. Bur., Tech. Comm. No. 7. Oxford. 131 pp.

———. and E. L. Stone Jr. 1974. White grubs: a description for foresters, and an evaluation of their silvicultural significance. Canadian For. Serv. Info. Report 0-X-212. Great Lakes For. Res. Centre. Sault Ste. Marie, Ontario. 21 pp.

Swain, Albert M. 1973. A history of fire and vegetation in northeastern Minnesota as recorded in lake sediments. *Quaternary Res.* 3:383–396.

Swain, Tony. 1977. Secondary compounds as protective agents. *Ann. Rev. Plant Physiol.* 28:479–501.

Swank, Wayne T., and Gray S. Henderson. 1976. Atmospheric input of some cations and anions to forest ecosystems in North Carolina and Tennessee. *Water Resources Res.* 12:541–546.

Swanson, Frederick J., George W. Lienkaemper, and James R. Sedell. 1976. History, physical effects, and management implications of large organic debris in western Oregon streams. USDA For. Serv. Gen. Tech. Report PNW–56. Pacific Northwest For. and Rge. Exp. Sta., Portland, Ore. 15 pp.

———, and Douglas N. Swanston. 1977. Complex mass-movement terrains in the western Cascade Range, Oregon. *Rev. Eng. Geol.* 3:113–124.

———, and George W. Lienkaemper. 1978. Physical consequences of large organic debris in Pacific Northwest streams. USDA For. Serv. Gen. Tech. Report PNW-69. Pacific Northwest For. and Rge. Exp. Sta., Portland, Ore. 12 pp.

———, R. L. Fredriksen, and F. M. McCorison. 1980. Material transfer in a western Oregon forested watershed. *In* R. L. Edmonds (ed.), *Analysis of Coniferous Forest Ecosystems in the Western United States.* US/IBP Synthesis Series. Dowden, Hutchinson and Ross, Inc., Stroudsburg, Pa.

———, J. R. Sedell, and F. J. Triska. 1980. Land-water interactions: the riparian zone. *In* R. L. Edmonds (ed.), *Analysis of Coniferous Forest Ecosystems in the Western United States.* US/IBP Synthesis Series. Dowden, Hutchinson and Ross, Inc., Stroudsburg, Pa.

Switzer, G. L., and L. E. Nelson. 1972. Nutrient accumulation and cycling in loblolly pine (*Pinus taeda* L.) plantation ecosystems: the first twenty years. *Soil Sci. Soc. Am. Proc.* 36:143–147.

Syrach-Larsen, C. 1956. *Genetics in Silviculture.* Trans. by Mark L. Anderson. Oliver and Boyd, London. 224 pp.

Takahara, S., A. Kawana, and I. Tange. 1955. Influence of light intensity and soil moisture on the shade-endurance in the leaves of Shirakashi seedlings, *Cyclobalanopsis myrsinaefolia* Oerst. *Bot. Mag. (Tokyo)* 68:212–215.

Talbot, Lee M., W. J. A. Payne, H. P. Ledger, Lorna D. Verdcourt, and Martha H. Talbot. 1965. The meat production potential of wild animals in Africa; a review of biological knowledge. Tech. Comm. Commonwealth Bur. Animal Breeding and Genetics No. 16. 42 pp.

Tamm, Carl Olof. 1950. Northern coniferous forest soils: a popular survey of the phenomena which determine the productive character of the forest soils of North Sweden. Transl. from the Swedish by Mark L. Anderson. Scrivener Press, Oxford. 253 pp.

———. 1964. Determination of nutrient requirements of forest stands. *Int. Rev. For. Res.* 1:115–170.

Tansley, A. G. 1935. The use and abuse of vegetational concepts and terms. *Ecology* 16:284–307.

———. 1949. *The British Islands and their Vegetation. I.* Cambridge Univ. Press, Cambridge. 484 pp.

Tappeiner, John C., II. 1969. Effect of cone production on branch, needle, and xylem ring growth of Sierra Nevada Douglas-fir. *For. Sci.* 15:171–174.

Tarrant, R. F. 1956a. Changes in some physical soil properties after a prescribed burn in young ponderosa pine. *J. For.* 54:439–441.

———. 1956b. Effect of slash burning on some physical soil properties. *For. Sci.* 2:18–22.

Taylor, Alan R. 1969. Lightning effects on the forest complex. *In Proc. Annual Tall Timbers Fire Ecology Conference, 1969*, pp. 127–150. Tall Timbers Res. Sta., Tallahassee, Fla.

———. 1974a. Ecological aspects of lightning in forests. *In Proc. Annual Tall Timbers Fire Ecology Conference*, 13:455–482. Tall Timbers Res. Sta., Tallahassee, Fla.

———. 1974b. Forest fire. *In* D. N. Lapedes (ed.), *McGraw-Hill Yearbook of Science and Technology*. McGraw-Hill Book Co., Inc., New York.

———. 1977. Lightning and trees. *In*. R. H. Golde (ed.), *Lightning*, Vol. 2, *Lightning Protection*. Academic Press, Inc., London.

Taylor, B. W. 1957. Plant succession on recent volcanoes in Papua. *J. Ecol.* 45:233–253.

Taylor, T. M. C. 1959. The taxonomic relationship between *Picea glauca* (Moench) Voss and *P. engelmannii* Parry. *Madroño* 15:111–115.

Tevis, L., Jr. 1956. Pocket gophers and seedlings of red fir. *Ecology* 37:379–381.

Thie, J., and G. Ironside (eds.). 1977. *Ecological (Biophysical) Land Classification in Canada*. Ecological land classification series, No. 1, Lands Directorate, Env. Canada, Ottawa. 269 pp.

Thilenius, John F. 1972. Classification of deer habitat in the ponderosa pine forest of the Black Hills, South Dakota. USDA For. Serv. Res. Paper RM–91. Rocky Mountain For. and Rge. Exp. Sta., Ft. Collins, Colo. 28 pp.

Thompson, Daniel Q., and Ralph H. Smith. 1970. The forest primeval in the Northeast—a great myth? *In Proc. Annual Tall Timbers Fire Ecology Conference*. 10:255–265. Tall Timbers Res. Sta., Tallahassee, Fla.

Thornthwaite, C. W. 1941. Atlas of climatic types in the United States, 1900–1939. U.S. Dept. Agric. Misc. Pub. 421. 7 pp.

———. 1948. An approach toward a rational classification of climate. *Geog. Rev.* 38:55–94.

———, and J. R. Mather. 1957. The water balance. Drexel Inst. Tech., Lab. Climatol., Pub. Climatol., Vol. 8(1). 86 pp.

Tingey, David T., Raymond G. Wilhour, and Carol Standley. 1976. The effect of chronic ozone exposures on the metabolite content of ponderosa pine seedlings. *For. Sci.* 22:234–241.

Torkildsen, G. B. 1950. Om årsakene til granens dårlige gjenvekst i einstapebestand. (The cause of poor reproduction of spruce in stands with an undergrowth of bracken.) *Blyttia* 8(4):160–164.

Toumey, J. W. 1947. *Foundations of Silviculture upon an Ecological Basis*, 2d ed. (rev. by Clarence F. Korstian). John Wiley & Sons, Inc., New York. 468 pp.

———, and R. Kienholz. 1931. Trenched plots under forest canopies. Yale Univ. Sch. Forestry Bull. 30. 31 pp.

Transeau, E. N. 1935. The prairie peninsula. *Ecology* 16:423–437.

Trappe, James M. 1977. Selection of fungi for ectomycorrhizal inoculation in nurseries. *Ann. Rev. Phytopathol.* 15:203–222.

———, and Robert D. Fogel. 1977. Ecosystematic functions of mycorrhizae. *In* J. K. Marshall (ed.), *The Belowground Ecosystem: A Synthesis of Plant-Associated Processes*. Range Sci. Dep. Sci. Ser. No. 26. Colo. State Univ., Ft. Collins.

Treshow, Michael. 1970. *Environment and Plant Response*. McGraw-Hill Book Co., Inc., New York. 422 pp.

Trimble, George R., Jr. 1977. A history of the Fernow Experimental Forest and the Parsons Timber and Watershed Laboratory. USDA For. Serv. Gen. Tech. Report NE–28. Northeastern For. Exp. Sta., Upper Darby, Pa. 46 pp.

———, and S. Weitzman. 1956. Site index studies of upland oaks in the northern Appalachians. *For. Sci.* 2:162–173.

Triska, F. J., J. R. Sedell, and S. V. Gregory. 1980. The coniferous forest stream; physical, chemical and biological interactions through space and time. *In* R. L. Edmonds (ed.), *Analysis of Coniferous Forest Ecosystems in the Western United States*. US/IBP Synthesis Series. Dowden, Hutchinson and Ross, Inc., Stroudsburg, Pa.

Troeger, R. 1960. Kiefernprovenienzversuche. I. Teil. Der grosse Kiefernprovenienzversuch im südwürttembergischen Forstbezirk Schussenried. *AFJZ* 131(3–4):49–59.

Trousdell, K. B., and M. D. Hoover. 1955. A change in ground-water level after clearcutting of loblolly pine in the Coastal Plain. *J. For.* 53:493–498.

———, Donald E. Beck, and F. Thomas Lloyd. 1974. Site index for loblolly pine in the Atlantic Coastal Plain of the Carolinas and Virginia. USDA For. Serv. Res. Paper SE–115. Southeastern For. Exp. Sta., Asheville, N.C. 11 pp.

Tryon, E. H., and Rudolfs Markus. 1953. Development of vegetation on century-old iron-ore spoil banks. West Va. Agr. Exp. Sta. Bull. 360. 63 pp.

———. J. O. Cantrell, and K. L. Carvell. 1957. Effect of precipitation and temperature on increment of yellow-poplar. *For. Sci.* 3:32–44.

Tschudy, Robert H., and Richard A. Scott (eds.). 1969. *Aspects of Palynology*. John Wiley & Sons, Inc., New York. 510 pp.

Tubbs, Carl H. 1965. Influence of temperature and early spring conditions on sugar maple and yellow birch germination in upper Michigan. USDA For. Serv. Res. Note LS–72. North Central For. Exp. Sta., St. Paul, Minn. 2 pp.

———. 1969. The influence of light, moisture, and seedbed on yellow birch regeneration. USDA For. Serv. Res. Paper NC–27. North Central For. Exp. Sta., St. Paul, Minn. 12 pp.

———. 1973. Allelopathic relationship between yellow birch and sugar maple seedlings. *For. Sci.* 19:139–145.

———. 1976. Effect of sugar maple root exudate on seedlings of northern conifer species. USDA For. Serv. Res. Note NC–213. North Central For. Exp. Sta., St. Paul, Minn. 2 pp.

Tucker, John M. 1961. Studies in the *Quercus undulata complex*. I. A preliminary statement. *Amer. J. Bot.* 48:202–208.

Tukey, H. B., Jr. 1962. Leaching of metabolites from above-ground plant parts, with special reference to cuttings used in propagation. *Proc. Plant. Prop. Soc.*, pp. 63–70.

———. 1969. Implications of allelopathy in agricultural plant science. *Bot. Rev.* 35:1–16.

Turekian, Karl K. (ed.). 1971. *The Late Cenozoic Glacial Ages*. Yale Univ. Press, New Haven. 606 pp.

Turreson, Gote. 1922a. The species and the variety as ecological units. *Hereditas* 3:100–113.

———. 1922b. The genotypical response of the plant species to the habitat. *Hereditas* 3:211–350.

———. 1923. The scope and import of genecology. *Hereditas* 4:171–176.

Udvardy, Miklos D. F. 1969. Birds of the coniferous forest. *In* Richard D. Taber (ed.), *Coniferous Forests of the Northern Rocky Mountains*. Center for Nat. Res., Missoula, Mont.

U.S. Corps of Engineers. 1956. Snow hydrology. Summary report of the snow investigations. North Pac. Div., U.S.C.E., Portland, Ore. 437 pp.

U.S. Department of Agriculture. 1941. *Climate and Man*. 1941 Yearbook of Agriculture. U. S. Govt. Printing Office, Washington, D. C. 1248 pp.

———. 1965. *Silvics of Forest Trees of the United States*. USDA For. Serv. Agr. Handbook No. 271. Washington, D.C. 762 pp.

———. 1973. Trees for polluted air. Misc. Publ. 1230. 12 pp.

———. 1974. *Seeds of Woody Plants in the United States*. USDA For. Serv. Agr. Handbook No. 450, Washington, D.C. 883 pp.

U.S. Department of Interior. 1967. Surface mining and our environment—a special report to the nation. U.S. Govt. Printing Office, Washington, D.C.

U.S. Tropical Forest Experimental Station. 1951. Micro-environment significant to tree-growth in limestone region. *Carib. For.* 12(1):7–8, 24–25.

Vaartaja, O. 1954. Temperature and evaporation at and near ground level on certain forest sites. *Canad. J. Bot.* 32:760–783.

Van der Hammen, T., T. A. Wijmstra, and W. H. Zagwijn. 1971. The floral record of the late Cenozoic of Europe. *In* Karl K. Turekian (ed.), *The Late Cenozoic Glacial Ages*. Yale Univ. Press, New Haven.

Van Eck, W. A., and E. P. Whiteside. 1963. Site evaluation studies in red pine plantations in Michigan. *Proc. Soil Sci. Soc. Amer.* 27:709–714.

Van Haverbeke, David F. 1968a. A taxonomic analysis of *Juniperus* in the central and northern Great Plains. *In* Ornamental tree and shrub improvement—the forester's role. *Proc. Sixth Cent. States For. Tree Imp. Conf.* Carbondale, Ill.

———. 1968b. A population analysis of *Juniperus* in the Missouri River

Basin. Univ. of Nebraska Studies New Series No. 38. 82 pp.
Van Hook, R. I., W. F. Harris, and G. S. Henderson. 1977. Cadmium, lead, and zinc distributions and cycling in a mixed deciduous forest. *Ambio* 6:281–286.
Van Wagner, C. E. 1970. Fire and red pine. *In Proc. Annual Tall Timbers Fire Ecology Conference,* 10:211–219. Tall Timbers Res. Sta., Tallahassee, Fla.
Vegis, A. 1964. Dormancy in higher plants. *Ann. Rev. Plant Phys.* 15:185–224.
Verrall, A. F., and T. W. Graham. 1935. The transmission of *Ceratostomella ulmi* through root grafts. *Phytopathology* 25:1039–1040.
Verry, Elon S., and D. R. Timmons. 1977. Precipitation nutrients in the open and under two forests in Minnesota. *Can. J. For. Res.* 7:112–119.
Vézina, P. E. 1961. Variations in total solar radiation in three Norway spruce plantations. *For. Sci.* 7:257–264.
———, and D. W. K. Boulter. 1966. The spectral composition of near ultraviolet and visible radiation beneath forest canopies. *Can. J. Bot.* 44:1267–1284.
Viereck, Leslie A. 1973. Wildfire in the taiga of Alaska. *Quaternary Res.* 3:465–495.
———, and Joan M. Foote. 1970. The status of *Populus balsamifera* and *P. trichocarpa* in Alaska. *Canad. Field-Nat.* 84:169–173.
Vince-Prue, Daphne. 1975. *Photoperiodism in Plants.* McGraw-Hill Book Co., Inc., New York. 444 pp.
Viosca, Percy, Jr. 1931. Spontaneous combustion in the marshes of southern Louisiana. *Ecology* 12:439–442.
Viro, P. J. 1956. The role of silica in the decomposition of forest humus. Sixieme Congres de la Science du Sol. Paris. Comm. 11 (45):723–726.
———. 1974. Effects of forest fire on soil. *In* T. T. Kozlowski and C. E. Ahlgren (eds.), *Fire and Ecosystems.* Academic Press, Inc., New York.
Visher, Stephen S. 1954. *Climatic Atlas of the United States.* Harvard Univ. Press. 403 pp.
Vogelman, H. W., Thomas Siccama, Dwight Leedy, and Dwight C. Ovitt. 1968. Precipitation from fog moisture in the Green Mountains of Vermont. *Ecology* 49:1205–1207.
Vogl, Richard J., and Calvin Ryder. 1969. Effects of slash burning on conifer reproduction in Montana's Mission Range. *Northwest Sci.* 43:135–147.
Voigt, G. K. 1960. Distribution of rainfall under forest stands. *For. Sci.* 6:2–10.
———. 1968. Variation in nutrient uptake by trees. *In Forest Fertilization.* Tennessee Valley Authority, Muscle Shoals, Ala.
Wagener, Willis W. 1961. Past fire incidence in Sierra Nevada forests. *J. For.* 59:739–748.
Wagenknecht, Egon, Alexis Scammoni, Albert Richter, and J. Lehmann. 1956. *Eberswalde 1953: Wege zu Standortgerechter Forstwirtschaft.* Neumann Verlag, Berlin. 523 pp.
Wagner, Frederic H. 1969. Ecosystem concepts in fish and game management. *In* George M. Van Dyne (ed.), *The Ecosystem Concept in Natural*

*Resource Management*. Academic Press, Inc., New York.
Wagner, W. H., Jr. 1968. Hybridization, taxonomy and evolution. *In* V. H. Heywood (ed.), *Modern Methods in Plant Taxonomy*. Academic Press, Inc., New York.
———. 1970. Biosystematics and evolutionary noise. *Taxon* 19:146–151.
Wahl, Eberhard W. 1968. A comparison of the climate of the eastern United States during the 1830's with the current normals. *Monthly Weather Rev*. 96:73–82.
———, and T. L. Lawson. 1970. The climate of the midnineteenth century United States compared to the current normals. *Monthly Weather Rev*. 98:259–265.
Wahlenberg, W. G. 1949. Forest succession in the southern Piedmont region. *J. For*. 47:713–715.
Wakeley, Philip C. 1954. Planting the southern pines. USDA For. Serv. Agr. Monogr. No. 18. 233 pp.
———, and J. Marrero. 1958. Five-year intercept as site index in southern pine plantations. *J. For*. 56:332–336.
Wallwork, J. A. 1959. The distribution and dynamics of some forest soil mites. *Ecology* 40:557–563.
Walter, Heinrich. 1973. *Vegetation of the Earth: In Relation to Climate and the Eco-Physiological Conditions*. Springer-Verlag, New York. 237 pp.
Ward, R. T. 1956. The beech forests of Wisconsin—changes in forest composition and the nature of the beech border. *Ecology* 37:407–419.
Wardle, Peter. 1968. Engelmann spruce (*Picea engelmannii* Engel.) at its upper limits in the Front Range, Colorado. *Ecology* 49:483–495.
Wareing, P. F. 1950. Growth studies in woody species. I. Photoperiodism in first-year seedlings of *Pinus silvestris*. II. Effect of day-length on shoot-growth in *Pinus silvestris* after the first year. *Physiol. Plant. (Copenhagen)* 3(3):258–276.
———. 1951. Growth studies in woody species. III. Further photoperiodic effects in *Pinus silvestris*. *Physiol. Plant. (Copenhagen)* 4(1):41–56.
———. 1956. Photoperiodism in woody plants. *Ann. Rev. Plant Phys*. 7:191–214.
———. 1959. Problems of juvenility and flowering in trees. *J. Linn. Soc. London, Bot*. 56:282–289.
———. 1969. Germination and dormancy. *In* M. B. Wilkins (ed.), *The Physiology of Plant Growth and Development*. McGraw-Hill Book Co., Inc., New York.
———, and L. W. Robinson. 1963. Juvenility problems in woody plants. *Rep. Forest Res*. pp. 125–127.
Waring, R. H. 1974. Structure and function of the coniferous forest biome organization. *In* R. H. Waring and R. L. Edmonds (eds.), *Integrated Research in the Coniferous Forest Biome*. Conif. For. Biome Bull. No. 5, Coniferous Forest Biome, US/IBP, Univ. Washington, Seattle, Wash.
———, and J. Major. 1964. Some vegetation of the California coastal redwood region in relation to gradients of moisture, nutrients, light, and temperature. *Ecol. Monogr*. 34:167–215.
———, K. L. Reed, and W. H. Emmingham. 1972. An environmental grid for

classifying coniferous forest ecosystems. *In* Jerry F. Franklin, L. J. Dempster, and Richard H. Waring (eds.), *Proc—Research on Coniferous Forest Ecosystems—A Symposium*. USDA For. Serv., Pacific Northwest For. and Rge. Exp. Sta., Portland, Ore.

———, and J. F. Franklin. 1979. The evergreen coniferous forests of the Pacific Northwest. *Science* 204:1380–1386.

Watson, E. S., D. C. McClurkin, and M. B. Huneycutt. 1974. Fungal succession on loblolly pine and upland hardwood foliage and litter in north Mississippi. *Ecology* 55:1128–1134.

Watt, A. S. 1947. Pattern and process in the plant community. *J. Ecol.* 35:1–22.

Watts, W. A. 1970. The full-glacial vegetation of northwestern Georgia. *Ecology* 51:17–33.

Wearstler, Kenneth A. Jr., and Burton V. Barnes. 1977. Genetic diversity of yellow birch seedlings in Michigan. *Can. J. Bot.* 55:2778–2788.

Weaver, Harold. 1951. Fire as an ecological factor in Southwestern ponderosa pine forests. *J. For.* 49:93–98.

———. 1974. Effects of fire on temperate forests: western United States. *In* T. T. Kozlowski and C. E. Ahlgren (eds.) *Fire and Ecosystems*. Academic Press, Inc., New York.

Weaver, John E., and Frederic E. Clements. 1938. *Plant Ecology*. 2d ed. McGraw-Hill Book Co., Inc., New York. 601 pp.

Webb, Charles D., and Robert E. Farmer, Jr. 1968. Sycamore seed germination: the effects of provenance, stratification, temperature and parent tree. USDA For. Serv. Res. Note SE–100. Southeastern For. Exp. Sta., Asheville, N.C. 6 pp.

Webb, D. A. 1954. Is the classification of plant communities either possible or desirable? *Bot. Tidsskr.* 51:362–370.

Webb, L. J. 1959. A physiognomic classification of Australian rain forests. *J. Ecol.* 47:551–570.

———. 1968. Environmental relationships of the structural types of Australian rain forest vegetation. *Ecology* 49:296–311.

———, and J. G. Tracey, W. T. Williams, and G. N. Lance. 1967a. Studies in the numerical analysis of complex rain-forest communities. I. A comparison of methods applicable to site/species data. *J. Ecol.* 55:171–191.

———, ———, ———, and ———. 1967b. Studies in the numerical analysis of complex rain-forest communities. II. The problem of species-sampling. *J. Ecol.* 55:525–538.

———, ———, ———, and ———. 1970. Studies in the numerical analysis of complex rain-forest communities. V. A comparison of the properties of floristic and physiognomic-structure data. *J. Ecol.* 58:203–232.

Weber, F. P. 1971. Applications of airborne thermal remote sensing in forestry. *In* Gerd Hildebrandt (ed.), *Application of Remote Sensors in Forestry*. Druckhaus Rombach & Co., Freiburg, i. Br., Germany.

Weetman, G. F., and B. Webber. 1972. The influence of wood harvesting on the nutrient status of two spruce stands. *Can. J. For. Res.* 2:351–369.

Weiser, C. J. 1970. Cold resistance and injury in woody plants. *Science* 169:1269–1278.

Wellner, C. A. 1948. Light intensity related to stand density in mature stands of the western white pine type. *J. For.* 46:16–19.

———. 1970. Fire history in the northern Rocky Mountains. *In Proc. Intermountain Fire Research Council and Symposium—the Role of Fire in the Intermountain West*, pp. 42–64. Univ. Montana, School of Forestry, Missoula, Mont.

Wells, Carol A., Dennis Whigham, and Helmut Lieth. 1972. Investigation of mineral nutrient cycling in a (sic) upland Piedmont forest. *J. Elisha Mitchell Sci. Soc.* 88:66–78.

Wells, Carol G. 1971. Effects of prescribed burning on soil chemical properties and nutrient availability. *In* Prescribed burning symposium proceedings. USDA For. Serv. Southeast. For. Exp. Sta., Asheville, N.C.

Wells, O. O. 1964a. Geographic variation in ponderosa pine. I. The ecotypes and their distribution. *Silvae Genetica* 13:89–103.

———. 1964b. Geographic variation in ponderosa pine. II. Correlations between progeny performance and characteristics of the native habitat. *Silvae Genetica* 13:125–132.

Wells, P. V. 1965. Scarp woodlands, transported grassland soils, and concept of grassland climate in the great plains region. *Science* 148:246–249.

Wendt, George E., Richard A. Thompson, and Kermit N. Larson. 1975. Land systems inventory Boise National Forest, Idaho. USDA For. Serv., Intermountain Region, Ogden, Utah. 54 pp. + map.

Went, Frits W. 1957. *The Experimental Control of Plant Growth.* The Ronald Press Co., New York. 343 pp.

Went, J. C. 1963. Influence of earthworms on the number of bacteria in the soil. *In* J. Doeksen and J. van der Drift (eds.), *Soil Organisms.* North-Holland Publishing Co., Amsterdam.

Werner, Hans. 1962. Untersuchungen über das Wachstum der Hauptholzarten auf den wichtigsten Standortseinheiten der Mittleren Alb. *Mitt. Vereins forstl. Standortsk. Forstpflz.* 12:3–52.

Werner, Jörg. 1964. Zur Frage der Wirkung von Fichtenmonokulturen auf staunässeempfindliche Böden. *Standort, Wald und Waldwirtschaft in Oberschwaben*, "Oberschwäbische Fichtenreviere." Stuttgart.

Wert, Steven L., Paul R. Miller, and Robert N. Larsh. 1970. Color photos detect smog injury to forest trees. *J. For.* 68:536–539.

Wertz, W. A., and J. F. Arnold. 1972. Land systems inventory. USDA For. Serv., Intermountain Region, Ogden, Utah. 12 pp.

———, and J. F. Arnold. 1975. Land stratification for land-use planning. *In* B. Bernier and C. H. Winget (eds.), *Forest Soils and Forest Land Management.* Laval Univ. Press, Quebec.

West, Neil E. 1968. Rodent-influenced establishment of ponderosa pine and bitterbrush seedlings in central Oregon. *Ecology* 49:1009–1011.

West, R. G. 1968. *Pleistocene Geology and Biology.* Longmans, Green & Co., Ltd., London. 377 pp.

White, Edwin H. 1974. Whole-tree harvesting depletes soil nutrients. *Can. J. For. Res.* 4:530–535.

White, J., and J. L. Harper. 1970. Correlated changes in plant size and number in plant populations. *J. Ecol.* 58:467–485.

White, T. C. R. 1969. An index to measure weather-induced stress of trees associated with outbreaks of psyllids in Australia. *Ecology* 50:905–909.

———. 1974. A hypothesis to explain outbreaks of looper caterpillars, with special reference to populations of *Selidosema suavis* in a plantation of *Pinus radiata* in New Zealand. *Oecologia* 16:279–301.

———. 1978. The importance of a relative shortage of food in animal ecology. *Oecologia* (Berl.) 33:71–86.

Whitehead, Donald R. 1965. Palynology and pleistocene phytogeography of unglaciated eastern North America. *In* H. E. Wright, Jr., and David G. Frey (eds.), *The Quaternary of the United States.* Princeton Univ. Press, Princeton, N.J.

———. 1969. Wind pollination in the angiosperms: evolutionary and environmental considerations. *Evolution* 23:28–35.

———. 1973. Late-Wisconsin vegetational changes in unglaciated eastern North America. *Quaternary Res.* 3:621–631.

Whitford, Philip C. 1976. Resprouting capacity of oak roots: a ten-year experiment. *Mich. Bot.* 15:89–92.

Whittaker, R. H. 1953. A consideration of climax theory: the climax as a population and pattern. *Ecol. Monogr.* 23:41–78.

———. 1954. Plant populations and the basis of plant indication. *Angewandte Pflanzensoziologie.* 1:183–206.

———. 1956. Vegetaion of the Great Smoky Mountains. *Ecol. Monogr.* 26:1–80.

———. 1960. Vegetation of the Siskiyou Mountains, Oregon and California. *Ecol. Monogr.* 30:279–338.

———. 1962. Classification of natural communities. *Bot. Rev.* 28:1–239.

———. 1966. Forest dimensions and production in the Great Smoky Mountains. *Ecology* 47:103–121.

———. 1967a. Ecological implications of weather modification. *In* R. Shaw (ed.), *Ground Level Climatology.* AAAS, Washington, D.C. Publ. 86.

———. 1967b. Gradient analysis of vegetation. *Biol. Rev.* 42:207–264.

———. 1970. The biochemical ecology of higher plants. *In* Ernest Sondheimer and John B. Simeone (eds.), *Chemical Ecology.* Academic Press, Inc., New York.

———. 1973. Direct gradient analysis. *Handb. Veg. Sci.* 5:9–51, W. Junk B.V., Publishers, The Hague.

———. 1975. *Communities and Ecosystems.* Macmillan Publ. Co., 2nd ed. 385 pp.

———, R. B. Walker, and A. R. Kruckeberg. 1954. The ecology of serpentine soils. *Ecology* 35:258–288.

———, and W. A. Niering. 1964. Vegetation of the Santa Catalina Mountains. I. Ecological classification and distribution of species. *J. Ariz. Acad. Sci.* 3:9–34.

———, and ———. 1965. Vegetation of the Santa Catalina Mountains. II. A gradient analysis of the south slope. *Ecology* 46:429–452.

———, and ———. 1968a. Vegetation of the Santa Catalina Mountains. III.

Species distributions and floristic relations on the north slope. *J. Ariz. Acad. Sci.* 5:3–21.
———, and ———. 1968b. Vegetation of the Santa Catalina Mountains, Arizona. IV. Limestone and acid soils. *J. Ecol.* 56:523–544.
———, and P. P. Feeny. 1971. Allelochemics: chemical interactions between species. *Science* 171:757–770.
———, and G. M. Woodwell. 1968. Dimension and production relations of trees and shrubs in the Brookhaven Forest, New York. *J. Ecol.* 56:1–25.
———, and ———. 1969. Structure, production and diversity of the oak-pine forest at Brookhaven, New York. *J. Ecol.* 57:157–176.
———, and H. G. Gauch. 1973. Evaluation of ordination techniques. *Handb. Veg. Sci.* 5:287–321, W. Junk B.V., Publishers, The Hague.
———, and Peter L. Marks. 1975. Methods of assessing terrestrial productivity. *In* Helmut Lieth and Robert H. Whittaker (eds.), *Primary Productivity of the Biosphere*, Ecological Studies 14. Springer-Verlag, New York.
Wicht, C. L. 1967. Summary of forests and evapotranspiration session. *In* William E. Sopper and Howard W. Lull (eds.), *Forest Hydrology*. Pergamon Press, Inc., New York.
Wiegert, Richard G. 1975. Simulation models of ecosystems. *Ann. Rev. Ecol. Syst.* 6:311–338.
Wiersma, J. H. 1962. Enkele quantitative aspecten van het exotenvraagstuk. *Ned. Bosbouw Tijdschrift* 34:175–184.
———. 1963. A new method of dealing with results of provenance tests. *Silvae Genetica* 12:200–205.
Wieslander, A. E., and R. Earl Storie. 1952. The vegetation-soil survey in California and its use in the management of wild lands for yield of timber, forage and water. *J. For.* 50:521–526.
———, and ———. 1953. Vegetational approach to soil surveys in wild areas. *Proc. Soil Sci. Soc. Amer.* 17:143–147.
Wilcox, James R. 1968. Sweetgum seed stratification requirements related to winter climate at seed source. *For. Sci.* 14:16–19.
Wilde, S. A. 1954. Reaction of soils: facts and fallacies. *Ecology* 35:89–92.
———. 1958. *Forest Soils: Their Properties and Relation to Silviculture.* The Ronald Press Co., New York. 537 pp.
———, E. C. Steinbrenner, R. S. Pierce, R. C. Dosen, and D. T. Pronin. 1953. Influence of forest cover on the state of the ground-water table. *Proc. Soil Sci. Soc. Amer.* 17:65–67.
Williams, W. T., G. N. Lance, L. J. Webb, J. G. Tracey, and J. H. Connell. 1969. Studies in the numerical analysis of complex rain-forest communities. IV. A method for the elucidation of small-scale forest pattern. *J. Ecol.* 57:635–654.
Wilson, Brayton F. 1966. Development of the shoot system of *Acer rubrum* L. Harvard Forest Paper No. 14. Harvard Univ., Petersham, Mass. 21 pp.
———. 1970. *The Growing Tree.* Univ. Mass. Press, Amherst. 152 pp.
Wilson, C. C. 1948. Fog and atmospheric carbon dioxide as related to apparent photosynthetic rate of some broadleaf evergreens. *Ecology* 29:507–508.

Wirsing, John M., and Robert R. Alexander. 1975. Forest habitat types on the Medicine Bow National Forest, southeastern Wyoming: preliminary report. USDA For. Serv. Gen. Tech. Report RM–12. Rocky Mountain For. and Rge. Exp. Sta., Ft. Collins, Colo. 11 pp.
Witkamp, M., and D. A. Crossley, Jr. 1966. The role of arthropods and microflora in breakdown of white oak litter. *Pedobiol.* 6:293–303.
Witter, J. A., and L. A. Waisanen. 1978. The effect of differential flushing times among trembling aspen clones on tortricid caterpillar populations. *Env. Ent.* 7:139–143.
Wolfe, Jack A. 1971. Tertiary climatic fluctuations and methods of analysis of Tertiary floras. *Palaeogeog., Palaeoclimatol., Palaeoecol.* 9:27–57.
———, and E. B. Leopold. 1967. Neogene and early Quaternary vegetation of northwestern North America and northeastern Asia. *In* D. M. Hopkins (ed.), *The Bering Land Bridge.* Stanford Univ. Press, Stanford, Calif.
Wolfe, J. N., R. T. Wareham, and H. T. Scofield. 1949. Microclimates and macroclimates of Neotoma, a small valley in central Ohio. Ohio Biol. Surv. Bull. 41. 267 pp.
Woods, D. B., and N. C. Turner. 1971. Stomatal response to changing light by four tree species of varying shade tolerance. *New Phytol.* 70:77–84.
Woods, F. W. 1953. Disease as a factor in the evolution of forest composition. *J. For.* 51:871–873.
———. 1957. Factors limiting root penetration in deep sands of the southeastern coastal plain. *Ecology* 38:357–359.
———, and Royal E. Shanks. 1959. Natural replacement of chestnut by other species in the Great Smoky Mountains National Park. *Ecology* 40:349–361.
Woodwell, G. M., and D. B. Botkin. 1970. Metabolism of terrestrial ecosystems by gas exchange techniques: The Brookhaven approach. *In* D. E. Reichle (ed.), *Analysis of Temperate Forest Ecosystems.* Springer-Verlag, New York.
———, R. H. Whittaker, W. A. Reiners, G. E. Likens, C. C. Delwiche, and D. B. Botkin. 1978. The biota and the world carbon budget. *Science* 199:141–146.
Wright, H. E., Jr. 1970. Vegetational history of the Central Plains. *In* Wakefield Dort, Jr., and J. Knox Jones, Jr. (eds.), *Pleistocene and Recent Environments of the Central Great Plains.* Univ. Press of Kansas, Lawrence, Kan.
———. 1971. Late Quaternary vegetational history of North America. *In* Karl K. Turekian (ed.), *The Late Cenozoic Glacial Ages.* Yale Univ. Press, New Haven.
———. 1976. The dynamic nature of Holocene vegetation. *Quaternary Res.* 6:581–596.
———, and David G. Frey (eds.). 1965. *The Quaternary of the United States.* Princeton Univ. Press, Princeton, N.J. 922 pp.
———, and M. L. Heinselman (eds.). 1973. The ecological role of fire in natural conifer forests of western and northern North America. *Quaternary Res.* 3:317–513.

Wright, Jonathan W. 1955. Species crossability in spruce in relation to distribution and taxonomy. *For. Sci.* 1:319–352.
———1962. *Genetics of Forest Tree Improvement.* FAO Forestry and For. Prod. Studies No. 16, Rome. 399 pp.
———. 1976. *Introduction to Forest Genetics.* Academic Press, Inc., New York. 463 pp.
———, and H. I. Baldwin. 1957. The 1938 International Union Scotch pine provenance test in New Hampshire. *Silvae Genetica* 6:2–14.
———, and W. Ira Bull. 1963. Geographic variation in Scotch pine. *Silvae Genetica* 12:1–25.
———, and Scott S. Pauley, R. Brooks Polk, Jalmer J. Jokela, and Ralph A. Read. 1966. Performance of Scotch pine varieties in the North Central Region. *Silvae Genetica* 15:101–110.
———, Louis F. Wilson, and William Randall. 1967. Differences among Scotch pine varieties in susceptibility to European pine sawfly. *For. Sci.* 13:175–181.
Wright, Richard F., Torstein Dale, Egil T. Gjessing, George R. Hendrey, Arne Henriksen, Merete Johannessen, and Ivar P. Muniz. 1976. Impact of acid precipitation on freshwater ecosystems in Norway. *In Proc. The First International Symposium on Acid Precipitation and the Forest Ecosystem.*, pp. 459–476. USDA For. Serv. Gen. Tech. Report NE–23. Northeastern For. Exp. Sta., Upper Darby, Pa.
Wright, T. W. 1955, 1956. Profile development in the sand dunes of Culbin Forest, Morayshire. *J. Soil Sci.* 6:270–283; 7:33–42.
Wuenscher, James E., and Theodore T. Kozlowski. 1971a. The response of transpiration resistance to leaf temperature as a desiccation resistance mechanism in tree seedlings. *Physiol. Plant.* 24:254–259.
———, and Theodore T. Kozlowski. 1971b. Relationship of gas-exchange resistance to tree-seedling ecology. *Ecology* 52:1016–1023.
Wulff, E. V. 1943. *An Introduction to Historical Plant Geography.* Chronica Botanica Co., Waltham, Mass. 223 pp.
Yaroshenko, P. D. 1946. O smenakh rastitel 'nogo pokrova. (On changes in vegetational cover.) *Bot. Zh. S.S.S.R.* 31(5):29–40.
Yocom, Herbert A. 1968. Shortleaf pine seed dispersal. *J. For.* 66:422.
Young, Harold E. 1964. The complete tree concept—a challenge and an opportunity. *Proc. Soc. Amer. For., 1964:* 231–233.
———. 1971. Biomass sampling methods for puckerbrush stands. *In Forest Biomass Studies.* Misc. Publ. 132, Life Sci. and Agr. Exp. Sta., Univ. Maine, Orono.
Youngberg, Chester T. 1959. The influence of soil conditions, following tractor logging, on the growth of planted Douglas-fir seedlings. *Proc. Soil Sci. Soc. Amer.* 23:76–78.
——— (ed.). 1965. *Forest-Soil Relationships in North America.* Proc. Second North Am. For. Soils Conf. Oregon State Univ. Press, Corvallis. 532 pp.
———, and Charles B. Davey (eds.). 1970. *Tree Growth and Forest Soils.* Proc. Third North Am. For. Soils Conf. Oregon State Univ. Press, Corvallis. 527 pp.

Zach, Lawrence W. 1950. A northern climax, forest or muskeg? *Ecology* 31:304–306.

Zachariae, G. 1962. Zur Methodik bei Geländeuntersuchungen in der Bodenzoologie. *Z. Pflanzenernähr. Düng. Bodenk.* 97:224–233.

Zahner, Robert. 1955. Soil water depletion by pine and hardwood stands during a dry season. *For. Sci.* 1:258–264.

———. 1956. Evaluating summer water deficiencies. U.S. For. Serv. Southern For. Exp. Sta., Occ. Paper 150. 18 pp.

———. 1957. Field procedures for soil-site classification of pine land in South Arkansas and North Louisiana. U.S. For. Serv. Southern For. Exp. Sta., Occ. Paper 155. 17 pp.

———. 1958a. Site-quality relationships of pine forests in southern Arkansas and northern Louisiana. *For. Sci.* 4:162–176.

———. 1958b. Hardwood understory depletes soil water in pine stands. *For. Sci.* 4:178–184.

———. 1962. Loblolly pine site curves by soil groups. *For. Sci.* 8:104–110.

———. 1967. Refinement in empirical functions for realistic soil-moisture regimes under forest cover. *In* William E. Sopper and Howard W. Lull (eds.), *Forest Hydrology.* Pergamon Press, Inc., New York.

———. 1968. Water deficits and growth of trees. *In* T. T. Kozlowski (ed.), *Water Deficits and Plant Growth II.* Academic Press, Inc., New York.

———, and F. W. Whitmore. 1960. Early growth of radically thinned loblolly pine. *J. For.* 58:628–634.

———, and Ned A. Crawford. 1965. The clonal concept in aspen site relations. *In* Chester T. Youngberg (ed.), *Forest-Soil Relationships in North America.* Oregon State Univ. Press, Corvallis.

———, and A. R. Stage. 1966. A procedure for calculating daily moisture stress and its utility in regressions of tree growth on weather. *Ecology* 47:64–74.

———, and J. R. Donnelly. 1967. Refining correlations of rainfall and radial growth in young red pine. *Ecology* 48:525–530.

Zak, B. 1965. Aphids feeding on mycorrhizae of Douglas-fir. *For. Sci.* 11:410–411.

Zavitkovski, J., and Michael Newton. 1968. Effect of organic matter and combined nitrogen on nodulation and nitrogen fixation in red alder. *In* J. M. Trappe, J. F. Franklin, R. F. Tarrant, and G. M. Hansen (eds.), *Biology of Alder.* Northwest Sci. Assoc. Fortieth Ann. Meeting Symp. Proc. pp. 209–223. Pacific Northwest For. and Rge. Exp. Sta., Portland, Ore.

Zimmermann, Martin H., and Claud L. Brown. 1971. *Trees: Structure and Function.* Springer-Verlag, New York. 336 pp.

Zinke, P. J. 1961. Forest site quality as related to soil nitrogen content. *Trans. 7th Int. Cong. Soil Sci.* 3:411–418.

———. 1967. Forest interception studies in the United States. *In* William E. Sopper and Howard W. Lull (eds.), *Forest Hydrology.* Pergamon Press, Inc., New York.

Zobel, Donald B. 1969. Factors affecting the distribution of *Pinus pungens,* an Appalachian endemic. *Ecol. Monogr.* 39:303–333.

———, W. A. McKee, G. M. Hawk, and C. T. Dyrness. 1974. Correlation of forest communities with environment and phenology on the H. J. Andrews Experimental Forest, Oregon. *In* R. H. Waring and R. L. Edmonds (eds.), *Integrated Research in the Coniferous Forest Biome*. Conif. For. Biome Bull. No. 5, Coniferous Forest Biome, US/IBP, Univ. Washington, Seattle, Wash.

———, ———, ———, and ———. 1976. Relationships of environment to composition, structure, and diversity of forest communities of the central western Cascades of Oregon. *Ecol. Monogr.* 46:135–156.

Zoltai, S. C. 1965. Forest sites of site regions 5S and 4S, Northwestern Ontario. Vol. I. Ontario Dept. Lands and For. Res. Rept. 65. 121 pp.

# Index

## CONVERSION FACTORS FOR SELECTED ENGLISH AND METRIC UNITS [1]

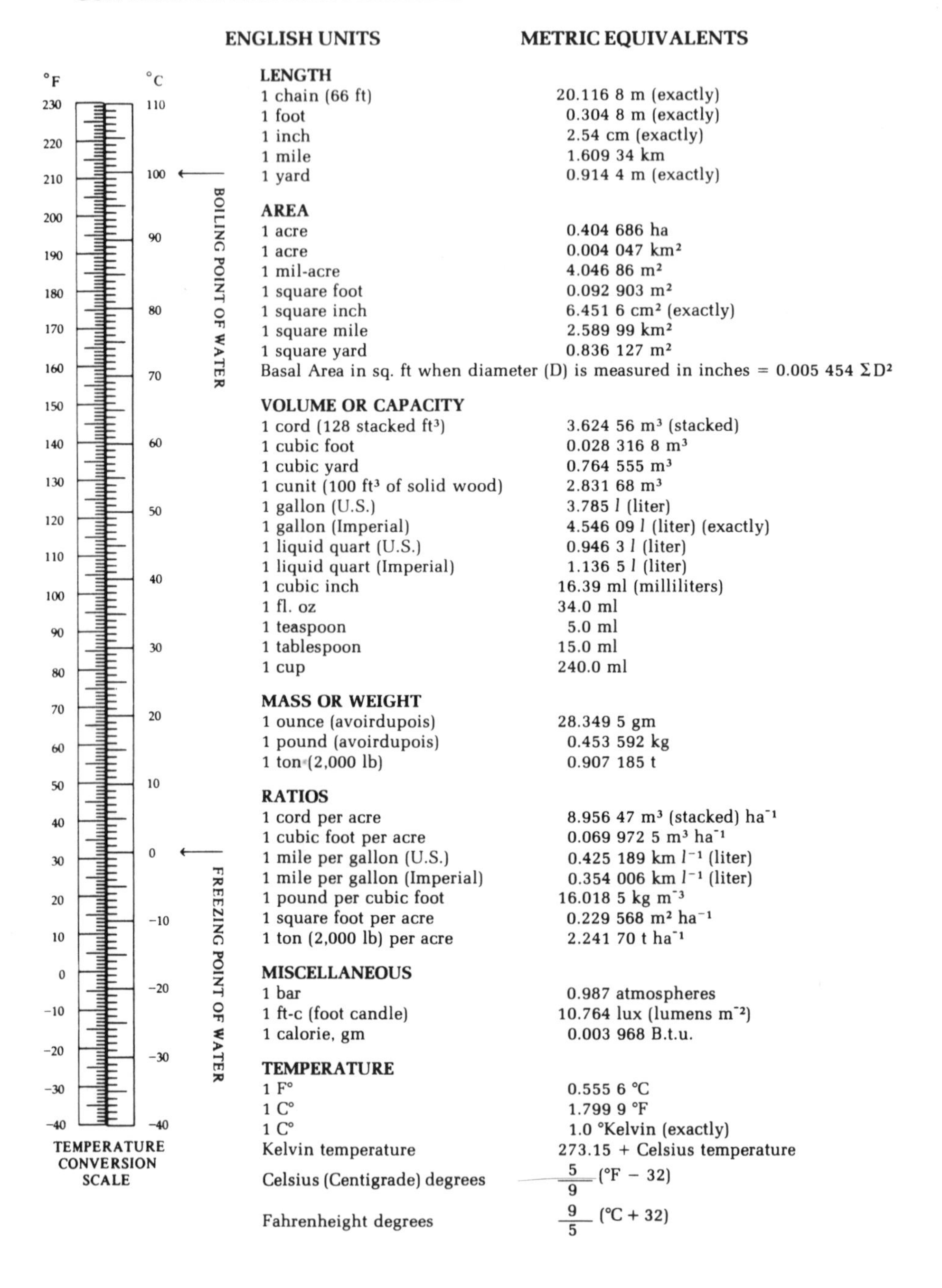


| ENGLISH UNITS | METRIC EQUIVALENTS |
|---|---|
| **LENGTH** | |
| 1 chain (66 ft) | 20.116 8 m (exactly) |
| 1 foot | 0.304 8 m (exactly) |
| 1 inch | 2.54 cm (exactly) |
| 1 mile | 1.609 34 km |
| 1 yard | 0.914 4 m (exactly) |
| **AREA** | |
| 1 acre | 0.404 686 ha |
| 1 acre | 0.004 047 $km^2$ |
| 1 mil-acre | 4.046 86 $m^2$ |
| 1 square foot | 0.092 903 $m^2$ |
| 1 square inch | 6.451 6 $cm^2$ (exactly) |
| 1 square mile | 2.589 99 $km^2$ |
| 1 square yard | 0.836 127 $m^2$ |
| Basal Area in sq. ft when diameter (D) is measured in inches | = 0.005 454 $\Sigma D^2$ |
| **VOLUME OR CAPACITY** | |
| 1 cord (128 stacked $ft^3$) | 3.624 56 $m^3$ (stacked) |
| 1 cubic foot | 0.028 316 8 $m^3$ |
| 1 cubic yard | 0.764 555 $m^3$ |
| 1 cunit (100 $ft^3$ of solid wood) | 2.831 68 $m^3$ |
| 1 gallon (U.S.) | 3.785 *l* (liter) |
| 1 gallon (Imperial) | 4.546 09 *l* (liter) (exactly) |
| 1 liquid quart (U.S.) | 0.946 3 *l* (liter) |
| 1 liquid quart (Imperial) | 1.136 5 *l* (liter) |
| 1 cubic inch | 16.39 ml (milliliters) |
| 1 fl. oz | 34.0 ml |
| 1 teaspoon | 5.0 ml |
| 1 tablespoon | 15.0 ml |
| 1 cup | 240.0 ml |
| **MASS OR WEIGHT** | |
| 1 ounce (avoirdupois) | 28.349 5 gm |
| 1 pound (avoirdupois) | 0.453 592 kg |
| 1 ton (2,000 lb) | 0.907 185 t |
| **RATIOS** | |
| 1 cord per acre | 8.956 47 $m^3$ (stacked) $ha^{-1}$ |
| 1 cubic foot per acre | 0.069 972 5 $m^3$ $ha^{-1}$ |
| 1 mile per gallon (U.S.) | 0.425 189 km $l^{-1}$ (liter) |
| 1 mile per gallon (Imperial) | 0.354 006 km $l^{-1}$ (liter) |
| 1 pound per cubic foot | 16.018 5 kg $m^{-3}$ |
| 1 square foot per acre | 0.229 568 $m^2$ $ha^{-1}$ |
| 1 ton (2,000 lb) per acre | 2.241 70 t $ha^{-1}$ |
| **MISCELLANEOUS** | |
| 1 bar | 0.987 atmospheres |
| 1 ft-c (foot candle) | 10.764 lux (lumens $m^{-2}$) |
| 1 calorie, gm | 0.003 968 B.t.u. |
| **TEMPERATURE** | |
| 1 F° | 0.555 6 °C |
| 1 C° | 1.799 9 °F |
| 1 C° | 1.0 °Kelvin (exactly) |
| Kelvin temperature | 273.15 + Celsius temperature |
| Celsius (Centigrade) degrees | $\frac{5}{9}$ (°F − 32) |
| Fahrenheight degrees | $\frac{9}{5}$ (°C + 32) |

## CONVERSION FACTORS FOR SELECTED ENGLISH AND METRIC UNITS [1]

| METRIC UNITS | ENGLISH EQUIVALENTS |
|---|---|
| **LENGTH** | |
| 1 cm (centimeter) | 0.393 701 inch |
| 1 km (kilometer) | 0.621 371 mile |
| 1 m (meter) | 0.049 709 7 chain (of 66 ft) |
| 1 m (meter) | 3.280 84 feet |
| 1 m (meter) | 1.093 61 yards |
| **AREA** | |
| 1 $cm^2$ (square centimeter) | 0.155 000 square inch |
| 1 ha (hectare) | 2.471 05 acres |
| 1 $km^2$ (square kilometer) | 0.386 102 square mile |
| 1 $km^2$ (square kilometer) | 247.10 acres |
| 1 $m^2$ (square meter) | 0.247 105 mil-acre |
| 1 $m^2$ (square meter) | 10.763 9 square feet |
| 1 $m^2$ (square meter) | 1.195 99 square yards |
| Basal Area in $m^2$ when diameter (D) is measured in cm | = 0.000 078 54 $\Sigma D^2$ |
| **VOLUME OR CAPACITY** | |
| 1 *l* (liter) | 0.219 969 gallon |
| 1 *l* (liter) | 1.056 7 liquid quarts (U.S.) |
| 1 ml (milliliter, or 1 cu. centimeter) | 0.061 024 cubic inch |
| 100 ml (milliliter) | 3.4 fl. oz (0.4 cup) |
| 1 $m^3$ (cubic meter) | 35.314 7 cubic feet |
| 1 $m^3$ (cubic meter) | 1.307 95 cubic yards |
| 1 $m^3$ (cubic meter) | 0.353 147 cunit (of 100 $ft^3$ of solid wood) |
| 1 $m^3$ (stacked) (stacked cubic meter) | 0.275 896 cord (of 128 stacked $ft^3$) |
| **MASS OR WEIGHT** | |
| 1 gm (gram) | 0.035 274 0 ounce (avoirdupois) |
| 1 kg (kilogram) | 2.204 62 pounds (avoirdupois) |
| 1 t (ton) | 1.102 31 tons (of 2,000 lb) |
| **RATIOS** | |
| 1 kg $m^{-3}$ (kilogram per cubic meter) | 0.062 428 pounds per cubic foot |
| 1 kg $ha^{-1}$ (kilogram per hectare) | 0.89 lb $acre^{-1}$ (pounds per acre) |
| 1 km $l^{-1}$ (kilometer per liter) | 2.351 90 miles per gallon (U.S.) |
| 1 km $l^{-1}$ (kilometer per liter) | 2.824 81 miles per gallon (Imperial) |
| 1 $m^2$ $ha^{-1}$ (square meter per hectare) | 4.356 00 square feet per acre |
| 1 $m^3$ $ha^{-1}$ (cubic meter per hectare) | 14.291 3 cubic feet per acre |
| 1 $m^3$ (stacked) $ha^{-1}$ (stacked cubic meter per hectare) | 0.111 651 cords per acre |
| 1 t $ha^{-1}$ (ton per hectare) | 0.446 090 tons (of 2,000 lb) per acre |

---

[1]Modified from Bowen (1974). SOURCE: Forest Management Institute, Canadian Forestry Service, Department of the Environment, Ottawa, Canada.